AF732366

Chromatographic Analysis of the Environment

Mass Spectrometry Based Approaches

FOURTH EDITION

Chromatographic Analysis of the Environment

Mass Spectrometry Based Approaches

FOURTH EDITION

Edited by
Leo M.L. Nollet
Dimitra A. Lambropoulou

CRC Press
Taylor & Francis Group
Boca Raton London New York

CRC Press is an imprint of the
Taylor & Francis Group, an informa business

CRC Press
Taylor & Francis Group
6000 Broken Sound Parkway NW, Suite 300
Boca Raton, FL 33487-2742

First issued in paperback 2019

ISBN-13: 978-1-4665-9756-3 (hbk)
ISBN-13: 978-0-367-86858-1 (pbk)

Library of Congress Cataloging-in-Publication Data

Names: Nollet, Leo M. L., 1948- editor. | Lambropoulou, Dimitra A., editor.
Title: Chromatographic analysis of the environment : mass spectrometry based approaches.
Description: Fourth edition / edited by Leo M. L. Nollet and Dimitra Lambropoulou. |
Boca Raton, FL : CRC Press, [2017]
Identifiers: LCCN 2016028978 | ISBN 9781466597563 | ISBN 9781315316208
Subjects: LCSH: Chromatographic analysis. | Environmental chemistry.
Classification: LCC QD79.C4 C48 2017 | DDC 628.5028/7--dc23
LC record available at https://lccn.loc.gov/2016028978

Visit the Taylor & Francis Web site at
http://www.taylorandfrancis.com

and the CRC Press Web site at
http://www.crcpress.com

This book is dedicated to all victims of the attacks in Brussels, Belgium, on March 22, 2016.

We hope it will compensate a very little bit for their sufferances and the sorrows of their families.

It is also dedicated to the Greek people, who are going through a very, very long period of uncertainty.

"When the going gets tough, the tough gets going."

Contents

Preface

Considerable analytical progress in chromatographic analysis has been made over the past decade, when hyphenated techniques involving highly efficient separation and sensitive detection have become the techniques of choice. Among them, methods based on chromatographic separation with mass spectrometric detection have opened new perspectives in terms of identification and are acknowledged as the major useful and authoritative methods for the determination of pollutants in the environment.

Since the publication of the third edition of this book in 2005, the focus of chromatographic research has changed again, and the variety of application fields that have been addressed in the literature shows that, due to their versatility, chromatographic mass spectrometric techniques have proven themselves successful in virtually any analytical challenge that makes them robust and effectively applicable alternatives in pollutant analysis. Obviously, in comparison with different detection methods, mass spectrometry or tandem mass spectrometry appears to be the technique of choice in environmental analysis, as it provides reliable results at the subnanogram per liter or per gram level, while the use of other specific detectors, such as flame-ionization detector, nitrogen–phosphorus detector, and electron capture detector, has been in gradual decline. Mass spectrometric analysis has increased exponentially during this time, and important progress has been made in terms of instrumentation in order to increase the throughput, mass resolving power, mass accuracy, tandem mass spectrometry capabilities, and resolution for mass spectrometry.

In this context, the main scope of the book is to outline those developments, which are particularly relevant to environmental analysis, especially when using chromatographic mass spectrometric techniques. Accordingly, the fourth edition has been largely changed and updated in comparison with the previous ones to address these new issues, the new lines of discussions, and new findings.

This edition contains six sections. In Section I, an overview of principles, instrumental aspects, and conceptual case studies in chromatographic analysis is given. The basic concepts and instrumentation of gas chromatography and liquid chromatography combined with mass spectrometry are discussed. This section includes also a chapter on thermal desorption.

The five chapters in Section II detail sample preparation methods for the determination of pollutants in air, in solid and complex environmental matrices, and in the aquatic environment. The reader finds much information on novel materials and methods in sample treatment and separation.

Important aspects, such as quality assurance, validation, proficiency testing, and reference methods, are the elements of the contents of the three chapters of Section III.

Sections IV, V, and VI enumerate chromatographic methods on air pollutants (Section IV), on environmental residues in different matrices (Section V, five chapters), and on transformation products of pharmaceutical and personal care products (Section VI). The five chapters of Section V essentially deal with residues with endocrine-disrupting capacities.

Most of the data are compiled in tables and figures to elucidate the text. Presenting important theoretical and practical aspects from sample collection to laboratory analysis, *Chromatographic Analysis of the Environment, Fourth Edition*, is a unique resource of chromatographic techniques, data, and references that are useful to all scientists involved in the analysis of environmental compounds.

The redaction of this book was not without obstacles and difficulties. These obstacles were largely due to severe health issues and other less severe issues of both editors. These health issues have now been overcome.

We are very thankful to all the contributors for their excellent input and especially for their time and patience.

I would like to give a kiss to Dimitra for her help, wisdom, and courage.

Leo M. L. Nollet
Dimitra A. Lambropoulou

Editors

Leo M. L. Nollet, PhD, earned an MS (1973) and a PhD degree (1978) in biology from Katholieke Universiteit Leuven, Leuven, Belgium. He is an editor and associate editor of numerous books. He edited for Marcel Dekker, New York—now CRC Press of Taylor & Francis—the first, second, and third editions of *Food Analysis by HPLC* and *Handbook of Food Analysis*. The last edition is a two-volume book. He also edited *Handbook of Water Analysis* (first, second, and third editions) and *Chromatographic Analysis of the Environment*, third edition (CRC Press).

With Fidel Toldrá, he coedited two books published in 2006 and 2007: *Advanced Technologies for Meat Processing* (CRC Press) and *Advances in Food Diagnostics* (Blackwell Publishing—now Wiley). With Michael Poschl, he coedited the book *Radionuclide Concentrations in Foods and the Environment*, also published in 2006 (CRC Press).

Dr. Nollet has also coedited with Y. H. Hui and other colleagues several books: *Handbook of Food Product Manufacturing* (Wiley, 2007), *Handbook of Food Science, Technology and Engineering* (CRC Press, 2005), *Food Biochemistry and Food Processing* (first and second editions; Blackwell Publishing, 2006 and 2012), and *Handbook of Fruits and Vegetable Flavors* (Wiley, 2010). In addition, he edited *Handbook of Meat, Poultry and Seafood Quality*, first and second editions (Blackwell Publishing, 2007 and 2012).

From 2008 to 2011, he published with Toldrá five books on animal products, namely, *Handbook of Muscle Foods Analysis*, *Handbook of Processed Meats and Poultry Analysis*, *Handbook of Seafood and Seafood Products Analysis*, *Handbook of Dairy Foods Analysis*, and *Handbook of Analysis of Edible Animal By-Products*. In 2011, also with Toldrá, he coedited for CRC Press two books: *Safety Analysis of Foods of Animal Origin* and *Sensory Analysis of Foods of Animal Origin*. In 2012, they both published *Handbook of Analysis of Active Compounds in Functional Foods*.

Coedited with Hamir Rathore, the book *Handbook of Pesticides: Methods of Pesticides Residues Analysis* was marketed in 2009, *Pesticides: Evaluation of Environmental Pollution* in 2012, and *Biopesticides Handbook* in 2015.

Other finished book projects include *Food Allergens: Analysis, Instrumentation, and Methods* (with Arjon van Hengel; CRC Press, 2011) and *Analysis of Endocrine Compounds in Food* (Wiley-Blackwell, 2011).

Dr. Nollet's recent projects include *Proteomics in Foods* with Toldrá (Springer, 2013) and *Transformation Products of Emerging Contaminants in the Environment: Analysis, Processes, Occurrence, Effects and Risks* with Dimitra A. Lambropoulou (Wiley, 2014).

This CRC series, Food Analysis & Properties, was edited by Dr. Nollet with Claudia Ruiz-Capillas and includes *Flow Injection Analysis of Food Additives* (CRC Press, 2015) and *Marine Microorganisms: Extraction and Analysis of Bioactive Compounds* (CRC Press, 2016).

Dimitra A. Lambropoulou, PhD, is currently assistant professor of environmental chemistry in the Environmental Pollution Control Laboratory, Department of Chemistry, Aristotle University of Thessaloniki. She earned her PhD in chemistry from University of Ioannina in 2002. Up till now, she has published over 75 publications in high-impact international peer-reviewed scientific journals. With Leo M. L. Nollet, she coedited a book published by John Wiley & Sons, Inc in 2014: *Transformation Products of Emerging Contaminants in the Environment: Analysis, Processes, Occurrence, Effects and Risks*. Her main research interests are the development and application of novel sample preparation techniques coupled to advanced mass spectrometry approaches in the field of environmental chemistry, design and application of new materials in analytical and separation sciences, high resolution mass spectrometry, occurrence, transport, fate and effects of emerging contaminants in the environment. She is also interested in the development of effective degradation and purification processes for the mineralization of organic micropollutants.

Contributors

Mohamed Abou-Elwafa Abdallah
Division of Environmental Health and Risk Management
School of Geography, Earth and Environmental Sciences
University of Birmingham
Birmingham, United Kingdom
and
Department of Analytical Chemistry
Faculty of Pharmacy
Assiut University
Assiut, Egypt

Adeola Adenugba
Faculty of Engineering
University of Regina
Regina, Canada

Charalampos Alexopoulos
General Chemical State Laboratory
Chemical Metrology Service
Athens, Greece

C. Marisa R. Almeida
Interdisciplinary Centre of Marine and Environmental Research
University of Porto
Porto, Portugal

M. Clara P. Basto
Interdisciplinary Centre of Marine and Environmental Research
and
Faculty of Sciences
University of Porto
Porto, Portugal

Dimitrios Bikiaris
Laboratory of Polymer Chemistry and Technology
Department of Chemistry
Aristotle University of Thessaloniki
Thessaloniki, Greece

Josep Caixach
Mass Spectrometry Laboratory/Organic Pollutants
Institute of Environmental Assessment and Water Research
Spanish Council for Scientific Research
Barcelona, Spain

Jun-Ji Cao
State Key Lab of Loess and Quaternary Geology
Institute of Earth Environment
Chinese Academy of Sciences
Xi'an, People's Republic of China

Judith C. Chow
Division of Atmospheric Sciences
Desert Research Institute
Reno, Nevada

Nuria Cortés-Francisco
Mass Spectrometry Laboratory/Organic Pollutants
Institute of Environmental Assessment and Water Research
Spanish Council for Scientific Research
and
Laboratori de l'Agència de Salut Pública de Barcelona
Barcelona, Spain

Adrian Covaci
Toxicological Center
Department of Pharmaceutical Sciences
University of Antwerp
Antwerp, Belgium

Alin C. Dirtu
Toxicological Center
Department of Pharmaceutical Sciences
University of Antwerp
Antwerp, Belgium
and
Department of Chemistry
"Al. I. Cuza" University of Iași
Iași, Romania

David Elustondo
Integrated Laboratory for Environmental Quality
Department of Chemistry and Soil Science
School of Sciences
University of Navarra
Pamplona, Spain

Eleni Evgenidou
Department of Chemistry
Aristotle University of Thessaloniki
Thessaloniki, Greece

Núria Fontanals
Departament de Química Analítica i Química Orgànica
Universitat Rovira i Virgili
Tarragona, Spain

Roberta Galarini
Center for Development and Validation of Methods
Istituto Zooprofilattico Sperimentale dell'Umbria e delle Marche
Perugia, Italy

Carlos Gonçalves
Interdisciplinary Centre of Marine and Environmental Research
University of Porto
Porto, Portugal

John V. Headley
Environment and Climate Change Canada
Saskatoon, Canada

Leonel Hernández-Mena
Centro de Investigación y Asistencia en Tecnología y Diseño del Estado de Jalisco
Guadalajara, Mexico

Steven Sai Hang Ho
Division of Atmospheric Sciences
Desert Research Institute
Reno, Nevada
and
Key Lab of Aerosol Chemistry and Physics
Institute of Earth Environment
Chinese Academy of Sciences
Xi'an, People's Republic of China

Yu Huang
Key Lab of Aerosol Chemistry and Physics
Institute of Earth Environment
Chinese Academy of Sciences
Xi'an, People's Republic of China

Elias Kakoulidis
General Chemical State Laboratory
Chemical Metrology Service
Athens, Greece

Barbara Kasprzyk-Hordern
Department of Chemistry
University of Bath
Bath, United Kingdom

Ioannis Konstantinou
Department of Chemistry
University of Ioannina
Ioannina, Greece

George Z. Kyzas
Laboratory of Polymer Chemistry and Technology
Department of Chemistry
Aristotle University of Thessaloniki
Thessaloniki, Greece

Dimitra A. Lambropoulou
Environmental Pollution Control Laboratory
Department of Chemistry
Aristotle University of Thessalonikis
Thessaloniki, Greece

Eugenia Lampi
General Chemical State Laboratory
Chemical Metrology Service
Athens, Greece

Esther Lasheras
Integrated Laboratory for Environmental Quality
Department of Chemistry and Soil Science
School of Sciences
University of Navarra
Pamplona, Spain

Alberto López-López
Centro de Investigación y Asistencia en Tecnología y Diseño del Estado de Jalisco
Guadalajara, Mexico

Dena McMartin
Faculty of Engineering
University of Regina
Regina, Canada

Simone Moretti
Center for Development and Validation of Methods
Istituto Zooprofilattico Sperimentale dell'Umbria e delle Marche
Perugia, Italy

Basil K. Munjanja
Department of Applied Chemistry
National University of Science and Technology
Bulawayo, Zimbabwe

Mario Murillo-Tovar
Centro de Investigación y Asistencia en Tecnología y Diseño del Estado de Jalisco
Guadalajara, Mexico

Jacek Namieśnik
Department of Analytical Chemistry
Chemical Faculty
Gdańsk University of Technology
Gdańsk, Poland

Leo M. L. Nollet
University College Ghent (Retired)
Ghent, Belgium

Myrsini Papageorgiou
Department of Chemistry
Aristotle University of Thessaloniki
Thessaloniki, Greece

Francisco Pena-Pereira
Analytical and Food Chemistry Department
Faculty of Chemistry
University of Vigo
Vigo, Spain
and
Department of Analytical Chemistry
Chemical Faculty
Gdańsk University of Technology
Gdańsk, Poland

Kerry Peru
Environment and Climate Change Canada
Saskatoon, Canada

Bruce Petrie
Department of Chemistry
University of Bath
Bath, United Kingdom

Ednei Gilberto Primel
Escola de Química e Alimentos
Universidade Federal do Rio Grande
Rio Grande, Brazil

Giorgio Saluti
Center for Development and Validation of Methods
Istituto Zooprofilattico Sperimentale dell'Umbria e delle Marche
Perugia, Italy

Vasilios G. Samaras
Water and Air Quality Laboratory
Department of Environment
University of the Aegean
Mytilene, Greece

Carolina Santamaría
Integrated Laboratory for Environmental Quality
Department of Chemistry and Soil Science
School of Sciences
University of Navarra
Pamplona, Spain

Jesús Miguel Santamaría
Integrated Laboratory for Environmental Quality
Department of Chemistry and Soil Science
School of Sciences
University of Navarra
Pamplona, Spain

Maria Augusta D. Sousa
Faculty of Pharmacy
University of Porto
Porto, Portugal

Sergiane Souza Caldas
Escola de Química e Alimentos
Universidade Federal do Rio Grande
Rio Grande, Brazil

Anna-Akrivi Thomatou
Department of Chemistry
University of Ioannina
Ioannina, Greece

Ramiro Vallejo-Rodríguez
Centro de Investigación y Asistencia en Tecnología y Diseño del Estado de Jalisco
Guadalajara, Mexico

Ana Laura Venquiaruti Escarrone
Instituto de Ciências Biológicas
Universidade Federal do Rio Grande
Rio Grande, Brazil

John G. Watson
Division of Atmospheric Sciences
Desert Research Institute
Reno, Nevada

Jian Zhen Yu
Department of Chemistry
Hong Kong University of Science and Technology
Kowloon, Hong Kong

Section I

Separation

1 Gas Chromatography–Mass Spectrometry

Basic Concepts and Instrumentation

Basil K. Munjanja

CONTENTS

1.1 INTRODUCTION

Gas chromatography–mass spectrometry (GC-MS) continues to play a pivotal role in environmental analysis of pollutants despite the increasing use of liquid chromatography–MS by most research groups. This is because of the high separation efficiency of GC and the wide range of mass analyzers that it can be coupled to simplify analysis of pollutants in complex environmental matrices. Examples of pollutants include nonpolar and thermally labile analytes such as polyaromatic hydrocarbons (PAHs), pesticides, polychlorinated biphenyls (PCBs), water disinfection by-products, polychlorinated dibenzo-*p*-dioxins and furans (PCDD/F), and endocrine disruptors, some of which are shown in Figure 1.1. In addition, GC-MS can also be applied in the analysis of polar substances such as pharmaceuticals and personal care products, only after derivatization (Subedi et al., 2011). To date the use of this technique in environmental analysis is on an upward trend because of the very high separation efficiency of GC coupled to the high sensitivity offered by MS. In view of this, it is critical that we discuss the basic concepts of GC-MS and all the essential aspects of instrumentation in this chapter. Moreover, we discuss recent advancements in sample introduction, chromatographic separation, mass analyzers, and the different applications of GC-MS in environmental analysis.

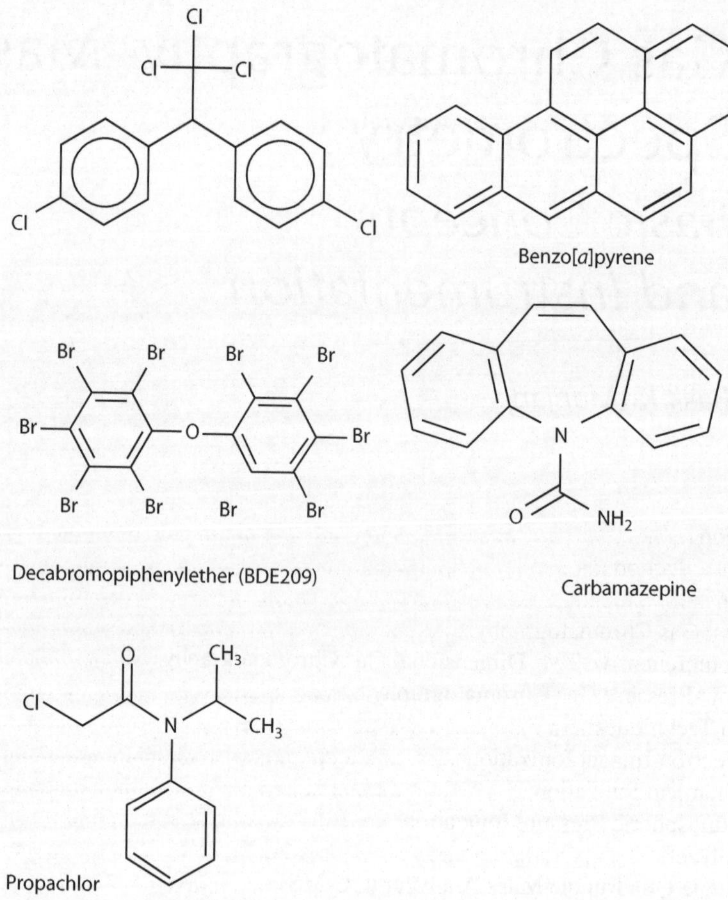

FIGURE 1.1 Chemical structures of some GC-MS-amenable organic compounds.

1.2 SAMPLE INTRODUCTION

The first stage in GC-MS is sample introduction. It is important to note that a good injection technique should not compromise the separation efficiency and should not induce a change in the sample composition. Hence, depending on the physical properties of the analyte such as physical state, chemical stability, and thermal degradation, one can choose between thermal desorption (TD) and large volume injection (LVI) techniques (Hird, 2008). TD techniques involve the removal of volatile (Ribes et al., 2007) or semivolatile analytes (Falkovich and Rudich, 2001) from solid matrices by heated gas flow and their subsequent extraction into a stream of inert gas and their transfer to the gas chromatograph in a small volume of concentrated vapor. TD techniques come in various forms, and their advantages are low costs and reduced labor requirements. However, their major drawback is the reduced sensitivity as the number of samples increases (Ho et al., 2011). Furthermore, other forms of TD suffer from the drawback of needing to modify the injector port and an additional transfer line. For this reason, in-injector port TD is the most common form used in environmental analysis, offering an additional advantage of high transfer efficiency because of the elimination of transfer lines between the sample and the analytical instrument. Thus, it has been widely used in the analysis of volatile organic compounds, explosives, and aerosol organics (Ho and Yu, 2004).

The use of unusually high injection volumes greater than 2 μm in LVI is an important feature that improves the sensitivity of the technique and simplifies the sample preparation step. In addition, LVI can serve as an interface for the automation of sample preparation steps such as solid-phase extraction (SPE) with GC (Hoh and Mastovska, 2008). The traditional LVI techniques such as on-column, programmed temperature vaporization (PTV) in solvent split mode had serious drawbacks with volatile analytes and high-molecular weight analytes (Li et al., 2009). However, newer modes of LVI, including direct sample introduction, splitless overflow, at-column and through-oven transfer adsorption–desorption overcome these drawbacks.

Some compounds may degrade in the injection port because either they lack thermal lability or they have relatively high polarity (Lambropolou and Hela, 2015). For such, an additional step of derivatization is done by silylation, acylation, or alkylation (Yang and Shin, 2013). For instance, in the analysis of seven pharmaceuticals and personal care products (PPCPs) in biosolids, after optimization of derivatization conditions, *N,O*-bis(trimethylsilyl)-trifluoroacetamide and 1% trimethylchlorosilane was used because it allowed effective detection of all polar compounds (Rice and Mitra, 2007). However, a major drawback of derivatization is that it is time consuming and some of the derivatization agents may damage the column.

1.3 SEPARATION

The column is the heart of the GC because separation takes place there. Earlier on, packed columns were used, but their use has since been discontinued because of their lower resolution when applied to the analysis of many compounds. They have since been replaced by the capillary column, which provides better resolution and increases the speed of analysis when coupled to a mass spectrometer. GC analysis can be one dimensional (1D), where one column is used, or two dimensional (2D), where two columns are used. In both cases, separation takes place based on the partition of the analyte between the mobile phase and the stationary phase. The mobile phase is usually nitrogen gas or helium, but the stationary phase can be polar (made from polyethylene glycol) or apolar (dimethylsiloxane, etc.). Thus, the selection of the stationary phase depends on the nature of the analytes. For instance, in the analysis of fatty acids such as methyl esters, polar stationary phases coated with polyethylene glycol (DB-WAX) or with bis(cyanopropyl) siloxane (e.g., BPX70) are used because they allow the differentiation of fatty acids with different carbon numbers, unsaturations, locations, and geometries of double bonds (Gu et al., 2011).

However, nowadays, a recent development is the use of ionic liquids as a new class of stationary films in capillary films with remarkable properties and benefits for GC-MS such as low volatility, high thermal stability, and selectivity toward certain chemical (Ballesteros-Gomez and Rubio, 2011). Initially, molten salts were first used as GC stationary phases. However, imidazolium-based ionic liquids that have higher viscosity, broader liquid range, and higher thermal stability have since replaced these. Other types of ionic liquids that are now used include dicationic ionic liquids, functionalized ionic liquids, and polymerized ionic liquids (Yao and Anderson, 2009). Ionic liquids were tested in 1D-MS and 2D-MS in the analysis of fatty acid methyl esters from algae. The ionic liquid stationary phases showed comparable resolution but lower column bleeding, with MS detection resulting in better sensitivity as compared to polyethylene glycol- and cyanopropyl-substituted polar stationary phases (Gu et al., 2011). It is worth noting that despite the unique selectivity that ionic liquids offer, they have not achieved the separation efficiency of polysiloxanes (Dorman et al., 2010). For this reason, different authors have proposed methods of combining polysiloxanes and ionic liquids based on their merits. For instance, Sun et al. (2010) synthesized an ionic liquid-bonded polysiloxane as a stationary phase, and it was used to construct an 8 m capillary column. The stationary phase showed good film-forming ability and high durability. In addition, it had stronger dispersive forces than the neat ionic liquid due to the presence of the polysiloxane skeleton. This accounted for its good selectivity and high separation efficiency for a wide range of analytes (Sun et al., 2010).

In addition to ionic liquids, other types of stationary phases such as graphene (Fan et al., 2015), graphene oxide (Feng et al., 2015), cyclotriveratrylene (CTV; Lv et al., 2015), and cyclodextrins (Shi et al., 2001; Grisales et al., 2009) are still being tried out by various research groups in a bid to increase the separation efficiency of GC. In most of these studies, high separation efficiency is being obtained

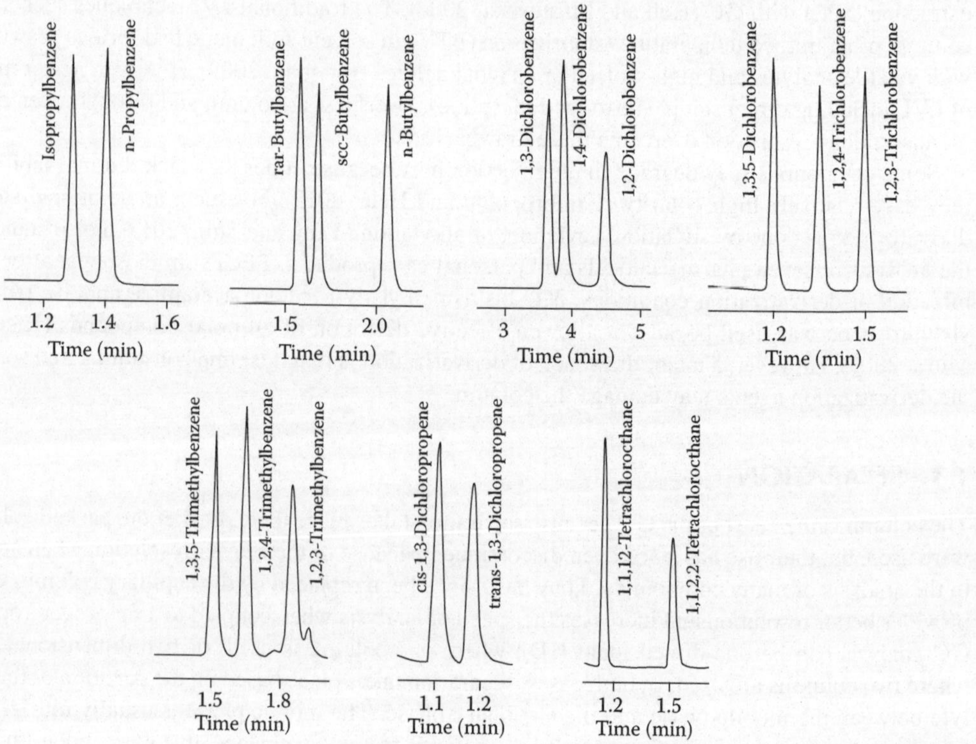

FIGURE 1.2 Chromatograms for GC separations of isomers on CTV capillary column. (Reproduced from *Journal of Chromatography A*, 1404, Lv, Q., Q. Zhang, M. Qi, H. Bai, Q. Ma et al., Cyclotriveratrylene as a new-type stationary phase for gas chromatographic separations of halogenated compounds and isomers, 89–94, Copyright (2015), with permission from Elsevier.)

for most organic compounds that compares very well with those of the conventional columns. Feng et al. (2015) explored the separation performance of grapheme oxide nanosheets as a stationary phase for the hydrogen bonding analytes such as alcohols and amines. Better separation compared to conventional columns was obtained, because the stationary phase interacts with the analytes by either hydrogen bonding, dipole–dipole, or dispersive interactions (Feng et al., 2015). In different studies, moreover, a remarkable feature of the new-generation stationary phases is their enhanced ability to resolve isomers of many organic compounds. For instance, Lv et al. (2015) investigated the potential of CTV as a stationary phase for GC separations. An important feature of the stationary phase was the excellent selectivity for halogenated compounds and positional and geometrical isomers (Lv et al., 2015), as shown in Figure 1.2. In a different study, Shi et al. (2001) found cyclodextrin phenyl carbamate to have good separation ability when it comes to disubstituted benzene isomers. Concisely, these developments in stationary phase will improve GC separations in the future, although their application in the environmental analysis of organic pollutants now is still scarce.

1.3.1 Fast Gas Chromatography

A major challenge for analytical chemists has always been to increase the sample throughput and minimize time wasted during analysis. Conventional capillary columns are as long as 30 m, thus increasing the analysis time. Thus, fast GC seeks to maintain the efficient separation of the GC process by altering the analysis time. Most of these changes have been done on the column itself. Fast GC utilizes a number of approaches outlined as follows:

- *Reduced column length and narrow internal diameter*: Most fast GC columns do not exceed 20 m in length, and their internal diameters vary from 0.10 to 0.18. The narrow internal diameters increase the separation efficiency by providing a higher signal-to-noise ratio, leading to higher sensitivity. Furthermore, less band broadening occurs in the narrow 1D columns because the analytes are diluted in a small volume of carrier gas (Banerjee and Utture, 2015).
- *Fast temperature programming*: This can be achieved using conventional ovens, resistive heating, or microwave ovens. To ensure that fast GC is achieved, the rate of temperature programming must be fast and so must the cooldown and equilibration times. This is where conventional ovens have a limitation because they can achieve rates of only 1–2°C s^{-1}, which is very slow (Mastovska and Lehotay, 2003). Resistive heating can achieve temperature programming rates of up to 20°C s^{-1} through the heating of a metal that encases the column and then determining the temperature by resistance measurements (Mastovska et al., 2001). Compared to the conventional oven, it offers rapid cooldown rates that result in higher sample throughput and very good repeatability of retention time is obtained. Furthermore, it also offers improved peak capacity and peak width compared to an isothermal separation (Reid et al., 2007). However, faster temperature programming leads to higher compound elution temperature, decreased separation efficiency, and greater thermal breakdown of susceptible analytes. For this reason, it is often combined with other techniques such as using a microbore column and a thin film of stationary phase in order to reduce the analysis time (Xu et al., 2008).
- *Altered stationary phase*: The use of a thin film of stationary phase ensures the rapid partitioning of analytes back into the carrier gas stream. This avoids band broadening. A column with a narrow internal diameter with a thin film of stationary phase has limited sample capacity compared to a conventional column. For this reason a smaller amount of sample is injected onto the column to avoid distorted peak shapes (Banerjee and Utture, 2015).
- *Higher flow rate of carrier gas*: A faster carrier gas flow rate causes the analytes to travel quickly through the column, leading to reduced analysis times. However, the carrier gas velocity should be optimized, as deviations might lead to reduced separation efficiency. Flow rates above the optimum value result in a reduced signal-to-noise ratio. On the other hand, lower velocities result in poor peak shapes and longer run times (Banerjee and Utture, 2015).
- *Microbore columns*: The specifications of the column are outlined in Table 1.1. Fast GC with microbore columns provides improved separation efficiency at reduced analysis times compared to conventional capillary columns (Húsková et al., 2009). However, their drawback is low sample capacity, which may cause band broadening, tailing, and ghost peaks (Kirchner et al., 2005).

TABLE 1.1
Classification of Capillary Columns

Category	Column Diameter Range (mm)	Standard Commercial Column Diameters (mm)	Max. Flow Rate ($mL\ min^{-1}$)
Megabore	≥0.5	0.53	≥660
Wide bore	≥0.3 to <0.5	0.32, 0.45	≥85 to <660
Narrow bore	≥0.2 to <0.3	0.20, 0.25, 0.28	≥17 to <86
Microbore	≥0.1 to <0.2	0.10, 0.15, 0.18	≥1 to <17
Sub-microbore	<0.1	Various	<1

Source: Reproduced from *Journal of Chromatography A*, 1000, Mastovska, K., and Lehotay, S.J., Practical approaches to fast-gas chromatography, 153 180, Copyright (2003), with permission from Elsevier.

1.3.2 Comprehensive Two-Dimensional Gas Chromatography

The use of multidimensional gas chromatography (MDGC) coupled to MS is on the increase. MDGC, also known as 2D-GC, is a separation technique that relies on two columns for enhanced resolving power. The main approaches use the same hardware, but operate differently, viz., heart-cutting 2D-GC and comprehensive two-dimensional gas chromatography (GC×GC). Of these two techniques, many reviews have been published on the latter, covering areas such as principles and developments in instrumentation (Adahchour et al., 2005; Mondello et al., 2008), detection techniques (Adahchour et al., 2006a), and applications (Adahchour et al., 2006b,c; Mitrevski et al., 2011). This shows that the applications of the latter are more than those of the former. This is partly because GC×GC allows solutc band concentration, which consequently enables detection of trace volatile substances with minimum interference from matrix components (Mondello et al., 2008).

Heart-cutting 2D-GC is famous for performing genius separations using the so-called heart-cutting technique, which is the passing of only a small fraction of the eluate from the primary column to the secondary column with the aid of an interface that switches on and off periodically. It follows that the technique offers high resolving power to a limited number of analytes (Seeley and Seeley, 2013). However, major drawbacks of this technique stem from the partial analysis of the sample in the two different dimensions and to the increased analysis time, thus not improving on the limitations of 1D-GC. Furthermore, the instrumentation is very complex and requires great skill to operate.

Liu and Philips (1991) overcame these limitations by producing a high-frequency heart-cutting separation of the entire sample and introduced GC×GC or 2D-GC. Compared to heart-cutting 2D-GC, where a fraction of the analyte from the primary column is separated in the secondary column, in GC×GC, the entire effluent of the primary column can pass to the secondary column, where it is also separated. The combination of the nonpolar primary column and the polar secondary column through the thermal modulator provides adequate separation efficiency for complex mixtures with a large number of compounds in a single analytical run (Seeley and Seeley, 2013). The high separation efficiency lies in the fact that the compounds are analyzed twice—that is, in both the primary and secondary columns. Moreover, the nonpolar primary column separates compounds by dispersive forces and not intermolecular forces, thus increasing the separation efficiency (Gorecki et al., 2006). The modulator transfers part of the effluent from the primary column in the form of a pulse. The time between the transfers is called the modulation period. Modulation results in cryofocusing of the narrow bands of each compound, leading to very narrow signals eluting from the secondary column (Banerjee and Utture, 2015). For this reason, 2D-GC is coupled to very fast detectors such as the time-of-flight (TOF) mass analyzers that were dominant in previous years. However, because of the high cost of TOF analyzers, many authors have recently made efforts to couple 2D-GC to quadrupole (Purcaro et al., 2010) and triple quadrupole (QqQ) mass analyzers (Tranchida et al., 2013). They have been successful because the mass analyzers are now capable of operating at high scan speeds that are comparable to those of the TOF ones. Most recently, Mitrevski and Marriot (2014) discovered that 2D-DC and a quadrupole time-of-flight (QTOF) mass analyzer complement each other in the analysis of isobaric compounds. For simple cases where QTOF mass spectrometry fails to differentiate isobaric compounds, 2D-GC is able to do so. In addition, where 2D-GC fails in separation, a QTOF mass analyzer makes use of mass filtering, and in the case of coeluting isobaric compounds, a QTOF mass analyzer can be used in tandem with mass analysis (tandem mass spectrometry [MS/MS] mode) (Mitrevski and Marriot, 2014).

A distinct advantage of 2D-GC over 1D-GC is the ease with which analyte identification can be carried out, as each peak is described by two retention times as compared to one in the latter. In addition, chemometric tools can also assist in peak identification, especially when chromatographic conditions such as oven temperature may change from run to run (Gorecki et al., 2006). A comprehensive review published by Pierce and Mohler describes the chemometrics used in terms of unsupervised, supervised, preprocessing, resolution, and image analysis algorithms (Pierce and

Mohler, 2012). In addition, they also discussed the factors that affect the conversion of data into useful information.

Many authors have reported the use of 2D-GC coupled to MS in environmental analysis of organic pollutants. Megson et al. used 2D GC coupled to time-of-flight mass spectrometry (TOFMS) to fingerprint PCBs in environmental samples (Megson et al., 2013). In a different study, chlorinated and brominated PAHs were analyzed in environmental samples such as soil by using 2D-GC coupled to high-resolution time-of-flight mass spectrometry (HR-TOFMS) with very high selectivity (Ieda et al., 2011). Another method was reported for ultratrace analysis of organochlorine pesticides in water by using 2D-GC coupled to HR-TOFMS (Ochiai et al., 2011). A review was published on the application of 2D-GC in the analysis of illicit drugs (Mitrevski et al., 2011). Another review was published to summarize some of the applications of 2D-GC/MS in the analysis of polybrominated diphenyl ethers (PBDEs) (Król and Namie, 2012). Lima Gomes et al. (2013) analyzed for steroids, caffeine, and methylparaben in water samples by using automated solid-phase microextraction (SPME) with 2D-GC–TOF-MS.

1.3.3 Low-Pressure Gas Chromatography

Low-pressure GC uses a short microbore column at the injector side, connected to a zero dead volume connector to a short megabore column (10 m × 0.53 mm) to be used with higher gas velocities. Injection is carried out at atmospheric pressure while the column is operated at lower pressures that are compatible with different mass analyzers (Ravindra et al., 2008). Consequently, the helium carrier gas becomes more viscous and behaves more like hydrogen, thereby shifting the optimal flow velocity from the Van Deemter equation, which increases the flow rates (Koesukwiwat et al., 2010). The increased flow rates result in reduced interaction with the active sites in the inlet (Banerjee and Utture, 2015). The advantages of low-pressure GC compared to those of conventional GC include shorter run times, enhanced sensitivity, reduced peak tailing, and increased ruggedness (due to wider column and thicker film) (Koesukwiwat et al., 2011). In addition, they also offer reduced degradation of thermally labile analytes and high sample loadability (Dirtu et al., 2008). However, the major drawback is that the high flow rates used may increase the pressure in the ion source to an intolerable limit, which may result in damage of the filament, especially when chemical ionization (CI) is used (Ravindra et al., 2008). Low-pressure GC has been applied in environmental analysis by using different mass analyzers such as single quadrupole (Hájková et al., 2007), ion trap (IT) (Joos et al., 2003; Ravindra et al., 2006), GC-QqQ (Gonzalez-Rodriguez et al., 2002; Fernandes et al., 2014), and TOF mass analyzers (Cajka et al., 2008; Lehotay et al., 2011).

1.4 IONIZATION TECHNIQUES

1.4.1 Electron Impact Ionization

In electron impact ionization (EI), the gaseous analyte is bombarded by a beam of electrons at 70 eV. Consequent ionization of molecules takes place by the loss of one electron. The use of 70 eV electrons results in transfer of sufficient energy, which results in structurally informative fragmentation and the generation of reproducible mass spectra. Consequently, because of this benefit, extensive libraries such as those of National Institute of Standards and Technology (NIST) and Wiley are compiled for 70 eV EI spectra (Banerjee and Utture, 2015). Thus, electron ionization is preferred for identifying unknowns, determining molecular structure, and confirming target the compound's identity through consistent ion abundance ratios (Banerjee and Utture, 2015). In EI, the ion source temperature is an important parameter that influences the extent of analyte ionization and fragmentation. It is usually set at a high value of approximately 220–240°C in order to increase ionization efficiency (Gómez et al., 2009; Prieto et al., 2010). However, it is important to note that the optimum value of ion source temperature varies with the instrument vendor.

1.4.2 Chemical Ionization

The extensive fragmentation of EI can also be a disadvantage, as it leaves no trace of the molecular ion, making it difficult to determine the molecular weight (Hajslova and Cajka, 2007). For this reason, CI, which is a soft ionization technique and does not produce extensive fragmentation, is used as an alternative. CI operates in two modes—namely, positive chemical ionization (PCI) and negative chemical ionization (NCI). In PCI, the ion source is filled with a reagent gas, such as methane, at a relatively high pressure—that is, ionized using EI to produce reagent ions. The sample molecules are ionized by the reagent gas, producing pseudomolecular ions that have less internal energy and are less prone to fragmentation than the molecular ions generated by EI. Thus, because of the reduced fragmentation in PCI, intense and more specific molecular ions are obtained and, consequently, unambiguous identification can be obtained. Furthermore, lower detection limits and increased sensitivity are obtained due to the reduction in background noise (Banerjee and Utture, 2015).

The principle of NCI is the same as that of electron capture detection, whereby a low-energy electron is captured by an electronegative sample molecule, forming a molecular anion that may undergo fragmentation, depending on its structure, thus generating a mass spectrum. For efficient ionization to be obtained, optimization of essential parameters (ionizing gas and its pressure, source temperature, electron energy, and emission current intensity of the beam) needs to be done (Amendola et al., 2002). It is worth noting that a relatively high pressure of the reagent gas is important in ensuring a sufficiently high number of ion–molecule collisions during the dwell time of the reactants in the ion source. On the other hand, extremely low electron energy and emission current decrease the ionization efficiency (Zhao et al., 2007). Furthermore, methane is commonly used as a reagent gas because it does not produce any negative ions, thus ensuring a low background signal (Worton et al., 2008). It follows that the advantages of NCI over CI and EI include improved sensitivity and high selectivity for analytes containing a halogen atom, a nitro group, or an extended aromatic ring. These include PAHs (Kamiya et al., 2015), PBDEs (Li et al., 2011), alkyl nitrates, and halocarbons (Worton et al., 2008). However, despite its great selectivity, a major disadvantage is that NCI is limited to a few chemical classes of compounds such as those listed earlier. In addition, it suffers from lack of commercial spectral libraries (Lambropolou and Hela, 2015). For this reason, another soft ionization technique, atmospheric pressure ionization (API), can also be used.

1.4.3 Atmospheric Pressure Ionization

Having been introduced by Horning et al. in 1977, this technique has been modified by several research groups to be what it is nowadays (Horning et al., 1977). For instance, McEwen and McKay (2005) and Schiewek et al. (2008) finally produced source designs that led to commercially available atmospheric pressure gas chromatography–mass spectrometry (AP-GC-MS) instruments. Ionization in the gas phase is generated using a corona needle by charge transfer or protonation using a reagent gas such as nitrogen, and, consequently, molecular and quasimolecular ions are generated. Charge transfer is carried out under dry conditions in the absence of a modifier, and it results in increased fragmentation. It is suitable for nonpolar compounds such as PCBs and chlorinated pesticides. On the contrary, proton transfer is carried out under wet conditions, in the presence of a modifier such as water, leading to better retention of the molecular ion. Likewise, it is excellent for compounds of moderate polarity such as organophosphates.

Portoles et al. (2010) investigated the ionization behavior of 100 GC-amenable pesticides in food by using AP-GC-QTOF-MS regarding the formation of either the molecular ion or the protonated molecule under charge transfer or protonated transfer. Thus, the presence of the molecular ion/protonated molecule allowed the rapid application of a screening method in a posttarget way, easily generating a list of compounds to be monitored by making use of their molecular formula. The presence of a QTOF mass analyzer made it possible to perform additional tandem mass analysis experiment to assist in the confirmation of the identity of detected compounds (Portoles et al., 2010).

In a different study, the same research group evaluated the potential of AP coupled to GC/MS/MS. For this purpose, 25 pesticides were selected to study their ionization behavior and to evaluate the advantages compared to EI. Portoles et al. (2012) found out that the molecular ion was present and highly abundant for most of the compounds with a noticeable difference in fragmentation compared to EI ionization as shown in Table 1.2. Moreover, the addition of the water as a modifier favored the formation of the protonated molecules such that the protonated molecules became the base peak of the spectrum for most compounds, leading to enhanced sensitivity using the QqQ mass analyzer as shown in Table 1.2. For instance, Figure 1.3 shows a comparison of buprofezin spectrum using EI, AP under charge transfer, and AP under proton transfer conditions.

Recently, Pintado-Herrera et al. (2014) explored the possibility of analyzing for regulated and emerging contaminants in water bodies by GC-AP-GC-TOF-MS. They compared the differences in

TABLE 1.2
Behavior of the Molecular Ions of Selected Pesticides under EI and APCI Modes and Single Quadrupole Mass Analysis

					APCI		APCI +		
t_r	Compound	Molecular Formula	M	EIM+	M+	MH+	H_2O MH+	EI Transitions	
4.70	Dichlorvos	$C_4H_7Cl_2O_4P$	220			+++	+++	185 > 93	109 < 79
5.97	Mevinphos	$C_7H_{13}O_6P$	224		++	+	++	127 > 109	192 > 127
6.96	Molinate	$C_9H_{17}NOS$	187	+		+++	+++	187 > 126	126 > 55
8.00	Dicrotophos	$C_8H_{16}NO_5P$	237	+	++	++	+++	127 > 109	127 > 95
8.24	Monocrotophos	$C_7H_{14}NO_5P$	223	+	++	++	+++	192 > 127	192 > 164
8.95	Terbufos	$C_9H_{21}O_2PS_3$	288				+	231 > 129	231 > 175
9.80	Phosphamidon	$C_{10}H_{19}ClNO_5P$	299		+	++	+++	127 > 109	264 > 127
9.76	Endosulfan ether	$C_9H_6Cl_6O$	340	+	+	++	++	272 > 237	239 > 204
9.94	Chlopyrifos methyl	$C_7H_{17}Cl_3NO_3PS$	321			++	+++	288 > 93	197 > 169
10.77	Chlopyrifos	$C_9H_{11}Cl_3NO_3PS$	349		+	+++	+++	199 > 171	316 > 260
10.85	Aldrin	$C_{12}H_8Cl_6$	362		+	+	++	261 > 191	263 > 193
11.39	Isodrin	$C_{12}H_8Cl_6$	362		++	++	++	193 > 157	195 > 123
11.56	Chlofernvinphos	$C_{12}H_{14}Cl_3O_4P$	358			++	+++	267 > 159	323 > 267
11.56	Oxychlordane	$C_{10}H_4Cl_8O$	420		+	+	++	185 > 121	235 > 141
11.56	Heptachlor epoxide B	$C_{10}H_5Cl_7O$	386		+	+	++	355 > 265	351 > 261
12.23	Endosulfan I	$C_9H_6Cl_6O_3S$	404			++	++	239 > 204	272 > 237
12.72	Buprofezin	$C_{16}H_{23}N_3OS$	305	+	++	+++	+++	105 > 77	172 > 115
12.73	Dieldrin	$C_{12}H_8Cl_6O$	378	+	++	++	++	263 > 193	261 > 191
13.10	Endrin	$C_{12}H_8Cl_6O$	378		++	++	++	263 > 193	261 > 191
13.36	Ethion	$C_9H_{22}O_4P_2S_4$	384		+	+++	+++	231 > 129	231 > 175
14.01	Endosulfan sulfate	$C_9H_6Cl_6O_4S$	420		++		++	274 > 239	274 > 237
15.63	Azinphos methyl	$C_{10}H_{12}N_3O_3PS_2$	317				+	160 > 77	160 > 132
15.66	Pyriproxyfen	$C_{20}H_{19}NO_3$	321			++	+++	131 > 96	136 > 78
16.04	Fenarimol	$C_{17}H_{12}Cl_2N_2O$	330	+	++	+++	+++	251 > 139	219 > 107
16.17	Azinphos ethyl	$C_{12}H_{16}N_3O_3PS_2$	345				+	160 > 132	160 > 77

Source: *Journal of Chromatography A*, 1260, Portoles, T., L. Cherta, J. Beltran, and F. Hernandez, Improved gas chromatography–tandem mass spectrometry determination of pesticide residues making use of atmospheric pressure chemical ionization, 183–192, Copyright (2012), with permission from Elsevier.

Note: MS/MS transitions commonly used under EI mode are shown. +, very small peak; ++, clearly identifiable peak (>20%); +++, base peak (or >80%).

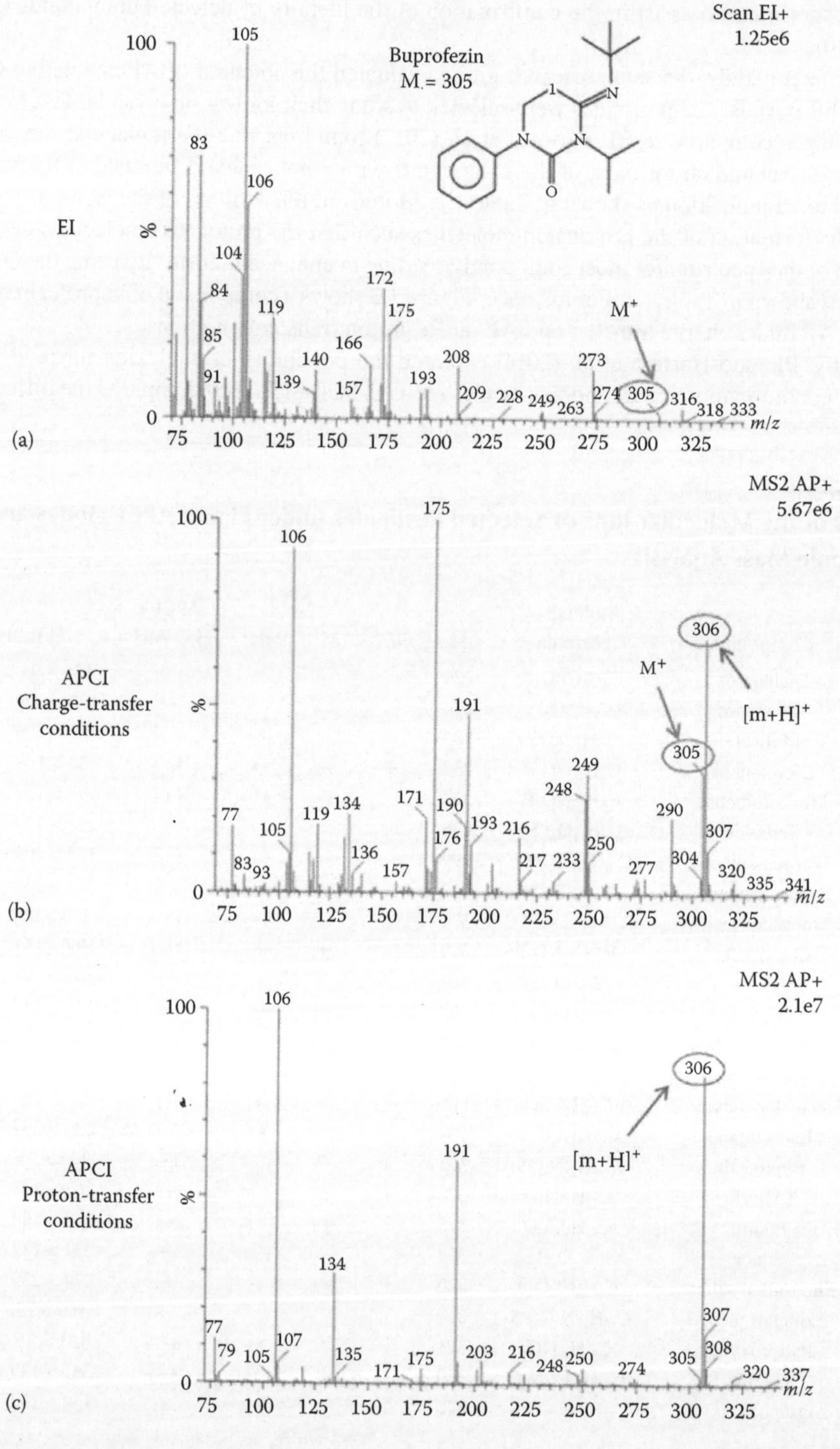

FIGURE 1.3 Comparison of buprofezin spectra using (a) an EI source, (b) an APCI source under charge transfer conditions, and (c) an APCI source under proton transfer conditions. (Reprinted from *Journal of Chromatography A*, 1260, Portoles, T., L. Cherta, J. Beltran, and F. Hernandez, Improved gas chromatography–tandem mass spectrometry determination of pesticide residues making use of atmospheric pressure chemical ionization, 183–192, Copyright (2012), with permission from Elsevier.)

ionization between AP-GC and EI. They discovered that extensive fragmentation was obtained for the macrocyclic fragrance Musk R1 as compared to AP-GC. In addition, the molecular ion at *m/e* 257 could not be detected, and some of the *m/z* ratios in the mass spectrum were common for other target and nontarget compounds, which could lead to false positives in environmental samples. On the contrary, when AP-GC is used, for Musk R1, the highest *m/z* was 257, but with the exception of other abundant fragment ions. Nevertheless, EI proved to be a better option for ionizing stable compounds such as PAHs. On the other hand, AP-GC provided better ionization for labile compounds such as fragrances. Lastly, because different cone voltages can be applied, progressive fragmentation of analytes can be carried out. This is necessary for some macrocyclic fragrances, which could not be determined by EI (Pintado-Herrera et al., 2014).

1.5 MASS ANALYZERS

The mass analyzer is the heart of the mass spectrometer because this is where mass analysis takes place. Mass analyzers can be classified according to accuracy, resolution, mass range, tandem analysis capabilities, and scan speeds.

- *Mass resolution* is the ability of a mass analyzer to separate ions of a similar mass. It can also be defined as the smallest difference between two equal magnitude peaks, so that the valley between them is a fraction of the peak height.
- *Mass resolving power* is the observed mass divided by the difference between two masses that can be separated.
- *Mass range* is the range of mass-to-charge ratios (*m/z*) over which a mass analyzer can separate and detect ions. Mass analyzers coupled to liquid chromatographic systems have increased mass ranges up to several orders.
- *Mass accuracy* is the deviation between measured mass (accurate mass) and calculated mass (exact mass) of an ion expressed as an error in millidaltons or parts per million (ppm). Unit-resolution mass analyzers provide a mass accuracy of approximately 0.1–0.2 Da. while high-resolution mass analyzers operate at a mass accuracy of less than 5 ppm.
- *Acquisition speed* is the time required for recording a mass spectrum or selected ions. The acquisition speed is expressed in daltons for low-resolution mass spectrometers and in hertz for high-resolution mass analyzers.
- *Tandem mass analysis* is the ability to carry out more than one stage of mass analysis either in space or in time.
- *Sensitivity* is the minimal signal-to-noise ratio at a given concentration of analyte.

1.5.1 Single Quadrupole Mass Analyzer (GC-MS)

The single quadrupole mass analyzer consists of four parallel rods of circular cross section that are connected in pairs and a combination of radio frequency and direct current voltage is applied between the rods. Ions will travel down the quadrupole between the rods, and for a given ratio of voltages, some will reach the detector, while others will collide with the rods and will not reach the detector (Hird, 2008). This process is called mass filtering, and it is wholly dependent on the voltage applied. A single quadrupole mass analyzer can be operated in either full-scan or selected ion-monitoring (SIM) mode. In full-scan mode, a wide range of ions is monitored, and this mode is particularly useful for identifying the components of a compound by using a mass spectrum. The latter, as its name suggests, monitors ions of a limited mass range, thus offering better sensitivity because monitoring of only a few ions takes place, thus increasing the acquisition time but compromising on the quality of the mass spectra (Hajslova and Cajka, 2007). However, in previous years this mode had the disadvantage of being complicated and difficult to maintain when the list of target analytes was increased. Instrument vendors such as Agilent have since solved this problem by use of

TABLE 1.3
Applications of GC/MS in Environmental Analysis

Group of Analytes	Environmental Matrix	Analytical Technique	Sensitivity	Reference
12 PCPs, 2 pharmaceuticals	Fish tissue	PLE-GC-EI-IT-MS	MDL: 1.2–38 ng g^{-1} MDL: 3.7–18 ng g^{-1}	Subedi et al. (2011)
Chlorinated and brominated PAHs	Soil	2D-GC-HR-TOFMS	LOD: 0.08–3.2 pg	Ieda et al. (2011)
16 OCPs	River water	SBSE-TD-2D-GC-HR-TOFMS	10–44 pg L^{-1}	Ochiai et al. (2011)
11 steroids, caffeine and methylparaben	Water	SPME-2D-GC-TOFMS	0.02–100 g L^{-1}	Lima Gomes et al. (2013)
2 OPPs, 2 fungicides	Soil	MAE-GC-EI-MS-SIM	0.10–0.12 ng g^{-1}	Merdassa, Liu, and Megersa (2013)
16 OCPs	Drinking water	SPME-GC-EI-MS (SIS) SPME-GC-EI-MS/MS	LOD: 0.2–6.6 ng L^{-1} LOD: 0.3–7.6 ng L^{-1}	Lara-Gonzalo et al. (2010)
10 OCPs	Sediments	MAE/HS-SPME GC-EI-IT-MS	LOD: 0.005–0.11 ng g^{-1}	Carvalho et al. (2008)
19 OCPs, 6 OPPs, 6 herbicides, 7 PCBs, 16 PAHs, 3 octyl/nonyl phenols, pentachlorobenzene	Water	SPE-GC-EI-QqQ-MS SPE-GC-NCI-MS	LOD: 1–150 ng L^{-1} LOQ: 25–250 ng L^{-1} LOD: 0.2–190 ng L^{-1} LOQ: 25–250 ng L^{-1}	Pitarch et al. (2007)
41 PBDEs	Fish	GC-QqQ-MS GC-HR-MS	LOD: 0.04–41 pg g^{-1} LOD: 5–85 pg g^{-1}	Mackintosh et al. (2012)
24 PAHs	Wastewater	SBSE-GC-EI-QqQ-MS	LOD: 0.002–0.01 μg L^{-1} LOQ: 0.005–0.100 μg L^{-1}	Barco-Bonilla et al. (2011)
40 pesticides	Airborne particulate matter	MAE-GPC-GC-EI-QqQ-MS	LOQ: 1.32–39.47 pg m^{-3}	Coscolla et al. (2011)
Chlorophenols, alkylphenols, nitrophenols, cresols	Soils	QuEChERS-GC-EI-QqQ-MS	LOD: 0.1–50 μg kg^{-1} LOQ: 1–100 μg kg^{-1}	Padilla-Sanchez et al. (2010)
17 pesticides	Environmental water and wastewater	DLLME-PTV-LVI-GC-QqQ-MS	0.5–18 ng L^{-1}	Carro et al. (2012)
44 pesticides, 13 PAHs	Wastewater	LLE-GC-QqQ-MS SPE-GC-QqQ-MS HS-SPME-GC-QqQ-MS	LOQ: 0.03–5.1 ng L^{-1} LOQ: 0.0–99 ng L^{-1} LOQ: 0.1–148.5 ng L^{-1}	Robles-Molina et al. (2013)
13 PhACs, 18 plasticizers, 8 PCPs, 9 acid herbicides, 8 triazines, 10 OPPs, 5 phenylureas, 12 OC biocides, 9 PAHs, 5 benzothiazoles and triazoles	River water	2D-GC-TOFMS	LOD: 0.5–100 ng L^{-1} LOQ: 2–185 ng L^{-1}	Matamoros, Jover, and Bayona (2010)

Note: DLLME, dispersive liquid–liquid microextraction; HS-SPME, headspace solid-phase microextraction; LLE, liquid–liquid extraction; LOD, limit of detection; LOQ, limit of quantification; MAE, microwave-assisted extraction; MDL, method detection limit; OC, organochlorine; OPPs, organophosphorus pesticide; PhACs, pharmaceuticals; PLE, pressurized liquid extraction; PTV-LVI, programmed temperature vaporization–large volume injection; SBSE, stir bar sorptive extraction; SIS, selected ion storage; SPE, solid-phase extraction; SPME, solid-phase microextraction.

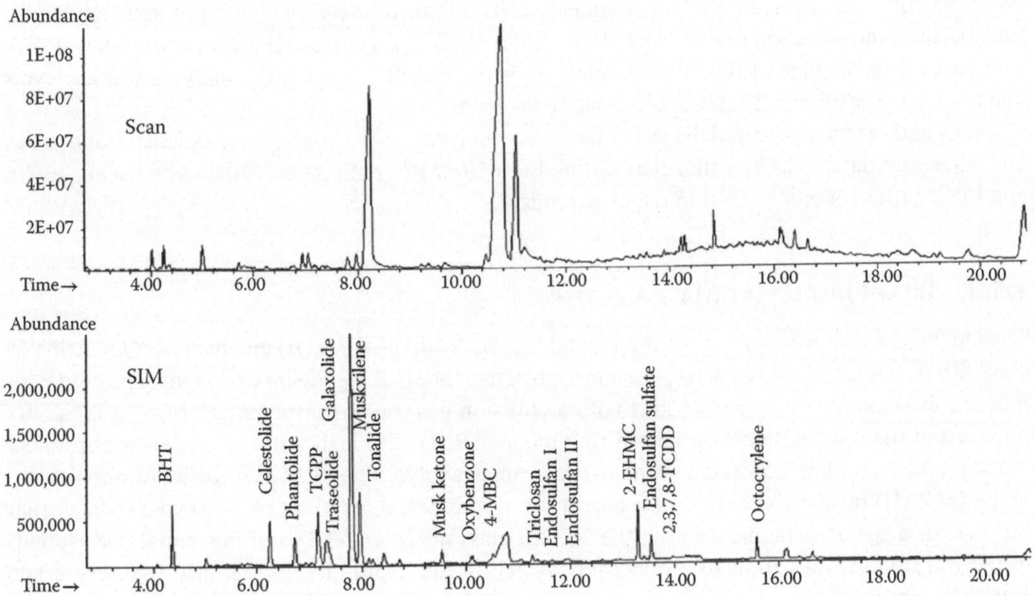

FIGURE 1.4 GC-MS chromatogram in the SIM/scan mode corresponding to a spiked wastewater effluent sample at 50 ng/L. (Reprinted from *Journal of Chromatography A*, 1216, Gómez, M. J., M. M. Gómez-Ramos, A. Agüera, M. Mezcua, S. Herrera et al., A new gas chromatography/mass spectrometry method for the simultaneous analysis of target and non-target organic contaminants in waters, 4071–4082, Copyright (2009), with permission from Elsevier.)

the retention time locking (Almeida et al., 2007). For this reason, many publications have reported on its use for quantitation purposes of different environmental pollutants such as PPCPs (Bisceglia et al., 2010), PBDEs (Gorga et al., 2013), PCBs (Zhou et al., 2010), and polycyclic aromatic hydrocarbons and pesticides (Borras et al., 2011; Merdassa et al., 2013; Tankiewicz et al., 2013), among others. Table 1.3 summarizes some of the applications of GC-MS in environmental analysis.

In some cases where target and nontarget analysis of contaminants is required, the SIM/scan mode is used. Gómez et al. (2009) demonstrated the use of this technique in the analysis of 934 organic contaminants as shown in Figure 1.4. The full-scan data were analyzed using Deconvolution Reporting Software, which identifies contaminants that are buried in the chromatogram by coextracted matrix components. Use of the retention time locking system was made for all compounds. The limit of detection obtained was lower in SIM mode than in full-scan mode (Gómez et al., 2009). In addition, by using the NIST and Automated Mass Spectral Deconvolution and Identification System (AMDIS) libraries, 12 new compounds were identified.

1.5.2 Ion Trap Mass Analyzer

From the column, the analytes are introduced to the IT mass analyzer through the transfer line. The IT mass analyzer consists of an entrance endcap, a ring electrode, and an exit endcap electrode. In between is a cavity where molecular and fragment ions are stored and stabilized. They travel in well-defined orbits governed by voltages applied between the ring electrode and the endcap electrodes. The carrier gas used is helium because it is light and does not degrade MS resolution.

An IT mass analyzer can be operated in full-scan, SIM, or MS/MS mode. The MS/MS mode provides higher sensitivity and selectivity; however, it requires careful optimization of parameters (Banerjee and Utture, 2015). In MS/MS mode, sample molecules are ionized in the ion source;

precursor ions are isolated and then fragmented. An IT mass analyzer is referred to as tandem in time because the same ion region is used for all the MS/MS processes (Hajšlová and Cajka, 2007.). The major disadvantage of the IT mass analyzer is the inability to quantify analytes at trace levels due to its low sensitivity (Lambropolou and Hela, 2015).

Many authors have reported the use of this mass analyzer for the analysis of pesticides (Gonçalves and Alpendurada, 2005; Carvalho et al., 2008; Lara-Gonzalo et al., 2010), PAHs (Leite et al., 2008), and PPCPs (Subedi et al., 2011) in the environment.

1.5.3 Triple Quadrupole Mass Analyzer

The QqQ mass analyzer consists of three quadrupoles, with the first (Q1) and the last (Q3) acting as mass filters, while the second is responsible for fragmentation of the precursor ion through collision-induced dissociation of precursor ions with a collision gas such as nitrogen or argon. It is mainly operated in the selected reaction-monitoring mode (SRM), although other modes such as product ion scan, precursor ion scan and neutral loss scan are available. The SRM (multireaction-monitoring mode [MRM]) involves selection of a parent ion in the first mass analyzer, followed by a similar process for a specific fragment ion in the second mass analyzer. The resulting signal corresponds to the transition from parent to product ion, which is free from any interference (Kotretsou and Koutsodimou, 2006).

The advantages of this mass analyzer are improved selectivity, improved sensitivity, wider dynamic linear range, and reduction in analysis time (Pitarch et al., 2007). Furthermore, the enhanced selectivity is important in the case of coeluting compounds where unambiguous identification and confirmation of coeluting peaks can be achieved through monitoring unique MRM transitions (Banerjee and Utture, 2015). However, homologous PBDEs and other matrix components can yield the same precursor and product ions (Mackintosh et al., 2012). This problem has been solved by the use of curved quadruples, which offer longer flight paths; hence, they can be used for high-resolution selection of *m/z*, which consequently allows for identification of compounds where coelution is likely to exist (Banerjee and Utture, 2015).

The application of GC-QqQ-MS was reported for the screening, quantification, and confirmation of 50 compounds belonging to different chemical classes, which are also included in the framework on European Water Policy (Pitarch et al., 2007). Another method was developed based on stir bar sorptive extraction capable of extracting simultaneously 24 PAHs from raw wastewater followed by GC-QqQ-MS quantification (Barco-Bonilla et al., 2011). Analysis of 40 pesticides in airborne particulate matter by using the same technique was also reported (Coscolla et al., 2011). Ultratrace analysis of 73 target organic environmental contaminants in fish and fish feed with high sensitivity and selectivity was carried out with the same technique (Kalachova et al., 2013b). The same research group proved GC-QqQ-MS to be an effective tool for ultratrace analysis of several brominated flame retardants in fish (Kalachova et al., 2013a). Another research group developed and optimized a multiresidue method for the determination of organochlorine pesticides in fish feed sample, and the performance parameters such as selectivity, precision, and recovery complied with European regulations (Nardelli et al., 2010). In a different study, a multiresidue method was developed and validated for the simultaneous analysis of 34 PAHs and phthalic acid esters (PAEs) in soil at trace levels. Limits of detection for PAEs were less than 0.84 $\mu g\ kg^{-1}$, and those for PAHs were less than 0.51 $\mu g\ kg^{-1}$ (Liao et al., 2010). Successful validation was carried out for 13 phenolic compounds in soil by using GC-QqQ-MS (Padilla-Sanchez et al., 2010). Carro et al. (2012) reported a solventless procedure for the determination of 17 pesticides in environmental water and wastewater using dispersive liquid–liquid microextraction coupled with GC-MS/MS with large volume PTV injection. Lastly, Robles-Molina et al. (2013) evaluated the analytical performance of three sample preparation techniques—namely, liquid–liquid extraction, SPE, and SPME for the GC-QqQ-MS determination of multiclass organic pollutants in wastewater.

1.5.4 Time-of-Flight Mass Analyzer

As its name suggests, the TOF analyzer is based on the movement of ions possessing the same energy but different masses traveling through a flight tube at different velocities. Thus, the lighter ones arrive before the heavier ones. Thus, the measurement of the TOF allows the determination of the mass. The cycle is repeated with the rate depending on the flight time with the highest mass to be recorded. Mass resolution is enhanced by the use of a reflectron, which is a series of ring rods with increasing voltage that create retarding fields. The higher energy ions reaching the reflectron area penetrate more deeply inside, and this results in the extension of the time until they are reflected. Due to this phenomenon, ions of the same *m/z* but with different initial energies hit the detector at almost the same time. The flight of the ions separated in a field-free region is proportional to the square root of the respective *m/z* value. GC-TOF-MS instruments can be further classified as high speed or high resolution.

High-speed TOF have a high spectral acquisition rate of 100–500 spectra per second, but provides only unit mass resolution. Hence, they allow the separation of overlapping peaks by using automated mass spectral deconvolution of overlapping signals. For this reason, they are ideal for fast GC analysis and GC×GC analysis. In the latter, they provide fast chromatographic separation in the second dimension, resulting in very narrow peaks with peaks of 50–600 ms at the baseline that can be constructed only by fast detectors.

HR-TOFMS instruments provide high resolution (>7000 full width at half maximum) with mass accuracy of ±5–10 ppm, but they have moderate spectral acquisition rates of up to 20 full spectra stored per second. Thus, due to the high resolution, there is less interference of the signal from matrix components. In addition, it can perform extracted ion chromatogram using a narrow mass window. This excludes a large amount of background noise, thus improving the signal-to-noise ratio (Portolés et al., 2007).

The QTOF mass analyzer is a hybrid analytical technique that can operate in both the MS (scan) and the MS/MS mode. It can be viewed as a QqQ system that has the last quadrupole replaced by a TOF analyzer. Compared to the QqQ mass analyzer, it has the capability of determining the accurate mass of the fragment ions generated in the collision cell, and this feature is important for structural elucidation of unknowns (Pico, 2008). The main advantage of this system is unambiguous identification provided by MS/MS (Portoles et al., 2010).

Ieda et al. (2011) described a method for the analysis of chlorinated and brominated PAHs in soil by GC×GC coupled to HR-TOFMS. The method allowed a highly selective group type of analysis in 2D chromatograms with very narrow mass windows (e.g., 0.02 Da) (Ieda et al., 2011). In a different study, GC coupled to HR-TOFMS was evaluated for the detection of PBDEs in sediments and fish by using EI and NCI. The method enhanced the detectability of the target analytes and enabled quantification of minor PBDE congeners and improved characterization of sample contamination patterns (Cajka et al., 2005). The advantages and limitations of GC×GC TOF-MS for the simultaneous screening of 97 contaminants in river water was demonstrated by Matamoros et al. (2010).

1.6 FUTURE TRENDS AND CONCLUSION

Many advances in terms of instrumentation have characterized GC-MS analyses, which include injection techniques, separation, ionization techniques, and mass analyzers (Ballesteros-Gomez and Rubio, 2011). All these advances coupled to software developments have increased the scope of GC-MS in the environmental analysis of pollutants. However, the complexity of matrices makes the identification of target analytes difficult. For this reason, GC×GC continues to be the method of choice in environmental analysis, as it can separate matrix components from target analytes. Furthermore, the recent advances in mass analyzers such as use of high-resolution and accurate mass measurement make the process of quantification, identification, and confirmation of analytes more accurate. Conclusively, GC-MS will continue to be a useful tool in environmental analysis of pollutants.

REFERENCES

Adahchour, M., Beens, J., Vreuls, R.J., Brinkman, U.A.T. 2006a. Recent developments in comprehensive two-dimensional gas chromatography (GC × GC): II. Modulation and detection. *Trends in Analytical Chemistry* 25: 540–553.

Adahchour, M., Beens, J., Vreuls, R.J., Brinkman, U.A.T. 2006b. Recent developments in comprehensive two-dimensional gas chromatography (GC × GC): III. Applications for petrochemicals and organohalogens. *Trends in Analytical Chemistry* 25: 726–741.

Adahchour, M., Beens, J., Vreuls, R.J., Brinkman, U.A.T. 2006c. Recent developments in comprehensive two-dimensional gas chromatography (GC × GC): IV. Further applications, conclusions and perspectives. *Trends in Analytical Chemistry* 25: 821–840.

Adahchour, M., Jover, E., Beens, J., Vreuls, R.J., Brinkman, U.A.T. 2005. Twin comprehensive two-dimensional gas chromatographic system: Concepts and applications. *Journal of Chromatography A* 1086: 128–134.

Almeida, C., Serodio, P., Florencio, M., Nogueira, J.M.F. 2007. New strategies to screen for endocrine-disrupting chemicals in the Portuguese marine environment utilizing large volume injection–capillary gas chromatography–mass spectrometry combined with retention time locking libraries (LVI-GC-MS-RTL). *Analytical and Bioanalytical Chemistry* 387: 2569–2583.

Amendola, L., Botre, F., Carollo, A., Longo, D., Zoccolillo, L. 2002. Analysis of organophosphorus pesticides by gas chromatography–mass spectrometry with negative chemical ionization: A study of the ionization conditions. *Analytica Chimica Acta* 461: 97–108.

Ballesteros-Gomez, A., Rubio, S. 2011. Recent advances in environmental analysis. *Analytical Chemistry* 83: 4579–4613.

Banerjee, K., Utture, S. 2015. Recent developments in gas chromatography mass spectrometry. In *Mass Spectrometry for the Analysis of Pesticide Residues and Their Metabolites*, 1st ed., eds. Tsipi, D., Botitsi, H., Economou, A. Wiley, Hoboken, NJ, pp. 91–112.

Barco-Bonilla, N., Romero-Gonzalez, R., Plaza Bolanos, P., Moreno Fernandez, J., Garrido Frenich, A. et al. 2011. Comprehensive analysis of polycyclic aromatic hydrocarbons in wastewater using stir bar sorptive extraction and gas-chromatography coupled to tandem mass spectrometry. *Analytica Chimica Acta* 693: 62–71.

Bisceglia, K.J., Yu, J.T., Coelhan, M., Bouwer, E.J., Roberts, A.L. 2010. Trace determination of pharmaceuticals and other wastewater-derived micropollutants by solid phase extraction and gas chromatography/mass spectrometry. *Journal of Chromatography A* 1217: 558–564.

Borras, E., Sanchez, P., Munoz, A., Tortajada-Genaro, L. 2011. Development of a gas chromatography–mass spectrometry method for the determination of pesticides in gaseous and particulate phases in the atmosphere. *Analytica Chimica Acta* 699: 57–65.

Cajka, T., Hajslova, J., Kazda, R., Poustka, J. 2005. Challenges of gas chromatography-high resolution time-of-flight mass spectrometry for simultaneous analysis of polybrominated diphenyl ethers and other halogenated persistent organic pollutants in environmental samples. *Journal of Separation Science* 28: 601–611.

Cajka, T., Hajslova, J., Lacina, O., Mastovska, K., Lehotay, S.J. 2008. Rapid analysis of multiple pesticides in fruit-based baby food using programmed temperature vaporiser injection-low-pressure-high resolution time-of-flight-mass spectrometry. *Journal of Chromatography A* 1186: 281–294.

Carro, A., Fernandez, S., Racamonde, I., Garcia-Rodriguez, S., Gonzalez, P. et al. 2012. Dispersive liquid–liquid microextraction coupled with programmed temperature vaporization–large volume injection gas chromatography–tandem mass spectrometry for multiclass pesticides in water. *Journal of Chromatography A* 1253: 134–143.

Carvalho, P., Rodrigues, P.N., Alves, F., Evangelista, R., Basto, M.C. et al. 2008. An expeditious method for the determination of organochlorine pesticides residues in estuarine sediments using microwave assisted pre-extraction and automated headspace solid-phase microextraction coupled to gas chromatography mass-spectrometry. *Talanta* 76: 1124–1129.

Coscolla, C., Castillo, M., Pastor, A., Yusa, V. 2011. Determination of 40 currently used pesticides in airborne particulate matter (PM 10) by microwave assisted extraction and gas chromatography coupled to triple quadrupole mass spectrometry. *Analytica Chimica Acta* 693: 72–81.

Dirtu, A., Ravindra, K., Roosens, L., Van Grieken, R., Neels, H. et al. 2008. Fast analysis of decabrominated diphenyl ether using low-pressure gas chromatography–electron-capture negative ionization mass spectrometry. *Journal of Chromatography A* 1186: 295–301.

Dorman, F., Whiting, J., Cochran, J., Gardea-Torresdey, J. 2010. Gas chromatography. *Analytical Chemistry* 82: 4775–4785.

Falkovich, A., Rudich, Y. 2001. Analysis of semivolatile organic compounds in atmospheric aerosols by direct sample introduction thermal desorption GC/MS. *Environmental Science & Technology* 35: 2326–2333.

Fan, J., Qi, M., Fu, R., Qu, L. 2015. Performance of graphene sheets as stationary phase for capillary gas chromatographic separations. *Journal of Chromatography A* 1399: 74–79.

Feng, Y., Hu, C., Qi, M., Fu, R., Qu, L. 2015. Separation performance of graphene oxide as stationary phase for capillary gas chromatography. *Chinese Chemical Letters* 26: 47–49.

Fernandes, V., Lehotay, S.J., Geis-Asteggiante, L., Kwon, H., Mol, H.G.J. et al. 2014. Analysis of pesticide residues in strawberries and soils by GC-MS/MS, LC-MS/MS and two-dimensional GC-time-of-flight MS comparing organic and integrated pest management farming. *Food Additives & Contaminants: Part A* 31: 262–270.

Gómez, M.J., Gómez-Ramos, M.M., Agüera, A., Mezcua, M., Herrera, S. et al. 2009. A new gas chromatography/mass spectrometry method for the simultaneous analysis of target and non-target organic contaminants in waters. *Journal of Chromatography A* 1216: 4071–4082.

Gonçalves, C., Alpendurada, M.F. 2005. Assessment of pesticide contamination in soil samples from an intensive horticulture area, using ultrasonic extraction and gas chromatography–mass spectrometry. *Talanta* 65: 1179–1189.

Gonzalez-Rodriguez, M., Garrido-Frenich, A., Arrebola, F., Martinez-Vidal, J. 2002. Evaluation of low-pressure gas chromatography linked to ion-trap tandem mass spectrometry for the fast trace analysis of multiclass pesticide residues. *Rapid Communications in Mass Spectrometry* 16: 1216–1224.

Gorecki, T., Panic, O., Oldridge, N. 2006. Recent advances in comprehensive two-dimensional gas chromatography (GC×GC). *Journal of Liquid Chromatography & Related Technologies* 29: 1077–1104.

Gorga, M., Martinez, E., Ginebreda, A., Eljarrat, E., Barceló, D. 2013. Determination of PBDEs, HBB, PBEB, DBDPE, HBCD, TBBPA and related compounds in sewage sludge from Catalonia (Spain). *Science of the Total Environment* 444: 51–59.

Grisales, J.O., Lebed, P.J., Keunchkarian, S., González, F.R., Castells, C.B. 2009. Permethylated β-cyclodextrin in liquid poly(oxyethylene) as a stationary phase for capillary gas chromatography. *Journal of Chromatography A* 1216: 6844–6851. doi:10.1016/j.chroma.2009.08.023.

Gu, Q., David, F., Lynen, F., Vanormelingen, P., Vyverman, W. et al. 2011. Evaluation of ionic liquid stationary phases for one dimensional gas chromatography mass-spectrometry and comprehensive two-dimensional gas chromatographic analyses of fatty acids in marine biota. *Journal of Chromatography A* 1218: 3056–3063.

Hájková, K., Pulkrabová, J., Schůrek, J., Hajšlová, J., Poustka, J. et al. 2007. Novel approaches to the analysis of steroid estrogens in river sediments. *Analytical and Bioanalytical Chemistry* 387: 1351–1363.

Hajslova, J., Cajka, T. 2007. Gas chromatography–mass spectrometry (GC-MS). In *Food Toxicants Analysis*, ed. Pico, Y. Elsevier, Amsterdam, pp. 419–473.

Hird, S. 2008. Analysis of pesticides by chromatographic techniques coupled with mass spectrometry. In *Analysis of Pesticides in the Environment*, ed. Tadeo, J. CRC Press, Boca Raton, FL, pp. 59–95.

Ho, S., Chow, J., Watson, J., Ng, L., Kwok, Y. et al. 2011. Precautions for in-injection port thermal desorption-gas chromatography/mass spectrometry (TD-GC/MS) as applied to aerosol filter samples. *Atmospheric Environment* 45: 1491–1496.

Ho, S., Yu, J. 2004. In-injection port thermal desorption and subsequent gas chromatography-mass spectrometric analysis of polycyclic aromatic hydrocarbon and *n*-alkanes in atmospheric air samples. *Journal of Chromatography A* 1059: 121–129.

Hoh, E., Mastovska, K. 2008. Large volume injection techniques in capillary gas chromatography. *Journal of Chromatography A* 1186: 2–15.

Horning, E.C., Carroll, D.I., Dzidic, I., Haegele, K.D., Lin, S., Oertli, C.U., Stillwell, R.N. 1977. Development and use of analytical systems based on mass spectrometry. *Clinical Chemistry* 23: 13–21.

Húsková, R., Matisová, E., Hrouzková, S., Svorc, L. 2009. Analysis of pesticide residues by fast gas chromatography in combination with negative chemical ionization mass spectrometry. *Journal of Chromatography A* 1216: 6326–6334.

Ieda, T., Ochiai, N., Miyawaki, T., Ohura, T., Horii, Y. 2011. Environmental analysis of chlorinated and brominated polycyclic aromatic hydrocarbons by comprehensive two-dimensional gas chromatography coupled to high-resolution time-of-flight mass spectrometry. *Journal of Chromatography A* 1218: 3224–3232.

Joos, P., Godoi, A., De Rong, R., de Zeeuw, J., Van Grieken, R. 2003. Trace analysis of benzene, toluene, ethylbenzene and xylene isomers in environmental samples by low-pressure gas chromatography-ion trap mass spectrometry. *Journal of Chromatography A* 985: 191–196.

Kalachova, K., Cajka, T., Sandy, C., Hajslova, J., Pulkrabova, J. 2013a. High throughput sample preparation in combination with gas chromatography coupled to triple quadrupole tandem mass spectrometry (GC-MS/MS): A smart procedure for (ultra)trace analysis of brominated flame retardants in fish. *Talanta* 105: 109–116.

Kalachova, K., Pulkrabova, J., Cajka, T., Drabova, L., Stupak, M. et al. 2013b. Gas chromatography-triple quadrupole tandem mass spectrometry: A powerful tool for the (ultra)trace analysis of multiclass environmental contaminants in fish and fish feed. *Analytical and Bioanalytical Chemistry* 405: 7803–7815.

Kamiya, Y., Ikemori, F., Ohura, T. 2015. Optimisation of pre-treatment and ionisation for GC/MS analysis for the determination of chlorinated PAHs in atmospheric particulate samples. *International Journal of Environmental Analytical Chemistry* 95: 1157–1168.

Kirchner, M., Matisová, E., Otrekal, R., Hercegova, A., de Zeeuw, J. 2005. Search on ruggedness of fast gas chromatography–mass spectrometry in pesticides residue analysis. *Journal of Chromatography A* 1084: 63–70.

Koesukwiwat, U., Lehotay, S.J., Leepipatpiboon, N. 2011. Fast, low-pressure gas chromatography triple quadrupole tandem mass spectrometry for analysis of 150 pesticide residues in fruits and vegetables. *Journal of Chromatography A* 1218: 7039–7050.

Koesukwiwat, U., Lehotay, S.J., Miao, S., Leepipatpiboon, N. 2010. High throughput analysis of 150 pesticides in fruits and vegetables using QuEChERS and low-pressure gas chromatography–time-of-flight mass spectrometry. *Journal of Chromatography A* 1217: 6692–6703.

Kotretsou, S., Koutsodimou, A. 2006. Overview of the applications of tandem mass spectrometry (MS/MS) in food analysis of nutritionally harmful compounds. *Food Reviews International* 22: 125–172.

Król, S., Namie, J. 2012. PBDEs in environmental samples: Sampling and analysis. *Talanta* 93: 1–17.

Lambropolou, D., Hela, D.G. 2015. Mass spectrometry. In *Handbook of Food Analysis*, 3rd ed., eds. Nollet, L., Toldrá, F. CRC Press, Boca Raton, FL, vol. 2, pp. 641–57.

Lara-Gonzalo, A., Sánchez-Uría, J.E., Segovia-Garcia, E., Sanz-Medel, A. 2010. Selected ion storage versus tandem MS/MS for organochlorine pesticides determination in drinking waters with SPME and GC-MS. *International Journal of Environmental Analytical Chemistry* 92: 856–867.

Lehotay, S., Koesukwiwat, U., van der Kamp, H., Leepipatpiboon, N. 2011. Qualitative aspects in the analysis of pesticide residues in fruits and vegetables using fast, low-pressure gas chromatography-time-of-flight mass spectrometry. *Journal of Agricultural and Food Chemistry* 59: 7544–7566.

Leite, N., Peralta-Zamora, P., Grassi, M. 2008. Multifactorial optimization approach for the determination of polycyclic aromatic hydrocarbons in river sediments by gas chromatography quadrupole ion trap selected ion storage mass spectrometry. *Journal of Chromatography A* 1192: 273–281.

Li, X., Huang, J., Zhang, W., Yu, G., Deng, S. et al. 2011. Determination of 41 polybrominated diphenyl ethers in soil using a pressurised solvent extraction and GC-NCI-MS method. *International Journal of Environmental Analytical Chemistry* 91: 37–41.

Li, Y., Whitaker, J., McCarty, C. 2009. New advances in large-volume injection gas chromatography-mass spectrometry. *Journal of Liquid Chromatography & Related Technologies* 32: 1644–1671.

Liao, C., Yang, P., Xie, Z., Zhao, Y., Cheng, X. et al. 2010. Application of GC–triple quadrupole MS in the quantitative confirmation of polycyclic aromatic hydrocarbons and phthalic acid esters in soil. *Journal of Chromatographic Science* 48: 161–166.

Lima Gomes, P., Barnes, B.B., Santos-Neto, Á.J., Lancas, F.M., Snow, N.H. 2013. Determination of steroids, caffeine and methylparaben in water using solid phase microextraction-comprehensive two dimensional gas chromatography–time of flight mass spectrometry. *Journal of Chromatography A* 1299: 126–130.

Liu, Z., Philips, J.B. 1991. Comprehensive two-dimensional gas chromatography using an on-column thermal modulator interface. *Journal of Chromatographic Science* 29: 227–231.

Lv, Q., Zhang, Q., Qi, M., Bai, H., Ma, Q. et al. 2015. Cyclotriveratrylene as a new-type stationary phase for gas chromatographic separations of halogenated compounds and isomers. *Journal of Chromatography A* 1404: 89–94.

Mackintosh, S.A., Pérez-Fuentetaja, A., Zimmerman, L.R., Pacepavicius, G., Clapsadl, M et al. 2012. Analytical performance of a triple quadrupole mass spectrometer compared to a high resolution mass spectrometer for the analysis of polybrominated diphenyl ethers in fish. *Analytica Chimica Acta* 747: 67–75.

Mastovska, K., Lehotay, S.J. 2003. Practical approaches to fast-gas chromatography. *Journal of Chromatography A* 1000: 153–180.

Mastovska, K., Hajslova, J., Godula, M., Krivankova, J., Kocourek, V. 2001. Fast temperature programming in routine analysis of multiple pesticide residues in food matrices. *Journal of Chromatography A* 907: 235–245.

Matamoros, V., Jover, E., Bayona, J.M. 2010. Part-per-trillion determination of pharmaceuticals, pesticides, and related organic contaminants in river water by solid-phase extraction followed by comprehensive two-dimensional gas chromatography time-of-flight mass spectrometry. *Analytical Chemistry* 82: 699–706.

McEwen, C.N., McKay RG. 2005. A combination atmospheric pressure LC/MS:GC/MS instrumentation. *Journal of the American Society for Mass Spectrometry* 16: 1730–1738.

Megson, D., Kalin, R., Worsfold, P.J., Gauchotte-Lindsay, C., Patterson, D.G. et al. 2013. Fingerprinting polychlorinated biphenyls in environmental samples using comprehensive two-dimensional gas chromatography with time-of-flight mass spectrometry. *Journal of Chromatography A* 1318: 276–283.

Merdassa, Y., Liu, J., Megersa, N. 2013. Development of a one-step microwave assisted-extraction method for simultaneous determination of organophosphorus pesticides and fungicides in soil by gas-chromatography mass spectrometry. *Talanta* 114: 227–234.

Mitrevski, B., Marriot, P. 2014. Evaluation of quadrupole time-of-flight mass spectrometry in comprehensive two-dimensional gas chromatography. *Journal of Chromatography A* 1362: 262–269.

Mitrevski, B., Wynne, P., Marriot, P. 2011. Comprehensive two-dimensional gas chromatography applied to illicit drug analysis. *Analytical and Bioanalytical Chemistry* 401: 2361–2371.

Mondello, L., Tranchida, P., Dugo, P., Dugo, G. 2008. Comprehensive two-dimensional gas chromatography–mass spectrometry: A review. *Mass Spectrometry Reviews* 27: 101–124.

Nardelli, V., Dell'Oro, D., Palermo, C., Centonze, D. 2010. Multi-residue method for the determination of organochlorine pesticides in fish feed based on a cleanup approach followed by gas chromatography-triple quadrupole tandem mass spectrometry. *Journal of Chromatography A* 1217: 4006–5003.

Ochiai, N., Ieda, T., Sasamoto, K., Takazawa, Y., Hashimoto, S. et al. 2011. Stir bar sorptive extraction and comprehensive two-dimensional gas chromatography coupled to high resolution time-of-flight mass spectrometry for ultra-trace analysis for organochlorine pesticides in river water. *Journal of Chromatography A* 1218: 6851–6860.

Padilla-Sanchez, J., Plaza Bolanos, P., Romero-Gonzalez, R., Garrido-Frenich, A., Martinez-Vidal, J. 2010. Application of a quick, easy, cheap, rugged and safe-based method for the simultaneous extraction of chlorophenols, alkylphenols, nitrophenols and cresols in agricultural soils, analyzed by using gas chromatography–triple quadrupole–mass spectrometry/ mass spectrometry. *Journal of Chromatography A* 1217: 5724–5731.

Pico, Y. 2008. New approaches in mass spectrometry. In *Comprehensive Analytical Chemistry*, 1st ed., ed. Pico, Y. Elsevier, Amsterdam, vol. 51, pp. 201–30.

Pierce, K., Mohler, R. 2012. A review of chemometrics applied to comprehensive two-dimensional separations from 2008–2010. *Separation & Purification Reviews* 41: 143–168.

Pintado-Herrera, M.G., González-Mazo, E., Lara-Martín, P.A. 2014. Atmospheric pressure gas chromatography–time-of-flight-mass spectrometry (APGC-ToF-MS) for the determination of regulated and emerging contaminants in aqueous samples after stir bar sorptive extraction (SBSE). *Analytica Chimica Acta* 851: 1–13.

Pitarch, E., Medina, C., Portolés, T., López, F.J., Hernández, F. 2007. Determination of priority organic micropollutants in water by gas chromatography coupled to triple quadrupole mass spectrometry. *Journal of Chromatography A* 583: 246–258.

Portoles, T., Cherta, L., Beltran, J., Hernandez, F. 2012. Improved gas chromatography–tandem mass spectrometry determination of pesticide residues making use of atmospheric pressure chemical ionization. *Journal of Chromatography A* 1260: 183–192.

Portolés, T., Pitarch, E., Francisco, J., Sancho, J.V., Hernandez, F. 2007. Methodical approach for the use of GC-TOF MS for screening and confirmation of organic pollutants in environmental water. *Journal of Mass Spectrometry* 42: 1175–1185.

Portoles, T., Sancho, J.V., Hernandez, F., Newton, A., Hancock, P. 2010. Potential of atmospheric pressure chemical ionization source in GC-QTOF MS for pesticide residue analysis. *Journal of Mass Spectrometry* 45: 926–936.

Prieto, A., Schrader, S., Moeder, M. 2010. Determination of organic priority pollutants and emerging compounds in wastewater and snow samples using multiresidue protocols on the basis of microextraction by packed sorbents coupled to large volume injection gas chromatography–mass spectrometry analysis. *Journal of Chromatography A* 1217: 6002–6011.

Purcaro, G., Tranchida, P.Q., Ragonese, C., Conte, L., Dugo, P. et al. 2010. Evaluation of a rapid-scanning quadrupole mass spectrometer in an apolar × ionic-liquid comprehensive two-dimensional gas chromatography system. *Analytical Chemistry* 82: 8583–8590.

Ravindra, K., Dirtu, A., Covaci, A. 2008. Low-pressure gas chromatography: Recent developments and trends. *TrAC Trends in Analytical Chemistry* 27: 291–302.

Ravindra, K., Godoi, A., Bencs, L., Van Grieken, R. 2006. Low-pressure gas chromatography–ion trap mass spectrometry for the fast determination of polycyclic aromatic hydrocarbons in air samples. *Journal of Chromatography A* 1114: 278–281.

Reid, V., McBrady, A., Synovec, R. 2007. Investigation of high-speed gas chromatography using synchronized dual-valve injection and resistively heated temperature programming. *Journal of Chromatography A* 1148: 236–243.

Ribes, A., Carrera, G., Gallego, E., Roca, X., Berenguer, J. et al. 2007. Development and validation of a method for air quality and nuisance odors monitoring of volatile organic compounds using multi-sorbent adsorption and gas/chromatography mass spectrometry thermal desorption system. *Journal of Chromatography A* 1140: 44–55.

Rice, S.L., Mitra, S. 2007. Microwave-assisted solvent extraction of solid matrices and subsequent detection of pharmaceuticals and personal care products (PPCPs) using gas chromatography–mass spectrometry. *Analytica Chimica Acta* 589: 125–132.

Robles-Molina, J., Gilbert-López, B., García-Reyes, J.F., Molina-Díaz, A. 2013. Comparative evaluation of liquid–liquid extraction, solid-phase extraction and solid-phase microextraction for the gas chromatography–mass spectrometry determination of multiclass priority organic contaminants in wastewater. *Talanta* 117: 382–391.

Schiewek, R., Lorenz, M., Glese, R., Brockmann, R., Benter, T., Gab, S., Schmitz, O.J. 2008. Development of a multipurpose ion source for LC-MS and GC-API MS. *Analytical and Bioanalytical Chemistry* 392: 87.

Seeley, J., Seeley, S. 2013. Multidimensional gas chromatography: Fundamental advances and new applications. *Analytical Chemistry* 85: 557–578.

Shi, X.Y., Wang, M., Chen, G.R., Fu, R.N., Gu, J.L. 2001. Synthesis and properties of new cyclodextrin phenyl carbamates as capillary gas chromatography stationary phases. *Analytica Chimica Acta* 445: 221–228.

Subedi, B., Mottaleb, M.A., Chambliss, C.K., Usenko, S. 2011. Simultaneous analysis of select pharmaceuticals and personal care products in fish tissue using pressurized liquid extraction combined with silica gel cleanup. *Journal of Chromatography A* 1218: 6278–6284.

Sun, X., Wu, C., Xing, J. 2010. Ionic liquid-bonded polysiloxane as stationary phase for capillary gas chromatography. *Journal of Separation Science* 33: 3159–3167.

Tankiewicz, M., Morrison, C., Biziuk, M. 2013. Multi-residue method for the determination of 16 recently used pesticides from various chemical groups in aqueous samples by using DI-SPME coupled with GC-MS. *Talanta* 107: 1–10.

Tranchida, P., Franchina, F., Zoccali, M., Panto, S., Sciarrone, D. et al. 2013. Untargeted and targeted comprehensive two-dimensional GC analysis using a novel unified high-speed triple quadrupole mass spectrometer. *Journal of Chromatography A* 1278: 153–159.

Worton, D., Mills, G., Oram, D., Sturges, W. 2008. Gas chromatography negative chemical ionization mass spectrometry: Application to the detection of alkyl nitrates and halocarbons in the atmosphere. *Journal of Chromatography A* 1201: 112–119.

Xu, F., Guan, W., Yao, G., Guan, Y. 2008. Fast temperature programming on a stainless-steel narrow-bore capillary column by direct resistive heating for fast gas chromatography. *Journal of Chromatography A* 1186: 183–188.

Yang, E., Shin, H. 2013. Trace level determinations of carbamate pesticides in surface water by gas chromatography mass spectrometry after derivatization with 9-xanthydrol. *Journal of Chromatography A* 1305: 328–332.

Yao, C., Anderson, J. 2009. Retention characteristics of organic compounds on molten salt and ionic liquid-based gas chromatography stationary phases. *Journal of Chromatography A* 1216: 1658–1712.

Zhao, Y., Yang, L., Wang, Q. 2007. Pulsed large volume injection gas chromatography coupled with electron capture negative ionization quadrupole mass spectrometry for simultaneous determination of typical halogenated persistent organic pollutants. *Journal of the American Society for Mass Spectrometry* 18: 1375–1386.

Zhou, Q., Sang, X., Wang, L., Ji, S., Ye, J. et al. 2010. Determination of polychlorinated biphenyls by GC/MS with ultrasound-assisted extraction from shellfish. *Frontiers of Agriculture in China* 4: 489–493.

2 LC-MS-Based Screening and Targeted Profiling Methods for Environmental Analysis

Focus on Pharmacologically Active Compounds in the Aqueous Environment and Critical Evaluation of Analytical Approaches

Barbara Kasprzyk-Hordern and Bruce Petrie

CONTENTS

2.1 INTRODUCTION

Chromatography coupled with mass spectrometry is widely used in environmental analysis due to its exceptional performance in both quantitative trace analysis and characterization of unknowns in environmental samples. This chapter aims to focus on pharmacologically active compounds (PACs) in the aqueous environment, current state of the art, and critical evaluation of existing analytical approaches.

PACs are environmental contaminants. Thousands of PACs are approved for human/veterinary use, although only a very small percentage of these compounds have been studied in the environment (<200 PACs), not to mention their active metabolites and degradation products. Human PACs enter the aqueous environment mainly through insufficiently treated sewage from households and hospitals, waste effluents from manufacturing processes, and runoff. Sludge from wastewater treatment plants (WWTPs) containing PACs can contribute to environmental contamination if used as fertilizer or transported to landfill. Veterinary PACs are discharged directly to the environment. PACs might enter the environment as unaltered bioactive compounds, or metabolites, or might undergo transformation during wastewater treatment to produce compounds of potential concern to humans and wildlife. Many of these compounds are ubiquitous and persistent in the environment. Additionally, they are continuously introduced into the environment and frequently quantified at parts per thousand (ppt)/parts per billion (ppb) levels.

The presence of PACs and their metabolites within environmental matrices was first reported in the literature during the 1970s. Hignite and Azarnoff (1977) quantified chlorophenoxyisobutyrate and salicylic acid at ppb concentrations in final effluent from a municipal WWTP. This study utilized ion-exchange chromatography for sample pretreatment and gas chromatography–mass spectrometry (GC-MS) for separation and quantitation. GC methods have since developed to incorporate a large number of compounds within a single analysis run (>50) with the use of suitable solid-phase extraction (SPE) media (Sebők et al., 2009; Andrási et al., 2011). However, such methods tend to require a run time of up to 1 h. The desire to reduce chromatographic run times and sample pretreatments, which have often been rate-limiting steps in such analysis, led to the introduction and widespread usage of liquid chromatography–mass spectrometry (LC-MS) methodologies. Their application arose from the development of robust ionization interfaces suitable for the coupling of LC to MS. This approach avoids the need for chemical derivatization with potentially toxic chemicals during sample preparation, a requirement of GC for polar compounds of low volatility and/or poor thermal stability. Furthermore, LC offers greater versatility in the manipulation of chromatographic separations over GC. For example, the liquid mobile phase of LC can be easily changed or modified with additives to influence separation and sensitivity. Consequently, LC-MS has become the tool of choice for the analysis of pollutants in environmental matrices.

2.2 LIQUID CHROMATOGRAPHY

2.2.1 Reversed-Phase Chromatography

Reversed-phase LC is the most widely used retention mechanism for separating a broad range of PACs and their metabolites (Baker and Kasprzyk-Hordern, 2011a; Ferrer and Thurman, 2012; Grabic et al., 2012) (Table 2.1). It relies upon a nonpolar stationary phase, which is commonly octadecyl C18, and a more polar mobile phase. A mobile phase gradient of water and an organic solvent such as methanol or acetonitrile are a common practice. This enables chromatography to be optimized to attain maximum resolution (without the loss of column efficiency). Although good separation of analytes is not required for detection by MS, it is necessary for environmental analysis to help limit the influence of remaining matrix within sample extracts on signal sensitivity. Moreover, conventional high-performance liquid chromatography (HPLC) has been superseded by ultraperformance liquid chromatography (UPLC) over the last decade, enabling the application of higher pressures

TABLE 2.1
Contemporary Multiresidue Methods for the Determination of PACs and Their Metabolites in Environmental Matrices by LC-MS/MS

Type of PAC	No. of Analytes	Matrix	Sampling Method	Sample Pretreatment	Sample Extraction	Glassware Pretreatment	LC	MS	MQL (ng L^{-1})	Ref.
					Reverse Phase Methods					
Steroid hormones, antimicrobials, preservatives, surfactants, human indicators, and metabolites	31	Aqueous: SW, EWW, IWW	–	Filtered (0.45 μm). Stored at 4°C	5 mL SW, 2 mL WW. Online HPLC Hypersil GOLD aQ; 20 mm × 2.1 mm, 12 μm. Gradient: 20 mM ammonium formate and 0.1% AA in H_2O and 20 mM ammonium formate and 0.1% AA in MeOH for positive ESI. Gradient: H_2O and MeOH for negative ESI	–	HPLC Purospher STAR RP-18 (C18); 125 mm × 2 mm, 5 μm. Gradient: 20 mM ammonium formate and 0.1% AA in H_2O and 20 mM ammonium formate and 0.1% AA in MeOH for positive ESI. Gradient: H_2O and MeOH for negative ESI	QqQ	0.28–109	Gorga et al. (2013)
Anticancer and associated drugs	10	Aqueous: IWW, HWW	Grab samples	Filtered (0.45 μm). Frozen at −20°C for maximum of 1 month. Modified to 0.1% w/w Na_2EDTA	50 mL. Oasis HLB (200 mg, 6 mL) elution with 10 mL MeOH	–	UPLC HSS T3; 50 mm × 2.1 mm, 1.7 μm. Gradient: 0.1% FA in H_2O and ACN	QqLIT	2.70–80	Ferrando-Climent et al. (2013)
Stimulants, analgesics, opioids, and metabolites	20	Particulate: sewage sludge	Grab samples	Transported at 4°C. Freeze-dried, finely ground, and stored at 4°C	0.5 g. Extracted PLE using MeOH/H_2O. Evaporated and diluted with H_2O. Extracted using polymeric Evolute ABN, elution with 3 mL MeOH and 3 mL 99:1 MeOH/FA	–	Purospher Star RP-18; 125 mm × 2 mm, 5 μm. Gradient: 20 mM FA/ammonium formate (pH 3.8) and ACN	QqLIT	0.30–23 ng g^{-1}	Mastroianni et al. (2013)
NSAIDs, lipid regulators, psychiatrics, antiepileptics, X-ray contrast media, plasticizer, artificial sweeteners, and metabolites	24	Aqueous: SW, IWW	Grab samples	Filtered (0.45 μm). Stored at 4°C and extracted within 24 h. pH adjusted to 2	250 mL WW, 500 mL SW. Chromabond HR-X (500 mg, 6 mL) elution with 10 mL MeOH	–	HPLC Agilent Zorbax SB-C18; 150 mm × 2.1 mm, 3.5 μm. Gradient: 2 mM NH_4Ac in H_2O and 2 mM NH_4Ac in MeOH	QqQ	0.50–30	Tran et al. (2013)

(Continued)

TABLE 2.1 (CONTINUED)
Contemporary Multiresidue Methods for the Determination of PACs and Their Metabolites in Environmental Matrices by LC-MS/MS

Type of PAC	No. of Analytes	Matrix	Sampling Method	Sample Pretreatment	Sample Extraction	Glassware Pretreatment	LC	MS	MQL (ng L^{-1})	Ref.
NSAIDs, analgesics, lipid regulators, psychiatrics, histamines, β-blockers, diuretics, antidiabetics, antihypertensives, statins, antiplatelets, opioids, anticoagulants, tranquilizers, antibiotics, and metabolites	100	Aqueous: SW	Grab samples	Stored at 4°C for up to 7 days before analysis	100 mL. Oasis HLB (200 mg, 6 mL) elution with 6 mL MeOH	–	HPLC Zorbax Eclipse VDB-C8; 150 mm × 4.6 mm, 3.5 μm. Gradient: 0.1% FA in H_2O and ACN	QTOF	1.00–1000	Ferrer and Thurman (2012)
Stimulants, analgesics, sedatives, opioids, and metabolites	24	Aqueous: EWW, IWW	Grab and 24-h composite (time proportional)	Processed within 6 h. Filtered (0.45 μm). pH adjusted to 4.5	500 mL EWW, 200 mL IWW. Oasis MCX elution into two seperate fractions (2 mL MeOH and 4 mL 95:5 MeOH/NH_4OH)	–	HPLC Nucleosil 100-3 C18 (125 mm × 2 mm, 3 μm). Gradient: 5 mM NH_4Ac in H_2O (pH 8.5) and 5 mM NH_4Ac in MeOH (pH 4.5)	QTOF	2.00–50	González-Mariño et al. (2012)
NSAIDs, analgesics, lipid regulators, psychiatrics, histamines, β-blockers, diuretics, antidiabetics, antihypertensives, statins, antiplatelets, opioids, hormones, anticoagulants, tranquilizers, antibiotics, and metabolites	90	Aqueous: SW, EWW	–	Filtered (0.45 μm). Acidified to pH 3 using sulfuric acid	100 mL. Oasis HLB (200 mg) elution with 5 mL MeOH and 3 mL ethyl acetate	–	HPLC C18 Hypersil GOLD aQ; 50 mm × 2.1 mm, 5 μm. Gradient: MeOH and ACN in H_2O (0.1% FA)	QqQ	0.10–78	Grabic et al. (2012)
NSAIDs, lipid regulator, antihypertensives, antiepileptics, diuretics, preservatives, and ultraviolet filters	17	Aqueous: SW, EWW	24-h composite	Stored at −18°C until analysis. Centrifuged at 4500 rpm	100 mL. Oasis HLB (6 mg) elution with 5 mL MeOH	–	UPLC BEH C18; 50 mm × 2.1 mm, 1.7 μm. Gradient: 0.01% FA in H_2O and MeOH	QqQ	2.00–280	Gracia-Lor et al. (2012)

(Continued)

TABLE 2.1 (CONTINUED)
Contemporary Multiresidue Methods for the Determination of PACs and Their Metabolites in Environmental Matrices by LC-MS/MS

Type of PAC	No. of Analytes	Matrix	Sampling Method	Sample Pretreatment	Sample Extraction	Glassware Pretreatment	LC	MS	MQL (ng L^{-1})	Ref.
NSAIDs, analgesics, lipid regulators, psychiatrics, histamines, β-blockers, diuretics, antidiabetics, antihypertensives, antiplatelets, anticoagulants, tranquilizers, antibiotics, and metabolites	81	Aqueous: SW, EWW, IWW	SW: grab samples; WW: 24-h composite	Filtered (0.45 µm). Modified to 0.1% w/w Na_2EDTA	100 mL SW, 50 mL EWW, 25 mL IWW. Automated Oasis HLB (200 mg, 6 mL) elution with 6 mL MeOH	–	UPLC HSS T3; 50 mm × 2.1 mm, 1.8 µm. Gradient: 10 mM NH_4Ac/FA in H_2O and MeOH for positive ESI UPLC BEH C18; 50 mm × 2.1 mm, 1.7 µm. Gradient: 5 mM NH_4Ac/NH_4OH in H_2O for negative ESI	QqLIT	0.70–140	Gros et al. (2012)
NSAIDs, analgesics, lipid regulators, psychiatrics, β-blockers, antiepileptics, pesticides, antibiotics, and metabolites	88	Aqueous: SW, EWW	24-h composite (time proportional)	Frozen at −20°C until analysis. Filtered (0.7 µm)	20 mL SW, 5 mL EWW. Online, sorbent mixture of Oasis HLB, Strata X-AW, Strata X-CW, and Isolute ENV+. Analytes eluted using 0.1% FA in MeOH and diluted with 5 mM NH_4Ac in H_2O	–	HPLC Atlantis T3; 150 mm × 3 mm, 3 µm. Gradient: 5 mM NH_4Ac in H_2O and 0.1% FA in MeOH	QqQ	2.30–186	Huntscha et al. (2012)
NSAIDs, analgesics, lipid regulators, psychiatrics, histamines, β-blockers, antibiotics, cardiovasculars, barbiturates, antidiabetics, antineoplastics, and metabolites	77	Aqueous: SW	–	7.5 mL aliquots frozen at −20°C. Filtered (0.45 µm). Modified to 0.1% FA and 0.02% TFA for positive ESI or 10 mM NH_4Ac for negative ESI	7.5 mL. Online TurboFlow™ column with loading mobile phase of 0.1% FA and 0.02% TFA for positive ESI or 10 mM NH_4Ac for negative ESI	–	BetaSil phenyl–hexyl; 50 mm × 3 mm, 3 µm. Gradient: 0.1% FA and 0.02% TFA in H_2O, 0.1% FA in H_2O, 0.3% FA and 0.02% TFA in H_2O, and 0.1% FA in ACN/acetone/isopropanol for positive ESI. Gradient: 0.5 mM NH_4Ac in MeOH/ACN, 0.1% FA in H_2O and 0.1% FA in ACN/acetone/isopropanol for negative ESI	QqQ	0.10–164	López-Serna et al. (2012)
Stimulants, hallucinogens, human indicators, opioids, morphine derivatives, benzodiazepines, antidepressants, anesthetics, drug precursors, and metabolites	65	Aqueous: SW, EWW, IWW	Grab samples	Filtered (0.7 µm), Acidified with HCl to pH 1.8–1.9 and stored in the dark at 4°C and extracted within 20 h	100–250 mL WW, 250–1000 mL SW. Oasis MCX (60 mg) elution with 3 mL 7% NH_4OH in MeOH	Silanized with 5% dimethylchlorosilane in toluene	UPLC BEH C18; 150 mm × 2.1 mm, 1.7 µm. Gradient: 79.7:20:0.3 H_2O/MeOH/AA (pH 2.9) and 99.7:0.3 MeOH/AA	QqQ	0.50–140	Baker and Kasprzyk-Hordern (2011a)

(Continued)

TABLE 2.1 (CONTINUED)
Contemporary Multiresidue Methods for the Determination of PACs and Their Metabolites in Environmental Matrices by LC-MS/MS

Type of PAC	No. of Analytes	Matrix	Sampling Method	Sample Pretreatment	Sample Extraction	Glassware Pretreatment	LC	MS	MQL (ng L^{-1})	Ref.
Stimulants, hallucinogens, opioids, morphine derivatives, benzodiazepines, antidepressants, anesthetics, drug precursors, and metabolites	60	Particulate: IWW	Grab samples	Stored in the dark at 4°C and processed within 18 h. Centrifuged at 25,931*g* and filtered. Particulates were homogenized, grinded with a pestle and mortar, and stored at −18°C	1 g dry weight. Extracted by PLE using $MeOH/H_2O$. Diluted in acidified H_2O (pH 1.8). Extracted using Oasis MCX (60 mg) elution with 3 mL 7% NH_4OH in MeOH	Silanized with 5% dimethylchlorosilane in toluene	UPLC BEH C18; 150 mm × 2.1 mm, 1.7 µm. Gradient: 79.7:20:0.3 H_2O/MeOH/AA (pH 2.9) and 99.7:0.3 MeOH/AA	QqQ	0.05–20.33 ng g^{-1}	Baker and Kasprzyk-Hordern (2011b)
NSAIDs, antibiotics, antiepileptics, and human indicators	15	Particulate: biosolids	–	Freeze-dried and stored at −20°C	0.5 g. Extracted by ASE using ACN and H_2O. pH adjusted to 3 and extracted using Oasis HLB. Elution by 5 mL $MeOH/H_2O$ containing 150 mg L^{-1} EDTA	–	Luna C18; 150 mm × 4.6 mm, 3 µm. Gradient: H_2O and 0.1% FA in ACN	QqQ	1.90–488 ng g^{-1}	Ding et al. (2011)
NSAIDs, analgesics, lipid regulators, statins, antidepressants, psychiatrics, and antibiotics	47	Aqueous: SW, EWW	WW: 24-h composite	Stored at −18°C	100 mL. Oasis HLB (60 mg) elution with 5 mL MeOH	–	UPLC HSS T3; 100 mm × 2.1 mm, 1.8 µm. Gradient: H_2O and MeOH (0.1 mM NH_4Ac, 0.01% FA)	QqQ	0.80–170	Gracia-Lor et al. (2011)
NSAIDs, analgesics, β-blockers, antiepileptics, antidepressants, bactericides, stimulants, and metabolites	14	Particulate: sediment, sludge	Grab	Stored at −20°C	1–10 g freeze-dried sludge. Extracted by ASE using MeOH and 0.1% FA in MeOH	Silanized	UPLC BEH C18; 50 mm × 2.1 mm, 1.7 µm. Gradient: $MeOH/H_2O$ containing 10 mM NH_4Ac	QqQ	1.00–50 ng g^{-1}	Langford et al. (2011)
NSAIDs, analgesics, lipid regulators, psychiatrics, histamines, β-blockers, antibiotics, cardiovasculars, barbiturates, diuretics, antidiabetics, anticancers, and metabolites	74	Aqueous: SW, EWW, IWW	–	Filtered (0.45 µm). Modified to 0.1% w/w Na_2EDTA	500 mL SW, 200 mL EWW, 100 mL IWW. Oasis HLB (60 mg, 3 mL) elution with 8 mL MeOH	–	UPLC BEH C18; 100 mm × 2.1 mm, 1.7 µm. Gradient: 0.1% FA in H_2O and ACN for positive ESI. Gradient: 10 mM NH_4Ac in H_2O and MeOH/ACN for negative ESI	QqQ	0.10–378	López-Serna et al. (2011)

(Continued)

TABLE 2.1 (CONTINUED)
Contemporary Multiresidue Methods for the Determination of PACs and Their Metabolites in Environmental Matrices by LC-MS/MS

Type of PAC	No. of Analytes	Matrix	Sampling Method	Sample Pretreatment	Sample Extraction	Glassware Pretreatment	LC	MS	MQL (ng L^{-1})	Ref.
NSAIDs, analgesics, lipid regulators, antibiotics, benzodiazepines, antidepressants, diuretics, cardiotonics, and antiulcer agents	23	Aqueous: SW, EWW, IWW	Grab samples	Filtered (0.45 μm) and stored at 4°C until extraction. Acidified to pH 2	50 mL IWW, 100 mL EWW. Oasis MAX (500 mg, 6 mL) elution with 6 mL MeOH	–	HPLC Pursuit UPS C18; 50 mm × 2.1 μm, 2.4 μm. Gradient: 10 mM FA in H_2O and MeOH	IT	3.00–219	Sousa et al. (2011)
NSAIDs, analgesics, lipid regulators, statins, antidepressants, antiulcer agents, psychiatric drugs, anxiolytics, cardiovasculars, and metabolites	20	Aqueous: SW, EWW, IWW	24-h composite	Frozen at −18°C until extraction. Centrifuged at 4500 rpm	100 mL. Oasis HLB (60 mg) eluted with 5 mL MeOH	–	UPLC BEH C18; 50 mm × 2.1 mm, 1.7 μm. Gradient: 0.1 mM NH_4Ac and 0.01% FA in H_2O and 0.1 mM NH_4Ac and 0.01% FA in MeOH	QqQ	0.22–44	Gracia-Lor et al. (2010)
Psychiatrics, antihistamines, angiotensins, β-blockers, cardiacs hormones, and metabolites	55	Aqueous: EWW, IWW	24-h composite	Stored at 4°C and extracted within 24 h. Filtered 1.6 μm	100 mL. Oasis HLB (200 mg, 6 mL) elution with 6 mL MeOH	–	UPLC BEH C18; 100 mm × 2.1 mm, 1.7 μm. Gradient: 10 mM FA/ ammonium formate (pH 3.5) and 0.1% FA in ACN in positive ESI. Gradient: H_2O and 90:10 ACN/MeOH in negative ESI	QqLIT	0.02–20	Huerta-Fontela et al. (2010)
NSAIDs, analgesics, lipid regulators, psychiatrics, histamines, β-blockers, antibiotics, cardiovasculars, barbiturates, diuretics, antidiabetics, anticancers, and metabolites	74	Aqueous: SW, EWW, IWW	–	Filtered (0.45 μm). Modified to 0.1% w/w Na_2EDTA	2.5 mL. Automated HySphere Resin GP	–	Purospher Star RP-18; 125 mm × 2.0 mm, 5 μm. Gradient: 0.1% FA in H_2O and ACN for positive ESI. Gradient: H_2O and ACN/MeOH	QqLIT	0.01–1278	López-Serna et al. (2010)
NSAIDs, analgesics, lipid regulators, stimulants, antihypertensives, X-ray contrast media, antibiotics, antihistamines, anticonvulsants, herbicides, and metabolites	46	Aqueous: SW, EWW	–	Stored at 4°C in the dark for 12–24 h	500 mL SW, 100 mL EWW. Oasis HLB (500 mg, 6 mL) elution with 4 mL MeOH and 4 mL ethyl acetate	–	HPLC Polaris C18-Ether; 150 mm × 2 mm, 3 μm. Gradient: 0.015% FA and 5% MeOH in H_2O and MeOH	QqQ	5.00–160	Nödler et al. (2010)

(Continued)

TABLE 2.1 (CONTINUED)
Contemporary Multiresidue Methods for the Determination of PACs and Their Metabolites in Environmental Matrices by LC-MS/MS

Type of PAC	No. of Analytes	Matrix	Sampling Method	Sample Pretreatment	Sample Extraction	Glassware Pretreatment	LC	MS	MQL (ng L^{-1})	Ref.
NSAIDs, antibiotics, tranquilizers, diuretics, anxiolytics, antiepileptics, contraceptives, antihypertensives, β-blockers, and analgesics	32	Particulate: soil, sediment	Grab samples	Sample dried at 30°C for 72 h and finely ground	20 g. Extracted by PLE using 0.1 M NH_4OH in H_2O and MeOH. Extraction using Oasis HLB (500 mg, 6 mL) and MAX (150 mg, 6 mL). Elution by MeOH and ethyl acetate	–	HPLC XTerra MS-C18; 100 × 2.1 mm, 3.5 μm. Gradient: formamide, FA, triethylamine, ACN, and MeOH in H_2O and formamide, FA, triethylamine, and H_2O in ACN and MeOH Atlantis HILIC; 100 mm × 2.1 mm, 2 μm. Gradient: formamide, FA, and ACN in H_2O, and formamide, FA, and H_2O in ACN	QqQ	0.30–7.1 ng g^{-1}	Pérez-Carrera et al. (2010)
					HILIC Methods					
Stimulants, analgesics, β-blockers, antibiotics, and metabolites	8	Aqueous: SW, EWW	–	pH adjusted to 4.5 and filtered (0.45 μm)	10 mL online SPE. HXLPP. Eluted with HILIC mobile phase	–	Fused-Core Ascentis Express HILIC; 50 mm × 2.1 mm, 2.7 μm. Gradient: 15 mM NH_4Ac/AA in H_2O (pH 4.5) and ACN	SIM	1.00–5.00	Fontanals et al. (2011)
Antidepressants, β-blockers, antidiabetics, and proton pump inhibitors	13	Aqueous: IWW	24-h composite (flow proportional)	pH adjusted to 2 and frozen at −20°C	Oasis HLB (200 mg, 6 mL) elution with 8 mL MeOH	–	Luna HILIC; 150 mm × 3 mm, 5 μm. Gradient: 5 mM NH_4Ac in H_2O and ACN/MeOH	QqQ	1.00–500	van Nuijs et al. (2010)
Stimulants and metabolites	9	Aqueous: IWW	24-h composite (flow proportional)	pH adjusted to 2 and frozen at −20°C	50 mL. Oasis MCX (60 mg) elution with 4 mL MeOH and 4 mL 5% NH_3 in MeOH	–	Luna HILIC; 150 mm × 3 mm, 5 μm. Gradient: 5 mM NH_4Ac in H_2O and ACN	QqQ	1.00–2.00	van Nuijs et al. (2009)

(Continued)

TABLE 2.1 (CONTINUED)
Contemporary Multiresidue Methods for the Determination of PACs and Their Metabolites in Environmental Matrices by LC-MS/MS

Type of PAC	No. of Analytes	Matrix	Sampling Method	Sample Pretreatment	Sample Extraction	Glassware Pretreatment	LC	MS	MQL (ng L^{-1})	Ref.
					Chiral Methods					
Analgesics, psychiatrics, antibiotics, cardiovascular drugs, β-blockers, and metabolites	18	Aqueous: SW, EWW, IWW	SW: grab samples; WW: 24-h composite	Filtered (0.7 μm) and frozen at −20°C until extraction	500 mL SW, 100 mL WW. Oasis HLB (60 mg, 3 mL), elution with MeOH	Silanized with 5% dimethylchlorosilane in toluene	Chirobiotic V, 250 mm × 2.1 mm, 5 μm. Isocratic; 4 mM NH_4Ac and 0.005% FA in MeOH	QqQ	0.23–28.3	López-Serna et al. (2013)
β-blockers, antidepressants, and stimulants	8	Aqueous: SW, EWW	Grab samples	Filtered (0.7 μm)	250 mL SW, 100 mL EWW. Oasis HLB (60 mg, 3 mL), elution with MeOH	Silanized with 5% dimethylchlorosilane in toluene	Chirobiotic V; 100 mm × 2 mm, 5 μm. Isocratic; 4 mM NH_4Ac and 0.005% FA in MeOH	QTOF	0.30–85.7	Bagnall et al. (2012)
Stimulants and antidepressants	9	Aqueous: EWW, IWW	Grab samples	Filtered (0.7 μm)	100 mL WW. Oasis HLB (60 mg, 3 mL), elution with MeOH	Silanized with 5% dimethylchlorosilane in toluene	CBH; 100 × 2 mm, 5 μm. Isocratic; 1 mM NH_4Ac in 90:10 H_2O/2-propanol	QqQ	2.40–10.1	Kasprzyk-Hordern et al. (2010)
β-blockers, selective serotonin reuptake inhibitors, and $β_2$-receptor agonists	10	Aqueous: EWW, IWW	Grab samples	Stored at 4°C and extracted within 24 h. Filtered (0.7 μm)	500 mL EWW, 100 mL IWW. Oasis HLB (60 mg, 3 mL) elution with 5 mL MeOH	–	Chirobiotic V; 250 mm × 4.6 mm, 5 μm. Isocratic; 20 mM NH_4Ac and 0.1% FA in 90:10 MeOH/H_2O	QqQ	1.00–25	MacLeod et al. (2007)

Note: AA, acetic acid; ACN, acetonitrile; ASE, accelerated solvent extraction; CBH, cellobiohydrolase; EDTA, ethylenediaminetetraacetic acid; ESI, electrospray ionization; EWW, effluent wastewater; FA, formic acid; HILIC, hydrophilic interaction liquid chromatography; HLB, hydrophilic–lipophilic balance; HLXPP, hyper-cross-linked precursor particles obtained by precipitation polymerization; HPLC, high-performance liquid chromatography; HWW, hospital wastewater; IT, ion trap; IWW, influent wastewater; LIT, linear ion trap; MCX, mixed-cation exchange; MeOH, methanol; MQL, method quantitation limit; MS, mass spectrometry; NH4Ac, ammonium acetate; NSAIDs, nonsteroidal anti-inflammatory drugs; PLE, pressurized liquid extraction; QqLIT, quadrupole linear ion trap; QqQ, triple quadrupole; QTOF, quadrupole time of flight; rpm, revolutions per minute; SIM, selected ion monitoring; SW, surface water; TFA, trifluoroacetic acid; UPLC, ultra-performance liquid chromatography; WW, wastewater.

(generally up to 15,000 psi). This allows columns packed with smaller particle sizes to reach their theoretical performance as described by the van Deemter equation (de Villiers et al., 2006). Other than improved sensitivity, run times are reduced, enabling the adequate separation of >50 PACs in under 10 min (Huerta-Fontela et al., 2010; López-Serna et al., 2011; Gros et al., 2012). Furthermore, methods have been developed which combine SPE and LC-MS into fully automated processes, further reducing total analysis times and sample volume requirements (López-Serna et al., 2012; Gorga et al., 2013). The application of fast user-friendly methods (off-line and automated) has resulted in a large volume of information on PAC occurrence and behavior in various environmental compartments to be gathered in recent years.

Multiresidue reversed-phase methods have successfully demonstrated that large numbers of PACs are ubiquitous to surface waters (Baker and Kasprzyk-Hordern, 2011a; Gracia-Lor et al., 2011, 2012; López-Serna et al., 2011, 2012; Ferrer and Thurman, 2012; Grabic et al., 2012; Gros et al., 2012; Huntscha et al., 2012; Gorga et al., 2013; Tran et al., 2013). Without adequate knowledge of their toxicity, either individually or synergistically, it is essential to limit the entry of these chemicals into the environment. This requires the validation and application of these methods to the most complex environmental matrices (e.g., raw sewage). For instance, Huerta-Fontela et al. (2010) successfully applied reversed-phase UPLC to measure 55 PACs and metabolites in influent and effluent wastewaters in six WWTPs. Such information is essential for assessing WWTP performance for the simultaneous removal of numerous PACs. Application of the method also demonstrated the advantage of being able to incorporate a large number of chemicals of varying physicochemical properties into a single run. For example, carbamazepine epoxide (a major metabolite of carbamazepine) was found to be in receiving sewage at concentrations of 880–4026 ng L^{-1}, whereas the parent compound carbamazepine ranged from <1.5 to 113 ng L^{-1} (Huerta-Fontela et al., 2010). Reversed-phase methods, which can successfully analyze metabolites along with parent chemicals, are essential to fate understanding.

To understand the mechanisms of removal and fate, determination in the particulate fraction of wastewater matrix is also needed (Petrie et al., 2013). However, there is a lack of methods that describe the analysis of a high number of PACs on particulate matter. In reality, this has not been restricted by difficulties with chromatographic separations but by the additional time required for sample collection and development of the extraction method. Baker and Kasprzyk-Hordern (2011b) developed and validated a method capable of quantifying 60 illicit drugs and pharmaceuticals from organic particulate matter. Numerous PACs were found sorbed by >10% of the total concentration to particulate matter of influent sewage. Specifically, fluoxetine was found sorbed by 39.2–73.9%, emphasizing the importance of undertaking particulate phase analysis for reporting PAC concentrations and for diagnosing their fate.

2.2.2 Hydrophilic Interaction Chromatography

Hydrophilic interaction chromatography (HILIC) is a mode of separating polar chemicals with conventional normal-phase silica-based polar stationary phases and mobile phases similar to reversed-phase chromatography (i.e., water and an organic solvent). Mobile phase conditions can be a gradient (similar to reversed-phase mode) with a higher starting percentage of the organic solvent or as an isocratic mixture (van Nuijs et al., 2009, 2010) (Table 2.1). This mode of separation is beneficial for very polar chemicals that elute very early or show no retention during reversed-phase chromatography. To demonstrate, the antidiabetic metformin has high polarity and shows no retention by conventional reversed-phase chromatography. However, HILIC has shown to retain and suitably separate this chemical (van Nuijs et al., 2010). Therefore, HILIC is a valuable complimentary method to reversed-phase chromatography. Furthermore, elution with a higher percentage of organic solvent during HILIC chromatography also results in greater ionization efficiencies and, consequently, sensitivity during MS detection (van Nuijs et al., 2011). Fontanals et al. (2011) and van Nuijs et al. (2009, 2010) successfully applied HILIC chromatography to determine drugs of abuse and pharmaceuticals and their metabolites in environmental matrices. van Nuijs et al. (2009) developed a method for the determination of five drugs of abuse and four metabolites in influent sewage. As previously discussed, the measurement of metabolites is necessary for fate evaluation. Their determination is also essential for the sewage epidemiology approach. The presence of metabolites in environmental matrices can indicate consumption of the corresponding parent chemical, whereas their absence indicates direct disposal (Baker et al., 2012). The hydrophilic nature of metabolites (which aids their excretion in urine) makes them ideal candidates for analysis by HILIC, illustrating its value for the analysis of environmental pollutants.

2.2.3 Chiral Chromatography

The concept that many PACs are chiral and exist as two or more enantiomers is often overlooked. This is despite knowledge that different enantiomers of the same chemical can exhibit very different toxicity (Kasprzyk-Hordern, 2010; Evans and Kasprzyk-Hordern, 2014). The challenge posed to the analyst is that these enantiomers cannot be distinguished from one another by MS detection. Therefore, to analyze each enantiomer, they must be separated chromatographically. However, their identical physicochemical properties make their separation difficult. Conventional reversed-phase chromatography and HILIC are unable to distinguish between enantiomers. Specialized chiral stationary phases are therefore required. These contain a chiral molecule bound to a solid support, most commonly silica. Separation relies on enantiomers of the same chemical exhibiting differing degrees of interaction with the stationary phase. This is a mechanism that is poorly understood; however, models that describe separation have been proposed (Davankov, 1997; Lämmerhofer, 2010). Methods capable of enantiomeric separation are well established for the pharmaceutical industry, but these tend to use ultraviolet detection. Unfortunately, these are often not applicable for environmental analysis, as interfacing to MS is not possible due to the use of incompatible mobile phases such as nonvolatile buffers or unsuitable additives.

The need to investigate the enantiomeric fate of pollutants in the environment has led to the development of chiral chromatography methods with suitable mobile phase compositions for coupling to MS (MacLeod et al., 2007; Kasprzyk-Hordern et al., 2010; Bagnall et al., 2012; López-Serna et al., 2013) (Table 2.1). The maximum back pressure of chiral columns is generally 2000 psi, limiting their operation to HPLC mode. Macromolecular and macrocyclic stationary phases have been applied to the analysis of pharmaceuticals in environmental matrices (Evans and Kasprzyk-Hordern, 2014). The most widely used macromolecular phase is cellobiohydrolase, a cellulase enzyme (Kasprzyk-Hordern et al., 2010; Bagnall et al., 2012). This has successfully separated nine different PACs (stimulants, antidepressants) in the aqueous phase of influent and effluent wastewaters (Kasprzyk-Hordern et al., 2010). The most popular of the macrocyclic stationary phases is vancomycin (MacLeod et al., 2007; Bagnall et al., 2012; López-Serna et al., 2013). This comprises 18 stereocenters (Ward and Farris, 2001) and has exhibited enantiomeric separation for a large number of chemicals (MacLeod et al., 2007; Bagnall et al., 2012; López-Serna et al., 2013). This includes analgesics, psychiatrics, antibiotics, cardiovascular drugs, β-blockers, selective serotonin reuptake inhibitors, β_2 receptor agonists, antidepressants, stimulants, and metabolites. However, including a large number of chemicals into a single multiresidue chiral method is difficult. Stationary phases that can successfully separate a larger number of chemicals of varying chemical class are now needed. Furthermore, stationary phases comprised smaller particle sizes (i.e., <2 μm) that can achieve the performance of UPLC would be beneficial. For further discussion on currently available chiral stationary phases and their application for environmental pollutants analysis, please refer to Evans and Kasprzyk-Hordern (2014). Application of chiral chromatography to the analysis of environmental pollutants has demonstrated enantiomer-selective biodegradation during secondary (biological) wastewater treatment (Kasprzyk-Hordern and Baker, 2012a) and in river waters (Bagnall et al., 2012). Kasprzyk-Hordern and Baker (2012b) also emphasized the value of chiral chromatography for sewage epidemiology of illicit drugs and drugs of abuse. It can be used to help identify the method of synthesis, route of administration, legality of use, and route of exposure to wastewater (i.e., excretion or direct disposal) (Kasprzyk-Hordern and Baker, 2012b). Due to the comparative infancy of method development in this area and the possibility of stereoselection during extraction solids, there are no methods reported in the literature for particulate analysis.

2.3 MASS SPECTROMETRY

Tandem mass spectrometry (MS/MS) is essential for the analysis of environmental pollutants due to the high sensitivity and specificity demanded for ppt-to-ppb determinations in such complex matrices. The selection of MS detector is critical for the analysis type sought (i.e., quantitative or qualitative). Targeted quantitative analysis traditionally uses triple quadrupoles (QqQs) and, increasingly, linear ion traps (LITs) due to their high sensitivity. In comparison, screening for unknowns and retrospective analysis tends to involve quadrupole time of flights (QTOFs) and, until more recently, Orbitrap technology.

2.3.1 Targeted Analysis with Triple Quadrupoles and Linear Ion Traps

The most common approach to targeted analysis is the detection using a QqQ (MacLeod et al., 2007; van Nuijs et al., 2009, 2010; Gracia-Lor et al., 2010, 2011, 2012; Kasprzyk-Hordern et al., 2010; López-Serna et al., 2010, 2012, 2013; Nödler et al., 2010; Pérez-Carrera et al., 2010; Baker and Kasprzyk-Hordern, 2011a,b; Ding et al., 2011; Langford et al., 2011; Grabic et al., 2012; Huntscha et al., 2012; Gorga et al., 2013; Tran et al., 2013) (Table 2.1). These MS/MS detectors offer excellent sensitivity for quantitative analyses. In brief, the molecular ion of the compound of interest is selected in the first Q. The second Q acts as a collision cell to fragment the ion and to focus product ions into the third Q (Farré et al., 2012). The third Q then selects for selected product ions. Alternatively, the third Q can be replaced with a LIT, which has the capacity to store and accumulate ions to improve signal-to-noise ratios. The use of LITs in the analysis of environmental pollutants is increasing notably (Huerta-Fontela et al., 2010; López-Serna et al., 2010; Gros et al., 2012; Ferrando-Climent et al., 2013; Mastroianni et al., 2013). The high selectivity and sensitivity of QqQs and LITs make them ideal for the trace determination of PACs in complex environmental matrices. Both QqQs and LITs have been successfully used to simultaneously determine >50 PACs in surface water and wastewater extracts (Baker and Kasprzyk-Hordern, 2011a,b; López-Serna et al., 2011; Grabic et al., 2012; Gros et al., 2012; Huntscha et al., 2012). Method quantitation limits achievable by such methods are <1 ng L^{-1} for numerous chemicals. These applications are ideal to gather environmental occurrence information and can support exposure-driven toxicity testing. Fully quantitative methods are also vital to monitor the performance of WWTPs. Such information can aid process selection or optimization to improve the removal of PACs during wastewater treatment.

2.3.2 Screening for Unknowns and Retrospective Analysis with QTOFs and Orbitraps

High sensitivity in full-scan mode, high mass resolution, and high accuracy make hybrid detectors such as QTOF ideally suited for nontarget screening and identification of unknowns. It relies on the acceleration of ions by an electric field. The time taken for these ions to reach the detector is reliant on their kinetic energy and, therefore, their mass-to-charge (*m*/*z*) ratio (Radjenović et al., 2009a). This high selectivity has seen QTOFs successfully applied to identify PACs, metabolites, and their degradation products in various environmental waters (Hernández et al., 2011; Masiá et al., 2013; Boix et al., 2014). The first such application was in 2001 by Bobeldijk et al. (2001). This investigation used MS-to-MS/MS switching to determine the exact mass determinations and calculated elemental compositions. Together with product ion spectra, this enabled the identification of three unknown chemicals including the antiepileptic carbamazepine. This method was successful at concentrations of 250 ng L^{-1} in surface water (Bobeldijk et al., 2001). Since then, there have been several QTOF applications to identify unknowns (namely, degradation products) in environmental matrices (Kosjek et al., 2009; Hernández et al., 2011; Masiá et al., 2013; Boix et al., 2014). It should be noted that without structure confirmation with a searchable mass spectral library, validation with a reference standard is needed. Improved sensitivity of modern QTOFs has lowered detection limits. However, the desire to use low sample volumes while achieving improved detection limits has seen their application restricted to mainly qualitative nontarget screening. Nevertheless, quantitative targeted analysis by QTOFs has been applied by several authors, achieving quantitation limits ranging broadly in the ppb-to-ppt range for a variety of chemicals in environmental waters (Bagnall et al., 2012; Ferrer and Thurman, 2012; González-Mariño et al., 2012).

Alternatively, Orbitrap technology is now available, which generally offers improved mass range, resolving power, and dynamic range (Farré et al., 2012). In brief, ions of a specific *m*/*z* ratio move in rings, oscillating along a central electrode. The frequency of these harmonic oscillations is independent of the ion velocity and inversely proportional to the square root of the *m*/*z* ratio (Radjenović et al., 2009a). Due to its more recent development and comparatively high cost, their use has been less widespread. Helbling et al. (2010) successfully identified degradation products (which were previously unreported) for five pharmaceuticals following exposure to activated sludge. The identification of unknowns, specifically degradation products, is essential. This can reveal pathways of biotransformation during biological wastewater treatment, which provides vital information for improving their removal during treatment. Furthermore, any structural change to a molecule may not result in loss of toxicity and environmental impact. This is a major limitation of exclusively using targeted quantitative analysis. Other than enabling identification of unknown chemicals, Orbitraps are suitable for quantitation at ppt levels (Bijlsma et al., 2012). For example, Cahill et al. (2012) achieved limits of quantitation ranging from 5.9 to 27 ng L^{-1} for five common acidic pharmaceuticals in wastewaters. The ability to quantify at trace concentrations and identify unknowns by using the same instrumentation is a huge advantage when investigating the fate and potential risk posed by PACs and their associated chemicals in the environment.

2.4 CRITICAL METHODOLOGICAL ASPECTS

2.4.1 Sample Collection and Preservation

2.4.1.1 Sampling Strategies

Sampling strategies for aquatic environmental pollutants including sample collection and preservation are hardly discussed in the literature and are very often ignored in research papers. This is surprising as inappropriate sample collection and preservation can lead to erroneous analytical measurements.

Collection of environmental samples with the objective of analyzing PACs is usually undertaken with the usage of grab sampling or time-, volume-, or flow-proportional composite sampling. A critical discussion of sample collection approaches for PACs in wastewater systems was undertaken by Ort et al. (2010). Sampling errors may result from an inappropriate choice of sampling mode and frequency. In the absence of a proper analysis of the system under investigation, the importance of short-term pollutant variations might also not be taken into account. Such a situation could lead to an overinterpretation of measured data, leading to incorrect conclusions (Ort et al., 2010). In order to reduce sampling uncertainty, the following should be taken into account: (i) goal of the study, (ii) choice of sampling mode, (iii) frequency of sampling, (iv) sampling location, and (v) sampled matrix. High-frequency grab sampling will provide detailed information on short-term variation in pollutant concentrations, but such an approach is not feasible in longer-term studies evaluating spatial and temporal occurrence of pollutants and their fate due to high cost and time commitment. Composite sampling mode is therefore recommended for screening purposes, but it must be remembered that it will not account for short-term variations in contaminant concentrations. In the case of river catchments and wastewater analysis, in order to account for variable flow rates, a flow-proportional sampling mode should be applied, as this is the only sampling mode that correctly weighs the individual subsamples, which are combined to form a composite sample and can be analyzed to obtain an average concentration. The volume- and time-proportional sampling modes (where either sampling volume or both frequency and sampling volume are constant) cannot provide a true average concentration because only the frequency changes, but the individual samples are not weighted properly according to the flow in the sewer or a river. Passive samplers have been recently employed to overcome some of the difficulties related to active sampling of PACs. There is, however, limited knowledge regarding their applicability in environmental analysis of PACs. Some of the drawbacks of the techniques include the following: varying flow velocities and clogging, which influence the uptake rate (Ort et al., 2010).

2.4.1.2 Stability of Analytes and Preservation Approaches

Time or flow-proportional 24 h composite sampling methods are widely used in environmental analysis for PACs which, as previously mentioned, do not account for short-term pollutant variations. Such approaches might also not account for errors resulting from low stability of analytes throughout the 24 h sampling period. This is because the majority of PACs are bioactive and may be metabolized or may undergo microbial degradation or other transformation reactions in an environmental matrix such as wastewater. Thus, depending on stability, quantifying a compound in wastewater or another environmental matrix that has been excreted several hours earlier may in fact lead to a significant over- or underestimation of the actual amount of residue originally present. As wastewater is characterized by a high level of biomass, microbial degradation within the 24 h sampling period should be expected.

The data for >60 PACs in wastewater reported by Baker and Kasprzyk-Hordern (2011c) show that from the perspective of stability, composite samplers are unsuitable with regard to some compounds (Table 2.2). For instance, concentration changes of 66% and 26% were reported for heroin and 6-acetylmorphine, respectively, after only 12 h in raw (unfiltered) wastewater at 2°C. Acidification of samples was found to improve stability for the majority of compounds. This is widely known to prevent bacterial activity and in turn preserves the sample (Kasprzyk-Hordern, Dinsdale et al., 2010). However, it has to be remembered that acidification can influence the partitioning of certain drugs between aqueous solution and (suspended) solids. If this is the case, both the aqueous solution and solids have to be analyzed for the presence of studied compounds (Baker and Kasprzyk-Hordern, 2011c). Similar results were observed by others (Castiglioni et al., 2006; van Nuijs et al., 2012; Gheorghe et al., 2008; Gonzalez-Marino et al., 2010). Gonzalez-Marino et al. (2010) reported using sodium azide (NaN_3, 0.2%) as a preserving agent NaN_3 (0.2%).

The key parameters affecting the stability of analytes include the following: level of biomass, type of analytes, storage temperature, pH, and the presence of suspended particulate matter. In the study by Baker and Kasprzyk-Hordern (2011b), pH was found to be the most important factor with a significant improvement in the stability of compounds overall when wastewater was acidified. The storage temperature of the samples was also found to affect stability, although to a lesser extent than pH. It is worth emphasizing here that refrigeration is a widely used approach to sample preservation during composite sampling. It is, however, clear from the results that it does not fully preserve environmental samples. For example, after 72 h in raw (unfiltered) wastewater at 2°C and 19°C, the percentages of compounds characterized by a significant concentration change were 54% and 69%, respectively. It is also important to note that the presence of suspended solids in the samples (unfiltered sample) facilitated a decrease in the stability of analytes, with the percentage of compounds reporting a concentration change in raw (unfiltered) wastewater at 19°C of 69% in comparison to filtered wastewater of 54%. These findings are of critical importance when assessing the validity of the usage of composite sampling in the analysis of PACs and indicate the need for reevaluation of current sampling approaches (Baker and Kasprzyk-Hordern, 2011c).

TABLE 2.2
Critical Methodological Aspects of LC-MS: Summary of Recoveries Obtained with Each Critical Parameter

Compound	Range of Recoveries Obtained with Each Parameter (%)		
	Analyte Losses due to (Bio)Degradation[a]	Analyte Losses due to Sorption to Glassware[b]	Analyte Losses during Evaporation (Effect of Temperature and Eluting Agent)[c]
Stimulants			
Cocaine	101 – 72	98 – 64	107 – 90
Benzoylecgonine	117 – 91	101 – 95	110 – 92
Norbenzoylecgonine	103 – 89	96 – 81	104 – 81
Norcocaine	100 – 69	99 – 73	103 – 90
Cocaethylene	100 – 81	95 – 70	108 – 90
Anhydroecgonine methyl ester	153 – 86	78 – 28	93 – 40
Ecgonidine	183 – 90	117 – 44	78 – 44
Amphetamine	188 – 89	88 – 16	96 – 51
Methamphetamine	112 – 86	89 – 23	95 – 59
Methcathinone	106 – 22	67 – 16	85 – 12
Benzylpiperazine	115 – 96	86 – 37	97 – 60
3-Trifluoromethylphenylpiperazine	139 – 50	90 – 37	100 – 70
Hallucinogens			
Methylendioxyamphetamine	110 – 89	99 – 41	102 – 83
3,4-Methylenedioxymethamphetamine	105 – 89	96 – 40	102 – 84
3,4-Methylenedioxy-*N*-ethylamphetamine	105 – 88	97 – 35	106 – 81
N-methyl-1,3-benzodioxolylbutanamine	99 – 78	93 – 27	97 – 75
1,3-Benzodioxolylbutanamine	103 – 62	92 – 27	96 – 73
Mescaline	99 – 73	96 – 76	112 – 82
LSD	103 – 80	93 – 87	108 – 90
2-Oxo-3-hydroxy LSD	183 – 86	96 – 80	108 – 92
Opioids and Morphine Derivatives			
Heroin	104 – 0	89 – 68	91 – 34
6-Acetylmorphine	99 – 23	118 – 91	164 – 91
Codeine	124 – 96	99 – 73	99 – 73
Norcodeine	105 – 90	97 – 68	101 – 68
Oxycodone	113 – 99	82 – 67	89 – 1
Oxymorphone	143 – 88	102 – 46	92 – 1
Morphine	189 – 91	110 – 58	103 – 58
Normorphine	114 – 93	104 – 57	93 – 57
Dihydrocodeine	109 – 92	100 – 77	103 – 77
Buprenorphine	104 – 83	89 – 78	96 – 81
Norbuprenorphine	117 – 92	97 – 88	102 – 88
Methadone	104 – 77	90 – 50	95 – 79
2-Ethylidene-1,5-dimethyl-3,3-diphenylpyrrolidine	104 – 28	89 – 16	97 – 80
2-Ethylidene-1,5-dimethyl-3,3-diphenylpyrrolidine	133 – 59	64 – 1	91 – 17

(Continued)

TABLE 2.2 (CONTINUED)
Critical Methodological Aspects of LC-MS: Summary of Recoveries Obtained with Each Critical Parameter

Compound	Range of Recoveries Obtained with Each Parameter (%)								
	Analyte Losses due to (Bio)Degradation[a]			Analyte Losses due to Sorption to Glassware[b]			Analyte Losses during Evaporation (Effect of Temperature and Eluting Agent)[c]		
Fentanyl	103	–	38	98	–	90	102	–	90
Norfentanyl	121	–	77	98	–	87	103	–	87
Propoxyphene	125	–	81	96	–	45	115	–	77
Norpropoxyphene	176	–	79	676	–	119	717	–	152
Tramadol	103	–	67	89	–	37	93	–	78
Nortramadol	111	–	37	98	–	81	109	–	87
Benzodiazepines									
Temazepam	119	–	27	90	-	86	106	–	79
Diazepam	97	–	56	97	-	88	108	–	82
Nordiazepam	150	–	78	91	-	76	102	–	70
Nitrazepam	101	–	0	80	-	66	95	–	44
7-Aminonitrazepam	136	–	98	94	-	58	117	–	73
Oxazepam	108	–	79	101	-	86	116	–	80
Chlordiazepoxide	113	–	49	107	-	86	113	–	87
Antidepressants									
Dosulepin	105	–	76	90	–	47	92	–	47
Amitriptyline	110	–	18	90	–	45	93	–	58
Nortriptyline	104	–	12	83	–	39	85	–	47
Fluoxetine	108	–	45	82	–	19	92	–	25
Norfluoxetine	109	–	44	74	–	19	81	–	27
Venlafaxine	118	–	66	100	–	61	104	–	91
Dissociative Anesthetics									
Phencyclidine	103	–	75	95	–	20	106	–	41
Ketamine	105	–	95	88	–	20	97	–	47
Norketamine	105	–	89	74	–	20	89	–	47
Others									
Methaqualone	107	–	72	94	–	21	109	–	63
Sildenafil	128	–	47	86	–	40	88	–	56
Ephedrine/pseudoephedrine	104	–	47	100	–	32	105	–	59
Norephedrine	105	–	16	114	–	33	122	–	31

Source: *Journal of Chromatography A*, 1218, Baker, D.R. and B. Kasprzyk-Hordern, Critical evaluation of methodology commonly used in sample collection, storage and preparation for the analysis of pharmaceuticals and illicit drugs in surface water and wastewater by solid phase extraction and liquid chromatography–mass spectrometry, 8036–8059, Copyright (2011), with permission from Elsevier.

Note: LSD, lysergic acid diethylamide.

[a] Stability of analytes in various storage conditions encountered during sample collection and storage (studied conditions: $t = 0–72$ h; $T = 2–19°C$; pH = 1.8–7.4).

[b] Evaporation of SPE extracts in silanized and nonsilanized vials.

[c] Evaporation of SPE extracts at temperatures of 20–50°C.

2.4.2 Sample Preparation

2.4.2.1 Solid-Phase Extraction of PACs from Aqueous Matrix and Key Phenomena

SPE is usually the method of choice for the analysis of PACs in the aqueous environment. SPE recoveries for PACs in environmental matrix can range from a small percentage to a well over 100% and can significantly deviate from expected full recovery. Usually, environmental matrix components and competitive sorption between analytes and interferences as well as actual SPE sorbents are responsible for ineffective extraction. However, as the SPE process is several steps long, each of these steps, and not the actual SPE sorbent, might introduce errors to the analytical protocol. If these are unaccounted for, it may result in misleading conclusions potentially being drawn, assigning low recoveries as a consequence of the SPE sorbent. A study by Baker and Kasprzyk-Hordern (2011c) emphasized the importance of key, often overlooked, parameters to the successful SPE of PACs in aqueous environmental samples with the usage of SPE-LC/MS: (i) effect of evaporation temperature and (ii) solvent with regard to SPE extracts, (iii) effect of silanizing glassware, (iv) SPE sorbent, (v) recovery of analytes during vacuum filtration through glass fiber filters and (vi) pre-LC-MS filter membranes, and (vii) low stability of analytes in environmental matrix at different storage times, temperatures, and pH values. Over 60 analytes were targeted, including stimulants, opioid and morphine derivatives, benzodiazepines, antidepressants, dissociative anesthetics, drug precursors, human urine indicators, and their metabolites (for detailed discussion, see Baker and Kasprzyk-Hordern [2011c]). Table 2.2 demonstrates that the most critical factors that affect recovery are low stability (previously discussed), interaction of analytes with glass surface, and the effect of the combination of different types of solvent and temperature on losses during the evaporation of SPE extracts.

2.4.2.2 Analyte Losses due to Sorption to Glassware

The possible loss of compounds due to adsorption onto glassware surfaces can be of significant importance in quantitative trace analysis (Middleditch, 1989). Untreated glassware contains silicate and silanol groups that act as ion-exchange and nucleophilic centers (Seed, 1997). Consequently, compounds may be lost due to their interactions with glass surface. Amines are especially prone to loss onto the slightly acidic surface of glass (Knapp, 1979). As the pH_{PZC} of silica is ~3, at $pH < pH_{PZC}$, silanols present on the surface of silica are dissociated and will be interacting with positively charged analytes (at pH < pKa of basic analytes). Therefore, at neutral, weakly acidic and weakly basic pH interactions of basic analytes with silanols will occur and should be taken into account to limit possible analyte losses. To prevent the loss of compounds, glassware used for trace analysis should be silanized. Silanization involves the reaction of polar Si–OH groups with chlorosilanes (e.g., dimethychlorosilane), which reduces the number of free silanol groups and, as a result, the possibility of interactions with basic analytes.

The results published by Baker and Kasprzyk-Hordern (2011c) and presented in Table 2.2 clearly indicate that the silanization of glassware provides significant improvements in SPE recovery of many basic analytes (e.g., amphetamines). Therefore, silanization of glassware should be considered when undertaking analysis of basic analytes. It has to be, however, emphasized that silanization of glassware introduces hydrophobicity on its surface, which, as a result, will increase binding of nonpolar molecules.

2.4.2.3 The Effect of Eluting Agent and Temperature on Loss during Evaporation

Evaporation of SPE extracts, before final reconstitution with solvents amenable to LC separation, is one of the key stages of each SPE protocol when preparing samples for reversed-phase, HILIC, or chiral chromatography. Both temperature and the solvent in which analytes are evaporated can have a significant impact on their recovery. In a study by Baker and Kasprzyk-Hordern (2011c), the highest recovery for almost all basic analytes was achieved at 20°C and was observed to decrease with an increase in temperature of evaporation. Furthermore, the impact of temperature was the most evident if methanol modified with ammonium hydroxide (as opposed to methanol only) was used as eluting agent. This is because the addition of ammonium hydroxide facilitates deprotonation of basic analytes (pH > pKa of basic analyte), which increases their volatility. The opposite situation would be observed in the case of acidic analytes, where higher volatility (and, as a result, evaporation losses) will be evident at low pH (pH < pKa of acidic analyte). In summary, solvents used for the elution of analytes from SPE sorbents can have a significant impact on their recovery. As the evaporation temperature of 40°C is usually selected for analytes dissolved either in pure methanol (commonly used in reversed-phase SPE) or in methanol modified with basic (e.g., ammonium hydroxide) or acidic (e.g., formic acid) modifiers (commonly used in mixed-mode SPE), the likely loss of compounds during the evaporation step is to be expected and should be taken into account.

2.4.2.4 Analyte Losses due to Sorption to Suspended Particulate Matter

Frequently, the analysis of PACs in aqueous environment includes only aqueous samples after the removal of suspended particulate matter (SPM) through the use of filtration or centrifugation. This is because either an assumption is made that analytes of interest do not sorb to SPM or there is no analytical method available to quantify analytes on the SPM material. Analysis of SPM is, however, critical in order to prevent underreporting of the levels of analyte that may be present in an aqueous matrix.

Extraction of environmental SPM is usually undertaken with the usage of pressurized liquid extraction (Díaz-Cruz et al., 2009; Nieto et al., 2010; Runnqvist et al., 2010), which are followed by SPE and then analysis using LC-MS/MS (Barron et al., 2008; Stein et al., 2008; Radjenović et al., 2009b; Vazquez-Roig et al., 2010). Baker and Kasprzyk-Hordern (2011b) utilized this approach to verify whether >60 PACs (stimulants, opioid and morphine derivatives, benzodiazepines, antidepressants, dissociative anesthetics, and their metabolites) sorb to SPM and if the extent of adsorption can lead to underreporting of PAC concentration in the aqueous environment. It is evident that for several compounds (such as antidepressants: dosulepin 64.7%, amitriptyline 50.3%, fluoxetine 73.9%, and norfluoxetine 89.4%), adsorption to SPM accounts for >50%, which, if not accounted for, could lead to significant underreporting of measured concentrations (Baker and Kasprzyk-Hordern, 2011b).

2.4.3 Data Acquisition and Quantification Approaches

In the most widely used LC and electrospray ionization (ESI)–QqQ analytical methodologies for trace quantitative analysis of PACs, data acquisition is carried out using multiple reaction monitoring (MRM) with the protonated molecular ion employed as the precursor and the most abundant transition product ion typically used for quantification, with a second transition used for confirmation. This is to fulfill European Union guidelines with four identification points for the confirmation of analytes with LC-MS/MS detection (European Commission, 2002). Additionally, the ratio between the two transitions is used as an identification criterion with values in environmental samples required to be within ±20–50% of those determined by spiked standards (European Commission, 2002; Baker and Kasprzyk-Hordern, 2011a).

There are, however, several problems associated with trace quantitative analysis in complex matrices. One of them is matrix-dependent signal suppression (or enhancement), which could potentially lead to under- or overestimation of analyte concentrations. Enantioselective ion suppression/enhancement in ESI-QqQ is of particular importance when undertaking chiral analysis, as it can lead to erroneous measurements (e.g., nonracemic composition for analytes, which are in actual fact racemic, or vice versa). For example, López-Serna et al. (2013) reported that in the case of several racemic chiral pharmaceuticals including albuterol, pindolol, and clenbuterol and especially propranolol, metoprolol, and timolol, increasingly complex environmental matrices resulted in enrichment factor (EF) values deviating from 0.5 (racemic mixture) indicating stereoselective signal suppression. For instance, EF for propranolol-d7 was 0.5 in the standard solution, 0.58 in surface water, and 0.66 in wastewater influent. In order to avoid matrix-related analytical errors, the usage of labeled analogues of standards (e.g., deuterated internal standards) should be considered as a necessary component of any analytical process (Evans and Kasprzyk-Hordern, 2014).

2.5 FUTURE OUTLOOK

Recent years have seen groundbreaking developments in analytical techniques for quantitative trace analysis of small molecules in complex environmental matrices. Significant advancements have been noted both in separation techniques (e.g., utilization of sub-2 µm particle size in LC) and in QqQ MS instrumentation, ultimately leading to higher accuracy, precision, and sensitivity of measurements as well as increased capability for multiresidue acquisition. Recent advances in high-resolution mass spectroscopy (HRMS, e.g., increased linearity range, speed, and sensitivity) such as QTOF and Orbitraps have led to increased interest in applications utilizing HRMS. Further developments in HRMS will lead to its more frequent application both in environmental analyses of unknowns and in targeted quantitative approaches.

REFERENCES

Andrási, N., Helenkár, A., Záray, G., Vasanits, A., Molnár-Perl, I. 2011. Derivatization and fragmentation pattern analysis of natural and synthetic steroids, as their trimethylsilyl (oxime) ether derivatives by gas chromatography mass spectrometry: Analysis of dissolved steroids in wastewater samples. *Journal of Chromatography A* 1218: 1878–1890.

Bagnall, J.P., Evans, S.E., Wort, M.T., Lubben, A.T., Kasprzyk-Hordern, B. 2012. Using chiral liquid chromatography quadrupole time-of-flight mass spectrometry for the analysis of pharmaceuticals and illicit drugs in surface and wastewater at the enantiomeric level. *Journal of Chromatography A* 1249: 115–129.

Baker, D.R., Kasprzyk-Hordern, B. 2011a. Multi-residue analysis of drugs of abuse in wastewater and surface water by solid-phase extraction and liquid chromatography-positive electrospray ionisation tandem mass spectrometry. *Journal of Chromatography A* 1218: 1620–1631.

Baker, D.R., Kasprzyk-Hordern, B. 2011b. Multi-residue determination of the sorption of illicit drugs and pharmaceuticals to wastewater suspended particulate matter using pressurised liquid extraction, solid phase extraction and liquid chromatography coupled with tandem mass spectrometry. *Journal of Chromatography A* 1218: 7901–7913.

Baker, D.R., Kasprzyk-Hordern, B. 2011c. Critical evaluation of methodology commonly used in sample collection, storage and preparation for the analysis of pharmaceuticals and illicit drugs in surface water and wastewater by solid phase extraction and liquid chromatography-mass spectrometry. *Journal of Chromatography A* 1218: 8036–8059.

Baker, D.R., Ocenášková, V., Kvicalova, M., Kasprzyk-Hordern, B. 2012. Drugs of abuse in wastewater and suspended particulate matter—Further developments in sewage epidemiology. *Environment International* 48: 28–38.

Barron, L., Tobin, J., Paull, B. 2008. Multi-residue determination of pharmaceuticals in sludge and sludge enriched soils using pressurized liquid extraction, solid phase extraction and liquid chromatography with tandem mass spectrometry. *Journal of Environmental Monitoring* 10: 353–361.

Bijlsma, L., Emke, E., Hernández, F., De Voogt, P. 2012. Investigation of drugs of abuse and relevant metabolites in Dutch sewage water by liquid chromatography coupled to high resolution mass spectrometry. *Chemosphere* 89: 1399–1406.

Bobeldijk, I., Vissers, J.P.C., Kearney, G., Major, H., van Leerdam, J.A. 2001. Screening and identification of unknown contaminants in water with liquid chromatography and quadrupole-orthogonal acceleration-time-of-flight tandem mass spectrometry. *Journal of Chromatography A* 929: 63–74.

Boix, C., Ibáñez, M., Zamora, T., Sancho, J.V., Niessen, W.M.A. et al. 2014. Identification of new omeprazole metabolites in wastewaters and surface water. *Science of the Total Environment* 468–469: 706–714.

Cahill, M.G., Dineen, B.A., Stack, M.A., James, K.J. 2012. A critical evaluation of liquid chromatography with hybrid linear ion trap-Orbitrap mass spectrometry for the determination of acidic contaminants in wastewater effluents. *Journal of Chromatography A* 1270: 88–95.

Castiglioni, S., Zuccato, E., Crisci, E., Chiabrando, C., Fanelli, R. et al. 2006. Identification and measurement of illicit drugs and their metabolites in urban wastewater by liquid chromatography-tandem mass spectrometry. *Analytical Chemistry* 78: 8421–8429.

Davankov, V.A. 1997. The nature of chiral recognition: Is it a three-point interaction? *Chirality* 9: 99–102.

de Villiers, A., Lestremau, F., Szucs, R., Gélébart, S., David, F. et al. 2006. Evaluation of ultra performance liquid chromatography: Part I. Possibilities and limitations. *Journal of Chromatography A* 1127: 60–69.

Díaz-Cruz, M.S., García-Galán, M.J., Guerra, P., Jelic, A., Postigo, C. et al. 2009. Analysis of selected emerging contaminants in sewage sludge. *TrAC Trends in Analytical Chemistry* 28: 1263–1275.

Ding, Y., Zhang, W., Gu, C., Xagoraraki, I., Li, H. 2011. Determination of pharmaceuticals in biosolids using accelerated solvent extraction and liquid chromatography/tandem mass spectrometry. *Journal of Chromatography A* 1218: 10–16.

European Commission. 2002. Commission decision (2002/657/EC): Implementing Council Directive 96/23/EC concerning the performance of analytical methods and the interpretation of results. *Official Journal of the European Communities* L221. Available from: http://eur-lex.europa.eu/en/index.htm. Accessed October 3, 2016.

Evans, S.E., Kasprzyk-Hordern, B. 2014. Application of chiral chromatography coupled with mass spectrometry in the analysis of chiral pharmaceuticals in the environment. *TrAC Trends in Environmental Analytical Chemistry* 1: e34–e51.

Farré, M., Kantiani, L., Petrovic, M., Pérez, S., Barceló, D. 2012. Achievements and future trends in the analysis of emerging organic contaminants in environmental samples by mass spectrometry and bioanalytical techniques. *Journal of Chromatography A* 1259: 86–99.

Ferrando-Climent, L., Rodriguez-Mozaz, S., Barceló, D. 2013. Development of a UPLC-MS/MS method for the determination of ten anticancer drugs in hospital and urban wastewaters, and its application for the screening of human metabolites assisted by information-dependent acquisition tool (IDA) in sewage samples. *Analytical and Bioanalytical Chemistry* 405: 5937–5952.

Ferrer, I., Thurman, E.M. 2012. Analysis of 100 pharmaceuticals and their degradates in water samples by liquid chromatography/quadrupole time-of-flight mass spectrometry. *Journal of Chromatography A* 1259: 148–157.

Fontanals, N., Marcé, R.M., Borrull, F. 2011. On-line solid-phase extraction coupled to hydrophilic interaction chromatography–mass spectrometry for the determination of polar drugs. *Journal of Chromatography A* 1218: 5975–5980.

Gheorghe, A., van Nuijs, A., Pecceu, B., Bervoets, L., Jorens, P.G. et al. 2008. Analysis of cocaine and its principal metabolites in waste and surface water using solid-phase extraction and liquid chromatography-ion trap tandem mass spectrometry. *Analytical and Bioanalytical Chemistry* 391: 1309–1319.

Gonzalez-Marino, I., Quintana, J.B., Rodríguez, I, Cela, R. 2010. Determination of drugs of abuse in water by solid-phase extraction, derivatisation and gas chromatography-ion trap-tandem mass spectrometry. *Journal of Chromatography A* 1217: 1748–1760.

González-Mariño, I., Quintana, J.B., Rodríguez, I., Gonzáez-Díez, M., Cela, R. 2012. Screening and selective quantification of illicit drugs in wastewater by mixed-mode solid-phase extraction and quadrupole-time-of-flight liquid chromatography-mass spectrometry. *Analytical Chemistry* 84: 1708–1717.

Gorga, M., Petrovic, M., Barceló, D. 2013. Multi-residue analytical method for the determination of endocrine disruptors and related compounds in river and waste water using dual column liquid chromatography switching system coupled to mass spectrometry. *Journal of Chromatography A* 1295: 57–66.

Grabic, R., Fick, J., Lindberg, R.H., Fedorova, G., Tysklind, M. 2012. Multi-residue method for trace level determination of pharmaceuticals in environmental samples using liquid chromatography coupled to triple quadrupole mass spectrometry. *Talanta* 100: 183–195.

Gracia-Lor, E., Sancho, J.V., Hernández, F. 2010. Simultaneous determination of acidic, neutral and basic pharmaceuticals in urban wastewater by ultra-high-pressure liquid chromatography-tandem mass spectrometry. *Journal of Chromatography A* 1217: 622–632.

Gracia-Lor, E., Sancho, J.V., Hernández, F. 2011. Multi-class determination of around 50 pharmaceuticals, including 26 antibiotics, in environmental and wastewater samples by ultra-high performance liquid chromatography-tandem mass spectrometry. *Journal of Chromatography A* 1218: 2264–2275.

Gracia-Lor, E., Martínez, M., Sancho, J.V., Peñuela, G., Hernández, F. 2012. Multi-class determination of personal care products and pharmaceuticals in environmental and wastewater samples by ultra-high performance liquid-chromatography-tandem mass spectrometry. *Talanta* 99: 1011–1023.

Gros, M., Rodríguez-Mozaz, S., Barceló, D. 2012. Fast and comprehensive multi-residue analysis of a broad range of human and veterinary pharmaceuticals and some of their metabolites in surface and treated waters by ultra-high-performance liquid chromatography coupled to quadrupole-linear ion trap tandem mass spectrometry. *Journal of Chromatography A* 1248: 104–121.

Helbling, D.E., Hollender, J., Kohler, H.-E., Singer, H., Fenner, K. 2010. High-throughput identification of microbial transformation products of organic micropollutants. *Environmental Science and Technology* 44: 6621–6627.

Hernández, F., Ibáñez, M., Gracia-Lor, E., Sancho, J.V. 2011. Retrospective LC-QTOF-MS analysis searching for pharmaceutical metabolites in urban wastewater. *Journal of Separation Science* 34: 3517–3526.

Hignite, C., Azarnoff, D.L. 1977. Drugs and drug metabolites as environmental contaminants: Chlorophenoxyisobutyrate and salicylic acid in sewage water effluent. *Life Sciences* 20: 337–341.

Huerta-Fontela, M., Galceran, M.T., Ventura, F. 2010. Fast liquid chromatography-quadrupole-linear ion trap mass spectrometry for the analysis of pharmaceuticals and hormones in water resources. *Journal of Chromatography A* 1217: 4212–4222.

Huntscha, S., Singer, H.P., McArdell, C.S., Frank, C.E., Hollender, J. 2012. Multiresidue analysis of 88 polar organic micropollutants in ground, surface and wastewater using online mixed-bed multilayer solid-phase extraction coupled to high performance liquid chromatography-tandem mass spectrometry. *Journal of Chromatography A* 1268: 74–83.

Kasprzyk-Hordern, B. 2010. Pharmacologically active compounds in the environment and their chirality. *Chemical Society Reviews* 39: 4466–4503.

Kasprzyk-Hordern, B., Baker, D.R. 2012a. Enantiomeric profiling of chiral drugs in wastewater and receiving waters. *Environmental Science and Technology* 46: 1681–1691.

Kasprzyk-Hordern, B., Baker, D.R. 2012b. Estimation of community-wide drugs use via stereoselective profiling of sewage. *Science of the Total Environment* 423: 142–150.

Kasprzyk-Hordern, B., Kondakal, V.V.R., Baker, D.R. 2010. Enantiomeric analysis of drugs of abuse in wastewater by chiral liquid chromatography coupled with tandem mass spectrometry. *Journal of Chromatography A* 1217 (27): 4575–4586.

Knapp, D.R. 1979. *Handbook of Analytical Derivatization Reactions*. Wiley, Hoboken, NJ.

Kosjek, T., Heath, E., Pérez, S., Petrović, M., Barceló, D. 2009. Metabolism studies of diclofenac and clofibric acid in activated sludge bioreactors using liquid chromatography with quadrupole—Time-of-flight mass spectrometry. *Journal of Hydrology* 372: 109–117.

Lämmerhofer, M. 2010. Chiral recognition by enantioselective liquid chromatography: Mechanisms and modern chiral stationary phases. *Journal of Chromatography A* 1217: 814–856.

Langford, K.H., Reid, M., Thomas, K.V. 2011. Multi-residue screening of prioritised human pharmaceuticals, illicit drugs and bactericides in sediments and sludge. *Journal of Environmental Monitoring* 13: 2284–2291.

López-Serna, R., Pérez, S., Ginebreda, A., Petrovic, M., Barceló, D. 2010. Fully automated determination of 74 pharmaceuticals in environmental and waste waters by online solid phase extraction-liquid chromatography–electrospray-tandem mass spectrometry. *Talanta* 83: 410–424.

López-Serna, R., Petrovic, M., Barceló, D. 2011. Development of a fast instrumental method for the analysis of pharmaceuticals in environmental and wastewaters based on ultra high performance liquid chromatography (UHPLC)–tandem mass spectrometry (MS/MS). *Chemosphere* 85: 1390–1399.

López-Serna, R., Petrovic, M., Barceló, D. 2012. Occurence and distribution of multi-class pharmaceuticals and their active metabolites and transformation products in the Ebro River basin (NE Spain). *Science of the total Environment* 440: 280–289.

López-Serna, R., Kasprzyk-Hordern, B., Petrovic, M., Barceló, D. 2013. Multi-residue enantiomeric analysis of pharmaceuticals and their active metabolites in the Guadalquivir River basin (South Spain) by chiral liquid chromatography coupled with tandem mass spectrometry. *Analytical and Bioanalytical Chemistry* 405: 5859–5873.

MacLeod, S.L., Sudhir, P., Wong, C.S. 2007. Stereoisomer analysis of wastewater-derived ß-blockers, selective serotonin re-uptake inhibitors, and salbutamol by high-performance liquid chromatography–tandem mass spectrometry. *Journal of Chromatography A* 1170: 23–33.

Masiá, A., Ibáñez, M., Blasco, C., Sancho, J.V., Picó, Y. et al. 2013. Combined use of liquid chromatography triple quadrupole mass spectrometry and liquid chromatography quadrupole time-of-flight mass spectrometry in systematic screening of pesticides and other contaminants in water samples. *Analytica Chimica Acta* 761: 117–127.

Mastroianni, N., Postigo, C., De Alda, M.L., Barcelo, D. 2013. Illicit and abused drugs in sewage sludge: Method optimization and occurrence. *Journal of Chromatography A* 1322: 29–37.

Middleditch, B.S. 1989. *Analytical Artifacts: GC, MS, HPLC, TLC, and PC*. Elsevier, Amsterdam.

Nieto, A., Borrull, F., Pocurull, E., Marcé, R.M. 2010. Pressurized liquid extraction: A useful technique to extract pharmaceuticals and personal-care products from sewage sludge. *TrAC Trends in Analytical Chemistry* 29: 752–764.

Nödler, K., Licha, T., Bester, K. 2010. Development of a multi-residue analytical method, based on liquid chromatography–tandem mass spectrometry, for the simultaneous determination of 46 micro-contaminants in aqueous samples. *Journal of Chromatography A* 1217 (42): 6511–6521.

Ort, C., Lawrence, M.G., Rieckermann, J., Joss, A. 2010. Sampling for pharmaceuticals and personal care products (PPCPs) and illicit drugs in wastewater systems: Are your conclusions valid? A critical review. *Environmental Science & Technology* 44: 6024–6035.

Pérez-Carrera, E., Hansen, M., León, V.M., Björklund, E., Krogh, K.A. et al. 2010. Multiresidue method for the determination of 32 human and veterinary pharmaceuticals in soil and sediment by pressurized-liquid extraction and LC-MS/MS. *Analytical and Bioanalytical Chemistry* 398: 1173–1184.

Petrie, B., McAdam, E.J., Scrimshaw, M.D., Lester, J.N., Cartmell, E. 2013. Fate of drugs during wastewater treatment. *TrAC Trends in Analytical Chemistry* 49: 145–159.

Radjenović, J., Petrovic, M., Barceló, D. 2009a. Complementary mass spectrometry and bioassays for evaluating pharmaceutical-transformation products in treatment of drinking water and wastewater. *TrAC Trends in Analytical Chemistry* 28: 562–580.

Radjenović, J., Jelić A., Petrović, M., Barceló, D. 2009b. Determination of pharmaceuticals in sewage sludge by pressurized liquid extraction (PLE) coupled to liquid chromatography-tandem mass spectrometry (LC-MS/MS). *Analytical and Bioanalytical Chemistry* 393: 1685–1695.

Runnqvist, H., Bak, S.A., Hansen, M., Styrishave, B., Halling-Sørensen, B. et al. 2010. Determination of pharmaceuticals in environmental and biological matrices using pressurised liquid extraction—Are we developing sound extraction methods? *Journal of Chromatography A* 1217: 2447–2470.

Sebok, A., Vasanits-Zsigrai, A., Helenkár, A., Záray, G., Molnár-Perl, I. 2009. Multiresidue analysis of pollutants as their trimethylsilyl derivatives, by gas chromatography–mass spectrometry. *Journal of Chromatography A* 1216: 2288–2301.

Seed, B. 1997. Silanizing glassware. In *Current Protocols in Immunology*. Wiley, Hoboken, NJ, pp. A.3K.1–A.3K.2.

Stein, K., Ramil M., Fink, G., Sander, M., Ternes, T.A. 2008. Analysis and sorption of psychoactive drugs onto sediment. *Environmental Science & Technology* 42: 6415–6423.

Tran, N.H., Hu, J., Ong, S.L. 2013. Simultaneous determination of PPCPs, EDCs, and artificial sweeteners in environmental water samples using a single-step SPE coupled with HPLC-MS/MS and isotope dilution. *Talanta* 113: 82–92.

van Nuijs, A.L.N., Tarcomnicu, I., Bervoets, L., Blust, R., Jorens, P.G. et al. 2009. Analysis of drugs of abuse in wastewater by hydrophilic interaction liquid chromatography-tandem mass spectrometry. *Analytical and Bioanalytical Chemistry* 395: 819–828.

van Nuijs, A.L.N., Tarcomnicu, I., Simons, W., Bervoets, L., Blust, R. et al. 2010. Optimization and validation of a hydrophilic interaction liquid chromatography-tandem mass spectrometry method for the determination of 13 top-prescribed pharmaceuticals in influent wastewater. *Analytical and Bioanalytical Chemistry* 398: 2211–2222.

van Nuijs, A.L.N., Tarcomnicu, I., Covaci, A. 2011. Application of hydrophilic interaction chromatography for the analysis of polar contaminants in food and environmental samples. *Journal of Chromatography A* 1218: 5964–5974.

van Nuijs, A.L.N., Abdellati, K., Bervoets, L., Blust, R., Jorens, P.G. et al. 2012. The stability of illicit drugs and metabolites in wastewater, an important issue for sewage epidemiology? *Journal of Hazardous Materials* 239: 19–23.

Vazquez-Roig, P., Segarra, R., Blasco C., Andreu, V., Picó Y. 2010. Determination of pharmaceuticals in soils and sediments by pressurized liquid extraction and liquid chromatography tandem mass spectrometry. *Journal of Chromatography A* 1217: 2471–2483.

Ward, T.J., Farris III, A.B. 2001. Chiral separations using the macrocyclic antibiotics: A review. *Journal of Chromatography A* 906: 73–89.

3 High-Resolution Mass Spectrometric Techniques for Structural Characterization and Determination of Organic Pollutants in the Environment

Nuria Cortés-Francisco and Josep Caixach

CONTENTS

3.1 INTRODUCTION

Mass spectrometry (MS) has a dynamic history with a continually progressing technology that has made significant advances into different fields and especially in environmental analysis (Gelpí, 2008, 2009). Since Aston's (1927) paper, magnetic sectors were the mass analyzers dominating the scene of high-resolution and accurate mass measurements.

The use of high-resolution mass spectrometry (HRMS) for structural characterization and determination of organic compounds was first applied by John Beynon in the 1950s. Beynon used a single magnetic sector for the identification of a wide variety of molecules based on accurate mass measurements. Beynon explained for the first time that if the mass of an ion from a chemical compound was determined with sufficient accuracy, the elemental composition of this compound could be determined (Beynon, 1954, 1960). In his work entitled *Mass Spectrometry and Its Applications to Organic Chemistry*, Beynon highlighted the importance of carrying out very accurate mass measurements to distinguish between different possible candidates when structural characterization and determination of an organic compound should be carried out. Based on his experience, Beynon already pointed out in his work different rules (e.g., isotopic pattern, ring plus double-bond equivalents [RDBE]) that could be applied to identify one molecule from a list of several candidates. Some of these assumptions are still in use nowadays.

At that moment, the availability of such measurements was limited due to the cost and complexity of instrumentation (Bristow, 2006). However, the panorama of MS early described by Brunée (1987) has been substantially modified (Makarov, 2009). HRMS is no longer restricted to magnetic sectors, and a wide variety of mass analyzers have been designed and improved for this purpose (Bristow, 2006).

The use of HRMS presents several advantages with regard to low-resolution or tandem mass spectrometry (MS/MS). In environmental analysis, the number of potential organic pollutants might be huge. These contaminants can be at very low concentrations and can be unknown or occur as a result of the degradation of the original contaminant (Hernández et al., 2012). High resolution helps to trace contaminants, as well as to identify analogues for which standards are not available (Krauss et al., 2010).

This chapter starts with a review of HRMS terminology, followed by the discussion of some parameters influencing accurate mass measurements and the identification of organic compounds when using HRMS. Section 3.5 focuses on the main mass analyzers capable of acquiring high-resolution mass data. Finally, some environmental applications are discussed. We have chosen only a few examples that illustrate the main advantages and/or requirements of using a high resolution, more than just giving a list of recent studies using HRMS applied to environmental analysis, which are available in some recent reviews (Krauss et al., 2010; Hernández et al., 2012; Lim et al., 2016).

3.2 HIGH-RESOLUTION MASS SPECTROMETRY TERMINOLOGY

As can be seen in the literature, there are many applications for which accurate mass measurement is used. However, in a study by Bristow and Webb (2003), a lack of understanding (particularly among newer users) has been detected. For this reason, guidance on undertaking key aspects of the methodology in order to obtain robust measurements and traceable data has been reported (Webb et al., 2004; Bristow, 2006). Moreover, concerning the number of publications quoting accurate mass data, in which some terms are used inconsistently, guidance for terminology and treatment of data has been also described in detail (Brenton and Godfrey, 2010) and the main definitions are summarized. Also, the International Union of Pure and Applied Chemistry (IUPAC) recommendations have been recently published (Murray et al., 2013). An example is given in Figure 3.1 to illustrate the main definitions.

- *Nominal mass:* The mass of an ion or molecule calculated using the mass of the most abundant isotope of each element rounded to the nearest integer value and multiplied by the number of atoms of each element.

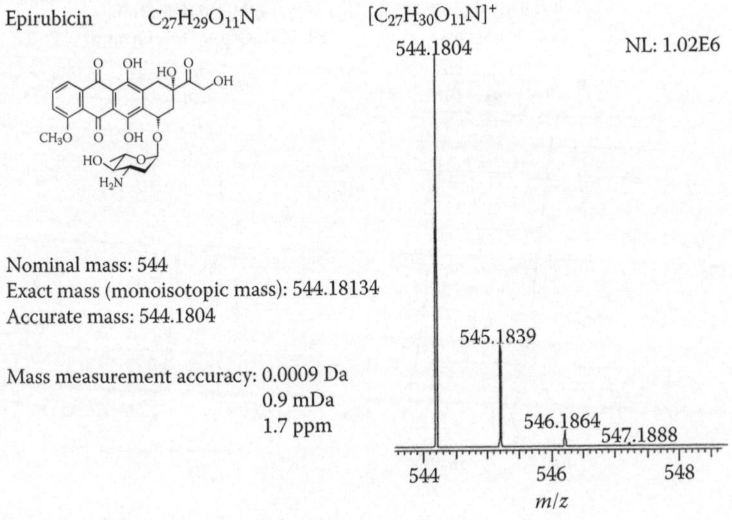

FIGURE 3.1 Nominal mass, exact mass, accurate mass, and mass measurement accuracy for the ion $C_{27}H_{30}O_{11}N^+$ of the neutral compound epirubicin. The mass spectra are also shown. NL, normalized level.

- *Accurate mass:* The experimentally determined mass of an ion measured to an appropriate degree of accuracy and precision used to determine, or limit the possibilities for, the elemental formula of the ion.
- *Exact mass:* The calculated mass of an ion whose elemental formula, isotopic composition, and charge state are known—that is, the theoretical mass. The IUPAC definition constricts the definition to using one isotope of each atom involved, usually the lightest isotope, but generalizes the definition to cover an ion or neutral molecule. The charge state is relevant, as the mass of the electron (0.00055 Da) or multiple charges may not be negligible in the context of mass measurement.
- *Monoisotopic mass:* The exact mass of an ion or molecule calculated using the mass of the most abundant isotope of each element.
- *Mass measurement accuracy:* The difference between the measured value (accurate mass) and the true value (exact mass) is the "accuracy" of the "accurate mass measurement," and it is suggested that the term *mass measurement accuracy* should be used to denote this difference (Brenton and Godfrey, 2010).

The mass measurement error (or accuracy) of a single reading will be the following:

$$\begin{aligned}\Delta m &= \left(m_{\text{Theoretical}} - m_{\text{Experimental}}\right) \text{in daltons} \\ &= \left(m_{\text{Theoretical}} - m_{\text{Experimental}}\right) \times 10^3 \text{ in millidaltons} \\ &= \left(\frac{m_{\text{Theoretical}} - m_{\text{Experimental}}}{m_{\text{Theoretical}}}\right) \times 10^6 \text{ in parts per million}\end{aligned} \tag{3.1}$$

If several measurements are recorded and an average result should be calculated, it is usually expressed as the root-mean-square (RMS) to prevent negative and positive values from cancelling each other. Another option is to calculate the average absolute mass measurement accuracy (Brenton and Godfrey, 2010).

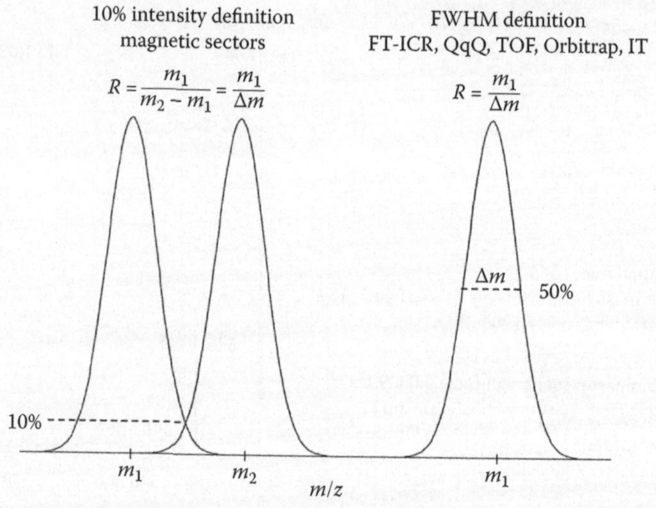

FIGURE 3.2 Definitions of resolution.

- *Mass measurement precision:* This is the repeatability of the measurement reflecting random errors. Random errors cause measurements to fall on either side of the average experimental measurement and affect the precision of the set of measurements. Although it is not usually reported in accurate mass measurements, precision is very important to know the uncertainty (U) of the mass measurement. Standard deviation may be used to quote the precision of the data set.
- *Mass resolution and mass resolving power:* It should be pointed out that the definition of resolution and resolving power has been a subject of controversy. In the last recommendations from IUPAC, resolution (R) is defined as

$$R = \frac{m}{\Delta m} \tag{3.2}$$

The m/z value at which the measurement was made should be reported as well as the definition and method of measurement of $\Delta(m/z)$. For instance, for magnetic sector instruments, two peaks of equal intensity are considered to be resolved when they are separated by a valley, which is 10% of the height of each peak. The definition used with quadrupole, Fourier transform ion cyclotron resonance (FT-ICR), ion trap (IT), time-of-flight (TOF), and FT Orbitrap mass spectrometers is based on a peak width measured at 50% peak height (full width half maximum [FWHM]), producing a value approximately double that calculated using the 10% valley definition (see Figure 3.2) (Bristow, 2006). The resolving power is the ability of an instrument or measurement procedure to distinguish between two peaks at m/z values differing by a small amount and expressed as the peak width in mass units (Murray et al., 2013).

3.3 FACTORS INFLUENCING THE ACCURACY AND PRECISION OF AN ACCURATE MASS MEASUREMENT

Ideally, we would like to obtain the best accuracy and precision, due to the fact that depending on the mass measurement accuracy (mDa or ppm) achieved, there will be a higher or lower number of potential elemental formulas for a compound. The better the accuracy, the less the ambiguity. But, can we measure the mass of a compound with such a great accuracy and precision?

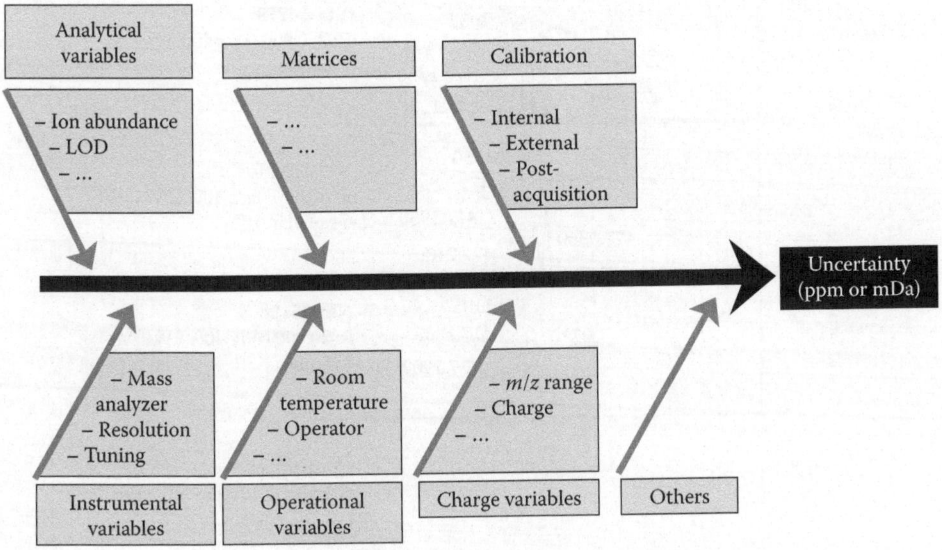

FIGURE 3.3 Factors that can influence the uncertainty (accuracy and precision) of a measurement.

The accuracy and the precision achieved depend on several factors as shown in Figure 3.3, such as resolution, interferences of the matrix, the mass analyzer, calibration protocol, and some operational factors, such as room temperature.

Webb et al. (2004) wrote a practice guide to provide users and suppliers of accurate mass instrumentation the essential steps in obtaining reliable data. They highlighted that in order to achieve high accuracy and precision, a number of key considerations when carrying out accurate mass measurements must be understood and optimized. These included tuning and peak shape, ion abundance, resolution, calibration, sample introduction, data manipulation, validation, and quality control checks. Some of these considerations should be general to any method development when using MS. The parameters that we have considered to be crucial for accurate mass measurements are discussed in the next sections.

3.3.1 Resolution

Regarding mass measurement accuracy, care should be taken when linking these ideas with high resolution. Low-resolution instruments have not traditionally been employed to record accurate mass measurements, but some very good results (accurate and precise measurements) have been obtained (Bristow, 2003; Balogh, 2004). These results showed that when there are no unresolved isobaric ions and a strict experimental protocol is used, accurate mass measurements can be independent from resolution. However, in everyday analysis when analyzing complex matrixes, such as in the field of environmental analysis, high resolution can help preventing false assignments. Moreover, high resolution is mandatory for the identification of nontarget compounds (transformation products of known compounds) or screening of unknowns. Definition of what is considered to be high resolution is difficult. Legislation applying to method development (European Commission, 2002) is still not updated. At the moment, we have only the old definition based on magnetic sector instruments (high resolution is 10,000 at 10% valley), which translates to 20,000–25,000 (FWHM). Fixing a value is always a controversial issue, due to commercial interests. Some reference guides state that a resolution higher than 20,000 in all mass ranges should be considered high resolution (European Commission, 2012); however, in complex mixtures such as natural organic matter (NOM), this is not enough, as shown in Figure 3.4. Even when $R = 20,000$ or 50,000 (FWHM) is used (Figure 3.4b

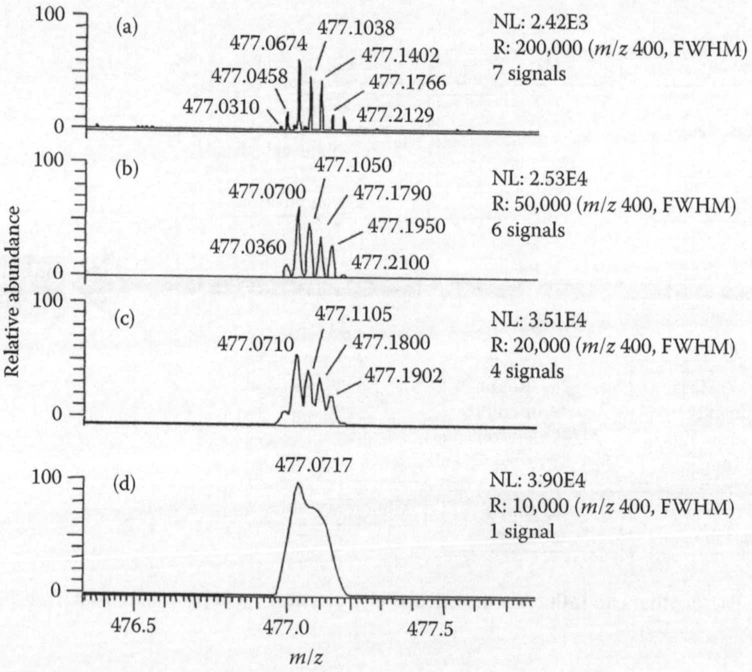

FIGURE 3.4 Mass spectrum of fulvic acids (zoom at *m/z* 477) from the International Humic Substance Society acquired with an FT-ICR mass analyzer at different resolutions (R). (a) R: 200,000, (b) R: 50,000, (c) R: 20,000, and (d) R: 10,000.

and c), all the signals are not resolved and the main consequence is that the elemental compositions that would be obtained by the accurate masses displayed in Figure 3.4b through d would be wrong and would never correspond to the real ones. This is an extreme case, but this is the effect that any interference can have on the accurate mass measurement of a compound. One may think that working in the highest resolution mode will solve all the problems; however, it should be taken into account that the resolution has two major effects on the mass spectra of compound: the width of the signal and the intensity. The ion abundance (or intensity) may also have an effect on accuracy and precision, and in addition, the scan rate would be slow when a higher resolution is achieved (see Section 3.5).

3.3.2 Tuning, Peak Shape, and Ion Abundance

The tuning of the instrument should be good enough to ensure that the peak shape is symmetric and we have enough ion abundance. If the optimization of the ion source (e.g., type of ion source, temperatures, voltages) and the mass analyzer (e.g., scan rate, resolution, ion accumulation) is not properly done, it may result in a low ion abundance and, therefore, in a worse mass accuracy. We strongly discourage to acquire data in centroid mode, instead of profile mode. Despite the inconvenience of memory that profile raw files may have, the peak shape can give us a clue to how properly or badly we are performing the accurate mass measurements. For instance, the peak shape can reveal if there is more than one compound under the envelope (Figure 3.4d).

3.3.3 Calibration Protocols

There are a number of methods for accurate mass measurement, which involve different approaches and instrumentation, but all involve calibration of the mass scale (Godfrey and

Brenton, 2012). Mass calibration consists of fitting the observed mass measurements to the exact mass of two or more different ions. The calibration process is crucial to achieving highly accurate mass measurements. Calibration may be external (i.e., reference masses from a mass spectrum of another analyte acquired under similar conditions) and/or internal (i.e., the reference masses are for ions of known elemental composition in the same mass spectrum as the analyte) (Marshall and Hendrickson, 2008).

In all mass spectrometers, external calibration is done automatically, with the infusion of a standard. The corrections to the mass axis are stored by the instrument and frozen in memory. A good external calibration is required to obtain good mass accuracy, and it only works satisfactorily if the instrument is stable (Webb et al., 2004). For some high-resolution mass spectrometers such as FT-ICR and FT Orbitrap, external calibration is very stable and no other calibration protocols might be applied. For TOF and magnetic sector mass analyzers, an additional internal calibration has been usually required, to prevent calibration drift (Cortés-Francisco et al., 2011).

In internal calibration, the *m/z* scale is calibrated using ions of known *m/z*, usually called lock masses, which are scanned during the same experiment as the compounds of interest. The lock masses can be introduced into the mass spectrometer mixed in the sample solution, infused from a separate source into the mass spectrometer, or teed in at the same time as the analytes. The main problem is that ion suppression and chemical interferences might be created between lock masses and analytes (Webb et al., 2004). With internal calibration, the mass range can be constantly recalibrated during acquisition, but an internal calibration can be also done a posteriori of the data acquisition, which is called postprocessing or postacquisition calibration. In postacquisition mode, no mass corrections are made to the peak in the scan; only external calibration stored in the instrument memory is used. The main purpose is to recalculate the data acquired with respect to the exact masses of a reference compound which was present in the ion source during the measurement. The reference compound serves as a "ruler" in calculating the accurate masses of the unknowns that are adjacent to the known compounds. The main advantage of postacquisition calibration is that the mass analyzers do not waste acquisition time looking for the lock mass and constantly recalibrating during acquisition, and more points per peak can be obtained in the analysis. Moreover, accuracy and precision are as good as with the common internal calibration (Cortés-Francisco et al., 2011). Internal calibration is typically twice accurate as external calibration (Marshall and Hendrickson, 2008).

3.4 ELEMENTAL FORMULA DETERMINATION

Once the accurate mass measurement of the ion has been recorded, the value is used to generate a list of elemental formulas. In most of the commercial software controlling the instruments, the elemental formula determination can be carried out manually. The user has to set a few minimum parameters to obtain a list of candidates: uncertainty of the data (mass tolerance), elements to be used, number of elements, and charge. Although these restrictions are set, the final list will contain many possible elemental formulas, and therefore, a number of key strategies should be employed to refine the list. Portolés et al. (2011) claimed that elucidation of unknowns cannot be achieved by following a standardized procedure, as both expertise and creativity are necessary in the process. We partially agree with this statement; however, a minimum of information should be included when reporting analysis of unknowns or simply when using HRMS. No one would ever submit a study for quantification without the quantification limits. Strict criteria should also be applied when judging an HRMS determination. However, with no doubt, HRMS is one of the areas of MS where expertise is specially needed.

3.4.1 Mass Measurement Accuracy and Precision: Uncertainty of the Data

As required by the *Journal of the American Society for Mass Spectrometry*, setting fixed acceptable error limits for accurate mass measurement is not recommended. It should be highlighted that it has always given more importance to accuracy rather than to precision. Precision has received limited coverage in the literature, but it is important, and both accuracy and precision should be considered when talking about uncertainty in the measurement and evaluation of mass spectrometers. As far as possible, it is necessary to know the uncertainty associated with an accurate mass measurement, as it is the limit of quantification associated to quantify. Only in this way can the list of theoretical candidates that should be taken into account be known.

It is important to understand the degree of accuracy and precision required relative to (i) the *m/z* of the ion that is measured; (ii) the intended use of the mass measurement, target and nontarget analysis (it is not the same confirming an identity from a known compound or looking for unknowns); and (iii) previous information that we have from the sample or the pollutants.

(i) The *m/z* of the ion that is measured: With increasing *m/z*, the number of formulas which will fit a measured molecular mass increases; therefore, an unambiguous result becomes impossible to obtain (Quenzer et al., 2002). In a previous study (Cortés-Francisco et al., 2011) considering mass range between *m/z* 100 and 600, it has been observed that there is less accuracy in mass measurements for higher *m/z* compounds than for lower ones for several mass analyzers tested and different calibration protocols applied. Regarding precision, the same behavior as for accuracy has been detected; precision of the measurements is worse at higher *m/z* values. It can be concluded that the uncertainty in the measurements is, in all cases, worse for higher *m/z* values. In contrast, a better mass measurement, in accuracy and precision terms, is needed for higher masses in order to have less candidates and the correct molecular formula (Webb et al., 2004). (ii) The intended use of the mass measurement: target or nontarget analysis: It is not the same confirming an identity from a known compound (where standards are available) or looking for unknowns. In some cases, sub-ppm or millidalton accuracy and precision for daily routine analysis have been considered adequate to discriminate between possible candidates, taking into account the nature of the compounds (*m/z* 100–600) (Cortés-Francisco et al., 2011). In many cases, additional information other than just the accurate mass measurement will be needed to obtain the correct elemental composition. This includes restrictions for the number of elements, isotopic pattern, RDBE, and nitrogen rule, which are discussed in the next section. The correct determination of the elemental formula will be definitively related to (iii) the previous information that we have from the sample or the pollutants.

3.4.2 Elements to Consider

For the formula generation, if we have no information of the kind of compound that we are looking for, we should consider all the possible elements in an organic compound. There are many elements to consider, and nobody is doing such. For example, in the seven golden rules study, Kind and Fiehn (2007) consider, in general, the following elements: C, H, N, O, P, S, F, Cl, Br, and Si. Logically, in the examples, they usually reduce the elements, for instance, C, H, N, O, S, P, and Si. That is what people usually do, based on previous information. Some studies report screening of unknowns, but the screening is focused on some derivates of well-known compounds. In these cases, the term *nontarget* would apply better.

3.4.3 Restrictions of Element Number

The restriction of element numbers is important to obtain a number of reasonable candidates that we can, afterward, analyze based on other strategies (e.g., isotopic pattern). The easy one is the calculation of maximum carbons: dividing the mass range (maximum nominal *m/z* of the spectra) through the element mass (12 for carbon). For the rest of elements, Kind and Fiehn (2007) recommend some restrictions. However, any previous information about the sample will help to set these restrictions in a more appropriate way. This is especially important for the analysis of tandem mass spectra, where no selection of the precursor is carried out ("all-ion fragmentation" [AIF] mode in Orbitrap or MS^E in TOF). Filtering the fragments of the spectra coming from the molecule in front of the rest coming from the matrix is easy if restrictions in elements and number of elements are done. For instance, if we assume that the molecular ion could be $[C_{27}H_{30}NO_{11}]^+$, we are going to restrict the elements as follows: $C \leq 27$, $H \leq 30$, $N \leq 1$, and $O \leq 11$.

3.4.4 Ion Type, Adduct Formation, and Charge: RDBE and Nitrogen Rule

If the ion type, the adducts formed in the ion source, and the charge are known (which usually the case), any formula giving an inappropriate RDBE value (meaning integer or noninteger) should give enough evidence to ignore incorrect formulas (Table 3.1) (Webb et al., 2004).

$$\mathrm{RDBE} = \mathrm{C} + \mathrm{Si} - \frac{1}{2}(\mathrm{H} + \mathrm{F} + \mathrm{Cl} + \mathrm{Br} + \mathrm{I}) + \frac{1}{2}(\mathrm{N} + \mathrm{P}) + 1 \tag{3.3}$$

It is true that the RDBE formula considers only the lowest valence state for each element and, sometimes, the true RDBE values are not calculated correctly. It is also suggested (Kind and Fiehn, 2007) to use it to detect formulas with an extremely high RDBE value, due to the fact that most of the compounds were found to have a RDBE <40 (exception of fullerenes).

The nitrogen rule (Table 3.2) is suggested for some authors (Webb et al., 2004; Bristow, 2006) to know if the molecule has an odd, even, or zero number of nitrogen atoms, based on the odd or even *m/z*. It has been shown that this assumption can be done only with nominal masses and when using accurate mass measurements for *m/z* <500.

3.4.5 Isotopic Pattern and Isotope Ion Ratio

From a list of feasible candidates, the isotopic pattern could help to eliminate the less probable formulas, especially not only for halogenated compounds, but also for Si and S compounds. This process is not easily done automatically by the software, and if it is done, the scientist should be skeptic on how the software is comparing the isotopic pattern and giving this percentage of similarity or isotopic fit. The first question that should be considered is, how many isotopic peaks (M+1, M+2, M+3) are taken into account for the comparison? How is the theoretical isotopic pattern simulated? Sensitivity and resolution can dramatically change the theoretical isotopic pattern. An example is shown in Figure 3.5. If a very high resolution is used (e.g., R = 300,000 FWHM) for the in silico

TABLE 3.1
Ion Type and RDBE Value Expected, Based on Equation 3.3

Ion Type	RDBE Value
Even electron ion, e.g., $[M+H]^+$ $[M-H]^-$	Noninteger
Odd electron ion, e.g., $M^{+\cdot}$	Integer

TABLE 3.2
Nitrogen Rule

Odd *m/z*	Even *m/z*	Number of Nitrogen Atoms
Even electron ion, e.g., $[M+H]^+$ $[M-H]^-$	Odd electron ion, e.g., $M^{+\cdot}$	N_0, N_2, N_4, ...
Odd electron ion, e.g., $M^{+\cdot}$	Even electron ion, e.g., $[M+H]^+$ $[M-H]^-$	N_1, N_3, N_5, ...

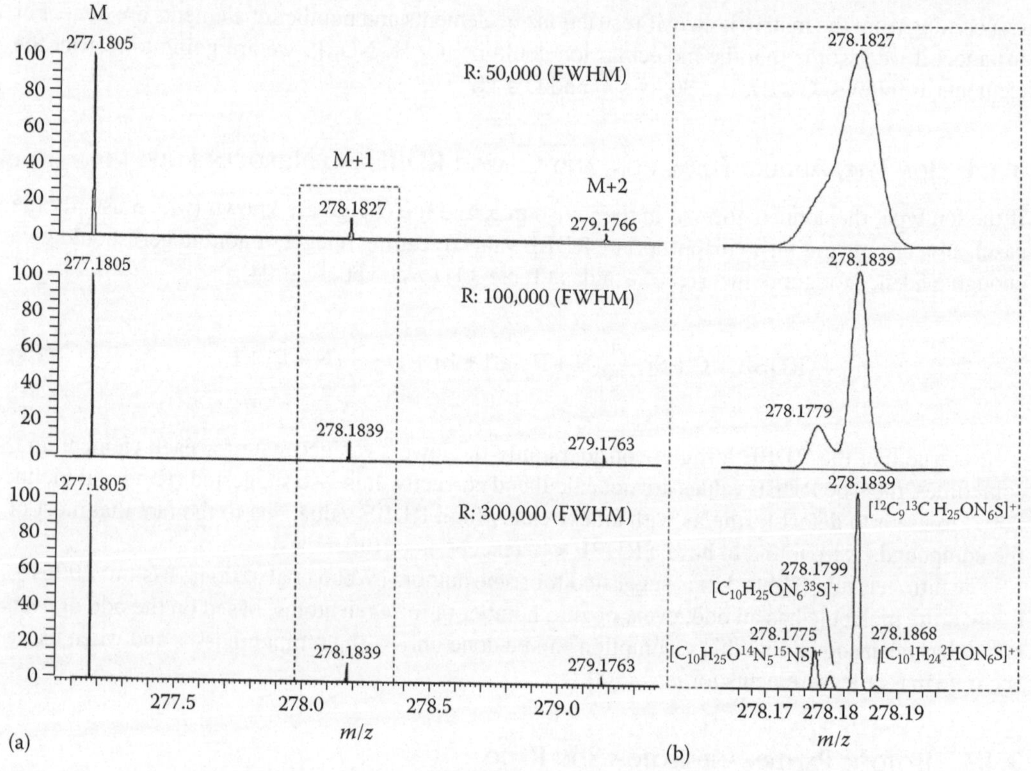

FIGURE 3.5 (a) Isotopic pattern simulation at R = 50,000, R = 100,000, and R = 300,000 (FWHM) of the ion $[C_{10}H_{25}ON_6S]^+$, where the molecular ion (M) and the isotopic ions (M+1 and M+2) are shown. (b) Zoom of M+1 (nominal mass 278), showing the different contributions of the natural abundant isotopes.

isotopic pattern, the M+1 peak can be so well resolved that the different contributions of the natural abundant isotopes can show a isotopic pattern very different from the experimental one. Simulation should be carried out under the same conditions that the experimental pattern is obtained.

If possible, an alternative to in silico isotopic pattern comparison will be the comparison of the isotopic pattern of standards (analyzed under identical experimental conditions) to the *m/z* of interest. This can be done by calculating the isotope ion ratio. The ratio is calculated as the ratio between the monoisotope ion (diagnostic ion) and the isotope ion (M+1, corresponding to the natural isotope ^{13}C). It is easier to perform it when the molecular ion has a relatively high *m/z* (*m/z* ≥500), and so, M+1 is around 50%. The calculation of the isotope ion ratio can be easily and automatically done by any software (Domènech et al., 2013).

3.4.6 Valences (Lewis and Senior Rules)

The Lewis and Senior rules measure the occupancy of s and p orbitals. The Lewis rule is based on the octet rule (elements share electrons with total occupancy). The Senior rule adds some criteria, so that the maximum and mixed valence states for each element are taken into account (Kind and Fiehn, 2007).

3.4.7 Element Ratios and Element Probability Check

When thousands of elemental formulas can be extracted from a spectrum, it is very useful to filter less probable candidates based on elemental ratios and element probability check. This protocol is usually applied in the characterization of NOM (Herzsprung et al., 2010) or metabolomics. A list of common element ratios is described in the study by Kind and Fiehn (2007), which can be very useful, especially for the H/C ratio.

3.4.8 Complementary Techniques to Obtain Structural Information

High resolution and high accuracy are powerful tools to confirm an identity, but they are insufficient for obtaining a unique structure of an unknown yet. Structural information is needed, and it can be complemented with MS/MS experiments and/or nuclear magnetic resonance (NMR), if possible (Webb et al., 2004; Cortés-Francisco et al., 2011).

3.4.9 Any Additional Information

Kind and Fiehn (2007) suggested the subtraction of the trimethylsilyl derivatization agent used for the analysis of some molecules by GC-MS to distinguish between candidates (rule 7). Generally speaking, any information that we can have from the sample or the pretreatment (derivatization step, pH extraction, purification, solubility) can help to restrict the parameters described in Sections 3.4.2 to 3.4.7 to obtain a shorter list of reasonable candidates. An example is given by Vichi et al. (2013) in a recent study applied for the determination of volatile thiols in a lipid matrix. Due to the fact that a very specific derivatization agent was used (ebselen), which contains selenium, restrictive criteria were set to generate reliable elemental formulas: $C \geq 13$, $O \geq 1$, $N = 1$, $Se = 1$, $S = 1$, $Na \leq 1$, and $RDBE \geq 8.5$. It was known that in this case, any nontargeted thiol will react with this agent, and so, a minimum number of elements and RDBE could be set. In addition, selenium has a characteristic isotopic pattern, so in this case, it was also very helpful for ranking candidates.

In some fields, the determination of the molecular formulas can be confirmed if any regular pattern between the different compounds of the sample is known. Such is the case for marine NOM, where regular patterns with mass differences of 14.0156 Da, which spanned the entire spectra, are the result of ions that differ from each other in the number of CH_2 groups. Other typical repeated spacing of 2.0156 Da is a result of ions that differ from each other in number of 2H, and moreover, mass distances arose from a replacement of CH_4 by O with a mass difference of 0.0364 Da. Van Krevelen plots and Kendrick mass defect (KMD) diagrams have been, until the moment, the best way adopted for marine chemistries and petroleomic analyzers to face this problem (Kim et al., 2003; Stenson et al., 2003). Van Krevelen diagrams are useful to classify substance classes in a first approach, and KMD is used to sort out homologous series of organic compounds.

For some other applications, such as metabolomics, a small subset of the mass spectrum signals can be used to build a mass difference network, based on metabolic pathways and reactions. This way, any *m/z* in the spectrum that cannot be part of this compositional space map could be considered to be not well assigned or not significant (Forcisi et al., 2013).

An example is given in the following to illustrate the different strategies that can be used to obtain an elemental formula from an accurate mass measurement.

For example, liquid chromatography (LC)–HRMS analysis (*R*: 50,000 FWHM) and accurate mass measurements of a wastewater sample after liquid–liquid extraction have been carried out. We want to identify a signal in the spectra (Figure 3.6) with accurate mass *m/z* 498.9302. We assume that no previous information is available (real unknown), and for this reason, we generate the list of candidates based on the parameters described in Table 3.3. When no additional information is available, it has been shown that setting very restrictive criteria will lead to an erroneous candidate (Cortés-Francisco et al., 2011).

If looking for unknowns, very general parameters have to be set and restrictions have to be done step by step, after careful observation of the spectra. The main drawback of setting very general parameters is that a long list of candidates is usually obtained. In this example, we get more than

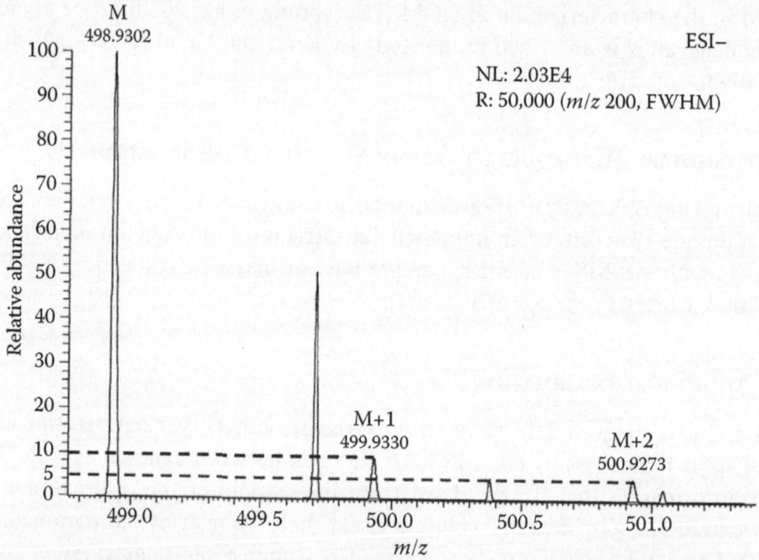

FIGURE 3.6 Experimental spectrum obtained from the analysis of a wastewater sample after liquid–liquid extraction, showing the signal that should be identified (*m/z* 498.9302) and the isotopic pattern (M+1 and M+2) and the intensities.

TABLE 3.3
Parameters Used to Generate the List of Candidates for the Identification of *m/z* 498.9302

Parameter			Comment
Uncertainty (accuracy and precision)	<2 ppm		Accuracy and precision obtained with $n = 7$ replicates of the calibration levels
Ion type/adduct formation/charge	M^-, charge = −1		Negative ESI mode
RDBE	Maximum 40		No fullerenes
Elements to consider	CHNOPSFClBrSi		A priori, if no additional information available
Restrictions of elements	Carbon	Maximum: 499/12 = 42 Minimum: 1	Maximum limit for a hypothetical molecule that consists exclusively of carbon
	Rest	Seven golden rules restrictions	A priori, if no additional information available

400 candidates and the correct hit is ranked in position 120. Based on the parameters defined in the text, we can work to refine the list and try to reduce the list (see Table 3.4).

In this case, detailed observation of the *isotopic pattern* helps to reduce the number of candidates. Halogenated compounds (containing Cl and Br) have a very characteristic pattern, but something similar happens with Si, due to the fact that ^{29}Si and ^{30}Si have natural abundances of 4.6% and 3.1%, respectively. These significant abundances will definitely contribute to the intensity of M+1 and M+2. Since for the compound of interest, M+1 is less than 10% and M+2 is around 5% (Figure 3.6), the isotopic pattern presents no special features. That is the reason why the principal hypothesis is that the main contribution for M+1 is due to ^{13}C. Without taking into consideration the halogenated compounds and Si, we reduce the number of candidates from more than 400 to 97. If we assume that M+1 is principally due to ^{13}C, we can recalculate the maximum number of carbons, reducing from 42 to 10 the number of carbon atoms.

TABLE 3.4
Parameters Used to Refine the List of Candidates for the Identification of *m/z* 498.9302

Parameter	Comment	Reset Parameters		Number of Candidates	Correct Hit in Position
Nitrogen rule	*m/z* 499 → N = 0 or even number of N atoms Hypothesis: N = 0	Elements to consider	CHOPSFClBrSi	>400	22nd
Isotopic pattern	No characteristic isotopic pattern from halogenated atoms	Elements to consider	CHOPSFSi	>400	9th
Isotopic pattern	Isotopic pattern Si: M+1 and M+2 more intense	Elements to consider	CHOPSF	97	3rd
Isotopic pattern: M+1	M+1 intensity: principally contribution of ^{13}C	Restrictions of elements	Maximum number of carbons: 10	73	1st
Isotopic pattern: M+2	M+2 intensity: principally contribution of ^{34}S and ^{18}O	Restrictions of elements	S = 2 and O = 0	No candidates	–
		Restrictions of elements	S = 1 and O = 1–5	2 candidates	1st
		Restrictions of elements	S = 0 and O = 22–27	1 candidate	Not the correct One
Lewis and Senior rules	Evaluate the 3 candidates	The 3 candidates fulfill the rules		3 candidates	–
Database	Search the 3 candidates	Only one candidate found		1 candidate	1st

With the elements in use (CHOPSF) and if we focus on M+2, we can see that the only elements that can contribute to M+2 are S and O (isotope natural abundances: ^{34}S, 4.3%, and ^{18}O, 0.2%). Based on the fact that M+2 is around 5%, we can have three different scenarios, as shown in Table 3.4. With these three options, we end up with three candidates (Figure 3.7).

The isotopic patterns of these compounds show very little differences, and there is no easy way to distinguish which is the correct option. At this point, we can check for the Senior and Lewis rules, as well as search in databases. Applying the software developed by Kind and Fiehn (2007), we checked that both the Lewis and Senior rules are fulfilled. However, when searching in databases, only the $[C8O3F17S]^-$ candidate is found. This formula corresponds to perfluorooctane sulfonate, which is an emerging pollutant that can be, for sure, present in a wastewater sample. This is the correct hit.

3.4.10 Database and Automatic Data Processing

Using HRMS enables the extraction of thousands of features from environmental samples. However, as it has been demonstrated previously, the identification of discriminating signals is the most laborious and time-consuming step of accurate mass measurements. It remains unsolved how to process all the information contained in a highly resolved mass spectrum in an automatic and fast way.

One of the reasons why GC-MS has dominated in environmental applications (for quantification and screening purpose) is due to the successful use of mass spectral libraries, such as that of the National Institute of Standards and Technology. Few attempts have been done to have the same for LC/MS. It has been a challenge due to solvent, voltage, and temperature effects in adduct formation and ionization efficiencies. Moreover, LC-MS/MS libraries have also been designed, but again, the different voltages applied for fragmentation in the cell change from one instrument to the other. Nowadays, MassBank is one of the free largest tandem mass spectra database available. Apart from

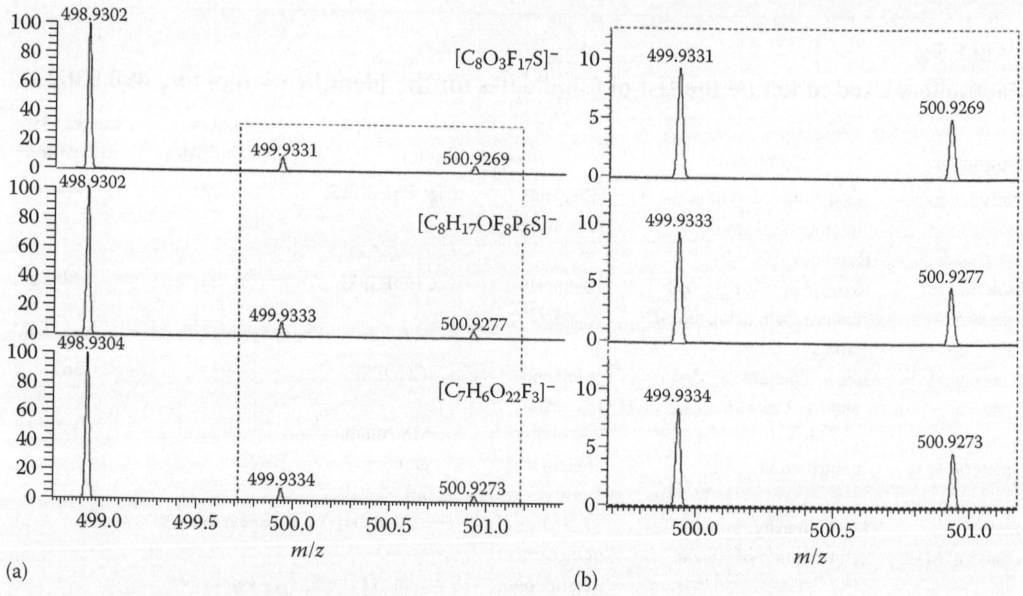

FIGURE 3.7 (a) Isotopic pattern simulation of the three final candidates. (b) Zoom of M+1 and M+2 (nominal mass 500) to show that no difference between the three molecules can be observed.

these spectral libraries, there are also compound databases such as PubChem, Reaxys, Chemspider, and KEGG, but usually they do not have spectral information.

At the moment, screening of HRMS fragments with several peaks within a reasonable time frame for automated batch processing and fast database queries is not available, although work is being done in this direction. For instance, MetFrag was designed to generate all possible topological fragments of a candidate compound in order to match the fragment mass with measured peaks (Wolf et al., 2010). More recently, MetFusion has been created to integrate the results from MassBank and MetFrag, so that the identification can be carried out with the combination of experimental spectral libraries and in silico fragmentation (Gerlich and Neumann, 2012).

3.5 INSTRUMENTATION

Identification and quantification of compounds require appropriate sample preparation (extraction, enrichment, purification) followed by instrumental analysis, generally coupling GC and/or LC to MS, depending on the nature of the compounds. Reasonably, GC or LC plays an important role in separation and consequent organic pollutant identification.

Apart from the chromatographic separation, the optimal ionization source should also be identified. There are many different commercially available ion sources for environmental analysis: electron ionization (EI), chemical ionization (CI), negative ion chemical ionization, and, more recently, atmospheric pressure chemical ionization (APCI) are used for volatile and semivolatile compounds eluted from GC; electrospray ionization (ESI), APCI, and atmospheric pressure photoionization (APPI) are used for thermally labile polar and nonpolar compounds eluted from LC. Each source varies in its ionization mechanism, and consequently, the analytical window for each is quite different.

The state of the art in the majority of GC-MS studies is the EI interface. EI is a positive hard ionization source which is highly reproducible. Fragmented spectra are usually obtained, and the identification of compounds is based on the use of libraries. However, the extensive fragmentation of the molecule provides no information about the molecular ion ($M^{+\cdot}$), so screening of nontarget compounds should be carried out based on the main fragment ion and the availability of the spectra in commercial libraries. Mass spectrometrists have ever been searching for ionization methods softer than EI, because molecular weight determination is of key importance for structure elucidation. The recent use of APCI with eluting compounds from GC permits to keep the molecular ion intact and facilitates the screening of unknowns based on their accurate mass measurement and molecular formula (Portolés et al., 2011). For LC-MS, ESI is now the ionization method of choice. With this ionization technique, the analysis of nonvolatile intact molecules can be performed (Kind and Fiehn, 2010). Compared to the other interfaces (APCI and APPI), ESI covers with great sensitivity the efficient ionization of a wide variety of chemical compounds. Moreover, as it is a soft ionization source, usually, molecular ions ($[M+H]^+/[M-H]^-$) are obtained, as no or little fragmentation occurs in the ion source. ESI is sensitive to changes in solvent composition, and so, molecular ions resulting from clusters with sodium or ammonium can occur. Moreover, multiple charged ions can also be formed, which enables the analysis of larger molecules such as peptides or polymers (Cole, 1997; Kebarle and Peschke, 2000; Cech and Enke, 2001).

Moreover, in the last decade, new approaches designed for the direct analysis without chromatographic separation have been introduced (Cooks et al., 2006). These approaches, pioneered by Cooks, enable fast analysis without any or with minimum sample preparation, which increases the sample throughput. Some of these new ionization techniques are desorption ESI (Takáts et al., 2004, 2005), direct analysis in real time (Gross, 2014), desorption APCI (Chen et al., 2007), and desorption APPI (Haapala et al., 2007).

In general, the choice of the ion source has no significant effect on mass measurement accuracy. However, it should be taken into account when selecting the ion source that mass measurement accuracy is directly proportional to mass spectral peak height-to-noise ratio (Chen et al., 1986). As a consequence, any improvement in ionization efficiency should translate into improved mass accuracy.

Once the best ion source has been determined and so ions are to be produced, those are guided through different mechanisms to the mass analyzer. The mass analyzer is the heart of the mass spectrometer and is the place where ions are separated based on their different *m/z* ratios.

This chapter focuses on the high-resolution mass analyzers (those capable of routine resolution >20,000 FWHM) for structural characterization and determination of organic pollutants in the environment. Those are double-focusing electrostatic/magnetic sector, FT-ICR, TOF, and FT-Orbitrap mass analyzers. Double-focusing electrostatic/magnetic sector instruments have been traditionally used coupled to GC with EI interface. TOF mass analyzers have been coupled to both GC and LC, with several different interfaces. The rest are usually coupled to LC with ESI source or eventually with any other atmospheric pressure ionization interfaces mentioned previously (Holčapek et al., 2012) but also since spring 2015 FT-Orbitrap is also available coupled to GC. Other analytical mass analyzers, such as quadrupole mass filter, IT, and triple quads, have important uses in environmental analysis but are not optimal for the highest resolution of complex matrixes.

The aim of this section is to give an overview of the main performance characteristics of the mass analyzers used in environmental analysis, and complete technical details of all the mass analyzers are out of the scope of this chapter. For this reason, no equations of motion will be included, as they can be found in articles and reviews dedicated exclusively to instrumentation (Marshall and Hendrickson, 2008; Xian et al., 2012).

3.5.1 Double-Focusing Electric/Magnetic Sector Instrument Mass Analyzer

Double-focusing electric/magnetic sector instruments, usually named magnetic sectors, are based on measurements of ion deflection (Marshall and Hendrickson, 2008). The term *double-focusing* refers to the fact that the combination of electrostatic and magnetic sectors focuses ions according to both direction and energy to provide higher resolution. The magnetic sector (B) exerts a force perpendicular to the ion motion to deflect ions according to their momentum. Higher mass ions are deflected less than lower mass ions, so ions are separated only by their masses. To obtain a spectrum of good resolution, the electrostatic sector (E) creates an electric field that exerts a force perpendicular to the ion motion to deflect ions according to their kinetic energy. Two geometries exist from these mass analyzers, depending on the order of the different sectors: EB or BE. Moreover, different combinations of magnetic and electrostatic sectors (multisectors) as well as combinations with some other mass analyzers (quadrupole, linear IT) were the first hybrid instruments that existed (Gross, 2011). Those instruments were replaced, due to space and technical requirements, and nowadays, the EBE configuration is the one that is most used.

Based on the sectors in the instrument, there are a number of different ways to acquire accurate mass measurements: peak matching, dynamic voltage scanning, and magnet scanning (Bristow, 2006). Briefly, the different scan modes differ whether the magnetic field is held constant (peak matching and dynamic voltage scanning) or the magnetic sector is scanned over a wide *m/z* range (magnet scanning). As observed in previous studies (Bristow et al., 2003), the accuracy is better when peak matching or dynamic voltage scanning is used; however, the *m/z* range in these modes is very narrow and lock mass reference is absolutely indispensable.

The coupling of high-resolution GC to HRMS has been an extraordinary tool due to its sensitivity, selectivity, specificity, and robustness. It has been, for several years, the reference instrumentation for analysis of many persistent organic pollutants, especially for dioxins, furans, and polybrominated diphenyl ethers (PBDEs), due to the demand for sensitivity as a consequence of strict regulations and extremely low levels of these compounds in different matrixes. As noted in several reviews, the evolution of magnetic sector in the last 10 years has been modest. Some improvements have been done on the amount of the sample injected (Li et al., 2009) and/or the simultaneous GC analysis (Medeiros and Simoneit, 2007); however, no improvements have been achieved on the mass analyzer itself. This technique needs a very expert operator, an expensive maintenance for optimum performance, a large space, and complementary accessories. For these reasons, magnetic sectors are no longer attractive to laboratories and users tend to replace this instrumentation with alternative techniques, such as TOF triple quads (Hernández et al., 2012) and Q-Orbitraps (Mol et al., 2016).

3.5.2 Fourier Transform Ion Cyclotron Resonance Mass Analyzer

Introduced in 1974 (Comisarow and Marshall, 1974), FT-ICR MS has evolved (Marshall, 2000; Marshall and Hendrickson, 2002) to become the highest-resolution broadband mass analysis technique (Marshall et al., 1998; Marshall and Hendrickson, 2008).

The principle of the FT-ICR analyzer is based on the circular oscillation that charged ions exhibit in the ICR cell once they are introduced into a homogeneous magnetic field. The circular motion defined by a cyclotron frequency is specific to each *m/z*. Linearity of detection and very high fidelity in the determination of frequency are inherent to FT instruments and allow very high mass accuracy (Scigelova et al., 2011). The cyclotron frequency varies as $(m/z)^{-1}$ (Marshall et al., 1998), and as a result, the ICR mass resolution varies as $(m/z)^{-1}$.

An FT-ICR mass analyzer is a pulsed detector. Because ion introduction is often temporally continuous (e.g., ESI), ions are typically accumulated externally during detection of ions from the preceding accumulation period (Senko et al., 1997). In FT-ICR, ions are externally accumulated in a multipole electric IT and simultaneously ejected toward the ICR cell. Ions should enter the ICR cell at low kinetic energies, so that they can be confined. Afterward, ions are coherently excited to a larger radius for a signal to be detected. The detection is carried out in the ICR cell wall where the detector plates are placed, in order to induce a measurable image current and transform it into mass spectra by using fast FT (Perry et al., 2008).

Resolution, mass accuracy, scan rate, *m/z* range, and dynamic range are proportional to magnetic field (B) strength. ICR mass resolution and scan rate increase proportionally to B, whereas mass accuracy, dynamic range, and upper *m/z* limit increases proportionally to B^2 (Marshall, 1996; Schaub et al., 2008).

As a consequence, FT-ICR instruments with a stronger magnetic field would increase resolution. A vacuum of 10^{-9} to 10^{-10} Torr is required to achieve high resolution, vacuum requirements being more stringent for higher-field magnets. Another way to increase the resolution is to allow for longer transient acquisitions. However, longer acquisition times might not be always practical, particularly in the case of coupling to LC. The magnetic field strength in commercially available FT-ICR instruments ranges from 7 to 15 T, and 21 T systems, this last one exclusive at the US National High Magnetic Field Laboratory and Pacific Northwest National Laboratory (Xian et al., 2012; Hendrickson et al., 2015).

Although FT-ICR instruments detect ions of a wide *m/z* range simultaneously, they present a limitation for the lower masses. Compared to modern TOF instruments that can detect ions starting from *m/z* 20 or the Orbitrap that can detect ions starting from *m/z* 50 (*m/z* 30 in the newest models), modern FT-ICR instruments are confined to a lower mass limit of *m/z* 125. As explained previously, the cyclotron frequency has an inverse nonlinear relationship with *m/z*, so very low masses are not easily excited for detection in the ICR cell. Moreover, in order to maintain the superior mass accuracy and resolution of FT-ICR mass analyzers, sampling rates >1.6 scans/s are shown to be necessary (Forcisi et al., 2013), which makes it incompatible with ultrahigh pressure liquid chromatography (UHPLC) or even HPLC. As a consequence, the coupling of LC to FT-ICR cannot be optimally performed.

The stored waveform inverse Fourier transform mode is used to excite the ions in the ICR cell for detection, as well as to perform ion isolation before MS/MS analysis in the ICR cell. To perform MS/MS experiments and carry out fragmentation in the ICR cell, gas has to be admitted and pressure in the cell has to be raised. However, for accurate mass measurements, enough time needs to be allowed for the gas to be dissipated and the ICR cell to return to the low pressure required for excitation and detection. It is much more practical to carry out collision-induced dissociation (CID) outside of the ultrahigh-vacuum region in an external analyzer or collision cell. In fact, the commercially available FT-ICR instruments are presented as a hybrid that consists of a linear IT or a quadrupole coupled to an FT-ICR mass analyzer. Progress is expected in the direction of higher-field magnets for FT-ICR, which will allow to significantly increase resolution at a given acquisition time.

FT-ICR mass analyzer presents an unsurpassed high accuracy and resolution. However, the large size, complexity, maintenance, and cost restrict the laboratory settings where it can be used. In many cases, their applications have involved extremely complex samples, such as molecular characterization of NOM (Hertkorn et al., 2008, 2013; Koch et al., 2008), aerosols (Schmitt-Kopplin et al., 2010), and diesel fuel (Hughey et al., 2001); the analysis of samples without pretreatment such as wine (Cooper and Marshall, 2001); and the application in areas such as metabolomics and proteomics (Witt et al., 2001).

3.5.3 Time-of-Flight Mass Analyzer

TOF mass analyzers are conceptually simple and are based on measurements of time of transit. Ions of the same initial position and velocity can be simultaneously accelerated (by a pulsed direct current electric field) and then allowed to fly freely (i.e., no external electric or magnetic fields) to a detector located some meters away, along the tube of flight (Marshall and Hendrickson, 2008). It will take a different time for each ion to reach the detector, depending on their *m/z*. Based on the way that they work, TOF mass analyzers are inherently fast and sensitive because all masses are measured simultaneously (i.e., the multiplex advantage), compared to scanning instruments (e.g., sectors and quadrupoles) that sequentially focus only one ion mass on the detector while all others are lost (Marshall and Hendrickson, 2008).

The TOF mass analyzer, initially projected by Stephan in 1946 (Mirsaleh-Kohan et al., 2008) and reduced to practice in 1948 (Cameron and Eggers, 1948), has evolved in the last 20 years, and since then, TOF has been considered a high-resolution mass analyzer and has been used for accurate mass measurements. For TOF, mass resolution is related to flight time, so increasing it, permitted to improve resolution. Delayed ion extraction, orthogonal ion introduction, and kinetic energy focusing by use of a reflectron introduced by Mamyrin in 1973 (Guilhaus, 1995; Mamyrin, 2001) were some of the main improvements carried out in TOF technology. Nowadays, manufacturers claim that with high-field pusher and dual-stage reflectrons, commercial TOF mass analyzers can reach a mass resolution of 40,000 (FWHM) (Pringle et al., 2007; Ow et al., 2010). Recent advances described TOF mass analyzers capable of attaining mass resolution of 50,000 or higher, with the multipass (Toyoda et al., 2003; Ioanoviciu, 2010; Shchepunov et al., 2010; Verentchikov and Yavor, 2011) and spiral designs (Satoh et al., 2007; Xian et al., 2012). The great speed of mass spectral acquisition is one of the main advantages of TOF analyzers, especially when coupled to GC and UHPLC. In fact, these fast scanning rates are what make this mass analyzer, at the eyes of some researchers and vendors, as suitable as higher-resolution mass analyzers, such as FT-ICR or Orbitrap. In fact, TOF mass analyzers may have more stable characteristics (better sensitivity and similar mass accuracy) than FT-ICR at high sampling rates (Forcisi et al., 2013).

For daily routine analysis, a broadband mass resolution of 10,000–40,000 and an RMS mass accuracy of 1–10 ppm may be attained. However, it should be highlighted that the mass accuracy may differ depending on the improved calibration routines. In most studies, internal calibration of mass spectrometers based on TOF technology has been reported (Wu and McAllister, 2003; Stroh et al., 2007; Kaufmann et al., 2008; Mol et al., 2008; Ortelli et al., 2009; Wang and Leung, 2009). The drift in the mass measurements in these mass analyzers may be important, mainly due to environmental factors, which leads to instability of the calibration. However, as instrumentation evolves extremely fast, new TOF mass analyzers utilize advanced temperature compensation systems, leading to better routine mass measurement accuracy.

Based on TOF technology, the development of the hybrid QTOF and tandem time-of-flight (TOF-TOF) mass spectrometer has also revolutionized the application of TOF to accurate mass measurement with the possibility of performing MS/MS experiments (Wolff et al., 2001; Ibañez et al., 2005; Wang and Leung, 2009). Although the hybrid models provide additional versatility, the analytical performance and fundamental principles of operation of TOF are the same. TOF-TOF is restricted to metabolomic and proteomic applications, and it is not so widely used in environmental analysis. Following the coherence of the present chapter, this hybrid mass spectrometer will not be described. Both TOF and QTOF have been widely used for target analysis and screening purposes in the environmental field (Hernández et al., 2007; Hayward and Wong, 2009; Portolés et al., 2011; Lebedev et al., 2013). However, QTOF permits the integration of MS/MS or MS^E experiments (acquisition is carried out at low and high collision energies within the same chromatographic run). The main advantages of QTOF are high sensitivity, mass resolution, and mass accuracy of the resulting tandem mass spectra in both precursor (MS) and product ion (MS/MS) modes (Chernushevich et al., 2001).

3.5.4 Fourier Transform Orbitrap Mass Analyzer

The Orbitrap, the newest FT mass analyzer, invented in 1999 (Makarov, 2000), has been widely distributed since its commercial introduction in 2005, with the launching of the linear trap quadrupole (LTQ) Orbitrap tandem mass spectrometer at the Annual American Society for Mass Spectrometry Conference in San Antonio, Texas (Hu et al., 2005).

The original roots of the Orbitrap stem from the principle of orbital trapping, the ability to trap charged particles in electrostatic fields, defined by Kingdon (1923) bears a similarity to two types of ion-trapping mass analyzers: the Paul trap (quadrupole IT) and the FT-ICR (Hu et al., 2005). Alexander Makarov used the term *Orbitrap* to describe the harmonic oscillations of the ions in this new mass analyzer.

The Orbitrap confines ions in an electrostatic quadrologarithmic potential well created between carefully shaped coaxial central and outer electrodes. In the absence of a superconducting magnet (Ham, 2008; Perry et al., 2008), ion stability is achieved only due to ions orbiting around an axial electrode. Orbiting ions perform harmonic oscillations along the electrode with a frequency proportional to $(m/z)^{-1/2}$. As a result, the mass resolution in the Orbitrap is proportional to $(m/z)^{-1/2}$ (Perry et al., 2008). In the Orbitrap, the axial frequency is used to derive the m/z ratio, since it is independent of the initial properties of the ions. It is this independence that is responsible for the HR and mass accuracy of the Orbitrap (Makarov, 2000). These oscillations are detected using image current detection and are transformed into mass spectra using fast FT, similarly to FT ICR.

Like the FT-ICR mass analyzer, the Orbitrap mass analyzer is a pulsed detector, but in this case, ions are collected in a "C" trap and injected simultaneously toward the Orbitrap (Perry et al., 2008). This method provides fast and uniform injection for large ion populations. Automatic gain control (AGC) target and maximum injection time are parameters typically found in ITs that should be also appropriately selected in the Orbitrap (Kaufmann et al., 2010). AGC target and maximum ion time are used to control the number of ions and the maximum time taken to fill the C trap, respectively.

The Orbitrap mass analyzer affords outstanding performance with respect to mass accuracy, resolution, sensitivity, and dynamic range. The Orbitrap has proven to be a robust mass analyzer that can routinely deliver high resolution and mass accuracy. Mass inaccuracy is typically <5 ppm for externally calibrated mass spectra and <2 ppm for internally calibrated mass spectra (Marshall and Hendrickson, 2008) but is demonstrated to be as low as 0.2 ppm under favorable conditions (Perry et al., 2008). In the commercial Orbitrap instruments, resolution can be switched in discrete steps. The resolution is controlled by varying the transient acquisition time (which changes the overall scan cycle time). To yield higher-resolution values, higher scan times should be employed, such as for a resolution of 100,000 (*m/z* 400 FWHM), some FT-Orbitrap models employ 1.9 s scan cycle time, whereas for a resolution of 60,000, they use 1 s.

As a result, some applications restricted until the moment to FT-ICR can be transferred to Orbitrap analyzers, such as proteomics, metabolomics, and NOM characterization. The main advantage is that the Orbitrap does not need such an expensive and delicate maintenance compared with an FT-ICR mass analyzer (Cortés-Francisco, 2011).

Since the first commercial LTQ Orbitrap, different Orbitrap-based mass spectrometers have been designed. For simplicity, it can be said that three different configurations can be found: the single-stage Orbitrap higher-energy collisional dissociation (HCD) mass spectrometer, the hybrid quadrupole-Orbitrap HCD mass spectrometer, and the hybrid linear IT-Orbitrap HCD mass spectrometer. Although the hybrid models provide additional versatility to MS/MS experiments, the analytical performance and fundamental principles of operation of the Orbitrap analyzers in all the instruments are identical (Perry et al., 2008). Currently, there are several commercial Orbitrap instruments and even instruments with the same configuration present in different technical specifications (resolution, fragmentation techniques, accuracy) depending on the specific model. The recent design of the Orbitrap mass analyzer consisted of increased field strength inside the Orbitrap, which has allowed to significantly increase resolution at a given acquisition time (Makarov et al., 2009). As described elsewhere (Forcisi et al., 2013), the Orbitrap is a formidable alternative to TOF and FT-ICR mass analyzers, as it closes the gap between the two mass analyzers.

Although the Orbitrap was first coupled to LC, a study presented already in 2010 a modification of an LTQ Orbitrap hybrid instrument for high mass accuracy and GC-HRMS. With the development of a new scan sequence (the nested scan), mass accuracy less than 1 ppm with internal calibration and resolution exceeding 100,000 was routinely achievable. In combination with isotopic distribution information, the GC-Orbitrap prototype facilitated the unambiguous determination of small-molecule elemental compositions and met the mass spectral requirements of challenging Environmental Protection Agency (EPA) methods (Peterson et al., 2010, 2014). Finally, the new GC-Q-Orbitrap was exhibited for the first time at the 63rd Annual American Society for Mass Spectrometry Conference. The GC-Orbitrap that is now commercially available is representing a significant step forward in the field of GC-HRMS, at the moment basically dominated by magnetic sectors and TOF analyzers.

3.5.5 Hybrid Mass Spectrometers

There have been enormous advances in high-resolution instrumentation in recent years, including hybrid mass spectrometers. Readers can refer to an in-depth review surrounding the development of hybrid instruments over the last 30 years (Glish and Burinsky, 2008). However, some last models may be missing; such is the case for the hybrid quadrupole-Orbitrap HCD mass spectrometer, as it was first commercially available in 2011. The technique requires two mass analyzers in series (or a single mass analyzer that can be used sequentially) to analyze the precursor and product ions (Kind and Fiehn, 2010).

Generally speaking, the hybrid instruments provide the possibility to acquire data in different modes of operation, which include precursor ion scans, product ion scans, neutral loss scans, and selected reaction monitoring. The different modes of operation are discussed by De Hoffmann (1996). It should be highlighted that the final quality of information obtained in each case may differ depending on the mass analyzers forming the hybrid instrument. CID is the most common technique to obtain tandem mass spectra. Precursor ion stability and internal energy under CID have been also previously discussed (Kertesz et al., 2009).

Ideally, last-generation hybrid TOF and Orbitrap mass analyzers have to obtain accurate masses at the MS^n level (Glish and Burinsky, 2008; Fjeldsted, 2009), without losing sensitivity, acquisition rate, resolution, and mass accuracy. Nowadays, with UHPLC, chromatographic peaks can be around 15 s or lower and it may be difficult to obtain full-scan spectra and tandem mass spectra (MS/MS) or MS^n scans (Staack and Hopfgartner, 2007; Bedair and Sumner, 2008) with enough points per peak.

Data-dependent acquisition mode can help to have enough time to perform additional MS/MS scans. The mass analyzers operate at full scan (or selected ion monitoring [SIM]) and by data-dependent mode, a fixed number of peaks selected from a survey scan using predetermined rules are selected, and the corresponding ions are subjected to MS/MS analysis (Murray, 2010). The predetermined rules may include a list of *m/z* precursors, reject list, dynamic exclusion, signal-to-noise threshold, loop count, and multiplex (Michalski et al., 2011; Kumar et al., 2013). The data-dependent mode may seem to be a semiautomatic way of performing MS/MS. However, meticulous optimization of some parameters is crucial, to prevent the mass analyzer to perform MS/MS of a solvent or contaminant ion, especially when ions are coeluting.

3.6 ENVIRONMENTAL APPLICATIONS

In environmental analysis, usually, the main goal of any target analysis is the identification and quantification of pollutants, and so, the main benefit of using HRMS is to have better detection limits and less interferences. Basically, the gain of an HRMS-based method compared to a triple quad is the fact that one is so much more sure of the fact that the compound is really "the compound."

However, for some applications, it is not only an advantage but also a real need. This is the case for the analysis of toxins, for which there are so many variety and no standards, NOM characterization, metabolites, and transformation products (disinfection by-products [DBPs], ozone derivative compounds).

The examples given in the following sections using GC-HRMS or LC-HRMS have been chosen to illustrate the advantages and need of using HRMS in front of other techniques.

3.6.1 GC-HRMS Applications

One of the reasons why GC-MS has dominated in environmental applications is due to the successful use of mass spectral libraries (Lebedev et al., 2013). Moreover, working in SIM mode, high sensitivity can be obtained. However, some analysis required lower detection limits and mass resolution, due to coelution in chromatography. Such is the case for PBDEs, chlorinated biphenyl congeners (polychlorinated biphenyls [PCBs]), dioxins, and furans. The EPA methods for the analysis of these compounds require the use of double-focusing electric/magnetic sector instruments (US EPA, 1994, 2007, 2008). Apart from some legal requirements, double-focusing electric/magnetic sector instruments have been used to solve some other environmental issues. Due to chromatographic coelution of some congeners of polychlorinated terphenyls, the use of high resolution was mandatory to resolve interferences in the detection and measurement (Caixach et al., 1994). With more than 670 compounds, toxaphene analysis was difficult due to the complexity of the mixture. No complete chromatographic resolution was achieved, and HRMS better than 10,000 (at 10% valley) was necessary (Santos et al., 1997). Another application in which double-focusing electric/magnetic sector instruments were used is the analysis of nitrosamines (Ontario Ministry of Environment, 2007; Planas et al., 2008). However, in this case, compared to dioxins, the use of magnetic sector is not required, and in fact, EPA proposed a new approach using GC-MS/MS and CI (US EPA, 2004) as well as some others that have been recently published (Yoon et al., 2012). Similar method detection limits have been achieved in the methods using HRMS and MS/MS. The fact that GC-MS/MS is more available than magnetic sectors will increase the number of methods quoting MS/MS, when similar performance methods can be achieved, as in this case.

Despite its low dynamic range and sensitivity, GC-TOF has been also reported for the quantitative analysis of pesticides, PBDEs, and PCBs (Hernández et al., 2007; Hayward and Wong, 2009) trying to be a substitute to magnetic sectors. The main advantage of developing a quantitative method for organic pollutants using GC-TOF is that full-scan data are acquired and the same injection can be used for screening purposes and elucidation of nontarget compounds. In this sense, GC-TOF can be a real alternative to double-focusing electric/magnetic sector instruments (Hernández et al., 2012). Several studies reported the use of TOF and QTOF analyzers for identifying contaminants and degradation products in the environment (Hernández et al., 2007; Portolés et al., 2011). In the study by Lebedev et al. (2013) for target and nontarget analysis of semivolatile compounds in snow samples, the use of GC-TOF compared to GC-MS identification (based only on mass spectral libraries) prevents from carrying out a false assignment. In the examples cited previously, the identification of the compounds is based on the elemental formula of several ions, in combination with the previous mass spectral list obtained from libraries or using any other software, except for the study of Portolés et al. (2011), which introduced several other rules (e.g., carbon filtering, maximum mass error based on previous experience, isotopic pattern for some halogens). The resolution that can be reached by the instruments used ranged from 7000 to 25,000 (FWHM), although it varies along the mass range, and for the moment, any resolution below 20,000 (FWHM) should not be considered high resolution. As the readers will notice, these references are older than the ones reported for LC-MS applications (see the next section), and in general, for the last 10 years, HRMS coupled to GC has offered less innovation. Recently, the GC-Q-Orbitrap was finally launched. Some distinguished research centers had the opportunity in 2015 to evaluate the beta version of the GC-Q-Orbitrap. In June 2016, it was finally commercially available. During this year, few applications using this new instrumentation have been published, but, of course, we can assure that more will be coming soon. Here we include the ones that are available today (Baldwin et al., 2016; Mol et al., 2016; Postigo et al., 2016).

3.6.2 LC-HRMS Applications

Contrary to what happens with GC-HRMS, in general, sensitivity has been criticized when using LC-HRMS in front of LC-MS/MS. Many studies claimed the limitations for instruments such as TOF and QTOF mass analyzers to achieve detection limits and linearity as good as those achieved by triple quadrupole mass analyzers (Hernández et al., 2004; Krauss et al., 2010). However, with the new QTOF and Orbitrap mass analyzers, these deficiencies have been overcome and detection limits are equal or better when using HRMS. For instance, the target analysis of cytostatic compounds in wastewater treatment plants has been carried out using both HRMS and MS/MS. In this case, the Orbitrap mass analyzer in full-scan and AIF modes has been used, acquiring at two spectra/s working with high resolution (*R*: 50,000 *m/z* 200, FWHM). Compared to a triple quad, we have obtained very good linearity and as good instrument detection limits, but better specificity, because both precursor and fragment ions are measured using HRMS (Gómez-Canela et al., 2013). In fact, hidden interferences from the matrices will contribute to the integrated peak for the target compound and there is a serious risk of having a case of overquantitation or false assignment (Scigelova et al., 2011; Kumar et al., 2013). Some readers will disagree with the statements proposed previously and will claim that better specificity in MS/MS can be obtained when monitoring more transitions per compound, without the necessity of buying new HRMS instrumentation. In many cases, HRMS is not absolutely mandatory, although it is of no doubt that when using HRMS for target analysis, interferences are better resolved and quantification can be carried out easily. In contrast, there are some environmental applications in which a high resolution is much more essential.

One of these fields that can take advantage of using HRMS is the determination of toxins. In a recent study, the determination of marine toxins by using HRMS has been carried out (Domènech et al., 2013). Quantification and confirmation criteria have been proposed and discussed, due to the fact that legislation has not been updated and several different strategies are found in literature (Mol et al., 2012), compared to MS/MS (Pitarch et al., 2007; Gerssen et al., 2010). Apart from the sensitivity of the developed method, it should be highlighted that the specificity of the technique permitted to resolve matrix interferences as shown in Figure 3.8.

Apart from sensitivity and selectivity, one of the main advantages of using HRMS is the fact that data acquisition is carried out in full-scan mode. We obtain the complete mass spectra of the molecular and fragment ions, when working in AIF with Orbitrap or MS^E with TOF, because no precursor selection is done. From these data, retrospective analysis can be conducted and compounds chemically related to the target analytes can be found.

The potential of using HRMS and the possibility to work in full-scan mode instead of MS/MS are given with an "in-house" example of algae extracts. A routine method was developed and validated using LC-MS/MS for the analysis of the toxin cylindrospermopsin (CYN) in algal bloom (Guzmán-Guillén et al., 2012). The same samples used in the study by Guzmán-Guillén et al. (2012) were voluntarily supplied by Dr. Ana Cameán and were analyzed using HRMS. Essentially, we received the extracts and worked on the optimization of the instrumental conditions. After LC separation with a Hypercarb column, MS analyses were carried out with an Orbitrap-Exactive HCD equipped with an ESI source. CYN is ionized in positive or negative mode, but during the method development, better

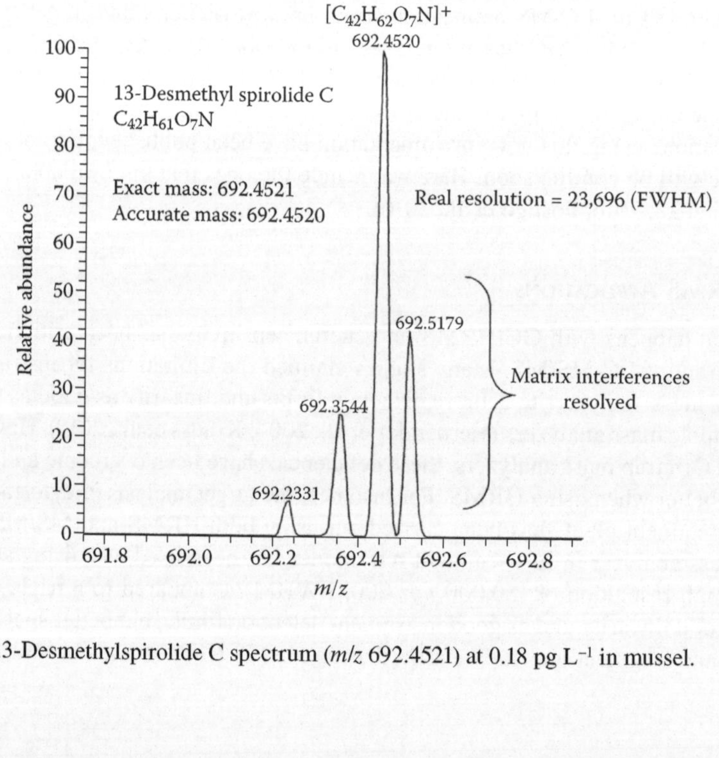

FIGURE 3.8 13-Desmethylspirolide C spectrum (m/z 692.4521) at 0.18 pg L^{-1} in mussel.

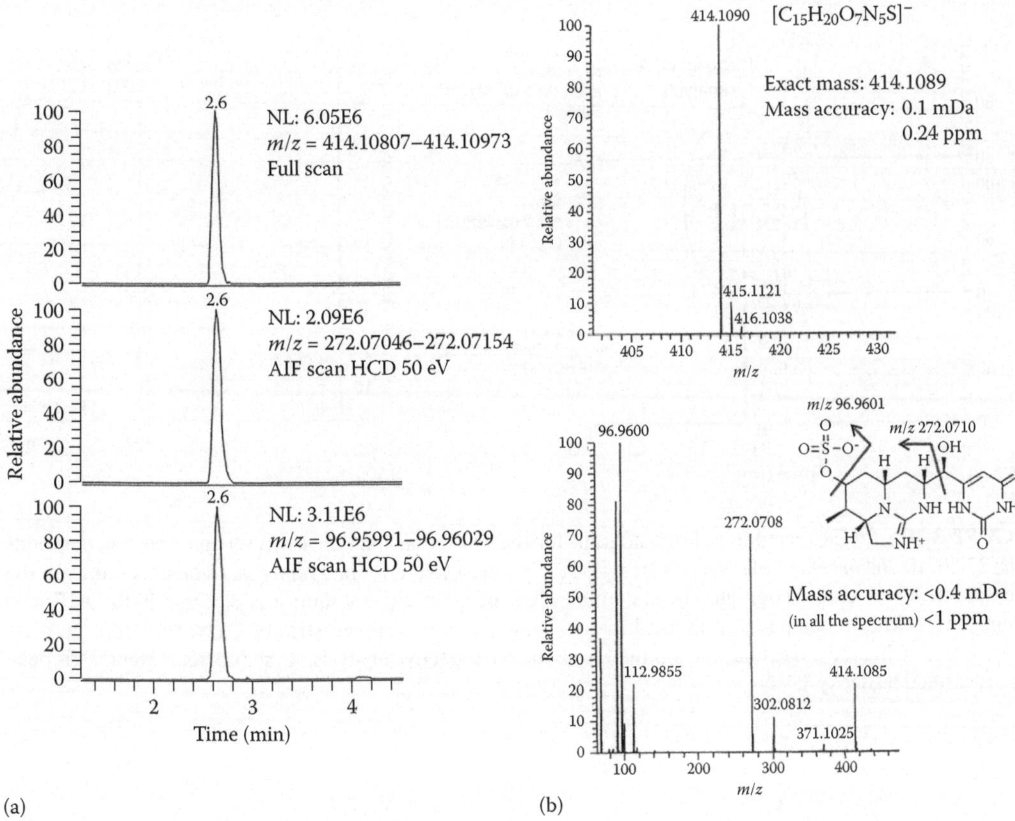

FIGURE 3.9 (a) Extracted ion chromatogram for the molecular ion *m/z* 414.1090 and two fragment ions (*m/z* 272.0710 and *m/z* 96.9601) of CYN (0.1 μg L^{-1} standard) with mass tolerance of 2 ppm (based on uncertainty of the measurements). (b) Full-scan spectrum and AIF spectrum showing the different fragments on the molecule for CYN.

limits of quantification in negative mode were obtained. Thus, the molecular ion (*m/z* 414.1090) for quantification and two fragment ions (*m/z* 272.0710 and *m/z* 96.9601) for confirmation were chosen (Figure 3.9). Two scan events were defined: full scan and AIF (HCD 50 eV) working in HR (*R*: 50,000 *m/z* 200, FWHM) with mass accuracy and precision better than 2 ppm.

CYN was detected and confirmed in the samples, due to the presence of the molecular ion and two fragment ions in the retention time (t_R: 2.6 min) of the standard (Figure 3.10). In addition, a chromatographic peak coming after the CYN (t_R: 2.9 min) was detected in the AIF scan event. The unknown compound presented the two fragment ion characteristics of CYN, so we assumed that it might be a chemically related compound. Retrospective analysis of the full-scan event revealed a molecular ion with *m/z* 398.1142.

This peak in the full-scan event corresponded to the *m/z* of CYN with a mass difference of 15.9948, which matched with an O atom. Based on the accurate mass measurement (*m/z* 398.1142) and the two fragment ions that corresponded to the same fragments as for CYN (*m/z* 96.9601 $[HO_4S]^-$ and *m/z* 272.0710 $[C_{10}H_{14}O_4N_3S]^-$), the new compound was identified as deoxy-cylindrospermopsin, a variety of cyanobacterial toxin already described in the literature (Dell'Aversano et al., 2004). A false-negative result was avoided, and not only CYN but also deoxy-CYN was found and quantified in several samples, as standards for CYN and deoxy-CYN are available (Figure 3.11). However, there are several toxins that have been reported, but there are no standards available. In this sense, HRMS is essential for preliminary identification.

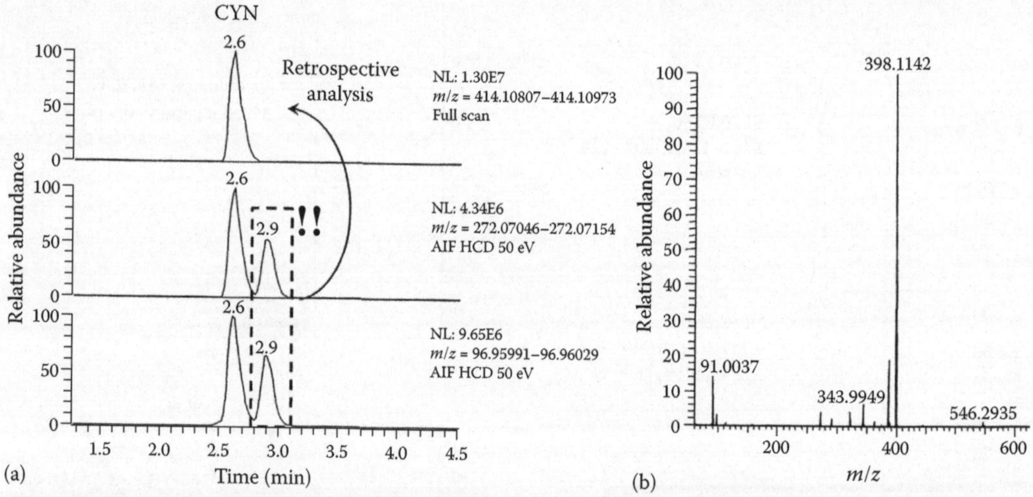

FIGURE 3.10 (a) Extracted ion chromatogram for the molecular ion *m/z* 414.1090 and two fragment ions (*m/z* 272.0710 and *m/z* 96.9601) of CYN (sample A) with mass tolerance of 2 ppm (based on uncertainty of the measurements). A chromatographic peak coming after the CYN (t_R: 2.9 min) was detected in the AIF scan event. The unknown compound presented the two fragment ion characteristics of CYN. (b) Full-scan mass spectrum showing the molecular ion obtained from the retrospective analysis of the full-scan event. This peak was identified as deoxy-CYN.

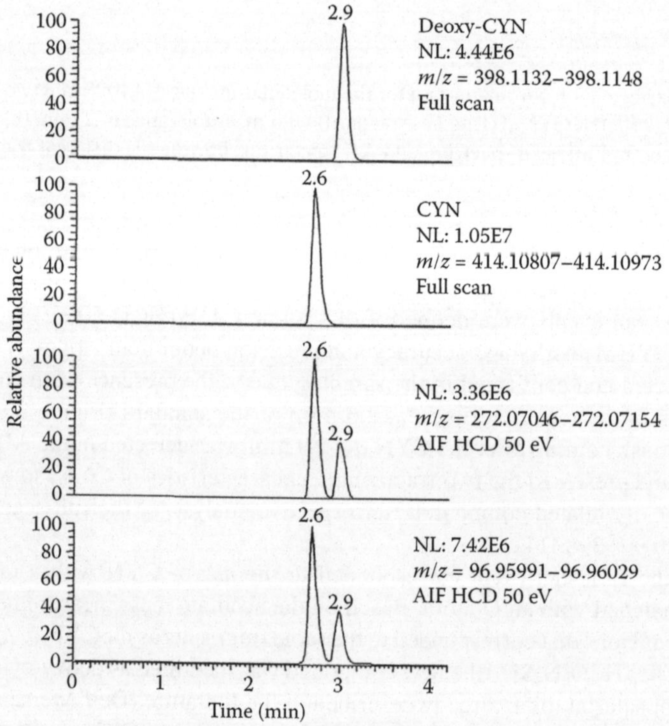

FIGURE 3.11 Extracted ion chromatogram for the molecular ion *m/z* 398.1142 (deoxy-CYN) and *m/z* 414.1090 (CYN) and the two characteristic fragment ions (*m/z* 272.0710 and *m/z* 96.9601) from a real sample. Mass tolerance of 2 ppm (based on uncertainty of the measurements).

In this example, a huge peak was easily detected in the AIF spectrum, so the elemental formula determination of this single compound was fast. As mentioned previously, the main bottleneck in the screening experiments is the identification of unknown compounds. This is the case in fields such as metabolomics and proteomics (and all the "omics" science), but we also have a similar challenge in environmental analysis with NOM characterization. Not surprisingly, the characterization of NOM has been one of the fields that have significantly progressed in the past few years, thanks to the evolution of HRMS. NOM characterization can consist of bulk parameters such as total organic carbon and dissolved organic carbon and spectroscopy techniques such as ultraviolet (UV) and visible absorption spectroscopy, specific UV absorbance, and excitation emission matrix fluorescence spectroscopy (Matilainen et al., 2011). However, high-field NMR spectroscopy and FT-ICR MS have been powerful discovery tools (Hertkorn et al., 2013). Based on these techniques, molecular-level NOM composition and structure exhibit far more variance than anticipated from more uniform bulk parameters, which are subjected to intrinsic averaging (Hertkorn et al., 2007).

The properties and the amount of dissolved organic matter (DOM) in natural waters can significantly affect the production of drinking water, as it may cause adverse aesthetic qualities such as color, taste, and odor. The use of HRMS has shown the effect of water treatment processes on DOM, such as activated carbon adsorption, ozonation, or membrane treatments (Herzsprung et al., 2012; Zhang et al., 2012a,b; Cortés-Francisco and Caixach, 2013). Based on HRMS, it has been shown that DOM can serve as main precursor of DBPs during chlorination (Hua and Reckhow, 2007; Richardson et al., 2007), as DOM molecules with a low degree of oxidation (low O/C) were found to be more reactive toward chlorine (Zhang et al., 2012b; Lavonen et al., 2013) and new DBPs have been characterized. Identification and formula determination are based on ultrahigh resolution (R > 100,000 FWHM), sub-ppm accuracy on accurate mass measurements, and posterior strategies to filter the list of candidates obtained—that is, elements in use, isotopic pattern, and element ratios. Due to the amount of information obtained from each analysis, the representation of the data is usually done based on van Krevelen diagrams, as mentioned previously. More recently, statistical analysis has been applied to compare the data obtained by FT-ICR and bring out hidden correlations (Sleighter et al., 2010).

REFERENCES

Aston, F.W. 1927. Letters to the Editor: The constitution of ordinary lead. *Nature* 120: 224.

Baldwin, S., Bristow, T., Ray, A., Rome, K., Sanderson, N., Sims, M., Cojocariu, C., Silcock, P. 2016. Applicability of gas chromatography/quadrupole-orbitrap mass spectrometry in support of pharmaceutical research and development. Rapid Communications in Mass Spectrometry 30(7): 873–880.

Balogh, M.P. 2004. Debating resolution and mass accuracy. *LCGC Europe* 17: 152–159.

Bedair, M., Sumner, L.W. 2008. Current and emerging mass spectrometry technologies for metabolomics. *TrAC Trends in Analytical Chemistry* 27: 238–250.

Beynon, J.H. 1954. Qualitative analysis of organic compounds by mass spectrometry. *Nature* 174: 735–37.

Beynon, J.H. 1960. *Mass Spectrometry and Its Applications to Organic Chemistry*, 1st en. Elsevier, London.

Brenton, A., Godfrey, A. 2010. Accurate mass measurement: Terminology and treatment of data. *Journal of the American Society for Mass Spectrometry* 21: 1821–1835.

Bristow, A.W.T. 2006. Accurate mass measurement for the determination of elemental formula—A tutorial. *Mass Spectrometry Reviews* 25: 99–111.

Bristow, A.W.T., Webb, K.S. 2003. Intercomparison study on accurate mass measurement of small molecules in mass spectrometry. *Journal of the American Society for Mass Spectrometry* 14: 1086–1098.

Brunnée, C. 1987. The ideal mass analyzer: Fact or fiction? *International Journal of Mass Spectrometry and Ion Processes* 76: 125–237.

Caixach, J., Rivera, J., Galceran, M.T., Santos, F.J. 1994. Homologue distributions of polychlorinated terphenyls by high-resolution gas chromatography and high-resolution mass spectrometry. *Journal of Chromatography A* 675: 205–211.

Cameron, A.E., Eggers, D.F. 1948. Ion velocitron. *Review of Scientific Instruments* 19: 605–607.

Cech, N.B., Enke, C.G. 2001. Practical implications of some recent studies in electrospray ionization fundamentals. *Mass Spectrometry Reviews* 20: 362–387.

Chen, L., Cottrell, C.E., Marshall, A.G. 1986. Effect of signal-to-noise ratio and number of data points upon precision in measurement of peak amplitude, position and width in Fourier transform spectrometry. *Chemometrics and Intelligent Laboratory Systems* 1: 51–58.

Chen, H., Zheng, J., Zhang, X., Luo, M., Wang, Z. et al. 2007. Surface desorption atmospheric pressure chemical ionization mass spectrometry for direct ambient sample analysis without toxic chemical contamination. *Journal of Mass Spectrometry* 42: 1045–1056.

Chernushevich, I.V., Loboda, A.V., Thomson, B.A. 2001. An introduction to quadrupole–time-of-flight mass spectrometry. *Journal of Mass Spectrometry* 36: 849–865.

Cole, B.R. 1997. *Electrospray Ionization Mass Spectrometry: Fundamentals, Instrumentation and Applications*. Wiley Interscience, Hoboken, NJ.

Comisarow, M.B., Marshall, A.G. 1974. Fourier transform ion cyclotron resonance spectroscopy. *Chemical Physics Letters* 25: 282–283.

Cooks, R.G., Ouyang, Z., Takáts, Z., Wiseman, J.M. 2006. Ambient mass spectrometry. *Science* 311: 1566–1570.

Cooper, H.J., Marshall, A.G. 2001. Electrospray ionization Fourier transform mass spectrometric analysis of wine. *Journal of Agricultural and Food Chemistry* 49: 5710–5718.

Cortés-Francisco, N., Caixach, J. 2013. Molecular characterization of dissolved organic matter through a desalination process by high resolution mass spectrometry. *Environmental Science & Technology* 47: 9619–9627.

Cortés-Francisco, N., Flores, C., Moyano, E., Caixach, J. 2011. Accurate mass measurements and ultrahigh-resolution: Evaluation of different mass spectrometers for daily routine analysis of small molecules in negative electrospray ionization mode. *Analytical and Bioanalytical Chemistry* 400: 3595–3606.

De Hoffmann, E. 1996. Tandem mass spectrometry: A primer. *Journal of Mass Spectrometry* 31: 129–137.

Dell'Aversano, C., Eaglesham, G.K., Quilliam, M.A. 2004. Analysis of cyanobacterial toxins by hydrophilic interaction liquid chromatography–mass spectrometry. *Journal of Chromatography A* 1028: 155–164.

Domènech, A., Cortés-Francisco, N., Palacios, O., Franco, J.M., Riobó, P. et al. 2013. Determination of lipophilic marine toxins in mussels: Quantification and confirmation criteria using high resolution mass spectrometry. *Journal of Chromatography A* 1328: 16–25.

European Commission. 2002. 2002/657/EC: Commission Decision of 12 August 2002 implementing Council Directive 96/23/EC concerning the performance of analytical methods and the interpretation of results. *Official Journal of the European Communities* L221.

European Commission. 2015. Directorate General Health and Consumer Protection, Guidance document on analytical quality control and method validation procedures for pesticides residues analysis in food and feed. SANTE/11945/2015.

Fjeldsted, J.C. 2009. Accurate mass measurements with orthogonal axis time-of-flight mass spectrometry. In *Liquid Chromatography Time-of-Flight Mass Spectrometry*, ed. Ferrer, I., Thurman, E.M. Wiley, New York 3–16.

Forcisi, S., Moritz, F., Kanawati, B., Tziotis, D., Lehmann, R. et al. 2013. Liquid chromatography–mass spectrometry in metabolomics research: Mass analyzers in ultra high pressure liquid chromatography coupling. *Journal of Chromatography A* 1292: 51–65.

Gelpí, E. 2008. From large analogical instruments to small digital black boxes: 40 years of progress in mass spectrometry and its role in proteomics; Part I: 1965–1984. *Journal of Mass Spectrometry* 43: 419–435.

Gelpí, E. 2009. From large analogical instruments to small digital black boxes: 40 years of progress in mass spectrometry and its role in proteomics; Part II: 1985–2000. *Journal of Mass Spectrometry* 44: 1137–1161.

Gerlich, M., Neumann, S. 2012. MetFusion: Integration of compound identification strategies. *Journal of Mass Spectrometry* 48: 291–298.

Gerssen, A., van Olst, E.H., Mulder, P.P., de Boer, J. 2010. In-house validation of a liquid chromatography tandem mass spectrometry method for the analysis of lipophilic marine toxins in shellfish using matrix-matched calibration. *Analytical and Bioanalytical Chemistry* 397: 3079–3088.

Glish, G.L., Burinsky, D.J. 2008. Hybrid mass spectrometers for tandem mass spectrometry. *Journal of the American Society for Mass Spectrometry* 19: 161–172.

Godfrey, A., Brenton, A. 2012. Accurate mass measurements and their appropriate use for reliable analyte identification. *Analytical and Bioanalytical Chemistry* 404: 1159–1164.

Gómez-Canela, C., Cortés-Francisco, N., Ventura, F., Caixach, J., Lacorte, S. 2013. Liquid chromatography coupled to tandem mass spectrometry and high resolution mass spectrometry as analytical tools to characterize multi-class cytostatic compounds. *Journal of Chromatography A* 1276: 78–94.

Gross, J.H. 2011. *Mass Spectrometry*, 2nd ed. Springer, Heidelberg.

Gross, J.H. 2014. Direct analysis in real time-a critical review on DART-MS. *Analytical and Bioanalytical Chemistry* 406: 63–80.

Guilhaus, M. 1995. Principles and instrumentation in time-of-flight mass spectrometry. *Journal of Mass Spectrometry* 30: 1519–1532.

Guzmán-Guillén, R., Prieto, A.I., Gustavo, A., Soria-Díaz, M.E., Cameán, A.M. 2012. Cylindrospermopsin determination in water by LC-MS/MS: Optimization and validation of the method and application to real samples. *Environmental Toxicology and Chemistry* 31: 2233–2238.

Haapala, M., Pól, J., Saarela, V., Arvola, V., Kotiaho, T. et al. 2007. Desorption atmospheric pressure photoionization. *Analytical Chemistry* 79: 7867–7872.

Ham, B.M. 2008. *Even Electron Mass Spectrometry with Biomolecule Applications*, 1st ed. Wiley Interscience, Hoboken, NJ.

Hayward, D.G., Wong, J.W. 2009. Organohalogen and organophosphorous pesticide method for ginseng root—A comparison of gas chromatography–single quadrupole mass spectrometry with high resolution time-of-flight mass spectrometry. *Analytical Chemistry* 81: 5716–5723.

Hendrickson, C.L., Quinn, J.P., Kaiser, N.K., Smith, D.F., Blakney, G.T., Chen, T., Marshall, A.G., Weisbrod, C.R., Beu, S.C. 2015. 21 Tesla Fourier transform ion cyclotron resonance mass spectrometer: A national resource for ultrahigh resolution mass analysis. *Journal of The American Society for Mass Spectrometry* 26(9): 1626–1632.

Hernández, F., Ibáñez, M., Sancho, J.V., Pozo, O.J. 2004. Comparison of different mass spectrometric techniques combined with liquid chromatography for confirmation of pesticides in environmental water based on the use of identification points. *Analytical Chemistry* 76: 4349–4357.

Hernández, F., Portolés, T., Pitarch, E., López, F.J. 2007. Target and nontarget screening of organic micropollutants in water by solid-phase microextraction combined with gas chromatography/high-resolution time-of-flight mass spectrometry. *Analytical Chemistry* 79: 9494–9504.

Hernández, F., Sancho, J.V., Ibáñez, M., Abad, E., Portolés, T. et al. 2012. Current use of high-resolution mass spectrometry in the environmental sciences. *Analytical and Bioanalytical Chemistry* 403: 1251–1264.

Hertkorn, N., Ruecker, C., Meringer, M., Gugisch, R., Frommberger, M. et al. 2007. High-precision frequency measurements: Indispensable tools at the core of the molecular-level analysis of complex systems. *Analytical and Bioanalytical Chemistry* 389: 1311–1327.

Hertkorn, N., Frommberger, M., Witt, M., Koch, B.P., Schmitt-Kopplin, P. et al. 2008. Natural organic matter and the event horizon of mass spectrometry. *Analytical Chemistry* 80: 8908–8919.

Hertkorn, N., Harir, M., Koch, B.P., Michalke, B., Schmitt-Kopplin, P. 2013. High-field NMR spectroscopy and FTICR mass spectrometry: Powerful discovery tools for the molecular level characterization of marine dissolved organic matter. *Biogeosciences* 10: 1583–1624.

Herzsprung, P., Hertkorn, N., Friese, K., Schmitt-Kopplin, P. 2010. Photochemical degradation of natural organic sulfur compounds CHOS from iron-rich mine pit lake pore waters—An initial understanding from evaluation of single-elemental formulae using ultra-high-resolution mass spectrometry. *Rapid Communications in Mass Spectrometry* 24: 2909–2924.

Herzsprung, P., von Tümpling, W., Hertkorn, N., Harir, M., Büttner, O. et al. 2012. Variations of DOM quality in inflows of a drinking water reservoir: Linking of van Krevelen diagrams with EEMF spectra by rank correlation. *Environmental Science & Technology* 46: 5511–5518.

Holčapek, M., Jirásko, R., Lísa, M. 2012. Recent developments in liquid chromatography–mass spectrometry and related techniques. *Journal of Chromatography A* 1259: 3–15.

Hu, Q., Noll, R.J., Li, H., Makarov, A., Hardman, M. et al. 2005. The Orbitrap: A new mass spectrometer. *Journal of Mass Spectrometry* 40: 430–443.

Hua, G., Reckhow, D.A. 2007. Characterization of disinfection byproduct precursors based on hydrophobicity and molecular size. *Environmental Science & Technology* 41: 3309–3315.

Hughey, C.A., Hendrickson, C.L., Rodgers, R.P., Marshall, A.G. 2001. Elemental composition analysis of processed and unprocessed diesel fuel by electrospray ionization Fourier transform ion cyclotron resonance mass spectrometry. *Energy Fuels* 15: 1186–1193.

Ibañez, M., Sancho, J.V., Pozo, O.J., Niessen, W., Hernández, F. 2005. Use of quadrupole time-of-flight mass spectrometry in the elucidation of unknown compounds present in environmental water. *Rapid Communications in Mass Spectrometry* 19: 169–178.

Ioanoviciu, D. 2010. The spiral main path electric deflector as a time-of-flight mass analyzer. *International Journal of Mass Spectrometry* 290: 145–147.

Kaufmann, A., Butcher, P., Maden, K., Widmer, M. 2008. Quantitative multiresidue method for about 100 veterinary drugs in different meat matrices by sub 2-μm particulate high-performance liquid chromatography coupled to time of flight mass spectrometry. *Journal of Chromatography A* 1194: 66–79.

Kaufmann, A., Widmer, M., Maden, K. 2010. Post-interface signal suppression, a phenomenon observed in a single-stage Orbitrap mass spectrometer coupled to an electrospray interfaced liquid chromatography. *Rapid Communications in Mass Spectrometry* 24: 2162–2170.

Kebarle, P., Peschke, M. 2000. On the mechanisms by which the charged droplets produced by electrospray lead to gas phase ions. *Analytica Chimica Acta* 406: 11–35.

Kertesz, T.M., Hall, L.H., Hill, D.W., Grant, D.F. 2009. CE50: Quantifying collision induced dissociation energy for small molecule characterization and identification. *Journal of the American Society for Mass Spectrometry* 20: 1759–1767.

Kim, S., Kramer, R.W., Hatcher, P.G. 2003. Graphical method for analysis of ultrahigh-resolution broadband mass spectra of natural organic matter, the Van Krevelen diagram. *Analytical Chemistry* 75: 5336–5344.

Kind, T., Fiehn, O. 2007. Seven Golden Rules for heuristic filtering of molecular formulas obtained by accurate mass spectrometry. *BMC Bioinformatics* 8: 105.

Kind, T., Fiehn, O. 2010. Advances in structure elucidation of small molecules using mass spectrometry. *Bioanalytical Reviews* 2: 23–60.

Kingdon, K.H. 1923. A method for the neutralization of electron space charge by positive ionization at very low gas pressures. *Physical Review* 21: 408–418.

Koch, B.P., Ludwichowski, K.-U., Kattner, G., Dittmar, T., Witt, M. 2008. Advanced characterization of marine dissolved organic matter by combining reversed-phase liquid chromatography and FT-ICR-MS. *Marine Chemistry* 111: 233–241.

Krauss, M., Singer, H., Hollender, J. 2010. LC-high resolution MS in environmental analysis: From target screening to the identification of unknowns. *Analytical and Bioanalytical Chemistry* 397: 943–951.

Kumar, P., Rúbies, A., Centrich, F., Granados, M., Cortés-Francisco, N. et al. 2013. Targeted analysis with benchtop quadrupole-orbitrap hybrid mass spectrometer: Application to determination of synthetic hormones in animal urine. *Analytica Chimica Acta* 780: 65–73.

Lavonen, E.E., Gonsior, M., Tranvik, L.J., Schmitt-Kopplin, P., Köhler, S.J. 2013. Selective chlorination of natural organic matter: Identification of previously unknown disinfection byproducts. *Environmental Science & Technology* 47(5): 2264–2271.

Lebedev, A.T., Polyakova, O.V., Mazur, D.M., Artaev, V.B. 2013. The benefits of high resolution mass spectrometry in environmental analysis. *Analyst* 138: 6946–6953.

Li, Y., Whitaker, J.S., McCarty, C.L. 2009. New advances in large-volume injection gas chromatography-mass spectrometry. *Journal of Liquid Chromatography & Related Technologies* 32: 1644–1671.

Lim, L., Yan, F., Bach, S., Pihakari, K., Klein, D. 2016. Fourier transform mass spectrometry: The transformation of modern environmental analyses. *International Journal of Molecular Sciences* 17(1): 104.

Makarov, A. 2000. Electrostatic axially harmonic orbital trapping: A high-performance technique of mass analysis. *Analytical Chemistry* 72: 1156.

Makarov, A. 2009. Invited presentation, *18th International Mass Spectrometry Conference, Bremen.*

Makarov, A., Denisov, E., Lange, O. 2009. Performance evaluation of a high-field Orbitrap mass analyzer. *Journal of the American Society for Mass Spectrometry* 20: 1391–1396.

Mamyrin, B.A. 2001. Time-of-flight mass spectrometry (concepts, achievements, and prospects). *International Journal of Mass Spectrometry* 206: 251–266.

Mamyrin, B.A., Karataev, V.I., Shmikk, D.V., Zagulin, V.A. 1973. The mass-reflectron, a new nonmagnetic time-of-flight mass spectrometer with high resolution. *Journal of Experimental and Theoretical Physics* 64: 82–89.

Marshall, A.G. 1996. Ion cyclotron resonance and nuclear magnetic resonance spectroscopies: Magnetic partners for elucidation of molecular structure and reactivity. *Accounts of Chemical Research* 29: 307–316.

Marshall, A.G. 2000. Milestones in Fourier transform ion cyclotron resonance mass spectrometry technique development. *International Journal of Mass Spectrometry* 200: 331–356.

Marshall, A.G., Hendrickson, C.L. 2002. Fourier transform ion cyclotron resonance detection: Principles and experimental configurations. *International Journal of Mass Spectrometry* 215: 59–75.

Marshall, A.G., Hendrickson, C.L. 2008. High-resolution mass spectrometers. *Annual Review of Analytical Chemistry* 1: 579–599.

Marshall, A.G., Hendrickson, C.L., Jackson, G.S. 1998. Fourier transform ion cyclotron resonance mass spectrometry: A primer. *Mass Spectrometry Reviews* 17: 1–35.

Matilainen, A., Gjessing, E.T., Lahtinen, T., Hed, L., Bhatnagar, A. et al. 2011. An overview of the methods used in the characterisation of natural organic matter (NOM) in relation to drinking water treatment. *Chemosphere* 83: 1431–1442.

Medeiros, P.M., Simoneit, B.R.T. 2007. Gas chromatography coupled to mass spectrometry for analyses of organic compounds and biomarkers as tracers for geological, environmental, and forensic research. *Journal of Separation Science* 30: 1516–1536.

Michalski, A., Damoc, E., Hauschild, J.-P., Lange, O., Wieghaus, A. et al. 2011. Mass spectrometry-based proteomics using Q Exactive, a high-performance benchtop quadrupole Orbitrap mass spectrometer. *Molecular & Cellular Proteomics* 10: M111.011015.

Mirsaleh-Kohan, N., Robertson, W.D., Compton, R.N. 2008. Electron ionization time-of-flight mass spectrometry: Historical review and current applications. *Mass Spectrometry Reviews* 27: 237–285.

Mol, H.G.J., Plaza-Bolaños, P., Zomer, P., de Rijk, T.C., Stolker, A.A.M. et al. 2008. Toward a generic extraction method for simultaneous determination of pesticides, mycotoxins, plant toxins, and veterinary drugs in feed and food matrixes. *Analytical Chemistry* 80: 9450–9459.

Mol, H.G.J., Tienstra, M., Zomer, P. 2016. Evaluation of gas chromatography - electron ionization - full scan high resolution Orbitrap mass spectrometry for pesticide residue analysis. *Analytica Chimica Acta* 935: 161–172.

Mol, H.J., Zomer, P., Koning, M. 2012. Qualitative aspects and validation of a screening method for pesticides in vegetables and fruits based on liquid chromatography coupled to full scan high resolution (Orbitrap) mass spectrometry. *Analytical and Bioanalytical Chemistry* 403: 2891–2908.

Murray, K.K. 2010. Glossary of terms for separations coupled to mass spectrometry. *Journal of Chromatography A* 1217: 3922–3928.

Murray, K.K., Boyd, R.K., Eberlin, M.N., Langley, G.J., Li, L. et al. 2013. Definitions of terms relating to mass spectrometry (IUPAC Recommendations 2013). *Pure and Applied Chemistry (IUPAC)* 85: 1515–1609.

Ontario Ministry of Environment. 2007. The Determination of *N*-nitrosamines in Water by Gas Chromatography–High Resolution Mass Spectrometry (GC/HRMS), NITROSO-E3388, Toronto.

Ortelli, D., Cognard, E., Jan, P., Edder, P. 2009. Comprehensive fast multiresidue screening of 150 veterinary drugs in milk by ultra-performance liquid chromatography coupled to time of flight mass spectrometry. *Journal of Chromatography B* 877: 2363–2374.

Ow, S.Y., Noirel, J., Salim, M., Evans, C., Watson, R. et al. 2010. Balancing robust quantification and identification for iTRAQ: Application of UHR-ToF MS. *Proteomics* 10: 2205–2213.

Perry, R.H., Cooks, R.G., Noll, R.J. 2008. Orbitrap mass spectrometry: Instrumentation, ion motion and applications. *Mass Spectrometry Reviews* 27: 661–699.

Peterson, A.C., McAlister, G.C., Quarmby, S.T., Griep-Raming, J., Coon, J.J. 2010. Development and characterization of a GC-enabled QLT-Orbitrap for high-resolution and high-mass accuracy GC/MS. *Analytical Chemistry* 82: 8618–8628.

Pitarch, E., Medina, C., Portolés, T., López, F.J., Hernández, F. 2007. Determination of priority organic micro-pollutants in water by gas chromatography coupled to triple quadrupole mass spectrometry. *Analytica Chimica Acta* 583: 246–58.

Peterson, A.C., Hauschild, J.P., Quarmby, S.T., Krumwiede, D., Lange, O., Lemke, R.A.S., Grosse-Coosmann, F., Horning, S., Donohue, T.J., Westphall, M.S., Coon, J.J. Griep-Raming, J. 2014. Development of a GC/Quadrupole-Orbitrap mass spectrometer, Part I: Design and characterization. *Analytical Chemistry* 86(20): 10036–10043.

Planas, C., Palacios, Ó., Ventura, F., Rivera, J., Caixach, J. 2008. Analysis of nitrosamines in water by automated SPE and isotope dilution GC/HRMS: Occurrence in the different steps of a drinking water treatment plant, and in chlorinated samples from a reservoir and a sewage treatment plant effluent. *Talanta* 76: 906–913.

Portolés, T., Pitarch, E., López, F.J., Hernández, F., Niessen, W.M.A. 2011. Use of soft and hard ionization techniques for elucidation of unknown compounds by gas chromatography/time-of-flight mass spectrometry. *Rapid Communications in Mass Spectrometry* 25: 1589–1599.

Postigo, C., Cojocariu, C.I., Richardson, S.D., Silcock, P.J., Barcelo, D. 2016. Characterization of iodinated disinfection by-products in chlorinated and chloraminated waters using Orbitrap based gas chromatography-mass spectrometry. *Analytical and Bioanalytical Chemistry* 408(13): 3401–3411.

Pringle, S.D., Giles, K., Wildgoose, J.L., Williams, J.P., Slade, S.E. et al. 2007. An investigation of the mobility separation of some peptide and protein ions using a new hybrid quadrupole/travelling wave IMS/oa-TOF instrument. *International Journal of Mass Spectrometry* 261: 1–12.

Quenzer, T.L., Robinson, J.M., Bolanios, B., Milgram, E., Greig, M.J. 2002. Automated accurate mass analysis using FT-ICR mass spectrometry. *Proceedings of the 50th Annual Conference on Mass Spectrometry and Allied Topics, Orlando, FL.*

Richardson, S.D., Plewa, M.J., Wagner, E.D., Schoeny, R., DeMarini, D.M. 2007. Occurrence, genotoxicity, and carcinogenicity of regulated and emerging disinfection by-products in drinking water: A review and roadmap for research. *Mutation Research/Reviews in Mutation Research* 636: 178–242.

Santos, F.J., Galceran, M.T., Caixach, J., Rivera, J., Huguet, X. 1997. Characterization of toxaphene by high resolution gas chromatography combined with high resolution mass spectrometry and tandem mass spectrometry. *Rapid Communications in Mass Spectrometry* 11: 341–348.

Satoh, T., Sato, T., Tamura, J. 2007. Development of a high-performance MALDI-TOF mass spectrometer utilizing a spiral ion trajectory. *Journal of the American Society for Mass Spectrometry* 18: 1318–1323.

Schaub, T.M., Hendrickson, C.L., Horning, S., Quinn, J.P., Senko, M.W. et al. 2008. High-performance mass spectrometry: Fourier transform ion cyclotron resonance at 14.5 Tesla. *Analytical Chemistry* 80: 3985–3990.

Schmitt-Kopplin, P., Gelencsér, A., Dabek-Zlotorzynska, E., Kiss, G., Hertkorn, N. et al. 2010. Analysis of the unresolved organic fraction in atmospheric aerosols with ultrahigh-resolution mass spectrometry and nuclear magnetic resonance spectroscopy: Organosulfates as photochemical smog constituents. *Analytical Chemistry* 82: 8017–8026.

Scigelova, M., Hornshaw, M., Giannakopulos, A., Makarov, A. 2011. Fourier Transform Mass Spectrometry. *Molecular & Cellular Proteomics* 10: M111.009431.

Senko, M.W., Hendrickson, C.L., Emmett, M.R., Shi, S.D.H., Marshall, A.G. 1997. External accumulation of ions for enhanced electrospray ionization Fourier transform ion cyclotron resonance mass spectrometry. *Journal of the American Society for Mass Spectrometry* 8: 970–976.

Shchepunov, V., Berdnikov, A., Lzumi, H., Giles, R., Gall, N. 2010. Doughnut multi-reflecting time-of-flight mass spectrometer. *Proceedings of the 58th ASMS Conference on Mass Spectrometry & Allied Topics, Salt Lake City, UT, May 23–27*.

Sleighter, R.L., Liu, Z., Xue, J., Hatcher, P.G. 2010. Multivariate statistical approaches for the characterization of dissolved organic matter analyzed by ultrahigh resolution mass spectrometry. *Environmental Science & Technology* 44: 7576–7582.

Staack, R.F., Hopfgartner, G. 2007. New analytical strategies in studying drug metabolism. *Analytical and Bioanalytical Chemistry* 388: 1365–1380.

Stenson, A.C., Marshall, A.G., Cooper, W.T. 2003. Exact masses and chemical formulas of individual Suwannee River fulvic acids from ultrahigh resolution electrospray ionization Fourier transform ion cyclotron resonance mass spectra. *Analytical Chemistry* 75: 1275–1284.

Stroh, J.G., Petucci, C.J., Brecker, S.J., Huang, N., Lau, J.M. 2007. Automated sub-ppm mass accuracy on an ESI-TOF for use with drug discovery compound libraries. *Journal of the American Society for Mass Spectrometry* 18: 1612–1616.

Takáts, Z., Wiseman, J.M., Gologan, B., Cooks, R.G. 2004. Mass spectrometry sampling under ambient conditions with desorption electrospray ionization. *Science* 306: 471–473.

Takáts, Z., Wiseman, J.M., Cooks, R.G. 2005. Ambient mass spectrometry using desorption electrospray ionization (DESI): Instrumentation, mechanisms and applications in forensics, chemistry, and biology. *Journal of Mass Spectrometry* 40: 1261–1275.

Toyoda, M., Okumura, D., Ishihara, M., Katakuse, I. 2003. Multi-turn time-of-flight mass spectrometers with electrostatic sectors. *Journal of Mass Spectrometry* 38: 1125–1142.

US EPA. 1994. Method 1613: Tetra- through Octa-chlorinated Dioxins and Furans by Isotopic Dilution HRGC/HRMS. EPA, Washington, DC.

US EPA. 2004. Method 521: Determination of Nitrosamines in Drinking Water by Solid Phase Extraction and Capillary Column Gas Chromatography with Large Volume Injection and Chemical Ionization Tandem Mass Spectrometry (MS/MS). EPA/600/R-05/054, EPA, Cincinnati, OH.

US EPA. 2007. Method 1614: Brominated Diphenyl Ethers in Water, Soil, Sediment, and Tissue by HRGC/HRMS. EPA, Washington, DC.

US EPA. 2008. Method 1668, Revision B: Method 1668B Chlorinated Biphenyl Congeners in Water, Soil, Sediment, Biosolids, and Tissue by HRGC/HRMS. EPA, Washington, DC.

Verentchikov, A.N., Yavor, M.I. 2011. Quasi-planar Multi-reflecting Time-of-Flight Mass Spectrometer. US Patent US 201101086729 A1.

Vichi, S., Cortés-Francisco, N., Caixach, J. 2013. Determination of volatile thiols in lipid matrix by simultaneous derivatization/extraction and liquid chromatography-high resolution mass spectrometric analysis: Application to virgin olive oil. *Journal of Chromatography A* 1318: 180–188.

Wang, J., Leung, D. 2009. Applications of ultra-performance liquid chromatography electrospray ionization quadrupole time-of-flight mass spectrometry on analysis of 138 pesticides in fruit- and vegetable-based infant foods. *Journal of Agricultural and Food Chemistry* 57: 2162–2173.

Webb, K., Bristow, T., Sargent, M., Stein, B. 2004. *Methodology for Accurate Mass Measurement of Small Molecules: Best Practice Guide*. LGC, London.

Witt, M., Fuchser, J., Baykut, G. 2001. Fourier transform ion cyclotron resonance mass spectrometry with NanoLC/microelectrospray ionization and matrix-assisted laser desorption/ionization: Analytical performance in peptide mass fingerprint analysis. *Journal of the American Society for Mass Spectrometry* 1: 553–561.

Wolf, S., Schmidt, S., Muller-Hannemann, M., Neumann, S. 2010. In silico fragmentation for computer assisted identification of metabolite mass spectra. *BMC Bioinformatics* 11: 148.

Wolff, J.C., Eckers, C., Sage, A.B., Giles, K., Bateman, R. 2001. Accurate mass liquid chromatography/mass spectrometry on quadrupole orthogonal acceleration time-of-flight mass analyzers using switching between separate sample and reference sprays: 2. Applications using the dual-electrospray ion source. *Analytical Chemistry* 73: 2605–2612.

Wu, J., McAllister, H. 2003. Exact mass measurement on an electrospray ionization time-of-flight mass spectrometer: Error distribution and selective averaging. *Journal of Mass Spectrometry* 38: 1043–1053.

Xian, F., Hendrickson, C.L., Marshall, A.G. 2012. High resolution mass spectrometry. *Analytical Chemistry* 84: 708–719.

Yoon, S., Nakada, N., Tanaka, H. 2012. A new method for quantifying *N*-nitrosamines in wastewater samples by gas chromatography-triple quadrupole mass spectrometry. *Talanta* 97: 256–261.

Zhang, H., Zhang, Y., Shi, Q., Ren, S., Yu, J. et al. 2012a. Characterization of low molecular weight dissolved natural organic matter along the treatment trait of a waterworks using Fourier transform ion cyclotron resonance mass spectrometry. *Water Research* 46: 5197–5204.

Zhang, H., Zhang, Y., Shi, Q., Hu, J., Chu, M. et al. 2012b. Study on transformation of natural organic matter in source water during chlorination and its chlorinated products using ultrahigh resolution mass spectrometry. *Environmental Science & Technology* 46: 4396–4402.

4 Application of Thermal Desorption–Mass Spectrometry for the Analysis of Environmental Pollutants

Steven Sai Hang Ho, Judith C. Chow, Jian Zhen Yu, John G. Watson, Jun-Ji Cao, and Yu Huang

CONTENTS

4.1 INTRODUCTION

Thermal desorption (TD) has a wide range of environmental applications, from monitoring personal exposure to toxic chemicals in the workplace to ambient and indoor air quality monitoring and to the determination of volatiles in soil and water matrices (DeLuca et al., 1990; Marbot, 1997; Coleman and Chung, 2002; Liu et al., 2002). Large numbers of samples are needed to relate ambient carbonaceous materials to pollution sources and to assess their environmental and health consequences. This demands fast, simple, and sensitive analytical methods. Solvent extraction (SE) followed by GC-MS analysis has been the most widely used approach for carbonaceous materials collected from sample matrices. Despite some documented advantages of using SE-GC-MS methods (e.g., selectivity, feasibility), it is labor and time consuming, which limits the number of samples that can be analyzed. It also requires large sample loadings that are not normally acquired in speciation networks.

TD is a well-established sample-introduction technique for GC analysis of nonpolar and low-polarity volatile organic compounds (VOCs) from soil, water, and air (DeLuca et al., 1990; Helmig and Arey, 1992; Hirvonen et al., 1994; Marbot, 1997; US EPA, 1997; Woolfenden, 1997; Lamanna and Goldstain, 1999; Sigman and Ma, 1999; Sunesson et al., 1999; Tran et al., 2000; Coleman and Chung, 2002; Liu et al., 2002). The TD technique can be used to analyze airborne particulate matter (PM) that has adverse effects on human morbidity and mortality (Vedal, 1997; Watson et al., 1997; Chow et al., 2006; Pope and Dockery, 2006), visibility (Chow et al., 2002; Watson, 2002), climate change (Jacobson, 2002), and materials (Davidson et al., 2000). These effects have sparked interest in the chemical and physical properties of PM, with increasing interest in its organic matter (OM) composition. The OM fraction is complex, comprising thousands of chemical constituents from numerous organic compound classes, such as alkanes (i.e., *n*-alkanes, hopanes, and steranes), aromatics (i.e., polycyclic aromatic hydrocarbons [PAHs]), alcohols (i.e., alkanols, phenols, and sugars), and carboxylic acids (i.e., alkanoic acids) (Rogge et al., 1993; Zheng et al., 1997; Mazurek, 2002; Fine et al., 2004). Various compounds within these classes are recognized as biologically toxic (Weyand and Bevan, 1986; Bayona et al., 1994; Hsiao et al., 2000; Pope and Dockery, 2006). Many are also used as organic markers in source/receptor models for source apportionment (Simoneit, 1985; Rogge et al., 1994; Schauer et al., 1999; Simoneit, 1999; Schauer and Cass, 2000; Watson and Chow, 2007). Simple and fast thermal methods are used to determine organic carbon (OC) and elemental carbon (EC) concentrations in thousands of ambient and source aerosols collected on quartz-fiber filters each year (Chow et al., 1993, 2001, 2004, 2005; Birch and Cary, 1996; Watson et al., 2005). Similar analytical simplicity can also be expected for the analysis of individual organic compounds using more specific detection methods for the thermally evolved organic compounds.

4.2 THERMAL DESORPTION MECHANISMS

4.2.1 VAPORIZATION

Vaporization is the conversion process by which a chemical in solid or liquid form is changed to the gas state by heating or reducing pressure. Most substances change their physical states in the order of solid to liquid to gas with increasing temperature. A few sublimate or change directly from solid to gas. The molecules in solid or liquid states bind with the intermolecular forces, which are fundamentally electrostatic interactions (e.g., ionic interactions, hydrogen bond, dipole–dipole interactions) or electrodynamic interactions (e.g., van der Waals forces) (Atkins, 2002). Vaporization is an endothermic process in which molecules gain kinetic energy as temperature increases. When molecular kinetic energy exceeds the enthalpy change of vaporization ($\Delta_v H$), which is the energy required to overcome intermolecular forces in the solid or the liquid state, individual molecules escape and enter the gas state (Atkins, 2002). This temperature is the boiling point at a constant pressure. Methane has a low standard enthalpy change of vaporization ($\Delta_v H^\theta = 8.19$ kJ mol^{-1}) because of weak van der Waals forces between its molecules (Sears et al., 1982). These forces result from the formation of momentary dipoles in uncharged nonpolar compounds (Seinfeld and Pandis, 1998). Higher $\Delta_v H$ are measured for formic acid and methanol ($\Delta_v H^\theta = 19.9$ and 37.4 kJ mol^{-1}, respectively) since the molecules bind with strong hydrogen (H) bonds (Stout and Fisher, 1941; Sears et al., 1982). The bond is an attractive interaction between an electronegative atom (e.g., oxygen [O] and nitrogen [N]) and an H atom bonded to another electronegative atom (McMurry, 1999). This explains why polar organic alcohols and acids have higher $\Delta_v H$ than alkanes with equivalent carbon numbers.

Apart from heating, vaporization occurs at reduced pressures. The boiling point of a substance is the temperature at which its vapor pressure is equal to the pressure of its environment (Atkins, 2002). When the pressure is reduced, vaporization takes place at a lower temperature. The boiling point of a substance decreases while the pressure decreases such that

$$T_B = \left(\frac{R\left[\ln(P) - \ln(101.325\,\text{kPa})\right]}{\Delta_v H} + \frac{1}{T_o} \right)^{-1}, \tag{4.1}$$

where T_B is the boiling point under pressure P in degrees Kelvin, R is the ideal gas constant (8.3143 m^3 Pa k^{-1} mol^{-1}), P is the vapor pressure at a given temperature in pascals, and T_o is the given temperature boiling point in degrees Kelvin under standard pressure (101.325 kPa) (Atkins, 2002). In a mixture, the vapor pressure of a solute is reduced, leading to the boiling point elevation of the solute. According to Raoult's law, the vapor pressure reduction is proportional to the mole fraction of the solute in the mixture. Deviations from Raoult's law are often observed as the interaction between solute and solvent is not ideal.

4.2.2 Thermal Stability of Organic Molecules

The thermal stability of an organic molecule is a combination of its thermodynamic and kinetic stabilities. The finite strength of chemical bonds puts an upper limit on the vibrational energy that molecules may possess without bond rupture, that is, decomposition. The Boltzmann energy distribution law dictates that at any temperature, a portion of the molecules will possess energy greater than the bond energy. The thermal degradation rate (k) of a molecule is expressed as

$$k = Ae^{-E/RT}. \tag{4.2}$$

A is the frequency factor related to the critical mode of vibration in the molecules and has the same dimension as k. For first-order kinetics, A is expressed in sec^{-1}. E is the activation energy supplied to the molecules to cause decomposition. For decomposition to occur, E needs to be greater than the dissociation energy of the weakest bond in the molecule; the weakest bond strength determines the upper limit of its thermal stability, if there are no other lower-energy decomposition pathways (Blake et al., 1961). Simple hydrocarbons, such as alkanes, decompose through the rupture of a C–C bond, with the formation of free radicals (Rice, 1931; Rice et al., 1932); however, the functional groups of many organic molecules can lead to the rearrangement of pathways and lower the thermal decomposition temperature (°K) (Johns et al., 1962a). The rearrangement pathways take place at temperatures far below the temperature required for straightforward bond rupture.

The thermal stability of a molecule is primarily determined by its chemical structure (Johns et al., 1962a,b). According to Equation 4.2, chemical structures that minimize A and maximize E lower the rate k, thereby increasing the stability of the molecules at a given temperature. Polybonding, in which an atom is bonded to more than one other atom in the molecule, has the effect of decreasing A, as the collision energy can be dissipated over more than one path (e.g., the thermal stability of graphite). The thermal stability of many inorganic solids is also related to this factor, rather than to their high bond strengths. More specifically, the structural features of compounds that favor high thermal stability include (1) all bonds in the molecule having high dissociation energies, (2) no easy paths to decomposition, (3) resonance structure, and (4) substituents that enhance resonance stabilization. Resonance structure moieties have the effect of increasing E in Equation 4.2, hence increasing the thermal stability (Johns et al., 1962a). Aromatic compounds have high thermal stability because of their resonance structure. For example, naphthalene in the vapor phase (at 1 atm) is thermally stable up to 620°C (Johns et al., 1962b). Nine PAHs (i.e., anthracene, phenanthrene, fluoranthene, chrysene, pyrene, triphenylene, benz[*a*]anthracene, benzo[*a*]pyrene, and dibenzo[*a*,*h*]anthracene) are thermally stable beyond 475°C (Madison and Roberts, 1957). In comparison, alkyl-PAHs have lower thermal stability. This follows from the introduction of lower dissociation energy bonds (i.e., alkyl C–C bonds) than the bonds in the aromatic rings. Madison and Roberts (1957) observed a slight thermal decomposition for 1-methyl naphthalene at 475°C. For the same reason, alkanes have lower thermal stabilities than PAHs. Hydrocarbons containing alkyl chains have been found to remain stable to about 350°C (Blake et al., 1961). Table 4.1 lists the chemical and physical properties of major organic classes and markers in aerosols.

Thermally stable molecules lack easy paths to decomposition (Johns et al., 1962b). Polar functional groups, such as $-NH_2$, –OH, –COOH, often provide lower energy decomposition in which bonds are broken and formed simultaneously (e.g., the loss of HX), and are therefore less thermally stable. For example, the thermal decomposition of alcohols involves the unimolecular elimination of water to form alkenes (Johnson, 1974). Schwarz et al. (1972) showed that polyhydroxylated compounds like sugars (e.g., mannitol, dulcitol, sorbitol, arabitol, xylitol, and ribitol) are stable only up to 200–250°C. The thermal stability of esters is lower than hydrocarbons. Esters of alcohols with β-hydrogen (e.g., methyl esters of fatty alcohols) are stable up to only ~285°C (Blake et al., 1961).

As shown in Equation 4.2, temperature (T) is the most important external factor affecting thermal stability. In addition, at a given temperature, decomposition generally occurs under high rather than low vapor pressure conditions due to greater collision frequencies (Johns et al., 1962a).

The presence of other molecules can lead to higher or lower stability by preventing or opening thermal degradation/reaction pathways, respectively. For example, in the petroleum industry, it has been known that the thermal stability of alkanes can be improved in the presence of hydrogen donors, for example, tetralin (Song et al., 1994). This is achieved by the interruption of the radical chain reaction through the H-donating action.

4.2.3 Pyrolysis

Pyrolysis converts an organic compound to one or more different compounds by thermal energy in an inert environment (absence of oxygen or catalysts) (McNaught and Wilkinson, 1997). Carbonization happens in extreme pyrolysis to form EC (McNaught and Wilkinson, 1997; Demirbas, 2004). Above a particular temperature, organic molecules gain sufficient vibrational energy to rupture their bonds and form free radicals that subsequently reform other lower-molecular weight compounds or combine with higher-molecular weight compounds (Napier and Norrish, 1967; Yuan et al., 1982; Merkulov et al., 1989; Buchanan et al., 1998; Fulle et al., 1998; Evans and Dellinger, 2006). Saturated alkanes break down C–H and C–C bonds to radicals that form smaller alkanes, alkenes, and hydrogen (Bradley, 1974; Rumyantsev and Oganesova, 1987; Watanabe et al., 2001). Simple aromatic compounds (e.g., benzene and toluene) produce aryl radicals, which attack other hydrocarbon molecules to give bi- and polyaryls (Smith, 1979; Brouwer et al., 1988; Hippler et al., 1990; Damm et al., 1999). The reaction causes the loss of H and X (any leaving group [e.g., F, NH_2, or OH]) from a compound with a configuration of H–C–C–X and results in the formation of a C=C bond (Boyd and Overton, 1972; Brown, 1980; Dominguez and Chuchani, 1992; Moss et al., 2005). This elimination also occurs by the loss of small molecules such as N_2, CO, CO_2, or SO_2, leading to the formation of reactive intermediates such as arynes, diradicals, carbenes, or nitrenes (Brown, 1980; Moss et al., 2005). The cleavage of particular C–C bonds in a carbon chain or ring always happens in pyrolysis. This includes the transfer of hydrogen during fragmentation that occurs by a cyclic process (Arnold et al., 1950, 1957; Jahngen et al., 2007). Pyrolysis can involve isomerization without elimination or fragmentation. These processes occur by way of intermediates such as diradicals from pericyclic reactions (List et al., 1971; Garnett et al., 1976; Morgan, 2000).

TABLE 4.1
Chemical and Physical Properties of Major Organic Classes and Markers in Aerosols

Organic Classes	Subgroups	Useful Marker Compounds	Molecular Weight	Melting Point (°C)	Boiling Point (°C)	Vapor Pressure at 25°C (mmHg)	Description (Bonds)	Thermal Stability	Amenability to TD	Amenability to SE
Aromatic	PAHs	Fluoranthene ($C_{16}H_{10}$)	202.6	110.8	375	1.23×10^{-8}	Aromatic ring, C–H	High	Yes	Yes
		Chrysene ($C_{18}H_{12}$)	228.3	255.8	448	6.23×10^{-9}				
		Benzo[*e*]pyrene ($C_{20}H_{12}$)	252.3	178.0	492	5.70×10^{-9}				
		Benzo[*a*]pyrene ($C_{20}H_{12}$)	252.3	176.5	495	1.80×10^{-14}				
		Indeno[1,2,3-*cd*]pyrene ($C_{22}H_{12}$)	276.3	162.5	536	1.00×10^{-10}				
		Benzo[*ghi*]perylene ($C_{22}H_{12}$)	276.3	278.3	500	1.00×10^{-10}				
		Picene ($C_{22}H_{14}$)	278.4	366–367	518–520	3.73×10^{-9}				
		Coronene ($C_{24}H_{12}$)	300.4	438.0	525	3.29×10^{-11}				
Alkanes	*n*-Alkanes	*n*-Nonacosane ($C_{29}H_{60}$)	408.8	63.7	440.8	3.24×10^{-14}	C–C, C–H	High	Yes	Yes
		n-Triacontane ($C_{30}H_{62}$)	422.8	62.3	449.7	2.74×10^{-15}				
		n-Hentriacontane ($C_{31}H_{64}$)	436.8	67.9	458	1.40×10^{-16}				
		n-Dotriacontane ($C_{32}H_{66}$)	450.9	69	467	8.49×10^{-18}				
		n-Tritriacontane ($C_{33}H_{68}$)	464.9	71–73	485.27[e]	4.02×10^{-19}				
		n-Tetratriacontane ($C_{34}H_{70}$)	478.9	72.6	496.87[e]	1.67×10^{-20}				
		n-Pentatriacontane ($C_{35}H_{72}$)	493.0	75.0	490	5.39×10^{-22}				
		n-Hexatriacontane ($C_{36}H_{74}$)	507.0	74–76	520.08[e]	1.34×10^{-23}				

(*Continued*)

TABLE 4.1 (CONTINUED)
Chemical and Physical Properties of Major Organic Classes and Markers in Aerosols

Organic Classes	Subgroups	Useful Marker Compounds	Molecular Weight	Melting Point (°C)	Boiling Point (°C)	Vapor Pressure at 25°C (mmHg)	Description (Bonds)	Thermal Stability	Amenability to TD	Amenability to SE
	Hopanes	17α(H)-22,29,30-trisnorhopane ($C_{27}H_{46}$)	370.7	112.79[d]	387.83[e]	2.09 × 10^{-6}[f]	Cyclic ring, C–C, C–H	High	Yes	Yes
		17α(H),21β(H)-30-norhopane ($C_{29}H_{50}$)	398.7	124.38[d]	407.67[e]	5.19 × 10^{-7}[f]				
		17α(H),21β(H)-hopane ($C_{30}H_{52}$)	412.8	127.07[d]	412.28[e]	3.91 × 10^{-7}[f]				
		17α(H),21β(H)-22R-homohopane ($C_{31}H_{54}$)	426.8	133.84[d]	423.88[e]	1.71 × 10^{-7}[f]				
	Steranes	ααα 20R-cholestane ($C_{27}H_{48}$)	372.7	115.54[d]	392.53[e]	8.79 × 10^{-6}[f]	Cyclic ring, C–C, C–H	High	Yes	Yes
		ααα 20R 24R-methyl-cholestane ($C_{28}H_{50}$)	386.7	122.93[d]	405.20[e]	9.19 × 10^{-7}[f]				
		ααα 20R 24R-ethyl-cholestane ($C_{29}H_{52}$)	400.7	130.33[d]	417.87[e]	3.84 × 10^{-7}[f]				
Esters	Phthalates	Dimethylphthalate ($C_{10}H_{10}O_4$)	194.2	0–2	282	1.65 × 10^{-3}	Aromatic ring with –COO–R_1, C–C, C–H (R_1 could be –CH_2 chain or aromatic ring)	Fair	Yes	Yes
		Diethyl phthalate ($C_{12}H_{14}O_4$)	222.2	–40.5	295	1.65 × 10^{-3}				
		Di-*n*-butyl phthalate ($C_{16}H_{22}O_4$)	278.3	–35	340	7.30 × 10^{-5}				
		Butyl benzyl phthalate ($C_{19}H_{20}O_4$)	312.4	<–35	370	8.25 × 10^{-6}				
		Bis(2-ethylhexyl)phthalate ($C_{24}H_{38}O_4$)	390.6	–50	386.9	9.75 × 10^{-6}				
		Di-*n*-octyl phthalate ($C_{24}H_{38}O_4$)	390.6	–25	220	2.60 × 10^{-6}				

(Continued)

TABLE 4.1 (CONTINUED)
Chemical and Physical Properties of Major Organic Classes and Markers in Aerosols

Organic Classes	Subgroups	Useful Marker Compounds	Molecular Weight	Melting Point (°C)	Boiling Point (°C)	Vapor Pressure at 25°C (mmHg)	Description (Bonds)	Thermal Stability	Amenability to TD	Amenability to SE
Alcohols	Sugars	Glycerol ($C_3H_8O_3$)	92.1	17.8	290	1.68×10^{-4}	–OH, C–C, C–H	Low	No	Yes
		Xylitol ($C_5H_{12}O_5$)	152.1	95–97	216	1.96×10^{-3}[f]				
		Levoglucosan ($C_6H_{10}O_5$)	162.1	170–180	313.77[e]	3.47×10^{-7}[f]				
		Glucose ($C_6H_{12}O_6$)	180.2	83.0	380.68[e]	1.82×10^{-8}				
		Fructose ($C_6H_{12}O_6$)	180.2	85.4[d]	340.99[e]	3.11×10^{-7}				
		Mannitol ($C_6H_{14}O_6$)	182.2	167.0	290–295	1.31×10^{-9}				
		Sorbitol ($C_6H_{14}O_6$)	182.2	93–97	389.71[e]	4.92×10^{-9}				
		Sucrose ($C_{12}H_{22}O_{11}$)	342.3	190–192	591.59[e]	5.15×10^{-17}				
		Cholesterol ($C_{27}H_{46}O$)	386.7	148.5	360	7.79×10^{-10}				
Carboxylic acids	Mono- and dicarboxylic acids	Oxalic acid ($C_2H_2O_4$)	90.0	189.0	190	2.34×10^{-4}	–COOH, C–C, C–H	Low	No	Yes
		Malonic acid ($C_3H_4O_4$)	104.1	135.6	140	1.50×10^{-3}				
		Succinic acid ($C_4H_6O_4$)	118.1	183–185	235	7.86×10^{-6}				
		Adpic acid ($C_6H_{10}O_4$)	146.1	152.0	337	7.30×10^{-2}[c]				
		Pimelic acid ($C_7H_{12}O_4$)	160.2	103–105	272[a]	2.08×10^{-5}				
		Suberic acid ($C_8H_{14}O_4$)	174.2	144.0	230[b]	1.69×10^{-8}				
		Palmitic acid ($C_{16}H_{32}O_2$)	256.4	62.0	271.4[b]	1.20×10^{-5}				
		Oleic acid ($C_{18}H_{34}O_2$)	282.3	13–14	460	5.46×10^{-7}				
		Stearic acid ($C_{18}H_{36}O_2$)	284.5	69.3	361	7.22×10^{-7}				

Source: Chow, J. C. et al. *Journal of Environmental Science and Health Part A*, 42, 1521–1541, 2007. With permission.

[a] Boiling point reported at the pressure of 100 mmHg.
[b] Boiling point reported at the pressure of 15 mmHg.
[c] Vapor pressure reported at the temperature of 18°C.
[d] Melting point was estimated with Gold and Ogle method (using EPI Suite developed by the EPA's Office of Pollution Prevention Toxics and Syracuse Research Corporation [SRC]).
[e] Boling point was estimated with adapted Stein and Brown method (using EPI Suite developed by the EPA's Office of Pollution Prevention Toxics and SRC).
[f] Vapor pressure was estimated with modified grain method (using EPI Suite developed by the EPA's Office of Pollution Prevention Toxics and SRC).

4.3 THERMOCHEMICAL PROPERTIES OF CARBONACEOUS MATERIALS

Carbonaceous materials can be thermally released to the gas phase either in chemically unaltered or altered forms. For carbonaceous aerosol, EC is thermally resistant, requiring oxidation to evolve into the gas phase in the form of CO or CO_2. OC comprises tens of thousands of organic compounds with diverse chemical structures. Depending on the chemical nature, some organic compounds are volatilized into the gas phase unaltered, while others are thermally labile, leading to degradation products released to the gas phase. As discussed earlier, oxygenated compounds are prone to thermal conversion during TD. *n*-Alkanols in a GC injection port at 250–350°C yielded *n*-aldehydes and 1-alkenes (Helmig et al., 1992; Hansen et al., 1995); levoglucosan partially degrades to levoglucosenone at 300°C (Williams et al., 2006); cholesterol spiked onto a quartz-fiber filter during TD-GC-MS analysis and partially converted to cholesta-3,5-diene. Thermal degradation products could be mistaken for compounds originally present in suspended particles. This highlights the importance of removing the thermally labile constituents or converting them to less sensitive functional groups before applying TD.

Thermal behaviors might be modified by other components coexisting in the particles. For example, *n*-alkanoic acid amides (C_{12}–C_{20}) and nitriles (C_{14}–C_{32}) were measured in the TD-GC-MS analysis of ambient aerosols collected in Augsburg, Germany (Schnelle-Kreis et al., 2005). Five nitriles were also reported in aerosols of a forest site in Germany (Helmig et al., 1990). These amides and nitriles are actually thermal reaction products from analogous acids and ammonium nitrate (NH_4NO_3) coexisting in the aerosol samples (Simoneit, 1992). Yu et al. (2002) observed changes in the thermal evolution pattern of levoglucosan in the presence and the absence of ammonium bisulfate (NH_4HSO_4). The presence of transition metal catalysts (e.g., Fe or Cu) in aerosols can also affect the conversion of OM during heating.

Matrix effects are little understood and may vary from sample to sample. Online instruments using TD minimize thermal decomposition due to reduced pressure and lower aerosol loading in the analysis zone. Laser desorption has been employed for online MS-based instruments (e.g., the studies of Murphy and Thomson [1995] and Noble and Prather [1996]). However, it experiences fluctuations in desorption/ionization conditions, which lead to a large pulse-to-pulse variation in ion signals that impede quantification.

4.4 THERMAL METHODS FOR GASEOUS ANALYSIS

VOC is a critical fraction to be studied due to the recognition of associated human health risks. Over 300 individual VOCs have been identified as crucial parameters in both indoor and outdoor pollution. The US Environmental Protection Agency (EPA) Compendium of Methods TO-15 is a standard offline technique for the collection and analysis of air samples in canisters (US EPA, 1999a). The sample VOCs condense in a cryogenic trap and are thermally desorbed into a laboratory GC-MS for detection. The advantage of active sampling is that it collects a large volume of air, and an extensive range of VOCs concentrations can be quantified at the same time. While canister samples have advantages, they are bulky, expensive, and require extensive training to operate. Canisters cannot be applied for the collection of semivolatile organic compounds.

In principle, any solid sample containing VOCs can be analyzed using some variation of TD. The VOCs are trapped in sorbent tubes or fibers coated with absorbents (e.g., Tenax TA, Chromosorb, Carboxen, activated carbon, or mixed absorbents) (US EPA, 1999b). During analysis, VOCs are thermally released from the adsorbent matrix in an external TD device and swept by a flow of inert gas into a secondary trap, usually a cold trap. The analytes are then refocused on the cold trap. Once the primary desorption step is complete, the cold trap is heated rapidly. The analytes are released as a tiny plug of vapor and introduced to the analyzer with minimal band spreading. Capillary cryofocusing can also substitute for the work done by the cold trap. VOCs are cryofocused at a low temperature (i.e., below 0°C) at the front of a GC column during the desorption step and then released onto the GC column by starting an oven temperature program. This two-stage desorption technique offers great versatility for monitoring trace level organic vapors and can save many hours of manual sample preparation. Schripp et al. (2014) adopted both the canister and sorbent sampling methods to investigate the variability in VOC concentrations at individual stores in the United States over the course of a year (Schripp et al., 2014). The main advantage of the adsorption method is that minimal amounts of background is emitted during sample collection, but the analytes lost in adsorption and/or desorption cannot be ignored (Meininghaus et al., 1999).

Passive sample substrates such as solid-phase micro-extraction (SPME) have excellent performances in long-term sampling for a few VOCs (e.g., benzene and toluene). This energy-free technique allows high selectivity sample collection when a proper absorbent-coated extraction head is chosen (Chai et al., 1993). The adsorbent media (either desorption tubes or SPME) can be used repeatedly, reducing sample costs.

The determination of polar species of small molecular size (e.g., formaldehyde and glyoxal) cannot be accomplished using the traditional TD technique. They are not easily desorbed due to their higher polarity, which likely gives rise to stronger binding with the absorbent matrix. A selective derivatizing agent can be coated on a suitable impurity-free sorbent inside a sampling or a desorption tube. During sample collection, the polar species in the air stream are trapped by the sorbent by reacting with the derivatizing agent. The resulting derivatives are thermostable at room temperature and atmospheric pressure and are suitable for GC analysis. An SPME with on-fiber derivatization supports the feasibility of coupling sorbent derivatization sampling with TD analysis. In the on-fiber derivatization/SPME technique, analytes in the gas phase diffuse onto the SPME fibers and subsequently react with the derivatization agent coated on the SPME fibers. The derivatized analytes are then thermally desorbed in a GC injector port by inserting the fiber into the injector port via a microsyringe. Pan et al. (1995, 1997), Pan and Pawliszyn (1997), Martos and Pawliszyn (1998), and Koziel et al. (2001) demonstrate that the combination of on-fiber derivatization and TD is a sensitive technique to sample and analyze airborne polar analytes. Ho and Yu (2002, 2004a) further established on-sorbent derivatization with *O*-(2,3,4,5,6-pentafluorobenzyl)hydroxylamine and 2,3,4,5,6-pentafluorophenyl hydrazine coupled with in-injection port TD for the measurement of airborne carbonyls.

4.5 THERMAL METHODS FOR AEROSOL ANALYSIS

The term *thermal extraction* (TE) refers to the direct removal of OM from the aerosol matrix by heating the sample; it has been used interchangeably with the term *thermal desorption*. More considerations are needed for the measurement of aerosol samples with TD approaches. Thermal conditions are optimized for the fast and complete vaporization of target compounds in their unaltered forms, along with a sensitive detection system for individual organic compounds (e.g., MS, GC-MS, and GC/flame ionization detector [FID]). Thermal conditions can also be selected to evolve nonvolatile and insoluble organic compounds as their pyrolysis products, and the parent compound identities are inferred from their thermal degradation products as in the application of pyrolysis GC for the determination of rubbers and plastics. TD coupled with a suitable detection system has the potential to become a simple and routine tool for the determination of organic aerosol chemical composition. With the perfection of the equipment, the procedures, and the data processing methods, it is possible to obtain data on currently acquired or archived samples with little additional cost and effort than is currently expended on OC and EC measurements.

4.5.1 Off-Line Thermal Desorption Analyses of Filters

4.5.1.1 Thermal Desorption–Mass Spectrometry

Schuetzle et al. (1973, 1975) and Cronn et al. (1977) first collected particles on glass-fiber filters and impaction plates, then introduced them directly into a high-resolution MS utilizing a temperature-programmed insertion tube (solid probe), which was heated from 20°C to 400°C at 20–30°C min^{-1} under a vacuum of 10^{-9} atm. Compounds were selectively removed from the aerosol matrix according to their volatility into the MS for analysis. The thermal desorption–mass spectrometry (TD-MS) generated mass thermograms, that is, the intensity of each ion current peak as a function of time or temperature. Compound identification was based on (1) a major characteristic ion and the temperature at which this signal was detected, (2) one or two additional ions at the same temperature, and (3) the plausibility of the existence of the component in the aerosol by vapor pressure calculation. Cronn et al. (1977) identified alkanes, alkenes, aromatics (i.e., alkyl benzenes, substituted naphthalenes, and PAHs), amines (i.e., alkyl piperidines), alcohols (i.e., phenol), and carboxylic acids. Schuetzle et al. (1975) identified a number of peroxides in aerosols collected in Pasadena, California, suggesting the low probability of decomposition of organic compounds with TD under high vacuum conditions. Nevertheless, there were signs of matrix effects and possible thermal decomposition. For example, C_3–C_7 dicarboxylic acids were found, while C_2 dicarboxylic acid was not found, inconsistent with results from other studies which showed that oxalic acid is more abundant than other dicarboxylic acids (Kawamura and Kaplan, 1987; Poore, 2000). This failure to detect oxalic acid in southern California reflects thermal decomposition. The matrix effect was demonstrated by the observation that four tested compounds (i.e., naphthalene, phenol, glutaric acid, and phenanthrene) spiked onto aerosol loaded substrates required 44–152°C higher temperatures to vaporize than those spiked onto a particle-free brass disk (Schuetzle et al., 1975). This is the effect of boiling point elevation due to the reduction in the vapor pressure of the substances in a mixture. The effect was more prominent for two polar compounds (i.e., phenol and glutaric acid) with hydrogen bonding.

Since TD takes place in a vacuum, organic compounds are vaporized at reduced temperatures. This in turn extends the analytical feasibility to compounds of lower thermal stability (e.g., peroxides) and higher molecular weight. Majer and Perry (1970) reported 12-ring hydrocarbons using TD-MS in soot as compared to a 7-ring PAH (coronene) using GC. However, organic compounds with similar thermal and mass fragmentation behaviors are not resolved by TD-MS.

4.5.1.2 Thermal Desorption Gas Chromatography–Mass Spectrometry

With capillary GC, organic compounds in aerosol samples are separated according to their physical (boiling point) and chemical (polarity) properties before quantification by MS. Chromatographic separation prior to FID or MS detection minimizes interferences from coexisting species.

External TD units use either conventional resistive heating (Hall et al., 1999; Hays et al., 2003) or Curie point heating (Neususs et al., 2000; Jeon et al., 2001) to thermally extract organic compounds. The in-injection port TD also has several variants. Falkovich and Rudich (2001) used a microvial to hold the filter strip inside a GC liner. Wauters et al. (1979) and Ho and Yu (2004b) adapted a conventional GC liner without modification. Helmig et al. (1990) altered the carrier gas lines, while Blanchard and Hopper (1997) modified the GC injector with a small T-connector, a three-way valve, and a needle valve. Welthagen et al. (2003) and Schnelle-Kreis et al. (2003, 2005) automated the insertion of GC injection liners preloaded with filter pieces and closing of the injector port.

The Curie point TD device uses radio frequencies to induce eddy currents in a ferromagnetic foil, in which filter strips are positioned, to heat the sample to the Curie point temperature of the foil. At the Curie point, the ferromagnetism of the metal changes to paramagnetism so that the foil remains at that constant temperature. The Curie point devices offer rapid heating. This might reduce decomposition as a result of volatilization being kinetically favored relative to the competitive decomposition processes (Beuhler et al., 1974).

External TD units require a transfer line and a connector to the GC injector and may consequently experience loss of organic compounds when the material of the transfer line is not inert or the line has cold spots. A contaminated line of inert material (due to deposition over time) can also become active. The line has to be inert and heated uniformly to a temperature above the desorption temperature. Residue (e.g., polar substances) may be trapped between the TD unit and the GC-MS, causing carryover (Greaves et al., 1985; Hays et al., 2003). TD performed in the GC injector is simple and eliminates the need for an external oven and transfer line. It requires no modifications or only a minor rearrangement of carrier gas lines or an injector cap. This development not only reduces the carryover, but also offers maximum transfer efficiency.

TD-GC-MS differs from SE-GC-MS primarily in the principle of extracting individual organic compounds (i.e., by heat or by solvent) from the filter. GC-MS analyses of the extracts are the same. Table 4.2 compares the attributes of the two methods.

The SE methods are better known for their selectivity, feasibility, and sample extract derivatization, fractionation, and cleanup. Solvent extracts generally pair better with off-line analytical instrumentation (e.g., ion chromatography [IC] and high-performance liquid chromatography [HPLC]; both transfer samples with liquid mobile phases) than TD, which has resulted in an expanded list of nonpolar and polar organic compounds (Kiss et al., 1997; Schauer and Cass, 2000; Fine et al., 2004; Odabasi et al., 2005). TD-GC-MS applied to aerosols has been largely limited to nonpolar PAHs and *n*-alkanes. Unlike SE, in which only a small portion of the extract (typically a small percentage) is injected, the entire sample mass from a filter portion is transferred to the TD-GC-MS. As a result, TD is at least an order of magnitude more sensitive than SE and requires considerably less sample mass. This attribute is particularly important when only micrograms of aerosols are available as on 25–47 mm diameter filters from speciation samplers (Chow, 1995; Watson and Chow, 2001). The high sensitivity of TD improves the temporal resolution of organic compounds. Since TD requires a small fraction (0.3–5.0 cm^2) of the filter for analysis (Falkovich and Rudich, 2001; Waterman et al., 2001a; Ho and Yu, 2004b; Ho et al., 2011), it can be applied to the remnants of archived quartz-fiber filters after ionic and carbon analyses without additional sampling.

4.5.1.3 Thermal Desorption with Two-Dimensional Gas Chromatography–Time-of-Flight Mass Spectrometry

Since the greater part of organics remain unresolved by TD-GC-MS, TD paired with comprehensive two-dimensional GC and TOFMS has been used to provide additional resolution. The compounds are first separated by boiling points with a nonpolar stationary phase column. As the compounds elute from this first column, they are refocused onto a shorter (1 m) polar stationary phase column for further separation according to polarity. Using this approach, Welthagen et al. (2003) detected ~15,000 compounds in particles collected in Germany, including categories of alkanes (i.e., *n*-alkanes, steranes, and hopanes), alkenes (i.e., terpenes), aromatic (i.e., alkylbenzenes, substituted aromatics, heterocyclic aromatics, PAHs, oxygenated PAHs, and branched PAHs), aldehydes, ketones (i.e., *n*-alkane-2-ones), carboxylic acids, nitriles (i.e., alkyl nitriles), and esters (Schnelle-Kreis et al., 2005). These methods can also identify oxygenated compounds (e.g., aldehydes, ketones, and carboxylic acids) ascribable to secondary organic aerosols (SOAs) (Hamilton et al., 2004, 2005). Many of the oxygenated species (e.g., 2-hexenal, furan) identified by Hamilton et al. (2004, 2005) were fairly volatile (vapor pressure: $>10^{-2}$ atm at 20°C). The presence of these volatile species needs further evaluation. It is possible that some of the oxygenated species detected are TD products of thermally labile compounds that were originally present in the aerosols. Using GC×GC-TOFMS to quantify organic compounds shows potential (Schnelle-Kreis et al., 2005).

4.5.1.4 Pyrolysis-Thermal Desorption–Gas Chromatography–Mass Spectrometry

The two most common pyrolysis techniques are the Curie point and controlled temperature programming pyrolysis. Pyrolysis is often used as a form of sample pretreatment for mass spectrometric

TABLE 4.2
Attributes of Thermal and Solvent Extraction Methods for the Analysis of Organic Compounds in Aerosols Collected on Filters

Variables	Thermal Desorption	Solvent Extraction
Extraction theory	Thermal behavior of analytes (e.g., boiling point, volatility)	Solubility of analytes in extraction solvent or solvent mixture (normally binary or tertiary)
Sample size and loading	Small punches (0.3–5.0 cm²) of filter depending on the design of the thermal desorption unit and the sample loading Can use remnants of archived quartz fiber samples from existing networks Requires >1 μg of OC[a] Medium collection flow rates (e.g., 5–30 L min⁻¹) provide adequate sample loading for analysis	Requires >50 μg of OC Combining samples into composites may be needed for low loading samples High collection flow rates (e.g., 100–1000 L min⁻¹) are required to increase the sample loading
Sample deposit homogeneity	Essential since only a fraction of filter is used	Not relevant when the entire filter is used for extractions
Solvent consumption	Low, a few microliters of internal standard containing solvent is used per sample	High, ranging from 10–300 ml per sample Common solvents used: Dichloromethane, methanol, 1-butanol, acetone, and *n*-hexane Generally, >90% cannot be recycled
Sample pretreatment	Spiking with internal standard, a few minutes of labor	Complicated steps (i.e., single or multiple volume reductions, extract transfer, and sample filtration) 1–20 h labor per sample
Analysis time[b]	1–2 h per sample	A few to tens of hours
Analytical instrument	MS GC/FID GC-MS	GC-MS GC/FID HPLC IC Capillary electrophoresis
Sample introduction volume	100%	0.5–25% of the solvent extract
Sample contamination	Low probability Potential carryover (usually negligible), transfer loss, molecular arrangement, or breakdown as temperature of ≥350°C Fragmentation of thermally labile compounds	High probability Possible contaminants from solvents and complicated extraction procedures Loss of volatile compounds during the extraction and pretreatment steps Possible carryover from injection port when high concentration samples are introduced
Sensitivity and MDLs	High sensitivity[a] *n*-Alkanes: 0.41–4.36 ng/sample[c] PAHs: 0.08–2.40 ng/sample[c]	Low sensitivity *n*-Alkanes: 37.7–125 ng/sample[c] PAHs: 10.3–47.9 ng/sample[c]
Suitability of filter matrix	Filters that can remain intact at TD temperatures, e.g., quartz fiber or Teflon-impregnated glass-fiber filters	Filters that do not dissolve in solvents, e.g., quartz fiber, glass fiber, or TIGF filters

Source: Chow, J. C. et al. *Journal of Environmental Science and Health Part A*, 42, 1521–1541, 2007. With permission.

[a] Calculation based on the detection limits of the organic compounds reported in the thermal desorption method (Hays et al., 2003).

[b] Including sample pretreatment and analytical separations.

[c] Data from the study by Ho and Yu (2004b).

investigations of complex materials (Labban et al., 2006). Pyrolysis-TD-GC-MS uses higher TD temperatures (often exceeding 400°C) than TD-GC-MS (Neususs et al., 2000; Blazso et al., 2003; Labban et al., 2006). As a result, pyrolysis-TD-GC-MS produces many more evolved products. The pyrolysis of resuspended soil samples at 740°C resolved 174 compounds (Labban et al., 2006), five times more than were obtained with TD for the same sample at 315°C. During pyrolysis, vaporized compounds include both volatile molecules (in the sample) and volatile fragments produced from the cracking of nonvolatilized molecules. The vaporized compounds are then collected, separated, and identified by GC-MS. In pyrolysis, polar compounds, either indigenous or those formed during thermal degradation, can readily undergo reactions such as decarboxylation and dehydration, consequently losing the functional group information. To overcome this, reactive pyrolysis mixes a derivatizing reagent with the sample to preserve the polar functional group information and also to improve their detection (Fabbri et al., 2002). Tetramethylammonium hydroxide and hexamethyldisilazane are the common methylation and silylation reagents, respectively, for thermally assisted derivatization (Fabbri et al., 2002).

Pyrolysis-TD-GC-MS has characterized biopolymers (e.g., lignins), soil particles, sediments, and other matrices containing macromolecules (e.g., the studies by Schulten and Leinweber [1996] and Leinweber and Schulten [1999]). Subbalakshmi et al. (2002) identified aerosol pyrolysis products and linked them to their parent molecules: (1) methoxyphenols originating from lignin; (2) furans, aldehydes, and ketones from carbohydrates; (3) pyrrole and indoles from proteins; and (4) hydrocarbons from lipid structures. Fabbri et al. (2002) applied conventional and reactive pyrolyses at 700°C to PM_{10} from an industrial area of Bologna, Italy, providing information on precursors (e.g., carbohydrates, lipids, synthetic rubbers, and conifer resins). Blazso et al. (2003) used thermally assisted methylation and silylation in their pyrolysis-TD-GC-MS analysis of size-segregated aerosol samples from a forest and a semiremote site in tropical Brazil, identifying fatty acids, alkadioic acids, hydroxybenzoic acids, levoglucosan, and other compounds of polysaccharide origin.

With pyrolysis-TD-GC-MS, the chemical nature of the sample is inferred from the molecular structure of its thermal degradation products. Compounds detected can be pyrolysis or evaporation products from one or more parent molecules. The identification of parent compounds requires extensive mapping to their pyrolysis products. Quantitative analysis is hindered by the presence of the large number of compounds detected and the effects of aerosol loading. However, pyrolysis-TD-GC-MS has the advantage of obtaining structural information for polymeric and insoluble organic compounds, which could not be otherwise obtained by conventional SE- or TD-GC-MS (Voorhees et al., 1988, 1991; Labban et al., 2006). Pyrolysis-TD-GC-MS also provides a rapid screening tool, since no sample pretreatment is required.

4.5.1.5 Parameters for Aerosol Analysis with TD-GC-MS and Pyrolysis-TD-GC-MS

Table 4.3 compares experimental parameters for 24 studies using TD-GC-MS and pyrolysis-TD-GC-MS. Important parameters include (1) type of sample (2) sample preparation, (3) TD unit, (4) analytical instrument, (5) TD temperature program, (6) TD time, (7) desorbent refocusing pathway, (8) GC column, (9) GC initial temperature, (10) GC oven temperature program, and (11) total analysis time. The quantitative significance of different parameter selections has not yet been determined.

TD temperature ramping and holding times varied by sample type, experimental objective, analytical equipment, and operation, which affected the analysis times. Both isocratic heating and gradient heating were applied in resistively heated TD devices, whereas rapid induction heating (~3 s) was used in Curie point pyrolysis. The TD of aerosols in off-line ovens and inside the GC injection port is completed in a matter of minutes, or seconds in the case of Curie-point pyrolysis. Isocratic desorption temperatures of 250–320°C have been used in a few studies (e.g., Greaves et al. [1985, 1987], Helmig et al. [1990], Veltkamp et al. [1996], Blanchard and Hopper [1997], Waterman et al. [2001a], and Blanchard et al. [2002]). For gradual heating, temperatures varied from ambient (~25°C) (Neususs et al., 2000; Jeon et al., 2001; Hays et al., 2003, 2004; Dong et al., 2004) to 175°C (Waterman et al., 2000). Temperature ramping in TD varied from 12°C min^{-1}

TABLE 4.3
Summary of Sample Key Parameters for TD-GC-MS and Pyrolysis-GC-MS

Reference	Sample Type	Sample Preparation	TD Unit	Analytical Instrument	TD Temperature Program	TD Time	Desorbent Refocusing Pathway	GC Column	Initial GC Temp.	GC Oven Temperature Program	Total Analysis Time
1. TD-GC-MS with Resistively Heated External Oven											
Greaves et al. (1985, 1987)[a] Veltkamp et al. (1996)	Aerosol sample and NIST SRM 1649	A 1.3 cm diameter piece of quartz filter was punched and placed on the top of a glass frit of a fritted glass sealing tube. The filter was firmly held on the position during sampling. The filter after collection was then directly analyzed by TD.	A cylindrical aluminum block containing a heating cartridge connected to a thermocouple	HP 5892A GC-MS in EI mode	Isocratic desorption at 254°C	15 min	Cold trapping onto capillary column head	Ultra Performance fused silica capillary (25 m × 0.32 mm × 0.52 μm); Ultra-2 (25 m × 0.32 mm × 0.17 μm)	*Ambient sample:* −50°C or −60°C *NIST standard:* 20°C	*Ambient sample:* −60°C to 0°C at 20°C min^{-1}, then to 300°C at 8°C min^{-1} *NIST standard:* 20–180°C at 20°C min^{-1}, then to 300°C at 8°C min^{-1}	*Ambient sample:* 55.5 min *NIST standard:* 45.5 min
Waterman et al. (2000)	NIST SRM 1649a	NIST urban dust was weighed, transferred into a glass-lined stainless steel GC liner containing glass wool, and spiked with internal standards.	External oven mounted on the top of the GC-MS system	HP 5890 GC/Fisons MD 800 MS, scan range: 40–520 amu	175–300°C at 50°C min^{-1}, held at 300°C for ~7.5 min	10 min	Cryogenic refocusing at −196°C during TE, ramped to 300°C in 20 s	Phenomenex ZB-5 (25 m × 0.25 mm × 25 μm)	40°C	40°C for 13 min, increased to 300°C at 5°C min^{-1}, held at 300°C for 25 min[c]	90 min
Waterman et al. (2001a)	NIST SRM 1649a	NIST urban dust standards (1–5 mg) were weighed and sealed into a glass tube filled with 120 mesh glass beads.	Same as above	HP 5890 GC/Fisons MD 800 MS, scan range: *m/z* 40–520	Isocratic desorption at 300°C	2 min	Same as above	Phenomenex ZB-5 (25 m × 0.25 mm × 25 μm)	40°C	40°C for 3 min, increased to 300°C at 5°C min^{-1}, held at 300°C for 33 min[c]	90 min

(Continued)

TABLE 4.3 (CONTINUED)
Summary of Sample Key Parameters for TD-GC-MS and Pyrolysis-GC-MS

Reference	Sample Type	Sample Preparation	TD Unit	Analytical Instrument	TD Temperature Program	TD Time	Desorbent Refocusing Pathway	GC Column	Initial GC Temp.	GC Oven Temperature Program	Total Analysis Time
Sidhu et al. (2001)	Aerosol collected on glass-fiber filters from combustion of alternative diesel fuel	Filter was sealed in a TD tube.	A stainless steel tube (0.635 cm O.D.) placed in a GC oven	Two GCs and one MS. The first GC is used as the TE unit. The second GC separates the desorbent.	30–300°C at 20°C min^{-1}	13.5 min	Cold trapping onto column head	DB-5MS (30 m × 0.25 μm)	−60°C	Not available	U[a]
Hays et al. (2003, 2007); Dong et al. (2004)	Aerosol collected from residential wood combustion, residential oil furnace, and fireplace appliance	A punch of filter was removed, spiked with internal standards, and placed into a glass desorption tube.	A glass tube placed in an external oven (TDS2 Gerstel Inc.)	Aglient 6890 GC/5793 MSD, scan range: 50–500 amu	25–300°C at 12°C min^{-1}	~23 min	Cryogenic focusing at −100°C in the inlet system, ramped to 300°C at 720°C min^{-1}	HP-5MS (30 m × 0.25 mm × 0.25 μm)	65°C	65°C for 10 min, 60–300°C at 10°C min^{-1}, held at 300°C for 41.5 min	99 min
2. Curie Point TD-GC-MS											
Jeon et al. (2001)	High-volume PM_{10} ambient samples collected along the US/Mexico border	A 1.5 × 18 mm^2 filter strip was cut and positioned inside a glass reaction tube lined with a ferromagnetic foil.	Curie point pyrolyzer	HP 5890 GC/5792 MSD	From ambient to 315°C	10 s	Cold trapping onto capillary column head	HP-5MS (25 m × 0.20 mm × 0.33 μm)	50°C	50–320°C at 3–15°C min^{-1}	U[a]
Neususs et al. (2000)	Ambient aerosol collected during the second aerosol characterization experiment	The filter strips were wrapped in a Pyrofoil with a Curie point of 590°C and placed into the pyrolyzer.	Curie point pyrolyzer	Fisons Trio 1000 GC-MS system	From ambient to 590°C	3.2 s	Cold trapping onto capillary column head	Chrompack CP-Sil-5 CB (30 m × 0.25 mm × 0.1 μm)	50°C	50°C for 2 min, 50–280°C at 10°C min^{-1}, held at 280°C for 10 min	35 min
3. In-Injection Port TE-GC-MS											
Helmig et al. (1990)	Aerosol samples collected on glass-fiber filters at a forest site	A small piece of sample filter was placed inside a glass liner.	GC injector port, with modified septum cap	Carlo Erba Mega 5160 GC/VG 250/70 SE MS, scan range: 45–400 amu	Isocratic desorption at 320°C	15 min	Cold trapping onto capillary column head	SE54 (15 m × 0.23 mm)	50°C	50–175°C at 7.5°C min^{-1}, then to 295°C at 10°C min^{-1}, held at 295°C for 5 min	47 min

(*Continued*)

TABLE 4.3 (CONTINUED)
Summary of Sample Key Parameters for TD-GC-MS and Pyrolysis-GC-MS

Reference	Sample Type	Sample Preparation	TD Unit	Analytical Instrument	TD Temperature Program	TD Time	Desorbent Refocusing Pathway	GC Column	Initial GC Temp.	GC Oven Temperature Program	Total Analysis Time
Hall et al. (1999)	NIST SRM 1649	A glass reaction tube with an internal volume of 30–40 μL was filled with 5 mg NIST urban dust standard. The tube was sealed before TD.	Microscale sealed vessel placed inside the injector port	HP 5890 GC/Fisons MD800 MS, scan range: 40–500 amu	45–300°C at 15°C min^{-1} and held at 300°C for 5 min	22 min	Cold trapping onto capillary column head	GC-5MS capillary column (25 m × 0.25 mm)	45°C	45°C for 1 min, 45–300°C at 6°C min^{-1}, held at 300°C for 17 min	82.5 min
Blanchard and Hopper (1997); Blanchard et al. (2002)	Aerosol samples collected on quartz and glass-fiber filters in Ontario	A 1 cm diameter section of the sample filter was punched and transferred into a GC liner spiked with internal standards.	A GC injection port was added with three minor components, including a small T-connector, a three-way valve, and a needle valve.	HP5890 GC/5972A MS in EI mode	Isocratic desorption at 300°C	15 min	Cold trapping onto capillary column head	HP-5MS (30 m × 0.25 mm × 0.25 μm)	30°C	30°C for 0.75 min, increased to 175°C at 7.5°C min^{-1}, then to 295°C at 10°C min^{-1}, held at 295°C for 5 min	71 min
Falkovich and Rudich (2001); Falkovich et al (2004); Graham et al. (2004)	NIST SRM 1649a; urban aerosols collected with an eight-stage impactor in Tel-Aviv, Israel	NIST standard or small pieces of filter sample were loaded into a disposable microvial (1.9 mm internal diameter; 12 mm long). The sample vials are placed by the vial holder in the GC liner.	Direct sample introduction device (ChromatoProbe, Varian Co.)	Varian Saturn 3400 GC-MS	120–350°C at a 200°C min^{-1}	4.15 min	Cold trapping onto capillary column head	DB-5MS column (30 m × 0.25 mm × 0.25 μm)	40°C	40°C for 4 min, 40–300°C at 5°C min^{-1}, held at 300°C for 4 min	64.2 min

(Continued)

TABLE 4.3 (CONTINUED)
Summary of Sample Key Parameters for TD-GC-MS and Pyrolysis-GC-MS

Reference	Sample Type	Sample Preparation	TD Unit	Analytical Instrument	TD Temperature Program	TD Time	Desorbent Refocusing Pathway	GC Column	Initial GC Temp.	GC Oven Temperature Program	Total Analysis Time
Ho and Yu (2004b); Yang et al. (2005)	Ambient aerosol samples collected on Teflon-impregnated glass-fiber filters in Hong Kong and on quartz filters at Nanjing, China	Two pieces of 1.45 cm² filter samples were spiked with deuterated internal standards, placed in the GC inject liner and held with glass wool.	Conventional GC injection port No modification of GC injector and liner	HP 5890 GC/5791 MSD, scan range: 50–650 amu	100–275°C within 7–8 min	7.0 min	Cold trapping onto capillary column head	HP-5MS (30 m × 0.25 mm × 0.25 μm)	30°C	30°C for 2 min, 30–120°C at 20°C min⁻¹, then to 300°C at 10°C min⁻¹, held at 300°C for 10 min	41.5 min
4. TD-GC × GC-MS											
Welthagen et al. (2003); Schnelle-Kreis et al. (2005)	Ambient samples in Augsburg, Germany	The filter strip was directly inserted into the DTD (direct thermal desorption) liner and subsequently loaded into a cold injection port.	Injection port Optic III with autoloader (ATAS-GL, Veldhoven, Netherlands)	Agilent 6890 GC/ LECO Pegasus III TOFMS with a LECO Pegasus 4D GC × GC moculator	50–350°C at 60°C min⁻¹	5 min	Cold trapping onto capillary column head	BPX5 (50 m × 0.22 mm × 0.25 μm) followed by BPX50 (1.5 m × 0.10 mm × 0.10 μm)	*GC × GC (2-D):* 60°C	60°C for 10 min, increased to 300°C at 1.5°C min⁻¹. The second column was kept at 5°C above the first column	175 min
Hamilton et al. (2004)	$PM_{2.5}$ aerosol collected in London	The filter strip (with 10 μg loading) was inserted into the GC liner and directly introduced into the GC injector.	Conventional GC injection port	Same as above, scan range: 20–350 amu	40–300°C at a 20°C min⁻¹	13 min	Cold trapping onto capillary column head	HP-5 (10 m × 0.18 mm × 0.18 μm) followed by DB17 (1.66 m × 0.1 mm × 0.1 μm)	40°C	40°C for 5 min, increased to 270°C at 3.5°C min⁻¹, held at 270°C for 10 min	93.7 min
Hamilton et al. (2005)	Secondary organic aerosol formed during the photooxidation of toluene with OH radicals	Same as above	Same as above	Same as above	Same as above	12.5 min	Cold trapping onto capillary column head	Same as above	70°C	70°C for 2 min, 70–250°C at 2.5°C min⁻¹, held at 270°C for 16 min	102.5 min

(Continued)

TABLE 4.3 (CONTINUED)
Summary of Sample Key Parameters for TD-GC-MS and Pyrolysis-GC-MS

Reference	Sample Type	Sample Preparation	TD Unit	Analytical Instrument	TD Temperature Program	TD Time	Desorbent Refocusing Pathway	GC Column	Initial GC Temp.	GC Oven Temperature Program	Total Analysis Time
5. In Situ Semicontinuous and Continuous TD Systems											
William et al. (2006)	In situ aerosol samples collected in Berkeley, California	The aerosols were collected into the collection-TD cell and thermally desorbed and transferred into the GC injector.	Collection-TD cell with conventional GC injection port	Agilent 6890 GC/5793 MSD, scan range: 29–550 amu	50–300°C at 30–40°C min^{-1}	10 min	Cold trapping onto capillary column head	Rtx-5MS (30 m × 0.25 mm × 0.25 μm)	45°C	45°C for 12 min, 45–300°C at 8.5°C min^{-1}, held at 300°C for ~7 min	59 min
6. Pyrolysis TD-GC-MS											
Voorhees et al. (1991)	$PM_{0.6}$ and $PM_{>0.45}$ collected on quartz filters in pristine the regions of Colorado	Filters were first extracted in sequence with methanol, dichloroetane, acetone, and hexane. The resulting insoluble material was introduced to a tube furnace.	A tube furnace directly interfaced to an GC-MS	Extrel Simulscan GC-MS, scan range: 35–450 amu	550°C for 10 min	10 min	Cold trapping onto capillary column head	DB-5 (25 m × 0.25 mm)	35°C	35–220°C at 8.5°C min^{-1}	31.7 min
Subbalakshmi et al. (2002)	Ambient aerosol collected on glass-fiber filters in Jakarta, Indonesia	Filter was introduced into a quartz capillary.	A pyroinjector	Agilent 6890 GC/5973 MS, scan range: 50–550 amu	450°C for 1 min	1 min	Cold trapping onto capillary column head	BP-5 (30 m × 0.22 mm × 1.0 μm)	40°C	40–290°C at 4°C min^{-1}	63.5 min

(Continued)

TABLE 4.3 (CONTINUED)
Summary of Sample Key Parameters for TD-GC-MS and Pyrolysis-GC-MS

Reference	Sample Type	Sample Preparation	TD Unit	Analytical Instrument	TD Temperature Program	TD Time	Desorbent Refocusing Pathway	GC Column	Initial GC Temp.	GC Oven Temperature Program	Total Analysis Time
Fabbri et al. (2002)	PM_{10} collected on glass-fiber filters in an industrial area in Italy	Filter was scraped with tweezers, and a weighted amount of the scraped material was inserted into a prepyrolyzed quartz tube.	A pyrolyzer directly connected to the GC injector port through an interface heated at 250°C	Varian 3400 GC/ Saturn II ion trap MS, scan range: 45–400 amu	700°C for 10 s	10 s	Cold trapping onto capillary column head	SPB-5 (30 m × 0.32 mm × 0.25 μm)	50°C	50°C for 2 min, 50–300°C at 5°C min^{-1}, held at 300°C for 5 min	57 min
Blazso et al. (2003)	$PM_{2.6}$ collected on quartz filters and size-segregated aerosol samples collected on Al foils in Brazil	The collection substrates were spiked with derivatization agents, dried, and introduced into a pyrolysis sample holder	A pyrolyzer	Agilent 6890 GC/5973 MS	400°C for 20 s	20 s	Cold trapping onto capillary column head	HP-5 (30 m × 0.25 mm)	50°C	50°C for 1 min, 50–300°C at 10°C min^{-1}, held at 300°C for 4 min	30.3 min
Labban et al. (2006)	PM_{10} of resuspended soil collected on quartz filters	Filters were placed in pyrofoil.	Curie point pyrolyzer	HP5890GC/5972 MS	740°C for 10 s	10 s	Cold trapping onto capillary column head	Rtx-5MS (30 m × 0.25 mm × 0.25 μm)	50°C	50–220°C at 10°C min^{-1}, 220–320°C at 30°C min^{-1}, held at 320°C for 5 min	25.5 min

Source: Chow, J. C. et al. *Journal of Environmental Science and Health Part A*, 42, 1521–1541, 2007. With permission.

[a] Total analysis time could not be determined because of insufficient experimental details.

(Hays et al., 2003, 2004) to 200°C min^{-1} (Falkovich and Rudich, 2001) without holding time and from 15°C min^{-1} (Hall et al., 1999) to 50°C min^{-1} (Waterman et al., 2000) with 5–7.5 min holding time. Maximum reported TD temperatures range from 275°C to 315°C (Jeon et al., 2001; Ho and Yu, 2004b) with holding times ranging from 0 to 10 min. These temperatures and holding times were optimized to ensure complete extraction and to minimize pyrolysis (Greaves et al., 1985).

The desorption temperature is directly related to the volatility and the thermal stability of the targeted analytes. At desorption temperatures of <275°C, the analytes are not completely extracted, especially for less volatile compounds (Ho and Yu, 2004b). Temperatures higher than 340°C may cause thermally labile compounds to decompose or pyrolyze (Cachier et al., 1989).

Helium (He) is the typical carrier gas. Hays and Lavrich (2007) suggested that the flow rate of He over the sample in the desorption unit may influence the desorption efficiency. Commercial TD units can be adjusted to different flows and maintain a stable gas stream into the GC column and the detector. However, the relationship between He gas flow and desorption efficiency has not been studied.

Thermally desorbed organic compounds are usually cryofocused, then heated and separated on an analytical column by their boiling points and polarities. Table 4.3 shows that initial temperatures in the GC oven vary from −60°C (Greaves et al., 1985, 1987) to 70°C (Hamilton et al., 2005), with the most common selection at 30–50°C. Higher initial temperatures may volatilize some of the OM during sample loading. Studies using higher initial temperatures did not quantify VOCs adsorbed onto the particles. Ramping rates in the GC oven vary from 1.5°C min^{-1} (Welthagen et al., 2003) to 100°C min^{-1} (Schnelle-Kreis et al., 2005), with the most common in the range of 5–20°C min^{-1}.

4.5.1.6 Intercomparison between TD- and SE-GC-MS

Hays and Lavrich (2007) tabulated compound classes from several TD-GC-MS studies. These include alkanes (i.e., *n*-alkanes, steranes, and hopanes), alkenes, aromatic (i.e., PAHs and furans), alcohols (i.e., phenols), amines, amides, esters (i.e., phthalates), carboxylic acids, aldehydes, ketones, and carboxylic acids. The quantification of PAHs by TD-GC-MS has been evaluated with Standard Reference Material 1649 (dusts collected in the urban Washington, DC, area in 1976–1977 and sieved to <125 μm) from the NIST (Greaves et al., 1985; Waterman et al., 2000, 2001b; Falkovich and Rudich, 2001; Sin et al., 2002) SE- and TD-GC-MS results were compared for PAHs and alkanes by Jeon et al. (2001) and Ho and Yu (2004b). These two methods agreed within a factor of 2.5 for the majority of PAHs and alkanes. Hansen et al. (1995) characterized the TD recoveries of three alcohols (i.e., tridecanol, octadecanol, and octacosanol), four carboxylic acids (i.e., nonanoic, dodecanoic, octadecanoic, and eicosanoic acid), two PAHs (i.e., 9-fluorenone and pyrene), and three source markers (nicotine, guaiacol, vanillin) at a TD temperature of 250°C. Only the two PAHs showed better than 75% recoveries with <55% for other compounds. Jeon et al. (2001) showed good correlation between SE- and TD-GC-MS for hopanes and steranes ($R^2 = 0.998$), PAHs ($R^2 = 0.978$), *n*-alkanes ($R^2 = 0.975$), and phenols ($R^2 = 0.940$) with the exception of *n*-alkanoic acids ($R^2 = 0.731$). Neususs et al. (2000) also demonstrated an agreement within 130 ± 30% for PAHs between SE- and Curie point TD-GC-MS, except for the lower concentrations of indeno[1,2,3]pyrene and benzo[g,h,i]perylene by the TD method. Neususs et al. (2000) suggested that transfer losses of the less volatile compounds might be the cause. Poor comparisons are probably due to partial decomposition during the TD process.

4.5.2 Continuous and Semicontinuous TD Aerosol Measurement Systems

TD has been applied in several first-generation semi-continuous and continuous instruments, typically coupled with MS (Jayne et al., 2000). Semicontinuous instruments integrate aerosol sample collection with TD-GC-MS-FID analysis (Williams, 2006), particle beam/MS (PB/MS) (Tobias and Ziemann, 1999, 2000), and chemical ionization/MS (CI/MS) (Voisin et al., 2003; Smith et al., 2004).

4.5.2.1 Thermal Desorption–Gas Chromatography–Mass Spectrometry–Flame Ionization Detector (TD-GC-MS-FID)

Williams et al. (2006) developed an in situ thermal desorption aerosol gas chromatography (TAG)/MS-FID to achieve a 30 min average organic speciation. TAG impacts particles into a thermal desorption cell (TDC) after a PM_2 cyclone and humidifier (to increase adhesion and minimize bounce). TD occurs under He in the TDC, which is temperature-programmed at 100°C to purge water from the system and then at 300°C to remove organic compounds followed by GC-MS-FID analysis. Desorbed analytes are focused onto the head of a GC column with the oven temperature set at 45°C. During the chromatographic separation step and subsequent MS or FID detection, the system is switched back to the sampling mode. Williams et al. (2006) concluded that a sampling time of 30 min (<1 m^3 of air in several cases) in urban environments (i.e., Berkeley, California) provides sufficient sample to speciate >100 organic compounds. With the current design, TAG provides <50% collection efficiency for particles smaller than ~65 nm, thereby underestimating ultrafine organic constituents.

4.5.2.2 Thermal Desorption–Particle Beam/Mass Spectrometer (TD-PB/MS)

TD-PB/MS identifies and quantifies organic species in chamber-produced SOA and diesel engine nanoparticles (Tobias et al., 1999, 2000; Sakurai et al., 2003). Particles are drawn into a high-vacuum chamber (6.6×10^{-11} atm), focused onto a narrow, low-divergence particle beam using aerodynamic lenses followed by impaction onto a metal (Mo) foil. The particles are either continuously vaporized by resistively heating the foil to a set temperature (e.g., 165°C or 250°C) for real-time analysis or heated using a linear temperature ramp as low as 1°C min^{-1} to achieve compound separation by volatility (Tobias et al., 1999). In both temperature ramping methods, the vapors are continuously mass-analyzed using a quadrupole MS. The optimal separation of target compounds is achieved by adjusting the temperature ramping rate. With the linear ramping temperature-programmed TD mode, separation was demonstrated for compounds having vapor pressures exceeding a range of a factor of 5. The mass spectra of the individual components are inferred from the time-dependent spectra for compound identification. The size-dependent chemical composition can also be obtained by passing the incoming aerosol through a differential mobility analyzer (Tobias et al., 1999).

TD-PB/MS provides higher time resolution and allows the study of aerosol chemistry dynamics than is possible with GC-MS analysis of filter samples. Sampling artifacts due to vapor adsorption on or particle volatilization from the filter are much reduced (Sakurai et al., 2003). Both laboratory experiments and theoretical calculations indicate that little evaporation occurs in the ~0.2 s transit time from the TD-PB/MS inlet to the vaporizer for alkanes down to ~C_{15}–C_{17}. Analytical artifacts caused by thermal decomposition are lower. The TD of unstable compounds, which can occur on a GC column, is reduced because vaporization during TD-PB/MS occurs in a vacuum (~1.3×10^{-5} atm) (Tobias et al., 2000). Thermally labile species such as α-alkoxylalkyl hydroperoxides and sulfuric acid (H_2SO_4) are detectable by TD-PB/MS (Tobias and Ziemann, 1999, 2000, 2001). Hydroperoxides can dehydrate into corresponding esters during GC analysis. A vaporization temperature of 165°C was demonstrated to completely evaporate hexadecanoic acid, tridecanoic acid, and oleic acid (Tobias and Ziemann, 2000). In comparison, TD with a GC inlet at 200°C only removed 70% of octadecane in ambient aerosols (Ho and Yu, 2004b).

TD-PB/MS captures ultrafine particles in the range of 20–500 nm in diameter and has quantified laboratory-generated SOA. Quantification with an estimated uncertainty of ±20% is achieved using calibration techniques that account for particle shape and transport efficiency (Tobias and Ziemann, 2000). The application of TD-PB/MS to ambient aerosols is still untested. Neither diesel particles nor chamber-generated SOA require prior knowledge of the types of products to be expected. The extraction of single compounds from the composite of the mass spectra of many compounds, in the case of ambient aerosols, is yet to be achieved.

4.5.2.3 Thermal Desorption–Chemical Ionization/Mass Spectrometer (TD-CI/MS)

TD-CI/MS is designed for real-time chemical analysis of ultrafine particles that are 6–20 nm in diameter using a highly efficient aerosol charger (Voisin et al., 2003; Smith et al., 2004). Particles are charged and collected on a Ni/Cr wire. The collected particles are heated to 300°C in an ionization chamber for particle evaporation, followed by chemical ionization by N_2 buffer gas ions such as H_3O^+, O_2^-, and CO_2^-. The buffer gas ions are produced by α-particles emitted from a radioactive material (i.e., americium 231 [^{231}Am]). The ions are detected with a triple quadrupole MS.

Online TD-CI/MS has been used in selective ion monitoring mode to size and identify inorganic ions (e.g., nitrate $\left[NO_3^-\right]$, sulfate $\left[SO_4^{2-}\right]$, and ammonium $\left[NH_4^+\right]$). The instrument has the potential to identify and detect OM in particle nuclei, but this has not yet been demonstrated.

4.5.2.4 Aerosol Mass Spectrometer (AMS)

The AMS integrates high-time resolution (5 s) sampling and sizing with MS detection of aerosol components. Particles passing through an aerodynamic lens are impacted onto a plate at 600°C in a vacuum ($<1.3 \times 10^{-6}$ atm). Flash volatilization converts PM constituents from condensed phase to gaseous form, which are introduced into either a quadrupole or a TOFMS detector. AMS also monitors inorganic components that can be evaporated, including NH_4^+, NO_3^-, SO_4^{2-}, chloride (Cl^-), and potassium (K^+) (Allan et al., 2003, 2004; Jimenez et al., 2003). Much progress has been made in interpreting the ion fragments. For example, fragment ions of *m/z* at 57 and 44 are used to separate OM into hydrocarbon-like organic compounds and oxygenated organic compounds, respectively (Zhang et al., 2005). However, the deconvolution of single mass spectra from multiple organic compounds remains a daunting challenge.

4.6 THERMAL DESORPTION METHOD LIMITATIONS

Not all types of substrates are suitable for TD (Helmig et al., 1990; Dong et al., 2004). Materials containing organic binders (e.g., Teflon with an operational temperature <~320°C) would decompose during the analysis. Sampling with quartz- or glass-fiber filters are the proper choices.

Compounds with similar volatilities limit the specificity of TD-MS without chromatographic separation. In addition, matrix effects associated with TD-MS analysis have been reported. Sakurai et al. (2003) found that the desorption profiles of diesel nanoparticles were influenced by loading due to adsorption onto diesel soot particles that accumulate on the vaporizer.

Carryover, transfer loss, analyte conversion, and degradation may interfere with some organic compounds. These analytical issues are inherently related to the presence of oxygenated organic compounds. Polar functional groups (oxygenated organic compounds, e.g., carboxylic acids and alcohols) adsorb on surfaces inside the analytical instruments (e.g., GC injection inlet), resulting in incomplete desorption and subsequent carryover. These groups are also responsible for analyte conversion and degradation under thermal stress. Consistently low recoveries (less than 30%) of carboxylic acids and alkanols have been documented (Hansen et al., 1995; Jeon et al., 2001; Hamilton et al., 2004). The presence of these compounds in aerosols is likely to impair quantification efforts and lead to compound misidentification. For example, even though Williams et al. (2006) identified polar levoglucosan, high minimum detection limits (MDLs) and low regression values in calibration raise uncertainties for its quantification. Highly polar compounds such as dicarboxylic acid have not yet been quantified with acceptable analytical uncertainties in TD analysis, although their boiling points are lower than the desorption temperature. They also affect the quantification of some thermally stable compounds when their degradation products are targeted for quantification. For example, *n*-alkanols degrade to their corresponding 1-alkenes, *n*-aldehydes, or possibly to *n*-alkane compounds (Helmig et al., 1992; Simoneit et al., 1992).

On-filter derivatization techniques such as thermally assisted alkylation using tetramethylammonium hydroxide or tetrabutylammonium hydroxide or silylation reagents are likely to be applicable to TD-GC-MS with minor filter pretreatment (Fabbri et al., 2002: Blazso et al., 2003). Online derivatization methods similar to those developed by Nimz and Morgan (1993), Docherty and Ziemann (2001), Blokker et al., (2002) and Orasche et al. (2011) appear to be feasible for measuring organic acids and alcoholic compounds. The derivatization reaction can proceed by filter spiking with reagent or by spiking inert material in the GC injector with reagent. Care must be taken with most derivatizing reagents because of their hazardous nature and ability to damage analytical instruments.

When TD methods are established for organic compounds, one must consider the possibility of the target compound undergoing thermal decomposition or being a decomposition product of other coexisting constituents. This can be achieved by evaluating the recovery by spiking the target compound onto aerosol matrices representative of the real samples. Comparison with other methods (e.g., SE-GC-MS) will also help to minimize the misidentification of thermal decomposition products as organic compounds originally present in the aerosols. Discrepancies between the independent methods need to be further examined. When a compound identified in the TD approach is absent from the analysis using an alternative method for similar samples, there must be a reason to expect that the compound in question is actually present in the aerosol sample and not an artifact, such as thermolysis product arising from the TD analytical procedure. Conversely, there should be a chemical/physical reason why a compound was not detected by one method or the other (Hansen et al., 1995). While method comparisons for an exhaustive list of organic compounds detected by the TD methods is desired, priority should be given to those species known to be source markers or of toxicological importance.

TD-MS has the potential to analyze unaltered, thermally labile compounds because TD takes place in a vacuum. This feature could be useful and should be explored for the analysis of species such as organic peroxides that have low thermal stability and are also sensitive to solvents.

As mentioned, few TD-GC-MS experiments use the temperature programming option for separating and analyzing organics. The GC-MS analysis of PM evolves incrementally by decreasing the GC column temperature and sample load, potentially increasing chromatographic resolution. In addition, they might also support individual compound identification with online MS measurements that employ specific temperature programming. This assumes that thermogram or individual particle mass spectra can be deconvoluted mathematically from temporally integrated ion chromatograms of the same sample. Nonetheless, this is surmised as a strong argument for the field deployment of both real-time and time-integrated aerosol sampling instruments at ambient monitoring stations.

Pyrolysis-TD-GC-MS can obtain polymeric and insoluble carbonaceous materials, which could not be obtained with any SE-based methods. However, work is needed to understand the relationship between parent molecules and their pyrolysis products before this unique advantage can be realized.

REFERENCES

Allan, J.D., Alfarra, M.R., Bower, K.N., Williams, P.I., Gallagher, M.W. et al. 2003. Quantitative sampling using an Aerodyne aerosol mass spectrometer 2. Measurements of fine particulate chemical composition in two U.K. cities. *Journal of Geophysical Research* 108: AAC2-1–AAC2-9.

Allan, J.D., Delia, A.E., Coe, H., Bower, K.N., Alfarra, M.R. et al. 2004. A generalised method for the extraction of chemically resolved mass spectra from aerodyne aerosol mass spectrometer data. *Journal of Aerosol Science* 35: 909–922.

Arnold, R.T., Danzig, M.J. 1957. Thermal decarboxylation unsaturated acids: II. *Journal of the American Chemical Society* 79: 892–894.

Arnold, R.T., Elmer, O.C., Dodson, R.M. 1950. Thermal decarboxylation unsaturated acids. *Journal of the American Chemical Society* 72: 4359–4361.

Atkins, P.W. 2002. *Physical Chemistry*, 7th ed., Oxford Univ. Press, New York.

Bayona, J.M., Casellas, M., Fernandez, P., Solanas, A.M., Albaiges, J. 1994. Sources and seasonal variability of mutagenic agents in the Barcelona city aerosol. *Chemosphere* 29: 441 450.

Beuhler, R.J., Flanlgan, E., Greene, L.T., Friedman, L.J. 1974. Proton transfer mass spectrometry of peptides: A rapid heating technique for underivatized peptides containing arginine. *Journal of the American Chemical Society* 96: 3990–3999.

Birch, M.E., Cary, R.A. 1996. Elemental carbon-based method for monitoring occupational exposures to particulate diesel exhaust. *Aerosol Science & Technology* 25: 221–241.

Blake, E.S., Hammann, W.C., Edwards, J.W., Reichard, T.E., Ort, M.R. 1961. Thermal stability as a function of chemical structure. *Journal of Chemical and Engineering Data* 6: 87–98.

Blanchard, P., Hopper, J.F. 1997. Concentrations and distributions of PAHs and n-alkanes in atmospheric aerosols at non-urban sites in Ontario, Canada. *Journal of Geophysical Research* 107: ICC-10-1–ICC-10-8.

Blanchard, P., Brook, J.R., Brazal, P. 2002. Chemical characterization of the organic fraction of atmospheric aerosol at two sites in Ontario, Canada. *Journal of Geophysical Research* 107: ICC10-1–ICC10-8.

Blazso, M., Janitsek, S., Gelencser, A., Artaxo, P., Graham, B. et al. 2003. Study of tropical organic aerosol by thermally assisted alkylation-gas chromatography mass spectrometry. *Journal of Analytical and Applied Pyrolysis* 68–9: 351–369.

Blokker, P., Pel, R., Akoto, L., Brinkman, U.A.T., Vreuls, R.J.J. 2002. At-line gas chromatographic-mass spectrometric analysis of fatty acid profiles of green microalgae using a direct thermal desorption interface. *Journal of Chromatography A* 959: 191–201.

Boyd, J., Overton, K.H. 1972. 1,3-Elimination with rearrangement during ester pyrolysis: A simple route to protoadamantene and 2,4-didehydroadamantane. *Journal of the Chemical Society, Perkin Transactions* 1: 2533–2539.

Bradley, J.N. 1974. A general mechanism for the high-temperature pyrolysis of alkanes: The pyrolysis of isobutane. *Proceedings of the Royal Society of London A* 337: 199–216.

Brouwer, L.D., Muller-Markgraf, W., Troe, J. 1988. Thermal decomposition of toluene: A comparison of thermal and laser photochemical activation experiments. *Journal of Chemical Physics* 92: 4905–4914.

Brown, R.F.C. 1980. *Pyrolytic Methods in Organic Chemistry: Application of Flow and Flash Vacuum Pyrolytic Techniques*. Academic, New York.

Buchanan, A.C. III, Britt, P.F., Skeen, J.T., Struss, J.A., Elam, C.L. 1998. Pyrolysis of silica-mobilized benzyl phenyl ether: Competing radical rearrangement pathways under restricted diffusion. *Journal of Organic Chemistry* 63: 9895–9903.

Cachier, H., Bremond, M.P., Buat-Ménard, P. 1989. Thermal separation of soot carbon. *Aerosol Science & Technology* 10: 358–364.

Chai, M., Arthur, C.L., Pawliszyn, J., Belardi, R.P., Pratt, K.F. 1993. Determination of volatile chlorinated hydrocarbons in air and water with solid-phase microextraction. *Analyst* 118: 1501–1505.

Chow, J.C. 1995. Critical review: Measurement methods to determine compliance with ambient air quality standards for suspended particles. *Journal of the Air & Waste Management Association* 45: 320–382.

Chow, J.C., Bachmann, J.D., Wierman, S.S.G., Mathai, C.V., Malm, W.C. et al. 2002. 2002 Critical review discussion—Visibility: Science and regulation. *Journal of the Air & Waste Management Association* 52: 973–999.

Chow, J.C., Watson, J.G., Chen, L.W.A., Arnolt, W.P., Moosmuller, H. et al. 2004. Equivalence of elemental carbon by thermal/optical reflectance and transmittance with different temperature protocols. *Environmental Science & Technology* 38: 4414–4422.

Chow, J.C., Watson, J.G., Chen, L.W.A., Paredes-Miranda, G., Chang, M.C.O. et al. 2005. Refining temperature measures in thermal/optical carbon analysis. *Atmospheric Chemistry & Physics* 5: 2961–2972.

Chow, J.C., Watson, J.G., Crow, D., Lowenthal, D.H., Merrifield, T.M. 2001. Comparison of IMPROVE and NIOSH carbon measurements. *Aerosol Science & Technology* 34: 23–34.

Chow, J.C., Watson, J.G., Mauderly, J.L., Costa, D.L., Wyzga, R.E. et al. 2006. 2006 critical review discussion—Health effects of fine particulate air pollution: Lines that connect. *Journal of the Air & Waste Management Association* 56: 1368–1380.

Chow, J.C., Watson, J.G., Pritchett, L.C., Pierson, W.R., Frazier, C.A. et al. 1993. The DRI thermal/optical reflectance carbon analysis system: Description, evaluation and applications in U.S. air quality studies. *Atmospheric Environment* 27A: 1185–1201.

Chow, J.C., Yu, J.Z., Watson, J.G., Ho, S.S.H., Bohannan, T.L. et al. 2007. The application of thermal methods for determining chemical composition of carbonaceous aerosols: A review. *Journal of Environmental Science and Health, Part A* 42: 1521–1541.

Coleman, W.M. III, Chung, H.L. 2002. Pyrolysis GC-MS analysis of Amadori compounds derived from selected amino acids with glucose and rhamnose. *Journal of Analytical and Applied Pyrolysis* 63: 49–366.

Cronn, D.R., Charlson, R.J., Knights, R.L., Crittenden, A.L., Appel, B.R. 1977. A survey of the molecular nature of primary and secondary components of particles in urban air by high-resolution mass spectrometry. *Atmospheric Environment* 11: 929–937.

Damm, M., Deckert, F., Hippler, H., Rink, G. 1999. Specific rate constants for the fragmentation of vibrationally excited benzyl radicals. *Physical Chemistry and Chemical Physics* 1: 81–90.

Davidson, C.I., Tang, W., Finger, S., Etyemezian, V., Striegel, M.F. et al. 2000. Soiling patterns on a tall limestone building: Changes over 60 years. *Environmental Science & Technology* 34: 560–565.

DeLuca, S., Sarver, E.W., Harrington, P.D., Voorhees, J.J. 1990. Direct analysis of bacterial fatty acids by curie-point pyrolysis tandem mass spectrometry. *Analytical Chemistry* 62: 1465–1472.

Demirbas, A. 2004. Effects of temperature and particle size on bio-char yield from pyrolysis of agricultural residues. *Journal of Analytical and Applied Pyrolysis* 72: 243–248.

Docherty, K.S., Ziemann, P.J. 2001. On-line, inlet-based trimethylsilyl derivatization for gas chromatography of mono- and dicarboxylic acids. *Journal of Chromatography A* 921: 265–275.

Dominguez, R.M., Chuchani, G. 1992. The direction of elimination in the pyrolysis kinetics of 3-bromobutyronitrile. *Reaction Kinetics and Catalysis Letters* 47: 247–254.

Dong, Y.J., Hays, M.D., Smith, N.D., Kinsey, J.S. 2004. Inverting cascade impactor data for size-resolved characterization of fine particulate source emissions. *Journal of Aerosol Science* 35: 1497–1512.

Evans, C.S., Dellinger, B. 2006. Formation of bromochlorodibenzo-p-dioxins and dibenzofurans from the high-temperature oxidation of a mixture of 2-chlorophenol and 2-bromophenol. *Environmental Science & Technology* 40: 3036–3042.

Fabbri, D., Pratis, S., Vassura, I. 2002. Molecular characterization of organic material in air fine particles (PM10) using conventional and reactive pyrolysis-gas chromatography-mass spectrometry. *Journal of Environmental Monitoring* 4: 210–215.

Falkovich, A.H., Rudich, Y. 2001. Analysis of semivolatile organic compounds in atmospheric aerosols by direct sample introduction thermal desorption GC/MS. *Environmental Science & Technology* 35: 2326–2333.

Falkovich, A.H., Schkolnik, G., Ganor, E., Rudich, Y., 2004. Adsorption of organic compounds pertinent to urban environments onto mineral dust particles. *Journal of Geophysical Research* 109: 1–19. doi:10.1029/2003JD003919.

Fine, P.M., Chakrabarti, B., Krudysz, M., Schauer, J.J., Sioutas, C. 2004. Diurnal variations of individual organic compound constituents of ultrafine and accumulation mode particulate matter in the Los Angeles basin. *Environmental Science & Technology* 38: 1296–1304.

Fulle, D., Dib, A., Kiefer, J.H., Zhang, Q., Yao, J. et al. 1998. Pyrolysis of furan at low pressures: Vibrational relaxation, unimolecular dissociation, and incubation times. *Journal of Physical Chemistry A* 102: 7480–7486.

Garnett, J.L., Johnson, W.D., Sherwood, J.E. 1976. Pyrolysis of 1-methylcyclohexene and methylenecyclohexane. *Journal of Chemistry* 29: 599–607.

Graham, B., Falkovich, A.H., Rudich, Y., Maenhaut, W., Guyon, P. et al. 2004. Local and regional contributions to the atmospheric aerosol over Tel Aviv, Israel: A case study using elemental, ionic and organic tracers. *Atmospheric Environment.* 38: 1593–1604.

Greaves, R.C., Barkley, R.M., Sievers, R.E. 1985. Rapid sampling and analysis of volatile constituents of airborne particulate matter. *Analytical Chemistry* 57: 2807–2815.

Greaves, R.C., Barkley, R.M., Sievers, R.E., Meglen, R.R. 1987. Covariations in the concentrations of organic compounds associated with springtime atmospheric aerosols. *Atmospheric Environment* 21: 2549–2561.

Hall, P.A., Watson, A.F.R., Garner, G.V., Hall, K., Smith, S. et al. 1999. An investigation of micro-scale sealed vessel thermal extraction-gas chromatography-mass spectrometry (MSSV-GC-MS) and micro-scale sealed vessel pyrolysis-gas chromatography-mass spectrometry applied to a standard reference material of an urban dust/organics. *Science of the Total Environment* 235: 269–276.

Hamilton, J.F., Webb, P.J., Lewis, A.C., Hopkins, J.R., Smith, S. et al. 2004. Partially oxidised organic components in urban aerosol using GCXGC-TOF/MS. *Atmospheric Chemistry & Physics* 4: 1279–1290.

Hamilton, J.F., Webb, P.J., Lewis, A.C., Reviejo, M.M. 2005. Quantifying small molecules in secondary organic aerosol formed during the photo-oxidation of toluene with hydroxyl radicals. *Atmospheric Environment* 39: 7263–7275.

Hansen, K.J., Hansen, B.N., Cravens, E., Sievers, R.E. 1995. Supercritical fluid extraction-gas chromatographic analysis of organic compounds in atmospheric aerosols. *Analytical Chemistry* 67: 3541–3549.

Hays, M.D., Lavrich, R.J. 2007. Developments in direct thermal extraction-GC/MS of fine aerosols. *Trends in Analytical Chemistry* 26: 88–102.

Hays, M.D., Smith, N.D., Dong, Y.J. 2004. Nature of unresolved complex mixture in size-distributed emissions from residential wood combustion as measured by thermal desorption-gas chromatography-mass spectrometry. *Journal of Geophysical Research* 109: D16S04-1–D16S04-13.

Hays, M.D., Smith, N.D., Kinsey, J., Dong, Y., Kariher, P. 2003. Polycyclic aromatic hydrocarbon size distributions in aerosols from appliances of residential wood combustion as determined by direct thermal desorption-GC/MS. *Journal of Aerosol Science* 34: 1061–1084.

Helmig, D., Arey, J. 1992. Organic chemicals in the air at Whitaker's Forest Sierra Nevada Mountains, California. *Science of the Total Environment* 112: 233–250.

Helmig, D., Bauer, A., Muller, J., Klein, W. 1990. Analysis of particulate organics in a forest atmosphere by thermodesorption GC/MS. *Atmospheric Environment* 24A: 179–184.

Helmig, D., Bauer, A., Muller, J., Klein, W. 1992. Discussion on analysis of particulate organics in a forest atmosphere by thermodesorption GC/MS. *Atmospheric Environment* 26A: 517–519.

Hippler, H., Reihs, C., Troe, Z.J. 1990. Elementary steps in the pyrolysis of toluene and benzy radicals. *Physical Chemistry* 167: 1–16.

Hirvonen, A., Pasanen, P., Tarhanen, J., Ruuskanen, J. 1994. Thermal desorption of organic compounds associated with settled household dust. *Indoor Air* 4: 255–264.

Ho, S.S.H., Chow, J.C., Watson, J.G., Ng, L.P.T., Kwok, Y. et al. 2011. Precautions in application of injection port-thermal desorption/mass spectrometry (TD-GC/MS) in analysis of aerosol loaded filters. *Atmospheric Environment* 45: 1491–1496.

Ho, S.S.H., Yu, J. 2002. Feasibility of collection and analysis of airborne carbonyls by on-sorbent derivatization and thermal desorption. *Analytical Chemistry* 74: 1232–1240.

Ho, S.S.H., Yu, J. 2004a. Determination of airborne carbonyls: Comparison of a thermal desorption/GC method with the standard DNPH/HPLC method. *Environmental Science and Technology* 38: 862–870.

Ho, S.S.H., Yu, J.Z. 2004b. In-injection port thermal desorption and subsequent gas chromatography-mass spectrometric analysis of polycyclic aromatic hydrocarbons and *n*-alkanes in atmospheric aerosol samples. *Journal of Chromatography A* 1059: 121–129.

Hsiao, W.L.W., Mo, Z.Y., Fang, M., Shi, X.M., Wang, F. 2000. Cytotoxicity of $PM_{2.5}$ and $PM_{2.5\text{-}10}$ ambient air pollutants assessed by the MTT and the Comet assays. *Mutation Research-Genetic Toxicology and Environmental Mutagenesis* 471: 45–55.

Jacobson, M.Z. 2002. Control of fossil-fuel particulate black carbon plus organic matter, possibly the most effective method of slowing global warming. *Journal of Geophysical Research* 107: ACH 16-1–ACH 16-22.

Jahngen, E., Mallett, J., O'Connor, R., Fischer, S. 2007. Mechanism of the decarboxylative rearrangement of a-(carbonyl) cyclopropane carboxylic acids to 2-substituted-4,5-dihydrofurans. *Arkivoc* ix: 135–149.

Jayne, J.T., Leard, D.L., Zhang, X., Davidovits, P., Smith, K.A. et al. 2000. Development of an aerosol mass spectrometer for size and composition analysis of submicron particles. *Aerosol Science & Technology* 33: 49–70.

Jeon, S.J., Meuzelaar, H.L.C., Sheya, S.A.N., Lighty, J.S., Jarman, W.M. et al. 2001. Exploratory studies of PM_{10} receptor and source profiling by GC/MS and principal component analysis of temporally and spatially resolved ambient samples. *Journal of the Air & Waste Management Association* 51: 766–784.

Jimenez, J.L., Jayne, J.T., Shi, Q., Kolb, C.E., Worsnop, D.R. et al. 2003. Ambient aerosol sampling using the aerodyne aerosol mass spectrometer. *Journal of Geophysical Research* 108: SOS13-1-SOS13-13.

Johns, I.B., McElhill, E.A., Smith, J.O. 1962a. Thermal stability of organic compounds. *I&EC Product Research and Development* 1: 2–6.

Johns, I.B., McElhill, E.A., Smith, J.O. 1962b. Thermal stability of some organic compounds. *Journal of Chemical and Engineering Data* 7: 277–281.

Johnson, W.D. 1974. Thermal decomposition of alcohols. 1. 2-methylbutan-2-ol (t-pentyl alcohol). *Australian Journal of Chemistry* 27: 1047–1052.

Kawamura, K., Kaplan, I.R. 1987. Motor exhaust emissions as a primary source for dicarboxylic acids in Los Angeles ambient air. *Environmental Science Technology* 21: 105–110.

Kiss, G., Gelencser, A., Krivacsy, Z., Hlavay, J. 1997. Occurrence and determination of organic pollutants in aerosol, precipitation, and sediment samples collected at Lake Balaton. *Journal of Chromatography A* 774: 349–361.

Koziel, J.A., Noah, J., Pawliszyn, J. 2001. Field sampling and determination of formaldehyde in indoor air with solid-phase microextraction and on-fiber derivatization. *Environmental Science & Technology* 35: 1481–1486.

Labban, R., Veranth, J.M., Watson, J.G., Chow, J.C. 2006. Feasibility of soil dust source apportionment by the pyrolysis-gas chromatography/mass spectrometry method. *Journal of the Air & Waste Management Association* 56: 1230–1242.

Lamanna, M.S., Goldstain, A.H. 1999. In-situ measurements of C_2-C_{10} VOCs above a Sierra Nevada ponderosa pine plantation. *Journal of Geophysical Research* 104: 21247–21262.

Leinweber, P., Schulten, H.R. 1999. Advances in analytical pyrolysis of soil organic matter. *Journal of Analytical and Applied Pyrolysis* 49: 359–383.

List, G.R., Evans, C.D., Selke, E., Glass, C.A., Hoffmann, R.L. et al. 1971. Pyrolysis of some acetoxynonenes. *Lipids* 6: 635–640.

Liu, N.A., Fan, W., Dobashi, R., Huang, L. 2002. Kinetic modeling of thermal decomposition of natural cellulosic materials in air atmosphere. *Journal of Analytical and Applied Pyrolysis* 63: 303–325.

Madison, J.J., Roberts, R.M. 1957. Pyrolysis of aromatic and related heterocyclics. *Industrial and Engineering Chemistry* 50: 237–250.

Majer, J.R., Perry, R. 1970. Mass spectrometric and chromatographic techniques for determination of polycyclic compounds. *Pure and Applied Chemistry* 24: 685–694.

Marbot, R. 1997. The selection of pyrolysis temperatures for the analysis of humic substances and related materials: 1. Cellulose and chitin. *Journal of Analytical and Applied Pyrolysis* 39: 97–104.

Martos, P.A., Pawliszyn, J. 1998. Sampling and determination of formaldehyde using solid-phase microextraction with on-fiber derivatization. *Analytical Chemistry* 70: 2311–2320.

Mazurek, M.A. 2002. Molecular identification of organic compounds in atmospheric complex mixtures and relationship to atmospheric chemistry and sources. *Environmental Health Perspectives* 110: 995–1003.

McMurry, J. 1999. *Organic Chemistry*, 5th ed. Thomson Brooks/Cole, Belmont, CA.

McNaught, A.D., Wilkinson, A. 1997. *IUPAC Gold Book Compendium of Chemical Terminology*, 2nd ed. Royal Society of Chemistry, Cambridge.

Meininghaus, R., Salthammer, T., Knöppel, H. 1999. Interaction of volatile organic compounds with indoor materials—A small-scale screening method. *Atmospheric Environment* 33: 2395–2401.

Merkulov, A.A., Ovsyannikov, A.A., Polak, L.S., Popov, V.T., Pustilnikov, V.Y. 1989. Initial stages of soot formation in thermal pyrolysis of acetylene: I. Mechanism for homogeneous pyrolysis of acetylene. *Plasma Chemistry and Plasma Processing* 9: 95–104.

Morgan, K.M. 2000. Reaction mechanisms: Part (II) Pericyclic reactions. *Annual Report on Progress Chemistry Section B* 96: 321–345.

Moss, R.A., Platz, M.S., Jones, M.J. 2005. *Reactive Intermediate Chemistry*. Wiley, Hoboken, NJ.

Murphy, D.M., Thomson, D.S. 1995. Laser ionization mass spectroscopy of single aerosol particles. *Aerosol Science & Technology* 22: 237–249.

Napier, I.M., Norrish, R.G.W. 1967. The photolysis and pyrolysis of nitromethane and methyl nitrite. *Proceedings of the Royal Society of London A* 299: 317–336.

Neususs, C., Pelzing, M., Plewka, A., Herrmann, H. 2000. A new analytical approach for size-resolved speciation of organic compounds in atmospheric aerosol particles: Methods and first results. *Journal of Geophysical Research* 105: 4513–4527.

Neususs, C., Weise, D., Birmili, W., Wex, H., Wiedensohler, A. et al. 2000. Size-segregated chemical, gravimetric and number distribution-derived mass closure of the aerosol in Sagres, Portugal during ACE-2. *Tellus B: Chemical and Physical Meteorology* 52: 169–184.

Nimz, E.L., Morgan, S.L. 1993. On-line derivatization for complex fatty acid mixtures by capillary gas chromatography/mass spectrometry. *Journal of Chromatographic Science* 31: 145–149.

Noble, C.A., Prather, K.A. 1996. Real-time measurement of correlated size and composition profiles of individual atmospheric aerosol particles. *Environmental Science Technology* 30: 2667–2680.

Odabasi, M., Ongan, O., Cetin, E. 2005. Quantitative analysis of volatile organic compounds (VOCs) in atmospheric particles. *Atmospheric Environment* 39: 3763–3770.

Orasche, J., Schnelle-Kreis, J., Abbaszade, G., Zimmermann, R. 2011. Technical note: In-situ derivatization thermal desorption GC-TOFMS for direct analysis of particle-bound non-polar and polar organic species. *Atmospheric Chemistry and Physics* 11: 8977–8993.

Pan, L., Pawliszyn, J. 1997. Derivatization/solid-phase microextraction: New approach to polar analytes. *Analytical Chemistry* 69: 196–205.

Pan, L., Adams, M., Pawliszyn, J. 1995. Determination of fatty acids using solid phase microextraction. *Analytical Chemistry* 67: 4396–4403.

Pan, L., Chong, J.M., Pawliszyn, J. 1997. Determination of amines in air and water using derivatization combined with solid-phase microextraction. *Journal of Chromatography A*: 249–260.

Poore, M.W. 2000. Oxalic acid in $PM_{2.5}$ particulate matter in California. *Journal of the Air & Waste Management Association* 50: 1874–1875.

Pope, C.A. III, Dockery, D.W. 2006. Health effects of fine particulate air pollution: Lines that connect. *Journal of the Air & Waste Management Association* 56: 709–742.

Rice, F.O. 1931. The thermal decomposition of organic compounds from the standpoint of free radicals: I. Saturated hydrocarbons. *Journal of the American Chemical Society* 53: 1959–1972.

Rice, F.O., Johnston, W.R., Evering, B.L. 1932. The thermal decomposition of organic compounds from the standpoint of free radicals: II. Experimental evidence of the decomposition of organic compounds into free radicals. *Journal of the American Chemical Society* 54: 3529–3543.

Rogge, W.F., Hildemann, L.M., Mazurek, M.A., Cass, G.R., Simoneit, B.R.T. 1994. Sources of fine organic aerosol: 6. Cigarette smoke in the urban atmosphere. *Environmental Science & Technology* 28: 1375–1388.

Rogge, W.F., Mazurek, M.A., Hildemann, L.M., Cass, G.R., Simoneit, B.R.T. 1993. Quantification of urban organic aerosols at a molecular level: Identification, abundance and seasonal variation. *Atmospheric Environment* 27A: 1309–1330.

Rumyantsev, A.N., Oganesova, E.Y. 1987. Mechanism of pyrolysis of *n*-alkanes: 1. *N*-butane. *Chemistry and Technology of Fuels and Oils* 23: 35–37.

Sakurai, H., Tobias, H.J., Park, K., Zarling, D., Docherty, K.S. et al. 2003. On-line measurements of diesel nanoparticle composition and volatility. *Atmospheric Environment* 37: 1199–1210.

Schauer, J.J., Cass, G.R. 2000. Source apportionment of wintertime gas-phase and particle-phase air pollutants using organic compounds as tracers. *Environmental Science & Technology* 34: 1821–1832.

Schauer, J.J., Kleeman, M.J., Cass, G.R., Simoneit, B.R.T. 1999. Measurement of emissions from air pollution sources: 2. C_1 through C_{30} organic compounds from medium duty diesel trucks. *Environmental Science & Technology* 33: 1578–1587.

Schnelle-Kreis, E., Sklorz, M., Peters, A., Cyrys, J., Zimmermann, R. 2005. Analysis of particle-associated semi-volatile aromatic and aliphatic hydrocarbons in urban particulate matter on a daily basis. *Atmospheric Environment* 39: 7702–7714.

Schnelle-Kreis, J., Welthagen, W., Sklorz, M., Zimmermann, R. 2005. Application of direct thermal desorption gas chromatography and comprehensive two-dimensional gas chromatography coupled to time of flight mass spectrometry for analysis of organic compounds in ambient aerosol particles. *Journal of Separation Science* 28: 1648–1657.

Schripp, T., Etienne, S., Fauck, C., Fuhrmann, F., Mark, L. et al. 2014. Application of proton-transfer-reaction-mass-spectrometry for indoor air quality research. *Indoor Air* 24: 178–189.

Schuetzle, D., Crittenden, A.L., Charlson, R.J. 1973. Application of computer controlled high resolution mass spectrometry to the analysis of air pollutants. *Journal of the Air Pollution Control Association* 23: 704–709.

Schuetzle, D., Cronn, D., Crittenden, A.L., Charlson, R.J. 1975. Molecular composition of secondary aerosol and its possible origin. *Environmental Science & Technology* 9: 838–845.

Schulten, H.R., Leinweber, P. 1996. Characterization of humic and soil particles by analytical pyrolysis and computer modeling. *Journal of Analytical and Applied Pyrolysis* 38: 1–53.

Schwarz, E.M., Grundstein, V.V., Ievins, A.F. 1972. Thermal investigation of polyols. *Thermal Analysis and Calorimetry* 4: 331–337.

Sears, F.W., Zermansky, M.W., Young, H.D. 1982. *University Physics*, 6th ed. Addison-Wesley, Boston.

Seinfeld, J.H., Pandis, S.N. 1998. *Atmospheric Chemistry and Physics: From Air Pollution to Climate Change*. Wiley, New York.

Sidhu, S., Graham, J., Striebich, R., 2001. Semi-volatile and particulate emissions from the combustion of alternative diesel fuels. *Chemosphere* 42(5–7): 681–690.

Sigman, M.E., Ma, C.Y. 1999. In-injection port thermal desorption for explosives vapor evidence analysis. *Analytical Chemistry* 71: 4119–4124.

Simoneit, B.R.T. 1985. Application of molecular marker analysis to vehicular exhaust for source reconciliations. *International Journal of Environmental Analytical Chemistry* 22: 203–233.

Simoneit, B.R.T. 1992. Discussion on analysis of particulate organics in a forest atmosphere by thermodesorption GC/MS. *Atmospheric Environment* 26A: 517–519.

Simoneit, B.R.T. 1999. A review of biomarker compounds as source indicators and tracers for air pollution. *Environmental Science and Pollution Research* 6: 159–169.

Sin, D.W.M., Fung, W.H., Lam, C.H. 2002. Measurement of carbonaceous aerosols: Validation and comparison of a solvent extraction–gas chromatographic method and a thermal optical transmittance method. *Analyst* 127: 614–622.

Smith, R.D. 1979. A direct mass spectrometric study of the mechanism of toluene pyrolysis at high temperatures. *Journal of Physical Chemistry* 83: 1553–1563.

Smith, J.N., Moore, K.F., McMurry, P.H., Eisele, F.L. 2004. Atmospheric measurements of sub-20 nm diameter particle chemical composition by thermal desorption chemical ionization mass spectrometry. *Aerosol Science & Technology* 38: 100–110.

Song, J., Zhan, B., Li, P. 1994. A study of SO_2 in fogwater of Qingdao. *Marine Science* 2: 66–69.

Stout, J.W., Fisher, L.H. 1941. The entropy of formic acid. *The Journal of Chemical Physics* 9: 163–168.

Subbalakshmi, Y., Patti, A.F., Lee, G.S.H., Hooper, M.A. 2002. Structural characterization of macromolecular organic material in air particulate matter using Py-GC-MS and solid state C-13-NMR. *Journal of Environmental Monitoring* 2: 561–565.

Sunesson, A.L., Sundgren, M., Levin, J.O., Eriksson, K., Carlson, R. 1999. Evaluation of two adsorbents of diffusive sampling and thermal desorption-gas chromatographic analysis of monoterpenes in air. *Journal of Environmental Monitoring* 1: 45–50.

Tobias, H.J., Ziemann, P.J. 1999. Compound identification in organic aerosols using temperature-programmed thermal desorption particle beam mass spectrometry. *Analytical Chemistry* 71: 3428–3435.

Tobias, H.J., Ziemann, P.J. 2000. Thermal desorption mass spectrometric analysis of organic aerosol formed from reactions of 1-tetradecene and O_3 in the presence of alcohols and carboxylic acids. *Environmental Science & Technology* 34: 2105–2115.

Tobias, H.J., Ziemann, P.J. 2001. Kinetics of the gas-phase reactions of alcohols, aldehydes, carboxylic acids, and water with the C_{13} stabilized criegee intermediate formed from ozonolysis of 1-tetradecane. *Journal of Physical Chemistry A* 105: 6129–6135.

Tobias, H.J., Kooiman, P.M., Docherty, K.S., Ziemann, P.J. 2000. Real-time chemical analysis of organic aerosols using a thermal desorption particle beam mass spectrometer. *Aerosol Science & Technology* 33: 170–190.

Tran, N.K., Steinberg, S.M., Johnson, B.J. 2000. Volatile aromatic hydrocarbons and dicarboxylic acid concentrations in air at an urban site in the southwestern U.S. *Atmospheric Environment* 34: 1845–1852.

US EPA. 1997. Compendium of methods for the determination of toxic organic compounds in ambient air, Method TO-17: Determination of volatile organic compounds in ambient air using active sampling onto sorbent tubes.

US EPA. 1999a. Compendium of Methods TO14a: Determination of Toxic Organic Compounds in Ambient Air: Determination of Volatile Organic Compounds (VOCs) in Ambient Air Using Specially Prepared Canisters with Subsequent Analysis by Gas Chromatography. Cincinnati, Ohio, 2nd ed. EPA, Washington, DC.

US EPA. 1999b. Compendium of Methods for the Determination of Toxic Organic Compounds in Ambient Air: Determination of Volatile Organic Compounds (VOCs) in Ambient Air using Specially Prepared Canisters with Subsequent Analysis by Gas Chromatography, 2nd ed. EPA, Washington, DC.

Vedal, S. 1997. Critical review—Ambient particles and health: Lines that divide. *Journal of the Air & Waste Management Association* 47: 551–581.

Veltkamp, P.R., Hansen, K.J., Barkley, R.M., Sievers, R.E. 1996. Chromatographic measurement of molecular markers of sources of atmospheric aerosol particles. *Environmental Geochemistry and Health* 18: 77–80.

Voisin, D., Smith, J.N., Sakurai, H., McMurry, P.H., Eisele, F.L. 2003. Thermal desorption chemical ionization mass spectrometer for ultrafine particle chemical composition. *Aerosol Science & Technology* 37: 471–475.

Voorhees, K.J., Schulz, W.D., Currie, L.A., Klouda, G.A. 1988. An investigation of the insoluble carbonaceous material in airborne particulates from vehicular traffic. *Journal of Analytical and Applied Pyrolysis* 14: 3–98.

Voorhees, K.J., Schulz, W.D., Kunen, S.M., Hendricks, L.J., Currie, L.A. et al. 1991. Analysis of insoluble carbonaceous materials from airborne particles collected in pristine regions of Colorado. *Journal of Analytical and Applied Pyrolysis* 18: 189–205.

Watanabe, M., Adschiri, T., Arai, K. 2001. Overall rate constant of pyrolysis of n-alkanes at a low conversion level. *Industrial and Engineering Chemistry Research* 40: 2027–2036.

Waterman, D., Horsfield, B., Hall, K., Smith, S. 2001b. Application of micro-scale sealed vessel thermal desorption-gas chromatography-mass spectrometry for the organic analysis of airborne particulate matter: Linearity, reproducibility and quantification. *Journal of Chromatography* 912: 143–150.

Waterman, D., Horsfield, B., Leistner, F., Hall, K., Smith, S. 2000. Quantification of polycyclic aromatic hydrocarbons in the NIST standard reference material (SRM1649a) urban dust using thermal desorption GC/MS. *Analytical Chemistry* 72: 3563–3567.

Waterman, D., Smith, S., Green, D., Horsfield, B., Hall, K. 2001a. The application of a thermal desorption GCMS technique for the organic analysis of airborne particulate matter. *Advances in Mass Spectrometry* 14: 887–888.

Watson, J.G. 2002. Visibility: Science and regulation. *Journal of the Air & Waste Management Association* 52: 628–713.

Watson, J.G., Chow, J.C. 2001. Ambient air sampling. In *Aerosol Measurement: Principles, Techniques, and Applications*, 2nd ed., ed. Baron, P., Willeke, K.E. Wiley, New York, pp. 821–844.

Watson, J.G., Chow, J.C. 2007. Receptor models for source apportionment of suspended particles. In *Introduction to Environmental Forensics*, 2nd ed., ed. Murphy, B., Morrison, R. Academic, New York, pp. 279–316.

Watson, J.G., Chow, J.C., Chen, L.W.A. 2005. Summary of organic and elemental carbon/black carbon analysis methods and intercomparisons. *Aerosol and Air Quality Research* 5: 65–102.

Watson, J.G., Wyzga, R.E., Burnett, R.T., Vostal, J.J., Romieu, I. et al. 1997. Ambient particles and health—Lines that divide: 1997 critical review discussion. *Journal of the Air & Waste Management Association* 47: 995–1008.

Wauters, E., Sandra, P., Verzele, M. 1979. Qualitative and semi-quantitative analysis of the non-polar organic fraction of air particulate matter. *Journal of Chromatography A* 170: 125–131.

Welthagen, W., Schnelle-Kreis, J., Zimmermann, R. 2003. Search criteria and rules for comprehensive two-dimensional gas chromatography-time-of-flight mass spectrometry analysis of airborne particulate matter. *Journal of Chromatography A* 1019: 233–249.

Weyand, E.H., Bevan, D.R. 1986. Benzo(*a*)pyrene disposition and metabolism in rats following intratracheal instillation. *Cancer Research* 46: 5655–5661.

Williams, B.J., Goldstein, A.H., Kreisberg, N.M., Hering, S.V. 2006. An in-situ instrument for speciated organic composition of atmospheric aerosols: Thermal desorption aerosol GC/MS-FID (TAG). *Aerosol Science & Technology* 40: 627–638.

Woolfenden, E. 1997. Monitoring VOCs in air using sorbent tubes followed by thermal desorption-capillary GC analysis: Summary of data and practical guidelines. *Journal of the Air & Waste Management Association* 47: 20–36.

Yang, H., Yu, J.Z., Ho, S.S.H., Xu, J.H., Wu, W.S. et al. 2005. The chemical composition of inorganic and carbonaceous materials in $PM_{2.5}$ in Nanjing, China. *Atmospheric Environment* 39: 3735–3749.

Yu, J.Z., Xu, J.H., Yang, H. 2002. Charring characteristics of atmospheric organic particulate matter in thermal analysis. *Environmental Science & Technology* 36: 754–761.

Yuan, W., Rabinovitch, B.S., Tosa, R. 1982. Vibrational energy transfer and pyrolysis of nitromethane by the variable encounter method. *Journal of Physical Chemistry* 86: 2796–2799.

Zhang, Q., Alfarra, M.R., Worsnop, D.R., Allan, J.D., Coe, H. et al. 2005. Deconvolution and quantification of hydrocarbon-like and oxygenated organic aerosols based on aerosol mass spectrometry. *Environmental Science & Technology* 39: 4938–4952.

Zheng, M., Wan, T.S.M., Fang, M., Wang, F. 1997. Characterization of the non-volatile organic compounds in the aerosols in Hong Kong—Identification, abundance and origin. *Atmospheric Environment* 31: 227–237.

Section II

Sample Preparation and Sampling Strategies

5 Sample Preparation Methods for Determination of Pollutants in Solid and Complex Environmental Matrices

Carlos Gonçalves, Maria Augusta D. Sousa, Vasilios G. Samaras, C. Marisa R. Almeida, and M. Clara P. Basto

CONTENTS

5.1 GENERAL INTRODUCTION

Current strategies for analyzing organic micropollutants in solid and complex environmental samples depend upon recently achieved analytical/instrumental breakthroughs. Sample pretreatment procedures are intended to provide concentrated target analytes in solution, isolated from the matrix, free of interferences, and ready for further instrumental analysis, therefore improving the method's sensitivity and selectivity (Albaseer et al., 2010). However, the evolution of instrumentation technology with ever more reliable equipments (more precise, accurate, and sensitive) also has an impact on the fundamental steps of an analytical process: sample preparation, chromatographic separation, and detection. Indeed, one of these steps tends to drive the remaining. When using a highly sensitive and selective detection system, sample preparation and chromatographic separation can be made less intensive (Pinto et al., 2010). Conversely, if such a system is not available, then the analytical problem can equally be overcome by using a more methodical sample preparation and a higher-resolution chromatographic separation. A good separation efficiency will also facilitate the tasks to be carried out during sample preparation and detection.

When dealing with environmental contaminants, the nature of the target analytes will determine to a great extent the entire analytical burden—for instance, whether the pollutant is polar or apolar or the fact that it has physicochemical properties similar to those of some interferences in the sample matrix will definitely play a role when planning the analytical steps to be carried out. Another important consideration is the degree of hazard the target pollutant poses to the environment or to human health: what levels can be tolerated without causing significant harm and what, if any, are the limits set by legislation intended to ensure acute and chronic safety. As the complexity of the matrix, the analyte characteristics, and the detection levels to be reached challenge the analytical capabilities, a more laborious methodology will have to be implemented to achieve the same degree of trueness of the results. Thus, as one may easily understand, analyzing sediments, wastewater, or sewage sludge will require more dedicated work than the analysis of a soil (or sand). Similarly, polar compounds typically require greater analytical effort and attention to all the potential interactions among the analytes and interferences than needed for analyzing less polar pollutants. As the Horwitz equation (Horwitz et al., 1980) well demonstrates, trace levels of pollutants cannot be determined with the same degree of precision (and accuracy) as major components, although the analytical instrumentation now available can provide more robust results. In addition, aged and weathered residues of several environmental pollutants tend to bind to the matrix. Residues bound to soil and sediments exhibit a different extraction behavior compared to the nonbound fractions. Therefore, a more exhaustive extraction procedure is required to release them from the matrix. However, as the extraction procedure becomes more intensive, more coextracted interferences are also expected (Salgado-Petinal et al., 2006a). Some authors highlighted the advantages of a mild extraction procedure (e.g., supercritical fluid extraction [SFE] for PAH analysis) that could give an indication of the bioavailable fraction of contaminants, as opposed to performing an exhaustive procedure that would provide information about the total amount of a given pollutant. The first technique would be more relevant from the ecotoxicological point of view (Björklund et al., 2000).

Despite the above trade-offs between the availability of laboratory instrumentation and the analytical challenges, a good sample preparation procedure is always needed (or at least recommended) to protect the analytical system from accumulation of dirty material, which will decrease the analytical performance in short term and reduce the lifetime of analytical columns and detectors (Salgado-Petinal et al., 2006a). As what happens in many areas, environmental research studies cope better with analytical uncertainty, aiming overall to demonstrate the usefulness of a certain analytical strategy, while routine work for compliance purposes (which will support decisions with a financial impact) seeks methodologies that are capable of ensuring the minimum chance of false positives and false negatives. In the latter, proven methodologies, although more traditional and laborious, are normally preferred. Having depicted as much as possible the potential scenarios encountered in the chromatographic analysis of environmental samples, one might understand why conventional analytical techniques are still used nowadays side by side with groundbreaking methodologies. The latter will eventually find their way into official standard methods once they have proved that they can provide the same or a higher degree of reliability in a more automated, miniaturized, faster, and environment-friendly way.

This chapter aims to provide a keen vision on the most recent analytical methodologies for environmental monitoring and assessment of solid and complex samples (soils, sediments, biota, sewage sludge, and wastewater) without neglecting the conventional ones when no alternatives have been developed or they do not meet the required figures of merit. The chapter is organized on the analytical strategies that have been pursued per solid and complex environmental matrix type. This chapter does not intend to be comprehensive but rather gives emphasis on the most innovative techniques and approaches.

The recently published Directive 2008/56/EC (European Commission, 2008), concerning marine quality strategy, and Directive 2013/39/EU (European Commission, 2013), amending Directive 2000/60/CE and revising the list of priority substances for which environmental quality standards are established, pose a great emphasis on the monitoring of organic pollutants in sediments and biota. Therefore, it is expected that sample preparation and chromatographic techniques for environmental matrices will play an important role in the implementation of the mentioned directives.

The list of 33 priority substances (plus 8 certain other pollutants) is the focus of studies aiming at ensuring that existing levels of contamination in biota and sediments will not significantly increase (European Union, 2010; European Commission, 2013). Some organic pollutants (such as pyrethroid pesticides) adsorb strongly onto soils and sediments, therefore not causing relevant concern for the water column. Nevertheless, it has been suggested that they may affect sediment-dwelling aquatic organisms and those that live just above the sediment (Albaseer et al., 2010). On the other hand, pollutants such as PBDEs can be slowly released from polluted sediments to the water phase and then their levels can be magnified in the trophic chain, particularly in the case of tetra- and penta-brominated congeners, impairing the health condition of aquatic organisms (Montes et al., 2010). Recently, investigation of organic pollutants in environmental matrices has expanded from persistent organic pollutants (POPs) to emerging contaminants (ECs), which are not necessarily new pollutants, but they are currently not included in routine monitoring programs. They may be candidates for future regulation in water, sediments, and biota, depending on research outcomes on their occurrence in different environmental compartments, their toxicity, their potential health effects, and public perception (Pintado-Herrera et al., 2013). Additionally, ECs do not need to persist in the environment to cause negative effects, as they are being continuously discharged through wastewater treatment plants (WWTPs) and anthropogenic activities (Pintado-Herrera et al., 2013).

To address the occurrence, distribution, and fate of these pollutants in the environment, efficient and robust analytical methods are necessary. Sample preparation followed by GC or LC separation and qualitative and quantitative analysis by MS has become the standard approach.

5.2 SAMPLE PREPARATION METHODS APPLIED TO SOLID AND COMPLEX SAMPLES

5.2.1 Preparatory Sample Handling

Prior to starting the procedures for extracting the analytes from the matrix, soil and sediment samples collected in the field are generally air-dried, ground, and sieved. Soil samples are usually sieved through a mesh with a size of 2 mm, while organic pollutants in sediments should be analyzed in the fraction with a particle size <63 μm, because hazardous substances are mainly contained in the finer sediment fractions with a large relative surface area. Furthermore, the fine fractions are the primary food source for biota (European Union, 2010). On the other hand, biota samples are normally subdivided, freeze-dried, and homogenized. Pretreatment of sewage sludge samples is usually performed by heating in an oven at low temperature or air- and freeze-drying. Autoclaving or addition of formaldehyde to minimize the microbial activity and addition of sodium ethylenediaminetetraacetic acid (Na_2EDTA) as a chelating agent to prevent bonding with metallic ions have also been suggested. Subsequently, samples are ground, sieved, and then stored in the dark at a low temperature (at 4°C to −20°C) before processing and analysis, normally within 48 h.

This preliminary sample preparation of solid matrices aims at obtaining samples in a dry state that can be stored for longer periods with minimal risk of hydrolysis and microbial degradation of the analytes. It also provides samples as homogeneous as possible from which representative aliquots can be taken for analysis. Due to the inhomogeneity of wastewaters discharged during the diurnal cycle, 24 h composite samples are usually collected, homogenized, eventually filtered, and analyzed. Caution must be taken during the 24 h collection period since chemical and microbiological degradation may be occurring; therefore, refrigeration of the container is highly recommended. Wastewater can be efficiently homogenized using the Ultra-Turrax apparatus. Needless to say, all processes that can potentially alter analyte concentration, such as photodecomposition, adsorption, vaporization, thermal decomposition, microbial action, and chemical reaction, should be minimized by proper preservation of the samples: normally keeping them in a cold (4°C) environment, in the absence of light. For instance, the stability of pyrethroids in samples with high moisture content is precarious, while sediments can be stored frozen for a month without significant losses. Acidity of the sample is also important to avoid pyrethroid epimerization (Albaseer et al., 2010). Nevertheless, analysis should always start as soon as possible after sampling (Albaseer et al., 2010).

5.2.2 Derivatization

Derivatization is usually incorporated into sample preparation methods to convert the analyte into a more volatile and thermally stable derivative and improve the selectivity and sensitivity for both GC and LC techniques. Derivatization can be performed before the extraction of analytes (benefiting extraction efficiency, chromatographic analysis, and detection), simultaneously performed with extraction, or applied to the organic solvent extract. Although a very large number of reactions have been reported (various types of silylation, alkylation, and acylation), in reality only a few are routinely used. Reagents are typically selected according to their reactivity with the analytes or the stability of their products. Silylation is the most widely used derivatization technique applied to pharmaceutically active compounds (PACs), phenolic compounds, natural and synthetic steroids. and personal care products (PCPs), the most employed derivatization reagents being *N,O*-bis(trimethylsilyl)trifluoroacetamide (BSTFA), *N*-methyl-*N*-(trimethylsilyl)trifluoroacetamide (MSTFA), and *N*-methyl-*N*-(*tert*-butyldimethysilyl)trifluoroacetamide (MTBSTFA). In some studies, the derivatization reaction was enhanced by the addition of a catalyst such as trimethylchlorosilane (TMCS), pyridine, and trimethylsilylimidazole (Morales et al., 2005; Martínez-Moral and Tena, 2011; Samaras et al., 2011; Azzouz and Ballesteros, 2012; Yu and Wu, 2012; Albero et al., 2014). The required reaction time depends on the chemical structure of the compounds and the reagent, ranging from a few minutes to a few hours. The reaction can be performed either at room temperature or under heating, while the optimum reagent volume must be appropriate to avoid partial derivatization of the molecules and extract dilution. Generally, the derivatization reaction is considered successful when it is performed in a short time and leads to the formation of a single derivative for each compound that is stable to the reaction conditions.

5.2.3 Fundamentals and Advantages of the Main Extraction and Cleanup Techniques

In general, sample preparation techniques for the determination of organic pollutants in environmental samples fall into three classes: (i) solvent extraction followed by solvent reduction, (ii) gas- or fluid-based extraction with fluid purging, and (iii) sorbent trapping with subsequent solvent elution or TD. Solvent extraction techniques include, for instance, ultrasonic-assisted extraction (UAE), microwave-assisted extraction (MAE), and pressurized liquid extraction (PLE), used as a replacement for the conventional Soxhlet extraction. Sorbent-trapping extraction techniques include, for instance, matrix solid-phase dispersion (MSPD), SPE, stir bar sorptive extraction (SBSE), and solid-phase microextraction (SPME). SFE belongs to the category of fluid-based techniques. According to Albaseer et al. (2010), the two first classes (i and ii) are mainly applied to the processing of soils and sediments (and also biota and sewage sludge), although sorbent-trapping techniques can still be used, namely, as part of a cleanup step. Aqueous and liquid samples are normally processed through a sorbent-trapping step, the most widely used being SPE on reversed-phase adsorbents.

Depending on the extraction technique selected, an optimization step precedes their application to real samples and should encompass critical parameters for the extraction efficiency, such as solvent polarity and volume, sorbent type, sample pH, extraction pressure, temperature, and time (Albaseer et al., 2010). For instance, the determination of pyrethroid pesticides in soils and sediments has mostly been carried out by UAE, MAE, and PLE. Recoveries around 80% or higher have been obtained employing apolar extraction solvents, such as toluene, ethyl acetate, hexane, and dichloromethane, sometimes in mixture with a polar solvent (Albaseer et al., 2010).

In the next sections, brief descriptions of the extraction principles of the most common techniques, highlighting any limitations and precautions that should be taken into consideration for their successful application, will be given. More detailed information can be found in the comprehensive review authored by Albaseer et al. (2010), the chapter by Pérez et al. (2011), and the several publications cited in this chapter.

5.2.3.1 Ultrasonic-Assisted Extraction

UAE has been used for the extraction of organic pollutants from soil, sediment, biota, and sewage sludge samples (Priego-Capote and de Castro, 2004; Martínez-Parreño et al., 2008; Ronan and McHugh, 2013). Ultrasonic radiation is a powerful tool for accelerating the partition of organic compounds from solid matrices to an extraction solvent. An ultrasonic bath is the most commonly available laboratory instrumentation for sonochemisty operating at a fixed frequency of 40 kHz. Ultrasonic probes have also been used in a few studies operating at a lower frequency of 20 kHz, referred to as focused ultrasound solid–liquid extraction (FUSLE) (Martínez-Moral and Tena, 2011). FUSLE is based on the application of high-power focused ultrasonic waves by using a titanium microtip immersed directly in the extraction mixture. The resulting cavitation bubbles produce high temperatures and pressures, which enable good extraction efficiency. UAE requires much cheaper equipments, available in most laboratories, when compared to MAE and PLE, providing good recoveries, particularly if the dynamic mode with solvent renewal is selected. It is applicable for a wide range of environmental samples and organic pollutants (Martínez-Parreño et al., 2008; Yamaguchi and Lee, 2010). pH, extraction solvent, number of extraction cycles, and extraction time are the most influential variables affecting the UAE procedure. The sample weight can also affect the extraction efficiency since the larger the sample amounts analyzed, the higher the amounts of matrix components extracted. Operation is also extremely easy, and the possibility of degradation of thermolabile analytes (which can occur when extraction is carried out at high temperatures and pressures, such as in PLE) is minimized in UAE (Albaseer et al., 2010). If kept under control, the slight heating effect can even be beneficial to improving the extraction efficiency. Furthermore, an ultrasonic bath allows a multiple sample extraction in one single batch; thus, UAE significantly reduces preparation time (Yamaguchi and Lee, 2010). In general, 10–30 min extractions are sufficient to allow quantitative yield and longer periods increase the risk of analyte degradation or readsorption onto the matrix (Albaseer et al., 2010). Further enrichment of the extract with SPE and a cleanup step (e.g., filtration, silica gel) to avoid interferences prior to chromatographic separation are commonly required for determination of trace analytes.

The capability of UAE to extract tightly bound pyrethroids and other pollutants from sediments and soils with excellent recoveries renders it an alternative to the more time-consuming Soxhlet technique (Domeno et al., 2006).

5.2.3.2 Microwave-Assisted Extraction

MAE has also been used as a replacement for Soxhlet extraction. Unlike classical heating methods where the solvent diffuses into the matrix and extracts the components by solubilization, microwaves heat the sample and the solvent mixture simultaneously without heating the vessel, causing the destruction of the matrix macrostructure and therefore enhancing both extraction speed and efficiency. Different parameters can be optimized to obtain the highest recoveries, such as solvent type and volume, temperature, pressure, extraction time, and microwave power. The choice of irradiation time and microwave power depends on the nature of the solvent, as well as on the type and the size of the sample matrix. Unlike classical techniques where a higher volume of solvent increases the extraction efficiency of the analytes, in MAE it may lead to lower recoveries due to inadequate mixing of the solvent with the matrix by the microwaves. Short irradiation times usually require an increased power for the efficient extraction of the analytes. In most cases, samples can be extracted with a single or a mixture of solvents with high dielectric constants to absorb the microwave energy efficiently. Less frequently, a mixture of solvents with low dielectric constants may be used to avoid the degradation of thermolabile compounds. For instance, Albaseer et al. (2010) brought special attention to the possible degradation or interconversion of pyrethroids extracted from soils and sediments whenever elevated temperatures and pressures are employed. MAE can also be performed with open vessels (focused MAE), at atmospheric pressure, or in closed vessels (pressurized MAE) under controlled pressure and temperature, with the last one having considerably more applications in literature (Dobor et al., 2010; Sanchez-Prado et al., 2010; Sturini et al., 2010; Raich-Montiu et al., 2011). Although, approximately quantitative extraction efficiencies have been obtained by MAE, the technique still requires considerable volumes of organic solvents. One appreciable advantage is the possibility of extracting several samples simultaneously (usually, up to 12) in a short cycle time (typically, from 10 to 30 min) (Sanchez-Prado et al., 2010). On the other hand, limited selectivity is a drawback of MAE, with consequent coextraction of significant amounts of interfering compounds, therefore demanding an additional cleanup of the obtained extracts (Salgado-Petinal et al., 2006a; Fernández-González et al., 2008; Albaseer et al., 2010). For that purpose, the combination of MAE and SPE (for cleanup) is currently the most widely used approach for the analysis of sewage sludge.

Recently, there have been reports on the combination of MAE with micellar media, named microwave-assisted micellar extraction (MAME), for the extraction of different organic compounds from solid samples. The process begins by loading the sample into the extraction vessel; then the surfactant is added, and the vessel is closed. Microwave radiation is applied, and extraction is initiated to heat up the extractants. After extraction finishes, samples are cooled down. Although in most situations a simple filtration of the extract suffices, some cases may require an additional cleanup step (Fernandez et al., 2001; Sosa Ferrera et al., 2004). The MAME methodology presented as advantages the use of reduced volumes of extractant, a lower cost, and a reduced toxicity in comparison to those of organic solvents.

A good review on the principles, influential parameters, and application of MAE on the determination of emerging pollutants in solid samples was published by Sanchez-Prado et al. (2010).

5.2.3.3 Pressurized Liquid Extraction

PLE, also known as accelerated solvent extraction (ASE), is the most recent extraction technique. The principle of PLE is based on the extraction of solid matrices with liquid solvents at elevated temperatures (higher than their boiling points) and pressures up to 200 bar. The decreased viscosity of solvents at high temperatures enhances the solubility of the target analytes and disrupts the strong analyte–matrix interactions, thus improving the extraction efficiency. In practice, the extraction cell which contains the sample is heated at a given temperature and filled up with the extraction solvent up to the desired pressure. For method development, proper operational parameters such as extraction solvent or mixture of solvents, temperature, static extraction time, and number of cycles need to be optimized. Other parameters such as pressure, purge time, and flush volume are usually not included for optimization because they do not significantly affect recoveries (Negreira et al., 2011). PLE is considered to tackle the limitations of former extraction techniques (namely, SFE) and has provided excellent results in the extraction of a wide range of pollutants from soils, sediments, biota, and sewage sludge (Radjenović et al., 2009; Albaseer et al., 2010; Gago-Ferrero et al., 2013). However, its remarkable efficiency at high extraction temperatures may also lead to coextraction of interfering species, giving a complex extract. Hence, a compromise between extraction efficiency and minimization of coextracted interfering compounds must be found, finely tuning the harshness of the extraction conditions (solvent composition, temperature, and pressure). Another aspect to bear in mind is the possibility of thermal instability of the analytes at high temperatures, leading to degradation or chemical conversion: for instance, the epimerization of pyrethroids should be checked, while carbamates are known to degrade (Albaseer et al., 2010). Frequently, a further cleanup of the extract cannot be avoided, particularly in sediments and soils with a high organic matter content and in biota (Fernández-González et al., 2008). The analysis of PBDEs has followed this strategy, with subsequent cleanup by adsorption chromatography, gel permeation chromatography (GPC), and/or oxidative treatments (Montes et al., 2010). Besides the recognized extraction efficiency, the main advantages of PLE are its high degree of automation and rapidness (Pintado-Herrera et al., 2013).

5.2.3.4 Pressurized Hot Water Extraction

Pressurized hot water extraction (PHWE), also called subcritical water extraction (SWE), is an emerging organic solvent-free technique which uses water as the solvent for the extraction of organic pollutants from solid matrices under increased temperature and pressure (using an ASE Dionex apparatus, for instance). Pressure should be high enough to keep water in the liquid state (Fernández-González et al., 2008). At temperatures between 100°C and 374°C, the polarity of water decreases markedly, water viscosity and surface tension decrease, and diffusivity increases, so water behaves similarly to some organic solvents (Concha-Graña et al., 2010; Pintado-Herrera et al., 2013). Water can effectively extract a wide range of analytes (e.g., PAHs, polychlorinated biphenyls [PCBs], and pesticides) from environmental samples, with results comparable to those of Soxhlet extraction (Rodil and Popp, 2006). But the PHWE technique is much faster and avoids the use of significant volumes of organic solvents. PHWE allows the extraction of analytes in a wide range of polarities (low to medium polarity) by changing the temperature and the pressure, easily providing class-selective extraction by temperature programming and/or by the addition of modifiers (Rodil and Popp, 2006; Pintado-Herrera et al., 2013). However, the dilute water extract cannot be directly analyzed by GC. Further processing by SPE, in cartridge or disk format, SPME or SBSE provides additional cleanup, concentration effect, and compatibility with GC, improving sensitivity and selectivity (Salgado-Petinal et al., 2006a; Yu and Hu, 2007; Concha-Graña et al., 2010; Pintado-Herrera et al., 2013; Lacroix et al., 2014).

5.2.3.5 Supercritical Fluid Extraction

One of the most promising extraction techniques in the early 1980s was SFE. Its development was prompted by a shift toward more environment-friendly techniques. Although several operational parameters need to be adjusted and despite almost only supercritical CO_2 has been used, several researchers succeeded on extracting pesticides (Gonçalves et al., 2006; Naeeni et al., 2011; Forero-Mendieta et al., 2012), PAHs (Bielská et al., 2013; Gonçalve et al., 2013), and dioxins (Hawthorne et al., 2012), among other pollutants, from soils and sediments, taking profit of the fluid high diffusivity in solid matrices (Albaseer et al., 2010). Nevertheless, the polarity of supercritical CO_2, equivalent to that of *n*-pentane, limits the type of analytes amenable to this technique, which, combined with the high cost of such dedicated equipment and matrix-dependent extraction yields, led to a gradual defeat of initial expectations (Rodil and Popp, 2006; Fernández-González et al., 2008; Martínez-Parreño et al., 2008; Sturini et al., 2010). Nevertheless, its applicability to more polar analytes can still be achieved if a polar cosolvent (e.g., methanol [MeOH]) is added. In such case, more interfering matrix substances will be coextracted, being the optimum modifier content a balance between extraction efficiency and selectivity (Gonçalves et al., 2006, 2013).

5.2.3.6 Matrix Solid-Phase Dispersion

MSPD is a simple, rapid extraction technique and does not require high solvent consumption or expensive equipment. In brief, MSPD extraction is performed by blending the matrix with an appropriate amount of sorbent in a glass mortar with a pestle, to produce a semidry homogeneous material. Then, the obtained mixed material is packed into a polypropylene syringe barrel, which may contain a cleanup sorbent, and the target analytes are isolated with a suitable elution solvent. Most of recent MSPD applications have utilized reversed-phase sorbents as dispersants, such as C18-bonded silica, and, less frequently, C8- and C30-bonded silica and primary and secondary amines (PSAs). MSPD has been applied mainly to the preparation of biological and semisolid samples such as vegetables and fruits, but it has been extended also to environmental samples—for example, sewage sludge and biota (Blanco et al., 2006; Sánchez-Brunete et al., 2007).

5.2.3.7 Solid-Phase Extraction

The first experimental applications of SPE date from approximately five decades ago (Liska, 2000). From then on, hundreds of scientific papers have described various developments and applications of SPE in environmental analysis (Hennion, 1991; Azzouz et al., 2010). Reference to SPE is done herein, as it is a very versatile technique. It can be used as an extraction and preconcentration tool for POPs and ECs in environmental matrices as well as for cleaning up of sample extracts. Normal phase (such as silica, alumina, and Florisil) and ion-exchange adsorbents are generally used for cleanup purposes, as their main function is to retain polar and ionizable interfering substances, while reversed-phase adsorbents can be used for preconcentration of analytes or cleanup, depending on the setup (Hennion, 1991).

The isolation process comprises the selective retention of analytes in a solid adsorbent, which in turn can be eluted with an organic solvent. Analytes are selectively trapped on the adsorbent, mainly through hydrogen bonds, van der Waals forces, hydrophobic effects, π-electron interactions, and cation- or anion-exchange processes (Olives et al., 2012).

The major goals of SPE are the extraction, concentration, and elimination of interfering components. Nevertheless, other advantageous uses of SPE have been identified: storage of analytes in the adsorbed phase (most useful for less stable compounds in the aqueous phase or volatile substances), the fractionation of the sample extract in different groups of compounds, and the derivatization of analytes in the cartridge (Hennion, 1991). However, SPE also has some drawbacks, such as the loss of more polar compounds during sample percolation, the coextraction of matrix interferences, clogging of filters, a noticeable time consumption for the extraction of large sample volumes, a significant consumption of organic solvents, an incomplete desorption of target analyses, the loss of volatile compounds, and the incomplete recovery of the dry extract (Pichon, 2000; Olives et al., 2012).

The most adequate adsorbents for the extraction of organic compounds are the reversed-phase sorbents, from the conventional alkyl-modified silica materials (C8 and C18 nonpolar phase) to the more recent polymeric sorbents, which improve the retention of polar compounds due to the high surface area and cross-links (e.g., poly(styrene-divinylbenzene)), and carbon materials (e.g., graphitized carbon black [GCB]) (Fontanals et al., 2005; Olives et al., 2012). Polymeric sorbents present different functional groups combined with a polymeric skeleton, allowing more than one kind of interaction between the analytes and the sorbent (hydrophilic, hydrophobic, and π–π interactions) (Olives et al., 2012).

Monolithic (particle-free) SPE is a recent strategy to deal with selectivity and sensitivity issues regarding LC-MS bioanalytical methods. Monolithic stationary phases are presented in the format of SPE disks (available in 15 different sorbent chemistries, such as C2, C8, and C18) (Gerhards et al., 2004) and 96-well plates, more easily automated. These materials present several advantages: low cost, mechanical robustness and high stability, no void volumes, easy control of the porous properties, high hydraulic permeability, and dominance of convection over the diffusion mechanism of mass exchange under dynamic conditions that allow the separation at extremely high flow rates (Farre et al., 2012). Extraction selectivity may also be increased using molecularly imprinted solid-phase extraction (MISPE). Molecularly imprinted polymers (MIPs) are synthetic polymers with a predetermined selectivity for a given analyte or group of structurally related compounds. They enclose specific recognition sites, which are complementary in shape, size, and functional group, to the analyte or analytes of interest (Farre et al., 2012; Olives et al., 2012). MIPs have already been synthesized for several emerging pollutants, including pharmaceuticals and endocrine-disrupting compounds (EDCs) (Zorita et al., 2008). Most recently, new nanomaterials (e.g., carbon nanotubes [NTs]) have been exploited as well, to integrate new extraction and cleanup technologies (Cai et al., 2003).

5.2.3.8 Solid-Phase Microextraction

SPME, a solvent-free extraction technique mostly used for the determination of organic micropollutants in aqueous samples, also has been useful in the sample preparation of solid samples. In particular, SPME involves the use of a fiber coated with a polymeric phase which extracts the analytes. The fiber is exposed to the sample until equilibrium is reached or for a rigorously controlled time period and the analytes are then thermally desorbed by transferring the fiber into the GC injection port. It provides fast and simple analyses, achieving extraction and concentration in the same analytical step with very low or no sample manipulation (Salgado-Petinal et al., 2006a; Pintado-Herrera et al., 2013). One of the strategies consists of wetting the soil matrix with water and then submitting the slurry to an elevated temperature to partition the analytes to the headspace, where the SPME fiber is placed (Gonçalves and Alpendurada, 2005; Salgado-Petinal et al., 2006a). The addition of water to a dry solid sample is required for improving the release of volatile and semivolatile analytes. A typical setup can involve moisturizing 0.5 g of the solid sample with 1–2 mL of water and extracting the pollutants with a polydimethylsiloxane (PDMS) fiber coating, exposed to the headspace at 100°C for 30–60 min (Salgado-Petinal et al., 2006a). The direct headspace solid-phase microextraction (HS-SPME) method is not appropriate for the extraction of nonpolar POPs from solid samples with high content of organic matter, such as sewage sludge (total organic carbon 34–42%) (Tölgyessy et al., 2013). Although it is not the most common strategy nowadays, another possibility would be to dip the SPME fiber into a water/soil or sediment slurry, which is heated for compound partitioning between the three-phase system soil/water/SPME adsorbent (Gonçalves and Alpendurada, 2005; Martínez-Parreño et al., 2008). Another procedure involves the extraction of the solid sample by either MAE, UAE, or PLE and then dilution of the extract with water, to which an SPME fiber is exposed by direct immersion into the aqueous solution (Gonçalves and Alpendurada, 2005). All these procedures afford simultaneously a cleanup effect, due to the selectivity of the SPME coating, and a further concentration factor (Albaseer et al., 2010). In the last strategy, an exhaustive extraction is accomplished by liquid–solid extraction (LSE) and complemented with a further selective

SPME step, originating chromatograms essentially free of interferences. A drawback of SPME dealing with the analysis of soils and sediments is the accumulation of matrix components other than the analyte in the fiber, which might lead to a decrease in SPME efficiency over time, limiting its lifetime. The fragile nature of the fused-silica fiber requires careful manipulation, avoiding any mechanical stress.

5.2.3.9 Stir Bar Sorptive Extraction

SBSE was introduced in the late 1990s by Baltussen et al. (1999). Since then, it has been applied successfully in the analysis of different types of micropollutants, mainly in water, soil, and sediment samples. This technique is based on the sorption of the analytes onto PDMS-coated stir bars by introducing directly the SBSE device into the aqueous sample, offering minimization of sampling handling, elimination of the use of organic solvents, and simplification of the analytical procedure. Most operating principles and influencing conditions are similar to those of the SPME technique. After SBSE, target analytes can be desorbed by either thermal desorption (TD) or liquid desorption (LD). While for SBSE-GC, TD is the most common technique, providing high sensitivity results, LD (often aided by sonication) has mostly been used for the determination of thermolabile and polar compounds, followed by LC separation. SBSE uses PDMS coating volumes much higher than in those used in SPME (20–125 μL vs. 0.5 μL), explaining its superior sensitivity, but at the cost of longer equilibration times and more difficult desorption (Rodil and Popp, 2006). For analyzing soil and sediment samples, most studies reported a previous extraction step with UAE, PLE, or MAE and finally dilution of the extract prior to the SBSE process.

5.2.3.10 Immunoaffinity Extraction

Immunoaffinity extraction (IAE) is a technique based on the affinity between antibodies and antigens, mainly applied as a cleanup technique in food and biological analyses. Although presenting high selectivity and allowing a better cleanup for polar compounds in complex matrices, its main drawbacks include small breakthrough volumes and low stability toward solvents, strong pH, and temperature (Farre et al., 2012). Its availability for a reduced number of pollutants and the high cost are major limitations.

5.2.3.11 Quick, Easy, Cheap, Effective, Rugged, and Safe

The quick, easy, cheap, effective, rugged, and safe technique, known by the acronym QuEChERS, gained popularity in recent years for the determination of organic pollutants in vegetables and biota samples, the name being explanatory of the claimed advantages. This technique uses both solvent extraction and sorbent-trapping processes. In general, this methodology consists of two steps: (i) extraction of the analytes from the matrix with a solvent by shaking or agitation in a vortex apparatus and (ii) dispersive solid-phase extraction (dSPE) cleanup. The original QuEChERS consisted of extraction of the sample with acetonitrile (ACN), adding $MgSO_4$ and NaCl to enhance the partition of the analytes and reduce polar interferences. After centrifugation, the extract was cleaned up by shaking with PSAs and $MgSO_4$. Later modifications of this procedure include the addition of buffers (acetic acid, anhydrous sodium acetate, or sodium citrate) to preserve base-sensitive analytes and adsorbents (C18 or GCB) and to reduce matrix interferences (Bragança et al., 2012). This procedure uses inexpensive and widely available apparatus in the laboratory. Furthermore, it is relatively fast and simple and provides an extract ready for analysis. The advantages of this procedure were thoroughly described by Bragança et al. (2012). QuEChERS has been adopted by official bodies as a standard method (AOAC 2007.01 and EN 15662).

5.2.3.12 Cloud-Point Extraction

Surfactants or surface-active agents are amphiphilic molecules, with a polar or hydrophilic head and a hydrophobic tail. Micelles are molecular aggregates of surfactant molecules, associated in aqueous solutions. They have a high capacity to solubilize solutes of different natures. However, there is a minimum concentration of surfactant needed for this aggregation to occur—the critical micellar concentration (Sosa Ferrera et al., 2004). Recently, this capacity of micelles to solubilize different compounds has been used during extraction and preconcentration of organic compounds from different samples, including environmental matrices, prior to their analysis through GC or LC (Eisert and Pawliszyn, 1997). The cloud-point temperature is a narrow temperature range over which a solution of a nonionic surfactant becomes turbid after heated. Above this temperature, phase separation occurs. This is a reversible phenomenon, and upon cooling, a single phase can be again obtained (Sosa Ferrera et al., 2004).

5.2.3.13 Soxhlet Extraction

Reference to Soxhlet extraction has to be made, as it is still used in the analysis of solid samples. It is mainly followed as a reference method for comparison purposes and due to the wide application by official standard methods (Wang et al., 2010). Even though efficient extractions may be achieved by this technique, the major drawbacks include long extraction time (up to 48 h), high solvent consumption, degradation of thermally labile compounds, and poor manual reproducibility during the sample preparation. Moreover, Soxhlet extraction is not selective enough to meet the analytical requirements needed for accurate multiresidue trace analysis, requiring further extensive cleanup and fractionation protocols.

5.3 SAMPLE PREPARATION METHODS FOR THE DETERMINATION OF ORGANIC POLLUTANTS IN SOIL

5.3.1 Introduction

Compared to sediments, soil has been a less-studied compartment regarding the levels of persistent, bioaccumulative, and toxic (PBT) compounds. Few monitoring studies have been published, despite the fact that several analytical methodologies have been developed (Rodil and Popp, 2006; Yu and Hu, 2007; Martínez-Parreño et al., 2008; Llorca-Pórcel et al., 2009; Martendal and Carasek, 2011; Raich-Montiu et al., 2011; Chitescu et al., 2012). Nevertheless, the European Commission stresses the need to protect the soil environment, inter alia, from chemical pollution, which has been identified as one of the threats that may limit the soil capacity to perform an array of vital functions for sustainable exploitation (Commission of the European Communities, 2006). Indeed, soil contamination has negative impacts on water and air quality, biodiversity, climate change, natural ecosystems, and the economy. It can also impair the health of animals and humans and threaten food and feed safety (Commission of the European Communities, 2006). Analytical science is expected to give an important contribution to soil characterization and protection and raise public awareness.

5.3.2 Extraction and Cleanup Protocols

In this section, the most recent sample preparation methods for the trace analysis of selected persistent and ECs in soils and their challenges and strengths will be discussed. A trend in recent years is the use of a combination of different extraction techniques, for instance, coupling solvent extraction and sorbent-trapping extraction, such as the use of SBSE or SPME after UAE, MAE, or PLE soil extraction. Table 5.1 comprises detailed information on representative analytical protocols and respective figures of merit.

Martínez-Parreño et al. (2008) proposed a method for determining a set of 51 POPs (organochlorine pesticides [OCPs], PCBs, PAHs, and PBDEs) in soil samples based on UAE, subsequent SBSE, and analysis by GC-MS. The use of a single internal standard for the 51 analytes proved insufficient, giving rise to linear regression coefficients (r^2) of 0.98 in the majority of cases. The authors emphasized that the 14 h extraction time was insufficient to reach the SBSE equilibrium

TABLE 5.1
Overview of Analytical Protocols Applied to the Determination of Selected Persistent and Emerging Pollutants in Soils

Matrix	Extraction Technique	Extraction and Cleanup Protocols	Compounds Determined	Analytical Technique	Recoveries	Precision (% RSD)	Analytical Thresholds	Ref.
Soil	UAE-SBSE	1 g of soil extracted with 15 mL MeOH for 30 min. Extract dissolved in 85 mL Milli-Q water. SBSE at 900 rpm and room temperature for 14 h. Stir bars thermally desorbed in TDS-2 connected to PTV system injector CIS-4.	PCBs, PAHs, PBDEs, OCPs	TD-GC-MS (SIM)	59–87%	10–18% r 10–20% R	LODs: 0.01–2.0 μg kg^{-1}	Martínez-Parreño et al. (2008)
Soil and sediment	UAE-SBSE	1 g of soil extracted with 15 mL MeOH by sonication for 30 min (40 kHz, 550 W). Extract mixed with 85 mL Milli-Q water and extracted by SBSE (900 rpm at room temperature for 6 h) with in situ derivatization (K_2CO_3 and acetic anhydride).	Chlorophenols, bisphenol A, 4-*tert*-octylphenol, 4-nonylphenol	TD-GC-MS (SIM)	73–88% soil 70–89% sediment	11–15% r 13–19% R	LODs: 0.2–1.7 μg kg^{-1} dw	Llorca-Pórcel et al. (2009)
Soils	UAE	5 g of soil extracted with 5 mL of ethyl acetate, three times 15 min. Direct analysis of the extracted concentrated down to 200 μL.	OCPs, OPPs, triazines, pyrethroids, other pesticides	GC-MS (SIM)	69–118%	3.9–16.8% r 7.5–20.1% R	LODs: 0.05–7.0 μg kg^{-1}	Gonçalves and Alpendurada (2005)
Soils and dust	UAE-SBSE	2 mL acetone, shaking with sample for 2 min and extracted for 20 min in an ultrasonic bath. Followed by 2 mL water containing 10% NaCl. Both transferred to 10 mL flask. 5 mL submitted to SBSE at 600 rpm for 10 min. Elution with 50 μL ethanol for 20 min in ultrasonic bath.	BFRs (TBBPA, PBDEs, and HBCD)	HPLC-UV	71–115% soil 56–118% dust	3.1–10.5% r 3.5–16.6% R	Not determined in soil	Yu and Hu (2007)

(*Continued*)

TABLE 5.1 (CONTINUED)
Overview of Analytical Protocols Applied to the Determination of Selected Persistent and Emerging Pollutants in Soils

Matrix	Extraction Technique	Extraction and Cleanup Protocols	Compounds Determined	Analytical Technique	Recoveries	Precision (% RSD)	Analytical Thresholds	Ref.
Soils	PSWE-SBSE	10 g soil samples extracted with water/ACN (75:25 v/v) at 130°C and 100 bar for three cycles of 10 min. 10 mL of extract submitted to SBSE for 180 min at a stirring speed of about 1000 rpm.	OCPs, chlorobenzenes	TD-GC-MS (SIM)	4.1–85.2% absolute 60–130% relative	3.2–13.0% except heptachlor 28.7% r <20% R	LODs: 0.002–4.7 µg kg^{-1}	Rodil and Popp (2006)
Soil and sludge	PLE	Aprox. 4.4 g soil or sludge in 5 mL ASE cell. Extraction with H_3PO_4 50 mM/ACN 1:1 at 65°C and 1500 psi during 10 min.	Fluoroquinolones, tetracyclines, bacteriostatics	LC-MS/MS	45–117% Oxitetracycline not suitable	10.6–25.1% r 11.6–33.4% IP	LODs: 0.9–17.6 µg kg^{-1} LOQs: 3–59 µg kg^{-1}	Pinto (2013)
Soil	PLE	30 g soil packed in 33 mL cell, extraction with acetone at 60°C for 5 min. Cleanup with C18 cartridges.	Natural estrogens	GC-MS (SIM) after derivatization with MSTFA	89–103%	<13%	LODs: 50 ng kg^{-1}	Beck et al. (2008)
Soil and plants	PLE	5 g of dried soil or 3 g of chopped plant. Extraction temperature 50°C; pressure 1500 psi; 2 cycles of 5 min each, static extraction; 50% flush volume; extraction solvent: acetone/citric acid 0.2 M (50:50) (pH 4.5) for soil and MeOH/citric acid 0.2 M (50:50) (pH 4.5) for plant material. 10 mL soil extract and 16.6 mL plant extract diluted to just 10% organic solvent content. Cleanup on Strata-X, 200 mg/6 mL.	Benzimidazole, tranquilizers, macrolides, sulfonamides, quinolones, penicillins, tetracyclines, NSAIDs, antiepileptics, lipid regulators, azole antifungals, polyene antifungals, mitotic inhibitors; azole biocides and fungicides	UHPLC-MS (Orbitrap HRMS)	70–100% for 68% of compounds, 40–70% for 21% of compounds, and <40% for 11% of compounds	Not determined	93% analytes detected at 10 µg kg^{-1} level and 7% at 50 µg kg^{-1}	Chitescu et al. (2012)

(Continued)

TABLE 5.1 (CONTINUED)
Overview of Analytical Protocols Applied to the Determination of Selected Persistent and Emerging Pollutants in Soils

Matrix	Extraction Technique	Extraction and Cleanup Protocols	Compounds Determined	Analytical Technique	Recoveries	Precision (% RSD)	Analytical Thresholds	Ref.
Soil and plants	UAE	Same aliquot as above. Extraction with 20 mL MeOH/citric acid 0.2 M (50:50) (pH 4.5) in ultrasound for 1 h, shaking for another 1 h. Same cleanup as above.	Same analytes as above	Same technique as above	10–50%	Not given	Not given	Chitescu et al. (2012)
Soil	MAE	1 g soil extracted with 5 mL ACN/water (90:10). Submitted to microwave irradiation (15 min, 120°C).	Avermectines (abamectine, doramectine, and ivermectine)	UPLC-MS/MS (APCI+)	>78.3% or >31.2%, depending on soil pH, for short aging period	1.5–7.4% r Similar IP	LODs: 18–25 μg kg^{-1}	Raich-Montiu et al. (2011)
Soil	MAE	0.5–1 g of soil. MAE cycle of 20 min, at 80°C with 8 mL 20% w/v $Mg(NO_3)_2$ and 2% v/v NH_3. Acidification with H_3PO_4 (1:5) before HPLC analysis.	Fluoroquinolones (ciprofloxacin, danofloxacin, enrofloxacin, flerofloxacin, levofloxacin, marbofloxacin, norfloxacin, and orbifloxacin)	HPLC-FLD	69–110%	0.9–2.4 % r 1.3–16.5% IP	LODs: 0.7–4.3 μg kg^{-1} LOQs: 2.0–13.0 μg kg^{-1}	Sturini et al. (2010)
Soil	MAE	Irradiation with 320 W of microwave power; 12 mL of ACN infused at 0.8 mL min^{-1} flow rate. Analytes trapped in alumina column and eluted with 2.0 mL of 0.3% acetic acid.	Sulfonamides (sulfadiazine, sulfameter, sulfamonomethoxine, and sulfaquinoxaline)	LC-MS/MS	82.6 ± 6.0% to 93.7 ± 5.5%	2.7–5.3% r 5.6–6.7% IP	LODs: 1.4–4.8 μg kg^{-1} LOQs: 4.6–16.0 μg kg^{-1}	Chen et al. (2009)

(Continued)

TABLE 5.1 (CONTINUED)
Overview of Analytical Protocols Applied to the Determination of Selected Persistent and Emerging Pollutants in Soils

Matrix	Extraction Technique	Extraction and Cleanup Protocols	Compounds Determined	Analytical Technique	Recoveries	Precision (% RSD)	Analytical Thresholds	Ref.
Soils, sediments and sludge	MAE	1 g samples extracted with 10 mL MeOH/water 3:2 under the action of microwave energy at 500 W for 6 min. Analytes trapped on Oasis HLB sorbent eluted with ethyl acetate, silylated at 70°C for 20 min, and determined by GC-MS.	18 pharmaceuticals (analgesics, antibacterials, antiepileptics, β-blockers, lipid regulators, and nonsteroidal anti-inflammatories), 1 PCP and 3 hormones	GC-MS (SIM) after silylation with BSTFA + 1% TMCS	91–101%	4.2–6.0% r 4.8–6.9% IP	LODs: 0.8–5.1 ng kg^{-1}	Azzouz and Ballesteros (2012)
Soil	MAE-SBME	1.0 g of soil extracted under MAE with 10 mL of water at 60°C for 20 min. Toluene was used in SBME.	PAHs	GC-MS (SIM)	80.9–106.8%	3.8–9.8% r	LODs: 0.03–0.25 μg kg^{-1} LOQs: 0.10–0.97 μg kg^{-1}	Guo and Lee (2013)
Soil	UAE-SBME (PVDF)	1.0 g of soil extracted with 10 mL of ultrapure water at 100 W for 30 min at 30–35°C. SBME with about 4 μL of 1-octanol for 20 min of extraction with stirring (400 rpm).	Chlorobenzenes (1,2,3-trichlorobenzene, 1,2,3,4-tetrachlorobenzene, hexachlorobenzene, and 1-chloro-4-nitrobenzene)	GC-MS (SIS)	93–105%	3.6–6.8% r	LODs: 0.7–27.3 μg kg^{-1} LOQs: 2.2–90.9 μg kg^{-1}	Wang et al. (2012)
Soil	MAE	1 g aliquot of soil submitted to 400 W irradiation (100% output) with 12 mL of acetone/hexane (2:1 v/v), at 160°C for 10 min.	OPPs and fungicides	GC-MS (SIM)	92.6–103.7%	2.5–12% r 0.3–8.5% IP	LODs: 0.10–0.12 μg kg^{-1} LOQs: 0.26–0.37 μg kg^{-1}	Merdassa et al. (2013)

(Continued)

TABLE 5.1 (CONTINUED)
Overview of Analytical Protocols Applied to the Determination of Selected Persistent and Emerging Pollutants in Soils

Matrix	Extraction Technique	Extraction and Cleanup Protocols	Compounds Determined	Analytical Technique	Recoveries	Precision (% RSD)	Analytical Thresholds	Ref.
Soil-mined areas and test fields	MAE	Samples extracted with 25 mL sodium phosphate buffer (0.1 M, pH 8). MAE was ramped to 80°C for 5 min (max. 500 W). The temperature was then held for 5 min (max. 900 W) followed by 10 min (max. 500 W).	TNT and related compounds	GC-NPD	76–104% TNT	3–18% r	LOD: 7 ng kg^{-1} TNT	Kjellström et al. (2008)
Soil	SAEAC	5 g of soil extracted in glass column with 5 mL of *n*-hexane for 15 min in an ultrasonic water bath (150 W and a frequency of 35 kHz) at room temperature. Extraction repeated in 2 cycles.	Cyclic and linear siloxanes	GC-MS (SIM)	87.7–108% and 84.9–107.6% for cyclic and linear siloxanes, respectively	4.5–8.9% r 7.6–10.4% IP	LODs: 0.4–1.1 µg kg^{-1} LOQs: 1.5–4.0 µg kg^{-1}	Sánchez-Brunete et al. (2010)
Soil	UAE-SDME	2.0 g portion of soil placed in a screw cap 5 mL glass vial. 3.0 µL of octanol as extraction solvent; sample dilution 1:1 with water (20% sodium chloride), 1:1 w/v, extraction at 60°C for 30 min; application of ultrasounds, without any pH adjustment.	OPPs (thionazin, sulfotep, dimethoate, disulfoton, and parathion)	GC-NPD	Enrichment factor: 1.4–12.7	2.1–6.9% R	LODs: 0.1–2.0 µg kg^{-1} LOQs: 0.3-6.6 µg kg^{-1}	Salemi et al. (2013)

(*Continued*)

TABLE 5.1 (CONTINUED)
Overview of Analytical Protocols Applied to the Determination of Selected Persistent and Emerging Pollutants in Soils

Matrix	Extraction Technique	Extraction and Cleanup Protocols	Compounds Determined	Analytical Technique	Recoveries	Precision (% RSD)	Analytical Thresholds	Ref.
Soil	UAE-SPE	Extraction of 1.0 g of soil with 30 mL dichloromethane, placed in ultrasonic bath for 30 min at room temperature. Extract dried under nitrogen at 30°C, and dissolved in 20 mL deionized water + 400 μL MeOHas modifier. A titania (TiO_2) nanotube/titanium (Ti) plate modified with *n*-octadecanethiol (C18) monolayer-protected Ag nanoparticles was used as the adsorbent in SPE of PAHs for 90 min.	PAHs	HPLC-FLD	70.32–115.51%	3.14–13.56% r 4.92–14.87% IP	LODs: 0.0015–0.4 μg kg^{-1}	Pan et al. (2013)
Soil	UAE-SPE	Extraction with 20 mL of acetone/water (70:30 v/v) in ultrasonic bath for 4 min. 1 mL extract passed through the C18 SPE cartridge. Derivatization with 4-chloro-3,5-dinitrobenzotrifluoride at 60°C for 10 min at pH 9.0 in H_3BO_3–$Na_2B_4O_7$ medium.	Tobramycin	HPLC-UV	78.0–91.0%	3.38–9.79 % r	LOD: 0.02 μg g^{-1}	He et al. (2011)

(*Continued*)

TABLE 5.1 (CONTINUED)
Overview of Analytical Protocols Applied to the Determination of Selected Persistent and Emerging Pollutants in Soils

Matrix	Extraction Technique	Extraction and Cleanup Protocols	Compounds Determined	Analytical Technique	Recoveries	Precision (% RSD)	Analytical Thresholds	Ref.
Soil and water samples	MISPE	Dinelli's method modified. 20-mL of sodium hydrogencarbonate solution (0.1 M, pH 7.8) was added to 10.0 g of spiked soil sample. The suspension was shaken for 1 h. The slurry was then centrifuged at 5000 rpm for 10 min. After the addition of 3 g of EDTA, the extract was passed through the MISPE cartridge. Cartridge washed with 2.0 mL CH_2Cl_2/ACN solution (93:7 v/v). Analytes eluted with 2.0 mL CH_2Cl_2/MeOH solution (90:10 v/v).	SUs: nicosulfuron, thifensulfuron-methyl, metsulfuron-methyl, sulfometuron-methyl, chlorsulfuron	LC-UV-DAD	71–139% in soil 96–107% in rainwater	4.3–6.5% r 1.8–9.4% R	LODs: 5–12 μg kg^{-1} in soil 2–14 ng L^{-1} in water	Zhu et al. (2002)
Soil	SFE	CO_2 with 15% w/w MeOH as modifier at 15 MPa, 318 K and 4 static stages of 10 min.	Organochlorine, organophosphorus, pyrethroid, and diverse pesticides	GC-μECD and NPD	<70%	Not determined	LOQs: 0.007–17.71 mg kg^{-1}	Forero-Mendiet et al. (2012)
Soil	SFE	Extraction of 2.3 g soil with 15 mL of CO_2 + 8% MeOH at 42 MPa and 120°C.	Organochlorine, organophosphorus, triazine, acetanilide, and other pesticides	GC-MS/MS	80.4–106.5%	0.7–10.8% r 4.2–15.7% IP	LODs: 0.1–3.7 μg kg^{-1} LOQs: 0.3–12 μg kg^{-1}	Gonçalves et al. (2006)
Soil and marine sediments	SFE	Extraction with supercritical CO_2 at 150 bar, 60°C, 10 min static and 30 min dynamic steps. 1.0 mL of ACN + 17 μL CCl_4 injected in 5 mL H_2O.	OPPs	DLLME and GC-FID	44.4–95.4%	3.1–7.5% r	LODs: 0.001–0.009 mg kg^{-1}	Naeeni et al. (2011)

(Continued)

TABLE 5.1 (CONTINUED)
Overview of Analytical Protocols Applied to the Determination of Selected Persistent and Emerging Pollutants in Soils

Matrix	Extraction Technique	Extraction and Cleanup Protocols	Compounds Determined	Analytical Technique	Recoveries	Precision (% RSD)	Analytical Thresholds	Ref.
Soil	SFE	5 min. static extraction followed by 40 min dynamic extraction at 182.4 bar and 338 K, CO_2 flow rate of 0.5 mL min^{-1}. 25 μL of MeOH added as modifier. 4.5 mL tetrahydrofuran as the collecting solvent. 30 mg of decanoic acid was added to the SFE collecting solvent, and the mixture was rapidly injected into 40 mL water.	SU herbicides (metsulfuron-methyl, bensulfuron-methyl, and chlorsulfuron)	SUPRAS-ME-HPLC-UV	28.3–52.3%	6.4–7.1% r 9.5–10.2% IP	LODs: 0.05–0.11 mg kg^{-1}	Asiabi et al. (2013)
Soil and sediments	QuEChERS	5 g soil extracted with 3 mL of water (pH 2.5) and 7 mL of ACN with 1% acetic acid. QuEChERS: 6 g of $MgSO_4$, 1.5 g of NaCl, 1.5 g of sodium citrate dehydrate, and 0.750 g of sodium citrate sesquihydrate	Ibuprofen, hydroxyibuprofen, and carboxyibuprofen	HPLC-FLD	79.5–101%	<3%	LODs <22.4 μg kg^{-1}	Bragança et al. (2012)

(Continued)

TABLE 5.1 (CONTINUED)
Overview of Analytical Protocols Applied to the Determination of Selected Persistent and Emerging Pollutants in Soils

Matrix	Extraction Technique	Extraction and Cleanup Protocols	Compounds Determined	Analytical Technique	Recoveries	Precision (% RSD)	Analytical Thresholds	Ref.
Soil	QuEChERS (and PLE)	2 g sample extracted with 15 mL of dichloromethane. Afterward, 4 g $MgSO_4$, 1 g NaCl, 1 g sodium citrate dehydrate, and 0.5 g disodium hydrogen citrate sesquihydrate are added, shaken, and centrifuged. (PLE: 2 g sample mixed with 4 g anhydrous Na_2SO_4, placed in a 34 mL stainless steel cell. Extraction carried out with dichloromethane at 40°C and 10 MPa; preheat time of 5 min, 10 min static extraction, and 100% flush volume.)	Chlcrobenzenes, chlorophenols, chlorinated hycrocarbons, and chlorinated olefins	GC-MS (SIM)	60–100% (ASE: 42–85%)	1.5–8.1% r 2.6–37.2% IP	LODs: 2–50 μg kg^{-1} LOQs: 7–170 μg kg^{-1} except chlorophenols (ASE: 58–540 μg kg^{-1} LODs, 194–1800 μg kg^{-1} LOQ)	Rouvière et al. (2012)
Soil	QuEChERS	2.5 g soil sample added with 1.5 mL ultrapure water and shaken. Ethyl acetate added, and the mixture shaken for 1 min. 1 g of magnesium sulfate added and shaken for 1 min.	Chloroform, 1,2-dichlorobenzene, and hexachlorobenzene	GC-ECD	62–93%	3.5–7.6% r	LODs: 0.15–2.2 μg kg^{-1}	Pinto et al. (2010)

(*Continued*)

TABLE 5.1 (CONTINUED)
Overview of Analytical Protocols Applied to the Determination of Selected Persistent and Emerging Pollutants in Soils

Matrix	Extraction Technique	Extraction and Cleanup Protocols	Compounds Determined	Analytical Technique	Recoveries	Precision (% RSD)	Analytical Thresholds	Ref.
Soil	SPME cold fiber	500 mg soil sample added with 12 mL of 194 g L^{-1} aqueous NaCl solution in a 15 mL vial. 80 min total extraction time with a sample temperature of 90°C (50 min in direct SPME mode with the coating at 90°C followed by 30 min in headspace mode with the coating at 30°C.)	PAHs and phthalate esters	GC-FID	21.7–113.8%	4–19% r	LODs: 0.2–2.3 µg g^{-1}	Martendal and Carasek (2011)
Soil	SPME cold fiber	100 mg sample + 5 µL diethylamine. Sample heated at 200°C. Extraction with PDMS fiber coating at 50°C for the first 35 min and 30°C for the last 10 min.	PAHs	GC-FID	–	–	–	Guo et al. (2013)

Note: APCI+, atmospheric pressure chemical ionization; BFRs, brominated flame retardants; CIS-4, cooled injection system; DAD, diode array detector; DLLME, dispersive liquid–liquid microextraction; dw, dry weight; ECD, electron capture detector; FID, flame ionization detector; FLD, fluorescence detection; HBCD, hexabromocyclododecane; IP, intermediate precision; LOD, limit of detection; LOQ, limit of quantification; MISPE, molecularly imprinted solid-phase extraction; NPD, nitrogen–phosphorus detector; NSAIDs, nonsteroidal anti-inflammatory drugs; OCPs, organochlorine pesticides, OPPs, organophosphorus pesticides; psi, pounds per square inch; PSWE, pressurized subcritical water extraction; PTV, programmed temperature vaporization; PVDF, polyvinylidene fluoride; r, repeatability; R, reproducibility; RSD, relative standard deviation; SDME, single-drop microextraction; SUPRAS-ME, supramolecular solvent microextraction; TBBPA, tetrabromobisphenol A; TDS-2, thermal desorption system; TNT, trinitrotoluene; µECD, microelectron capture detector.

for apolar compounds and longer extraction would be needed to reach 100% recoveries, as performed by Camino-Sánchez et al. (2011). This apparent disadvantage of SBSE, which derives from the fact that around 50- to 200-fold bigger volumes of PDMS are used compared to those used in SPME, is in part tackled by performing simultaneous UAE and SBSE to a maximum batch of 20 samples (Rodil and Popp, 2006; Martínez-Parreño et al., 2008). The reader should also be aware that the extract aqueous solution submitted to SBSE had 15% MeOH, and this can limit the recoveries of the more polar compounds (log K_{ow} <3.17)—for example, pesticides. The same authors propose a similar method for the determination of chloro- and alkylphenols (APs) in soils (Llorca-Pórcel et al., 2009). Detection after derivatization with acetic anhydride was greatly improved, resulting in limits of detection (LODs) roughly 10-fold lower (at the sub-microgram per kilogram level). The selectivity, evaluated by the cleanliness of the chromatograms and the absence of matrix effects, is also improved owing to the inherent properties of SBSE and the derivatization reaction (Llorca-Pórcel et al., 2009).

In contrast to the majority of works, which propose SBSE methodologies based on commercial PDMS stir bars, Yu and Hu (2007) developed three new adsorbents (polydimethysiloxane-β-cyclodextrin [PDMS-β-CD], carbowax [CW]-PDMS-poly(vinyl alcohol) [PVA], and PDMS-PVA) by using sol–gel chemistry for the determination of brominated flame retardants (BFRs) in soils and dust. BFRs are persistent and lipophilic compounds with a tendency to bioaccumulate and biomagnify through the food chain; therefore, they are among the most investigated pollutants in soil, sediments, and biota (Salgado-Petinal et al., 2006a; Montes et al., 2010). Besides the fact that PDMS-β-CD-coated stir bars gave the best response for the target analytes, especially for brominated diphenyl ether-99 (BDE-99), BDE-100, and BDE-153, the most remarkable feature of this new coating was the possibility to attain the equilibrium in just 10 min. The reproducibility of the stir bars preparation process was <16% relative standard deviation (RSD), both within batch and batch to batch (Yu and Hu, 2007). Although the authors did not estimate the detection limits of the method in solid matrix, presumably they are quite high (lowest quantified amount in soil was 0.12 mg kg^{-1}) given the lack of sensitivity of UV detectors and the LD variant applied to the SBSE bar. Nevertheless, the concept of coating PDMS-β-CD on a single side provided durable stir bars able to withstand the frictional forces at a high stirring speed and that could be reused for at least 100 times (Yu and Hu, 2007).

Rodil and Popp (2006) proposed a method based on SWE coupled with SBSE-GC-MS, with TD, for the determination of OCPs (hexachlorocyclohexanes, cyclodienes, and DDX [dichlorodiphenyldichloroethylene, dichlorodiphenyldichloroethane, dichlorodiphenyltrichloroethane]) and chlorobenzenes in soil samples. The extract obtained under SWE is perfectly suited for SBSE, which is an efficient procedure for the enrichment of semivolatile organic compounds from aqueous samples (Rodil and Popp, 2006). Nevertheless, the determined optimum extraction time to approach the equilibrium was 180 min. Although the addition of 25% ACN as SWE modifier proved to be mandatory for efficient extraction of OCPs and chlorobenzenes from soil, the authors also observed that the presence of ACN in the solution affected negatively the sorption of all analytes in the SBSE polymer, namely the most polar ones: hexachlorocyclohexanes, endrin aldehydes, and endosulfan sulfate (Rodil and Popp, 2006). Despite the fact that the authors highlighted the selectivity of SWE as a major advantage in comparison with UAE and PLE, they did observe a significant matrix effect on the SBSE process. The use of isotope-labeled internal standards spiked in the extraction cell was then a required tool to compensate for matrix effects, benefiting accuracy and precision, with final regression correlation coefficients between 0.9993 and 0.9999 (Rodil and Popp, 2006).

Although successful, only a limited number of references can be found reporting the analysis of organic pollutants in soil samples using SPME: pesticides (Peruga et al., 2014); PAHs (Martendal and Carasek, 2011; Rocha et al., 2011), PCBs (Wu et al., 2014); benzene, toluene, ethylbenzene, and xylenes (Niño et al., 2014), chlorobenzenes (Ghasemi and Sillanpää, 2014), and phthalate esters (Martendal and Carasek, 2011). This fact can be attributed to some limitations in terms of fiber stability, analyte volatility, and matrix effects (Martendal and Carasek, 2011; Đurović and Đorđević, 2012; Higashikawa et al,, 2013). As it is general knowledge, SPME is an equilibrium process and recoveries are often low due to the insufficient volatility of the analytes or difficult release from the sample matrix. An option is to increase the temperature of the sample; however, this also decreases the distribution coefficients of the analytes between the fiber coating and the sample matrix. To overcome this, the cold fiber SPME concept was developed: the sample matrix is heated to a high temperature, accelerating the desorption of the analytes from matrix components while the fiber coating is simultaneously kept cool (Guo et al., 2013). Martendal and Carasek (2011) used this strategy in combination with different SPME sampling modes in the analysis of PAHs and phthalate esters in soil. The SPME fiber was immersed in the soil slurry for 50 min with sample heating at 90°C followed by 30 min in headspace mode, cooling the coating at 30°C. This approach allowed an efficient extraction of both volatiles and semivolatiles; however, the recovery of heavier PAHs remained difficult, probably due to a stronger binding to the soil matrix (Martendal and Carasek, 2011). Despite this, the overall detection limits of the method are not impressive due to the use of low sensitivity detection (flame ionization detector [FID]) and the tiny sample-to-aqueous phase ratio (500 mg to 12 mL). Based on the outcomes of this work, Guo et al. (2013) introduced the programmed coating temperature SPME system for the analysis of PAHs in the headspace mode. The addition of diethylamine (5 μL) as modifier to the soil was revealed crucial for weakening the affinity of heavier PAHs to the matrix, improving their extraction. Quantitative recoveries were obtained for naphthalene, acenaphthylene, acenaphthene, fluorene, and phenanthrene while recoveries of fluoranthene, pyrene, benzo[*b*]fluoranthene, and benzo[*a*]pyrene ranged from 60% to 75% due to stronger interaction with the matrix (Guo et al., 2013).

Besides being the technique of choice for POP analysis, PLE is currently also the preferred technique for the determination of emerging compounds in soil. Emerging pollutants, namely, antibiotics, can reach the environment through several pathways. Massive quantities are used worldwide for human and animal therapy, after which they are excreted to wastewater or simply disposed once the shelf life expires, without environment precautions (Azzouz and Ballesteros, 2012; Bragança et al., 2012). Animal and fish farming rely heavily on drugs mostly administered as food additives to prevent diseases, promote growth, or fight parasites (Azzouz and Ballesteros, 2012). Besides being found in wastewater and surface water, which are the most impacted compartments, ECs can also find their way into the terrestrial compartment. They enter the soil after amendment of agricultural fields with animal manure and organic fertilizers and adsorption in sewage sludge (Beck et al., 2008; Raich-Montiu et al., 2011; Bragança et al., 2012; Chitescu et al., 2012). Some pharmaceuticals have physicochemical properties that render them persistent and bioaccumulable pollutants in the environment (Raich-Montiu et al., 2011; Azzouz and Ballesteros, 2012), while antibiotics, in particular, can alter the natural microbial communities in soils, impairing their recycling functions and selecting resistant strains of microorganisms that may become harmful to human health and/or may inhibit plant growth (Chen et al., 2009; Chitescu et al., 2012). Moreover, pharmaceuticals can leach from soil and reach groundwater and surface water. Nevertheless, there is a great lack of knowledge on the levels, fate, and effects of natural estrogens and antibiotics in soils, although it is highly likely that these substances are widespread.

The work by Azzouz and Ballesteros (2012) is a good source for a wide array of references on extraction and cleanup strategies suitable for the determination of pharmaceuticals in soils, sediments, and sludge. It can be used in complement with this chapter. A study was conducted on the determination of the levels of fluoroquinolones (enrofloxacin, ciprofloxacin, sarafloxacin, difloxacin, norfloxacin, and ofloxacin), oxytetracycline, trimethoprim, and sulfamethoxazole in soils and sewage sludge by PLE and LC-MS/MS (Pinto, 2013). The best extraction solvent was demonstrated to be a mixture of 50 mM H_3PO_4 and ACN 1:1, where the aqueous phase contributes to wet the matrix and release the analytes from the binding sites through the action of the acidic pH. Moreover, these analytes are quite hydrophilic. The recoveries of fluoroquinolones were in the range of 45–77%, and those of bacteriostatics were between 105% and 117%. The method is not suitable for the analysis of oxytetracycline (apparent recovery 8%), in part due to a very intense ionization suppression and the need of a quelating agent (Pinto, 2013). An appreciable matrix effect was observed in the analysis of dehydrated sewage sludge (relative recoveries 30–42%) when compared with soil (recoveries 82–97%). The most intense matrix effect was observed for the first eluting analytes, with considerable ionization suppression of trimethoprim, ofloxacin, and ciprofloxacin and signal enhancement of norfloxacin up to 130% (sludge) and 145% (soil) (Pinto, 2013). Both the analyzed sewage sludge samples contained levels of ciprofloxacin and norfloxacin between 137 and 231 μg kg^{-1}, while one soil contained all the investigated fluoroquinolones at levels between 28 and 226 μg kg^{-1}. To minimize the matrix effects, namely, ionization suppression in the electrospray ion source, a cleanup or dilution of the extract is mandatory, allowing to detect all compounds of interest at the microgram per kilogram level (Pinto, 2013).

Beck et al. (2008) developed a method for the determination of 17α-estradiol (17α-E2) and 17β-E2, estrone (E1), and estriol (E3) by also employing PLE, followed by SPE cleanup with C18 cartridges and final determination by GC-MS after derivatization with MSTFA. Acetone or ethyl acetate gave equally good extraction efficiency, whereas ACN and MeOH gave lower recoveries, particularly of E1, which is very much affected by polarity changes. Cleanup and further concentration with C18 adsorbent allowed combining several extracts to a sample equivalent of 500 g and therefore achieved LODs at the nanogram per kilogram level, corresponding to the levels typically found in cropland and grazed pasture soils (Beck et al., 2008). Chitescu et al. (2012) also preferred PLE with acetone as the most suitable extraction technique for the determination of 42 analytes (including pharmaceuticals, azole biocides, and fungicides) in soils and plants tissue compared to UAE. The LC-MS analysis was performed in a ultrahigh pressure liquid chromatography (UHPLC)-Exactive Orbitrap MS (at 50,000 FWHM), which presented additional capabilities for screening of nontarget substances and retrospective target and untargeted analysis, besides a significant sensitivity and selectivity improvement (Chitescu et al., 2012). The low recoveries obtained in this work were in agreement with those reported in other works (Pinto, 2013), demonstrating the difficulty of analyzing polar pharmaceuticals in both the extraction and electrospray ionization (ESI) steps (Chitescu et al., 2012). The authors concluded that PLE afforded better recoveries than UAE and allowed detecting all compounds of interest, most of them at the 10 μg kg^{-1} level (Chitescu et al., 2012). Furthermore, PLE was faster and fully automated, offering better and repeatable results.

MAE has also been applied successfully in the determination of several classes of pharmaceuticals: avermectines (Raich-Montiu et al., 2011), sulfonamides (Chen et al., 2009), and fluoroquinolones (Sturini et al., 2010) in soils. The analysis of avermectines revealed particularly demanding, and MAE under optimum conditions could give only an estimate of the extractable/bioavailable fraction (Raich-Montiu et al., 2011). Recoveries were soil dependent and heavily dependent on the aging period. Aging of these compounds in soil for 1 month decreased recoveries from around 90% to about 40–50%, whereas at 2.5 months, they could be as low as just 14–18% (Raich-Montiu et al., 2011). Soil pH seemed to be determinant in the sorption process, being stronger in acid soils, while organic matter might be indirectly implicated, as it shows different functional groups at different pH values. No ionization suppression was observed for the avermectines in LC atmospheric pressure chemical ionization mode after a cleanup step by SPE on Oasis HLB cartridges. The analysis of avermectins is highly relevant, as they persist in manure for long-term periods and manure is spread on fields as fertilizer. Furthermore, avermectins can be toxic to nontarget organisms such as the springtails (Raich-Montiu et al., 2011).

Sturini et al. (2010) developed a true "organic solvent-free" MAE-based method for the determination of human-use and veterinary fluoroquinolones in soil that uses a 20% w/v $Mg(NO_3)_2$ aqueous solution as leaching/chelating agent, binding the deprotonated carboxylic acid group of the analytes. Recoveries were very much influenced by Mg(II) percentage and, in a lesser extent, by irradiation time, while temperature did not influence the extraction process (Sturini et al., 2010). Further comparison between MAE and UAE demonstrated that a single MAE gave a greater yield when using a lower Mg(II) concentration than UAE in three sequential steps. Therefore, the procedure can be classified as simple, fast, highly selective, and sensitive. A further refinement of the MAE technique is proposed by Chen et al. (2009) for the determination of sulfonamides in soils, called dynamic microwave-assisted extraction, coupled online with SPE. The analytical setup consisted of a continuous extraction system infusing ACN through the sample cell under microwave irradiation. Neutral alumina was used as SPE sorbent for matrix cleanup, which allowed online introduction of the organic extract into SPE. The volume and flow of the extraction solvent showed a more significant effect on extraction yields compared to microwave power (Chen et al., 2009). As in the case of the avermectines, the recoveries of sulfonamides in soil decreased with aging—from 86.3–101.9% in the day after spiking to 37.6–47.5% 4 weeks after. Nevertheless, the authors were not able to clarify if this fact was due to a rapid formation of nonextractable residues or transformation of the sulfonamides with increasing standing time (Chen et al., 2009).

Azzouz and Ballesteros (2012) proposed a MAE-based method for the determination of pharmaceuticals, PCPs, and hormones in soils, sediments, and sludge in a continuous setup, although the concept is not a synonym of online coupling of all system components. MAE, SPE, and GC-MS steps are offline combined, and just the SPE processing is automated, where an Oasis HLB adsorbent is inserted in a loop configuration for repeated use. The authors selected a medium irradiation power (500 W) for 6 min to avoid degradation of some analytes under extreme conditions. The diluted extract with water was cleaned up and concentrated by SPE with 60 mg of Amberlite XAD-2 or XAD-4, LiChrolut EN, reversed-phase C18, or Oasis HLB, the last one providing the best results. Owing to several concentration steps by solvent evaporation, the final method affords LODs roughly 1000 times lower than those of similar methods (ng kg^{-1} level). Consequently, all agricultural soils, pond and river sediments, and sewage sludges displayed levels of pharmaceuticals and personal care products (PPCPs) and hormones that would remain undetectable using alternative methods (Azzouz and Ballesteros, 2012).

The efforts toward simplification, automation, and online operation of MAE proceeded with the works of Guo and Lee (2013), which coupled MAE with solvent bar microextraction (SBME). In SBME the extraction solvent is confined within a short hollow fiber membrane (e.g., Accurel Q 3/2 polypropylene) to form a solvent bar, which is placed in the sample solution for extraction (Guo and Lee, 2013). The solvent bar can tumble and move freely, and in the present case, it was added to MAE vessels during extraction. Therefore, the PAHs extracted from solid samples into the water were simultaneously cleaned up, extracted, and concentrated by SBME. The overall method is fast, one step, convenient, and environmentally benign (less than 10 μL of organic solvent was used), and the extract is directly amenable to GC analysis (Guo and Lee, 2013). A similar procedure was proposed by Wang et al. (2012) for the determination of chlorobenzenes in soil except that UAE was used for partitioning the analytes into water and the SBME process was not performed simultaneously with UAE but in sequence.

Wang et al. (2010) compared the performance of Soxhlet extraction, MAE, and PLE for the determination of both PCBs and PBDEs in soil and fish. The two most recent techniques gave comparable or better extraction results than Soxhlet extraction (up to 176%, especially for some PCB congeners with higher K_{ow} values). This improved efficiency is attributed by the authors to the use of high temperatures and pressures that favor the extraction of PCBs from solid matrices. Nevertheless, the MAE conditions must be carefully adjusted depending on the matrix to avoid the degradation of higher brominated BDEs and obtaining optimum efficiency. In general, PLE provided more robust and repeatable results compared to the other techniques. MAE was demonstrated to be also suitable for the determination of various pesticides and fungicides in soil (Merdassa et al., 2013) and trinitrotoluene and related compounds in soils of mined areas and test fields after cleanup on ABS Elut Nexus SPE (Kjellström et al., 2008), among many other applications.

Although UAE is sometimes replaced with PLE and MAE, some recent innovations regarding this technique can still be found. For instance, Sanchéz-Brunete et al. (2010) developed a method for the determination of cyclic and linear siloxanes in soil using the proprietary sonication-assisted extraction in small columns (SAESC). Siloxanes are very persistent in the environment and are widely employed in consumer products such as paints, cosmetics, and medical products. They enter WWTPs and are distributed in soil by land application of the sludge (Sanchéz-Brunete et al., 2010).

Siloxanes are suspected to have endocrine-disrupting properties and cause effects on the reproduction, raising concerns about their effect on humans and the environment (Sanchéz-Brunete et al., 2010). Procedural contamination was the main difficulty encountered by the authors due to the widespread use of silicones in consumer goods and bleed from GC septa and capillary columns. SAESC is performed in a glass column of 10 cm × 20 mm internal diameter (i.d.), where the samples are placed between filter paper and anhydrous sodium sulfate layers. The packed column is submitted to ultrasonication, and the extract is drained, already filtered, and is ready for analysis. By using this procedure, decamethylcyclopentasiloxane and dodecamethylcyclohexasiloxane were found in higher concentrations in industrial soils than in agricultural and sludge-amended soils (Sanchéz-Brunete et al., 2010).

Single-drop microextraction (SDME) is a popular technique for the analysis of organic pollutants in water and was adapted by Salemi et al. (2013) for determining organophosphorus pesticides (OPPS) in soils after UAE. SDME is quick, is easy to perform, is inexpensive, and uses very little solvent; thus, it is being considered an environment-friendly approach. It is compatible with GC analysis of volatile or semivolatile substances, since after extraction in the headspace, the microdrop is retracted into the microsyringe and injected for analysis. The authors used ultrasonic energy, temperature, and salting out to accelerate the extraction and partitioning of the pesticides into the gas phase, although the equilibrium state required more than 40 min to be attained (Salemi et al., 2013). Although the enrichment factors were not superb, the whole extraction phase was injected and analyzed, resulting in figures of merit comparable to those of other approaches.

Another innovation regarding UAE is the combination with new SPE adsorbents. Pan et al. (2013) proposed a method for determination of PAHs in soil whose main innovation is the use of a titania (TiO_2) NTs/titanium (Ti) plate modified with *n*-octadecanethiol (C18) monolayer-protected Ag nanoparticles as the adsorbent in SPE cleanup/concentration. The UAE extract was dissolved in 20 mL water with 400 μL of MeOH as organic modifier, and the TiO_2 NTs/Ti plate adsorbent was exposed for 90 min under stirring. After desorption with ACN and HPLC–fluorescence detection (FLD) analysis, the achieved LODs were about 100-fold lower than those in alternative methods; however, the extraction time for SPE with the new adsorbent required a long contact period (90 min). The efficiency of the new adsorbent is explained by its large surface area and hydrophobic nature (Pan et al., 2013).

MISPE is also among the SPE-based techniques used for environmental analysis. Zhu et al. (2002) selectively analyzed traces of five sulfonylureas (SUs) (nicosulfuron, thifensulfuron-methyl, metsulfuron-methyl, sulfometuron-methyl, and chlorsulfuron) in water and soil samples, using MISPE coupled to reversed-phase HPLC–diode array detector (DAD). An MIP was synthesized using metsulfuron-methyl herbicide as a template, 2-(trifluoromethyl)acrylic acid as a functional monomer, divinylbenzene (DVB) as a cross-linker, and dichloromethane as a porogen. The stability of these MIPs was tested, and it was confirmed that their performance did not vary even after 200 enrichment and desorption cycles. SU recoveries from soil samples ranged from 71% to 139%, with LODs between 5 and 12 μg kg^{-1}, while in rainwater samples, the recovery rates were between 96% and 107%, with LODs as low as 3–14 ng L^{-1}. In this work, the newly developed MISPE method proved to be highly selective for the enrichment of SU herbicides from environmental samples, offering several practical advantages over other SPE materials, such as C18. The equilibrium between MIP and metsulfuron-methyl was almost established within 1 min. On the other hand, the nonspecific binding of various SU herbicides on the polymer was very low.

Besides some technical drawbacks already mentioned, SFE is still used nowadays. It has been used in the conventional configuration (Forero-Mendieta et al., 2012), with results similar to those obtained before (Gonçalves et al, 2006). It has been also combined with other advanced techniques (Naeeni et al., 2009; Asiabi et al., 2013) or used as a tool to access the extractability/bioavailability of pollutants from soils and sediments, taking advantage of its selectivity and nonexhaustive extraction character

(Hawthorne et al., 2012; Bielská et al., 2013). Coupling of SFE and dispersive liquid–liquid microextraction provided very low detection limits (ng kg^{-1} range) due to the excellent preconcentration factors achieved by the combined techniques, in the range of 67–144. The main advantages of this strategy include low cost, environmental friendliness, short analysis time, clean chromatograms, and the lack of need to evaporate the collecting organic solvent (Naeeni et al., 2011). A similar strategy was developed for the determination of SU pesticides in soil by using supramolecular solvent microextraction (SUPRAS-ME) for enrichment of the SFE extract (Asiabi et al., 2013). SUPRASs are nanostructured liquids generated from amphiphiles through a sequential self-assembly process occurring at molecular scale and nanoscale. A SUPRA solvent made up of decanoic acid assemblies dispersed in tetrahydrofuran and water provided preconcentration factors in the range of 174–322; however, the method's LODs are not impressive due to the use of UV detection, which is known for its low sensitivity and selectivity.

The QuEChERS technique, initially developed for the determination of pesticides in fruits and vegetables, has also been demonstrated as a useful methodology for the determination of pharmaceuticals and degradation products (Bragança et al., 2012), aromatic organochlorines (Rouvière et al., 2012), and other chlorinated compounds in soils (Pinto et al., 2010). Bragança et al. (2012) obtained the best results for the determination of ibuprofen and degradation products in soil and sediment by combining the solvent indicated in the AOAC 2007.01 method (1% acetic acid in ACN) and the QuEChERS composition (including citrate salts) used in the EN 15662 method. Rouvière et al. (2012) proposed a method for the determination of 34 organochlorine compounds in soil by using a non-water-miscible solvent, dichloromethane, in QuEChERS. It proved to be a better extraction solvent than ACN, especially for chlorinated hydrocarbons and olefins, and for chlorobenzenes. Furthermore, it provided clean extracts, avoiding the need for purification by dSPE with PSA. Moreover, this method was compared with PLE and proved to be more appropriate since in the latter technique volatile compounds are lost, resulting in low recoveries (Rouvière et al., 2012). Pinto et al. (2010) introduced a simplified version of QuEChERS for the determination of chlorinated compounds, replacing ACN with ethyl acetate and not performing a dSPE cleanup step. The satisfactory results obtained were definitely influenced by the use of a highly selective and sensitive electron capture detector (ECD).

In conclusion, the selection of the extraction technique for soil analysis is highly dependent on the analyte type and matrix composition. In general, organic matter and clay contents are key factors that affect the binding strength of the analytes. Aging of the pollutants and particle size fractions may also influence the harshness of the extraction conditions required. A sample preparation approach that is not sensitive to matrix composition provides more robust results for quantitative determinations, while moderate and selective techniques might supply information on the bioavailability of the pollutants to living organisms and leaching potential (Björklund et al., 2000; Hawthorne et al., 2012; Bielská et al., 2013).

5.4 SAMPLE PREPARATION METHODS FOR THE DETERMINATION OF ORGANIC POLLUTANTS IN SEDIMENTS

5.4.1 INTRODUCTION

In recent years, special attention has been devoted to the study of organic pollutants in sediments, as this compartment represents the final sink for many dangerous substances (Salgado-Petinal et al., 2006a). River sediments act as integrators of many pollution inputs (industrial, agricultural, and domestic) within a catchment, revealing historical contamination rather than being a witness of acute episodes. Concentrations of certain xenobiotics measured in sediments are often greater than those in surface water, which indicates that sediments accumulate these substances and constitute a secondary source for the aquatic systems (Klosterhaus et al., 2013). Assays performed with *Chironomus tentans* demonstrated that a significant proportion (i.e., 94%) of the synthetic hormone 17α-ethinylestradiol (EE2) bioaccumulated came from the sediment. Thus, this source of contamination appears to be important for deposit feeders, such as *C. tentans*, which can ingest daily sediments several times their body weight (Dussault et al., 2009). Furthermore, there is an expected increase in the studies dedicated to the levels and fate of such pollutants in sediments as the European legislation puts special emphasis on the monitoring of certain substances with significant potential for accumulation for which sediment and biota should be matrices of concern (European Commission, 2008; European Union, 2010; Tölgyessy et al., 2013).

5.4.2 Extraction and Cleanup Protocols

The most recent and innovative sample preparation methods for the trace analysis of selected persistent and ECs in sediments will be discussed in this section, emphasizing their challenges and strengths. Table 5.2 comprises detailed information on the analytical protocols adopted and respective figures of merit.

A promising method is proposed by Camino-Sánchez et al. (2011), which includes an efficient and exhaustive extraction of marine sediments by PLE followed by selective cleanup and further concentration by using SBSE. With such powerful sample pretreatment coupled with the selectivity and sensitivity afforded by a QqQ LC-MS/MS system, LODs in the nanogram per kilogram range were obtained (termed *ultratrace*) in the determination of 86 priority pollutants, including PCBs, PAHs, PBDEs, and pesticide families (Camino-Sánchez et al., 2011). The strategy is similar to that followed by Martínez-Parreño et al. (2008), but it provides, in general, better sensitivity due to the fact that a 10-fold higher amount of sample was used and a more sensitive detection mode was employed (MS/MS instead of MS in SIM mode). Although superb sensitivity was obtained with the mentioned strategy, the system requires dedicated and expensive accessories and components to achieve an automated procedure (e.g., ASE 300, MPS autosampler, TD unit, and CIS-4 programmed temperature vaporization [PTV] cryofocusing injection port). Speed of analysis is not among the advantages of this procedure since each extraction takes 24 h with interruption for salt addition at 12 h (not compatible with normal working hours). However, this drawback was minimized by processing several (nine) samples simultaneously by using different SBSE bars in a multiposition stirrer. Additionally, the productivity of the method benefited from the inclusion of 86 analytes in a single chromatographic run. The method requires considerable amounts of solvent (50 mL per sample) and NaCl (60 g per sample) for salting out. The authors failed to prove the equivalence of the different SBSE bars efficiency to allow the use of the same calibration; however, since isotope-labeled internal standards were used, it is to be believed that this issue does not have any negative impact on the accuracy of the results. The authors justified the need for a method with detection limits in the nanogram per kilogram range in order to improve knowledge on not yet regulated contaminants in sediments, which would enable to take measures to prevent, limit, or reduce the impact of environmental contamination (Camino-Sánchez et al., 2011).

Given the considerable solvent consumption and sample handling time typically required by LSE coupled with SBSE methods discussed so far, Tölgyessy et al. (2013) proposed a miniaturized alternative involving UAE and subsequent SBSE-TD-GC-MS, which the authors advocated as the main improvement. The authors succeeded in analyzing PCBs and PBDEs in freshwater sediments by using a small aliquot of 0.2 g, extracted in two cycles with 1.2 and 1 mL of MeOH. As the whole process is miniaturized, SBSE is performed in just 5 mL of extract aqueous dilution, differently from previously cited works, fast equilibration times on the PDMS adsorbent are achieved (Tölgyessy et al., 2013). Sorptive techniques such as SBSE avoid the cleanup of extracts and reduce the use of organic solvents (just 2.2 mL in this methodology), thus constituting an environment-friendly approach (Tölgyessy et al., 2013).

Fernandez et al. (2001) reported the development of a MAME methodology using nonionic surfactants (polyoxyethylene 10 lauryl ether [POLE]) and oligoethylene glycol monoalkyl ether (Genapol X-080) as extractants for the analysis of PCBs in two marine sediment samples, collected

TABLE 5.2
Overview of Analytical Protocols Applied to the Determination of Selected Persistent and Emerging Pollutants in Sediments

Matrix	Extraction Technique	Extraction and Cleanup Protocols	Compounds Determined	Analytical Technique	Recoveries	Precision (% RSD)	Analytical Thresholds	Ref.
Marine sediments	PLE-SBSE	Aliquots of 10.0 ± 0.1 g of dried sample. Extraction with MeOH at 80°C for 10 min, pressure of 1000 psi, preheat time of 5 min. SBSE at 500 rpm for 24 h and 30% NaCl.	PCBs, PAHs, PBDEs, OCPs, OPPs, urons, triazines	TD-GC-MS/MS	63–19%	<35 % R	LODs: 0.001–0.30 μg kg^{-1} LOQs: 0.014–1.0 μg kg^{-1}	Camino-Sánchez et al. (2011)
River sediments	UAE-SPME	Sediment aliquot of 1.0 g extracted with 5 mL of MeOH in ultrasound for 30 min. SPME of 360 μL of extract dissolved in 18 mL of water for 60 min at 60°C.	PAHs	SPME-GC-MS	86–106%	11–27% IP	LODs: 0.70–17 μg kg^{-1}	Gonçalves et al. (2016)
Freshwater sediments	UAE-SBSE	0.2 g sediment sample extracted with MeOH (1.2 and 1.0 mL) in two 5 min extraction cycles, in 1.5 mL microtest tubes. SBSE for 1 h at 900 rpm.	PCBs, PBDEs	TD-GC-MS (SIM)	62.8–91.5%	3.6–15.0% r	LOQs: 0.33–2.3 μg kg^{-1}	Tölgyessy et al. (2013)
Marine sediments	MAME and Soxhlet (comparative study)	MAME: 10 mL of surfactant solution (0.02 M) added to the sediment and submitted to MAE, under acidic pH, at 40% power. Extracted solution separated and 2× 5 mL of surfactant solution added to the sediment. Final extracted solution used for posterior analysis. Soxhlet: 2 g of sediments extracted with 34 mL of acetone/*n*-hexane (10:7 v/v) for 2 h at 50°C.	PCBs	HPLC-FLD	MAME: 35–103% Soxhlet: 55–98%	MAME: 0.1–1% 0.22–0.93% with POLE and 0.13–0.79% with Genapol X-080	MAME LODs: 0.6–4.9 μg L^{-1} in the sediment extract	Fernandez et al. (2001)

(Continued)

TABLE 5.2 (CONTINUED)
Overview of Analytical Protocols Applied to the Determination of Selected Persistent and Emerging Pollutants in Sediments

Matrix	Extraction Technique	Extraction and Cleanup Protocols	Compounds Determined	Analytical Technique	Recoveries	Precision (% RSD)	Analytical Thresholds	Ref.
River and marine sediments	SPME	0.25–0.5 g sample weighed in 22 mL SPME glass vessels. Addition of 5 mL water, 0.5 mL of H_2SO_4 (dilution 1:3 water), heating at 100°C, SPME with polyacrylate fiber in the headspace for 40 min.	PBDEs	GC-MS/MS	76–111%	1.2–14.0% r	LOQs: 0.01–0.15 μg kg^{-1}	Montes et al. (2010)
Sediments, soil, and sewage sludge	SPME	0.5 g solid samples moisturized with 2 mL water, placed in a 100°C bath, and stirred at 700 rpm. Extraction with PDMS 100 μm for 60 min.	PBBs, PBDEs	GC-MS/MS	82–104% sediment 96–105% soil	3.9–16% r 3.6–12% IP	LODs: 0.005–0.214 μg kg^{-1} sediment 0.014–0.625 μg kg^{-1} soil	Salgado-Petinal et al. (2006a)
Sediments	PHWE-SBSE	2 g of dried and milled sediment + 18 g of siliceous earth, placed in a 22 mL steel extraction cell. Extraction with water containing 10% MeOH as organic modifier. PHWE at 100°C and 1500 psi in three 5 min cycles (preheating time 6 min, flush volume 60%, and purge time 60 s). Extract acidified with phosphoric acid (pH 2), stirred with SBSE at 900 rpm during 8 h at room temperature and in darkness conditions. LD with 200 μL of ethyl acetate under sonication during 30 min.	PPCPs	Derivatization with MTBSTFA followed by GC-MS	2.2–140%	<19% r	LODs: 0.01–5 μg kg^{-1}	Pintado-Herrera et al. (2013)

(Continued)

TABLE 5.2 (CONTINUED)
Overview of Analytical Protocols Applied to the Determination of Selected Persistent and Emerging Pollutants in Sediments

Matrix	Extraction Technique	Extraction and Cleanup Protocols	Compounds Determined	Analytical Technique	Recoveries	Precision (% RSD)	Analytical Thresholds	Ref.
Marine sediments	PHWE-SPME	0.5 g of sediment extracted by PHWE at 2000 psi and 200°C and using 10% MeOH as organic modifier (static time 10 min.) SPME with 65 μm PDMS/DVB fiber at a temperature of 60°C for 60 min.	Parent and alkylated PAHs	GC-MS (SIM)	SRM 1944: 55–120% SRM 1941b: 32–110%	3.3–9.6% r	LODs: 0.4–15 μg kg^{-1} LOQs: 1.2–51 μg kg^{-1}	Fernández-González et al. (2008)
Marine sediments	PHWE-SPME	Extraction at 1500 psi and 150°C with MeOH/water (5:95) for 5 min static time. 18 mL of the obtained extract submitted to SPME with 65 μm PDMS/DVB fiber at 60°C for 45 min.	28 OCPs	GC-MS (SIM)	80–115%	5–18% r 7–21% IP	LODs: 0.11–16 μg kg^{-1} LOQs: 0.24–22 μg kg^{-1}	Concha-Graña et al. (2010)

Note: DVB, divinylbenzene; PBBs, polybrominated biphenyls; POLE, polyoxyethylene 10 lauryl ether.

in different areas of the Gran Canaria Island. In general, the results obtained using Genapol X-080 were better than those obtained using POLE, probably due to a greater interaction between the Genapol X-080 micelles and the different analytes. The results achieved demonstrated that MAME recovery rates were satisfactory and similar to those obtained with conventional Soxhlet extraction.

Differently from the method discussed in the preceding paragraph, Gonçalves et al. (2016) developed a methodology with the major aim to investigate the role of different salt marsh plants on remediation of PAH-contaminated sediments. PAHs are a group of organic pollutants known by their carcinogenicity, mutagenicity, and bioaccumulation potential originated from background processes as well as fossil fuel spills and combustion. They are highly lipophilic and tend to deposit in sediments, soils, and biota. The procedure was based on UAE of sediments using MeOH and further concentration using SPME (100 μm PDMS fiber) followed by GC-MS analysis in SIM mode. This procedure turned out to be fairly straightforward and used equipment readily available in analytical laboratories, with a productivity of 20 samples per day. The LODs achieved were in the low nanogram per kilogram range and were sufficient to perform a general characterization of the sediments and assess the influence of the root system of different types of salt marsh plants and sediment characteristics on the PAH levels accumulated in the respective rhizosphere. Such an analytical setup proved to be very consistent (evaluating the results of quality control standards), due to the use of deuterated internal standards which compensated for the matrix effects observed for five- and six-ring PAHs and the progressive accumulation of matrix coextractives in the fiber adsorbent (Gonçalves et al., 2016). The influence of MeOH content (aliquot of extract) in the SPME efficiency was evaluated until 5%, in view of improving the sensitivity of the method, and it did not show any detrimental effect (Gonçalves et al., 2016).

The substantial matrix effect observed in the determination of some organic pollutants in sediments by using SPME, which could be confirmed by the authors of this chapter, was tackled by Montes et al. (2010) through chemical treatment of the matrix by adding potassium permanganate and sulfuric acid. The amount of reagents was related to the organic matter content of the sediment (20–40 mg $KMnO_4$/mg carbon) and allowed further extraction with an SPME polyacrylate (PA) fiber in the headspace at 100°C for 40 min. Recoveries of PBDEs from the sediment water slurry were dramatically improved, and the method benefited from the selectivity of direct SPME in the headspace mode coupled with analysis by MS/MS. The resulting method affords automation capabilities and is much simpler, as the typical prior organic solvent extraction step is circumvented by combining an oxidative sample treatment with simultaneous extraction, concentration, and cleanup by HS-SPME (Montes et al., 2010). Despite the remarkable advantages noticed by the authors, the use of strong oxidant conditions limits the lifetime of the SPME fiber to about 50 extractions, after which a 30% reduction in extraction efficiency was noted. This work followed a previous one in which the authors demonstrated that the sample matrix type has a remarkable influence on the extraction efficiency, which is linked to its organic carbon content. In that work, a marine sandy sediment allowed the best extraction efficiency, a river sediment and soil gave comparable efficiencies, and impracticable extraction was observed in sewage sludge (Salgado-Petinal et al., 2006a). In order to obtain reliable results, the authors opted by using a standard addition calibration methodology in real river and harbor sediments.

Pintado-Herrera et al. (2013) proposed a method for determining pharmaceuticals (analgesics, antipyretics, antifungals, neuroleptics, β-blockers, lipid regulators, antidepressives, and estrogens) and PCPs (UV filters, APs, fragrances, biocides, and phthalates) in sediments, among other matrices, based on SWE and SBSE, appropriately termed "environmentally friendly analysis of emerging contaminants." The work has the merit of advancing knowledge on the determination of PPCPs in soil and sediments, for whose matrices there is a great lack of data on ECs, although levels have been measured typically between 0.5 and 10 μg kg^{-1} (Pintado-Herrera et al., 2013). LD of SBSE bars was preferred versus TD, as most of the PPCPs under analysis required prior derivatization to be suitable for analysis by GC-MS. As acetic anhydride does not react with acidic pharmaceuticals, MTBSTFA was preferred. Temperature is considered a critical parameter on the SWE efficiency;

therefore, it was optimized between 50°C and 150°C, as the set of compounds to be analyzed ranged from very polar to semipolar and belonged to different chemical families. To adjust the subcritical water properties, a good balance needed to be found between the extraction of polar compounds, which require lower temperatures, and apolar ones, which are favored by higher temperatures (Pintado-Herrera et al., 2013). Use of temperatures above 150°C could produce degradation, volatilization, or increase the coextraction of matrix interferences that would make subsequent analysis difficult (Pintado-Herrera et al., 2013). The optimal extraction temperature was considered to be 100°C.

Often researchers feel the need to add a small amount of an organic solvent (frequently, MeOH) to the extraction vessel during SBSE, in order to decrease the losses of hydrophobic analytes adsorbed onto the vial glass walls, improving extraction efficiency this way (Rodil and Popp, 2006; Pintado-Herrera et al., 2013). When SBSE follows a previous extraction by SWE, then the organic modifier added to water solvent (here, 10% MeOH) with the aim to improve the SWE efficiency from the solid matrix has the double function of benefiting also SBSE efficiency if added at a low percentage that does not prejudice the partition of analytes (Pintado-Herrera et al., 2013).

Although using an approach different from the previous methodologies, Fernández-González et al. (2008) claimed to have developed a fully automated, environment-friendly, simple, selective, and sensitive method for the analysis of PAHs in sediment samples. The aim was to analyze 27 alkylated and nonalkylated PAHs, enabling to characterize spilled oils and relate them to the origin sources. The coupling of PHWE and SPME is very straightforward, but compared to the approach proposed by Gonçalves et al. (2016), a dedicated and expensive extraction system was necessary to perform PHWE (Fernández-González et al., 2008). In line with the PHWE principles, the maximum operating pressure of the equipment (2000 psi) was used to keep water in a liquid state at 200°C, conditions required to extract analytes with log P values ranging from 3.4 to 7.2. Moreover, the authors were forced to add MeOH to improve recoveries in PHWE, especially those of the heavier PAHs, but at the expense of decreasing the SPME efficiency (Fernández-González et al., 2008). Thus, the authors selected 10% MeOH as a compromise. These observations are in agreement with the experiments of Gonçalves et al. (2016), which did not exceed 2% MeOH, in practice. The LODs obtained by Fernández-González et al. (2008) are comparable with those of Gonçalves et al. (2016) when using half of the sediment amount.

Concha-Graña et al. (2010) applied a very similar approach for the determination of 28 OCPs, which can still be found in the environment and biota due to their persistence, although most of them have been banned two to three decades ago. Since they are, generally, more polar than PAHs (log P 2.23–6.89), the extraction temperature (150°C), pressure (1500 psi), and percentage of modifier (5% MeOH) were lower than those used for the analysis of PAHs, also by PHWE-SPME-GC/MS. Thus, sorbent-trapping extraction techniques can be an option for the determination of different organic compounds in sediments, namely, as a cleanup step after a solvent partition technique.

The total measure of pollutants in sediments does not always give an accurate picture of the threat posed to aquatic organisms, as their bioavailability depends on a number of sediment characteristics. Some authors have exploited the potential of SPME for investigating the bioavailability of pyrethroids, chlorpyrifos, and PCBs in sediments (Cui et al., 2013; Harwood et al., 2013). Harwood et al. (2013) used PDMS SPME fibers and Tenax beads to quantify the bioavailable fraction of pyrethroids in contaminated field sediments that were toxic to benthic invertebrates. Tenax provided a more sensitive method than SPME, and a correlation between Tenax-extractable concentrations and mortality was observed (Harwood et al., 2013). Cui et al. (2013) used PDMS fibers loaded with stable isotope-labeled hydrophobic organic compounds as a strategy to circumvent the need to reach equilibrium in conventional SPME for free/dissolved pollutant determination, which would take weeks or months in soil or sediments. This kinetic calibration approach ($K_{adsorption} = K_{absorption}$) may be used under nonequilibrium conditions, allowing to accurately estimate the concentration of the free pollutants (PCB-52, PCB-153, bifenthrin, and *cis*-permethrin), thus assessing their bioavailability. Cui et al. (2013) reckoned SPME as a biomimetic tool ideally suited for measuring the bioavailability of hydrophobic organic compounds given its small size and nondepletive nature.

5.5 SAMPLE PREPARATION METHODS FOR THE DETERMINATION OF ORGANIC POLLUTANTS IN BIOTA

5.5.1 INTRODUCTION

Determination of organic micropollutants in biota/biological tissues is a current topic of concern. The Environmental Quality Standards Directive enforces the determination of the so-called priority substances in biota species, and such analysis is also required for ecotoxicological risk assessment with the ultimate aim of ensuring environmental quality, biodiversity safeguard, and sustainability (European Union, 2010; European Commission, 2013). The pollutant categories identified to pose risk to or via the biota, namely, toxicity, bioaccumulation, and biomagnifications, are PBDEs, PAHs, certain pesticides, polychlorinated dibenzo-*p*-dioxins/furans (PCDD/Fs), mercury and its compounds, hexabromocyclododecane (HBCD), and perfluorooctane sulfonic acid (PFOS) and its derivatives (European Commission, 2013). Environmental evidence of the persistence, bioaccumulation, and worldwide distribution of PBDEs, particularly in biota and solid samples, has led to the ban of most applications of penta- (tetra- to hexabrominated congeners) and octa-mix (hepta- to nonabrominated species) solutions in many countries (Montes et al., 2010), the same happening with some OCPs. Conversely, other pollutants such as PAHs and PCDD/F are biogenically or unintentionally produced, and their levels should be controlled and reduced.

Biological samples are among the most complex matrices containing a great number of potential interfering molecules (e.g., lipids) with low concentrations of target contaminants (Lacroix et al., 2014). Most sample pretreatment techniques (extraction and cleanup) for biota analysis are rather similar to those used for soils or sediments. Several good review papers have been published specifically addressing biota samples (Fidalgo-Used et al., 2007; Andreu and Picó, 2012; Huerta et al., 2012; Valsecchi et al., 2013) or solid environmental samples in general (van Leeuwen and de Boer, 2008; Bester, 2009; Wille et al., 2011; Aufartová et al., 2012; Król et al., 2012; von der Kammer et al., 2012), and these should be considered in complement to this section. Some reviews focused on a group of environmental pollutants, for example, PBDEs (Król et al., 2012), musk fragrances (Bester, 2009), pharmaceuticals (Huerta et al., 2012), steroid estrogens (Aufartová et al., 2012), currently used pesticides (Andreu and Picó, 2012), engineered nanomaterials (von der Kammer et al., 2012), and perfluorinated compounds (PFCs) (Valsecchi et al., 2013), while others covered several substances classified as POPs (Fidalgo-Used et al., 2007; van Leeuwen and de Boer, 2008) or ECs (Wille et al., 2011).

5.5.2 EXTRACTION AND CLEANUP PROTOCOLS

A wide array of extraction techniques has been employed for biota sample preparation, including noninstrumental (ion pair extraction, alkaline digestion, QuEChERS, mechanical shaking, and MSPD [Andreu and Picó, 2012; Huerta et al., 2012; Valsecchi et al., 2013]) and instrumental (SFE, MAE, PLE, and UAE [Nerín et al., 2002; Fidalgo-Used et al., 2007; van Leeuwen and de Boer, 2008; Wille et al., 2011; Huerta et al., 2012; Valsecchi et al., 2013]). Cleanup has been performed mainly employing GPC, column chromatography on silica, Florisil, alumina ($Al_2O_3{\cdot}H_2O$, of activity IV), chemical treatment with sulfuric acid, or selective extraction techniques such as SPME and SBSE (Fidalgo-Used et al., 2003, 2007; Huerta et al., 2012). Lipids are important interferences in the analysis of biological tissues; thus, these components must be removed. For the analysis of PCBs and pesticides, treatment with sulfuric acid is usually sufficient. Nevertheless, sulfuric acid destroys

PAHs; therefore, it cannot be used in the cleanup of such extracts. Saponification with KOH can be used instead.

Given the good assortment of compilations of the methods for the analysis of biota found in the literature, only recent applications and their distinctive features will be discussed in this section. The most recent and innovative approaches will be detailed, emphasizing their challenges and strengths. Table 5.3 comprises detailed information on the analytical protocols and their respective figures of merit.

Aufartová et al. (2012) reviewed green microextraction techniques for the determination of steroid hormones in biological and environmental samples, namely, SPME with MIPs, in-tube solid-phase microextraction (IT-SPME), SBSE, microextraction in packed syringe, and liquid-phase microextraction (LPME). Similar to soils and sediments, the SPME fiber can be exposed to the aqueous solution of the biota extract for analyte enrichment and cleanup.

Fidalgo-Used et al. (2003) determined OCPs in fish tissue by combining conventional Soxhlet extraction and SPME (100 µm PDMS fiber) over the extract, while Carrasco et al. (2007) determined methylmercury in dogfish muscle by employing alkaline digestion with KOH followed by propylation with $NaBPr_4$, HS-SPME (100 µm PDMS fiber) sampling, and, finally, GC-pyrolysis-atomic fluorescence spectrometry determination. Lacroix et al. (2014) proposed a method for the determination of 21 PAHs in reduced aliquots of marine biota samples (fish muscle, mussel, and oyster tissues) using alkaline digestion with KOH combined with SBSE and TD-GC-MS/MS (SBSE-GC-MS/MS), building on the approach adopted by Pfannkoch et al. (2010), which used QuEChERS followed by SBSE. Carryover of the SBSE coating is an aspect to take in consideration when highly contaminated samples are analyzed, particularly for high-molecular weight analytes (e.g., PAHs), for which amounts between 1% and 6% were found in a second desorption. Also, matrix effects was a phenomenon of concern, since response ratios in matrix-matched calibrations of 0.77–1.21 in fish muscle and 0.51–1.47 in mussel tissue were obtained (Lacroix et al., 2014). The authors attributed them to competition between analytes and interfering molecules during extraction, especially in lipid-rich mussel tissue. Besides these drawbacks commonly found in adsorptive techniques, and corroborated by the present authors, the sensitivity of the method merits to be highlighted as well as the accuracy and precision, owing to the use of six deuterated internal standards/surrogates and matrix-matched calibrations. It should be mentioned that the operating pH of the saponification did not have any negative impact on the SBSE efficiency, being even beneficial (Lacroix et al., 2014).

Solvent extraction techniques have been also used for biota analysis. Ronan and McHugh (2013) described a method for the determination of natural and synthetic estrogens in mussels, which were ultrasonic-extracted with MeOH. The lipid coextractives were removed by agitation with *n*-hexane, and the methanolic extract dissolved in water was further cleaned up with Oasis HLB SPE cartridges.

Gu et al. (2014) implemented an analytical method for the determination of 15 EDCs constituted mainly by phthalates, APs, and bisphenol A (BPA) in seafood that uses UAE with ACN instead of *n*-hexane, as this solvent showed a high background level of diethylhexylphthalate. PSA was preferred as cleanup sorbent over Florisil and Oasis HLB, as better recoveries were obtained for all analytes in all matrices. Nevertheless, due to the different polarities of the analytes (and ionization), PSA had to be eluted in two steps: with 5 mL of ACN (phthalic acid esters [PAEs] and APs) followed by 6 mL of formic acid–MeOH (3:97 v/v) (monoalkyphthalate ester [MPE] and BPA). GC-MS/MS and LC-MS/MS were used for separation and detection, as the analytes required different analytical conditions (type of chromatography, mobile phase, additives, etc.) for optimum resolution and sensitivity (Gu et al., 2014). Matrix-matched calibrations were used to compensate for the huge matrix effect (especially for APs) under LC-ESI-MS/MS.

Although the use of organotin compounds was banned several years ago, they still represent a threat to aquatic organisms. Analytical methodologies used are generally based on extraction with organic solvents and acid mixtures, sometimes assisted by complexing agents such as tropolone, hydrolysis with tetramethylammonium hydroxide or enzymatic digestion with protease and lipase.

TABLE 5.3
Overview of Analytical Protocols Applied to the Determination of Selected Persistent and Emerging Pollutants in Biota

Matrix	Extraction Technique	Extraction and Cleanup Protocols	Compounds Determined	Analytical Technique	Recoveries	Precision (% RSD)	Analytical Thresholds	Ref.
Biota (fish muscle, mussel, and oyster)	Alkaline digestion	200 mg mussel or 300 mg fish tissue added with 0.8 or 0.7 mL pure water, respectively. 10 mL of an ethanolic solution of KOH 0.05 g mL^{-1} was added. Samples are digested for 3 h at 80°C. Addition of 100 mL water and SBSE for 2 h at 700 rpm.	21 PAHs	SBSE-TD-GC-MS/MS	94–117%	0.35–12.1% r	LOQs: 1–10 μg kg^{-1} in fish muscle LOQs: 0.5–10 μg kg^{-1} in mussel tissue	Lacroix et al. (2014)
Biota (mussels)	UAE-SPE	10 mL of CH_3OH added to a 50 mL centrifuge tube containing 0.5 g of freeze-dried tissue and extracted by ultrasonication for 1 min. The CH_3OH fraction was reduced to dryness and resuspended with 9 mL deionized water. Cleanup on Oasis HLB.	E1, 17β-E2, and 17α-EE2	LC/ESI-MS/MS	88–100%	3.2–8.8% r	LODs: 0.3–0.9 μg kg^{-1} LOQs: 0.6–1.7 μg kg^{-1}	Ronan and McHugh (2013)
Fish, mollusk and prawn	UAE	Homogenized fresh sample (2 g) weighed in a 15 mL glass centrifuge tube. 2 mL of ultra-pure water added into the tube and shaken in a vortex for 30 s. 10 mL of ACN was added, the tube was shaken for 1 min and extracted in ultrasonic bath for 15 min. 1 g NaCl was added, the tube was shaken and centrifuged. Cleanup by SPE with PSA (1 g/6 mL) cartridge.	15 EDCs: 5 PAEs, MPEs), 4 APs, and bisphenol A	LC-MS/MS and GC-MS/MS	60–127% for most compounds	<24% and <15% for majority of analytes: r <25% and up to 43% for DEHP (due to background contamination): IP	LOD: 0.015–2.2 μg kg^{-1} (ww)	Gu et al. (2014)

(*Continued*)

TABLE 5.3 (CONTINUED)
Overview of Analytical Protocols Applied to the Determination of Selected Persistent and Emerging Pollutants in Biota

Matrix	Extraction Technique	Extraction and Cleanup Protocols	Compounds Determined	Analytical Technique	Recoveries	Precision (% RSD)	Analytical Thresholds	Ref.
Biota (mussel and oyster)	UAE-MIP	0.1 g sample extracted with 5 mL of 1:1 MeOH/acetic acid. Extraction under sonication for 30 s at 20–25% amplitude. Extract derivatized with sodium tetraethylborate. MIP cleanup: 2 mL of extract plus 10 mL of water loaded in a cartridge containing 50 mg MIP. Eluted with 6 mL of 0.3 M HCl in MeOH.	Monobutyltin, dibutyltin, and tributyltin	GC-FPD	89–100% mussels 88–120% oysters	5–6% mussels 10–20% oysters Both repeat	LOD: 3 μg kg^{-1} LODs: 10 μg kg^{-1}	Gallego-Gallegos et al. (2006)
Fish tissue	PLE	1 g of Florisil placed at the bottom of the cells. 1 g of freeze-dried fish mixed with Florisil. PLE in two cycles of 5 min of static time at100°C and 1500 psi using ethyl acetate/DCM (1:1 v/v). 25 mL extract diluted to 200 mL with HPLC water further purified by Isolute C18 (500 mg, 3 mL) cartridges.	8 UV filters and transformation products	LC-MS/MS	70–112% except OD-PABA (36–42%)	5–14% r	LOD: 0.1–6.0 μg kg^{-1} dw LOQs: 0.3–20.0 μg kg^{-1} dw	Gago-Ferrero et al. (2013)
Fish homogenate and organs	PLE	1 g of fish homogenate was mixed with Hydromatrix and placed in a 22 mL stainless steel extraction cell. 2 g of neutral aluminum oxide placed in the cell as a lipid retainer. PLE conditions: MeOH as extraction solvent; oven temperature, 50°C; pressure, 1500 psi; 5 min heat-up time; three static cycles; static time, 5 min. GPC: Enviro-Prep column, mobile phase DCM/MeOH (90:10 v/v), at 5 mL min^{-1} flow rate, collected fraction 13.5–26.5 min.	20 pharmaceuticals and metabolites (NSAIDs, diuretics, psychiatric drugs, antihistaminics, antihelmintics, β2-adrenergic, receptor agonists, and β-blockers)	LC-MS/MS	81–116% for fish homogenate 81–108% for fish liver 80–106% for fish muscle	2.1–21% r for fish homogenate 3.9–20% r for fish liver 1.1–20% r for fish muscle	LODs: 0.03–0.50 μg kg^{-1} for fish homogenate 0.01–0.42 μg kg^{-1} for fish muscle 0.08–0.98 μg kg^{-1} for fish liver	Huerta et al. (2013)

(Continued)

TABLE 5.3 (CONTINUED)
Overview of Analytical Protocols Applied to the Determination of Selected Persistent and Emerging Pollutants in Biota

Matrix	Extraction Technique	Extraction and Cleanup Protocols	Compounds Determined	Analytical Technique	Recoveries	Precision (% RSD)	Analytical Thresholds	Ref.
Fish fillet (sea bass, black sea bass, sole, and sea trout)	PLE	1 g of sample mixed with diatomaceous earth and transferred into the 33 mL extraction cell. PLE with acetone/*n*-hexane (1:1 v/v), one cycle at 100°C (5 min heating and 5 min static time), and 10 MPa. Flush with solvent (60% of the cell volume); purge time with nitrogen 90 s. Cleanup with EXtrelut NT3 cartridge containing 3 mL concentrated sulfuric in tandem with a Bond Elut SI silica cartridge.	18 PCBs	GC-MS (SIM)	80–110%	1.7–7.7% r 2.1–16% R	LODs: 0.4–1.1 μg kg^{-1} LOQs: 1.3–3.8 μg kg^{-1}	Ottonello et al. (2014)
Fish (eel, salmon), egg, and mozzarella cheese	PLE	10 g of sample mixed with 6 g of Isolute, transferred to a 125 mL extraction cell. PLE with hexane/acetone (1:1 v/v) at 100°C and 1500 psi in three 5 min cycles. Heating and static time was 5 min with a flush volume of 40% and a purge time of 300 s. Cleanup with acid silica (PowerPrep) followed by silica SPE cartridge.	HBCD isomers and TBBPA	LC-MS/MS	104% for the sum of HCBD isomers	4% for α-HBCD, 033% for β-HBCD, and 11% for γ-HBCD R	LODs: μg kg^{-1} for HBCD and TBBPA except for egg TBBPA (0.1 μg kg^{-1})	ten Dam et al. (2012)

(*Continued*)

TABLE 5.3 (CONTINUED)
Overview of Analytical Protocols Applied to the Determination of Selected Persistent and Emerging Pollutants in Biota

Matrix	Extraction Technique	Extraction and Cleanup Protocols	Compounds Determined	Analytical Technique	Recoveries	Precision (% RSD)	Analytical Thresholds	Ref.
Birds and fish organs and tissues (liver and kidney of 17 birds; plasma, liver, and brain of 7 cyprinoid fish)	Enzymatic digestion, UAE, and LLE	0.5 g of liver, kidney, and brain; 1 mL of plasma. 2 mL of acetate buffer and β-glucuronidase/sulfatase (10 μL) incubated for 14 h at 37°C. Ice-cold ACN (5 mL) was added, shaken for 1 min, and ultrasonicated for 10 min. The extract was added to 5% NaCl solution (50 mL), 1 M acetic acid (200 μL), and MTBE (15 mL) and was vigorously shaken for 20 min. Cleanup onto activated silica gel (3 g). GPC with cyclohexane/ethyl acetate (3:1 v/v) as mobile phase for liver and brain samples. Final cleanup with Oasis HLB cartridges	14 pharmaceuticals and 3 PCPs	LC-MS/MS	Absolute recoveries: 48–88% Relative recoveries: 90–130%	2.0–8.7% r 3.5–11% R	LODs: 0.0092–3.2 μg kg^{-1} ww LOQs: 0.020–8.7 μg kg^{-1} ww	Tanoue et al. (2014)

(Continued)

TABLE 5.3 (CONTINUED)
Overview of Analytical Protocols Applied to the Determination of Selected Persistent and Emerging Pollutants in Biota

Matrix	Extraction Technique	Extraction and Cleanup Protocols	Compounds Determined	Analytical Technique	Recoveries	Precision (% RSD)	Analytical Thresholds	Ref.
Mussels	QuEChERS	Freeze-dried mussels (2 g) in a 50 mL centrifuge tube. Rehydrated with 10 mL of ultrapure water and vortexed for 30 s. Addition of 10 mL of ACN and agitation for 1 min. 4 g Na_2SO_4 (anhydrous), 1 g NaCl, 1 g $Na_3Cit{:}2H_2O$ and 0.5 g $Na_2HCit{:}3H_2O$ added and agitated. 2.5 mL of the upper ACN layer transferred into a 15 mL tube containing cleanup sorbent (750 mg Na_2SO_4, 125 mg PSA, and 125 mg C18). 50 µL of formic acid added and shaken for 1 min.	Carbamazepine, oxcarbazepine, and 6 transformation products	LC-MS/MS Orbitrap Exactive	67–110%	5–11% r 7–18% R	LODs: 0.1–0.3 µg kg^{-1} LOQs: 0.2–1.0 µg kg^{-1} MDLs: 0.5–1.5 µg kg^{-1}	Bueno et al. (2013)
	PLE	Bottom of the PLE 66 mL cells: 7 g of Florisil and 50 g of glass beads (1 mm diameter, Assistent) between glass fiber filters. 3 g homogenized mussel placed in extraction cells and mixed with glass beads and finally covered with 15 g of glass beads. Extraction carried out with 70 mL water/acetone (3:2 v/v). Extraction conditions: cell heating time (5 min), static time (10 min), pressure (1500 psi), temperature (80°C), purging time (200 s), flushing volume. (60%) and cycles (2). Extracts were cleaned by SPE using Oasis HLB cartridges and eluted with 6 mL MeOH.			4–67%			

(Continued)

TABLE 5.3 (CONTINUED)
Overview of Analytical Protocols Applied to the Determination of Selected Persistent and Emerging Pollutants in Biota

Matrix	Extraction Technique	Extraction and Cleanup Protocols	Compounds Determined	Analytical Technique	Recoveries	Precision (% RSD)	Analytical Thresholds	Ref.
Mussels	QuEChERS	2 g freeze-dried mussels weighed in a 50 mL polypropylene centrifuge tubes. Mussels rehydrated with 10 mL of Milli-Q water and vortexed for 30 s. Addition of 10 mL of ACN and agitation for 2 min. Addition of the salt mixture (4 g anhydrous Na_2SO_4, 1 g NaCl, 1 g $Na_3Cit{:}2H_2O$, and 0.5 g $Na_2HCit{:}3H_2O$). Agitation for 1 min and centrifugation. 2.5 mL of the upper layer transferred to tube containing the cleanup mixture (750 mg Na_2SO_4, 125 mg Bondesil C18, and 125 mg PSA silica bulk). 50 µL of formic acid added and the tubes agitated for 1 min.	UV filters (EHMC, octocrylene, OD-PABA), UV stabilizers (UV-P and UV-326), and musk fragrances (galaxolide, celestolide, cashmeran, and musk ketone)	GC-MS/MS	81–126% except UV-326, (33–50%)	3–16% r 5–26% R	LODs: 5–50 µg kg^{-1} LOQs: 0.5–50 µg kg^{-1}	Groz et al. (2014)

(*Continued*)

TABLE 5.3 (CONTINUED)
Overview of Analytical Protocols Applied to the Determination of Selected Persistent and Emerging Pollutants in Biota

Matrix	Extraction Technique	Extraction and Cleanup Protocols	Compounds Determined	Analytical Technique	Recoveries	Precision (% RSD)	Analytical Thresholds	Ref.
Fish (*Luciobarbus sclateri* and *Cyprinus carpus*) and sediments	QuEChERS	1 g lyophilized fish or sediment. QuEChERS with the following composition: • For fish samples, $MgSO_4$ (6 g) and NaCl (1.5 g) were used for extraction, and PSA (50 mg), $MgSO_4$(150 mg), C18(50 mg), and active coal (15 mg) for cleanup. • For sediment samples, $MgSO_4$ (6 g), NaCl (1.5 g), trisodium citrate dehydrate ($Na_3C_6H_5O_7{\cdot}2H_2O$) (1.5 g), and disodium hydrogen citrate sesquihydrate ($HOC(COOH)(CH_2COONa)_2{\cdot}1.5H_2O$) (0.75 g) were used for phase separation and PSA (50 mg), $MgSO_4$ (150 mg), and C18 (50 mg) for cleanup.	50 currently used pesticides and degradation products	LC-MS/MS	>40% in biota and sediments	–	LODs: 0.08–3.75 μg kg^{-1} in biota 0.03–1.67 μg kg^{-1} in sediments	Masiá et al. (2013a)

(Continued)

TABLE 5.3 (CONTINUED)
Overview of Analytical Protocols Applied to the Determination of Selected Persistent and Emerging Pollutants in Biota

Matrix	Extraction Technique	Extraction and Cleanup Protocols	Compounds Determined	Analytical Technique	Recoveries	Precision (% RSD)	Analytical Thresholds	Ref.
Fish, sediment, fruits, and bees	QuEChERS	1 g of lyophilized sediment, fish or bees, and 10 g fruits. Lyophilized samples were moisturized with 7.5 mL water and extracted with 10 mL ACN. Vigorous agitation by hand for 1 min. A mixture of 4 g anhydrous $MgSO_4$, 1 g NaCl, 1 g trisodium citrate dehydrate, and 0.5 g disodium hydrogen citrate sesquihydrate was added. Cleanup: 1 mL of the ACN phase was transferred to a 15 mL centrifuge tube containing 125 mg PSA, 750 mg anhydrous $MgSO_4$, and 15 mg of GCB, and the tube shaken for 2 min.	54 pesticides	UPLC-LTQ Orbitrap	Recoveries between 71% and 100% 82% for pesticides in fish 98% for pesticides in honeybees 94% for pesticides in fruits 89% for pesticides in sludge	<22%	LODs equal for all pesticides 10 μg kg^{-1} for sediment and fish 20 μg kg^{-1} for honeybees 1 μg kg^{-1} for fruits	Farré et al. (2014)
Mussels and tellins	Micro-MSPD	Lyophilized samples (100 mg) placed in a glass mortar and blended with 0.4 g C18 phase for 5 min. The mixture was transferred to a 3 mL SPE tube containing 0.1 g Florisil phase. Tubes flushed with 1.2 mL ACN (collected extract of about 0.8 mL). Extracts were diluted with 2.6 mL of water. 2.0 mL of the resulting solutions and 100 μL of deionized water were loaded into the capillary column (70 cm × 0.32 mm i.d)., 3 μm film thickness, column coated with 95% methyl-/5% phenyl-polysiloxane. Finally, elution with mobile phase.	PAHs	In-tube SPME and HPLC-FLD	52–100% for MSPD 18–28% for in-tube SPME 13–27% overall Real samples: 10–28%	2–10% r 17–28% IP	LODs: 0.05–0.6 μg kg^{-1} (dw)	Campíns-Falcó et al. (2008)

(Continued)

TABLE 5.3 (CONTINUED)
Overview of Analytical Protocols Applied to the Determination of Selected Persistent and Emerging Pollutants in Biota

Matrix	Extraction Technique	Extraction and Cleanup Protocols	Compounds Determined	Analytical Technique	Recoveries	Precision (% RSD)	Analytical Thresholds	Ref.
Mussels and cockles	Micro-MSPD	MSPD procedure same as above. Extracts were diluted with 2.6 mL of water and submitted to SPME by immersing a PDMS-DVB fiber for 45 min under continuous stirring.	OCPs and PDBEs	SPME and GC-ECD	55–130% relative recoveries	1–23% r	LODs: 0.3–7.1 μg kg^{-1} (ww) LOQs: 1.0–23 μg kg^{-1} (ww)	Moliner-Martinez et al. (2009)
Several mollusks (mussels, cockles, clams)	MSPD	0.5 g of freeze-dried sample, 0.5 g of a PSA as solid support, a cleanup sorbent combination of 1.75 g of Florisil (deactivated with 5% Milli-Q water), 1.75 g of acidified silica (10% w/w H_2SO_4) and 0.5 g of silica, and 10 mL DCM as elution solvent were used.	15 halogenated flame retardants (9 PBDEs + 9 novel HFRs)	GC-NCI-MS	70–120% except for DEHTBP in clams (46%)	1–23% r	LODs: 0.003–0.07 μg kg^{-1} (dw) except for decabromodiphenyl ethane (0.6 μg kg^{-1} [dw])	Villaverde-de-Sá et al. (2013)

Note: DCM, dichloromethane; DEHP, diethylhexyl phthalate; DEHTBP, bis(2-ethylhexyl)-3,4,5,6-tetrabromo-phthalate; EHMC, 2-ethyl-hexyl-4-trimethoxycinnamate; E1, estrone; E2, estradiol; EE2, ethynylestradiol; HFRs, halogenated flame retardants; FPD, flame photometric detector; LTQ, linear ion trap; MDL, method detection limit; MPE, monoalkyphthalate esters; MTBE, methyl *tert*-butyl ether; Na_2HCit, sodium hydrogen citrate; Na_3Cit, sodium citrate; NCI, negative chemical ionization; OD-PABA, 2-ethylhexyl 4-dimethylaminobenzoate; PAEs, phthalic acid esters; TBBPA, tetrabromobisphenol A; ww, wet weight.

Gallego-Gallegos et al. (2006) highlighted matrix effects as one of the main factors compromising the analytical performance, which is especially true for ESI-based mass spectrometric methods. Indeed, despite their high selectivity, ionization suppression (or enhancement) is a frequent phenomenon and the major source of bias in the results if not appropriately compensated for with often expensive, isotope-labeled internal standards. Pronounced matrix effects were observed for monobutyltin in mussels and for mono-, di-, and tributyltin in oyster. Therefore, the authors proposed a cleanup of the biological tissue extract with a tributyltin-targeted MIP (Gallego-Gallegos et al., 2006). When MIP cleanup was employed, an external standard calibration provided good results on certified reference materials avoiding the need for a tedious standard addition calibration (Gallego-Gallegos et al., 2006).

PLE seems to be the technique of choice for the preparation of biota samples in the latest years. Analytical methodologies have been developed for the determination of HBCD and tetrabromobisphenol A (TBBPA) (Dam et al., 2012), PCBs (Ottonello et al., 2014), UV filters (Gago-Ferrero et al., 2013), and multiclass pharmaceuticals in fish tissues (Huerta et al., 2013). Gago-Ferrero et al. (2013) developed and validated a method for the determination of eight extensively used UV filters and transformation products in different fish species from the Guadalquivir River basin. UV filters are of increasing environmental concern since they exhibit estrogenic activity and have been detected in seawater, wastewater, sediments, and biological fluids (e.g., milk, semen) (Gago-Ferrero et al., 2013). Biomagnification in predator fish species has been reported. The analytical work was based on a PLE methodology, in-cell cleanup with Florisil, SPE cleanup with Isolute C18 cartridges, and LC-MS/MS determination. The desired selectivity of the method was achieved using a mixture of 1:1 dichloromethane/ethyl acetate as extraction solvent, yet capable of extracting UV filters and transformation products, and up to 100°C extraction temperature can be used. The cleanup was performed by diluting the extract in water and percolating the solution through a reversed-phase adsorbent. Confirming the complexity of fish samples analysis, even after a two-step purification process, ionization suppression ranged from 15% to 65%, therefore requiring the use of isotopically labeled internal standards for compensating for the matrix effects (Gago-Ferrero et al., 2013).

Despite their higher hydrophilicity than that of conventional PBT compounds, the potential of certain pharmaceuticals and/or their metabolites to accumulate in tissues of aquatic organisms (crustaceans, mollusks, and fish), as a consequence of their chronic exposure, has been highlighted by several studies. Thus, Huerta et al. (2013) developed a method based on PLE, GPC cleanup, and LC-MS/MS for the determination of 20 PACs and metabolites from seven therapeutic families in fish homogenate, muscle, and liver. PLE was the most efficient technique, while QuEChERS provided acceptable recoveries (>40%) for most of the analytes and UAE could recover only 5 out of the 20 compounds. PLE afforded better overall recoveries, lower RSD (%), and significantly lesser matrix interferences during analysis when compared to QuEChERS (Huerta et al., 2013). To address the removal of lipid content from extracts, which is recognized as one of the most important sources of matrix effects, three techniques were tested: purification with neutral aluminum oxide placed at the bottom of the extraction cell, freezing-lipid filtration, and defatting by using hexane, with the first approach selected as the most efficient precleanup method. The main cleanup was conducted with a single GPC step, as the use of Florisil reduced the recoveries to a great extent and SPE on Oasis HLB followed by GPC also gave poor recoveries, namely, for β-blockers. Despite employing two sequential cleanup steps, as done in the previous study, ionization suppression as high as 83% was observed in liver extract (48% lipid content). The authors concluded that, as expected, ionization suppression (carbamazepine, clopidogrel, and venlafaxine) or enhancement (diclofenac) was more obvious as the percentage of lipids in the sample increased (Huerta et al., 2013). For an accurate determination of pharmaceuticals in fish samples, the authors recommend the use of the internal sample calibration method (internal standard calibration on matrix-matched extracts), especially in those cases where the analogue isotopically labeled standards are not available. For an optimum degree of accuracy, the authors demonstrated that internal sample calibration curves in an appropriate matrix, that is, extract of the same species or at least species with similar characteristics (lipid content), must be followed. The developed method allowed to quantify, for the first time, nine pharmaceutical compounds in fish from Mediterranean rivers (Ebro, Llobregat, Jucar, and Guadalquivir), all subjected to intensive anthropogenic activity (Huerta et al., 2013).

Ottonelo et al. (2014) used a different approach for the determination of PCBs in fish samples, which was dictated by the properties of the analytes (highly apolar) and the analytical technique (GC-MS). Indeed, GC-MS can be considered a more robust technique concerning the influence of the matrix on the accuracy of the results compared to LC-MS. The nature of the extraction solvent in PLE did not show a significant impact in the PCB recoveries, and a mixture of acetone/hexane (1:1 v/v) was selected due to practical considerations (lower toxicity and cost). The extract was subjected to a cleanup for fat digestion on an EXtrelut NT 3 cartridge, where the adsorbent (diatomaceous earth) acted as a solid support for 3 mL of concentrated sulfuric acid. A silica cartridge was placed in tandem with the former for further removal of polar interferences. During processing of fish samples, a cleanup is needed to prevent the deterioration of the chromatographic column and contamination of the ion source of the mass spectrometer by the coextracted fat, which causes poor analytical performance. Despite the ban of PCBs in the 1970s, they are still detected in the fat tissue of edible fish species; therefore, methods such as the one developed by Ottonelo et al. (2014), with a total extraction time of about 12 min per sample and usage of 13 mL of solvent for cleanup, simplify the analytical tasks and reduce costs.

Dam et al. (2012) proposed a method for the determination of HBCD and TBBPA, two of the most used BFRs worldwide, in eel, salmon, cheese, and by egg using PLE followed by acid silica cleanup and quantitative analysis by LC-MS/MS. HBCD is of environmental concern due to its high lipophilicity, accumulating in sediments and in the fatty tissue of aquatic organisms. The European Commission has classified it as PBT. Although much less bioaccumulative and more easily metabolized and eliminated from organisms, TBBPA still has been found in significant concentrations in sediments and aquatic species. Both substances are suspected of impairing the endocrine system. Due to the physicochemical characteristics of the compounds, suitable solvents also extract a great deal of lipids. The paper of Dam et al. (2012) is a good source of suitable extraction and cleanup protocols applied to analyze BFRs, where GPC, sulfuric acid oxidation, and SPE are highlighted as the preferred cleanup techniques. The authors tested these approaches and adopted an automated procedure (PowerPrep) integrating cleanup with a high-capacity disposable acid silica column followed by a Sep-Pak Plus silica cartridge for achieving satisfactory cleanup before LC-MS/MS analysis (Dam et al., 2012).

Tanoue et al. (2014) also addressed the determination of PPCPs in birds and fish tissues, using techniques other than PLE. The authors developed a comprehensive method capable of determining free and protein-bound PPCPs in the livers and kidneys of birds and in the plasma, livers, and brains of fish with selective removal of fats, proteins, pigments, other high-molecular weight compounds, and salts coextracted from these complex matrices. The methodology involved enzymatic hydrolysis with β-glucuronidase/sulfatase for deconjugation and/or deconnection of the protein-bound analytes, UAE and liquid–liquid partition followed by sequential cleanup using silica gel chromatography, GPC (facultative) and Oasis HLB purification, and analysis via ultrahigh-performance liquid chromatography (UHPLC) with MS/MS (Tanoue et al., 2014). Enzymatic hydrolysis gave better PPCP recoveries than simple extraction with acid ACN, while GPC cleanup was needed only for liver and brain samples, whose lipidic fraction could not be appropriately separated by silica gel chromatography. Even after such extensive sample pretreatment, ionization suppression on the ESI source was considerable, reaching 88%. Therefore, the method involved the use of isotope-labeled internal standards for compensating for such matrix effects. By using this strategy, acceptable internal standard-corrected recovery rates with low RSDs (<15%) were obtained for all biological samples tested, demonstrating the robustness of the method. The authors advocated that the limits of quantitation achieved were lower than those reported in previous studies of biota samples. Despite this judicious job on proper sample pretreatment for biota samples, the final protocol is labor intensive and consumes considerable volumes of organic solvents, particularly if a GPC step is included.

The application of the method to real samples revealed that nonsteroidal anti-inflammatory drugs (NSAIDs), antibacterial agents, and psychotropic agents were frequently detected in the fish tissues, which is a threat to these organisms and to fish-eating birds (top predators of the freshwater food web) (Tanoue et al., 2014).

QuEChERS has also been frequently applied to the analysis of biota samples. Bueno et al. (2013) investigated the occurrence and levels of carbamazepine (once the most ubiquitous PAC in the aquatic environment), oxcarbazepine, and six transformation products in marine mussels (*Mytilus galloprovincialis*) by using this technique. After rehydration of the freeze-dried sample, the authors extracted the mussels with ACN and partitioning was achieved with a mixture of citrate salts (Na_2SO_4, NaCl, sodium citrate:$2H_2O$, sodium hydrogencitrate:$3H_2O$), followed by cleanup with Na_2SO_4/PSA/C18. The mixture Na_2SO_4/PSA/C18 provided equally good results than the new material (Z-Sep Plus) and Z-Sep/C18, but it was preferred for the analysis of transformation products. The Z-Sep material contains zirconium atoms, which act as a Lewis acid, while the phosphate groups in phospholipids act as a strong Lewis base, strongly binding with zirconium atoms (Bueno et al., 2013). After cleanup, four compounds showed no matrix effect (<20%) in LC-HRMS, while three analytes underwent a medium effect (50–20%), even though the authors used matrix-matched calibration curves to compensate for the matrix effect and avoid any under-/overestimation during quantification. The authors compared the QuEChERS procedure with an optimized PLE method and concluded that the former was more advantageous in terms of extraction time (~1.5 h per eight samples vs. ~2.5 h per sample in PLE), organic solvent consumption (10 mL vs 41 mL in PLE), extraction efficiency, sample amount, use of laboratory glass material, and simpler analytical protocol.

Groz et al. (2014) applied the aforementioned QuEChERS method for the determination of three UV filters (2-ethyl-hexyl-4-trimethoxycinnamate, octocrylene, and octyldimethyl *p*-aminobenzoic acid), two UV stabilizers (UV-P and UV-326), and four musks (galaxolide, celestolide, cashmeran, and musk ketone) in marine mussels. Musks and UV filters have been recently recognized as important organic contaminants of the aquatic environment, due to their lipophilic properties ($\log K_{ow} > 4.30$), thus being prone to bioaccumulation. Among the three cleanup sorbents tested, the mixture Na_2SO_4/PSA/C18 provided better cleanup results than Z-Sep and $CaCl_2$/PSA for all compounds (Groz et al., 2014). The addition of 2% formic acid was important to disrupt compound–protein bindings. Matrix effect was negligible for all analytes (<25%) except for UV-P, which presented a high signal enhancement (>200%). Although fats are not freely soluble in ACN, mussels contain about 15% of lipids and a certain quantity is normally coextracted. Filtration in *n*-heptane was used for separating these substances. The authors demonstrated with this method that bathing activities in summer seasons are partially responsible for mussel contamination, besides the WWTP discharges (Groz et al., 2014).

Masiá et al. (2013a) applied QuEChERS methods for the determination of 50 currently used pesticides in sediment (original method) and biota (citrate-buffered method) of the Guadalquivir River basin for over 2 years (2010–2011). Farré et al. (2014) also used QuEChERS as a polyvalent method for the determination of pesticides in fruits, fish, bees, and sediments. Subsequent analysis by ultraperformance liquid chromatography (UPLC)–HRMS LTQ Orbitrap enabled the simultaneous detection/identification of not only the target 54 pesticides but also of nontarget pesticides, pharmaceuticals, drugs of abuse, and mycotoxins, as well as their metabolites. The QuEChERS method demonstrated potential for extracting a broad range of compounds; thus, when coupled with HRMS, with acquisition in scan and data-dependent MS/MS spectra, it is suitable for application in screening and retrospective analysis (Farré et al., 2014). The degree of ionization suppression and enhancement of the UHPLC/ESI LTQ-Orbitrap MS method was comparable to that observed in other mass analyzers. In the absence of isotope-labeled standards, matrix-matched standard calibration curves were required to compensate for matrix effects so as to improve the method's quantitative accuracy, especially for fat-rich matrices.

Campíns-Falcó et al. (2008) and Moliner-Martinez et al. (2009) developed miniaturized MSPD methods for the determination of PAHs and OCPs and PBDEs in biota, respectively. While in the first work, MSPD was coupled to in-tube in-valve SPME and HPLC-FLD analysis, the second employed SPME with a PDMS-DVB fiber and GC-ECD/FID analysis. Campíns-Falcó et al. (2008) studied the smallest amount of sample that could produce reproducible results, that is, 100 mg. With this principle in mind, all the remaining steps could be miniaturized, still providing good overall sensitivity, as almost the entire extract was further concentrated by IT-SPME and injected for analysis. The amounts of C18 (dispersant) and Florisil (fat retainer) used were in the phase-to-sample ratios of 4:1 and 1:1, respectively. The MSPD tubes were flushed with 1.2 mL of ACN. The mussel and tellin extracts collected after MSPD were diluted with water and injected into a 70 cm capillary column coated with 5% phenyl–95% PDMS mounted in a switching valve and later diverted to the analytical column. The miniaturized MSPD procedure coupled with IT-SPME is simple and rapid (total analysis time of ≈20 min), and the consumption of organic solvents and extractive phases was drastically reduced. The method is according to green chemistry principles and can be automated. By comparing the features of the developed method with others published in the literature, encompassing Soxhlet extraction, UAE, MAE, PLE, and MSPD, the authors demonstrated that much lower volumes of solvents were used and the sensitivity was maximized (lowest LODs among those reported). The information given in table 5.5 of the paper of Campíns-Falcó et al. (2008) is worth checking as a compilation of methods for the determination of PAHs in biota samples. Following the same MSPD procedure, Moliner-Martinez et al. (2009) developed a method for the analysis of OCPs and PBDEs in mussels and cockles. The authors tested several dispersant/fat retainer agents (Florisil/silica, C18, and C18/Florisil), concluding that C18/Florisil provided better recoveries and compatibility with the subsequent SPME step. This methodology benefited from the same advantages already mentioned. Villaverde-de-Sá et al. (2013) also adopted an MSPD procedure for the determination of 15 halogenated flame retardants (HFRs)—nine BDEs and six novel HFRs—in different kinds of mollusks, followed by GC-MS with negative chemical ionization (NCI). A detailed study of the MSPD conditions was performed, including different solid support materials (C18, silica, Florisil, alumina, and PSA), activation/deactivation of Florisil, different contents of sulfuric acid in silica, and different volumes of elution solvent. Better recoveries of some analytes and less residue (mostly fat) were obtained with PSA as the dispersing agent. For the cleanup step, acidified silica impregnated with 10% of H_2SO_4 and Florisil deactivated with 5% of Milli-Q water were used in combination, allowing a fat content below 0.5% in the extract (Villaverde-de-Sá et al., 2013). Nevertheless, a signal enhancement for some analytes (up to threefold) could not be avoided. These matrix effects could not be compensated for the novel HFRs either by the use of internal standards or by the inclusion of analyte protectors (sorbitol or olive oil). Thus, calibration through standard addition was required for correct quantification. It is quite evident that the inclusion of an additional selective/cleanup technique such as SPME, adopted in two previous miniaturized procedures, simplifies the chromatographic determination. The authors highlighted, once again, the advantages of the MSPD procedure, including low solvent and sample intake, simplicity, reduced cost, and integration of extraction and cleanup into a single step (Villaverde-de-Sá et al., 2013).

Simon et al. (2010) developed a dedicated cleanup procedure for the purification of extracts of biota prior to biotesting of the endocrine-disrupting activity. The quality of the extraction and cleanup steps was identified as a key factor for successful bioassay measurements (Simon et al., 2010). The stepwise cleanup included the following:

- Dialysis through membranes (polyethylene lay-flat membranes)
- GPC cleanup, performed on two polystyrene–diphenylbenzene columns (PLgel)
- Normal-phase HPLC, for removal of lipids with smaller molecular size, such as cholesterol, some fatty acids, and triglycerides, which were still present in the GPC extracts

Even after such extensive cleanup, some small lipids remained in the extracts, but the purpose was to determine endocrine-disrupting activity by using bioassays, and LC-MS/MS would probably not be affected by this residual fraction (0.2% lipids) (Simon et al., 2010).

SFE has not been exploited recently in the analysis of biota, and one of the most representative works remains the determination of OCPs and some metabolites in frogs, developed by Nerín et al. (2002).

According to the above discussion, determination of organic pollutants in biota can require very time-consuming and labor-intensive extraction and cleanup procedures, combining two or more extraction techniques/steps to eliminate coextracted lipids normally present in these types of matrices.

5.6 SAMPLE PREPARATION METHODS FOR THE DETERMINATION OF ORGANIC POLLUTANTS IN SEWAGE SLUDGE

5.6.1 Introduction

Municipal wastewater management strategies are currently characterized by two aspects. The first concerns the ongoing effort to improve the effluent quality by upgrading the existing technology and the application of new efficient treatment systems. The second relates to the growing awareness of the problems associated with the sludge produced by the various processes. These problems include increased sludge production, sustainable sludge disposal, and the risk associated with the environment and human health (Kelessidis and Stasinakis, 2012). Nowadays, more than 50,000 WWTPs operating in the European Union produce 8.1 Mt of biosolids per year. The implementation of Directive 91/271/EC by the new member states is expected to cause a significant increase in annual sludge production, surpassing 13 Mt by 2020 (Kelessidis and Stasinakis, 2012). The main disposal methods so far are agricultural application, disposal in landfills, incineration, and composting. Among them, the application of sewage sludge as fertilizer to agricultural lands has become the most widespread disposal method since it recycles the nutrients and improves soil quality. However, considering sludge as a by-product of the wastewater treatment process, it accumulates numerous xenobiotic organic compounds, pathogens, and heavy metals, which can cause adverse effects on biological life and the soil ecosystem (Díaz-Cruz et al., 2009). Systematic investigation on the fate and long-term monitoring surveys of synthetic organic compounds has received little attention until now, due to the lack of regulation (Díaz-Cruz et al., 2009).

During the past 10 years, several research papers have been published on the occurrence of micropollutants in sewage sludge. Among them, surfactants, PCPs, synthetic musk fragrances, PACs, estrogens, organophosphates (OPs), and BFRs, polychlorinated naphthalenes, PFCs (perfluorooctane sulfonic acid [PFOS], perfluorooctanoic acid), PBDEs, organotins, PAEs and PDMSs, benzothiazoles, illicit drugs, and UV filters have been classified as the most frequently detected contaminants. The concentration levels of these compounds have been reviewed recently by some authors, ranging from a few nanograms per kilogram (estrogens, PFOS, PACs) to several grams per kilogram dry weight (dw) (linear alkylbenzene sulfonates) (Díaz-Cruz et al., 2009; Clarke and Smith, 2011). Due to the growing concerns regarding their toxicity and environmental persistence, this, in turn has resulted in extensive research on the development of new analytical methods and monitoring efforts to provide reliable data on their concentrations levels in sludge (Clarke and Smith, 2011).

5.6.2 Current Trends in Sample Preparation of Sewage Sludge

Based on current methodologies for the determination of trace contaminants in sewage sludge, the typical steps of sample preparation consist of sludge stabilization, homogenization, extraction for isolation and preconcentration of the target analytes, purification of the final extracts, and, occasionally, derivatization. Analysis is then carried out by chromatographic methods coupled with MS or hybrid MS (MS/MS, QTOF, Orbitrap) for the identification and quantification of residues. From an analytical perspective, sewage sludge is a challenging matrix owing to the problems derived from (i) matrix interferences (surfactants, colloids, sulfur, and lipids) observed during quantitative analysis, (ii) the presence of trace levels of analytes, and (iii) strong interactions of analytes (hydrophobic and electrostatic) with organic matter.

Among the classical extraction methods, UAE is currently one of the most used techniques, because the equipment required is widely available and the extraction can be performed using a small volume of solvent and shortened extraction times compared to Soxhlet/Soxtec extraction and mechanical agitation. Current trends are toward less laborious and automated extraction methods providing greater selectivity and good precision. PLE and MAE are currently the most commonly employed novel techniques. The extraction is usually performed at elevated temperatures and pressures, while extraction times vary from 5 to 30 min depending on the specific requirements of the method. Alternative extraction methods such as SFE, MSPD, SPME, QuEChERS, and FUSLE proved to be efficient in other research areas, but did not reach widespread application in sludge analysis, being limited to a few studies (Table 5.4). The general trend in sludge analysis is characterized by developing and validating multiresidue methodologies for the simultaneous determination of analytes with widely different polarities (acidic, neutral, and basic). Sample preparation is analyte dependent and should take into consideration the physicochemical characteristics of the target compounds, such as polarity, hydrophobicity, and charge, usually by following factorial experimental designs (Doehlert, Box–Behnken, Plackett–Burman designs) for method optimization (Morales et al., 2005; Llop et al., 2010; Negreira et al., 2011; Dorival-García et al., 2012; Benítez-Villalba et al., 2013). The most recent methods for sludge analysis encompass simultaneous extraction in a single step by using PLE or MAE followed by SPE as a cleanup step. Filtration of the sample extract prior to the concentration procedure and proper pH adjustment of the dissolved solutes in the water phase has also been reported as a common practice (Barron et al., 2008; Lillenberg et al., 2009; Ordoñez et al., 2013). Extraction of organic acids is usually performed at pH values ranging two units below the pK_a of the acid. Similarly, a base in neutral form needs to be adjusted to two units above the pK_b. Further sample cleanup may be required depending on the instrumentation employed. In most cases, the cleanup step after the extraction is crucial in order to remove organic and inorganic coextractives that might interfere with analyte chromatographic separation and detection, resulting in high detection limits and decreased reproducibility. Purification of the eluates can be performed by normal-phase SPE with polar sorbents such as silica gel, alumina, and charcoal; GPC and treatment with copper powder; or a combination of them. Among them, SPE is clearly the most preferred technique since it enables the isolation, preconcentration, and separation of numerous solutes, since elution conditions are selective for the retention and/or recovery of the analytes. A wide range of SPE sorbents with different properties has been reported in the literature that enable interactions based on adsorption, H bonding, polar and nonpolar interactions, cation/anion exchange, or size exclusion. Sorbent selection is generally based on the analyte (nonpolar, polar, or ionized) and whether it is ionized (strong or weak acid/base), while sorbent conditioning and solvent elution can be varied for method development. Another widely adopted practice suggests the combination of two selected SPE adsorbents with different properties operated either in series or in parallel in order to classify target residues into two groups for further analysis. Similarly, selective derivatization before analysis by GC is followed. Online SPE coupled to LC-MS/MS has also been reported by a number of authors, as it offers minimum sample preparation, automation, and high selectivity (Montesdeoca-Esponda et al., 2013; Vega-Morales et al., 2013).

In recent studies, both matrix-matched calibration and standard calibration curves were plotted in order to evaluate the matrix effects in terms of recovery rates, standard deviations, and measurement uncertainties, where the addition of stable isotopes was used to compensate for these effects during quantification (García-Galán et al., 2013; Albero et al., 2014). Radjenović et al. (2009) reported strong variations in the recovery values of 31 PACs depending on the sludge type (primary, secondary, Koch hollow fiber membrane bioreactors [MBRs], Kubota flat sheet MBRs, and treated sludge), attributing this effect to different compositions of sludge (e.g., fatty acids, salts, amines). Similarly, Martín et al. (2010) concluded that the sludge matrix components can strongly influence the extraction efficiency of 16 PACs and proposed the need for validating the method in

TABLE 5.4
Overview of Analytical Protocols Applied to the Determination of Selected Persistent and Emerging Pollutants in Sewage Sludge

Matrix	Extraction Technique	Extraction and Cleanup Protocols	Compounds Determined	Analytical Technique	Recoveries	Precision (% RSD)	Analytical Thresholds	Ref.
Sewage sludge	PLE	1 g of sample extracted with dichloromethane at 100°C. Cleanup: filtration.	4 abused drugs and their metabolites	LC-MS/MS	25–65%	<14% r <20% R	LODs: 0.5–10 μg kg^{-1}	Arbeláez et al. (2014)
Digested sludge	PLE	1.0 g of sludge mixed with sand. Extraction with 50:50 v/v MeOH/H_2O at 60°C. Cleanup: SPE, adjustment to pH 5.5, Oasis HLB cartridge.	27 PACs	LC-MS/MS	2–130%	<16.5%	LODs: 2–580 μg kg^{-1}	Barron et al. (2008)
Sewage sludge	PLE	4–6 g sample mixed with 10 g diatomaceous earth. Extraction with ethyl acetate at 90°C and 150 bar. Cleanup: silica SPE and size exclusion chromatography.	15 PCPs, plasticizers, and flame retardants	GC-MS (PTV splitless mode)	36–114%	5–23%	LODs: 3–30 μg kg^{-1}	Chen and Bester (2009)
Primary and secondary sewage sludge	PLE	0.5 g of sample mixed with sand. Extraction with aqueous 500 mM formate buffer (pH 3.5) at 80°C. Cleanup: SPE, adjustment to pH 5.5, Oasis HLB cartridge.	6 artificial sweeteners	LC-MS/MS	72–105%	<16%	LODs: 0.3–16 μg kg^{-1}	Ordoñez et al. (2013)
Sewage sludge and soil samples	PLE	2 g of sample mixed with Hydromatrix dispersing agent. Extraction with ACN/water (25:75 v/v) at 50°C. Cleanup: filtration and SPE, Oasis HLB cartridge.	22 sulfonamides, including 5 acetylated metabolites	LC-MS/MS	60–130%	<23%	LODs: 0.03–2.23 μg kg^{-1}	Montesdeoca-Esponda et al. (2013)

(Continued)

TABLE 5.4 (CONTINUED)
Overview of Analytical Protocols Applied to the Determination of Selected Persistent and Emerging Pollutants in Sewage Sludge

Matrix	Extraction Technique	Extraction and Cleanup Protocols	Compounds Determined	Analytical Technique	Recoveries	Precision (% RSD)	Analytical Thresholds	Ref.
Sewage sludge	PLE	0.5 g of sample mixed with 5 g activated Al_2O_3. Extraction with a mixture of MeOH/water (9:1 v/v) at 50°C. Cleanup: SPE, polymeric Evolute ABN cartridge.	20 abused and illicit drugs and their metabolites	LC-ESI-(QqLIT) MS/MS	26–89%	14%	LODs: <3.3 μg kg^{-1} for all compounds except cannabinoids (LODs: 8.2–22.5 μg kg^{-1})	Mastroianni et al. (2013)
Anaerobic digested sludge	PLE	1 g of sample mixed with Hydromatrix. Extraction with a MeOH/water mixture (1:2 v/v) at 100°C. Cleanup: SPE, Oasis HLB cartridge and filtration.	43 PACs	LC-QqLIT-MS	40.3–215%	<13% r <15% R	LODs: 0.01–8.84 μg kg^{-1}	Jelic et al. (2009)
Sewage sludge	PLE	9 ± 1 g (ww) of sample mixed 1:1 with sand. Extraction with 0.35% phosphoric acid and ACN mixture (1:1 v/v) adjusted to pH 2.50 with 0.01 M citric acid monohydrate. Cleanup: SPE, adjustment to pH 2, Oasis HLB cartridge.	7 fluoroquinolones, tetracyclines, and sulfonamides	LC-MS	43–96%	–	LODs: 0.1–160 μg kg^{-1}	Lillenberg et al. (2009)
Sewage sludge	PLE	1 g of sample mixed with Hydromatrix. Extraction with MeOH–water (1:2 v/v) at 100°C. Cleanup: SPE, Oasis HLB cartridge and filtration.	31 PACs	LC-MS/MS	3.2–123.6%	0.1–15.3% r 0.9–17.4% IP	LODs: 0.35–159.8 μg kg^{-1} for treated sludge 0.15–113.8 μg kg^{-1} for primary and secondary sludge 0.17–258.5 μg kg^{-1} for the sludge from the membrane bioreactor s	Radjenović et al. (2009)

(Continued)

TABLE 5.4 (CONTINUED)
Overview of Analytical Protocols Applied to the Determination of Selected Persistent and Emerging Pollutants in Sewage Sludge

Matrix	Extraction Technique	Extraction and Cleanup Protocols	Compounds Determined	Analytical Technique	Recoveries	Precision (% RSD)	Analytical Thresholds	Ref.
Primary and biological sludge	PLE	Cells were filled (bottom to top) with 1 g of diatomaceous earth, 0.5 g of graphitized carbon, 0.5 g of diatomaceous earth, and 0.5 g of sludge, previously homogenized with 2 g of diatomaceous earth. Extraction with *n*-hexane/dichloromethane (80:20 v/v) at 75°C. Cleanup: SPE cartridge containing 0.5 g of PSA.	8 UV filters	GC-MS	73–112%	6–13%	LOQs: 17–61 μg kg^{-1}	Negreira et al. (2011)
Sewage sludge	PLE	1 g of sample mixed with activated Al_2O_3. Extraction with mixture of water (pH 7) and MeOH (1:1) at 100°C. Cleanup: filtration.	10 PCPs (including UV filters, preservatives, and antimicrobials)	UHPLC-MS/MS	<72% except for 2,4-dihydroxy-benzophenone (UV filter) (30%)	<8% r <13% R	LODs: 8 μg kg^{-1}	Nieto et al. (2009)
Sewage sludge	PLE	1 g of sample extracted with MeOH/acetone (1:1) followed by two 3 min cycles with water (pH 7):MeOH (1:1) at 75°C. Cleanup: filtration.	11 estrogens and conjugated estrogens	LC-MS/MS	>81%	<6% r <9% R	LODs: <26 μg kg^{-1} except for α-E2, E2, EE2, and E2-17α in the range 150–175 μg kg^{-1}	Nieto et al. (2008)
Sewage sludge	PLE	500 mg of sample mixed with quartz. Extraction with MeOH–H_2O (50/50, v/v) at 100°C. Cleanup: SPE, adjustment to pH 4, Oasis HLB cartridge.	11 sulfonamides, macrolides, and trimethoprim	LC-MS/MS	78–142%	2–8%	LOQs: 2–51 μg kg^{-1}	Göbel et al. (2005)
Sewage sludge	PLE	1.0 g of sewage sludge extracted with ethyl acetate at 100°C. No cleanup: direct injection.	BPA and its chlorinated derivatives	LC-MS/MS	97.7–100.6%	<6% both r and IP	LODs: 4–8 μg kg^{-1}	Dorival-García et al. (2012)

(*Continued*)

TABLE 5.4 (CONTINUED)
Overview of Analytical Protocols Applied to the Determination of Selected Persistent and Emerging Pollutants in Sewage Sludge

Matrix	Extraction Technique	Extraction and Cleanup Protocols	Compounds Determined	Analytical Technique	Recoveries	Precision (% RSD)	Analytical Thresholds	Ref.
Treated sludge (including thickening, dewatering, and anaerobic digestion)	PHWE	0.5 g of sample extracted with 0.01 M NaOH at 120°C. No cleanup: direct injection.	4 NSAIDs	HF-LPME-LC-ESI-MS	101–109%	<11.4 r <13.1 R	LODs: 0.4–3.7 μg kg^{-1}	Saleh et al. (2011)
Anaerobic digested and activated sludge	PHWE	50 mg of sample mixed with quartz sand. Extraction with water (pH 9.4) at 200°C. Cleanup: SPE, Carbograph 4 cartridge.	Surfactants	LC-MS	>87%	<8%	–	Bruno et al. (2002)
Mixture of primary and secondary sludge	PHWE	1 g of sample mixed with 3 g of diatomaceous earth. Extraction with water at 80°C. Cleanup: SPE, Oasis HLB cartridge and filtration.	13 benzothiazole, benzotriazole, and benzenesulfonamide derivates	LC-Orbitrap-HRMS	>80%	<10% both r and R	LODs: 0.25–25 μg kg^{-1}	Zhou et al. (2012)
Sewage sludge	SPLE	0.5 g of sample mixed with 4 g of Florisil. Extraction with dichloromethane at 130°C. Cleanup: Florisil as cleanup sorbent eluted with dichloromethane.	BPA and APs	GC-MS (50 mL of BSTFA at 50°C for 15 min)	81–105%	<6%	LODs: 10.7 μg kg^{-1} for BPA 1.2–41.6 μg kg^{-1} for APs	Martínez-Moral and Tena (2011)
Compost	PLE	1 g of sample extracted with MeOH at 120°C. No cleanup: extracts directly injected in the chromatograph.	5 parabens	UPLC-MS/MS	94.6–103.7%	0.5–4.7% r 1.8–6.4% IP	LODs: 3–4 μg kg^{-1}	Benítez-Villalba et al. (2013)
Compost	USE	1 g of sample extracted with 5 mL of MeOH. Vortexed for 2 min and sonicated for 15 min at 75% amplitude, two extraction cycles.	5 parabens	UPLC-MS/MS	94.0–105.0%	1.6–3% r 1.1–4.6% IP	LODs: 3–7 μg kg^{-1}	Benítez-Villalba et al. (2013)

(*Continued*)

TABLE 5.4 (CONTINUED)
Overview of Analytical Protocols Applied to the Determination of Selected Persistent and Emerging Pollutants in Sewage Sludge

Matrix	Extraction Technique	Extraction and Cleanup Protocols	Compounds Determined	Analytical Technique	Recoveries	Precision (% RSD)	Analytical Thresholds	Ref.
Sewage sludge	MAE (household microwave oven)	1 g of sample extracted with 3:2 MeOH/water. Cleanup: SPE, adjustment to pH 7, Oasis HLB cartridge, and filtration.	18 PACs	GC-MS (70 μL of BSTFA + 1% TMCS at 70°C for 20 min)	91–101%	<7%	LODs: 0.8–5.1 ng kg^{-1}	Azzouz and Ballesteros (2012)
Sewage sludge	MAE	0.03–0.3 g of dried sample extracted with 1:1 hexane/acetone with the addition of 1 mL of water at 120°C. Cleanup: SPE, Oasis HLB cartridge.	NP and NPEO	HPLC-FLD	61.4% for NPEO and 91.4% for NP	<5% r	LODs: 1.82 mg g^{-1} for NPEO and 2.86 mg g^{-1} for NP	Fountoulakis et al. (2005)
Raw sludge samples from the thickener	MAE	1 g of sample extracted with 5 mL MeOH. Cleanup: filtration, online SPE, two Oasis HLB cartridges working in parallel.	11 EDCs	UHPLC-MS/MS	>77%	<9%	LODs: 0.1–0.7 μg kg^{-1}	Vega-Morales et al. (2013)
Primary and activated sludge	MAE	0.5 g of activated sludge samples extracted with 50 mL H_2O from 60°C to 100°C. Cleanup: DME (0.5 g neutral alumina as sorbent and 0.25 g aluminum sulfate as electrolyte). SPE, Oasis HLB cartridge.	4 NSAIDs	GC-MS (hexamethyl-disilazane + trifluoroacetic acid and hydroxylamine HCl in pyridine)	80–105 ± 15%	<18%	LODs: 15–22 μg kg^{-1}	Dobor et al. (2010)
Primary, secondary, raw, and digested sludge	MAE	0.25–0.5 g sample mixed with 15 g of anhydrous Na_2SO_4. Extraction with 30 mL of hexane/acetone (3:1 v/v) from room temperature to 130°C. Cleanup: silica gel column with celite and anhydrous Na_2SO_4.	8 PBDEs	GC-NCI-MS	Dry sludge: 95–108% Wet sludge: 74–113%	Dry sludge: <8.8% Wet sludge: <8.4%	LODs: 1.8–6.1 μg kg^{-1} for dry sludge 1.4–7.7 μg kg^{-1} for wet sludge	Shin et al. (2007)

(Continued)

TABLE 5.4 (CONTINUED)
Overview of Analytical Protocols Applied to the Determination of Selected Persistent and Emerging Pollutants in Sewage Sludge

Matrix	Extraction Technique	Extraction and Cleanup Protocols	Compounds Determined	Analytical Technique	Recoveries	Precision (% RSD)	Analytical Thresholds	Ref.
Sewage sludge	MAE	1 g of sample extracted with ACN. Cleanup: filtration, online SPE, two columns working in parallel.	7 BUVSs	LC-MS/MS	46.1–83.9%	7.8–15.5%	LODs: 53.3–146 ng kg^{-1}	Montesdeoca-Esponda et al. (2013)
Mixture of primary and secondary sludge (80:20)	MAE	0.5 g of sample extracted with 30 mL of a 1:1 acetone/MeOH mixture at 130°C. Cleanup: SPE, adjustment to pH 2.5, Oasis HLB cartridge.	Triclosan and 2 *metabolites* (2,4-DCP and 2,4,6-TCP)	GC-MS/MS (50 μL of MTBSTFA for 5 min at room temperature)	78.3–106.6%	<13%	LOQs: 0.4–0.8 μg kg^{-1}	Morales et al. (2005)
Digested sludge	MAE	5.0 g of sample extracted with 25 mL MeOH at 105°C. 10 mL of the resulting extract was concentrated in a heater at 80°C, 1 h.	Linear alkylbenzene sulfonate (C10:0–C14:0)	LC-FLD		1.3–3.2%	LODs: 3.3–5.4 mg kg^{-1}	Cantarero et al. (2011)
Sewage sludge	UAE	1 g of sample extracted with 5 mL of MeOH containing 1% v/v formic acid for 20 min (three times). Cleanup: SPE, HLB cartridge followed by ENVI-Carb cleanup cartridge.	4 EDCs and 10 PPCPs	GC-MS (100 μL of MTBSTFA at 70°C for 60 min)	57.9–103.1%	1.3–9.5%	LODs: 4.7–39 μg kg^{-1}	Yu and Wu (2012)
Sewage sludge	UAE	0.1 g of sample extracted with 20 mL + 2 × 10 mL ethyl acetate/cyclohexane (5:2 v/v) for 10 min. Cleanup: Florisil cartridges.	27 flame retardants (OFFRs, NBFRs, and PBDEs)	GC-EI-MS/MS	OPFRs: 64–131% PBDEs: 90–121% NBFRs: 92–140%	OPFRs: ≤14% PBDEs: ≤20% NBFRs: ≤13%	LODs: 3.7–575 μg kg^{-1}	Cristale and Lacorte (2013)

(Continued)

TABLE 5.4 (CONTINUED)
Overview of Analytical Protocols Applied to the Determination of Selected Persistent and Emerging Pollutants in Sewage Sludge

Matrix	Extraction Technique	Extraction and Cleanup Protocols	Compounds Determined	Analytical Technique	Recoveries	Precision (% RSD)	Analytical Thresholds	Ref.
Sewage sludge and sediment samples	UAE	0.1 g of sludge and 1 g of sediment, vortexed for 2 min and ultrasonicated for 15 min with ACN/ultrapure water (5:3, v/v) Cleanup: SPE, adjustment to pH 7, Oasis HLB cartridge.	19 PACs, steroid hormones, and PCPs	UHPLC-MS/MS (dervivatized with 200 μL of dansyl chloride at 60°C for 5 min)	63–119%	15%	LODs: 0.1–3 μg kg^{-1}	Yu et al. (2011)
Dewatered secondary sludge	UAE	40 mg of sample extracted with a mixture of $MeOH/H_2O$ (2.5:1.5 mL) at 50°C for 30 min. Cleanup: SPE, adjustment to pH 2.5, C18 cartridge.	5 EDCs and 4 NSAIDs	GC-MS (50 μL of BSTFA 1% TMCS with 10 μL of pyridine at 70°C for 20 min)	91.8–117%	4–12% r 5.4–14% IP	LODs: 15.0–32.9 μg kg^{-1}	Samaras et al. (2011)
Sewage sludge	UAE	0.5 g of sample sonicated for 15 min with 5 mL MeOH/ McIlvaine buffer, 50:50 v/v, pH = 3, two extraction cycles. No cleanup: direct injection.	13 quinolones	LC-ESI-MS/MS	97.9–104.8%	<7% both r and IP	LODs: 2–5 μg kg^{-1}	Dorival-García et al. (2013)
Activated sludge	UAE	0.5 g of sample extracted with 10 mL of DCM/MeOH (1:9 v/v). Cleanup: C18 cartridge.	HBCD, TBBPA (BPA, mono-BBPA, di-BBPA, and tri-BBPA)	LC-QqLIT-MS	39–120%	<13%	LODs: 1.4–66 μg kg^{-1}	Guerra et al. (2010)

(Continued)

TABLE 5.4 (CONTINUED)
Overview of Analytical Protocols Applied to the Determination of Selected Persistent and Emerging Pollutants in Sewage Sludge

Matrix	Extraction Technique	Extraction and Cleanup Protocols	Compounds Determined	Analytical Technique	Recoveries	Precision (% RSD)	Analytical Thresholds	Ref.
Primary, secondary, and anaerobically digested dehydrated sludge and compost	UAE	1 g of primary or secondary sludge, 1.5 g of digested sludge, or 2 g of compost extracted with 5 and 2 mL of MeOH and 2 mL of acetone. In each extraction step, sample was vigorously shaken for 30 s and ultrasonicated for 15 min. Cleanup: SPE, adjustment to pH 2, Oasis HLB cartridge.	16 PACs	HPLC	Primary sludge: 61.5–107% Secondary sludge: 47.8–100% Digested sludge: 41.1–115% Compost: 81–108%	–	LODs: Primary sludge: 3.21–355 μg kg^{-1} Secondary sludge: 3.09–350 μg kg^{-1} Digested sludge: 1.39–249 μg kg^{-1} Compost: 0.40–166 μg kg^{-1}	Martín et al. (2010)
Dewatered sewage sludge	LSE-UAE	100 mg of dewatered sludge extracted with 7.5 mL of 1% v/v acetic acid and 1.5 mL of MeOH, LSE (3 × 7.5 mL) by vortex followed by UAE for 15 min. Cleanup: SPE, adjustment to pH 4, Oasis HLB cartridge.	18 PFCAs, PFASs, and PFSAs	LC-ESI-MS/MS	Absolute recoveries: 71–115% except for the long-chain PFCs (26–39%)	<15%	LODs: 0.15–1.5 μg kg^{-1} dw	Arvaniti et al. (2014)
Dehydrated sludge	FUSLE	0.5 g of sample and 8 mL of ACN as extraction solvent. Ultrasonic irradiation power of 65% for only 20 s, two cycles. Cleanup: filtration.	PFCAs and PFOS	UPLC-QTOF-MS	69–104%	<8%	LODs: 0.2 μg kg^{-1}	Martínez-Moral and Tena (2013)
Sewage sludge	UAE	0.5 g of sample extracted three times with 10 mL MeOH/formic acid (96:4 v/v), each period 15 min. Cleanup: SPE, adjustment to pH 2, Oasis HLB cartridge.	Veterinary PACs		13–91%	4–21%	LODs: 0.08–0.3 mg g^{-1}	Carvalho et al. (2013)

(Continued)

TABLE 5.4 (CONTINUED)
Overview of Analytical Protocols Applied to the Determination of Selected Persistent and Emerging Pollutants in Sewage Sludge

Matrix	Extraction Technique	Extraction and Cleanup Protocols	Compounds Determined	Analytical Technique	Recoveries	Precision (% RSD)	Analytical Thresholds	Ref.
Nondigested sludge	MSPD	0.5 g of lyophilized samples mixed and dispersed with 2 g diatomaceous earth in a glass mortar with a pestle for 5 min. Cleanup: polypropylene syringe containing 1 g of PSA.	9 BUVSs	GC-QTOF-MS	70–111%	2–13%	LODs: 2–10 μg kg^{-1}	Casado et al. (2013)
Sewage sludge	MSPD	0.5 g of freeze-dried sludge samples were mixed and dispersed with 2 g of C18 as cleanup sorbent. Cleanup: PSA.	8 OPs	LC-ESI-QTOF-MS	69–123%	<13%	LODs: 2–50 μg kg^{-1}	Celano et al. (2014)
Sewage sludge	MSPD	1 g lyophilized sample mixed with 2 g of C18 and 1 g anhydrous sodium sulfate. Mixture blended with a pestle and transferred to Florisil column. Analytes eluted with 10 mL ethyl acetate/MeOH (90:10, v/v) and extract concentrated to 1 mL in evaporator. 100 μL extract analyzed by GC-MS/MS.	Parabens and two chlorinated by-products	GC-MS/MS	80–125%	–	LODs: 0.1–2.0 μg kg^{-1}	Albero et al. (2012)
Sewage sludge	MSPD-UAE	2 g of lyophilized sample blended with alumina deactivated with water at 3% (3 g) and 3 g activated copper powder, extracted with 2 × 3 mL dichloromethane assisted by ultrasound. Cleanup: filtration, SPE, C18 columns.	16 OCPs	GC-MS	84.5–104.1%	<13%	LODs: 0.03–0.7 μg kg^{-1}	Sánchez-Brunete et al. (2008)

(*Continued*)

TABLE 5.4 (CONTINUED)
Overview of Analytical Protocols Applied to the Determination of Selected Persistent and Emerging Pollutants in Sewage Sludge

Matrix	Extraction Technique	Extraction and Cleanup Protocols	Compounds Determined	Analytical Technique	Recoveries	Precision (% RSD)	Analytical Thresholds	Ref.
Sewage sludge	MSPD-UAE	2 g of lyophilized sludge were placed in a glass mortar containing 3 g of alumina. Cleanup: SPE, C18 columns.	27 PAHs	GC-MS	94.2–101.3%	<14%	LODs: 0.03–0.45 μg kg^{-1}	Sánchez-Brunete et al. (2007)
Sewage sludge	MSPD	0.2 g of sample was soaked in 1 mL of saturated methanolic potassium hydroxide solution in a glass mortar, then 1 g of Florisil and 0.5 g of anhydrous sodium sulfate were added, and the mixture was thoroughly blended with a pestle. Cleanup: packed column containing a layer of 1 g Florisil and 2 g of silica.	17 PAHs	HPLC-FLD	76.3–103.6%	<6%	LODs: 0.0001–0.005 mg kg^{-1}	Pena et al. (2008)
Pelletized sewage sludge and poultry manure samples	QuEChERS	0.5 g of sample extracted with 2 mL of water and 3 mL of ACN, shaken in a vortex for 1 s, followed by UAE for 15 min. Afterward, $MgSO_4$ (800 mg) and NaOAc (300 mg) were added to the mixture, which was hand shaken for 1 min, vortexed for 30 s. and finally centrifuged for 15 min. Cleanup: dSPE cleanup (Sovirel tube containing 500 mg $MgSO_4$, 200 mg PSA, and 200 mg C18). Further cleanup of the extracts was carried out using Evolute ABN cartridges.	13 natural and synthetic hormones	GC-MS/MS (100 μL of MSTFA/TMIS/DTE [1000:2:5 v/v/w] at room temperature for 5 min)	76–124%	≤16%	LODs: 0.5–3.0 μg kg^{-1}	Albero et al. (2014)

(Continued)

TABLE 5.4 (CONTINUED)
Overview of Analytical Protocols Applied to the Determination of Selected Persistent and Emerging Pollutants in Sewage Sludge

Matrix	Extraction Technique	Extraction and Cleanup Protocols	Compounds Determined	Analytical Technique	Recoveries	Precision (% RSD)	Analytical Thresholds	Ref.
Drinking water treatment sludge	QuEChERS	10 g of sample hand-shaken for 15 s with 10 mL ACN, acidified with 100 μL acetic acid, partitioning of the analytes into the organic phase was performed by adding 4 g $MgSO_4$ and 1 g NaCl. Cleanup: dSPE cleanup (2 mL aliquot of the extract was added to 150 mg $MgSO_4$ and 50 mg of chitin).	12 pesticides, PACs, and PCPs	LC-ESI-MS/MS	50–120%	≤15%	LODs: 1–50 μg kg^{-1}	Cerqueira et al. (2014)
Sewage sludge (limed, digested, dried, liquid, and composted)	QuEChERS	2 g of sample extracted added with 10 mL 0.1 M EDTA and 10 mL ACN + 1% v/v acetic acid, with 30 s vortex shaking. Cleanup: dSPE cleanup (150 mg PSA and 900 mg $MgSO_4$ extract were filtered through a 0.45 μm PTFE syringe filter).	136 PACs and hormones	LC-TOF-MS	15–131%	<20%	LODs: 1–2500 μg kg^{-1}	Peysson and Vulliet (2013)
Sewage sludge	SFE	PAHs: 1 g of sample and 0.5 g of deionized water, pure CO_2 at 80°C. PCBs: 1 g of sample, pure CO_2 at 80°C. OCPs: 3 g of sample, pure CO_2 at 50°C. All mixed with copper powder. Cleanup: PAHs and PCBs: double-layer column of silica gel (bottom layer) and naphaluminum oxide (top layer). OCPs: Extrelut NT 3 (diatomaceous earth) column, SPE (C18) and Florisil cleanup.	8 PAHs, 7 PCBs, 12 OCPs	GC-MS	PAHs: 104–125% PCBs: 75–106% OCPs: 56–121%	–	–	Berset and Holzer (1999)

(Continued)

TABLE 5.4 (CONTINUED)
Overview of Analytical Protocols Applied to the Determination of Selected Persistent and Emerging Pollutants in Sewage Sludge

Matrix	Extraction Technique	Extraction and Cleanup Protocols	Compounds Determined	Analytical Technique	Recoveries	Precision (% RSD)	Analytical Thresholds	Ref.
Sewage sludge	PHWE-HS-SPME	PHWE: 1 g of sample mixed with water at pH 4, extraction at 100°C for 15 min. No cleanup: direct injection.	10 aliphatic primary amines	GC-IT-MS/MS (the fiber was introduced in PFBAY at 40°C for 15 min)	–	<19% r <24% R	LODs: 9–135 μg kg^{-1}	Llop et al. (2010)
Dewatered sludge	HS-SPME	65 μm PDMS/DVB fiber was exposed for 45 min in the headspace of 0.25 g sewage sludge sample mixed with 0.5 mL of water stirred at 750 rpm at 80°C. No cleanup: direct injection.	Macrocyclic musk fragrances	GC-MS	–	1–9% r 6–15% IP	LODs: 10–25 ng kg^{-1}	Vallecillos et al. (2013)
Dewatered sludge	MA-HS-SPME	5 g of sample mixed with 20 mL H_2O (containing 3 g NaCl) extracted by a PDMS-DVB fiber placed in the headspace when the extraction slurry was microwave irradiated at 80 W for 5 min. No cleanup: direct injection.	6 synthetic polycyclic musks	GC-MS	85–96%	<14% r <16% IP	LODs: 0.04–0.1 μg kg^{-1}	Wu and Ding (2010)

Note: 2,4-DCP, 2,4-dichlorophenol; 2,4,6-TCP, 2,4,6-trichlorophenol; BUVs, benzotriazole ultraviolet stabilizers; DTE, 1,4-dithyoerythritol; E2, estradiol; EE2, ethinylestradiol; MA-HS-SPME, microwave-assisted HS-SPME; NBFRs, novel brominated flame retardants; NP, nonylphenol; NPEO, nonylphenol ethoxylate; OPFRs, organophosphorus flame retardants; PFASs, perfluoroalkyl sulfonates; PFBAY, pentafluorobenzaldehyde; PFCAs, perfluorocarboxylic acids; PFCs, perfluorinated compounds; PFSAs, perfluoroalkyl sulfonamides; PTFE, polytetrafluoroethylene; QqLIT, quadrupole linear ion trap; TMIS, trimethyliodosilane.

terms of recovery, precision, and LODs in each type of sludge. Indeed, extraction recoveries were highly dependent on the type of sludge (primary, secondary, and anaerobically, dehydrated sludge and compost samples). Moreover, the influence of sample size on the extraction efficiency is also a crucial parameter. Samaras et al. (2011) studied the effect of the extracted biomass on the recoveries of selected EDCs and NSAIDs with the use of UAE. The extraction of higher amounts of biomass affected the extraction efficiency and yielded lower absolute recoveries for most of the hydrophobic EDCs. In contrast, acidic NSAIDs (diclofenac, ketprofen, ibuprofen, and naproxen) with lower log K_{ow}, ranging between 3.4 and 3.5, resulted just in slightly lower recoveries. Table 5.4 gives an overview of typical sample preparation methods and figures of merit reported for the trace analysis of selected persistent and ECs in sewage sludge. Soxhlet/Soxtec extraction and mechanical agitation are not included in this review, as they are considered non-environment-friendly methods and their application has declined in the last few years in favor of novel extraction techniques such as PLE and MAE.

5.6.3 Extraction and Cleanup Protocols

As mentioned previously, UAE is still popular because the equipment needed is widely available and the extraction can be performed with a small volume of solvent, typically 5–25 mL, within an extraction time between 10 and 30 min. Several methodologies based on UAE have been reported in the past 3 years to determine PAHs, parabens, quinolones, EDCs, PACs, steroid hormones, PPCPs, flame retardants and PFCs in sewage sludge (Samaras et al., 2011; Yu et al., 2011; Dorival-García et al., 2012; Yu and Wu, 2012; Benítez-Villalba et al., 2013; Carvalho et al., 2013; Cristale and Lacorte, 2013; Arvaniti et al., 2014).

Recently, Martínez-Moral and Tena (2013) presented a method for the determination of perfluorocarboxylic acids and PFOS in sewage sludge based on FUSLE avoiding SPE cleanup steps. Two extraction cycles with an ultrasonic irradiation power of 65% for only 20 s were enough for the quantitative extraction of PFCs. The whole method FUSLE-UPLC-MS/MS provided LODs lower than 0.2 μg kg^{-1}, excellent repeatability, intermediate precision with RSDs below 8%, and good recovery values between 69% and 104%. In previous comparative studies, FUSLE was not selected to quantitatively recover BPA and APs (Martínez-Moral and Tena, 2011) and 14 benzotriazole (BT), benzothiazole, and benzenesulfonamide (BSA) derivatives from sewage sludge, and extraction was performed with a selective pressurized liquid extraction (SPLE) method and PHWE (Herrero et al., 2014), respectively.

Arvaniti et al. (2014) provided compensation for variations in the extraction percentages of 18 PFCs by UAE with the use of internal standards and adequate LODs ranging from 0.15 to 1.5 μg kg^{-1} (dw) in sewage sludge were obtained. Zhou et al. (2012) reported recoveries by UAE in the range of 50–150% for 31 antibiotics in manure samples at pH 3, while 28 and 25 compounds were in the same recovery range (50–150%) at pH 4.5 and 7, respectively. In this study, different recoveries were observed among two solid matrices, manure, and sediment.

Among the novel extraction techniques, PLE is the most often applied in sewage sludge analysis. Commonly applied temperatures fall in the range between 70°C and 100°C to extract different groups of compounds including artificial sweeteners (Ordoñez et al., 2013), sulfonamides (García-Galán et al., 2013) abused and illicit drugs (Mastroianni et al., 2013; Arbeláez et al., 2014), antibiotics (sulfonamides, macrolides, and fluoroquinolones) (Göbel et al., 2005), natural and synthetic estrogens (Nieto et al., 2008), PACs (Barron et al., 2008; Jelic et al., 2009; Radjenović et al., 2009), PCPs (Nieto et al., 2009), and steroid hormones (Nieto et al., 2008; Combalbert et al., 2010). Lillenberg et al. (2009) reported a decrease in the recovery of antibiotics at temperatures exceeding 120°C, resulting either from thermal degradation or from a loss in method selectivity due to the more exhaustive extraction of interfering matrix components. Higher extraction temperatures (170°C) were reported by Meesters and Schroder (2002) to determine 4-nonylphenol (4-NP) and BPA in digested sludge with recoveries of 100% and precision less than 1.5% RSD for both compounds. In terms of solvent selection, nonpolar and polar aprotic organic solvents have been reported. The mixture of MeOH:H_2O has been applied most, with high recoveries for several groups of PACs (Barron et al., 2008; Radjenović et al., 2009; Mastroianni et al., 2013) and antibiotics (Göbel et al., 2005) due to their high polarity. In a few studies, less polar solvents such as ethyl acetate and dichloromethane were used to extract BPA and APs (Martínez-Moral and Tena, 2011), drugs of abuse (morphine, codeine, cocaine, and methadone) (Arbeláez et al., 2014) and PCPs, plasticizers, and flame retardants (Chen and Bester, 2009). The static extraction time and the number of cycles are related parameters, which also have a strong influence on extraction efficiency. The static extraction time is usually set at 3–15 min to ensure contact between the analytes and the solvent, while the number of cycles ranges from 1 to 4. Lillenberg et al. (2009) applied five continuous extractions to the same sample with identical extraction conditions. The recoveries were considered negligible (lower than 2%) during the fourth cycle. On the other hand, one extraction cycle was considered as the best compromise by Mastroianni et al. (2013), taking into account the multiresidue character of the methodology for the determination of 20 abused and illicit drugs and their metabolites. Contributions of the second and third cycles to the total extractable amounts were below 17% and 5%, respectively.

Lately, a new approach described in the literature as an efficient way of improving sample cleanup is the addition and mixing of the sample with different sorbents (Florisil, silica, alumina, and graphitized carbon) or reagents (potassium hydroxide, Na_2EDTA, aluminum oxide, and copper) within the PLE cell. This approach, also known as SPLE, was followed by a number of authors for the extraction of APs and BPA (Martínez-Moral and Tena, 2011), PAHs (Pena et al., 2010), artificial sweeteners (Ordoñez et al., 2013), and UV filters (Negreira et al., 2011).

A variant of PLE, known as PHWE, includes the use of water as extraction solvent (Bruno et al., 2002; Llop et al., 2010; Saleh et al., 2011; Ordoñez et al., 2013; Herrero et al., 2014). A PHWE method with an SPE cleanup followed by LC-Orbitrap HRMS has been developed for the simultaneous determination of benzothiazole, BT, and BSA derivatives (Herrero et al., 2014). In this study, different extraction temperatures and water pH values (3, 7, and 9) were tested. Maximum extraction efficiency was obtained at 80°C for all tested compounds, while thermal degradation of BTs and BSAs was observed at 120°C. The absolute recoveries obtained were similar for the three pH values, but the matrix effect was slightly higher at a pH of 9. Similarly, the pH of water was the least influential factor for the extraction of the nine Environmental Protection Agency (EPA) *N*-nitrosamines based on PHWE followed by HS-SPME and GC-CI-MS/MS (Llop et al., 2010).

Recently, Saleh et al. (2011) described hollow-fiber liquid-phase microextraction (HF-LPME) as a preconcentration and cleanup step after PHWE for the determination of four NSAIDs in sewage sludge. Optimization of HF-LPME was carried out by testing the donor phase pH (1–6), ionic strength (addition of NaCl, 0–20% w/v), and extraction time (30–270 min). The optimum conditions (2 h extraction at pH 1.5 without salt addition) decreased the matrix effect (−8.9% to +14.6%) in comparison with other published values obtained using SPE for cleanup and produced high enrichment factors, in the range of 947–1213 times.

The MAE technique has also been successfully applied to the extraction of a wide range of pollutants covering different physicochemical properties in sewage sludge and other solid environmental matrices (Fountoulakis et al., 2005; Morales et al., 2005; Dobor et al. 2010; Cantarero et al., 2011 Shin et al., 2011; Azzouz and Ballesteros, 2012; Montesdeoca-Esponda et al., 2013; Vega-Morales et al., 2013). A MAE method and online PE cleanup (Waters Sep-Pak C18 cartridges) combined with UHPLC-MS/MS was developed by Vega-Morales et al. (2011) for the extraction of various E2-mimicking compounds (NP, NP_{1-7}EO, OP, OP_1-7EO, BPA, EE2, E2, and E3) from sludge samples. The optimum operational parameters were achieved by using a 2^3 factorial design where the influence of three variables (at two levels)—power (100 and 300 W), time (5 and 15 min), and volume (5 and 10 mL)—on the recovery of the analytes were examined. Under optimum conditions (5 mL of MeOH, 300 W, and 10 min), the method provided recoveries between 71.7% and 103.1%, with RSD lower than 11.1% and LODs ranging from 0.6 to 3.5 μg kg^{-1}. A MAE method followed by GC-MS analysis has been successfully developed for the determination of polycyclic (ESI mode) and nitro musks (NCI mode) in sewage sludge samples. Adequate low LOD values from 27 to 41 μg kg^{-1} for the polycyclic musks and approximately 4 μg kg^{-1} for the nitro musks were reported after a cleanup step with a silica gel column (Svoboda et al., 2007).

The water content of the sludge matrix has been reported to be crucial for efficient MAE, due to the high dipole moment of the water molecules, which results in efficient heating of the sample. The addition of 1 mL of water to the dried sample was found to enhance the extraction efficiency of NP and nonylphenol ethoxylates (NPEOs) (Fountoulakis et al., 2005). The moisture content has also been reported to yield higher recoveries for NPs when using UAE as the extraction technique (Gatidou et al., 2007; Samaras et al., 2011). Satisfactory recoveries of four acidic NSAIDs (ibuprofen, ketoprofen, naproxen, and diclofenac) by using MAE followed by dispersive matrix extraction (DME)–SPE as cleanup step and water as extraction solvent were also reported by Dobor et al. (2010). However, without the application of DME, the SPE sorbent was contaminated by the colloidal fraction and LOQs ranged higher by about one order of magnitude.

MSPD was applied for the first time to sewage sludge samples for the extraction of bromophenols and halogenated bisphenols (Blanco et al., 2006). A few studies followed for the multianalysis of 27 PAHs (Sánchez-Brunete et al., 2007), 17 PAHs (Pena et al., 2008), and 16 OCPs in sewage sludge (Sánchez-Brunete et al., 2007). To the best of our knowledge, only two research groups published new methods based on MSPD followed by hybrid techniques (QTOF-MS) in the past 2 years. Celano et al. (2014) presented an MSPD method followed by LC-QTOF-MS analysis for the determination of eight OPs. In particular, 0.5 g of freeze-dried sludge samples were mixed and dispersed with 2 g of C18 in a glass mortar, with a pestle, for 5 min. Then, the blend was transferred to a polypropylene syringe containing 1 g of PSA as cleanup sorbent. Analytes were recovered passing ACN through the packed syringe. The extract was evaporated and diluted with ultrapure water (1:1) before injection in the LC-QTOF-MS system. In this study, it was proved that replacement of C18 by PSA sorbents leads to cleaner extracts. Similarly, Casado et al. (2013) developed a method based on MSPD followed by GC-QTOF-MS for the determination of nine BT UV stabilizers. In this study, lyophilized samples were mixed and dispersed with diatomaceous earth and then transferred to a polypropylene syringe containing 1 g of PSA as the cleanup sorbent and ethyl acetate was used as extraction solvent. The extract was concentrated with N_2 before injection into the chromatographic system. The applicability of both MSPD methods provided the advantage of decreasing the amount of coextracted substances, enabling direct injection of the extract in the chromatographic systems. Albero et al. (2012) developed a rapid method for the determination of seven parabens and two chlorinated by-products in sewage sludge by MSPD-GC-MS/MS. Samples collected in various urban, industrial, and rural zones in Madrid were analyzed with the optimized method, and the results showed that methylparaben was detected in 95% of the samples, in ranges between 5.1 and 26.2 μg kg^{-1}, and propylparaben in 74% of WWTPs, at levels of up to 44.1 μg kg^{-1}. The analytical procedure also presented good recovery rates (80–125%) and low LODs, ranging from 0.1 to 2.0 μg kg^{-1}. In conclusion, the results obtained by MSPD are promising and show that its application should most likely expand in the near future.

An opposite trend is followed by SFE with a few developed methods in the analysis of PAHs (Miège et al., 2003) and polycyclic and nitro musks in sludge samples (Lee et al., 2003; Smyth et al., 2007), probably due to the development of PLE and MAE, which have become widely accepted extraction techniques. SFE has been reported as an appropriate technique for extracting thermolabile analytes in environmental analysis, as it allows the possibility to perform extractions at moderate temperatures (~40°C). However, the main drawback of SFE is the difficulty of extracting samples that have high lipid and moisture content. In this case further cleanup is required or rather low recoveries are obtained. In the late 1990s, Berset and Holzer (1999) presented an SFE method for the analysis of PAHs, PCBs, and OCPs in sewage sludge. Most PCB extracts could be measured directly after SFE, whereas for PAHs a short cleanup improved the final separation by high-resolution capillary GC. Due to the amount of coextracted compounds and the low levels of OCPs, extracts obtained by SFE were subjected to a multistep cleanup by diatomaceous earth in an Extrelut column followed by C18 on an SPE cartridge and Florisil before analysis.

The application of SPME is rather infrequent in sewage sludge analysis mainly due to the strong dependence of this technique on matrix. Recently, HS-SPME coupled with UAE, PLE, and MAE has been successfully applied to volatile and semivolatile organic compounds (Salgado-Petinal et al., 2006b; Llop et al., 2010; Wu and Ding, 2010). Wu and Ding (2010) presented a microwave-assisted HS-SPME method for the extraction of six synthetic polycyclic musks. The dewatered sample, mixed with 20 mL deionized water, was extracted by a PDMS-DVB fiber placed in the headspace when the extraction slurry was microwave irradiated at 80 W for 5 min. The LODs of this method ranged from 0.04 to 0.1 μg kg^{-1} (fresh weight). Similarly, Llop et al. (2010) combined PHWE and HS-SPME-GC-MS/MS for the analysis of 10 amines. The entire process took approximately 1 h with reported LODs in the range between 9 and 135 μg kg^{-1} (dw). Salgado-Petinal et al. (2006b) coupled UAE of sludge sample and HS-SPME-GC-MS/MS of the obtained extract for the analysis of eight polybrominated biphenyls and PBDEs. The proposed method exhibited good performance in terms of linearity and precision, with recoveries exceeding 92% and LODs in the sub-microgram per kilogram level. On the contrary, Vallecillos et al. (2013) presented a direct HS-SPME-GC-MS method for the determination of macrocyclic musk fragrances in sewage sludge. The optimal HS-SPME conditions were achieved when a PDMS/DVB 65 μm fiber was exposed for 45 min in the headspace of 0.25 g sewage sludge samples mixed with 0.5 mL of water stirred at 750 rpm at 80°C. Method detection limits were found between 10 and 25 ng kg^{-1} (dw).

SBSE has also been seldom applied in the analysis of sewage sludge and mainly in combination with TD coupled to GC-MS, after in situ derivatization. Tan et al. (2008) first reported a method to analyze a wide variety of sex steroid hormones, phthalates, APs, and tamoxifen in sludge samples. The protocol was based on the extraction of 1 g of wet solids/sludge with 10 mL of Milli-Q water adjusted at pH >11 with sodium carbonate and acetic anhydride as the derivatization agent. SBSE was performed at room temperature for 5 h at 500 rpm. The stir bar was then placed in a glass TD tube for delivery into the GC system. Although studied LODs ranged up to the low microgram per kilogram level, recoveries of the target compounds were reported between 3% and 60%. Similarly, Casas Ferreira et al. (2011) reported extraction efficiencies of 5 parabens, triclosan, and methyl-triclosan between 14% and 43% in sludge samples. In both studies, the efficiency of SBSE was strongly affected by the complexity of the matrices. In general, the need for tight control of the extraction conditions (not only equilibrium time and agitation speed but also pH adjustment and ionic strength of the matrix), the strong dependence of SBSE on the matrix, and the limited spectrum of analyte polarities for the available stationary phases make this technique not suitable for sewage sludge analysis.

QuEChERS extraction is a relatively new, simple method, with only three articles reported so far on sludge analysis (Peysson and Vulliet, 2013; Albero et al., 2014; Cerqueira et al., 2014). QuEChERS multiresidue extraction followed by a dSPE cleanup and LC-TOF-MS analysis has been successfully applied for the analysis of 136 substances, including 119 PACs and 17 hormonal steroids. The analytes were first extracted from the matrix with a mixture of EDTA, acidified ACN (acetic acid 1%), and heptane, which was subsequently purified by the weak anion-exchanger sorbent PSA. The use of PSA improved recoveries and reduced matrix effects when compared to graphitized carbon and C18 sorbents. The developed method was applied to various types of sludge (limed, digested, dried, liquid, and composted), with LODs ranging from 1 to 2500 μg kg^{-1} (Peysson and Vulliet, 2013). Due to the diversity of chemical classes among the selected compounds, the results varied greatly with recoveries from 15% to 131% and intraday precision (RSD) from 1% to 88%. In a latter study, the use of chitin obtained from shrimp shell waste, as an alternative sorbent in the dSPE step to the commonly used C18, PSA, and GCB, has been employed for the simultaneous extraction of different organic contaminants (simazine, atrazine, clomazone, tebuconazole, amitriptyline, caffeine, diclofenac, ibuprofen, methylparaben, propylparaben, triclocarban, and BPA) in drinking water treatment sludge by employing LC-ESI-MS/MS (Cerqueira et al., 2014). Albero et al. (2014) presented a QuEChERS method for the analysis of 13 natural and synthetic hormones in biosolids and poultry manure. The incorporation of labeled standards allowed an accurate quantification with LODs comparable to those of other authors. The optimized method needed minimal amounts of organic solvent (<10 mL) and time (2 h) per sample.

5.6.4 Comparative Studies

A number of comparative studies carried out between novel and conventional techniques showed that the performance of PLE and MAE was more effective than those of traditional methods such as Soxhlet extraction, UAE, mechanical agitation, and vortex agitation in terms of recoveries, accuracy, and sensitivity (Bruno et al., 2002; Benítez-Villalba et al., 2013; Dorival-García et al., 2013). Furthermore, both methods offered several advantages over the classic methods such as reduction in extraction time and in solvent consumption.

Benítez-Villalba et al. (2013) reported no statistically significant difference between the extraction efficiencies of UAE and PLE for the analysis of parabens in compost from sewage sludge. However, PLE provided slightly higher recoveries, easy operation, shorter analysis time, and higher degree of automation. Although the highest signal suppression was observed in PLE extracts by direct injection, its extraction efficiency was still greater, demonstrating that temperature and pressure are crucial parameters affecting the extraction of the studied parabens. Dorival-García et al. (2013) compared the extraction efficiencies of three different extraction techniques (UAE, MAE, and PLE) for the determination of 13 acid quinolone antibiotics. By following a factorial experimental design, 0.5 g of dry sewage sludge samples treated with a mixture of MeOH/McIlvaine's buffer (50:50 v/v) at pH 3 were selected as optimum extraction conditions. MAE and PLE provided the highest recoveries (>90%) in comparison with UAE (~80%). UAE was clearly not as efficient as MAE and PLE, especially for acidic quinolones. MAE was slightly better than PLE, with a maximum difference in recoveries of 6%. However, no significant differences were observed in relation to sensitivity, selectivity, accuracy, and precision. Mean values of signal suppression of the three extraction techniques were compared using least significant difference multiple range test with a 95.0% confidence level. Higher signal suppression was significantly observed in MAE and PLE extracts than in UAE etxracts, probably due to the more drastic extraction conditions of these two automated techniques (Dorival-García et al., 2013). Bruno et al. (2002) used a PHWE method at a neutral pH for the simultaneous extraction of several classes of surfactants. The optimized procedure was found to be more efficient in extracting anionic surfactants and acidic metabolites when compared to Soxhlet extraction. Similarly, Fountoulakis et al. (2005) reported better extraction efficiency of NP and NPEOs when MAE was compared to Soxhlet extraction and UAE. In general, MAE and PLE have been proved to constitute efficient and reproducible methods for the extraction of a wide range of organic pollutants in sewage sludge, with LODs ranging in the low nanogram per kilogram level.

5.7 SAMPLE PREPARATION METHODS FOR THE DETERMINATION OF ORGANIC POLLUTANTS IN WASTEWATER

5.7.1 Introduction

Environmental contamination by organic micropollutants, such as PACs, PCPs, BFRs, and phthalates, among others, often persistent and bioaccumulable, is mainly of anthropogenic origin (Mompelat et al., 2009). Moreover, it is consensual that the most significant entry route of these pollutants into the aquatic environment is discharge from WWTPs. Since there is no unit specifically designed to remove these compounds, the treatment processes used by most WWTPs seem to be inefficient, and, along with the treated wastewater, these pollutants are released into the aquatic environment (Castiglioni et al., 2006). The situation is particularly critical in a scenario of climate change and water scarcity, mainly in Mediterranean countries, where a considerable flux of surface water streams is composed of wastewater discharges. Afterward, surface waters can contaminate groundwater, even after soil or bank filtration (Jux et al., 2002). Thus, there is a need for analyzing these organic compounds in wastewater. In recent years, special attention has been devoted to the study of organic pollutants in influent and effluent wastewaters, assessing the station removal efficiency, while efforts have been carried out to develop more efficient treatment processes. The analysis of wastewater is especially challenging, as many contaminant sources are conveyed to wastewater, aside from the huge load of organic matter, which makes these matrices very complex. Furthermore, some compounds, such as natural and synthetic estrogens, have endocrine-disrupting properties at concentrations in the low nanogram per liter range, close to the detection capabilities of current instruments. Developments in extraction and cleanup procedures are welcome to allow increasing the knowledge of the behavior and fate of POPs and ECs in wastewater.

5.7.2 Extraction and Cleanup Protocols

Over the past few decades, several research works have been conducted on the use of SPE for the extraction of a variety of compounds from wastewater samples. For instance, Sousa et al. (2011) described the development and validation of an SPE method, with simultaneous cleanup capability, for the concurrent determination of 23 pharmaceutical compounds belonging to different therapeutic classes in wastewater. An exhaustive study of different combinations of adsorbents (normal phase, ion exchange, and mixed composition) and elution solvents was performed, exploiting the physical and chemical properties of the analytes versus interferences. The best results were achieved with the mixed-anion exchange Oasis mixed-anion exchange (MAX cartridges, which provided high recovery rates (>60%) and reasonably low detection limits (1–20 ng L^{-1}), considering such highly complex matrices. Table 5.5 comprises detailed information on the analytical protocols and respective figures of merit.

Castiglioni et al. (2005) also developed and validated a two-step extraction methodology for the simultaneous determination of 30 PACs in urban wastewater. The first group of compounds was extracted with Oasis mixed-cation exchange (MCX) cartridges, acidifying the sample at pH 1.5–2.0, while LiChrolut EN adsorbent was selected for the second group, at pH 7.0. In both cases, reversed-phase LC-MS/MS analysis followed. This optimized method presented recovery rates >70% for most pharmaceuticals, with a method overall variability <8% RSD. Moreover, 19 of the 30 PACs studied were detected in concentrations between 0.5 and 2000 ng L^{-1}.

Benito-Pena and coworkers (2006) selected a mixed-mode SPE cartridge type for the simultaneous preconcentration of eight β-lactam antibiotics in wastewater: penicillin G, amoxicillin, ampicillin, penicillin V, oxacillin, cloxacillin, dicloxacillin, and nafcillin. Initially, two SPE cartridges were compared for sample cleanup and concentration: a reversed-phase cartridge (Bond Elut C18) and a strong polymeric mixed-mode anion exchanger (Oasis MAX). Separation was conducted by LC in both situations, followed by UV-DAD detection. The final results demonstrated that Oasis MAX cartridges provided the best recovery rates (28–114%), as well as a good method reproducibility, even for high conductivity samples. Masiá et al. (2013b) proposed a multiresidue method for the analysis of pesticides included in the Water Framework Directive 2000/60/EC (WFD), by coupling IT-SPME with UPLC-MS/MS. The optimized method exhibited good linearity over the dynamic range, LODs between 0.025 and 2.5 μg L^{-1}, and interday variation coefficients <32%. The method was applied to water samples from different sources, from tap water to river water and wastewater, and these compounds were quantified in the low ppb range. It was further demonstrated that the online enrichment of the analytes allowed a minimal sample manipulation.

TABLE 5.5
Overview of Analytical Protocols Applied to the Determination of Selected Persistent and Emerging Pollutants in Wastewater

Matrix	Extraction Technique	Extraction and Cleanup Protocols	Compounds Determined	Analytical Technique	Recoveries	Precision (% RSD)	Analytical Thresholds	Ref.
Wastewater	SBSE	Water acidified with phosphoric acid (pH 2), PDMS SBSE stirred at 900 rpm for 8 h at room temperature and in the dark. 200 µL of ethyl acetate for LD by sonication for 30 min.	PPCPs	Derivatization with MTBSTFA followed by GC-MS	2.2–87.7%	<1.2% r	LODs: 0.2–222 ng L^{-1}	Pintado-Herrera et al. (2013)
Wastewater: influents and effluents	SPE	Oasis MAX cartridges loaded with 50 or 100 mL influent or effluent wastewater, respectively (adjusted to pH 3, filtered 0.45 µm). Cartridges washed and eluted with MeOH (6 mL), and eluates evaporated and reconstituted in 250 µL MeOH/H_2O (25:75).	23 PACs	LC-MS/MS	>60%	2–22% r	LODs: 1–20 ng L^{-1} (except ibuprofen)	Sousa et al. (2011)
Wastewater: effluents	SPE	Preconditioned cartridges loaded with 500 mL sample (Oasis MCX pH 2; LiChrolut EN pH 7). Cartridges washed and eluted (Oasis MCX with 2 mL MeOH, 2 mL 2% ammonia solution in MeOH, and 0.2% NaOH in MeOH; LiChrolut EN with 3 mL MeOH and 3 mL ethyl acetate). Eluates evaporated and reconstituted in 100 µL 0.01% acetic acid in Milli-Q water (pH 3.5).	PACs	LC-MS/MS	>70%	1–17% r	LODs: 0.2–5 ng L^{-1}	Castiglioni et al. (2005)
Wastewater: industrial and influent and effluent urban sewage water	SPE	Preconditioned Oasis MAX cartridges loaded with 250 mL sample (adjusted to pH 7.5). Cartridges washed and eluted with 2 × 1 mL 0.05 M TBA in MeOH, eluates filtered (0.22 µm), and diluted onefold with Milli-Q water.	β-Lactam antibiotics	LC-UV-DAD	28–114%	0.3–6% r	LODs: 2.2–25.6 µg L^{-1}	Benito-Pena et al. (2006)

(*Continued*)

TABLE 5.5 (CONTINUED)
Overview of Analytical Protocols Applied to the Determination of Selected Persistent and Emerging Pollutants in Wastewater

Matrix	Extraction Technique	Extraction and Cleanup Protocols	Compounds Determined	Analytical Technique	Recoveries	Precision (% RSD)	Analytical Thresholds	Ref.
Wastewater: two influents and two effluents Four river waters and two coastal waters	IT-SPME	GC TRB-5 capillary (40 cm × 0.32 mm i.d.), coated with 5% diphenyl–95% PDMS (3 μm film thickness) for IT-SPME Surface waters and wastewaters were vacuum filtered through 1 μm paper fiber filter, followed with 0.7 μm glass fiber filter. 4 mL water samples passed through the capillary. Desorption with 40 μL MeOH.	Pesticides	UPLC-MS/MS	8–92%	3.3–26% r	LODs: 0.025–2.5 μg L^{-1}	Masiá et al. (2013b)
Wastewater: influent and effluent from sewage treatment plant	SPME with CW-DVB fiber	25 mL sample adjusted to pH 4.5 and 10% KCl w/v, after which it was exposed for 20 min (with stirring) to a CW/TPR-coated SPME fiber. The fiber was then statically desorbed in 100 μL MeOH for 30 min.	Sulfonamide antibiotics	LC-MS/MS	SPME: 40–112%		SPME LODs: 9–55 ng L^{-1}	Balakrishnan et al. (2006)
	SPE	Preconditioned MCX sorbent cartridges loaded with 500 mL sample (adjusted to pH 3, filtered 0.45 μm). Cartridges washed and eluted with MeOH/5% NH_4OH (2 mL), and eluates evaporated and reconstituted in 250 μL NH_4OAc (20 mM).			SPE: 3–115%		SPE LODs: 3–9 ng L^{-1}	
Wastewater: influent and effluent (grab and composite)	SPME	CW/TPR fiber immersed in 1.5 mL sample and desorption into 1.5 mL MeOH. Extract evaporated to dryness and reconstituted with 30% ACN/70% H_2O; final volume 75 μL.	Sulfonamide, trimethoprim, and macrolide antibiotics	HPLC-MS/MS			SPME LOQs: 16–1380 ng L^{-1}	McClure and Wong (2007)
	SPE	Preconditioned Oasis HLB cartridges loaded with 100 mL influent and 500 mL effluent. Cartridges washed and eluted with 6 mL MeOH. Eluate filtered (0.22 μm), evaporated, and reconstituted in 1 mL MeOH.					SPE LOQs: 0.86–15 ng L^{-1}	

(Continued)

TABLE 5.5 (CONTINUED)
Overview of Analytical Protocols Applied to the Determination of Selected Persistent and Emerging Pollutants in Wastewater

Matrix	Extraction Technique	Extraction and Cleanup Protocols	Compounds Determined	Analytical Technique	Recoveries	Precision (% RSD)	Analytical Thresholds	Ref.
Wastewater: influents and effluents	SBSE with LD	SBSE: 10 mm long glass-encapsulated magnetic stir bar, coated with 24 µL PDMS. 50 mL water sample at pH 5 and room temperature, stirred at 900 rpm for 180 min. Stir bar removed, rinsed with 2 mL Milli-Q water, dried, and placed in a vial with 1 mL ACN, for 15 min at 30°C. Stir bar removed and stripping solvent evaporated. Dried residue redissolved in 200 µL ACN/H_2O (1:1).	6 PCPs	UPLC-ESI-MS/MS	25–72%	<15% r	5–10 ng L^{-1}	Pedrouzo et al. (2010)
Wastewater: influents and effluents of domestic sewage treatment plants	SBSE with LD (10 mm long glass-encapsulated magnetic stir bar, coated with 32 µL EG Silicone)	SBSE: 10 mm long glass-encapsulated magnetic stir bar, coated with 32 µL EG Silicone. 50 mL water sample at pH 5, stirred for 4 h at room temperature. Stir bar removed, introduced in a vial with 1 mL MeOH, and placed in an ultrasonic bath for 15 min. Extracts diluted in Milli-Q water (1:1 v/v).	PPCPs	LC-ESI-MS/MS	1–80%	<17%	5–10 ng L^{-1}	Gilart et al. (2013)
Wastewater	IAE	After SPE (LiChrolut EN and C18 cartridges loaded with 1 L sample), extracts eluted with acetone, evaporated to dryness, and reconstituted in 25 mL of 5% MeOH/water. Purification with immunosorbent cartridges. Extract washed with 5 mL Milli-Q water and eluted with 4 mL 70% MeOH/water, followed by 1 mL Milli-Q water. Eluate evaporated to dryness and reconstituted in 200 µL of 25% ACN in Milli-Q water.	Steroid estrogens E2 and E1	LC/ESI-MS	>90%	<5%	LOD E2: 0.18 ng L^{-1} LOD E1: 0.07 ng L^{-1}	Ferguson et al. (2001)

Note: CW/TFR, carbowax–template resin; MAX, mixed-anion exchange; MCX, mixed-cation exchange; TBA, tetra-*n*-butylammonium hydrogen sulfate.

Two studies have comparatively tested SPME versus SPE techniques to extract a similar group of antibiotic compounds from wastewater samples. First, Balakrishnan et al. (2006) analyzed 10 sulfonamide antibiotics, employing extraction by SPME. Of the five different SPME fiber assemblies tested, the CW/DVB fiber showed the best results, allowing to overcome the confounding matrix effects. On the other hand, for the SPE procedure, carried out with MCX cartridges, higher LODs, and poorer sensitivity were observed. Later on, McClure and Wong (2007) also developed and optimized an SPME methodology, using CW-template resin fibers, for the simultaneous analysis of sulfonamides, trimethoprim, and macrolide antibiotics, prior to quantification by LC-ESI-MS/MS. Nevertheless, in this case, the SPE technique allowed obtaining lower LOQs. Antibiotics were quantified in both influent and effluent wastewater samples.

SBSE is another commonly used microextraction technique in wastewater analysis. Pedrouzo et al. (2010) developed an SBSE method with LD and UPLC-ESI-MS/MS for the analysis of six PCPs in environmental waters: four UV filters (2,2-dihydroxy-4-methoxybenzophenone, benzophenone-3, octocrylene, and octyldimethyl-*p*-aminobenzoic acid) and two antimicrobial agents (triclocarban and triclosan). The extraction process using PDMS-coated stir bars took place over 180 min, to allow reaching the equilibrium. The method was sensitive enough to determine these compounds, at trace levels, in environmental samples. Benzophenone-3 was quantified in levels between 75 and 127 ng L^{-1} in influent sewage water and fell below 25 ng L^{-1} in effluent sewage water. Method detection limits ranged from 5 to 10 ng L^{-1} for influent and effluent sewage water (Table 5.5). This method's performance was compared to that of an SPE procedure developed in a previous study, using Oasis HLB and Bond Elut Plexa sorbents. The fact that overall recoveries and LOQs were similar in both SBSE and SPE methodologies, along with the decrease in the sample volume, high selectivity, and low ionic suppression allowed by the SBSE technique, showed that SBSE was effective for the determination of most apolar compounds in 3 h extraction.

More recently, Gilart et al. (2013) optimized SBSE methods coupled with LC-MS/MS, using two new commercially available polar coatings: PA with polyethylene glycol (PEG) (Acrylate Twister) and PEG-modified silicone (EG Silicone Twister). LD was performed by immersion of the stir bar in 1 mL MeOH, placed in an ultrasonic bath for 15–60 min. These new materials were compared to the classic coating based on PDMS during the extraction of a group of PPCPs, including UV filters, from wastewater samples. The final results showed that the PEG modified silicone coating extracted some of the polar compounds more efficiently while improving the sorption on nonpolar compounds compared to the other two coatings. The validated method presented good linearity, precision, and low limits of quantification.

As mentioned before, IAE is a highly selective sample preparation technique, allowing the removal of interferences present in complex matrices such as wastewater. For the determination of steroid estrogens in wastewater, Fergunson et al. (2001) developed a new method based on IAE coupled to LC-ESI-MS. The optimized method showed quite low detection limits for both steroid estrogens, β-E2 and E1, not to mention their high recovery percentages. On the opposite, the nonimmunogen (but structurally related) analyte EE2 was not retained (recovery <2%), illustrating the extreme high selectivity of the immunoextraction purification step. Both the E1 and E2 estrogens were detected in the analyzed wastewaters, in concentrations ranging from 0.77 to 18 ng L^{-1}. Like in the majority of analytical method development studies, deuterium-labeled surrogate analogues were also used to automatically correct for losses of analytes during extraction and/or sample preparation procedures, as well as to account for variations in instrument response between injections. Moreover, the authors also concluded that the significant sensitivity increase of the developed method in comparison to that attained by other studies concerning estrogen analysis by HPLC-ESI-MS was owed to the considerable reduction in sample matrix extraction provided by the IAE sample preparation. To finalize, it should be mentioned that emerging techniques such as MSPD or SPME have so far been limited to a restricted number of studies.

5.8 SAMPLE PREPARATION METHODS FOR THE DETERMINATION OF ORGANIC POLLUTANTS IN MISCELLANEOUS MATRICES

MAE is a versatile technique and has been applied to extract PAHs in airborne particulate matter, wood samples, marine sediment, muscles of polluted fish, and sewage sludge. PLE was also selected for extracting POPs, such as PAHs, PCBs, and pesticides from urban dust, soil, and marine sediment (Yamaguchi and Lee, 2010). Yamaguchi and Lee (2010) compared an UAE procedure for the determination of PAHs in solid environmental matrices (herein, urban dust SRM 1649a) with MAE using the same solvent system. Overall, the performance of both techniques did not differ statistically, although slightly better recoveries were achieved by the latter for higher-molecular weight PAHs, that is, benzo[*k*]fluoranthene (BkF), benzo[*a*]pyrene (BaP), indeno[1,2,3-cd]pyrene (InP), dibenzo[*a*,*h*]anthracene (DahA), and benzo[*g*,*h*,*i*]perylene (BghiP). On the other hand, lower recoveries of low-molecular weight PAHs were obtained, probably due to their evaporation into the headspace during the heating process, with consequent losses of some compounds (Yamaguchi and Lee, 2010). Mixtures of polar and nonpolar solvents (e.g., 2:3 acetone/*n*-hexane) showed higher recoveries than those obtained by single solvents (Yamaguchi and Lee, 2010). More detailed information on the analytical protocol and respective figures of merit can be found in Table 5.6.

Ryno et al. (2006) compared different extraction techniques (Soxhlet extraction, MAE, and PLE) for the characterization of PAH composition in environmental air samples and in diesel exhaust particulate matter. MAE was the weakest technique, since about 10-fold lower values were obtained for the volatile PAHs, namely, naphthalene, acenaphthene, fluorene, phenanthrene, anthracene, and fluoranthene. PLE was the best option, since it operates in a closed vessel and provides fast, reproducible, and efficient extractions (Ryno et al., 2006). A novel extraction and cleanup procedure including microwave-assisted digestion *and concomitant micro-solid-phase extraction* (MAD-μSPE) was developed by Basheer et al. (2008) for the analysis of OCPs and PCBs in human ovarian cancer tissue samples. This procedure is mentioned herein as an inspiration for future application in the determination of POPs in environmental samples (biota, sediments, and sewage sludge), as it proved promising in the analysis of wastewaters (Basheer et al., 2006) and during the optimization process with fish tissue. The μSPE device consisted of the sorbent material (20 mg of mixture HySep-A and C18, at 1:1 ratio) enclosed within a polypropylene flat-sheet Accurel membrane envelope (2 cm × 0.5 cm and 0.2 μm pore size). Microwave digestion with aqueous NaOH and extraction and cleanup by μSPE were carried out simultaneously. Above and below 60°C, a decrease in analyte enrichment was observed. Excellent sensitivity and recoveries were obtained, despite the fact that MAD-μSPE is an equilibrium extraction technique. The μSPE device could be reused for more than 20 analyses (Basheer et al., 2008).

5.9 CONCLUSIONS

Sample preparation is still the most critical and time-consuming step within the overall analytical procedure for the accurate determination of environmental pollutants in solid and complex matrices, which is particularly true in the case of biota samples. One common feature to all matrices dealt within this chapter is the fact that, generally, not only one extraction technique is applied, but rather the analytical protocols involve the combination of two or more extraction techniques, in an effort to improve the amount of analyte recovered from the solid matrix, as well as to reduce the interferences caused by coextracted compounds. From the literature review carried out in this chapter, two main trends are highlighted:

- The use of hyphenated techniques, composed of an exhaustive extraction (e.g., PLE, UAE, MAE, or PHWE) followed by a simultaneous enrichment and cleanup step (e.g., SPME, SBSE), is frequently employed, affording ultimate sensitivity and selectivity, especially when combined with MS. For instance, the determination of PBDEs in environmental samples has followed this approach mainly.

TABLE 5.6
Overview of Analytical Protocols Applied to the Determination of Selected Persistent and Emerging Pollutants in Miscellaneous Matrices

Matrix	Extraction Technique	Extraction and Cleanup Protocols	Compounds Determined	Analytical Technique	Recoveries	Precision (% RSD)	Analytical Thresholds	Ref.
Solid samples, e.g., urban dust	UAE-SBSE	100 mg sample extracted with 40 mL of acetone/hexane (2:3 v/v), 20 min ultrasonic extraction. PDMS SBSE in 30% alcohol/water solution for 4 h extraction.	16 US EPA PAHs	TD-GC/MS (SIM)	63.3–122% 60.5–125.7%	Not determined	LODs: 1.59–16.7 μg kg^{-1}	Yamaguchi and Lee (2010)
Ovarian cancer tissue	MAD-μSPE (propylene membrane envelope)	Tissues were mixed with 10 mL of 2 M sodium hydroxide solution for microwave digestion and extraction. The μSPE device was added to the vessel. Tissue digestion and extraction were allowed to proceed for 35 min with a power of 1200 W, 60°C. The analytes were desorbed from the μSPE device by ultrasonication in 100 μL of hexane for 20 min.	11 OCPs and 5 PCBs	GC-MS	68.2–95.3% with RSDs 6–15%	0.14–12.7%	LODs: 0.001–0.009 μg kg^{-1}	Basheer et al. (2008)

Note: MAD-μSPE, microwave-assisted digestion and concomitant micro-solid-phase extraction.

- The use of isotope-labeled internal standards and/or surrogates is a consensual procedure to obtain precise and accurate results in the analysis of solid and complex samples. Matrix-matched calibration is a frequently used option when such internal standards are not available. The concept of internal sample calibration allows the ultimate reliability of analytical results in complex environmental samples. Quantification of pollutants prone to matrix effects under LC-ESI-MS/MS analysis, for example, pharmaceuticals in biota samples with diverse lipid contents, benefits with this strategy.

Soil and sediment samples can be analyzed with a relatively simple integrated cleanup step, while biota and sewage sludge samples require an extensive procedure for the removal of the lipidic fraction (biota) and complex organic matter coextracted along with the analytes under the vigorous conditions required to disrupt the analyte–matrix bonds. In the latter case, sulfuric acid cleanup, GPC, and a sequence of normal-phase and/or ion-exchange adsorbents are required, sample preparation accounting for the lengthiest step.

PLE, MAE, and UAE require a relatively short extraction time and a small amount of solvent. However, judicious cleanup of the extracts obtained from complex matrices is necessary due to the low selectivity of the extraction procedure. PLE is nowadays the technique of choice due to its efficiency, rapidity, and unattended operation. PHWE has emerged as a very effective and adjustable technique, besides its inherent environmental friendly features, in agreement with the principles of green analytical chemistry. Great potential is foreseen in the near future, as PHWE is amenable for direct coupling with a sorbent-trapping technique, such as SPE, SPME, or SBSE for further enrichment and cleanup of the aqueous extracts, providing excellent sensitivity and accuracy.

In general, SBSE methods following PHWE, PLE, UAE, or MAE provide LODs 10- to 100-fold lower than those of SPME-based methods, but the extraction time increases from 1 to around 8 h and require a TD unit in conjunction with a CIS PTV injector. SBSE has been mainly performed with PDMS stir bars of different thicknesses and lengths, with a few exceptions: combined PDMS-β-CD was applied for the analysis of BFRs in soil and dust samples; PA and PEG, and PEG-silicone (recently available) were used to determine PPCPs in wastewater.

Adsorbent-based extraction techniques have several convenient features comparable to those of solvent-based procedures: minimization of the use of organic solvents, reduction of sample manipulation steps, reduction of sample preparation time, and reuse of the adsorbent, polymer fiber, or bar after extraction. Nevertheless, solid samples almost invariably require prior solvent extraction, preferably carried out in a closed vessel under elevated temperature and pressure. Reduction of the sample size provides the means for reduction of all the subsequent steps, as exemplified by the micro-MSPD methodologies, as long as representativeness of the sample aliquot is not endangered.

A recent trend, exploiting the full capabilities of modern analytical equipments and broad-scope extraction protocols, seems to be the screening/quantitation of multiple contaminants (which can reach hundreds) in one run. The use of Orbitrap mass spectrometers has seen a huge increment in recent years and supports the mentioned strategy by allowing target and nontarget screening of an endless number of environmental pollutants, with superior confirmation capabilities and possibility of retrospective analysis of acquired data.

Last but not least, it should be mentioned that the degree of confidence in the analytical results of organic micropollutants in solid and complex matrices has drastically improved in recent years with the application of thoughtful sample preparation strategies and the use of sophisticated instruments, allowing to perform more meaningful and relevant environmental studies.

5.10 FUTURE TRENDS

Analytical methodologies for the determination of organic micropollutants in soils exist to a similar extent as for sediments; however, less interest has been devoted to monitoring soil quality. This topic should be addressed in the coming years in order to achieve the objectives of the thematic strategy for soil protection. Contamination of the soil compartment with PACs and other ECs is raising a huge concern, due to residues from livestock production and dispersion of sewage sludge. Moreover, progress is still needed in the determination of environmental pollutants in biota samples, as these matrices constitute a considerable challenge in terms of sample preparation and both the WFD and the Marine Strategy Framework Directive demand that biota be monitored in a routine basis.

The simultaneous analysis of several classes of organic pollutants in a single run appears to be the future direction. The use of broad-scope extraction and cleanup protocols will expand further in the near future, as high-resolution mass spectrometers (mainly Orbitraps) are "invading" environmental and food analytical laboratories. Solventless extraction, miniaturization, automation, and selectivity should be sought as driving principles when developing innovative sample preparation methodologies.

Above all, a good array of analytical methodologies is required for the determination of organic pollutants in all environmental matrices (wastewater, sewage sludge, soil, sediments, and biota), being crucial in studying their sources, transport, pathways, persistence, and ecological and human effects. Systematic monitoring studies that could provide comprehensive overviews of environmental contamination should deserve more insights; as such, routine studies are normally seen as less enchanting when compared to demonstrating the feasibility of innovative analytical approaches.

ACKNOWLEDGMENTS

The authors acknowledge partial funding by the Project ECORISK (NORTE-07-0124-FEDER-000054), cofinanced by ON.2, under the National Strategic Reference Framework, through the European Regional Development Fund.

LIST OF ACRONYMS

4-NP	4-nonylphenol
ACN	acetonitrile
AP	alkylphenol
ASE	accelerated solvent extraction
BDE	brominated diphenyl ether
BFR	brominated flame retardant
BPA	bisphenol A
BSA	benzenesulfonamide derivative
BSTFA	*N,O*-bis(trimethylsilyl)trifluoroacetamide
BT	benzotriazole
CW	carbowax
DAD	diode array detector
DME	dispersive matrix extraction
DVB	divinylbenzene
dSPE	dispersive solid-phase extraction
E1	estrone
E2	estradiol
E3	estriol
EC	emerging contaminant
ECD	electron capture detector
EDC	endocrine-disrupting compound
EE2	ethinylestradiol
ESI	electrospray ionization
FID	flame ionization detector
FLD	fluorescence detection
FUSLE	focused ultrasound solid–liquid extraction

FWHM	full with at half maximum
GC	gas chromatography
GCB	graphitized carbon black
Genapol X-080	oligoethylene glycol monoalkyl ether
GPC	gel permeation chromatography
HBCD	hexabromocyclododecane
HF-LPME	hollow-fiber liquid-phase microextraction
HFR	halogenated flame retardant
HRMS	high-resolution mass spectrometry
HS-SPME	headspace solid-phase microextraction
IAE	immunoaffinity extraction
IT-SPME	in-tube solid-phase microextraction
LC	liquid chromatography
LC-MS	liquid chromatography–mass spectrometry
LD	liquid desorption
LOD	limit of detection
LOQ	limit of quantification
LPME	liquid-phase microextraction
LSE	liquid–solid extraction
MAD-μSPE	microwave-assisted digestion and concomitant micro-solid-phase extraction
MAE	microwave-assisted extraction
MAME	microwave-assisted micellar extraction
MAX	mixed-anion exchange
MBR	membrane bioreactor
MCX	mixed-cation exchange
MeOH	methanol
MIP	molecularly imprinted polymer
MISPE	molecularly imprinted solid-phase extraction
MPE	monoalkyphthalate ester
MS	mass spectrometry
MSPD	matrix solid-phase dispersion
MSTFA	*N*-methyl-*N*-(trimethylsilyl)trifluoroacetamide
MTBSTFA	*N*-methyl-*N*-(*tert*-butyldimethysilyl)trifluoroacetamide
NCI	negative chemical ionization
NPEO	nonylphenol ethoxylate
NSAID	nonsteroidal anti-inflammatory drug
OCP	organochlorine pesticide
OP	octylphenol
OPs	organophosphates
PA	polyacrylate
PAC	pharmaceutically active compound
PAE	phthalic acid ester
PAH	polycyclic aromatic hydrocarbon
PBDE	polybrominated diphenyl ether
PBT	persistent, bioaccumulative, and toxic
PCB	polychlorinated biphenyl
PCDD/F	polychlorinated dibenzo-*p*-dioxin/furan
PCP	personal care product
PDMS	polydimethylsiloxane
PDMS-β-CD	poly(dimethysiloxane)-β-cyclodextrin
PEG	polyethylene glycol

PFAS	perfluoroalkyl sulfonate
PFC	perfluorinated compound
PFCA	perfluorocarboxylic acid
PFOS	perfluorooctane sulfonic acid
PHWE	pressurized hot water extraction
PLE	pressurized liquid extraction
POLE	polyoxyethylene 10 lauryl ether
POP	persistent organic pollutant
PPCPs	pharmaceuticals and personal care products
PSAs	primary and secondary amines
PVA	poly(vinyl alcohol)
QTOF	quadrupole-time-of-flight
QuEChERS	quick, easy, cheap, effective, rugged, and safe
RSD	relative standard deviation
SAESC	sonication-assisted extraction in small columns
SBME	solvent bar microextraction
SBSE	stir bar sorptive extraction
SDME	single drop microextraction
SFE	supercritical fluid extraction
SPE	solid-phase extraction
SIM	selective ion monitoring
SPLE	selective pressurized liquid extraction
SPME	solid-phase microextraction
SU	sulfonylurea
SUPRAS-ME	supramolecular solvent microextraction
SWE	subcritical water extraction
TBBPA	tetrabromobisphenol A
TD	thermal desorption
TMCS	trimethylchlorosilane
UAE	ultrasonic-assisted extraction
UPLC	ultraperformance liquid chromatography
WFD	Water Framework Directive
WWTP	wastewater treatment plant

REFERENCES

Albaseer, S.S., Rao, R.N., Swamy, Y.V., Mukkanti, K. 2010. An overview of sample preparation and extraction of synthetic pyrethroids from water, sediment and soil. *Journal of Chromatography A* 1217: 5537–5554.

Albero, B., Perez, R.A., Sanchez-Brunete, C., Tadeo, J.L. 2012. Occurrence and analysis of parabens in municipal sewage sludge from wastewater treatment plants in Madrid (Spain). *Journal of Hazardous Materials* 239–240: 48–55.

Albero, B., Sánchez-Brunete, C., Miguel, E., Aznar, R., Tadeo, J.L. 2014. Rapid determination of natural and synthetic hormones in biosolids and poultry manure by isotope dilution GC-MS/MS. *Journal of Separation Science* 37(7): 811–819.

Andreu, V., Picó, Y. 2012. Determination of currently used pesticides in biota. *Analytical and Bioanalytical Chemistry* 404: 2659–2681.

Arbeláez, P., Borrull, F., Marcé, R.M., Pocurull, E. 2014. Simultaneous determination of drugs of abuse and their main metabolites using pressurized liquid extraction and liquid chromatography–tandem mass spectrometry. *Talanta* 125: 65–71.

Arvaniti, O.S., Asimakopoulos, A.G., Dasenaki, M.E., Ventouri, E.I., Stasinakis, A.S. et al. 2014. Simultaneous determination of eighteen perfluorinated compounds in dissolved and particulate phases of wastewater, and in sewage sludge by liquid chromatography-tandem mass spectrometry. *Analytical Methods* 6(5): 1341–1349.

Asiabi, H., Yamini, Y., Moradi, M. 2013. Determination of sulfonylurea herbicides in soil samples via supercritical fluid extraction followed by nanostructured supramolecular solvent microextraction. *Journal of Supercritical Fluids* 84: 20–28 .

Aufartová, J., Mahugo-Santana, C., Sosa-Ferrera, Z., Santana-Rodríguez, J.J., Nováková, L. et al. 2012. Determination of steroid hormones in biological and environmental samples using green microextraction techniques: An overview. *Analytica Chimica Acta* 704: 33–46.

Azzouz, A., Ballesteros, E. 2012. Combined microwave-assisted extraction and continuous solid-phase extraction prior to gas chromatography–mass spectrometry determination of pharmaceuticals, personal care products and hormones in soils, sediments and sludge. *Science of the Total Environment* 419: 208–215.

Azzouz, A., Souhail, B., Ballesteros, E. 2010. Continuous solid-phase extraction and gas chromatography–mass spectrometry determination of pharmaceuticals and hormones in water samples. *Journal of Chromatography A* 1217(17): 2956–2963.

Balakrishnan, V., Terry, K., Toito, J. 2006. Determination of sulfonamide antibiotics in wastewater: A comparison of solid phase microextraction and solid phase extraction methods. *Journal of Chromatography A* 1131(1–2): 1–10.

Baltussen, E., Sandra, P., David, F., Cramers, C. 1999. Stir bar sorptive extraction (SBSE), a novel extraction technique for aqueous samples: Theory and principles. *Journal of Microcolumn Separations* 11: 737–747.

Barron, L., Tobin, J., Paull, B. 2008. Multi-residue determination of pharmaceuticals in sludge and sludge enriched soils using pressurized liquid extraction, solid phase extraction and liquid chromatography with tandem mass spectrometry. *Journal of Environmental Monitoring* 10(3): 353–361.

Basheer, C., Alnedhary, A.A., Rao, B.S.M., Valiyaveettil, S., Lee, H.K. 2006. Development and application of porous membrane-protected carbon nanotube micro-solid-phase extraction (μ-SPE) combined with gas chromatography/mass spectrometry. *Analytical Chemistry* 78(8): 2853–2858.

Basheer, C., Narasimhan, K., Yin, M., Zhao, C., Choolani, M. et al. 2008. Application of micro-solid-phase extraction for the determination of persistent organic pollutants in tissue samples. *Journal of Chromatography A* 1186: 358–364.

Beck, J., Totsche, K.U., Kogel-Knabner, I. 2008. A rapid and efficient determination of natural estrogens in soils by pressurised liquid extraction and gas chromatography–mass spectrometry. *Chemosphere* 71: 954–960.

Benítez-Villalba, J.C., Zafra-Gómez, A., Dorival-García, N., Camino-Sánchez, F.J., Cantarero, S. et al. 2013. Ultra-performance liquid chromatography MS/MS method for the determination of parabens in compost from sewage sludge: Comparison of the efficiency of two extraction techniques. *Journal of Separation Science* 36(16): 2635–2645.

Benito-Pena, E., Partal-Rodera, A., Leon-Gonzalez, M., Moreno-Bondi, M. 2006. Evaluation of mixed mode solid phase extraction cartridges for the preconcentration of beta-lactam antibiotics in wastewater using liquid chromatography with UV-DAD detection. *Analytica Chimica Acta* 556(2): 415–422.

Berset, J.D., Holzer, R. 1999. Quantitative determination of polycyclic aromatic hydrocarbons, polychlorinated biphenyls and organochlorine pesticides in sewage sludges using supercritical fluid extraction and mass spectrometric detection. *Journal of Chromatography* 852(2): 545–558.

Bester, K. 2009. Analysis of musk fragrances in environmental samples. *Journal of Chromatography A* 1216: 470–480.

Bielská, L., Smídová, K., Hofman, J. 2013. Supercritical fluid extraction of persistent organic pollutants from natural and artificial soils and comparison with bioaccumulation in earthworms. *Environmental Pollution* 176: 48–54.

Björklund, E., Nilsson, T., Bøwadt, S., Pilorz, K., Mathiasson, L. et al. 2000. Introducing selective supercritical fluid extraction as a new tool for determining sorption/desorption behavior and bioavailability of persistent organic pollutants in sediment. *Journal of Biochemical and Biophysical Methods* 43: 295–311.

Blanco, E., Casais, M.C., Mejuto, M.C., Cela, R. 2006. Approaches for the simultaneous extraction of tetrabromobisphenol A, tetrachlorobisphenol A, and related phenolic compounds from sewage sludge and sediment samples based on matrix solid-phase dispersion. *Analytical Chemistry* 78(8): 2772–2778.

Bragança, I., Plácido, A., Paíga, P., Domingues, V.F., Delerue-Matos, C. 2012. QuEChERS: A new sample preparation approach for the determination of ibuprofen and its metabolites in soils. *Science of the Total Environment* 433: 281–298.

Bruno, F., Curini, R., Corcia, A.D., Fochi, I., Nazzari, M. et al. 2002. Determination of surfactants and some of their metabolites in untreated and anaerobically digested sewage sludge by subcritical water extraction followed by liquid chromatography–mass spectrometry. *Environmental Science and Technology* 36(19): 4156–4161.

Bueno, M.J.M., Boillot, C., Fenet, H., Chiron, S., Casellas, C. et al. 2013. Fast and easy extraction combined with high resolution-mass spectrometry for residue analysis of two anticonvulsants and their transformation products in marine mussels. *Journal of Chromatography A* 1305: 27–34.

Cai, Y., Jiang, G., Liu, J., Zhou, Q. 2003. Multi-walled carbon nanotubes packed cartridge for the solid-phase extraction of several phthalate esters from water samples and their determination by high performance liquid chromatography. *Analytica Chimica Acta* 494(1–2): 149–156.

Camino-Sánchez, F.J., Zafra-Gómez, A., Pérez-Trujillo, J.P., Conde-González, J.E., Marques, J.C. et al. 2011. Validation of a GC–MS/MS method for simultaneous determination of 86 persistent organic pollutants in marine sediments by pressurized liquid extraction followed by stir bar sorptive extraction. *Chemosphere* 84: 869–881.

Campíns-Falcó, P., Verdú-Andrés, J., Sevillano-Cabeza, A., Molins-Legua, C., Herráez-Hernández, R. 2008. New micromethod combining miniaturized matrix solid-phase dispersion and in-tube in-valve solid-phase microextraction for estimating polycyclic aromatic hydrocarbons in bivalves. *Journal of Chromatography A* 1211: 13–21.

Cantarero, S., Zafra-Gómez, A., Ballesteros, O., Navalón, A., Vílchez, J.L. et al. 2011. Matrix effect study in the determination of linear alkylbenzene sulfonates in sewage sludge samples. *Environmental Toxicology and Chemistry* 30(4): 813–818.

Carrasco, L., Díez, S., Bayona, J.M. 2007. Methylmercury determination in biota by solid-phase microextraction: Matrix effect evaluation. *Journal of Chromatography A* 1174: 2–6.

Carvalho, P.N., Pirra, A., Basto, M.C.P., Almeida, C.M.R. 2013. Multi-family methodologies for the analysis of veterinary pharmaceutical compounds in sediment and sludge samples: Comparison among extraction techniques. *Analytical Methods* 5: 6503–6510.

Casado, J., Rodríguez, I., Carpinteiro, I., Ramil, M., Cela, R. 2013. Gas chromatography quadrupole time-of-flight mass spectrometry determination of benzotriazole ultraviolet stabilizers in sludge samples. *Journal of Chromatography A* 1293: 126–132.

Castiglioni, S., Bagnati, R., Calamar, D., Fanelli, R., Zuccato, E. 2005. A multiresidue analytical method using solid-phase extraction and high-pressure liquid chromatography tandem mass spectrometry to measure pharmaceuticals of different therapeutic classes in urban wastewaters. *Journal of Chromatography A* 1092(2): 206–215.

Castiglioni, S., Bagnati, R., Fanelli, R., Pomati¸ F., Calamari, D. et al. 2006. Removal of pharmaceuticals in sewage treatment plants in Italy. *Environmental Science and Technology* 40(1): 357–363.

Celano, R., Rodríguez, I., Cela, R., Rastrelli, L., Piccinelli, A.L. 2014. Liquid chromatography quadrupole time-of-flight mass spectrometry quantification and screening of organophosphate compounds in sludge. *Talanta* 118: 312–320.

Cerqueira, M.B.R., Caldas, S.S., Primel, E.G. 2014. New sorbent in the dispersive solid phase extraction step of quick, easy, cheap, effective, rugged, and safe for the extraction of organic contaminants in drinking water treatment sludge. *Journal of Chromatography A* 1336: 10–22.

Chen, X., Bester, K. 2009. Determination of organic micro-pollutants such as personal care products, plasticizers and flame retardants in sludge. *Analytical and Bioanalytical Chemistry* 395(6): 1877–1884.

Chen, L., Zeng, Q., Wang, H., Su, R., Xu, Y. et al. 2009. On-line coupling of dynamic microwave-assisted extraction to solid-phase extraction for the determination of sulfonamide antibiotics in soil. *Analytica Chimica Acta* 648: 200–206.

Chitescu, C.L., Oosterink, E., Jong, J.D., Stolker, A.A.M.L. 2012. Ultrasonic or accelerated solvent extraction followed by U-HPLC-high mass accuracy MS for screening of pharmaceuticals and fungicides in soil and plant samples. *Talanta* 88: 653–662.

Clarke, B.O., Smith, S.R. 2011. Review of "emerging" organic contaminants in biosolids and assessment of international research priorities for the agricultural use of biosolids. *Environment International* 37: 226–247.

Combalbert, S., Pype, M.-L., Bernet, N., Hernandez-Raquet, G. 2010. Enhanced methods for conditioning, storage, and extraction of liquid and solid samples of manure for determination of steroid hormones by solid-phase extraction and gas chromatography–mass spectrometry. *Analytical and Bioanalytical Chemistry* 398(2): 973–984.

Commission of the European Communities. 2006. Communication from the Commission to the Council, the European Parliament, the Economic and Social Committee and the Committee of the Regions, "Thematic strategy for soil protection," COM(2006) 231 final, Brussels, 12 pp.

Concha-Graña, E., Fernández-González, V., Grueiro-Noche, G., Muniategui-Lorenzo, S., López-Mahía, P. et al. 2010. Development of an environmental friendly method for the analysis of organochlorine pesticides in sediments. *Chemosphere* 79: 698–705.

Cristale, J., Lacorte, S. 2013. Development and validation of a multiresidue method for the analysis of polybrominated diphenyl ethers, new brominated and organophosphorus flame retardants in sediment, sludge and dust. *Journal of Chromatography A* 1305: 267–275.

Cui, X., Bao, L., Gan, J. 2013. Solid-phase microextraction (SPME) with stable isotope calibration for measuring bioavailability of hydrophobic organic contaminants. *Environmental Science and Technology* 47: 9833–9840.

Díaz-Cruz, M.S., García-Galán, M.J., Guerra, P., Jelic, A., Postigo, C. et al. 2009. Analysis of selected emerging contaminants in sewage sludge. *TrAC Trends in Analytical Chemistry* 28(11): 1263–1275.

Dobor, J., Varga, M., Yao, J., Chen, H., Palkó, G. et al. 2010. A new sample preparation method for determination of acidic drugs in sewage sludge applying microwave assisted solvent extraction followed by gas chromatography-mass spectrometry. *Microchemical Journal* 94(1): 36–41.

Domeno, C., Blasco, M., Sánchez, C., Nerín, C. 2006. A fast extraction technique for extracting polycyclic aromatic hydrocarbons (PAHs) from lichens samples used as biomonitors of air pollution: Dynamic sonication versus other methods. *Analytica Chimica Acta* 569: 103–112.

Dorival-García, N., Zafra-Gómez, A., Navalón, A., Vílchez, J.L. 2012. Improved sample treatment for the determination of bisphenol A and its chlorinated derivatives in sewage sludge samples by pressurized liquid extraction and liquid chromatography–tandem mass spectrometry. *Talanta* 101: 1–10.

Dorival-García, N., Zafra-Gómez, A., Camino-Sánchez, F.J., Navalón, A., Vílchez, J.L. 2013. Analysis of quinolone antibiotic derivatives in sewage sludge samples by liquid chromatography–tandem mass spectrometry: Comparison of the efficiency of three extraction techniques. *Talanta* 106: 104–118.

Đurović, R.D., Đorđević, T.M. 2012. Effects of soil composition on solid phase microextraction determination of triazine and organophosphorus pesticides. *Journal of Environmental Science and Health, Part B: Pesticides, Food Contaminants, and Agricultural Wastes* 47: 851–857.

Dussault, E.B., Balakrishnan, V.K., Borgmann, U., Solomon, K.R., Sibley, P.K. 2009. Bioaccumulation of the synthetic hormone 17α-ethinylestradiol in the benthic invertebrates *Chironomus tentans* and *Hyalella azteca*. *Ecotoxicology and Environmental Safety* 72: 1635–1641.

Eisert, R., Pawliszyn, J. 1997. Automated in-tube solid-phase microextraction coupled to high-performance liquid chromatography. *Analytical Chemistry* 69(16): 3140–3147.

European Commission. 2008. Directive 2008/56/EC of the European Parliament and of the Council of 17 June 2008 establishing a framework for community action in the field of marine environmental policy (Marine Strategy Framework Directive). *Official Journal of the European Communities* L164/19–40.

European Commission. 2013. Directive 2013/39/EU of the European Parliament and of the Council of 12 August 2013 amending Directives 2000/60/EC and 2008/105/EC as regards priority substances in the field of water policy. *Official Journal of the European Communities* L226/1–17.

European Union. 2010. Guidance Document No. 25—Guidance on chemical monitoring of sediment and biota under the Water Framework Directive. Common implementation strategy for the Water Framework Directive (2000/60/EC). Technical Report-2010-04, 82 pp.

Farre, M., Kantiani, I., Petrovic, M., Perez, S., Barcelo, D. 2012. Achievements and future trends in the analysis of emerging organic contaminants in environmental samples by mass spectrometry and bioanalytical techniques. *Journal of Chromatography A* 1259(0): 86–99.

Farré, M., Picó, Y., Barceló, D. 2014. Application of ultra-high pressure liquid chromatography linear ion-trap orbitrap to qualitative and quantitative assessment of pesticide residues. *Journal of Chromatography A* 1328: 66–79.

Ferguson, P., Iden, C., McElroy, A., Brownawell, B. 2001. Determination of steroid estrogens in wastewater by immunoaffinity extraction coupled with HPLC–electrospray-MS. *Analytical Chemistry* 73(16): 3890–3895.

Fernandez, A., Sosa Ferrera, Z., Rodriguez, J. 2001. Application of microwave-assisted extraction using micellar media to the determination of polychlorinated biphenyls in marine sediments. *Analytica Chimica Acta* 433(2): 237–244.

Fernández-González, V., Concha-Grãna, E., Muniategui-Lorenzo, S., López-Mahía, P., Prada-Rodríguez, D. 2008. Pressurized hot water extraction coupled to solid-phase microextraction–gas chromatography–mass spectrometry for the analysis of polycyclic aromatic hydrocarbons in sediments. *Journal of Chromatography A* 1196–1197: 65–72.

Ferreira, A.M.C., Möder, M., Laespada, M.E.F. 2011. Stir bar sorptive extraction of parabens, triclosan and methyl triclosan from soil, sediment and sludge with in situ derivatization and determination by gas chromatography–mass spectrometry. *Journal of Chromatography A* 1218(25): 3837–3844.

Fidalgo-Used, N., Centineo, G., Blanco-González, E., Sanz-Medel, A. 2003. Solid-phase microextraction as a clean-up and preconcentration procedure for organochlorine pesticides determination in fish tissue by gas chromatography with electron capture detection. *Journal of Chromatography A* 1017: 35–44.

Fidalgo-Used, N., Blanco-González, E., Sanz-Medel, A. 2007. Sample handling strategies for the determination of persistent trace organic contaminants from biota samples. *Analytica Chimica Acta* 590: 1–16.

Fontanals, N., Marce, R., Borrull, F. 2005. New hydrophilic materials for solid-phase extraction. *TrAC Trends in Analytical Chemistry* 24(5): 394–406.

Forero-Mendieta, J.R., Castro-Vargas, H.I., Parada-Alfonso, F., Guerrero-Dallos, J.A. 2012. Extraction of pesticides from soil using supercritical carbon dioxide added with methanol as co-solvent. *Journal of Supercritical Fluids* 68: 64–70.

Fountoulakis, M., Drillia, P., Pakou, C., Kampioti, A., Stamatelatou, K. et al. 2005. Analysis of nonylphenol and nonylphenol ethoxylates in sewage sludge by high performance liquid chromatography following microwave-assisted extraction. *Journal of Chromatography A* 1089(1–2): 45–51.

Gago-Ferrero, P., Díaz-Cruz, M.S., Barceló, D. 2013. Multi-residue method for trace level determination of UV filters in fish based on pressurized liquid extraction and liquid chromatography–quadrupole-linear ion trap–mass spectrometry. *Journal of Chromatography A* 1286: 93–101.

Gallego-Gallegos, M., Liva, M., Olivas, R.M., Cámara, C. 2006. Focused ultrasound and molecularly imprinted polymers: A new approach to organotin analysis in environmental samples. *Journal of Chromatography A* 1114: 82–88.

García-Galán, M.J., Díaz-Cruz, S., Barceló, D. 2013. Multiresidue trace analysis of sulfonamide antibiotics and their metabolites in soils and sewage sludge by pressurized liquid extraction followed by liquid chromatography–electrospray-quadrupole linear ion trap mass spectrometry. *Journal of Chromatography A* 1275: 32–40.

Gatidou, G., Thomaidis, N.S., Stasinakis, A.S., Lekkas, T.D. 2007. Simultaneous determination of the endocrine disrupting compounds nonylphenol, nonylphenol ethoxylates, triclosan and bisphenol A in wastewater and sewage sludge by gas chromatography–mass spectrometry. *Journal of Chromatography A* 1138(1–2): 32–41.

Gerhards, P., Sadjadi, S., Motyka, R. 2004. Rapid HPLC column and mobile phase screening for fast LC/MS/MS method development. Application Note A02106, Varian 1–3.

Ghasemi, E., Sillanpää, M. 2014. Optimization of headspace solid phase microextraction based on nanostructured ZnO combined with gas chromatography–mass spectrometry for preconcentration and determination of ultra-traces of chlorobenzenes in environmental samples. *Talanta* 130: 322–327.

Gilart, N., Miralles, N., Marce, R., Borrull, F., Fontanals, N. 2013. Novel coatings for stir bar sorptive extraction to determine pharmaceuticals and personal care products in environmental waters by liquid chromatography and tandem mass spectrometry. *Analytica Chimica Acta* 774: 51–60.

Göbel, A., Thomsen, A., McArdell, C.S., Alder, A.C., Giger, W. et al. 2005. Extraction and determination of sulfonamides, macrolides, and trimethoprim in sewage sludge. *Journal of Chromatography A* 1085: 179–189.

Gonçalves, C., Alpendurada, M.F. 2005. Assessment of pesticide contamination in soil samples from an intensive horticulture area, using ultrasonic extraction and gas chromatography–mass spectrometry. *Talanta* 65: 1179–1189.

Gonçalves, C., Carvalho, J.J., Azenha, M.A., Alpendurada, M.F. 2006. Optimization of supercritical fluid extraction of pesticide residues in soil by means of central composite design and analysis by gas chromatography–tandem mass spectrometry. *Journal of Chromatography A* 1110: 6–14.

Gonçalves, C., Pinto, M.d.R., Alpendurada, M.F. 2013. Benefits of a binary modifier with balanced polarity for an efficient supercritical fluid extraction of PAHs from solid samples followed by high performance liquid chromatography. *Journal of Liquid Chromatography and Related Technologies* 24(19): 2943–2959.

Gonçalves, C., Teixeira, C., Basto, M.C.P., Almeida, C.M.R. 2016. PAHs levels in Portuguese estuaries and lagoons: Salt marsh plants as potential agents for the containment of PAHs contamination in sediments. *Regional Studies in Marine Science* 7: 211–221.

Groz, M.P., Bueno, M.J.M., Rosain, D., Fenet, H., Casellas, C. et al. 2014. Detection of emerging contaminants (UV filters, UV stabilizers and musks) in marine mussels from Portuguese coast by QuEChERS extraction and GC-MS/MS. *Science of the Total Environment* 493: 162–169.

Gu, Y.-y., Yu, X.-j., Peng, J.-f., Chen, S.-b., Zhong, Y.-y. et al. 2014. Simultaneous solid phase extraction coupled with liquid chromatography tandem mass spectrometry and gas chromatography tandem mass spectrometry for the highly sensitive determination of 15 endocrine disrupting chemicals in seafood. *Journal of Chromatography B* 965: 164–172.

Guerra, P., Eljarrat, E., Barceló, D. 2010. Simultaneous determination of hexabromocyclododecane, tetrabromobisphenol A, and related compounds in sewage sludge and sediment samples from Ebro River basin (Spain). *Analytical and Bioanalytical Chemistry* 397: 2817–2824.

Guo, J., Jiang, R., Pawliszyn, J. 2013. Determination of polycyclic aromatic hydrocarbons in solid matrices using automated cold fiber headspace solid phase microextraction technique. *Journal of Chromatography A* 1307: 66–72.

Guo, L., Lee, H.K. 2013. Microwave assisted extraction combined with solvent bar microextraction for one-step solvent-minimized extraction, cleanup and preconcentration of polycyclic aromatic hydrocarbons in soil samples. *Journal of Chromatography A* 1286: 9–15.

Harwood, A.D., Landrum, P.F., Weston, D.P., Lydy, M.J. 2013. Using SPME fibers and Tenax to predict the bioavailability of pyrethroids and chlorpyrifos in field sediments. *Environmental Pollution* 173: 47–51.

Hawthorne, S.B., Chai, Y., Grabanski, C.B., Davis, J.W., Wilken, M. et al. 2012. Investigating differential binding of polychlorinated dibenzo-*p*-dioxins/dibenzofurans in soil and soil components using selective supercritical fluid extraction. *Chemosphere* 88: 261–269.

He, S., Chen, Q., Sun, Y., Zhu, Y., Luo, L., Li, J. et al. 2011. Determination of tobramycin in soil by HPLC with ultrasonic-assisted extraction and solid-phase extraction. *Journal of Chromatography B* 879: 901–907.

Hennion, M. 1991. Sample handling strategies for the analysis of non-volatile organic compounds from environmental water samples. *TrAC Trends in Analytical Chemistry* 10(10): 317–323.

Herrero, P., Borrull, F., Marcé, R.M., Pocurull, E. 2014. A pressurised hot water extraction and liquid chromatography-high resolution mass spectrometry method to determine polar benzotriazole, benzothiazole and benzenesulfonamide derivates in sewage sludge. *Journal of Chromatography A* 1355: 53–60.

Higashikawa, F.S., Cayuela, M.L., Roig, A., Silva, C.A., Sánchez-Monedero, M.A. 2013. Matrix effect on the performance of headspace solid phase microextraction method for the analysis of target volatile organic compounds (VOCs) in environmental samples. *Chemosphere* 93: 2311–2318.

Horwitz, W., Kamps, L.R., Boyer, K.W. 1980. Quality assurance in the analysis of foods and trace constituents. *Journal of the Association of Official Analytical Chemists* 63(6): 1344–1354.

Huerta, B., Rodríguez-Mozaz, S., Barceló, D. 2012. Pharmaceuticals in biota in the aquatic environment: Analytical methods and environmental implications. *Analytical and Bioanalytical Chemistry* 404: 2611–2624.

Huerta, B., Jakimska, A., Gros, M., Rodríguez-Mozaz, S., Barceló, D. 2013. Analysis of multi-class pharmaceuticals in fish tissues by ultra-high-performance liquid chromatography tandem mass spectrometry. *Journal of Chromatography A* 1288: 63–72.

Jelic, A., Petrovíc, M., Barceló, D. 2009. Multi-residue method for trace level determination of pharmaceuticals in solid samples using pressurized liquid extraction followed by liquid chromatography/quadrupole–linear ion trap mass spectrometry. *Talanta* 80(1): 363–371.

Jux, U., Baginski, R., Arnold, H., Kronke, M., Seng, P. 2002. Detection of pharmaceutical contaminations of river, pond, and tap water from Cologne (Germany) and surroundings. *International Journal of Hygiene and Environmental Health* 205(5): 393–398.

Kelessidis, A., Stasinakis, A.S. 2012. Comparative study of the methods used for treatment and final disposal of sewage sludge in European countries. *Waste Management* 32(6): 1186–1195.

Kjellström, A., Brantlind, M., Eldsäter, C. 2008. Optimized microwave extraction for trace detection of 2,4,6-trinitrotoluene in soil samples. *Chemosphere* 71: 1701–1708.

Klosterhaus, S.L., Grace, R., Hamilton, M.C., Yee, D. 2013. Method validation and reconnaissance of pharmaceuticals, personal care products and alkylphenols in surface waters, sediments, and mussels in an urban estuary. *Environment International* 54: 92–99.

Król, S., Zabiegała, B., Namieśnik, J. 2012. PBDEs in environmental samples: Sampling and analysis. *Talanta* 93: 1–17.

Lacroix, C., Cuff, N.L., Receveur, J., Moraga, D., Auffret, M. et al. 2014. Development of an innovative and "green" stir bar sorptive extraction–thermal desorption–gas chromatography–tandem mass spectrometry method for quantification of polycyclic aromatic hydrocarbons in marine biota. *Journal of Chromatography A* 1349: 1–10.

Lee, H.-B., Peart, T.E., Sarafin, K. 2003. Occurrence of polycyclic and nitro musk compounds in Canadian sludge and wastewater samples. *Water Quality Research Journal of Canada* 38(4): 683–702.

Lillenberg, M., Yurchenko, S., Kipper, K., Herodes, K., Pihl, V. et al. 2009. Simultaneous determination of fluoroquinolones, sulfonamides and tetracyclines in sewage sludge by pressurized liquid extraction and liquid chromatography electrospray ionization–mass spectrometry. *Journal of Chromatography A* 1216(32): 5949–5954.

Liska, I. 2000. Fifty years of solid-phase extraction in water analysis—Historical development and overview. *Journal of Chromatography A* 885(1–2): 3–16.

Llop, A., Borrull, F., Pocurull, E. 2010. Pressurised hot water extraction followed by simultaneous derivatization and headspace solid-phase microextraction and gas chromatography-tandem mass spectrometry for the determination of aliphatic primary amines in sewage sludge. *Analytica Chimica Acta* 665(2): 231–236.

Llorca-Pórcel, J., Martínez-Parreño, M., Martínez-Soriano, E., Valor, I. 2009. Analysis of chlorophenols, bisphenol-A, 4-*tert*-octylphenol and 4-nonylphenols in soil by means of ultrasonic solvent extraction and stir bar sorptive extraction with in situ derivatisation. *Journal of Chromatography A* 1216: 5955–5961.

Martendal, E., Carasek, E. 2011. A new approach based on a combination of direct and headspace cold-fiber solid-phase microextraction modes in the same procedure for the determination of polycyclic aromatic hydrocarbons and phthalate esters in soil samples. *Journal of Chromatography A* 1218: 1707–1714.

Martín, J., Santos, J.L., Aparicio, I., Alonso, E. 2010. Multi-residue method for the analysis of pharmaceutical compounds in sewage sludge, compost and sediments by sonication-assisted extraction and LC determination. *Journal of Separation Science* 33(12): 1760–1766.

Martínez-Moral, M.P., Tena, M.T. 2011. Focused ultrasound solid–liquid extraction and selective pressurised liquid extraction to determine bisphenol A and alkylphenols in sewage sludge by gas chromatography–mass spectrometry. *Journal of Separation Science* 34(18): 2513–2522.

Martínez-Moral, M.P., Tena, M.T. 2013. Focused ultrasound solid-liquid extraction of perfluorinated compounds from sewage sludge. *Talanta* 109: 197–202.

Martínez-Parreño, M., Llorca-Pórcel, J., Valor, I. 2008. Analysis of 51 persistent organic pollutants in soil by means of ultrasonic solvent extraction and stir bar sorptive extraction GC-MS. *Journal of Separation Science* 31: 3620–3629.

Masiá, A., Campo, J., Vázquez-Roig, P., Blasco, C., Picó, Y. 2013a. Screening of currently used pesticides in water, sediments and biota of the Guadalquivir River Basin (Spain). *Journal of Hazardous Materials* 236P: 95–104.

Masiá, A., Moliner-Martinez, Y., Munoz-Ortuno, M., Pico, Y., Campíns-Falco, P. 2013b. Multiresidue analysis of organic pollutants by in-tube solid phase microextraction coupled to ultra-high performance liquid chromatography–electrospray-tandem mass spectrometry. *Journal of Chromatography A* 1306: 1–11.

Mastroianni, N., Postigo, C., de Alda, M.L., Barcelo, D. 2013. Illicit and abused drugs in sewage sludge: Method optimization and occurrence. *Journal of Chromatography A* 1322: 29–37.

McClure, E., Wong, C. 2007. Solid phase microextraction of macrolide, trimethoprim, and sulfonamide antibiotics in wastewaters. *Journal of Chromatography A* 1169(1–2): 53–62.

Meesters, R.J.W., Schröder, H.F. 2002. Simultaneous determination of 4-nonylphenol and bisphenol A in sewage sludge. *Analytical Chemistry* 74(14): 3566–3574.

Merdassa, Y., Liu, J.-f., Megersa, N. 2013. Development of a one-step microwave-assisted extraction method for simultaneous determination of organophosphorus pesticides and fungicides in soils by gas chromatography–mass spectrometry. *Talanta* 114: 227–234.

Miège, C., Dugay, J., Hennion, M.-C. 2003. Optimization, validation and comparison of various extraction techniques for the trace determination of polycyclic aromatic hydrocarbons in sewage sludges by liquid chromatography coupled to diode-array and fluorescence detection *Journal of Chromatography A* 995(1–2): 87–97.

Moliner-Martinez, Y., Campíns-Falcó, P., Molins-Legua, C., Segovia-Martínez, L., Seco-Torrecillas, A. 2009. Miniaturized matrix solid phase dispersion procedure and solid phase microextraction for the analysis of organochlorinated pesticides and polybrominated diphenylethers in biota samples by gas chromatography electron capture detection. *Journal of Chromatography A* 1216: 6741–6745.

Mompelat, S., Le Bot, B., Thomas, O. 2009. Occurrence and fate of pharmaceutical products and by-products, from resource to drinking water. *Environment International* 35(5): 803–814.

Montes, R., Rodríguez, I., Cela, R. 2010. Solid-phase microextraction with simultaneous oxidative sample treatment for the sensitive determination of tetra- to hexa-brominated diphenyl ethers in sediments. *Journal of Chromatography A* 1217: 14–21.

Montesdeoca-Esponda, S., Sosa-Ferrera, Z., Santana-Rodríguez, J.J. 2013. Microwave-assisted extraction combined with on-line solid phase extraction followed by ultra-high-performance liquid chromatography with tandem mass spectrometric determination of benzotriazole UV stabilizers in marine sediments and sewage sludges. *Journal of Separation Science* 36: 781–788.

Morales, S., Canosa, P., Rodríguez, I., Rubí, E., Cela, R. 2005. Microwave assisted extraction followed by gas chromatography with tandem mass spectrometry for the determination of triclosan and two related chlorophenols in sludge and sediments. *Journal of Chromatography A* 1082: 128–135.

Naeeni, M.H., Yamini, Y., Rezaee, M. 2011. Combination of supercritical fluid extraction with dispersive liquid–liquid microextraction for extraction of organophosphorus pesticides from soil and marine sediment samples. *Journal of Supercritical Fluids* 57: 219–226.

Negreira, N., Rodríguez, I., Rubí, E., Cela, R. 2011. Optimization of pressurized liquid extraction and purification conditions for gas chromatography–mass spectrometry determination of UV filters in sludge. *Journal of Chromatography A* 1218: 211–217.

Nerín, C., Batlle, R., Sartaguda, M., Pedrocchi, C. 2002. Supercritical fluid extraction of organochlorine pesticides and some metabolites in frogs from National Park of Ordesa and Monte Perdido. *Analytica Chimica Acta* 464: 303–312.

Nieto, A., Borrull, F., Pocurull, E., Marcé, R.M. 2008. Determination of natural and synthetic estrogens and their conjugates in sewage sludge by pressurized liquid extraction and liquid chromatography–tandem mass spectrometry. *Journal of Chromatography A* 1213(2): 224–230.

Nieto, A., Borrull, F., Marcé, R.M., Pocurull, E. 2009. Determination of personal care products in sewage sludge by pressurized liquid extraction and ultra high performance liquid chromatography-tandem mass spectrometry. *Journal of Chromatography A* 1216(30): 5619–5625.

Niño, L.R., Grosseli, G.M., Mozeto, A.A., Fadini, P.S. 2014. Optimization study of BTEX extraction from soils spiked with two kinds of Brazilian diesel by HS SPME using Box–Behnken experimental design and multi-response analysis. *Analytical Methods* 6(10): 3247–3253.

Olives, A., Gonzalez-Ruiz, V., Martin, M. 2012. Isolation and quantitative methods for analysis of non-steroidal anti-inflammatory drugs. *Anti-inflammatory & Anti-allergy Agents in Medicinal Chemistry* 11: 65–95.

Ordoñez, E.Y., Quintana, J.B., Rodil, R., Cela, R. 2013. Determination of artificial sweeteners in sewage sludge samples using pressurised liquid extraction and liquid chromatography-tandem mass spectrometry. *Journal of Chromatography A* 1320: 10–16.

Ottonello, G., Ferrari, A., Magi, E. 2014. Determination of polychlorinated biphenyls in fish: Optimisation and validation of a method based on accelerated solvent extraction and gas chromatography–mass spectrometry. *Food Chemistry* 142: 327–333.

Pan, D., Wang, J., Chen, C., Huang, C., Cai, Q. et al. 2013. Ultrasonic assisted extraction combined with titanium-plate based solid phase extraction for the analysis of PAHs in soil samples by HPLC-FLD. *Talanta* 108: 117–122.

Pedrouzo, M., Borrull, F., Marce, R., Pocurull, E. 2010. Stir-bar-sorptive extraction and ultra-high-performance liquid chromatography–tandem mass spectrometry for simultaneous analysis of UV filters and antimicrobial agents in water samples. *Analytical and Bioanalytical Chemistry* 397(7): 2833–2839.

Pena, M.T., Casais, M.C., Mejuto, M.C., Cela, R. 2008. Development of a matrix solid-phase dispersion method for the determination of polycyclic aromatic hydrocarbons in sewage sludge samples. *Analytica Chimica Acta* 626(2): 155–165.

Pena, M.T., Casais, M.C., Mejuto, M.C., Cela, R. 2010. Development of a sample preparation procedure of sewage sludge samples for the determination of polycyclic aromatic hydrocarbons based on selective pressurized liquid extraction. *Journal of Chromatography A* 1217(4): 425–435.

Pérez, S., Farré, M., Gonçalves, C., Alpendurada, M.F., Barceló, D. 2011 Green analytical chemistry in the determination of organic pollutants in the environment. In *Challenges in Green Analytical Chemistry*, ed. delaGuardia, M., Garrigues, S. Royal Society of Chemistry, London, pp. 223–283.

Peruga, A., Beltrán, J., López, F., Hernández, F. 2014. Determination of methylisothiocyanate in soil and water by HS-SPME followed by GC-MS-MS with a triple quadrupole. *Analytical and Bioanalytical Chemistry* 406: 5271–5282.

Peysson, W., Vulliet, E. 2013. Determination of 136 pharmaceuticals and hormones in sewage sludge using quick, easy, cheap, effective, rugged and safe extraction followed by analysis with liquid chromatography-time-of-flight-mass spectrometry. *Journal of Chromatography A* 1290: 46–61.

Pfannkoch, E.A., Stuff, J.R., Whitecavage, J.A. 2010. High throughput method for the determination of PAHs in seafood by QuEChERS-SBSE-GC-MS. Gerstel Application Note, 6 Part B: 1.

Pichon, V. 2000. Solid-phase extraction for multiresidue analysis of organic contaminants in water. *Journal of Chromatography A* 885(1–2): 195–215.

Pintado-Herrera, M.G., González-Mazo, E., Lara-Martín, P.A. 2013. Environmentally friendly analysis of emerging contaminants by pressurized hot water extraction–stir bar sorptive extraction–derivatization and gas chromatography–mass spectrometry. *Analytical and Bioanalytical Chemistry* 405: 401–411.

Pinto, J.A.G. 2013. Determination of quinolones/fluoroquinolones in soil and sewage sludge from a wastewater treatment plant through ASE and LC-MS/MS. Graduation report. Faculty of Sciences, University of Porto, Porto, Portugal. Unpublished results, 63 pp.

Pinto, C.G., Laespada, M.E.F., Martín, S.H., Ferreira, A.M.C., Pavón, J.L.P. et al. 2010. Simplified QuEChERS approach for the extraction of chlorinated compounds from soil samples. *Talanta* 81: 385–391.

Priego-Capote, F., de Castro, M.D.L. 2004. Analytical uses of ultrasound: I. Sample preparation. *TrAC Trends in Analytical Chemistry* 23(9): 644–653.

Radjenović, J., Jelić, A., Petrović, M., Barceló, D. 2009. Determination of pharmaceuticals in sewage sludge by pressurized liquid extraction (PLE) coupled to liquid chromatography–tandem mass spectrometry (LC-MS/MS). *Analytical and Bioanalytical Chemistry* 393(6–7): 1685–1695.

Raich-Montiu, J., Prat, M.D., Granados, M. 2011. Extraction and analysis of avermectines in agricultural soils by microwave assisted extraction and ultra high performance liquid chromatography coupled to tandem mass spectrometry. *Analytica Chimica Acta* 697: 32–37.

Rocha, M.J., Ferreira, P.C., Reis, P.A., Cruzeiro, C., Rocha, E. 2011. Determination of polycyclic aromatic hydrocarbons in coastal sediments from the Porto region (Portugal) by microwave-assisted extraction, followed by SPME and GC-MS. *Journal of Chromatographic Science* 49: 695–701.

Rodil, R., Popp, P. 2006. Development of pressurized subcritical water extraction combined with stir bar sorptive extraction for the analysis of organochlorine pesticides and chlorobenzenes in soils. *Journal of Chromatography A* 1124: 82–90.

Ronan, J.M., McHugh, B. 2013. A sensitive liquid chromatography/tandem mass spectrometry method for the determination of natural and synthetic steroid estrogens in seawater and marine biota, with a focus on proposed Water Framework Directive Environmental Quality Standards. *Rapid Communications in Mass Spectrometry* 27: 738–746.

Rouvière, F., Buleté, A., Cren-Olivé, C., Arnaudguilhem, C. 2012. Multiresidue analysis of aromatic organochlorines in soil by gas chromatography–mass spectrometry and QuEChERS extraction based on water/dichloromethane partitioning: Comparison with accelerated solvent extraction. *Talanta* 93: 336–344.

Ryno, M., Rantanen, L., Papaioannou, E., Konstandopoulos, A.G., Koskentalo, T. et al. 2006. Comparison of pressurized fluid extraction, Soxhlet extraction and sonication for the determination of polycyclic aromatic hydrocarbons in urban air and diesel exhaust particulate matter. *Journal of Environmental Monitoring* 8: 488–493.

Saleh, A., Larsson, E., Yamini, Y., Jönsson, J.Å. 2011. Hollow fiber liquid phase microextraction as a preconcentration and clean-up step after pressurized hot water extraction for the determination of non-steroidal anti-inflammatory drugs in sewage sludge. *Journal of Chromatography A* 1218(10): 1331–1339.

Salemi, A., Rasoolzadeh, R., Nejada, M.M., Vosough, M. 2013. Ultrasonic assisted headspace single drop micro-extraction and gas chromatography with nitrogen-phosphorus detector for determination of organophosphorus pesticides in soil. *Analytica Chimica Acta* 769: 121–126.

Salgado-Petinal, C., Garcia-Chao, M., Llompart, M., Garcia-Jares, C., Cela, R. 2006a. Headspace solid-phase microextraction gas chromatography tandem mass spectrometry for the determination of brominated flame retardants in environmental solid samples. *Analytical and Bioanalytical Chemistry* 385: 637–644.

Salgado-Petinal, C., Llompart, M., García-Jares, C., García-Chao, M., Cela, R. 2006b. Simple approach for the determination of brominated flame retardants in environmental solid samples based on solvent extraction and solid-phase microextraction followed by gas chromatography–tandem mass spectrometry. *Journal of Chromatography A* 1124: 139–147.

Samaras, V.G., Thomaidis, N.S., Stasinakis, A.S., Lekkas, T.D. 2011. An analytical method for the simultaneous trace determination of acidic pharmaceuticals and phenolic endocrine disrupting chemicals in wastewater and sewage sludge by gas chromatography–mass spectrometry. *Analytical and Bioanalytical Chemistry* 399(7): 2549–2561.

Sánchez-Brunete, C., Miguel, E., Tadeo, J.L. 2007. Analysis of 27 polycyclic aromatic hydrocarbons by matrix solid-phase dispersion and isotope dilution gas chromatography-mass spectrometry in sewage sludge from the Spanish area of Madrid. *Journal of Chromatography A* 1148(2): 219–227.

Sánchez-Brunete, C., Miguel, E., Tadeo, J.L. 2008. Determination of organochlorine pesticides in sewage sludge by matrix solid-phase dispersion and gas chromatography-mass spectrometry. *Talanta* 74(5): 1211–1217.

Sánchez-Brunete, C., Miguel, E., Albero, B., Tadeo, J.L. 2010. Determination of cyclic and linear siloxanes in soil samples by ultrasonic-assisted extraction and gas chromatography–mass spectrometry. *Journal of Chromatography A* 1217: 7024–7030.

Sanchez-Prado, L., Garcia-Jares, C., Llompart, M. 2010. Microwave-assisted extraction: Application to the determination of emerging pollutants in solid samples. *Journal of Chromatography A* 1217: 2390–2414.

Shin, M., Svoboda, M.L., Falletta, P. 2007. Microwave-assisted extraction (MAE) for the determination of polybrominated diphenylethers (PBDEs) in sewage sludge. *Analytical and Bioanalytical Chemistry* 387: 2923–2929.

Simon, E., Lamoree, M.H., Hamers, T., Weiss, J.M., Balaan, J. et al. 2010. Testing endocrine disruption in biota samples: A method to remove interfering lipids and natural hormones. *Environmental Science and Technology* 44: 8322–8329.

Smyth, S.A., Lishman, L., Alaee, M., Kleywegt, S., Svoboda, L. et al. 2007. Sample storage and extraction efficiencies in determination of polycyclic and nitro musks in sewage sludge. *Chemosphere* 67(2): 267–275.

Sosa Ferrera, Z., Sanz, C., Santana, C., Rodriguez, J. 2004. The use of micellar systems in the extraction and pre-concentration of organic pollutants in environmental samples. *TrAC Trends in Analytical Chemistry* 23(7): 469–479.

Sousa, M.A., Gonçalves, C., Cunha, E., Hajslova, J., Alpendurada, M.F. 2011. Cleanup strategies and advantages in the determination of several therapeutic classes of pharmaceuticals in wastewater samples by SPE–LC-MS/MS. *Analytical and Bioanalytical Chemistry* 399(2): 807–822.

Sturini, M., Speltini, A., Maraschi, F., Rivagli, E., Profumo, A. 2010. Solvent-free microwave-assisted extraction of fluoroquinolones from soil and liquid chromatography–fluorescence determination. *Journal of Chromatography A* 1217: 7316–7322.

Svoboda, M.L., Yang, J.-J., Falletta, P., Lee, H.-B. 2007. A microwave-assisted extraction method for the determination of synthetic musks in sewage sludge. *Water Quality Research Journal of Canada* 42(1): 11–19.

Tan, B.L.L., Hawker, D.W., Müller, F.F., Tremblay, L.A., Chapman, H.F. 2008. Stir bar sorptive extraction and trace analysis of selected endocrine disruptors in water, biosolids and sludge samples by thermal desorption with gas chromatography–mass spectrometry. *Water Research* 42(1): 404–412.

Tanoue, R., Nomiyama, K., Nakamura, H., Hayashi, T., Kim, J.-W. et al. 2014. Simultaneous determination of polar pharmaceuticals and personal care products in biological organs and tissues. *Journal of Chromatography A* 1355: 193–205.

ten Dam, G., Pardo, O., Traag, W., van der Lee, M., Peters, R. 2012. Simultaneous extraction and determination of HBCD isomers and TBBPA by ASE and LC-MSMS in fish. *Journal of Chromatography B* 898: 101–110.

Tölgyessy, P., Vrana, B., Šilhárová, K. 2013. An improved method for determination of polychlorinated biphenyls and polybrominated diphenyl ethers in sediment by ultrasonic solvent extraction followed by stir bar sorptive extraction coupled to TD–GC–MS. *Chromatographia* 76(3–4): 177–185.

Vallecillos, L., Pocurull, E., Borrull, F. 2013. A simple and automated method to determine macrocyclic musk fragrances in sewage sludge samples by headspace solid-phase microextraction and gas chromatography-mass spectrometry. *Journal of Chromatography A* 1314: 38–43.

Valsecchi, S., Rusconi, M., Polesello, S. 2013. Determination of perfluorinated compounds in aquatic organisms: A review. *Analytical and Bioanalytical Chemistry* 405: 143–157.

van Leeuwen, S.P.J., de Boer, J. 2008. Advances in the gas chromatographic determination of persistent organic pollutants in the aquatic environment. *Journal of Chromatography A* 1186: 161–182.

Vega-Morales, T., Sosa-Ferrera, Z., Santana-Rodríguez, J.J. 2011. Determination of various estradiol mimicking compounds in sewage sludge by the combination of microwave-assisted extraction and LC-MS/MS. *Talanta* 85: 1825–1834.

Vega-Morales, T., Sosa-Ferrera, Z., Santana-Rodríguez, J.J. 2013. The use of microwave assisted extraction and on-line chromatography-mass spectrometry for determining endocrine-disrupting compounds in sewage sludges. *Water, Air, and Soil Pollution* 224(3): 1486. DOI: 10.1007/s11270-013-1486-4.

Villaverde-de-Sá, E., Valls-Cantenys, C., Quintana, J.B., Rodil, R., Cela, R. 2013. Matrix solid-phase dispersion combined with gas chromatography–mass spectrometry for the determination of fifteen halogenated flame retardants in mollusks. *Journal of Chromatography A* 1300: 85–94.

von der Kammer, F., Ferguson, P.L., Holden, P.A., Masion, A., Rogers, K.R. et al. 2012. Analysis of engineered nanomaterials in complex matrices (environment and biota): General considerations and conceptual case studies. *Environmental Toxicology and Chemistry* 31(1): 32–49.

Wang, P., Zhang, Q., Wang, Y., Wang, T., Li, X. et al. 2010. Evaluation of Soxhlet extraction, accelerated solvent extraction and microwave-assisted extraction for the determination of polychlorinated biphenyls and polybrominated diphenyl ethers in soil and fish samples. *Analytica Chimica Acta* 663: 43–48.

Wang, L., Wang, L., Chen, J., Du, W., Fan, G. et al. 2012. Ultrasonic-assisted water extraction and solvent bar microextraction followed by gas chromatography–ion trap mass spectrometry for determination of chlorobenzenes in soil samples. *Journal of Chromatography A* 1256: 9–14.

Wille, K., Brabander, H.F.D., Wulf, E.D., Caeter, P.V., Janssen, C.R. et al. 2012. Coupled chromatographic and mass-spectrometric techniques for the analysis of emerging pollutants in the aquatic environment. *TrAC Trends in Analytical Chemistry* 35: 87–108.

Wu, S.-F., Ding, W.-H. 2010. Fast determination of synthetic polycyclic musks in sewage sludge and sediments by microwave-assisted headspace solid-phase microextraction and gas chromatography–mass spectrometry. *Journal of Chromatography A* 1217: 2776–2781.

Wu, Y.-Y., Yang, C.-X., Yan, X.-P. 2014. Fabrication of metal–organic framework MIL-88B films on stainless steel fibers for solid-phase microextraction of polychlorinated biphenyls. *Journal of Chromatography A* 1334: 1–8.

Yamaguchi, C., Lee, W.-Y. 2010. A cost effective, sensitive, and environmentally friendly sample preparation method for determination of polycyclic aromatic hydrocarbons in solid samples. *Journal of Chromatography A* 1217: 6816–6823.

Yu, C., Hu, B. 2007. Novel combined stir bar sorptive extraction coupled with ultrasonic assisted extraction for the determination of brominated flame retardants in environmental samples using high performance liquid chromatography. *Journal of Chromatography A* 1160: 71–80.

Yu, Y., Wu, L. 2012. Analysis of endocrine disrupting compounds, pharmaceuticals and personal care products in sewage sludge by gas chromatography–mass spectrometry. *Talanta* 89: 258–263.

Yu, Y., Huang, Q., Cui, J., Zhang, K., Tang, C. et al. 2011. Determination of pharmaceuticals, steroid hormones, and endocrine-disrupting personal care products in sewage sludge by ultra-high-performance liquid chromatography–tandem mass spectrometry. *Analytical and Bioanalytical Chemistry* 399: 891–902.

Zhou, L.-J., Ying, G.-G., Liu, S., Zhao, J.-L., Chen, F. et al. 2012. Simultaneous determination of human and veterinary antibiotics in various environmental matrices by rapid resolution liquid chromatography–electrospray ionization tandem mass spectrometry. *Journal of Chromatography A* 1244: 123–138.

Zhu, Q., Degelmann, P., Niessner, R., Knopp, D. 2002. Selective trace analysis of sulfonylurea herbicides in water and soil samples based on solid-phase extraction using a molecularly imprinted polymer. *Environmental Science and Technology* 36(24): 5411–5420.

Zorita, S., Boyd, B., Jonsson, S., Yilmaz, E., Svensson, C. et al. 2008. Selective determination of acidic pharmaceuticals in wastewater using molecularly imprinted solid-phase extraction. *Analytica Chimica Acta* 626(2): 147–154.

6 Application of Novel Materials in Sample Treatment and Separation

Cleanup and Chromatographic Improvements

Núria Fontanals

CONTENTS

6.1 INTRODUCTION

At present, environmental analysis is focused on monitoring the source, fate, effects, and risks associated with contaminants, with a particular shift toward polar ones. Therefore, in recent years, one of the greatest challenges in analytical chemistry has been to develop fast and efficient methods for the trace analysis of target and nontarget organic compounds in complex matrices. In order to achieve this aim, hyphenated chromatography–MS techniques have already been well established for the determination of contaminants in environmental samples. However, one of the different challenges that remains is dealing with multiresidue methods in a reasonably short analysis time, as well as handling complex samples, that might occur in form of signal suppression or enhancement (known as the matrix effect [ME]) during ionization in MS-based detectors (Farré et al., 2012; Richardson, 2014).

Several strategies have been proposed to eliminate the problems resulting from the ME. The most effective strategy lies in sample cleanup, which helps to reduce interferents. However, the risk of analyte loss is still an issue that arises (Caban et al., 2012).

In this respect, significant progress has been achieved in the development of novel materials for sorptive extraction techniques, with selectivity becoming an additional requirement during its development. Of the different sorptive extraction techniques, SPE is the most widely used technique for liquid samples or liquid extracts. However, considerable progress has been made in solid-phase microextraction (SPME) or stir bar sorptive extraction (SBSE). The selectivity of a material in sample preparation can be toward a target analyte or a group of analytes with related properties. Hence, the target analyte or analytes can interact with the selective material, while matrix interferences or nonrelated analytes are less retained or even distinguished. To achieve selectivity, all parameters involved in the material development play an important role. In addition, the conditions used during the sample preparation process should be carefully optimized, enhancing their ability to support an effective cleanup (usually composed of organic solvent) without analyte losses.

Regarding chromatographic separation, liquid chromatography (LC) appears to be the most common approach since it is able to separate quite complicated mixtures of analytes with different properties, including polar analytes. Moreover, progress in LC column technology has been made in recent years in order to achieve multiresidue analysis and, in some cases, deal with the separation of highly polar compounds in a competitive manner (in terms of sensitivity gained) for subsequent MS detection.

Summing up, there have been significant advances over recent years in the development of materials for both sample preparation and chromatographic separation in environmental analysis. These two issues are addressed in the current chapter, with a particular emphasis on material development.

6.2 MATERIALS FOR SORPTIVE EXTRACTION TECHNIQUES

Since the early 1990s, one of the main areas of interest in sample preparation, and specifically in sorptive extraction techniques, has been the development of high-capacity materials. These materials have a greater specific surface area (e.g., hyper-cross-linked sorbents), higher hydrophilicity, or both (Fontanals et al., 2005). The enhanced morphological and chemical properties of these materials are suitable for the effective retention of the increasing number of moderately and highly polar organic contaminants present in the environment. However, they also retain interfering compounds. This situation has led to the emergence of selective materials.

Molecularly imprinted polymers (MIPs), as well as mixed-mode polymeric materials, can be categorized as selective materials. The common feature between MIPs and mixed-mode sorbents is their ability to retain the target analytes after an effective cleanup step that involves organic solvent (that washes out the matrix interferences), representing a great step forward in terms of sample preparation when dealing with complex samples.

MIPs and mixed-mode materials are described in the following two sections. In both cases, the fundamentals and preparation are firstly outlined briefly, and the main part is dedicated to the application of these selective materials in environmental fields and their improvements for cleaning up during the sample preparation step. Most of the application examples in this chapter relate to SPE, since it has become the most widely applied option for preconcentrating target analytes and cleaning the matrix with environmental samples. However, examples with other sorptive extraction techniques, such as SPME and SBSE, are also illustrated.

6.2.1 Molecularly Imprinted Polymers

The selectivity of MIPs arises from the synthetic procedure used, whereby functional monomers (that provide the functional groups that interact with the target compound) and a cross-linker agent (that enhances the morphological properties of the material) are polymerized in the presence of the target compound (template) that controls the sizes and functionality of the cavities during the polymerization. Once polymerization has taken place, the template is removed and binding sites are established with the shape, size, and functionalities that complement the target compound. The resulting MIPs are remarkably robust, relatively cheap, and straightforward to prepare. Therefore, their behavior imitates the interactions established by natural receptors to retain a target molecule selectively (e.g., antibody–antigen) but without the associated limitations on stability. In recent years, the field of MIPs has grown rapidly and become very broad, including a variety of fields, such as sensors, catalysis, and drug delivery, among others (Vasapollo et al., 2011; Schirhagl, 2013). However, ever since Sellergren (1994) first reported their use in SPE, the so-called molecularly imprinted solid-phase extraction (MISPE) is, by far, the most advanced technical application of MIPs. MISPE has been applied for a range of types of analytes from several different matrices (Beltran et al., 2010; Hu et al., 2013; Martín-Esteban, 2013). In this case, the applications of MISPE are reviewed as a part of a method that includes MS as the detection system in environmental analysis, in which multiresidue methods are also of great interest for determining as many analytes as possible in the same analysis. Thus, there is growing interest in MIPs that show cross-reactivity (the retention of target molecule, as well as other structurally related compounds). In view of this scenario, the application of MIPs in environmental analysis is shifting toward an effective cleanup step that reduces the ME. Table 6.1 illustrates some of the representative examples of this. For instance, ciprofloxacin was the template in the synthesis of MIP. However, in its application for the analysis of environmental waters, it was observed that this MIP was class selective for fluoroquinolones, but not for other antibiotics (Chen et al., 2010). With respect to the ME, in this study, different aqueous samples (river, lake, and influent and effluent sewage waters) were analyzed, with either water, methanol (MeOH), or acetonitrile (ACN) being used as a cleanup solvent. According to the results, ACN was the only solvent suitable for removing all the interferences and eliminating the ME while maintaining complete recoveries. The ME was also reduced in a study (Lopez-Nogueroles et al., 2013) in which a MIP for extracting nitro musk was used to analyze ultrapure, river, sea, and effluent sewage water samples. In addition, the ME obtained with this MIP was lower than that obtained with Oasis HLB (a highly retentive sorbent), which was attributed to the effective cleanup step consisting of 1 mL of MeOH/H_2O (80:20 v/v).

Another approach for enhancing class selectivity is multitemplate MIP, consisting of using more than one template during the synthesis. A MIP was developed using 16 PAHs as templates, and it showed superior recoveries than other MIPs for PAHs (Lai et al., 2004; Krupadam et al., 2010) when seawater samples were analyzed. However, the binding capacity of the MIP was still affected by matrices with a high organic and inorganic content (Song et al., 2012). Duan et al. (2013) adopted a similar strategy when they prepared a MIP using five pharmaceuticals as templates. The multitemplate MIP showed high selectivity for the templates, but not for the interferences. In addition, it showed good performance and negligible ME when MISPE was applied as a cleanup step in the extraction of 10 g of sediment sample.

Class-selective MIPs have recently been commercialized by Supelco, Biotage, and Polyintell for different groups of compounds such as estrogens (Affinilute MIP-Estrogens), nonsteroidal anti-inflammatory drugs (Affinilute MIP-NSAIDs or SupelMIP NSAIDs), amphetamines (SupelMIP Amphetamine), among others. However, only some of them have been applied in environmental analysis. For instance, SupelMIP NSAIDs were evaluated in the analysis of sewage water (Zorita et al., 2008b; Gilart et al., 2012), and the cleanup step performed was able to achieve effective cleaning that provided no significant signal recovery differences when analyzing complex samples and lower ME in comparison to Oasis HLB. These results are in agreement with Gros et al. (2008), whose study demonstrated a decrease in the ME provided by SupelMIP β-blocker in comparison to Oasis HLB for the extraction of β-blockers from sewage samples. Affinilute MIP-Estrogen outperformed a C18 cartridge in the determination of a group of estrogens from 100 mL of river and tap water, with the recoveries being acceptable and the ME insignificant (Lucci et al., 2011).

The selectivity provided by MIPs has led to the direct coupling of MISPE-MS and MISPE-MS/MS. Recently, SupelMIP NSAIDs directly coupled to MS/MS were applied in the detection and quantification of NSAIDs from 10 mL of influent and 50 mL of effluent sewage without chromatographic separation (Gilart et al., 2013). Even so, the signal suppression achieved (higher than 70%) was attributed to the absence of chromatographic separation in the method. Therefore, inclusion of chromatographic separation or enhancement of the selectivity of the MIP should be further evaluated to deal with the ME in the direct coupling of MISPE-MS/MS.

TABLE 6.1
In-House and Commercial Molecularly Imprinted Materials Applied in Extraction Techniques in Methods with MS Detection in Environmental Analysis

	Technique	Template	Class Selectivity	Sample	Washing Step	Matrix Effect	Ref.
In-house	On-line MISPE-LC-MS/MS	Cyclobarbital	Phenobarbital, amobarbital, and phenytoin	River	Aqueous ammonium acetate	n.d.	Hoshina et al. (2009)
	Off-line MISPE/LC-MS/MS	Carbamazepine	No	Sewage	5 mL water + 2 mL ACN	Low	Beltran et al. (2007)
	Off-line MISPE/LC-MS/MS	Ciprofloxacin	Fluoroquinolones	River and sewage	2 × 3 mL ACN	88%	Chen et al. (2010)
	Off-line MISPE/GC-MS/MS	16 PAHs	16 PAHs	Sea	2 mL MeOH/water (10:90)	Low	Song et al. (2012)
	Off-line MISPE/LC-MS/MS	NSAIDs	NSAIDs	Lake, sewage, and sediments	2 mL DCM/ACN (94:6)	No	Duan et al. (2013)
Commercial	Off-line MISPE/LC-MS/MS	Affinilute MIP-NSAIDs	15 drugs	Sewage	5 mL water + 5 mL ACN/water (40:60)	20–95%	Gilart et al. (2012)
	Off-line MISPE/MS/MS	Affinilute MIP-NSAIDs	NSAIDs	Sewage	5 mL water + 5 mL ACN/water (40:60)	50–100%	Gilart et al. (2013)
	Off-line MISPE/LC-MS/MS	Affinilute MIP-Estrogens	7 estrogens	River and tap	4 mL ACN/water (80:20) + 2 mL water	Low	Lucci et al. (2011)
	Off-line MISPE/LC-MS/MS	MIP4SPE-β-blockers	8 β-blockers	Sewage	2 mL water + 1 mL ACN + 1 mL DCM	Low	Gros et al. (2008)
In-house	MI-SPME-GC-MS	BDE-209	PBDE	Sewage	No	n.d.	Li et al. (2009)
	MI-SPME-GC-MS	DBP	Phthalates	Reservoir water	No	n.d.	He et al. (2010)
	MI-SPME-GC-MS	Ametryn	Triazines	Tap	No	n.d.	Djozan et al. (2009)
	MI-SBSE-LC-MS	(S)-citalopram	Enantioselective	Water	Dipping for 15 min in toluene/DCM (1:3)	n.d.	Gomez-Caballero et al. (2011)

Note: BDE-209, decabromodiphenylether; DBP, dibutylphthalate; DCM, dichloromethane; n.d., no data; NSAIDs, nonsteroidal anti-inflammatory drugs; PAHs, polyaromatic hydrocarbons; PBDE, polybromodiphenylether.

Extraction of organic contaminants from solid samples, mainly sludge and soil, has been performed in recent years, primarily by ultrasonic extraction and pressurized liquid extraction (PLE). Due to the complexity of the matrices extracted, these strategies often also lead to the extraction of unwanted interferences, which are usually removed using an external cleanup step based on SPE. As mentioned previously, the effectiveness of this cleanup step is quite dependent on the sorbent used in SPE. For instance, 10 mL aliquots of PLE extracts from soil samples were further purified using a commercial MIP selective for triazines. The recovery results obtained in this procedure were equivalent to the recoveries obtained in ultrapure water, which might imply, in this case, that the MIP worked for selectively retaining the target analytes despite the presence of interferences. Subsequently, the method based on PLE/MISPE/LC-MS/MS was applied to analyze different types of environmental samples such as sludge-fertilized agricultural soil and sludge samples, and some triazines were found at nanogram per gram concentration levels (García-Galán et al., 2010).

Another strategy used in PLE involves performing internal cleanup with the addition of sorptive material into the cell so that the sample matrix components are adsorbed to the material while the targets are being extracted. This strategy is more straightforward (i.e., it is based on a single extraction step rather than two steps involving PLE followed by SPE). However, to date, its application has been investigated only with classical sorbents, such as alumina, Florisil, and silica-based sorbents, which sometimes helps to clean the extracts from the solid matrix (Runnqvist et al., 2010).

Some very recent studies also used a quick, easy, cheap, effective, rugged, and safe (QuEChERS) technique for the extraction of environmental solid samples (Salvia et al., 2012; Berlioz-Barbier et al., 2014), although this extraction technique was already well established for food samples. In that case, the cleanup step of the QuEChERS extract is performed by dispersive SPE (dSPE), which differs from SPE as the materials are placed in an Eppendorf tube and put into contact with the sample and analytes by means of agitation. To date, the materials used in dSPE include classical sorbents such as primary secondary amine, graphitized carbon black, alumina, and magnesium sulfate and their combinations in different proportions, as they are commercially available, together with the kits for QuEChERS extraction (Wilkowska and Biziuk, 2011). Thus, the field is open to exploit more selective materials, such as MIPs, for cleaning extracts from solid samples.

MIPs have also been applied in a variant of SPE known as magnetic solid-phase extraction, in which typically magnetic nanoparticles (MNPs) are added to sample matrices containing target analytes. After the extraction procedure, the MNPs are easily isolated from the solution by applying an external magnetic field, which simplifies and accelerates the isolation process. In recent years, a significant number of functionalized MNPs have been prepared and recently reviewed by Xie et al. (2014), which also included MIPs. For example, Xia et al. (2013) prepared a MIP coated with MNPs that are selective for a group of structurally similar estrogens, of which the recognition availability was not significantly affected by the presence of matrix interferences. Similarly, Chen et al. (2013) reported a signal enhancement/reduction of around 20–30% for the target analytes (sulfonamides and their acetylated metabolites) when 500 mL of lake or river water was percolated through a magnetic MIP, in comparison to Oasis HLB.

SPME and SBSE are based on establishing a single partitioning equilibrium of analytes between the sample and the solid material (coating). The SPME format is a material coating on an extraction fiber, while the SBSE format is a material that coats a magnetic stir bar. Thus, these formats restrict the morphology and chemistry of the coatings to those that can be deposited onto the fiber and stir bar, respectively. Originally, these techniques were based on polydimethylsiloxane as a coating, but some alternative (more polar) materials have become commercially available and prepared in-house for SPME and, recently, for SBSE. In a trend similar to SPE materials, molecularly imprinted polymer-coated solid-phase microextraction (MI-SPME) fibers and stir bar sorptive extraction molecularly imprinted polymer coatings (MI-SBSE) have been introduced and applied in several different fields (Xu et al., 2013; Gilart et al., 2014a). With respect to environmental field analysis, different MI-SPME coatings have been prepared in-house by using sol–gel (Li et al., 2009) or monolithic (Turiel et al., 2007; Djozan et al., 2009; Tan et al., 2009; He et al., 2010) approaches to extract different analytes. However, only a few of them have been included in a method that used MS for detection, none of which evaluated the ME or even included a cleanup step in the procedure (Table 6.1 provides the details). Only two MI-SPME/LC-diode array detection methods for the determination of bisphenol A (Tan et al., 2009) and triazines (Turiel et al., 2007) included a cleanup step, based on ACN/H_2O (25:75 v/v) and toluene, respectively. MI-SBSE coatings have also been prepared using sol–gel (Zhu et al., 2006) or monolithic technology (Hu et al., 2010, 2011; Yang et al., 2010; Chen et al., 2011; Gomez-Caballero et al., 2011; Sheng et al., 2012), in which the MIP was first chemically bonded to a glass bar where the polymer grew afterward. Similarly to MI-SPME, only one MI-SBSE coating selective for citalopram followed by LC-MS included a cleanup step, consisting of dipping the stir bar for 15 min in a solution of toluene/dichloromethane (3:1 v/v). However, the best signal was attained when no cleaning was performed, which was attributed to the existence of nonspecific interactions with the polymer backbone (Gomez-Caballero et al., 2011). Other authors demonstrated the selectivity of the coating by adding nonstructurally related compounds and evaluating their affinity. In all cases, though, they were highly susceptible to interfering compounds when analyzing complex matrices. In view of these outcomes with MI-SPME and MI-SBSE coatings, further research into selective coating development is highly advisable.

6.2.2 Mixed-Mode Polymeric Materials

Ion-exchange sorbents are a good alternative for extracting ionic or ionizable compounds, since they are based on ionic interactions. The first ion-exchange sorbents were silica based and chemically modified with an ion-exchange group. Over the last decade, mixed-mode polymeric sorbents have been developed, which combine a polymeric skeleton with ion-exchange groups. Thus, these mixed-mode sorbents display ion-exchange interactions (specific) with the charged analytes and reversed-phase interactions (nonspecific) with the uncharged ones, and they can be classified into four groups (combinations of cationic and anionic, strong and weak) depending upon the ion-exchange group that modifies the polymer skeleton. The main difference between the strong and weak ion-exchange materials relates to the chargeability of the materials. Thus, strong ion-exchange materials always remain charged, whereas in the case of weak ion-exchange materials, they may be charged or not depending on the pH. Therefore, it is important to select the suitable material and follow a suitable SPE protocol to enhance the ionic interactions with the analytes of interest. Figure 6.1 summarizes the recommended protocol for each of the four types of mixed-mode materials. The specific interactions (controlled by the pH in the loading and elution step), as well as the effective cleanup (which always involves adding pure organic solvent), are the key to achieving selectivity and getting rid of interferences with mixed-mode materials.

In contrast to MIPs, to the best of our knowledge, mixed-mode materials have been applied only as SPE sorbents. Therefore, all the details hereafter relate to SPE. In SPE, despite being relatively new, mixed-mode materials have been applied in various fields to extract different types of analytes in a selective manner from complex matrices, such as environmental, foodstuff, and biological samples, among others (Fontanals et al., 2010, 2014a). Due to the widespread use of these materials, several mixed-mode materials are commercially available. Figure 6.2 classifies the commercially available mixed-mode materials, along with their manufacturers, into the four groups. In order to improve the retention features, our research group has pioneered the preparation of mixed-mode hyper-cross-linked polymers (with their own high specific surface area and microporous structure that lead to enhanced reversed-phase interactions). Figure 6.2 also summarizes these in-house mixed-mode materials.

Table 6.2 gives an overview of the applications in which mixed-mode materials are included in a method with MS detection to analyze environmental samples. For instance, Oasis MCX sorbent is widely applied to the extraction of drugs of abuse (since most of them are basic analytes and are retained ionically with this strong cation-exchange [SCX] sorbent). However, there are differences in the SPE protocol and its cleanup step and, therefore, in the ME encountered. Thus, some authors

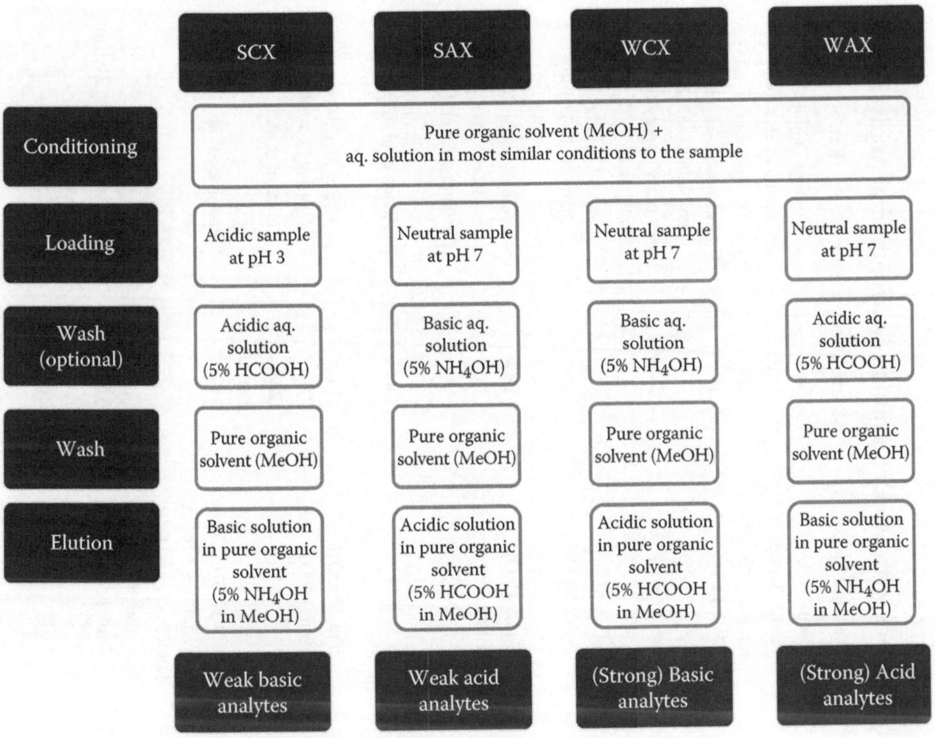

FIGURE 6.1 SPE protocols and type of analytes recommended for each type of mixed-mode ion-exchange polymeric sorbent. In brackets, an example of most usual conditions.

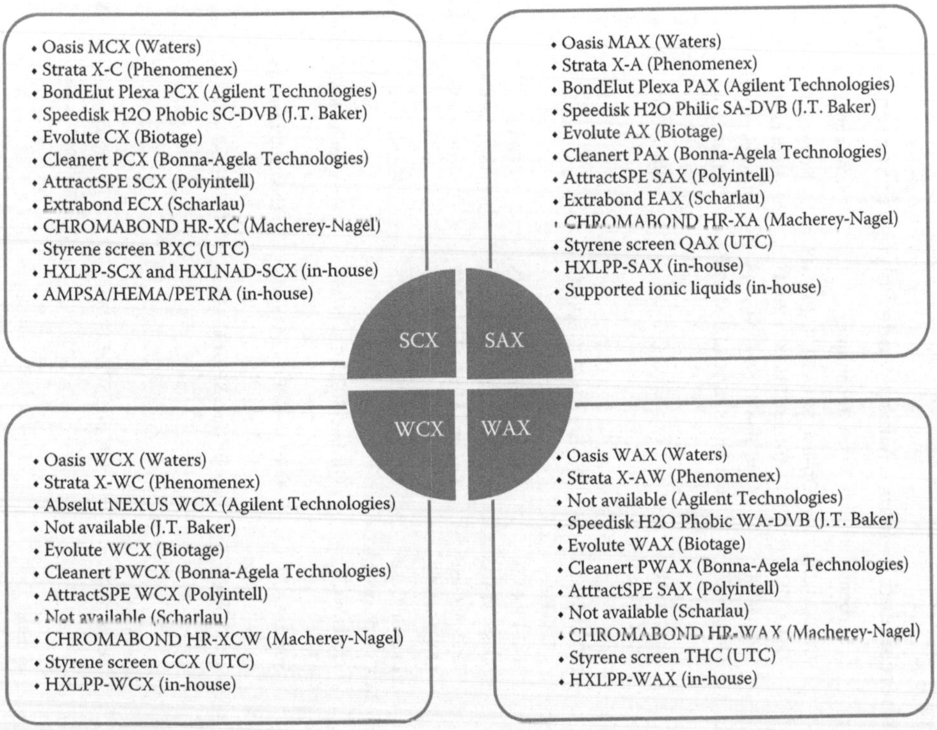

FIGURE 6.2 Classification of the mixed-mode polymeric sorbents commercially available and in-house prepared.

TABLE 6.2
Commercial and In-House Mixed-Mode Polymeric Sorbents Applied in SPE in Methods with MS Detection in Environmental Analysis

	Sorbent	Technique	Analytes	Sample	Washing Step	Matrix Effect	Ref.
SCX	Oasis MCX	Off-line SPE/HILIC-MS/MS	Neurotoxin	Sewage	2 mL MeOH	n.d.	Combes et al. (2013)
	Oasis MCX	Off-line SPE/LC-MS/MS	Herbicides	Soil	10 mL MeOH	n.d.	Nanita et al. (2008)
	Strata-X-C	Off-line SPE/LC-MS/MS	Pharmaceuticals	Sewage	3 mL basic MeOH	No	Van De Steene et al. (2006)
	Oasis MCX	Off-line SPE/HILIC-MS/MS	Illicit drugs	Sewage	Water	High	Gheorghe et al. (2008); van Nuijs et al. (2009)
	Tandem Oasis MCX + HLB	Off-line SPE/LC-MS/MS	Illicit drugs	Sewage	Aqueous + MeOH	Low	Baker and Kasprzyk-Hordern (2011); González-Mariño et al. (2011)
	AMPSA/HEMA/PETRA	Off-line SPE/LC-MS/MS	Drugs	Sewage	5 mL MeOH	Low	Gilart et al. (2014b)
	HXLPP-SCX	Off-line SPE/LC-MS/MS	Drugs	Sewage	5 mL MeOH	Low	Fontanals et al. (2014b)
SAX	Oasis MAX	Off-line SPE/LC-MS/MS	Organophosphorus	Sewage	10 mL water	Low	García-López et al. (2010)
	Tandem Oasis MAX + HLB	Off-line SPE/LC-MS/MS	Benzotriazoles	River and sewage	No	Low	Carpinteiro et al. (2012)
	Tandem Oasis MAX + Plexa SAX	Off-line SPE/LC-MS/MS	Herbicides	Storm water	No	n.d.	Zhang et al. (2014)
WCX	Oasis WCX	Off-line SPE/LC-MS/MS	Fluoroquinolones	Sewage	5 mL MeOH	Low	Lee et al. (2007b)
					2 mL MeOH	Low	Zorita et al. (2008a)
	Strata-X-CW	Off-line SPE/LC-MS/MS	Therapeutic drugs	River	No	Cleaner extracts	Scheurer et al. (2009)
WAX	Oasis WAX	Off-line SPE/LC-MS	Estrogens	River sediments	3 mL 5% NH_4OH in water/MeOH (85:15) + 3 mL 5% NH_4OH in water/THF (80:20)	n.d.	Matejicek et al. (2007)
	Oasis WAX	Off-line SPE/LC-MS	Fluorescence whiting agents	Water and biota	4 mL MeOH	n.d.	Chen et al. (2006); Taniyasu et al. (2005)

Note: n.d., no data.

have adopted a cleanup step based on aqueous solution (Gheorghe et al., 2008; Bijlsma et al., 2009; van Nuijs et al., 2009; Pedrouzo et al., 2011), while others have included a methanolic solution in this step (Baker and Kasprzyk-Hordern, 2011; González-Mariño et al., 2011). To a reasonable extent, the ME was lower when the cleanup step was based on MeOH. However, some authors did not include it as the analytes retained by reversed-phase interactions were eluted with MeOH (Pedrouzo et al., 2011). González-Mariño et al. (2011) compared the ME that resulted from Oasis HLB and Oasis MCX with and without the cleanup step with MeOH when a group of drugs of abuse was extracted from effluent sewage. Figure 6.3 represents the % ME (% ME = 100% means that no effect was encountered) for each of the studied analytes. As can be seen in this figure, when the cleanup step was not included with Oasis MCX or Oasis HLB (cases B and C, respectively), the ME was as low as 10% in the case of morphine; meanwhile, when it was included (case A), the MEs were all above 60%. In view of these results, Oasis MCX with methanolic cleanup was the chosen extraction method in this study. In any case, these studies further dealt with the ME by adding isotopically labeled internal or surrogate standard (Van De Steene et al., 2006; Bijlsma et al., 2009; Baker and Kasprzyk-Hordern, 2011; González-Mariño et al., 2011; Pedrouzo et al., 2011; Wilkowska and Biziuk, 2011). However, in more complex samples such as influent sewage, the samples were needed to be diluted up to five times to quantify the analytes properly if the cleanup step was not performed (Lee et al., 2007a; Bijlsma et al., 2009).

Four in-house SCX polymers (namely, 2-acrylamide-2-methylpropane sulphonic acid AMPSA/hydroxyethylmethacrylate [HEMA]/pentaerythritol triacrylate [PETRA], sulfonated HEMA/divinylbenzene, and sulfonated hyper-cross-linked precursor particles obtained by precipitation polymerization [HXLPP-SCX] or by nonaqueous dispersion polymerization [HXLNAD-SCX]) were also evaluated in terms of ME and recovery when a group of therapeutic and illicit drugs was determined by SPE/LC-MS/MS from influent and effluent sewage samples. The low ME encountered in these studies was due to the cleanup step performed (with 5 mL of MeOH), which simplifies the complexity of the matrix (Fontanals et al., 2014b; Gilart et al., 2014b).

Oasis MAX sorbent was compared to Oasis HLB for the extraction of a group of benzotriazoles and benzothiazoles from aqueous matrices, which were further determined by LC-MS/MS. Both sorbents were used with the same SPE protocol, which did not contain a cleanup step. However, Oasis MAX was selected since it provided clear extracts with lower complexity, which is attributed to the retention of the acidic interferences in the matrix with the amino moieties existing in the structure of the MAX sorbent (Carpinteiro et al., 2012). Zhang et al. (2014) tested Oasis MAX and Bond Elut PAX in the extraction of a broad group of herbicides from environmental waters. Both sorbents

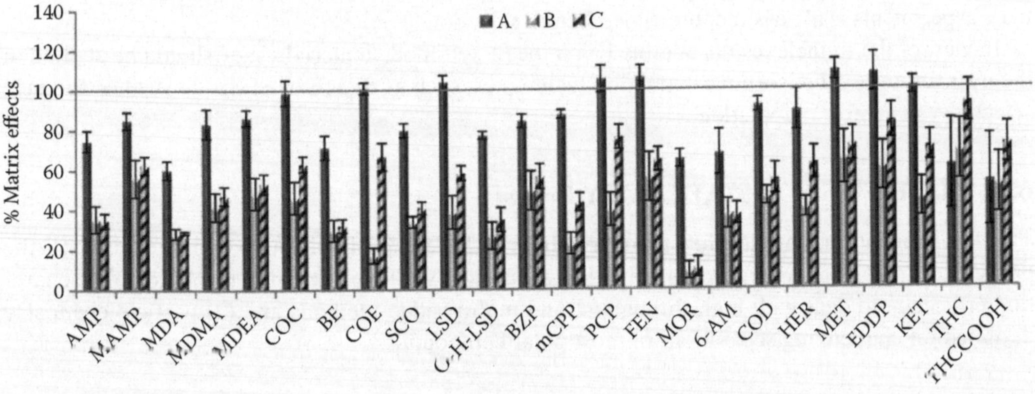

FIGURE 6.3 Percentage of ME in effluent wastewater depending on the SPE protocol: (A) Oasis MCX including cleanup step; (B) Oasis MCX, without cleanup step; and (C) Oasis HLB, without cleanup step. (Reprinted with permission from González-Mariño et al. (2011), 1708–1717. Copyright 2011 American Chemical Society.)

enabled the complete extraction of all of the herbicides as neutral, and basic compounds were eluted first with the methanolic fraction (the fraction intended to be the cleanup step), and the acidic compounds were subsequently eluted with acidified MeOH. Therefore, a cleanup step that gets rid of the undesirable compounds was, in fact, not included in this protocol. Nonetheless, the authors found that the ME can be tackled depending on the type and percentage of acid incorporated during the elution step. Thus, of the three eluents tested (1.5% trifluoroacetic acid [TFA], 2% TFA, and 5% HCOOH in MeOH), it was concluded that using TFA as an acidic additive may induce the elution of matrix interferences and the subsequent ion suppression in the LC-MS/MS system. Therefore, 5% HCOOH in MeOH was selected to avoid the elution of matrix interferences as much as possible.

Oasis WCX sorbent was selected over Oasis HLB and Oasis MCX due to the requirements of high recovery and cleanness of the extract in two studies (Lee et al., 2007b; Zorita et al., 2008a) for the extraction of a group of fluoroquinolones from sewage. Oasis WCX was also tested in the online SPE-LC-MS approach to determine a group of illicit drugs from sewage (Fontanals et al., 2013). This study reported a reduction in the ME when the mixed-mode sorbent involved an effective cleanup step based on MeOH.

With respect to the Strata-X brand, few applications have been reported, especially in environmental analysis. For instance, Strata-X-WC was used in the extraction of a group of antidiabetic drugs from environmental waters by LC-MS/MS (Scheurer et al., 2009). The complete recoveries and the absence of the ME were attributed to the selectivity of the mixed-mode sorbent, as well as the small volume of the sample percolated (10 mL). Strata-X-AW was integrated into a passive sampling device that effectively extracted a wider group of perfluorinated alkyl carboxylates and sulfonates from water than the corresponding nonmixed-mode sorbent (Kaserzon et al., 2012).

Oasis WAX sorbent was compared to Oasis HLB and Oasis MAX for the determination of estrogens and their conjugates. The results were similar for the three sorbents when the targets were extracted from ultrapure water samples. However, Oasis WAX showed the best results when the targets were extracted from river sediments, which might be attributed to the effectiveness of the cleanup step performed (Matejicek et al., 2007).

Other studies compared the performance of the four types of mixed-mode sorbents (i.e., Oasis MCX, MAX, WCX, and WAX) for the determination of basic/neutral therapeutic and illicit drugs in surface and river waters. Oasis MCX was selected since it provided the best recovery. However, as the cleanup step was based only on aqueous solution, the ME remained, even though several internal and surrogate standards were added (Kasprzyk-Hordern et al., 2007). Oasis MCX, Oasis HLB, and Supel MIP-Amphetamine (commercial MIP selective for amphetamines) were also compared. Regarding the ME, MIP clearly provided better performance than any of the Oasis sorbents, which is the reason why it was selected (González-Mariño et al., 2009). Thus, the choice of one material or another in sample preparation is not straightforward, and conducting a series of preliminary experiments is always recommended before selection.

In view of the suitable results obtained with the mixed-mode materials, they should be applied in the near future to other sorptive extraction techniques, such as SPME or SBSE, to further enhance selectivity in sample preparation.

6.3 MATERIALS FOR STATIONARY PHASES

The developments of materials for stationary phases (SPs) can be divided into two categories: (1) the particle technology approach to achieving fast separation while maintaining resolution and separation efficiency to deal with high-throughput and multiresidue methods and (2) particle chemistry designed for challenging separations of most polar compounds.

6.3.1 Fast and Highly Efficient Separations

In the last few decades, three approaches have been developed and commercialized to improve throughput and efficiency in LC: the use of monolithic columns, the use of ultrahigh-pressure liquid chromatography (UHPLC) using sub-2 μm particle columns, and the use of superficially porous particles and high-temperature liquid chromatography (HTLC). Except HTLC, improvements in throughput and efficiency in all of the cases depend on the column material. In the case of monolithic columns, their main features result in low backpressure and good mass transfer that enable higher flow rates (up to 10 mL min^{-1}) than particle-packed columns, which results in enhanced throughput. However, these high flow rates limit its connection to MS detectors. These limitations have influenced their application, and only a few studies have applied them in environmental analysis, which, to the best of our knowledge, aimed to evaluate the chromatographic parameters that resulted from the monolithic column (Ghanem and Ikegami, 2011; Kurganov, 2013; Urban and Jandera, 2013). The other two approaches concerning particle technology have been more applied in environmental field, and they are described in the following two sections.

6.3.1.1 Sub-2 μm Particles

The benefits of using smaller particle sizes in the performance of LC were previously reported by Knox and Saleem (1969) some time ago. Later, in 2004, after solving some technical problems, columns packed with totally porous sub-2 μm particles became commercially available. With sub-2 μm particle technology, the throughput results in a ninefold increase, while maintaining a similar kinetic efficiency to the conventional 5 μm particle packing. However, reducing the particle size by a factor of 3 results in an increase in the backpressure by a factor of 27 (600–1000 bar), which makes it unfeasible to work with a conventional LC instrument. To address this problem, UHPLC instruments emerged that can withstand pressures beyond 600 bar. In addition, the UHPLC system must be designed with the ability to work with columns of reduced diameter (up to 2.1 mm internal diameter [i.d.]) in ultrahigh modes, so that the frictional heating and/or the solvent compressibility is limited.

Another feasibility issue with the UHPLC-MS coupling is the elevated acquisition rate, as very narrow peaks are produced (2–4 s wide on average) in UHPLC compared to those in LC (10 s on average). Regarding quantitative proposals, at least 10–15 acquisition points per peak are recommended to define the chromatographic peak correctly and achieve suitable performance (Rodriguez-Aller et al., 2013). In this context, the determination of 37 multiclass pesticides in wastewater was performed using UHPLC coupled with a fast-acquisition QqQ mass analyzer, which allowed decreasing dwell times or ionization switching time without apparent sensitivity losses. In this study, it was possible to acquire three selected reaction monitoring (SRM) transitions per compound at a dwell time of 10 ms, with a total of 117 SRM transitions acquired, distributed along eight functions in order to reach an acceptable data point rate for UHPLC peaks (5–7 s), in the chromatographic analysis time of 10 min (Barco-Bonilla et al., 2010).

Despite the previously mentioned limitations, nowadays, a wide variety of columns packed with totally porous sub-2 μm particles (more than 100 different columns from about 15 suppliers) are commercially available (Fekete et al., 2014). These have led to numerous applications in different fields wherein UHPLC-MS has been used. Of these, Table 6.3 summarizes some representative examples of the inclusion of sub-2 μm particle-packed columns in environmental fields. It should be noted that, although there are many methods that use UHPLC instruments described in the literature, most of them used conventional columns or chromatographic conditions not pertinent of ultrafast separations. Therefore, only applications that, in fact, used ultrafast separation conditions were focused on. For instance, Kasprzyk-Hordern et al. (2007) selected an Acquity ethylene-bridged hybrid (BEH) C18 column (100 mm × 1 mm, 1.7 μm) that enables high-speed analysis (15 min) and low mobile phase rates (0.07 mL min^{-1}) in the separation of 28 pharmaceutical compounds. Moreover, it enabled direct introduction into the ion source from LC without the need for splitting. The same author selected the same column but longer (150 mm × 1 mm, 1.7 μm) for the separation of 65 compounds (illicit drugs, their metabolites, and internal standards) in less than 23.1 min using a flow rate of 0.04 mL min^{-1} (Baker and Kasprzyk-Hordern, 2011). Gros et al. (2013) used a sub-2 μm column in the determination of a group of 53 residue antibiotics and some of their metabolites in

TABLE 6.3
Examples of Applications of Columns Packed with Sub-2 μm Particles in Environmental Analysis

Column	Instrument	Analytes	Sample	Analysis Time	Ref.
Acquity HSS T3 (50 mm × 2.1 mm, 1.8 μm) Acquity BEH C18 (50 mm × 2.1 mm, 1.7 μm)	UHPLC-QqLIT	81 pharmaceuticals and metabolites	Sewage and river	6 min	Gros et al. (2012)
Acquity BEH C18 (50 mm × 2.1 mm, 1.7 μm)	UHPLC-QqLIT	55 pharmaceuticals and hormones	Sewage	9 min	Huerta-Fontela et al. (2010)
Acquity HSS T3 (50 mm × 2.1 mm, 1.8 μm)	UHPLC-QqLIT	53 antibiotics and their metabolites	Sewage, hospital sewage, and river	2.5 min	Gros et al. (2013)
Agilent Zorbax Eclipse PAHs (50 mm × 2.1 mm, 1.8 μm)	UHPLC-QqQ	16 PAHs	Water	–	Cai et al. (2009)
Acquity BEH C18 (100 mm × 1 mm, 1.7 μm)	UHPLC-QqQ	28 pharmaceuticals and illicit drugs	River	15.5 min	Kasprzyk-Hordern et al. (2007)
Acquity BEH C18 (150 mm × 1 mm, 1.7 μm)	UHPLC-QqQ	65 illicit drugs and metabolites	Sewage	23.1 min	Baker and Kasprzyk-Hordern (2011)
Acquity HSS T3 (100 mm × 2.1 mm, 1.8 μm)	UHPLC-QqQ	37 pesticides	Sewage and surface water	10 min	Marín et al. (2009)
Acquity BEH C18 (100 mm × 1 mm, 1.7 μm)	UHPLC-QqQ	43 polar pesticides	Sewage	9 min	Barco-Bonilla et al. (2010)

environmental waters with UHPLC–quadrupole linear ion trap (QqLIT). The authors claimed that higher resolution and sensitivities and reduced analysis time were achieved with this determination when using the UHPLC system. In addition, they further justified its selection by the reduction in the ME produced by isobaric coeluting sample compounds, thanks to the enhanced chromatographic resolving power offered by UHPLC in comparison to conventional LC.

6.3.1.2 Superficially Porous Particles

The most recent approach to improving chromatographic performance is the use of superficially porous particles, also called fused-core particles or core–shell particles. They consist of a solid fused core and a thin porous shell. Today, these columns are commercialized by different suppliers, with particle dimensions differing slightly. In chronological order, released in 2007, HALO (Advanced Material Technology) is a column packed with 2.7 μm superficially porous particles that consist of 1.7 μm of silica core particles coated with 0.5 μm of porous silica. Poroshell (Agilent Technologies) and Ascentis Express (Sigma-Aldrich) have the same dimensions, and very recently, Ascentis Express has also been presented in a 5 μm particle size format. However, in the case of this larger particle, there is no information available about the dimensions of the core and the shell. In 2009, Phenomenex marketed Kinetex columns with core–shell technology in the 2.6 μm particle size and later in the 1.7 or 1.3 μm formats. They consist, respectively, of 1.9, 1.24, and 0.9 μm and a nonporous core layered with 0.35, 0.23, and 0.2 μm porous shells. In a similar fashion, in 2011, other vendors released similar sub-3 μm superficially porous particles, such as Accucore (Thermofisher), Cortects (Waters), Nucleoshell (Macherey-Nagel), and Sunshell (Nacalia).

The solid fused core of silica, which is impenetrable to the analytes, results in a faster mass transfer, shorter diffusion path, and less band broadening. These properties enable the columns to exhibit efficiencies and throughput similar to the sub-2 μm particles. In addition, the more ordered particle distribution also leads to reasonable column pressures (operates at half or third pressure compared to fully porous sub-2 μm particles), without the need for resourcing UHPLC systems (Guillarme et al., 2010). In this context, the column manufacturers advertised that UHPLC performance can be obtained in LC when using these superficially porous particles. However, this assumption was brought into question by Fekete et al. (2014), who considered different cases and drew the conclusion that only superficially porous particles packed in columns of 4.6 mm i.d. can be applied without loss in column efficiency in conventional LC systems, and in the case of columns with 3 and 2.1 mm i.d., the LC systems must be optimized.

In any case, applications of superficially porous particles are expanding rapidly, and this expansion is expected to be much faster in the near future as a range of chemistry support products have recently become available. Regarding environmental analysis, the number of examples is still limited. For example, Kinetex C18 (100 mm × 2.1 mm, 2.6 μm) was selected for the determination of 15 pharmaceuticals and four of their metabolites in sewage because it provided narrow peaks that improved sensitivity with an analysis time of 9 min (Tarcomnicu et al., 2011). Gilart et al. (2014b) also used an Ascentis Express C18 column (100 mm × 4.6 mm, 2.7 μm) for the separation of a group of 17 illicit and therapeutic drugs that also attained narrow peaks. However, the analysis time was 16 min, which can presumably be attributed to the low flow rate applied in a column with such a large inner diameter.

6.3.2 Chemistry of Stationary Phases

Although reversed-phase liquid chromatography (RPLC) with alkyl-bonded (mainly, C18 and C8) silica is the most common SP in environmental analysis, when dealing with challenging separations, other SPs might be helpful and have been increasingly applied in recent years, such as polar-embedded reversed phase, fluorinated reversed phase, phenyl–hexyl, and also those used in hydrophilic interaction liquid chromatography (HILIC). It is also worth mentioning that alkylsiloxane-bonded silica materials have been improved in recent years in order to enhance endcapping or to improve the bonding of the alkyl groups, among others, resulting in better LC performance. Figure 6.4 groups these improvements in SPs and displays their chemical structure.

Waters commercializes Acquity BEH columns based on BEH technology, which enables its application across a much wider pH range (1–12) and also provides versatile and robust separation due to the intrinsic chemical stability. Later, Acquity HSS T3 became available, which uses a trifunctional C18 alkyl phase bonded at a ligand density together with the T3 endcapping process, which is much more effective than traditional trimethylsilane (TMS) endcapping. This unique combination of bonding and endcapping provides superior polar compound retention and aqueous mobile phase compatibility (Gros et al., 2013). In fact, the Acquity HSS T3 column was selected over the Acquity BEH C18 column for the separation of 53 antibiotics since, in combination with an optimized mobile phase, it provided the best peak shape (Gros et al., 2013). Other companies also have released similar columns with higher ligand bonding density and extreme endcapping. Examples of these are Zorbax Eclipse XDB (Agilent), Hypersil GOLD aQ (Thermo Scientific), Atlantis T3 (Waters), XBridge (Waters), Nucleodur EC (Macherey-Nagel), among others. It should be mentioned that these SPs have progressively been marketed in different particle technologies (e.g., totally porous, sub-2 µm, and superficially porous).

Pentafluorophenyl (PFP) and phenyl–hexyl silica SPs offer alternative selectivity to alkyl-bonded silica through other interactions such as π–π, dipole–dipole, and hydrogen bond interactions. However, the application of such SPs in environmental analyses is still limited. For instance, PFP was selected over C18, C8, or phenyl–hexyl column for the determination of antifouling agent in marine sediment, as the PFP column provided the best combination of retention time and peak shape (Nichols et al., 2014). A superficially porous PFP SP was used to achieve the separation of a group of perfluoralkyl carboxylate, sulfonate, and sulfonamide isomers (Benskin et al., 2012) in landfill leachate samples. In fact, this SP is reported to have displayed enhanced retention and

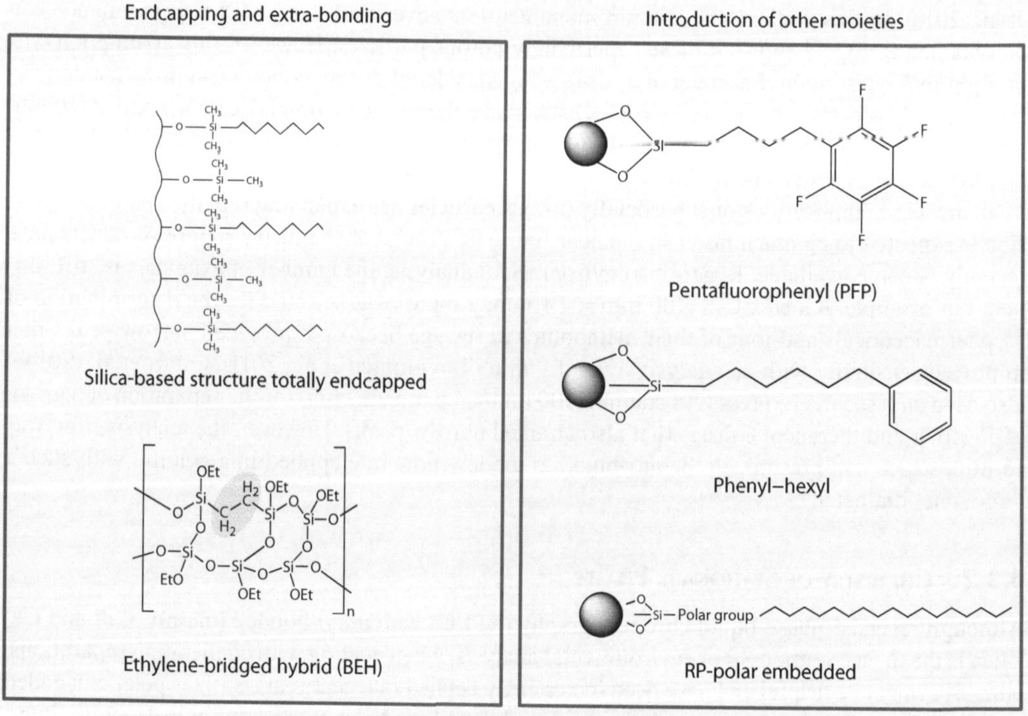

FIGURE 6.4 Examples of enhancements of the alkyl-bonded silica-based materials to improve the RPLC separation of polar compounds.

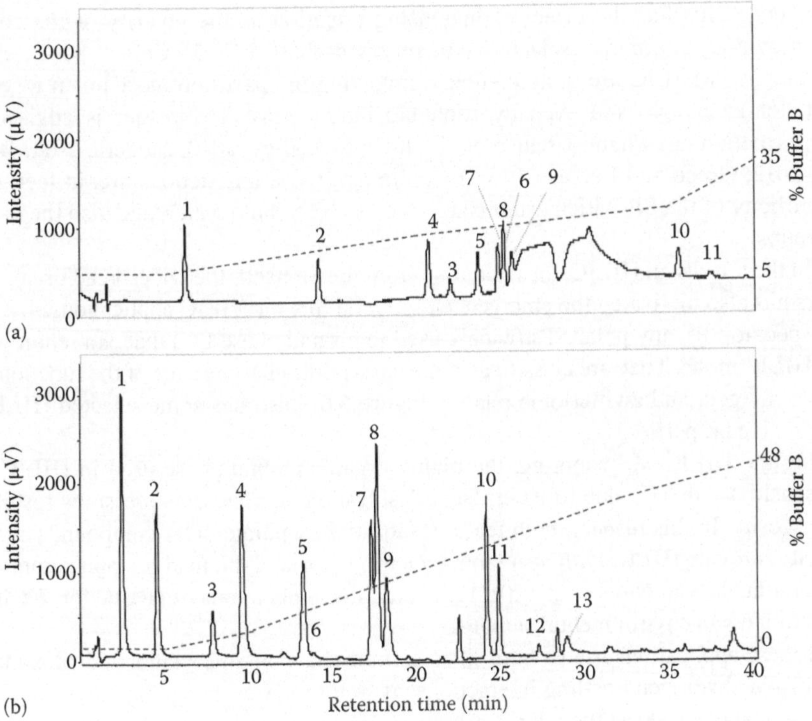

FIGURE 6.5 Chromatograms of standard phenol solution analyzed under optimized conditions with (a) C18 reverse phase column and (b) RP-Amide column. The analytes are gallic acid (1), 3,4-dihydroxybenzoic acid (2), dihydrocaffeic acid (3), 4-hydroxybenzoic acid (4), vanillic acid (5), 2,3-dihydroxybenzoic acid (6), caffeic acid (7), syringic acid (8), 2,4-dihydroxybenzoic acid (9), ferulic acid (10), lawsone (11), juglone (12), and ellagic acid (13). (Reprinted from Restivo et al., *Plos One*, 9, e88762–e88772, 2014. With permission.)

selectivity for the separation of halogenated compounds and also for positional isomers and non-planar molecules (Núñez et al., 2012).

Polar-embedded RPs are another class of SPs that offer an alternative in terms of enhancing retention and selectivity for polar compounds. One example is amide-functionalized (RP-Amide) silica SP (which contains a polar amide group on the bonded alkyl chain). The polar amide moieties can interact with the residual silanol groups in the phase through electrostatic and/or hydrogen bonding interactions and also add retention increments for polar compounds (Benhaim and Grushka, 2010). RP-Amide SP has been used in some instances to determine compounds from environmental samples. The separation of a group of phenolic acids and derivatives was conducted with a classical C18 column and an Ascentis Express RP-Amide column. The two columns differ completely in terms of retention, as can be seen in Figure 6.5. The RP-Amide column separated several peaks that almost coeluted with the C18 column (Restivo et al., 2014). Similar comparison (C18 over RP-Amide) was performed in the separation of a group of polyether ionophores. In that case, the retention factors for both columns were similar, but there were significant differences in selectivity, resulting in very different elution profiles for the two columns. Eventually, RP-Amide was selected, which, under optimized conditions, provided a chromatographic separation in less than 5 min with good resolution (Herrero et al., 2012).

6.3.2.1 Hydrophilic Interaction Liquid Chromatography

HILIC mode was first employed by Alpert (1990), but it has since gained popularity, as it is well suited for the separation of polar and/or ionizable compounds. HILIC is a method in which a polar SP is used in combination with hydrophobic mobile phase (which contains a high percentage of organic solvent). This has the effect of increasing retention as the polarity of the compounds increases, providing alternative selectivity to reversed-phase LC (McCalley, 2010). The retention mechanisms in HILIC are quite complex, involving the partition mechanism together with adsorption, ion-exchange, and even hydrophobic interactions. The reader is advised to read detailed information elsewhere (Chauve et al., 2010; McCalley, 2010; Jandera, 2011; Buszewski and Noga, 2012; Greco and Letzel, 2013). In summary, these interactions are, in fact, related to the hydrophilicity of the SP, which establishes its capacity to hold water, and also the presence of charged groups.

Although the novelty in HILIC does not arise from the SP itself, the widespread use of HILIC in recent years has also motivated the emergence of several SPs for HILIC applications.

With respect to SPs, any polar SPs (widely used in normal phase LC) that can retain water may be used in HILIC mode. They are classified on the basis of the charged state of the functional groups into neutral, charged, and zwitterionic phases. Figure 6.6 illustrates some selected HILIC phases with their relative properties.

Regarding the HILIC-MS coupling, the highly organic mobile phase used in HILIC provides enhanced sensitivity in MS, due to its efficient desolvation and low backpressures resulting from their low viscosity. In this respect, both features (ability to separate polar compounds and increasing sensitivity) provide HILIC with a suitable scenario to expand the field of application. As regard environmental fields, van Nuijs et al. (2011) reviewed the applications of HILIC for the analysis of polar contaminants in environmental samples.

Some of these applications used a bare silica column that contains silanol and siloxane bridges, which can display hydrogen bonding interactions, as well as ion-exchange interactions when working at pH levels above 4–5, as the silanol groups are deprotonated. These features mean that bare silica columns are most commonly used for the separation of neutral and basic compounds. For example, Peru et al. (2006) developed a HILIC-MS/MS method for the detection and quantification of a group of (basic) veterinary drugs from liquid manure and rainfall runoff. HILIC-MS/MS surpassed the already established RPLC methods, since the compounds were well separated from the matrix interferences, resulting in lower ME. In addition, the high amount of organic solvent in

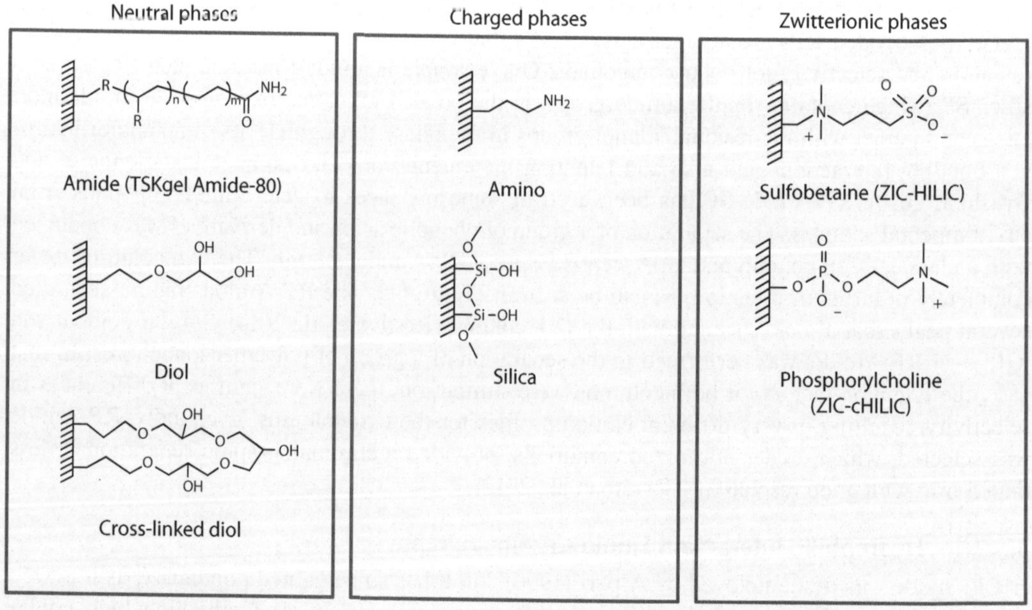

FIGURE 6.6 Classification of the polar stationary phases used in HILIC separations.

the mobile phase leads to a high electrospray ionization efficiency, resulting in increased sensitivity. A significant increase in sensitivity for a group of (basic) illicit drugs was also found when the HILIC-MS/MS method was compared to RPLC-MS/MS (Gheorghe et al., 2008).

Zwitterionic columns (see Figure 6.6) generally contain both a strongly acidic and basic functionality bonded to the silica surface, providing a net surface charge equal to zero. However, the different spatial disposition of the charge groups or the length of the chain is responsible for the chargeability of the phase. Therefore, it may display ion-exchange interactions. In addition, the ionic functionalities strongly absorb water and polar (hydrogen bonding and dipole–dipole) interactions are of primary importance. In this respect, the chromatographic properties of zwitterionic phases significantly differ from those of other HILIC phases (Jandera, 2011). Salas et al. (2014) compared a bare silica column (Atlantis HILIC, Waters) with a zwitterionic column (Syncronis, Thermo Scientific) for the determination of a group of sweeteners from environmental water. Under optimal conditions, the zwitterionic column was the only column able to separate all of the studied analytes, whereas with the bare silica column, some of the charged analytes were poorly retained. A similar zwitterionic column (ZIC-HILIC, Merck) was selected over C18 and RP-Amide phases for the separation of a group of iodinated X-ray contrast media. The authors selected this column, as it overcame certain limitations of the other columns, such as the early elution of the compounds (leading to the ME) and the high aqueous content of the mobile phase used (Echeverría et al., 2013).

The introduction of these novel SPs is becoming increasingly attractive in the environmental field, and they are expected to become well established in the future.

6.4 CONCLUSIONS

The development of materials that improve selectivity in the sample preparation process is a growing research topic in environmental analysis. This is widely demonstrated by the range of successful applications where these materials are used.

Materials focusing on improvements in particle technologies and in phase chemistry are also a cutting-edge issue in environmental analysis.

Despite all of the research in this field over years, further developments can be expected in materials that continue to improve the cleaning of the samples and the chromatographic separation.

ACKNOWLEDGMENT

The author thanks the Ministry of Science and Innovation (CTQ2011-24179) for financial support.

REFERENCES

Alpert, A.J. 1990. Hydrophilic-interaction chromatography for the separation of peptides, nucleic acids and other polar compounds. *Journal of Chromatography A* 499: 177–196.

Baker, D.R., Kasprzyk-Hordern, B. 2011. Multi-residue analysis of drugs of abuse in wastewater and surface water by solid-phase extraction and liquid chromatography–positive electrospray ionisation tandem mass spectrometry. *Journal of Chromatography A* 1218: 1620–1631.

Barco-Bonilla, N., Romero-González, R., Plaza-Bolaños, P., Garrido Frenich, A., Martínez Vidal, J.L. 2010. Analysis and study of the distribution of polar and non-polar pesticides in wastewater effluents from modern and conventional treatments. *Journal of Chromatography A* 1217: 7817–7825.

Beltran, A., Borrull, F., Marcé, R.M., Cormack, P.A.G. 2010. Molecularly-imprinted polymers: Useful sorbents for selective extractions. *TrAC Trends in Analytical Chemistry* 29: 1363–1375.

Beltran, A., Caro, E., Marce, R.M., Cormack, P.A.G., Sherrington, D.C., Borrull, F. 2007. Synthesis and application of a carbamazepine-imprinted polymer for solid-phase extraction from urine and wastewater. *Analytica Chimica Acta* 597: 6–11.

Benhaim, D., Grushka, E. 2010. Characterization of Ascentis RP-Amide column: Lipophilicity measurement and linear solvation energy relationships. *Journal of Chromatography A* 1217: 65–74.

Benskin, J.P., Ikonomou, M.G., Woudneh, M.B., Cosgrove, J.R. 2012. Rapid characterization of perfluoralkyl carboxylate, sulfonate, and sulfonamide isomers by high-performance liquid chromatography–tandem mass spectrometry. *Journal of Chromatography A* 1247: 165–170.

Berlioz-Barbier, A., Vauchez, A., Wiest, L., Baudot, R., Vulliet, E., Cren-Olivé, C. 2014. Multi-residue analysis of emerging pollutants in sediment using QuEChERS-based extraction followed by LC-MS/MS analysis. *Analytical and Bioanalytical Chemistry* 406: 1259–1266.

Bijlsma, L., Sancho, J.V., Pitarch, E., Ibáñez, M., Hernández, F. 2009. Simultaneous ultra-high-pressure liquid chromatography-tandem mass spectrometry determination of amphetamine and amphetamine-like stimulants, cocaine and its metabolites, and a cannabis metabolite in surface water and urban wastewater. *Journal of Chromatography A* 1216: 3078–3089.

Buszewski, B., Noga, S. 2012. Hydrophilic interaction liquid chromatography (HILIC)—A powerful separation technique. *Analytical and Bioanalytical Chemistry* 402: 231–247.

Caban, M., Migowska, N., Stepnowski, P., Kwiatkowski, M., Kumirska, J. 2012. Matrix effects and recovery calculations in analyses of pharmaceuticals based on the determination of β-blockers and β-agonists in environmental samples. *Journal of Chromatography A* 1258: 117–127.

Cai, S.-S., Syage, J.A., Hanold, K.A., Balogh, M.P. 2009. Ultra performance liquid chromatography–atmospheric pressure photoionization–tandem mass spectrometry for high-sensitivity and high-throughput analysis of U.S. Environmental Protection Agency 16 priority pollutants polynuclear aromatic hydrocarbons. *Analytical Chemistry* 81: 2123–2128.

Carpinteiro, I., Abuin, B., Ramil, M., Rodríguez, I., Cela, R. 2012. Simultaneous determination of benzotriazole and benzothiazole derivatives in aqueous matrices by mixed-mode solid-phase extraction followed by liquid chromatography-tandem mass spectrometry. *Analytical and Bioanalytical Chemistry* 402: 2471–2478.

Combes, A., El Abdellaoui, S., Sarazin, C., Vial, J., Mejean, A., Ploux, O., Pichon, V. 2013. Validation of the analytical procedure for the determination of the neurotoxin β-*N*-methylamino-L-alanine in complex environmental samples. *Analytica Chimica Acta* 771: 42–49.

Chauve, B., Guillarme, D., Cléon, P., Veuthey, J.-L. 2010. Evaluation of various HILIC materials for the fast separation of polar compounds. *Journal of Separation Science* 33: 752–764.

Chen, C., Yang, L., Zhou, J. 2011. Trace bensulfuron-methyl analysis in tap water, soil, and soybean samples by a combination of molecularly imprinted stir bar sorption extraction and HPLC-UV. *Journal of Applied Polymer Science* 122: 1198–1205.

Chen, H., Zhang, Y., Gao, B., Xu, Y., Zhao, Q. et al. 2013. Fast determination of sulfonamides and their acetylated metabolites from environmental water based on magnetic molecularly imprinted polymers. *Environmental Science and Pollution Research* 20: 8567–8578.

Chen, H.-C., Wang, S.-P., Ding, W.-H. 2006. Determination of flourescent whitening agents in environmental waters by solid-phase extraction and ion pair liquid chromatography-tandem mass spectrometry. *Journal of Chromatography A* 1102: 135–142.

Chen, L., Zhang, X., Xu, Y., Du, X., Sun X. et al. 2010. Determination of fluoroquinolone antibiotics in environmental water samples based on magnetic molecularly imprinted polymer extraction followed by liquid chromatography–tandem mass spectrometry. *Analytica Chimica Acta* 662: 31–38.

Djozan, D., Mahkam, M., Ebrahimi, B. 2009. Preparation and binding study of solid-phase microextraction fiber on the basis of ametryn-imprinted polymer: Application to the selective extraction of persistent triazine herbicides in tap water, rice, maize and onion. *Journal of Chromatography A* 1216: 2211–2219.

Duan, Y.-P., Dai, C.-M., Zhang, Y.-L., Ling, C. 2013. Selective trace enrichment of acidic pharmaceuticals in real water and sediment samples based on solid-phase extraction using multi-templates molecularly imprinted polymers. *Analytica Chimica Acta* 758: 93–100.

Echeverría, S., Herrero, P., Borrull, F., Pocurull, E., Fontanals, N. 2013. Performance of zwitterionic hydrophilic liquid chromatography in a method to determine a group of iodinated X-ray contrast media *Journal of Separation Science* 36: 3688–3695.

Farré, M., Kantiani, L., Petrovic, M., Pérez, S., Barceló, D. 2012. Achievements and future trends in the analysis of emerging organic contaminants in environmental samples by mass spectrometry and bioanalytical techniques. *Journal of Chromatography A* 1259: 86–99.

Fekete, S., Kohler, I., Rudaz, S., Guillarme, D. 2014. Importance of instrumentation for fast liquid chromatography in pharmaceutical analysis. *Journal of Pharmaceutical and Biomedical Analysis* 87: 105–119.

Fontanals, N., Borrull, F., Marce, R.M. 2005. New hydrophilic materials for solid-phase extraction. *TrAC Trends in Analytical Chemistry* 24: 394–406.

Fontanals, N., Cormack, P.A.G., Marcé, R.M., Borrull, F. 2010. Mixed-mode ion-exchange polymeric sorbents: Dual-phase materials that improve selectivity and capacity. *TrAC Trends in Analytical Chemistry* 29: 765–779.

Fontanals, N., Borrull, F., Marcé, R.M. 2013. On-line weak cationic mixed-mode solid-phase extraction coupled to liquid chromatography–mass spectrometry to determine illicit drugs at low concentration levels from environmental waters. *Journal of Chromatography A* 1286: 16–21.

Fontanals, N., Borrull, F., Marcé, R.M. 2014a. Mixed-mode sorbents in solid-phase extraction. In *Advanced Separations by Specialized Sorbents*, ed. Dragan, E.S. CRC Press, Boca Raton, FL, Vol. 10, pp. 245–283.

Fontanals, N., Miralles, N., Abdullah, N., Davies, A., Gilart, N., Cormack, P.A.G. 2014b. Evaluation of strong cation-exchange polymers for the determination of drugs by solid-phase extraction–liquid chromatography–tandem mass spectrometry. *Journal of Chromatography A* 1343: 55–62.

García-Galán, M.J., Díaz-Cruz, M.S., Barceló, D. 2010. Determination of triazines and their metabolites in environmental samples using molecularly imprinted polymer extraction, pressurized liquid extraction and LC–tandem mass spectrometry. *Journal of Hydrology* 383: 30–38.

García-López, M., Rodríguez, I., Cela, R. 2010. Mixed-mode solid-phase extraction followed by liquid chromatography-tandem mass spectrometry for the determination of tri- and di-substituted organophosphorus species in water samples. *Journal of Chromatography A* 1217: 1476–1484.

Ghanem, A., Ikegami, T. 2011. Recent advances in silica-based monoliths: Preparations, characterizations and applications. *Journal of Separation Science* 34: 1945–1957.

Gheorghe, A., van Nuijs, A, Pecceu, B., Bervoets, L., Jorens, P. et al. 2008. Analysis of cocaine and its principal metabolites in waste and surface water using solid-phase extraction and liquid chromatography-ion trap tandem mass spectrometry. *Analytical and Bioanalytical Chemistry* 391: 1309–1319.

Gilart, N., Marcé, R.M., Borrull, F., Fontanals, N. 2012. Determination of pharmaceuticals in wastewaters using solid-phase extraction-liquid chromatography-tandem mass spectrometry. *Journal of Separation Science* 35: 875–882.

Gilart, N., Marcé, R.M., Fontanals, N., Borrull, F. 2013. A rapid determination of acidic pharmaceuticals in environmental waters by molecularly imprinted solid-phase extraction coupled to tandem mass spectrometry without chromatography. *Talanta* 110: 196–201.

Gilart, N., Marcé, R.M., Borrull, F., Fontanals, N. 2014a. New coatings for stir-bar sorptive extraction of polar emerging organic contaminants. *TrAC Trends in Analytical Chemistry* 54: 11–23.

Gilart, N., Cormack, P.A.G., Marcé, R.M., Fontanals, N., Borrull, F. 2014b. Selective determination of pharmaceuticals and illicit drugs in wastewaters using a novel strong cation-exchange solid-phase extraction combined with liquid chromatography–tandem mass spectrometry. *Journal of Chromatography A* 1325: 137–146.

Gomez-Caballero, A., Guerreiro, A., Karim, K., Piletsky, S., Goicolea, M.A. et al. 2011. Chiral imprinted polymers as enantiospecific coatings of stir bar sorptive extraction devices. *Biosensors & Bioelectronics* 28: 25–32.

González Mariño, I., Quintana, J.B., Rodríguez, I., Rodil, R., González-Peñas, J. et al. 2009. Comparison of molecularly imprinted, mixed-mode and hydrophilic balance sorbents performance in the solid phase extraction of amphetamine drugs from wastewater samples for liquid chromatography-tandem mass spectrometry determination. *Journal of Chromatography A* 1216: 8435–8441.

González-Mariño, I., Quintana, J.B., Rodríguez, I., González-Díez, M., Cela, R. 2011. Screening and selective quantification of illicit drugs in wastewater by mixed-mode solid-phase extraction and quadrupole-time-of-flight liquid chromatography–mass spectrometry. *Analytical Chemistry* 84: 1708–1717.

Greco, G., Letzel, T. 2013. Main interactions and influences of the chromatographic parameters in HILIC separations. *Journal of Chromatography Science* 51: 684–693.

Gros, M., Pizzolato, T.-M., Petrovic, M., de Alda, M.J.L., Barceló, D. 2008. Trace level determination of [beta]-blockers in waste waters by highly selective molecularly imprinted polymers extraction followed by liquid chromatography-quadrupole-linear ion trap mass spectrometry. *Journal of Chromatography A* 1189: 374–384.

Gros, M., Rodríguez-Mozaz, S., Barceló, D. 2012. Fast and comprehensive multi-residue analysis of a broad range of human and veterinary pharmaceuticals and some of their metabolites in surface and treated waters by ultra-high-performance liquid chromatography coupled to quadrupole-linear ion trap tandem mass spectrometry. *Journal of Chromatography A* 1248: 104–121.

Gros, M., Rodríguez-Mozaz, S., Barceló, D. 2013. Rapid analysis of multiclass antibiotic residues and some of their metabolites in hospital, urban wastewater and river water by ultra-high-performance liquid chromatography coupled to quadrupole-linear ion trap tandem mass spectrometry. *Journal of Chromatography A* 1292: 173–188.

Guillarme, D., Ruta, J., Rudaz, S., Veuthey, J.-L. 2010. New trends in fast and high-resolution liquid chromatography: A critical comparison of existing approaches. *Analytical and Bioanalytical Chemistry* 397: 1069–1082.

He, J., Lv, R., Zhan, H., Wang, H., Cheng, J., Lu, K. et al. 2010. Preparation and evaluation of molecularly imprinted solid-phase micro-extraction fibers for selective extraction of phthalates in an aqueous sample. *Analytica Chimica Acta* 674: 53–58.

Herrero, P., Borrull, F., Pocurull, E., Marcé, R.M. 2012. Novel amide polar-embedded reversed-phase column for the fast liquid chromatography–tandem mass spectrometry method to determine polyether ionophores in environmental waters. *Journal of Chromatography A* 1263: 7–13.

Hoshina, K., Horiyama, S., Matsunaga, H., Haginaka, J. 2009. Molecularly imprinted polymers for simultaneous determination of antiepileptics in river water samples by liquid chromatography-tandem mass spectrometry. *Journal of Chromatography A* 1216: 4957–4962.

Hu, Y., Li, J., Hu, Y., Li, G. 2010. Development of selective and chemically stable coating for stir bar sorptive extraction by molecularly imprinted technique. *Talanta* 82: 464–470.

Hu, Y., Li, J., Li, G. 2011. Synthesis and application of a novel molecularly imprinted polymer-coated stir bar for microextraction of triazole fungicides in soil. *Journal of Separation Science* 34: 1190–1197.

Hu, Y., Pan, J., Zhang, K., Lian, H., Li, G. 2013. Novel applications of molecularly-imprinted polymers in sample preparation. *TrAC Trends in Analytical Chemistry* 43: 37–52.

Huerta-Fontela, M., Galceran, M.T., Ventura, F. 2010. Fast liquid chromatography–quadrupole-linear ion trap mass spectrometry for the analysis of pharmaceuticals and hormones in water resources. *Journal of Chromatography A* 1217: 4212–4222.

Jandera, P. 2011. Stationary and mobile phases in hydrophilic interaction chromatography: A review. *Analytica Chimica Acta* 692: 1–25.

Kaserzon, S.L., Kennedy, K., Hawker, D.W., Thompson, J., Carter, S. et al. 2012. Development and calibration of a passive sampler for perfluorinated alkyl carboxylates and sulfonates in water. *Environmental Science & Technology* 46: 4985–4993.

Kasprzyk-Hordern, B., Dinsdale, R.M., Guwy, A.J. 2007. Multi-residue method for the determination of basic/neutral pharmaceuticals and illicit drugs in surface water by solid-phase extraction and ultra performance liquid chromatography-positive electrospray ionisation tandem mass spectrometry. *Journal of Chromatography A* 1161: 132–145.

Knox, J.H., Saleem, M. 1969. Kinetic conditions for optimum speed and resolution in column chromatography. *Journal of Chromatography Science* 7: 614–622.

Krupadam, R.J., Khan, M.S., Wate, S.R. 2010. Removal of probable human carcinogenic polycyclic aromatic hydrocarbons from contaminated water using molecularly imprinted polymer. *Water Research* 44: 681–688.

Kurganov, A. 2013. Monolithic column in gas chromatography. *Analytica Chimica Acta* 775: 25–40.

Lai, J.-P., Niessner, R., Knopp, D. 2004. Benzo[*a*]pyrene imprinted polymers: Synthesis, characterization and SPE application in water and coffee samples. *Analytica Chimica Acta* 522: 137–144.

Lee, H.-B., Sarafin, K., Peart, T.E. 2007a. Determination of [beta]-blockers and [beta]2-agonists in sewage by solid-phase extraction and liquid chromatography-tandem mass spectrometry. *Journal of Chromatography A* 1148: 158–167.

Lee, H.-B., Peart, T.E., Svoboda, M.L. 2007b. Determination of ofloxacin, norfloxacin, and ciprofloxacin in sewage by selective solid-phase extraction, liquid chromatography with fluorescence detection, and liquid chromatography-tandem mass spectrometry. *Journal of Chromatography A* 1139: 45–52.

Li, M.K.-Y., Lei, N.-Y., Gong, C., Yu, Y., Lam, K.-H. et al. 2009. An organically modified silicate molecularly imprinted solid-phase microextraction device for the determination of polybrominated diphenyl ethers. *Analytica Chimica Acta* 633: 197–203.

Lopez-Nogueroles, M., Lordel-Madeleine, S., Chisvert, A., Salvador, A., Pichon, V. 2013. Development of a selective solid phase extraction method for nitro musk compounds in environmental waters using a molecularly imprinted sorbent. *Talanta* 110: 128–134.

Lucci, P., Núñez, O., Galceran, M.T. 2011. Solid-phase extraction using molecularly imprinted polymer for selective extraction of natural and synthetic estrogens from aqueous samples. *Journal of Chromatography A* 1218: 4828–4833.

Marín, J.M., Gracia-Lor, E, Sancho, J.V, López, F.J., Hernández, F. 2009. Application of ultra-high-pressure liquid chromatography-tandem mass spectrometry to the determination of multi-class pesticides in environmental and wastewater samples. Study of matrix effects. *Journal of Chromatography A* 1216: 1410–1420.

Martín-Esteban, A. 2013. Molecularly-imprinted polymers as a versatile, highly selective tool in sample preparation. *TrAC Trends in Analytical Chemistry* 45: 169–181.

Matejicek, D., Houserova, P., Kuban, V. 2007. Combined isolation and purification procedures prior to the high-performance liquid chromatographic-ion-trap tandem mass spectrometric determination of estrogens and their conjugates in river sediments. *Journal of Chromatography A* 1171: 80–89.

McCalley, D.V. 2010. The challenges of the analysis of basic compounds by high performance liquid chromatography: Some possible approaches for improved separations. *Journal of Chromatography A* 1217: 858–880.

Nanita, S.C., Pentz, A.M., Grant, J., Vogl, E., Devine, T.J. et al. 2008. Mass spectrometric assessment and analytical methods for quantitation of the new herbicide aminocyclopyrachlor and its methyl analogue in soil and water. *Analytical Chemistry* 81: 797–808.

Nichols, D., Jordan, T., Kerr, N. 2014. Determination of tributyltin in marine sediment and waters by pressurised solvent extraction and liquid chromatography-tandem mass spectrometry. *Analytical and Bioanalytical Chemistry* 406: 2993–2998.

Núñez, O., Gallart-Ayala, H., Martins, C.P.B., Lucci, P. 2012. New trends in fast liquid chromatography for food and environmental analysis. *Journal of Chromatography A* 1228: 298–323.

Pedrouzo, M., Borrull, F., Pocurull, E., Marcé, R.M. 2011. Drugs of abuse and their metabolites in waste and surface waters by liquid chromatography-tandem mass spectrometry. *Journal of Separation Science* 34: 1091–1101.

Peru, K.M., Kuchta, S.L., Headley, J.V., Cessna, A.J. 2006. Development of hydrophilic interaction chromatography-mass spectrometry assay for spectinomycin and lincomycin in liquid hog manure supernatant and run off from cropland. *Journal of Chromatography A* 1107: 152–158.

Restivo, A., Degano, I., Ribechini, E., Colombini, M.P. 2014. Development and optimisation of an HPLC-DAD-ESI-QToF method for the determination of phenolic acids and derivatives. *Plos One* 9: e88762–e88772.

Richardson, S.D. 2014. Water analysis: Emerging contaminants and current issues. *Analytical Chemistry* 86: 2813–2848.

Rodriguez-Aller, M., Gurny, R., Veuthey, J.-L., Guillarme, D. 2013. Coupling ultra high-pressure liquid chromatography with mass spectrometry: Constraints and possible applications. *Journal of Chromatography A* 1292: 2–18.

Runnqvist, H., Bak, S.A., Hansen, M., Styrishave, B., Halling-Sørensen, B. et al. 2010. Determination of pharmaceuticals in environmental and biological matrices using pressurised liquid extraction—Are we developing sound extraction methods? *Journal of Chromatography A* 1217: 2447–2470.

Salas, D., Borrull, F., Marcé, R.M., Fontanals, N. 2014. Study of the retention behavior of iodinated X-ray contrast agents in hydrophilic interaction liquid chromatography, comparing bare silica and zwitterionic stationary phases. *Journal of Separation Science* 37: 1111–1117.

Salvia, M.-V., Vulliet, E., Wiest, L., Baudot, R., Cren-Olivé, C. 2012. Development of a multi-residue method using acetonitrile-based extraction followed by liquid chromatography–tandem mass spectrometry for the analysis of steroids and veterinary and human drugs at trace levels in soil. *Journal of Chromatography A* 1245: 122 133.

Scheurer, M., Sacher, F., Brauch, H.-J. 2009. Occurrence of the antidiabetic drug metformin in sewage and surface waters in Germany. *Journal of Environmental Monitoring* 11: 1608–1613.

Schirhagl, R. 2013. Bioapplications for molecularly imprinted polymers. *Analytical Chemistry* 86: 250–261.

Sellergren, B. 1994. Direct drug determination by selective sample enrichment on an imprinted polymer. *Analytical Chemistry* 66: 1578–1582.

Sheng, N., Wei, F., Zhan, W., Cai, Z., Du, S. et al. 2012. Dummy molecularly imprinted polymers as the coating of stir bar for sorptive extraction of bisphenol A in tap water. *Journal of Separation Science* 35: 707–712.

Song, X., Li, J., Xu, S., Ying, R., Ma, J. et al. 2012. Determination of 16 polycyclic aromatic hydrocarbons in seawater using molecularly imprinted solid-phase extraction coupled with gas chromatography-mass spectrometry. *Talanta* 99: 75–82.

Tan, F., Zhao, H., Li, X., Quan, X., Chen, J. et al. 2009. Preparation and evaluation of molecularly imprinted solid-phase microextraction fibers for selective extraction of bisphenol A in complex samples. *Journal of Chromatography A* 1216: 5647–5654.

Taniyasu, S., Kannan, K., So, M.K., Gulkowska, A., Sinclair, E. et al. 2005. Analysis of fluorotelomer alcohols, fluorotelomer acids, and short- and long-chain perfluorinated acids in water and biota. *Journal of Chromatography A* 1093: 89–97.

Tarcomnicu, I., van Nuijs, A.L.N., Simons, W., Bervoets, L., Blust, R. et al. 2011. Simultaneous determination of 15 top-prescribed pharmaceuticals and their metabolites in influent wastewater by reversed-phase liquid chromatography coupled to tandem mass spectrometry. *Talanta* 83: 795–803.

Turiel, E., Tadeo, J.L., Martin-Esteban, A. 2007. Molecularly imprinted polymeric fibers for solid-phase microextraction. *Analytical Chemistry* 79: 3099–3104.

Urban, J., Jandera, P. 2013. Recent advances in the design of organic polymer monoliths for reversed-phase and hydrophilic interaction chromatography separations of small molecules. *Analytical and Bioanalytical Chemistry* 405: 2123–2131.

Van De Steene, J.C., Mortier, K.A., Lambert, W.E. 2006. Tackling matrix effects during development of a liquid chromatographic-electrospray ionisation tandem mass spectrometric analysis of nine basic pharmaceuticals in aqueous environmental samples. *Journal of Chromatography A* 1123: 71–81.

van Nuijs, A., Tarcomnicu, I., Bervoets, L., Blust, R., Jorens, P. et al. 2009. Analysis of drugs of abuse in wastewater by hydrophilic interaction liquid chromatography-tandem mass spectrometry. *Analytical and Bioanalytical Chemistry* 395: 819–828.

van Nuijs, A.L.N., Tarcomnicu, I., Covaci, A. 2011. Application of hydrophilic interaction chromatography for the analysis of polar contaminants in food and environmental samples. *Journal of Chromatography A* 1218: 5964–5974.

Vasapollo, G., Del Sole, R., Mergola, L., Lazzoi, M.R., Scardino, A. et al. 2011. Molecularly imprinted polymers: Present and future prospective. *International Journal of Molecular Sciences* 12: 5908–5945.

Wilkowska, A., Biziuk, M. 2011. Determination of pesticide residues in food matrices using the QuEChERS methodology. *Food Chemistry* 125: 803–812.

Xia, X., Lai, E.C., Örmeci, B. 2013. Duo-molecularly imprinted polymer-coated magnetic particles for class-selective removal of endocrine-disrupting compounds from aqueous environment. *Environmental Science and Pollution Research* 20: 3331–3339.

Xie, L., Jiang, R., Zhu, F., Liu, H., Ouyang, G. 2014. Application of functionalized magnetic nanoparticles in sample preparation. *Analytical and Bioanalytical Chemistry* 406: 377–399.

Xu, J., Zheng, J., Tian, J., Zhu, F., Zeng, F. et al. 2013. New materials in solid-phase microextraction. *TrAC Trends in Analytical Chemistry* 47: 68–83.

Yang, L., Zhao, X., Zhou, J. 2010. Selective enrichment and determination of nicosulfuron in water and soil by a stir bar based on molecularly imprinted polymer coatings. *Analytica Chimica Acta* 670: 72–77.

Zhang, P., Bui, A., Rose, G., Allinson, G. 2014. Mixed-mode solid-phase extraction coupled with liquid chromatography tandem mass spectrometry to determine phenoxy acid, sulfonylurea, triazine and other selected herbicides at nanogram per litre levels in environmental waters. *Journal of Chromatography A* 1325: 56–64.

Zhu, X., Cai, J., Yang, J., Su, Q., Gao, Y. 2006. Films coated with molecular imprinted polymers for the selective stir bar sorption extraction of monocrotophos. *Journal of Chromatography A* 1131: 37–44.

Zorita, S., Larsson, L., Mathiasson, L. 2008a. Comparison of solid-phase sorbents for determination of fluoroquinolone antibiotics in wastewater. *Journal of Separation Science* 31: 3117–3121.

Zorita, S., Boyd, B., Jönsson, S., Yilmaz, E., Svensson, C. et al. 2008b. Selective determination of acidic pharmaceuticals in wastewater using molecularly imprinted solid-phase extraction. *Analytica Chimica Acta* 626: 147–154.

7 Advances in Sample Preparation for Molecular Imprinting in Environmental Applications

George Z. Kyzas, Dimitrios Bikiaris, and Dimitra A. Lambropoulou

CONTENTS

7.1 INTRODUCTION

Molecular imprinting is a modern technique and not quite common. It is a specific enough technique according to which the selective binding/isolation of target species (molecules or ions/metals) from a mixture can be achieved (Whitcombe et al., 1995). The ability to selectively recognize a target molecule in a vast pool of similar molecules is essential to biological and chemical processes. This process is called molecular recognition, and it is an event that occurs everywhere in nature. It occurs when two molecules are both geometrically and chemically complementary; that is, when they can both "fit together" spatially and bind to each other by using noncovalent forces, including hydrogen bonds, electrostatic interactions, hydrophobic interactions, and weak metal coordination (Chen et al., 2002). Therefore, a specific technology has been bloomed and redesigned in the last 40 years, which is called molecular imprinting. The whole process is based on adsorption/rebinding/separation technology, which is already one of the most successful techniques for pollutant removal (Srinivasan, 2011; Balogh et al., 2012; Jiang et al., 2012; Alahmadi et al., 2013; Mobarak et al., 2013; Radenović et al., 2013; Sasaki et al., 2013). The targets-for-isolation species (usually called templates or template molecules) are compounds of high-added value, which their isolation from mixture can lead to future recovery and exploitation.

Some of the basic application areas of molecular imprinting include separation processes (chromatography, SPE, membrane separations), artificial antibodies, and sensor recognition elements (Bergemann et al., 1999; Chen et al., 2011). Molecular imprinting presents wide recognition due to some properties such as stability and easy and low-cost preparation (Liu and Wulff, 2004). It is worth noting the impressive increase number of research groups which employ this technique since the increasing number of publications can confirm it. The simplicity of use, the relatively low cost, and the broad range of possible guest molecules (small organic molecules, ions but also biological macromolecules) have since led to the important development of this technique, as illustrated by the increasing number of publications over recent years (Figure 7.1).

As mentioned previously, the major applications of MIPs belong to the analytical field with targets (templates) of environmental pollutants. Therefore, special mention should be done for the existing analytical instrumentation for the determination of the compounds/species under separation. In general, the target molecules (analytes) are determined using techniques of chromatography, which are coupled to common detectors (UV, fluorescence). More recently, MS or MS/MS was applied for the same purpose. But in the case of MS, the direct injections of crude/nonmodified sample extracts are allowed but not recommended, because matrix components can inhibit or even enhance the analyte ionization, hampering accurate quantification. For this reason, sample preparation is even now a procedure of the whole chain of analytical process that is under investigation.

The main targets of sample preparation are the following: (i) removal of potential interferents, (ii) analyte preconcentration, (iii) converting (if this is necessary) the analyte into a more suitable form for detection or separation, and (iv) providing a robust and reproducible method which will be independent of variations in the sample matrix. Recently, other targets (more specific) have been addressed about sample preparation: (i) using smaller initial sample sizes, (ii) improving the selectivity in extraction, (iii) facilitating the automation, and (iv) minimizing the amount of glassware and organic solvents to be used (Smith, 2003).

Based on the previous discussion, the common and "traditional" liquid–liquid extraction is now not applicable and cannot "satisfy" the recent requirements. Therefore, analytical laboratories and industries have substituted it with other new extraction techniques. Nowadays, the most used techniques are the (i) SPE, (ii) SPME, (iii) SBSE, (iv) matrix solid-phase dispersion (MSPD), (v) microsolid-phase extraction, (vi) liquid-phase microextraction, etc.

Until now, many research and review articles were published emphasizing the importance of sample preparation in the whole analytical process. A very interesting special issue ("Advances in Sample Preparation—Part I and Part II") was published in the *Journal of Chromatography A* (Elsevier) showing some updates on this topic (Ramos and Smith, 2007a,b). However, all of these techniques do not present such high selectivity, and, therefore, further optimization of typical steps

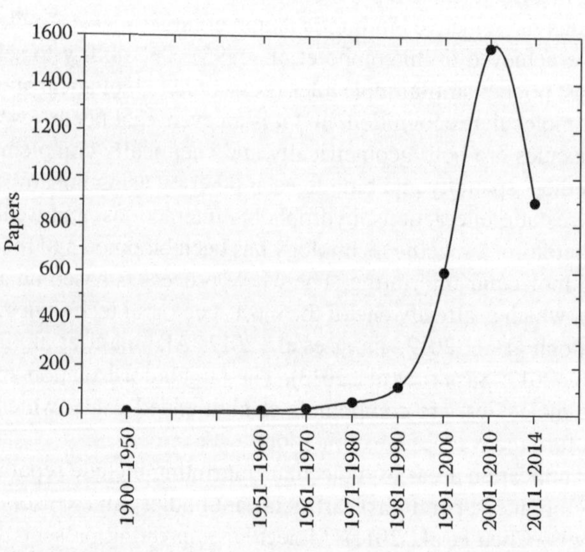

FIGURE 7.1 Works published using the term *molecular imprinted polymers* (data after search in Scopus).

needs to be realized. Unfortunately, even after careful optimization (conditioning, sample loading, washing and elution during the extraction, and cleanup of target analytes), some matrix components are co-eluted with target analytes, making it difficult to reach limits of detection in line with the stringent regulations nowadays.

In the near past, antibodies immobilized on an adequate support, namely, immunosorbents, were proposed as an alternative for use in SPE applications (Martín-Esteban et al., 1997; Pichon et al., 1999) in order to overcome the aforementioned drawbacks associated with typical nonspecific sorbents. Different immunosorbents have been employed for the determination of pesticides, drugs, and polyaromatic hydrocarbons, among others, showing an excellent degree of cleanup, owing to the inherent selectivity of the antibodies used. However, the obtainment of antibodies is difficult, time consuming, and expensive, so it is difficult to guarantee its success. Also, it is important to say that after the antibodies have been obtained, they need to be immobilized on an adequate support, which may result in poor antibody orientation or even complete denaturation.

The use of MIPs as selective sorbent materials allows customized sample treatment prior to the final determination. Thus, their use in SPE, so-called MISPE, is, by far, the most advanced technical application of MIPs (Caro et al., 2006; Pichon and Haupt, 2006; Tamayo et al., 2007; Haginaka, 2009; Lasáková and Jandera, 2009). An increasing interest in the combination of MIPs with other sample-preparation techniques is presented nowadays (i.e., polymeric membrane extraction, SBSE, and MSPD).

The scope of this work is to highlight the updates in MIPs for sample preparation (especially for SPE and SPME), describing the different approaches employed in selected applications in environmental targets (i.e., pharmaceuticals, biomedical, biological, etc.).

7.2 SYNTHESIS OF MOLECULARLY IMPRINTED POLYMERS

The preparation of MIPs starts by positioning the functional monomers around a template molecule. The monomers interact with sites on the template via interactions that can be reversibly covalent or noncovalent (hydrogen, ionic, van der Waals, π–π, etc.). Then, they are polymerized and cross-linked around the template in order to fix their position and to "freeze" the geometry of the pores in the network. The template molecule is then extracted, leaving a polymer with functional sites capable of molecular recognition.

The first step is the contact between the functionalized monomers and the template molecule, which leads to the formation of a complex. Its structure and stability will then determine the behavior of the future MIP. The interactions involved must be sufficiently strong to remain intact during the polymerization stage but sufficiently labile to enable template extraction and reinsertion of guest molecules in the later stages. These interactions must therefore occur rapidly and be reversible. It is crucial to optimize the choice of the different components of the system at this point.

Different types of strategies can be distinguished depending on whether the bonding between template and host is covalent or not: (i) covalent interactions, (ii) noncovalent interactions, and (iii) double approach: covalent plus noncovalent.

The second step is the polymerization step of the monomers around the complex formed in the first step (Piletsky et al., 2002). Owing to its ease of use, radical polymerization is the most frequent. The crucial question is to determine how to carry out this polymerization/cross-linking step with minimum disturbance of the complex already in place. Choices must be made; for instance, in radical polymerization, the radicals can be generated at 60°C (a,a′-azobutyronitrile as initiator; denoted as AIBN) or 45°C (azobis valeronitrile as initiator; denoted as ABDV), which could cause heat destabilization of the complex or at 4°C with low-temperature photochemical radical production (AIBN, 360 nm). Comparative studies on recognition specificity have shown that the photochemical approach gives the most specific materials (O'Shannessy et al., 1989).

Another crucial point of the preparation of MIPs is the extraction step. The proportion of extraction of the template molecules interacting with the MIP via easily hydrolyzable covalent bonds or noncovalent linkages is estimated to be about 90%. The remaining molecules are trapped in highly cross-linked zones. This problem is exacerbated with macromolecules where steric hindrance lowers the efficiency of extraction. In addition, the synthesis of MIPs requires large quantities of guest molecule (50–500 μmol per gram of polymer). Therefore, when a pure template molecule is difficult or expensive to obtain, reaching quantitative template extraction yields can be primordial. Extraction conditions must then be optimized to obtain yields of over 99%. The extraction step uses an appropriate solvent. It often proves to be long and the actual process involved is dependent on the system in question. Thus, automation of the washing steps for industrial applications still remains problematic. Extraction of the template leaves a three-dimensional material in which the cavity shapes and functional group locations are complementary to the guest molecule.

Solvent plays an important role in the formation of the porous structure of MIPs, which are a subset of a larger class known as macroporous polymers. The morphological properties of porosity and surface area are determined by the type of solvent, referred to as "porogen," used in the polymerization. Many solvents were tested in literature for the preparation of MIPs, presenting their main differentiation in the phase. The main organic solvents used are dimethyl sulfoxide, dimethylformamide, acetonitrile, toluene, and chloroform, while deionized water is also used as solvent (Spivak et al., 1997). Moreover, it is notable to report that 70–98% of the final mass of a prepared MIP is composed typically of its cross-linker. Numerous cross-linkers have been tested for the preparation of a rigid MIP such as divinyl benzene (DVB), ethylene glycol dimethacrylate, ethylenebis-acrylamide, and trimethylolpropane trimethacrylate (TRIM). Wulff et al. (1987) and Sibrian-Vazquez and Spivak (2004) have systematically studied the effect of cross-linkers on the recognition properties of MIPs.

Another important point is that the synthesis is related to the template used. For example, the synthesis of MIP and the binding forces between template and polymeric matrix can be strongly influenced from the valence of the template. Therefore, other reagents are selected in order to form a polymeric matrix for monovalent metal as templates and other for others.

After the preparation of MIPs, a full characterization is necessary to examine the morphological and functionalized properties of the prepared materials. The common techniques of MIP characterization are spectroscopy (nuclear magnetic resonance, Fourier transform infrared spectroscopy), swelling tests, and Brunauer–Emmett–Teller analysis (Sellergren, 1994a,b).

In summary, two strategies for molecular imprinting have been established based on whether the template is associated with interactive monomers using covalent bonds or noncovalent interactions. Of the two strategies, the noncovalent approach has been used more extensively for three reasons: (i) noncovalent methodology is easier because it does not require synthetic steps toward the prepolymer complex and interactions between monomers and template are easily obtained when all components are mixed in solution; (ii) removal of the template is generally much easier, and usually, it is accomplished by continuous extraction; and (iii) a greater variety of functionality can be introduced into the MIP binding site using noncovalent methods. An example for MIP preparation with noncovalent bonding for dye removal is given (Figure 7.2).

Prior to its use in experiments, an MIP is usually evaluated to check its recognition properties for a target. Chromatographic evaluation and equilibrium batch rebinding experiments are the methods most commonly used to investigate the selectivity of the imprinted materials. Chromatographic evaluation allows measurement of capacity factors (k') and imprinting factors of MIPs. These values are obtained from the retention time of the template molecule on a chromatographic column packed with the MIP and a second column packed with the nonimprinted polymer (NIP, which is synthesized just the same as an MIP but without the target compound). If the MIP is imprinted, then the analyte should be retained more strongly on the MIP than on the NIP because of the selective interactions. In some studies (Caro et al., 2003), the selectivity of the MIP was also probed using compounds structurally related to the template. If the MIP retains these compounds almost as well or better than the template, this indicates that the MIP shows cross-reactivity (Caro et al., 2003).

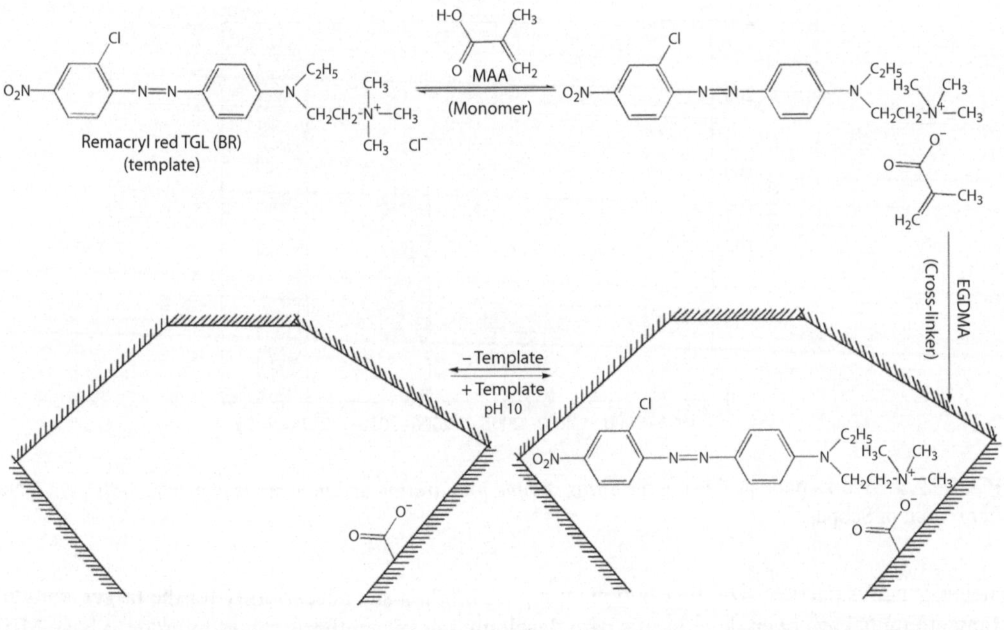

FIGURE 7.2 Schematic representation of the imprinting process for basic dye. (Reprinted from Kyzas, G. Z., and D. N. Bikiaris, *Advances in Materials Science and Engineering*, 2014, 93263, 2014. With permission.)

For equilibrium batch rebinding experiments, a known mass of template in solution is added to a vial containing a fixed mass of polymer. Once the system has come to equilibrium, the concentration of free template in the solution is measured and the mass of template adsorbed to the MIP calculated (Brüggemann et al., 2004). Some of these experiments are based on radioligand binding, which is a very highly sensitive method to study the population of binding sites with the strongest binding characteristics (Shea and Sasaki, 1991). Typically, the sample is incubated with the radioligand for several hours and a centrifugation step is used to sediment the polymer particles. The radioactivity in the supernatant is then measured.

7.3 SAMPLE PREPARATION FOR MOLECULARLY IMPRINTED POLYMERS

Many works for MIPs were published in literature with the topic "sample preparation" (Figure 7.3). The impressive increase of the related published works in the last 5 years is noteworthy (2011–2015). This implies the need for further consideration of sample preparation.

7.3.1 Solid-Phase Extraction for Molecularly Imprinted Polymers (Molecularly Imprinted Solid-Phase Extraction)

SPE is one of the most promising and important techniques for MIPs. This is confirmed by the fact that many industries and companies attempt (and some have achieved this) to make commercially available cartridges packed with MIPs, which are appropriate for the target analyte determination in different samples. Many papers were also published regarding the applicability of MIPs for SPE. A drawback in literature is that the studies published only suggest (or describe) the need for using different templates for different applications. A little number of works suggests new trends for limiting the inherent drawbacks in preparing and using MIPs (template bleeding, tedious synthesis procedure, and slow mass transfer).

As it was described in Section 7.2, the bulk polymerization was the first designing plan/strategy for synthesis of MIPs. After extraction and sieving, the MIP particles obtained invariably possessed

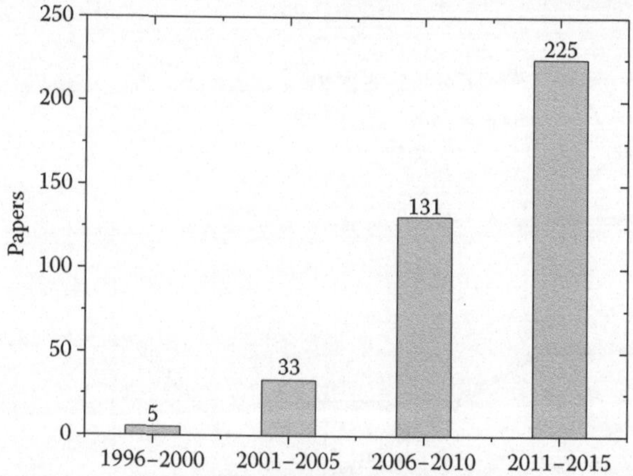

FIGURE 7.3 Works published using the terms *sample preparation* and *molecular imprinted polymers* (data after search in Scopus).

a heterogeneous particle size distribution with poor binding site accessibility for the target analyte. Many attempts have been done in this path developing novel synthesis routes in order to have MIP microspheres with suitable properties (physical characteristics, size, porosity, pore volume, and surface area). Only multistep swelling and polymerization (Hosoya et al., 1996) and precipitation polymerization (Wang et al., 2003) have been used for the preparation of MIPs for SPE of a wide variety of analytes in biological samples (Haginaka et al., 1999; Haginaka and Sanbe, 2000; Sanbe and Haginaka, 2003) and environmental samples (Cacho et al., 2003; Tamayo et al., 2003, 2005).

Various MISPE modes have been assayed such as (i) batch SPE technique, where MIPs are incubated with the sample; (ii) conventional off-line SPE technique, where MIPs are packed into cartridges; and (iii) several modes of online SPE technique. The former has been changed and displaced by the conventional off-line SPE format. Andersson et al. (1997) revealed for first time that the use of a close structural analogue of analyte as imprint species prevented the problems associated with template bleeding, which, until now, is one of the most serious and common problems of using MIPs.

Off-line protocols: Off-line MISPE protocols do not have any difference from other SPE procedures. In a typical procedure, a small amount (15–500 mg) of MIPs is packed into polyethylene (PET) cartridges. Then, after conditioning, loading, and washing, analytes are eluted, ideally free of co-extractives, and the elution extract is further analyzed by LC, GC, or capillary electrophoresis.

A huge development of off-line MISPE methods has been observed recently for the determination of many analytes in environmental samples (river water, groundwater, wastewater, tap water, seawater, and soil extracts), biofluids (urine, plasma, serum, and blood), and food samples. In general, the sample is loaded onto the MIP cartridge in a low-polarity solvent, since in such media, specific interactions are maximized, and after multiple washings for the removal of compounds, which have been nonspecifically bound to the polymeric matrix, analytes are eluted with a solvent able to disrupt the typical noncovalent interactions between the analyte and the imprinted polymer. Aqueous samples can also be directly loaded onto MIP cartridges. However, in this case, MIPs behave like a reversed-phase sorbent, and thus, both target analytes and matrix components are retained through nonspecific interactions. Then, a washing solvent able to remove matrix components and to redistribute nonspecifically bound analytes to the selective imprints is introduced.

Online protocols: In a typical procedure, a small precolumn packed with MIPs (typically about 50 mg) is placed in the loop of a six-port injection valve. After loading the sample and washing out interfering compounds, the analytes are eluted by the mobile phase and then separated in the analytical column.

In-line protocols: Extraction, enrichment, separation, and determination of target analytes can be achieved in one single step by the direct coupling of an MIP column in-line with the detection system thanks to the high selectivity provided by MIPs.

Advanced batch protocols: The batch MISPE technique has been displaced by other SPE techniques/modes. However, batch MISPE has reemerged using magnetic MIPs (Chen et al., 2009; Ji et al., 2009; Zhang et al., 2009), which, after extraction, can be separated from the media by a simple magnet, avoiding filtration and/or centrifugation. Based on literature, this procedure has been successfully used for the extraction of triazines from soil, soybean, lettuce, and millet (Zhang et al., 2009); tetracycline antibiotics from egg and tissue (Chen et al., 2009); and bisphenol A in environmental water and milk (Ji et al., 2009). Another approach is the molecularly imprinted microsolid phase extraction (MIMSPE) (Feng et al., 2009). The MIMSPE device consists of MIPs enclosed within a polypropylene flat-sheet membrane envelope. The membrane presents a porous structure that is ideal for holding submicrosphere MIP particles not suitable for MISPE since they cannot be hold back by the typical frits used in SPE cartridges. A successful example of the application of MIMSPE is the selective extraction of phenolic compounds from tap, river, and sewage waters (Feng et al., 2009) and ochratoxin A from coffee, grape juice, and urine (Lee et al., 2012).

7.3.1.1 Molecularly Imprinted Solid-Phase Extraction for Environmental Applications

The first MISPE approach for MIPs was done by Sellergren (1994a), who determined pentamidine in urine samples. The number of works published with MISPE environmental applications is large and can be divided into subtopics. One of the most important applications is the selective extraction of analytes from environmental samples (Andersson, 2001; Caro et al., 2006; Baggiani et al., 2007; Pichon and Chapuis-Hugon, 2008; Lok and Son, 2009).

Environmental samples (mainly environmental waters, sediments, and soils) have been applied to MISPE for the extraction of pesticides, pharmaceuticals, and phenolic compounds (Table 7.1). An initial off-line MISPE work describes an environmental application dealt with the extraction and enrichment of six chlorotriazines from groundwater and sediment samples using a terbuthylazine-imprinted polymer (Ferrer et al., 2000). An aliquot of 100 mL of water sample was percolated, and after a cleanup step with dichloromethane, the matrix components were removed, whereas target analytes were selectively retained with recoveries higher than 80% (except for propazine, with a recovery of 53%) after percolating 100 mL of water sample.

Despite these results, subsequent works have concluded that small amounts of water on the MIP have a negative effect on the retention of analytes, making it necessary to include an exhaustive drying step before the cleanup. Besides, matrix effects due to the presence of divalent cations in real water (such as calcium ions) have been observed. An ion-exchange interaction between these cations and the hydrogen of the carboxyl groups of the MIP occurs, thus preventing subsequent interactions (by H-bonding) between the triazines and the MIP (Chapuis et al., 2003). However, this matrix effect can be easily avoided by introducing an acidic washing step after the percolation of the water sample. Alternatively, as mentioned previously, it is possible to perform a two-step SPE procedure by firstly loading water samples onto conventional sorbents (i.e., C18, DVD) and then retained analytes eluted with an organic solvent suitable for further loading on the MIP (Turiel et al., 2001; Pap et al., 2002).

MISPE has been successfully applied, although in a lesser extent, to the cleanup of soil sample extracts for the determination of triazines (Turiel et al., 2001; Chapuis et al., 2004), monosulfuron (Dong et al., 2004), and fluoroquinolones (Turiel et al., 2007a).

Especially for pharmaceutical application, the MISPE examples are numerous. Table 7.2 summarizes pharmaceutical applications of MISPE, which are divided into two categories: one is to extract a target compound(s) in pharmaceutical preparations and herbal medicines or herbs selectively (Hu et al., 2003, 2005; Mena et al., 2003; Karasová et al., 2005; Nakamura et al., 2005; Karasová et al., 2006; Lin et al., 2006; Liu et al., 2006; Djozan and Baheri, 2007; Farrington and Regan, 2007; Rezaei et al., 2008). The other is to extract a compound(s) structurally related to a

TABLE 7.1
Relevant MISPE Applications in Environmental Analysis

Analytes	Template/Monomer/ Cross-Linker	Porogen	Sample	MISPE Mode
Triazines	Terbutylazine/MAA/EDMA	Dichloromethane	Groundwater and sediments	Off-line
Triazines	Terbutylazine/MAA/EDMA	Dichloromethane	Waters (industrial effluent)	Off-line
Triazines	Terbutylazine/MAA/EDMA	Toluene	Waters	Online (SPE + MIP)
Triazines	Propazine/MAA/EDMA	Toluene	Tap, groundwater soil	Off-line (SPE + MIP)
Triazines	Terbutylazine/MAA/EDMA Ametryn/MAA/EDMA	Dichloromethane	Soil and grape juice	Off-line
Monosulfuron	Monosulfuron/MAA/EDMA	Dimethylformamide	Soil	Off-line
Fluoroquinolones	Ciprofloxacin/MAA/EDMA	Methanol	Soil	Off-line and in-line
4-Nitrophenol	4-Nitrophenol/4-VP/EDMA	Acetonitrile	River water	Online
Triazines	Terbuthylazine/MAA/EDMA	Toluene	River water	Online (RAM + MIP)
Chlorinated phenoxy acids	2,4,5-Trichlorophenoxyacetic acid/4-VP/EDMA	Methanol/water	River water	Off-line
Bisphenol A	Bisphenol A/4-VP/EDMA	Toluene	River water	Online
Bisphenol A	Bisphenol A-d16/4-VP/EDMA	Toluene/poly(vinyl alcohol)	River water	Off-line
Bisphenol A	Bisphenol A/4-VP/TRIM	Acetonitrile	Tap and lake water	In-line
Methylthiotriazines	Irgarol/TFMAA/EDMA	Toluene	River water	Online
Penicillin G	Penicillin G procaine salt/1-(4-vinylphenyl)-3-(3,5-bis(trifluoromethyl)phenyl)urea + methacrylamide/EDMA	Acetonitrile	Tap and river water	Off-line
Nonylphenol and its ethoxylated derivatives	Nonylphenol/MAA/EDMA	Toluene/acetonitrile	Environmental solids	Off-line
Diclofenac	Diclofenac (2-VP/EDMA)	Toluene	Surface and wastewater	Off-line
Acidic pharmaceuticals	Not available (commercial MIP)		Wastewater	Off-line
Atrazine and metabolites	Terbutylazine/MAA/EDMA	Dichloromethane	Soils	Off-line
Fluoroquinolones	Enrofloxacin/1-(4-vinylphenyl)-3-(3,5-bis(trifluoromethyl)phenyl) urea + methacrylamide/EDMA	Acetonitrile	Environmental waters	Off-line
Ethynyl estradiol	Ethynyl estradiol/MAA/EDMA	Acetonitrile	River water	Online
Benzimidazole fungicides	Thiabendazole/MAA/DVB	Toluene/acetonitrile	Environmental waters	Online

(Continued)

TABLE 7.1 (CONTINUED)
Relevant MISPE Applications in Environmental Analysis

Analytes	Template/Monomer/ Cross-Linker	Porogen	Sample	MISPE Mode
Polycyclic aromatic hydrocarbons	PAH solution/MAA/EDMA	Acetonitrile	Coastal sediments, atmospheric particulates, and industrial effluents	Off-line
Benzimidazole compounds	Thiabendazole/MAA/DVB	Toluene/acetonitrile	Waters	Off-line and online
Antiepileptics	Cyclobarbital/4-VP/EDMA	Toluene	River water	Online (RAM-MIP)
Phenolic compounds	2,4,6-Trichlorophenol/MAA/ EDMA	Acetonitrile	Tap, river, and sewage waters	Off-line

Source: *Analytica Chimica Acta*, 668, Turiel, E., and A. Martín-Esteban, Molecularly imprinted polymers for sample preparation: A review, 87–99, Copyright (2010), with permission from Elsevier.

Note: 2-VP, 2-vinylpyridine; 4-VP, 4-vinylpyridine; DVB, divinylbenzene; EDMA, ethylene glycol dimethacrylate; HEMA, 2-hydroxyethyl methacrylate; MAA, methacrylic acid; TFMAA, 2-(trifluoromethyl)-acrylic acid; TRIM, trimethylolpropane trimethacrylate.

TABLE 7.2
Relevant MISPE Applications in Pharmaceutical Applications

Template	Target Molecules	Preparation Method	MISPE Method
Amobarbital	Phenobarbital	Suspension	On-column, LC
Atropine	Atropine, scopolamine	Multistep swelling	Online, LC
Chloramphenicol	Chloramphenicol	Bulk	Offline, voltammetry
Diacetylmorphine	Diacetylmorphine	Monolith	SPME, GC, GC-MS
(*E*)-piceatannol	(*E*)-piceatannol, butein, quercetin	Bulk	Offline, LC-MS
Esculetin	Esculetin and analogues	Bulk	Offline, LC
Harman	Harmaline and harmine	Bulk	Offline, LC-MS
Ibuprofen	Ibuprofen	Bulk	Offline, LC
Piroxicam	Piroxicam	Bulk	Off-line, UV
Protocatechuic acid	Phenolic acids	Bulk	Offline, LC
Quercetin	Quercetin, kaempferol	Bulk	Offline, LC
Quercetin	(*E*)-piceatannol, butein	Bulk	Offline, LC
trans-Resveratrol	*trans*-Resveratrol, emodin	Bulk	Offline, LC
Sinomenine	Sinomenine	Bulk	Offline, LC
Sulfamethoxazole	Sulfamethoxazole	Monolith	On-column, LC
Trimethoprim	Trimethoprim	Bulk, suspension	Offline, LC

Source: Haginaka, J.: Molecularly imprinted polymers as affinity-based separation media for sample preparation. *Journal of Separation Science*. 2009. 32. 1548–1565. Copyright Wiley-VCH Verlag GmbH & Co. KGaA. Reproduced with permission.

template molecule from crude extracts of herbal medicines or herbs (Xie et al., 2001, 2002, 2003; Zhu and Xu, 2003; Zhuang et al., 2008).

An MIP for atropine was prepared using a multistep swelling and polymerization method. The MIP produced was applied for the determination of tropane alkaloids (atropine and scopolamine) in a commercial gastrointestinal drug by a column-switching LC system, consisting of the MIP as a precolumn and a cation-exchange column as an analytical column. Since the amount of atropine included in the commercial gastrointestinal drug was high, the leaked template molecule, atropine, did not interfere the assay of atropine. When injecting the sample solution directly to the analytical column, large amounts of interfering components are eluted at the peak of scopolamine. However, by online MISPE, tropane alkaloids (atropine and scopolamine) could be determined accurately.

A combination of an MIP column and an MS detector afforded affinity extraction and online identification of antitumor components from the herb extracts (Xie et al., 2002). An MIP for harman, a structural analogue of antitumor components, was prepared. Two antitumor components, harmine and harmaline, were extracted and identified from the methanol extract of *Peganum nigellastum* seeds. Furthermore, an MIP for quercetin, an active inhibitor of epidermal growth factor receptor (EGFR), was prepared. The crude extract of *Caragana jubata*, a Chinese traditional Tibetan medicine, was loaded onto the MIP, and two novel active EGFR inhibitors, (*E*)-piceatannol and butein, were found to be selectively retained on the MIP (Zhu and Xu, 2003). When an MIP for (*E*)-piceatannol was used as affinity media, two different compounds besides the template itself were specifically recognized by the MIP, which were identified to be butein and quercetin (Zhu et al., 2003).

7.3.1.2 Molecularly Imprinted Solid-Phase Microextraction for Environmental Applications

Although the combination of molecular imprinting technology with SPME is still in its infancy, several papers have been published demonstrating the selectivity provided by imprinted fibers, allowing the determination of target analytes at trace levels in different kinds of samples.

The simplest way to desorb analytes after MIP-SPME is performing elution by immersing the fiber in a small volume of a solvent able to disrupt the specific interactions taking place at imprinted cavities. The first example was reported by Koster et al. (2001) for the determination of brombuterol in urine samples. Analytes were extracted from urine for 45 min using an MIP-coated fiber of ~75 μm. After a washing step, brombuterol was statically eluted for 5 min in a small volume (200 μL) of a methanol/acetic acid (90:10) mixture and further analyzed by HPLC with electrochemical detection. In a similar manner, the selective extraction of triazines from soils and vegetables was performed by immersing propazine-imprinted monolith in toluene-based sample extracts for 1 h under low stirring. Then, to remove nonspecific interactions, fibers were immersed in toluene for 15 min under stirring and subsequently air-dried for 5 min and analytes were finally desorbed with 200 μL of methanol by sonication for 5 min. The full methanolic extracts were diluted with 200 μL of water and further analyzed by HPLC with UV detection. The high selectivity provided by the propazine-imprinted fiber allows the detection of target analytes at very low concentration levels, which would be extremely difficult without performing any cleanup (Turiel et al., 2007b).

When off-line elution is performed, only a small fraction of the total volume used for the elution is injected in the chromatographic system. Thus, in order to improve detection limits, the online combination of SPME to HPLC is a useful alternative. This approach has been followed by Li et al. (2009) for the SPME-HPLC analysis of prometryn in soybean and corn (Hu et al., 2007b); triazines in soybean, corn, lettuce, and soils (Hu et al., 2007a), tetracyclines in chicken feed, chicken muscle, and milk (Hu et al., 2008); and μ-blockers in urine and plasma (Hu et al., 2009) using MIP-coated fibers. After extraction and washing steps, the fibers were immersed into the chamber of the specially designed SPME-HPLC six-way injection valve, the analytes being eluted statically for a fixed time with a suitable solvent or directly by the mobile phase used for the chromatographic separation of target analytes. Regardless of the methodology used, clean chromatograms were obtained reaching detection limits at very low concentration levels.

As mentioned previously, most of the reported imprinted fibers are thermally stable up to temperatures ranging from 250°C to 300°C, thus making them suitable for SPME-GC, which is the simplest, cheapest, and more environmental friendly approach. This methodology has been used for the determination of diacetylmorphine and analogous compounds in street heroin (Djozan and Baheri, 2007); triazines in water, rice, onion, and maize (Djozan and Ebrahimi, 2008; Djozan et al., 2009); and polybrominated diphenyl ethers in municipal wastewaters (Li et al., 2009) with high sensitivity and selectivity. Although few are the applications reported, until now, the thermal stability of imprinted fibers will surely help to the development of MIP-based SPME and to its implementation in analytical laboratories in the near future.

ACKNOWLEDGMENTS

The support for this study was received from the Greek Ministry of Education and Religious Affairs through Operational Program "Education and Lifelong Learning" of the National Strategic Reference Framework (NSRF)—Research Funding Program "Excellence II (Aristeia II)" under the title "Advanced microextraction approaches based on novel nanopolymers to measure pharmaceuticals, personal care products and their transformation products in the aquatic environment" (No. 4199), which is gratefully appreciated.

REFERENCES

Alahmadi, S.M., Mohamad, S., Jamil Maah, M. 2013. Preparation of organic-inorganic hybrid materials based on MCM-41 and its applications. *Advances in Materials Science and Engineering* 2013: 634863.

Andersson, L.I. 2001. Selective solid-phase extraction of bio- and environmental samples using molecularly imprinted polymers. *Bioseparation* 10: 353–364.

Andersson, L.I., Paprica, A., Arvidsson, T. 1997. A highly selective solid phase extraction sorbent for preconcentration of sameridine made by molecular imprinting. *Chromatographia* 46: 57–66.

Baggiani, C., Anfossi, L., Giovannoli, C. 2007. Solid phase extraction of food contaminants using molecular imprinted polymers. *Analytica Chimica Acta* 591: 29–39.

Balogh, A.G., Baba, K., Cohen, D.D., Elliman, R.G., Ensinger, W. et al. 2012. Modification, synthesis, and analysis of advanced materials using ion beam techniques. *Advances in Materials Science and Engineering* 2012: 431297.

Bergemann, C., Müller-Schulte, D., Oster, J., À Brassard, L., Lübbe, A.S. 1999. Magnetic ion-exchange nano- and microparticles for medical, biochemical and molecular biological applications. *Journal of Magnetism and Magnetic Materials* 194: 45–52.

Brüggemann, O., Visnjevski, A., Burch, R., Patel, P. 2004. Selective extraction of antioxidants with molecularly imprinted polymers. *Analytica Chimica Acta* 504: 81–88.

Cacho, C., Turiel, E., Martín-Esteban, A., Pérez-Conde, C., Cámara, C. 2003. Clean-up of triazines in vegetable extracts by molecularly-imprinted solid-phase extraction using a propazine-imprinted polymer. *Analytical and Bioanalytical Chemistry* 376: 491–496.

Caro, E., Marcé, R.M., Cormack, P.A.G., Sherrington, D.C., Borrull, F. 2003. On-line solid-phase extraction with molecularly imprinted polymers to selectively extract substituted 4-chlorophenols and 4-nitrophenol from water. *Journal of Chromatography A* 995: 233–238.

Caro, E., Marcé, R.M., Borrull, F., Cormack, P.A.G., Sherrington, D.C. 2006. Application of molecularly imprinted polymers to solid-phase extraction of compounds from environmental and biological samples. *TrAC Trends in Analytical Chemistry* 25: 143–154.

Chapuis, F., Pichon, V., Lanza, F., Sellergren, S., Hennion, M.C. 2003. Optimization of the class-selective extraction of triazines from aqueous samples using a molecularly imprinted polymer by a comprehensive approach of the retention mechanism. *Journal of Chromatography A* 999: 23–33.

Chapuis, F., Pichon, V., Lanza, F., Sellergren, B., Hennion, M.C. 2004. Retention mechanism of analytes in the solid-phase extraction process using molecularly imprinted polymers: Application to the extraction of triazines from complex matrices. *Journal of Chromatography B: Analytical Technologies in the Biomedical and Life Sciences* 804: 93–101.

Chen, B., Piletsky, S., Turner, A.P.F. 2002. Molecular recognition: Design of keys. *Combinatorial Chemistry and High Throughput Screening* 5: 409–427.

Chen, L., Liu, J., Zeng, Q., Wang, H., Yu, A. et al. 2009. Preparation of magnetic molecularly imprinted polymer for the separation of tetracycline antibiotics from egg and tissue samples. *Journal of Chromatography A* 1216: 3710–3719.

Chen, L., Xu, S., Li, J. 2011. Recent advances in molecular imprinting technology: Current status, challenges and highlighted applications. *Chemical Society Reviews* 40: 2922–2942.

Djozan, D., Baheri, T. 2007. Preparation and evaluation of solid-phase microextraction fibers based on monolithic molecularly imprinted polymers for selective extraction of diacetylmorphine and analogous compounds. *Journal of Chromatography A* 1166: 16–23.

Djozan, D., Ebrahimi, B. 2008. Preparation of new solid phase micro extraction fiber on the basis of atrazine-molecular imprinted polymer: Application for GC and GC/MS screening of triazine herbicides in water, rice and onion. *Analytica Chimica Acta* 616: 152–159.

Djozan, D., Mahkam, M., Ebrahimi, B. 2009. Preparation and binding study of solid-phase microextraction fiber on the basis of ametryn-imprinted polymer: Application to the selective extraction of persistent triazine herbicides in tap water, rice, maize and onion. *Journal of Chromatography A* 1216: 2211–2219.

Dong, X., Wang, N., Wang, S., Zhang, X., Fan, Z. 2004. Synthesis and application of molecularly imprinted polymer on selective solid-phase extraction for the determination of monosulfuron residue in soil. *Journal of Chromatography A* 1057: 13–19.

Farrington, K., Regan, F. 2007. Investigation of the nature of MIP recognition: The development and characterisation of a MIP for ibuprofen. *Biosensors and Bioelectronics* 22: 1138–1146.

Feng, Q., Zhao, L., Lin, J.M. 2009. Molecularly imprinted polymer as micro-solid phase extraction combined with high performance liquid chromatography to determine phenolic compounds in environmental water samples. *Analytica Chimica Acta* 650: 70–76.

Ferrer, I., Lanza, F., Tolokan, A., Horvath, V.V., Sellergren, B. et al. 2000. Selective trace enrichment of chlorotriazine pesticides from natural waters and sediment samples using terbuthylazine molecularly imprinted polymers. *Analytical Chemistry* 72: 3934–3941.

Haginaka, J. 2009. Molecularly imprinted polymers as affinity-based separation media for sample preparation. *Journal of Separation Science* 32: 1548–1565.

Haginaka, J., Sanbe, H. 2000. Uniform-sized molecularly imprinted polymers for 2-arylpropionic acid derivatives selectively modified with hydrophilic external layer and their applications to direct serum injection analysis. *Analytical Chemistry* 72: 5206–5210.

Haginaka, J., Takehira, H., Hosoya, K., Tanaka, N. 1999. Uniform-sized molecularly imprinted polymer for (S)-naproxen selectively modified with hydrophilic external layer. *Journal of Chromatography A* 849: 331–339.

Hosoya, K., Yoshizako, K., Shirasu, Y., Kimata, K., Araki, T. et al. 1996. Molecularly imprinted uniform-size polymer-based stationary phase for high-performance liquid chromatography structural contribution of cross-linked polymer network on specific molecular recognition. *Journal of Chromatography A* 728: 139–147.

Hu, S.G., Wang, S.W., He, X.W. 2003. An amobarbital molecularly imprinted microsphere for selective solid-phase extraction of phenobarbital from human urine and medicines and their determination by high-performance liquid chromatography. *Analyst* 128: 1485–1489.

Hu, S.G., Li, L., He, X.W. 2005. Solid-phase extraction of esculetin from the ash bark of Chinese traditional medicine by using molecularly imprinted polymers. *Journal of Chromatography A* 1062: 31–37.

Hu, X., Hu, Y., Li, G. 2007a. Development of novel molecularly imprinted solid-phase microextraction fiber and its application for the determination of triazines in complicated samples coupled with high-performance liquid chromatography. *Journal of Chromatography A* 1147: 1–9.

Hu, X., Hu, Y., Li, G. 2007b. Preparation and characterization of prometryn molecularly imprinted solid-phase microextraction fibers. *Analytical Letters* 40: 645–660.

Hu, X., Pan, J., Hu, Y., Huo, Y., Li, G. 2008. Preparation and evaluation of solid-phase microextraction fiber based on molecularly imprinted polymers for trace analysis of tetracyclines in complicated samples. *Journal of Chromatography A* 1188: 97–107.

Hu, X., Pan, J., Hu, Y., Li, G. 2009. Preparation and evaluation of propranolol molecularly imprinted solid-phase microextraction fiber for trace analysis of β-blockers in urine and plasma samples. *Journal of Chromatography A* 1216: 190–197.

Ji, Y., Yin, J., Xu, Z., Zhao, C., Huang, H. et al. 2009. Preparation of magnetic molecularly imprinted polymer for rapid determination of bisphenol A in environmental water and milk samples. *Analytical and Bioanalytical Chemistry* 395: 1125–1133.

Jiang, G., Chen, T., Yang, Q. 2012. Photocatalytic materials. *Advances in Materials Science and Engineering* 2012: 186948.

Karasová, G., Lehotay, J., Sádecká, J., Skačáni, I., Lachová, M. 2005. Selective extraction of derivates of *p*-hydroxybenzoic acid from plant material by using a molecularly imprinted polymer. *Journal of Separation Science* 28: 2468–2476.

Karasová, G., Lehotay, J., Kłodzinska, E., Gadzała-Kopciuch, R., Buszewski, B. 2006. Comparison of several extraction methods for the isolation of benzoic acid derivatives from *Melissa officinalis*. *Journal of Liquid Chromatography and Related Technologies* 29: 1633–1644.

Koster, E.H.M., Crescenzi, C., Den Hoedt, W., Ensing, K., De Jong, G.J. 2001. Fibers coated with molecularly imprinted polymers for solid-phase microextraction. *Analytical Chemistry* 73: 3140–3145.

Kyzas, G.Z., Bikiaris, D.N. 2014. Molecular imprinting for high-added value metals: An overview of recent environmental applications. *Advances in Materials Science and Engineering* 2014: 93263.

Lasáková, M., Jandera, P. 2009. Molecularly imprinted polymers and their application in solid phase extraction. *Journal of Separation Science* 32: 799–812.

Lee, T.P., Saad, B., Khayoon, W.S., Salleh, B. 2012. Molecularly imprinted polymer as sorbent in micro-solid phase extraction of ochratoxin A in coffee, grape juice and urine. *Talanta* 88: 129–135.

Li, M.K.Y., Lei, N.Y., Gong, C., Yu, Y., Lam, K.H. et al. 2009. An organically modified silicate molecularly imprinted solid-phase microextraction device for the determination of polybrominated diphenyl ethers. *Analytica Chimica Acta* 633: 197–203.

Liu, J.Q., Wulff, G. 2004. Functional mimicry of the active site of carboxypeptidase A by a molecular imprinting strategy: Cooperativity of an amidinium and a copper ion in a transition-state imprinted cavity giving rise to high catalytic activity. *Journal of the American Chemical Society* 126: 7452–7453.

Lin, L.Q., Zhang, J., Fu, Q., He, L.C., Li, Y.C. 2006. Concentration and extraction of sinomenine from herb and plasma using a molecularly imprinted polymer as the stationary phase. *Analytica Chimica Acta* 561: 178–182.

Liu, X., Ouyang, C., Zhao, R., Shangguan, D., Chen, Y. et al.2006. Monolithic molecularly imprinted polymer for sulfamethoxazole and molecular recognition properties in aqueous mobile phase. *Analytica Chimica Acta* 571: 235–241.

Lok, C.M., Son, R. 2009. Application of molecularly imprinted polymers in food sample analysis—A perspective. *International Food Research Journal* 16: 127–140.

Martín-Esteban, A., Fernández, P., Cámara, C. 1997. Immunosorbents: A new tool for pesticide sample handling in environmental analysis. *Fresenius' Journal of Analytical Chemistry* 357: 927–933.

Mena, M.L., Agüí, L., Martinez-Ruiz, P., Yáñez-Sedeño, P., Reviejo, A.J. et al. 2003. Molecularly imprinted polymers for on-line clean up and preconcentration of chloramphenicol prior to its voltammetric determination. *Analytical and Bioanalytical Chemistry* 376: 18–25.

Mobarak, Y., Bassyouni, M., Almutawa, M. 2013. Materials selection, synthesis, and dielectrical properties of PVC nanocomposites. *Advances in Materials Science and Engineering* 2013: 149672.

Nakamura, M., Ono, M., Nakajima, T., Ito, Y., Aketo, T. et al. 2005. Uniformly sized molecularly imprinted polymer for atropine and its application to the determination of atropine and scopolamine in pharmaceutical preparations containing Scopolia extract. *Journal of Pharmaceutical and Biomedical Analysis* 37: 231–237.

O'Shannessy, D.J., Ekberg, B., Mosbach, K. 1989. Molecular imprinting of amino acid derivatives at low temperature (0°C) using photolytic homolysis of azobisnitriles. *Analytical Biochemistry* 177: 144–149.

Pap, T., Horváth, V., Tolokán, A., Horvai, G., Sellergren, B. 2002. Effect of solvents on the selectivity of terbutylazine imprinted polymer sorbents used in solid-phase extraction. *Journal of Chromatography A* 973: 1–12.

Pichon, V., Chapuis-Hugon, F. 2008. Role of molecularly imprinted polymers for selective determination of environmental pollutants—A review. *Analytica Chimica Acta* 622: 48–61.

Pichon, V., Haupt, K. 2006. Affinity separations on molecularly imprinted polymers with special emphasis on solid-phase extraction. *Journal of Liquid Chromatography and Related Technologies* 29: 989–1023.

Pichon, V., Bouzige, M., Miège, C., Hennion, M.C. 1999. Immunosorbents: Natural molecular recognition materials for sample preparation of complex environmental matrices. *TrAC Trends in Analytical Chemistry* 18: 219–235.

Piletsky, S.A., Piletska, E.V., Karim, K., Freebairn, K.W., Legge, C.H. et al. 2002. Polymer cookery: Influence of polymerization conditions on the performance of molecularly imprinted polymers. *Macromolecules* 35: 7499–7504.

Radenović, A., Malina, J., Sofilić, T. 2013. Characterization of ladle furnace slag from carbon steel production as a potential adsorbent. *Advances in Materials Science and Engineering* 2013: Article ID 198240.

Ramos, L., Smith, R.M. 2007a. Special issue: Advances in sample preparation—Part I. *Journal of Chromatography A* 1152: 1–286.

Ramos, L., Smith, R.M. 2007b. Special issue: Advances in sample preparation—Part II. *Journal of Chromatography A* 1153: 1–310.

Rezaei, B., Mallakpour, S., Rahmanian, O. 2008. A selective solid-phase extraction and preconcentration method with using molecularly imprinted polymer for piroxicam in pharmaceutical sample. *Analytical Letters* 41: 1818–1831.

Sanbe, H., Haginaka, J. 2003. Restricted access media-molecularly imprinted polymer for propranolol and its application to direct injection analysis of β-blockers in biological fluids. *Analyst* 128: 593–597.

Sasaki, H., Kobashi, Y., Nagai, T., Maeda, M. 2013. Application of electron beam melting to the removal of phosphorus from silicon: Toward production of solar-grade silicon by metallurgical processes. *Advances in Materials Science and Engineering* 2013: Article ID 857196.

Sellergren, B. 1994a. Direct drug determination by selective sample enrichment on an imprinted polymer. *Analytical Chemistry* 66: 1578–1582.

Sellergren, B. 1994b. Imprinted dispersion polymers: A new class of easily accessible affinity stationary phases. *Journal of Chromatography A* 673: 133–141.

Shea, K.J., Sasaki, D.Y. 1991. An analysis of small-molecule binding to functionalized synthetic polymers by 13C CP/MAS NMR and FT-IR spectroscopy. *Journal of the American Chemical Society* 113: 4109–4120.

Sibrian-Vazquez, M., Spivak, D.A. 2004. Characterization of molecularly imprinted polymers employing crosslinkers with nonsymmetric polymerizable groups. *Journal of Polymer Science, Part A: Polymer Chemistry* 42: 3668–3675.

Smith, R.M. 2003. Before the injection—Modern methods of sample preparation for separation techniques. *Journal of Chromatography A* 1000: 3–27.

Spivak, D., Gilmore, M.A., Shea, K.J. 1997. Evaluation of binding and origins of specificity of 9-ethyladenine imprinted polymers. *Journal of the American Chemical Society* 119: 4388–4393.

Srinivasan, R. 2011. Advances in application of natural clay and its composites in removal of biological, organic, and inorganic contaminants from drinking water. *Advances in Materials Science and Engineering* 2011: Article ID 872531.

Tamayo, F.G., Casillas, J.L., Martin-Esteban, A. 2003. Highly selective fenuron-imprinted polymer with a homogeneous binding site distribution prepared by precipitation polymerisation and its application to the clean-up of fenuron in plant samples. *Analytica Chimica Acta* 482: 165–173.

Tamayo, F.G., Casillas, J.L., Martin-Esteban, A. 2005. Evaluation of new selective molecularly imprinted polymers prepared by precipitation polymerisation for the extraction of phenylurea herbicides. *Journal of Chromatography A* 1069: 173–181.

Tamayo, F.G., Turiel, E., Martín-Esteban, A. 2007. Molecularly imprinted polymers for solid-phase extraction and solid-phase microextraction: Recent developments and future trends. *Journal of Chromatography A* 1152: 32–40.

Turiel, E., Martín-Esteban, A. 2010. Molecularly imprinted polymers for sample preparation: A review. *Analytica Chimica Acta* 668: 87–99.

Turiel, E., Martín-Esteban, A., Fernández, P., Pérez-Cond, C., Cámara, C. 2001. Molecular recognition in a propazine-imprinted polymer and its application to the determination of triazines in environmental samples. *Analytical Chemistry* 73: 5133–5141.

Turiel, E., Martín-Esteban, A., Tadeo, J.L. 2007a. Molecular imprinting-based separation methods for selective analysis of fluoroquinolones in soils. *Journal of Chromatography A* 1172: 97–104.

Turiel, E., Tadeo, J.L., Martin-Esteban, A. 2007b. Molecularly imprinted polymeric fibers for solid-phase microextraction. *Analytical Chemistry* 79: 3099–3104.

Wang, J., Cormack, P.A.G., Sherrington, D.C., Khoshdel, E. 2003. Monodisperse, molecularly imprinted polymer microspheres prepared by precipitation polymerization for affinity separation applications. *Angewandte Chemie* 115.

Whitcombe, M.J., Rodriguez, M.E., Villar, P., Vulfson, E.N. 1995. A new method for the introduction of recognition site functionality into polymers prepared by molecular imprinting—Synthesis and characterization of polymeric receptors for cholesterol. *Journal of the American Chemical Society*. 117: 7105–7111.

Wulff, G., Heide, B., Helfmeier, G. 1987. Enzyme-analogue built polymers, 24 On the distance accuracy of functional groups in polymers and silicas introduced by a template approach. *Reactive Polymers, Ion Exchangers, Sorbents* 6: 299–310.

Xie, J., Zhu, L., Luo, H., Zhou, L., Li, C. et al. 2001. Direct extraction of specific pharmacophoric flavonoids from gingko leaves using a molecularly imprinted polymer for quercetin. *Journal of Chromatography A* 934: 1–11.

Xie, J., Zhu, L., Xu, X. 2002. Affinitive separation and on-line identification of antitumor components from *Peganum nigellastrum* by coupling a chromatographic column of target analogue imprinted polymer with mass spectrometry. *Analytical Chemistry* 74: 2352–2360.

Xie, J., Chen, L., Li, C., Xu, X. 2003. Selective extraction of functional components derived from herb in plasma by using a molecularly imprinted polymer based on 2,2-bis(hydroxymethyl)butanol trimethacrylate. *Journal of Chromatography B: Analytical Technologies in the Biomedical and Life Sciences* 788: 233–242.

Zhang, Y., Liu, R., Hu, Y., Li, G. 2009. Microwave heating in preparation of magnetic molecularly imprinted polymer beads for trace triazines analysis in complicated samples. *Analytical Chemistry* 81: 967–976.

Zhu, L., Xu, X. 2003. Selective separation of active inhibitors of epidermal growth factor receptor from *Caragana jubata* by molecularly imprinted solid-phase extraction. *Journal of Chromatography A* 991: 151–158.

Zhu, L., Chen, L., Xu, X. 2003. Application of a molecularly imprinted polymer for the effective recognition of different anti-epidermal growth factor receptor inhibitors. *Analytical Chemistry* 75: 6381–6387.

Zhuang, X., Dong, X., Ma, S., Zhang, T. 2008. Selective on-line extraction of trans-resveratrol and emodin from *Polygonum cuspidatum* using molecularly imprinted polymer. *Journal of Chromatographic Science* 46: 739–742.

8 Sample Preparation Methods for Determination of Pollutants in Air Samples

Francisco Pena-Pereira and Jacek Namieśnik

CONTENTS

8.1 INTRODUCTION

Air pollutants are those substances that are known or suspected to cause serious environmental and health hazards. A wide range of primary air contaminants can be released directly from both natural and anthropogenic sources. Besides, secondary air pollutants can be formed in the atmosphere via oxidation or photochemical reactions of primary air pollutants. Air pollution monitoring pursues the ambitious objective of improving the overall quality of the environment (Gurjar et al., 2010). In this regard, the US EPA has reported a comprehensive list of 187 hazardous air pollutants (HAPs), which includes compounds from very volatile to nonvolatile organic and inorganic

(EPA, 2013). Figure 8.1 shows the types of HAPs considered by the US EPA according to their nature and volatility. Almost 90% of the regulated HAPs are of organic nature, from which volatile organic compounds (VOCs) and semivolatile organic compounds comprise about three-fourths of all listed HAPs. Many compounds that could be present in air and suspected to be hazardous are, however, nonregulated. The identification of harmful ambient air contaminants, as well as the determination of their levels, is of paramount concern, as several environmental problems are associated with the presence of certain levels of HAPs. Nevertheless, the presence of pollutants in air is not simply a matter of environmental significance. In fact, both occupational and household exposures to a wide range of harmful chemicals make the assessment of indoor air quality necessary. In addition, the presence of certain VOCs in breath air has been associated with health diseases, so breath air analysis can significantly contribute to the early diagnosis of diseases and malfunctions of bodily organs.

Direct chromatographic separation and subsequent detection of HAPs are not always possible due to the low concentrations of pollutants in air samples. Furthermore, the complexity of air matrices can make their analysis a challenging task. Thus, sample pretreatment is generally required prior to the determination of HAPs in air. Sample preparation techniques ought to meet several requirements, including a significant retention and enrichment of target analytes, removal or minimization of coexisting compounds, compatibility of the extract with the analytical instrumentation, minimum consumption of toxic reagents and solvents, negligible waste generation, and expeditiousness and economy. Together with the improvements in analytical instrumentation, the development of sample preparation techniques has significantly contributed to the current state of air analysis. There is a wide body of literature on both the theoretical and experimental aspects of sampling and sample preparation for HAP monitoring, being highly recommended for interested readers (Harper, 2000; Górecki and Namieśnik, 2002; Michulec et al., 2005; Wang and Austin, 2006a; Partyka et al., 2007; Seethapathy et al., 2008; Wardencki et al., 2008; Ras et al., 2009; Król et al., 2010, 2011; Marć et al., 2013).

In this chapter, the main sample preparation techniques employed for HAP monitoring in air samples are described and comparatively evaluated. Derivatization strategies and desorption techniques employed are also discussed. Finally, selected applications of sample preparation techniques in outdoor, indoor, and breath air monitoring are provided.

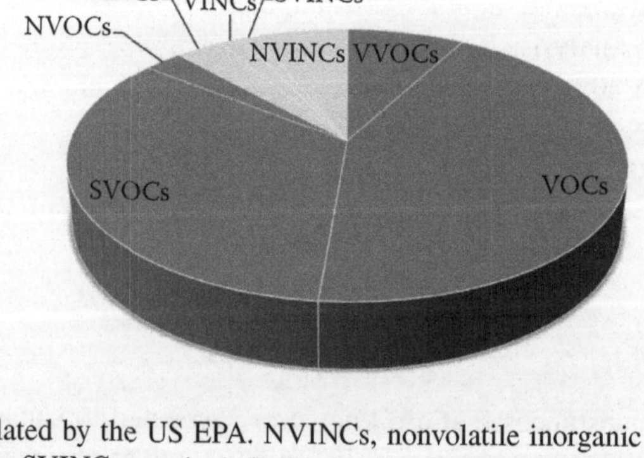

FIGURE 8.1 HAPs regulated by the US EPA. NVINCs, nonvolatile inorganic compounds; NVOCs, nonvolatile organic compounds; SVINCs, semivolatile inorganic compounds; SVOCs, semivolatile organic compounds; VINCs, volatile inorganic compounds; VOCs, volatile organic compounds; VVINCs, very volatile inorganic compounds; VVOCs, very volatile organic compounds. (From Wardencki, W., and J. Namieśnik, *New Horizon and Challenges in Environmental Analysis and Monitoring*, Chem. Fac. Gdańsk University of Technology, Gdańsk, 2003, pp. 446–463. With permission.)

8.2 SAMPLE PREPARATION TECHNIQUES

Analyte enrichment techniques can be grouped into three general categories (Michulec et al., 2005; Partyka et al., 2007):

- Dynamic enrichment, which is based on the forced passing of a gaseous sample through a system containing an appropriate sorbent or adsorbent material where analytes can be efficiently retained. Release of the enriched analytes is subsequently carried out by liquid desorption or thermal desorption (TD) for their determination, mainly by chromatographic techniques.
- Passive enrichment, which enables the preconcentration of target compounds into appropriate trapping systems by means of diffusion or permeation.
- Denudation enrichment, which consists of a combination of dynamic passing of a gaseous sample and diffusion of the analyte molecules from the gas stream into a retaining medium by means of the different forces involved in denudation processes, as explained in Section 8.2.3.

Depending on the sampling technique employed, several phenomena can be exploited to achieve an efficient mass transfer and enrichment of target compounds (Figure 8.2).

In this section, sample preparation techniques commonly applied in the determination of HAPs are described and briefly discussed. Advantages and limitations of these techniques are provided in Table 8.1.

8.2.1 Cryogenic Enrichment

Cryogenic enrichment is typically performed for preconcentrating trace-level organics from air samples. In cryogenic enrichment, air samples are passed at a controlled flow rate through a cooled

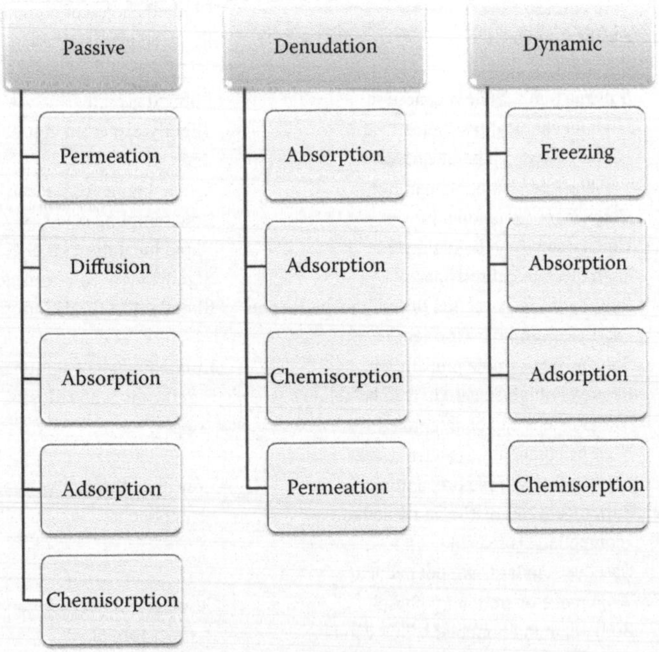

FIGURE 8.2 Physicochemical phenomena involved in analyte enrichment.

TABLE 8.1
Advantages and Limitations of Sample Preparation Techniques Employed for Air Samples

Sample Preparation Technique	Strengths	Weaknesses
Cryogenic enrichment	Simplicity Compatible with other sample preparation techniques	Reduced trapping efficiency Water vapor management is crucial Artifact formation due to trapped ozone
Impingers	Simplicity Large sample volumes can be used Selectivity can be tuned by appropriate selection of solvents and reagents	Time consuming Large consumption of solvents Reduced enrichment Not suitable for determination of instantaneous concentrations Additional sample preconcentration may be needed
Denuders	Physical speciation can be performed High selectivity Valid for polar and highly reactive analytes	Preparation and maintenance of denuders are time consuming Not suitable for determination of instantaneous concentrations
OTTs	Solvent-free sample preparation High mechanical resistance Very volatile analytes can be enriched at room temperature Complete removal of water from the OTT	Limited sample capacity Low sample flow rates Reduced enrichment of highly polar compounds
SPME	Solvent-free sample preparation High enrichment factors can be obtained Unitary steps can be integrated Selectivity can be tuned by appropriate selection of SPME fibers Easily automated Portability	Limited sorption capacity Fragility of SPME fibers Limited lifetime of the fibers Batch-to-batch variation of SPME fibers Limited choice of commercial SPME fibers Reduced precision
INCAT	Solvent-free sample preparation High mechanical resistance Unitary steps can be integrated	Limited sample capacity Slightly dispersed elution zones of the analytes
NTD	Solvent-free sample preparation Convenient for automation Unitary steps can be integrated High mechanical resistance Simultaneous extraction of free molecules and particles can be performed	Limited range of stationary phases Low sampling capacity Rapid breakthrough of NTD Significant needle-to-needle variation Slightly dispersed elution zones of the analytes
SPDE	Solvent-free sample preparation Relatively high sensitivity can be achieved Unitary steps can be integrated High mechanical resistance	Limited range of stationary phases Significant analyte losses during the aspiration step
GPE	Solvent-free sample preparation Enrichment of volatile and reactive compounds is feasible Gaseous standards are not needed	Low breakthrough volumes for weakly retained compounds
Single- and multibed adsorption systems	Wide range of stationary phases Both polar and nonpolar VOCs can be enriched	Displacement effects can occur Artifact formation Incomplete desorption of highly adsorbed compounds

(Continued)

TABLE 8.1 (CONTINUED)
Advantages and Limitations of Sample Preparation Techniques Employed for Air Samples

Sample Preparation Technique	Strengths	Weaknesses
Membrane-based techniques	High selectivity High enrichment power Reduced solvent consumption Convenient for automation Feasible for continuous monitoring	Time consuming Reduced enrichment of highly polar compounds

U-shaped tube, empty or packed with a suitable material (typically glass beads), immersed in a cryogenic fluid. After sampling, the retained compounds are transferred to the chromatographic column by slowly heating the trap at controlled temperature and purged with an inert gas.

Cryogenic enrichment is performed with cryogenic fluids, liquid argon, liquid nitrogen, and liquid oxygen being commonly used. From the three different cryogens, liquid argon is considered the best option because it allows the oxygen to pass through, even though its higher cost prevents more widespread use (Wang and Austin, 2006b).

Cryogenic enrichment is employed in a variety of reference methods for ambient air analysis, such as US EPA TO-3 and TO-12 (EPA, 1999). However, cryotrapping is prone to certain limitations, as shown in Table 8.1. In particular, cryogenic enrichment provides relatively poor trapping efficiency. In addition, the lack of selectivity of this sample preparation technique can give rise to certain problems in both the enrichment of VOCs and their subsequent separation by GC, especially in the case of ozone and water. On the one hand, ozone is enriched together with VOCs in the cryogenic trap, then leading to artifact formation due to the reactivity of certain VOCs with ozone during the sample pretreatment step (Helmig, 1997). On the other hand, water vapor collected on the trap can lead to adverse effects on sample preparation, separation, and detection of VOCs. Thus, condensed water can give rise to analyte losses and blockage of cryogenic traps through ice formation. Besides, water can be transferred to the GC column when heating the trap, then giving rise to problems in the chromatographic separation of enriched VOCs, namely, peak distortion, baseline drift, retention time shifts, and damage of the stationary phase. In addition, water can extinguish the flame when working with an FID and damage the MS ion source due to ionization of water (Wang and Austin, 2006b). Water management is therefore crucial when using cryotrapping systems. In this regard, several strategies have been reported to remove water from gas samples, including desiccants, adsorbents, cryocondensation, and permeation (Ras et al., 2009).

8.2.2 Impingers

Impinger extraction is based on the absorption of volatile analytes into an appropriate liquid solvent. Air samples are pumped into a flask containing the extractant by means of a tube with its outlet placed near the bottom of the flask. The air bubbles produced in the extractant improve the extraction kinetics owing to the increase on the interface surface. The lower the size of air bubbles, the faster the mass transfer process. Thus, the size of bubbles has an impact on the design of impingers (Camel and Caude, 1995).

The selectivity of this sample preparation technique can be improved by appropriate extraction with derivatization. A wide range of combinations of solvents and reagents can therefore provide the required selectivity by using impingers. Furthermore, impingers can be arranged in series with different extractants with the aim of separating different groups of analytes. Three methodologies involving impingers have been reported by the US EPA for the determination of aldehyde and ketones (TO-5), phosgene (TO-6), and phenols (TO-8) in ambient air (EPA, 1999).

Some drawbacks are, however, associated with the use of impingers, such as the limited enrichment attainable due to the large amounts of solvents employed. Thus, additional sample preconcentration may be needed prior to chromatographic separation to determine low concentrations of VOCs. This enrichment is mainly carried out by evaporation to dryness and subsequent reconstitution into a small volume of solvent, which can contribute to analyte losses in the case of nonderivatized highly volatile compounds. In addition, the chromatographic peaks of target analytes may be overlapped with the extraction solvent peak.

8.2.3 Denuders

Denudation is a convenient sample preparation technique for the enrichment and isolation of gaseous compounds present in air samples reported as early as 1971 (Fish and Durham, 1971). In denudation, gas and particulate phases are separated on the basis of the difference in vapor and particle diffusivities. Thus, VOCs are collected by a retaining medium at the inner wall of the device, while solid particles and aerosols practically do not contact the walls under laminar flow conditions. Two forces are involved in the process, one of them parallel to the direction of the gas flow that results from the forced flow of gas and the other perpendicular to denuder walls as a result of the radial diffusion (Kloskowski et al., 2002, 2007). The process continues until equilibrium conditions between

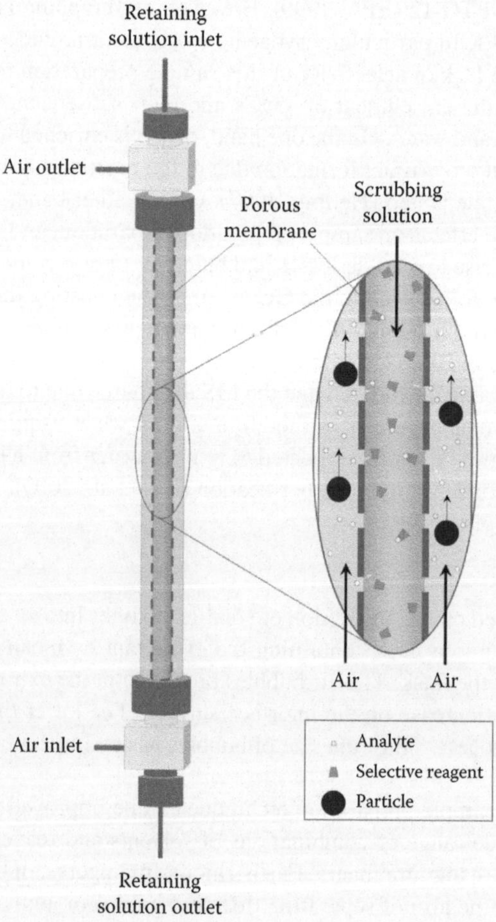

FIGURE 8.3 Scheme of an annular diffusion denuder.

the gas and the retaining medium are attained. Cyclones and highly efficient impactors are usually employed at the denuder inlet to minimize the contact of solid particles with the retaining medium. Particles and aerosols can subsequently be collected by placing a filter (or set of filters) at the outlet of the tube.

A variety of denuder designs can be used, including cylindrical, annular, honeycomb-type, or parallel plate-type denuders. All these configurations can be used with liquid extractants under the denomination of wet denuders that offer the possibility to perform continuous measurements when combined with chromatographic techniques (Kloskowski et al., 2002). A schematic of an annular diffusion denuder can be shown in Figure 8.3.

Unlike alternative sample preparation techniques, denuders allow carrying out physical speciation analysis, which means that analytes present in the gaseous phase are separated from solid particles that can be present in atmospheric samples. This fact is of crucial relevance in the determination of gaseous analytes in atmospheric samples, since target compounds can be present in both phases, and therefore, significant errors (positive and negative artifacts) can be produced (Król et al., 2010). The preparation and maintenance of denuders are, however, time consuming. Besides, long sampling times are commonly needed, so relevant information can be missed when dealing with the HAP monitoring in air as average concentrations are obtained.

8.2.4 Open Tubular Traps

Open tubular trapping (OTT) is a sorptive technique used for the isolation and enrichment of target compounds in air firstly reported in the mid-1980s (Burger and Munro, 1986). Preconcentration of target compounds is performed in OTT by passing gas samples at relatively low flow rates through a capillary tube coated on the inside wall with an appropriate trapping medium (Figure 8.4).

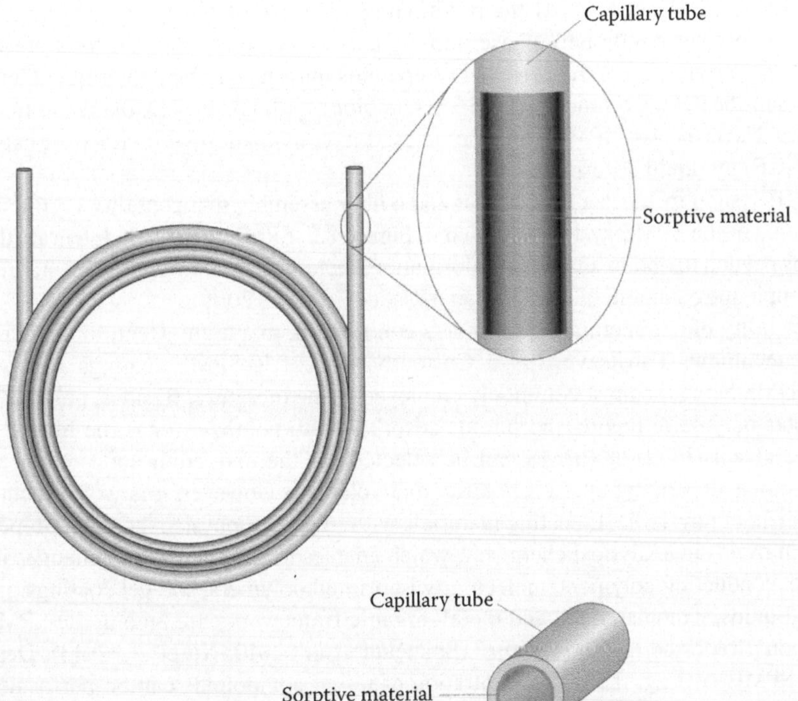

FIGURE 8.4 Schematic illustration of the OTT technique.

Volatile compounds are therefore retained by the stationary phase by diffusion at room temperature. Exhaustive and nonexhaustive methods can be employed in OTT. In the former case, sorption is performed as an equilibrium process, while in the latter, the sample volume should not exceed the breakthrough volume for complete enrichment of analytes (Kataoka et al., 2009). After enrichment for a prescribed time, analytes are transferred to the GC column for analysis mainly by TD, even though liquid desorption can also be performed.

Different OTTs have been reported in the literature. The first reported OTTs were prepared by immobilizing an SE-30 layer, activated carbon, or a porous organic polymer supported on immobilized SE-30 (Burger and Munro, 1986). Developments on the preparation of OTTs have enabled the application of OTTs coated with a thick film of stationary phase, which showed excellent performance in the breakthrough mode (Kloskowski et al., 2004; Pettersson et al., 2004). OTTs have been mainly prepared with materials showing excellent thermal resistance such as poly(dimethylsiloxane) (PDMS) as retaining medium. A piece of a commercial GC column can also be employed as OTT, thus broadening the applicability of this sample preparation technique. For instance, a piece of a GC column coated with PDMS has been employed for the enrichment of VOCs (Dudek et al., 2002a).

OTT can be considered as a solvent-free enrichment technique when combined with TD. The large stationary phase volume employed in OTT allows the determination of very low amounts of target compounds. However, the efficient enrichment of highly polar compounds is still a challenging subject to be solved in OTT, even though certain attempts have been made to increase the polarity of the stationary phase (Bonn et al., 2009). Besides, long sampling times may be needed when dealing with large sample volumes.

8.2.5 Solid-Phase Microextraction

SPME is a miniaturized solvent-free sample preparation technique that has become widely employed for the preconcentration of target compounds from liquid, solid, and gaseous samples (Arthur and Pawliszyn, 1990). SPME consists of the partitioning of target analytes between the sample and a thin layer of a polymeric stationary phase immobilized in a fused silica fiber for a prescribed time and subsequent introduction of the enriched compounds into an analytical instrument for their separation and quantification. Commonly, SPME is combined with GC by TD, thanks to their excellent compatibility. Nevertheless, SPME can also be used in combination with HPLC or capillary electrophoresis (CE) by liquid desorption.

The SPME system includes a fiber holder and a fiber assembly that contains a retractable SPME fiber. A scheme of the SPME system is shown in Figure 8.5. SPME enables to integrate the different unitary steps needed to obtain analytical information, including sampling, extraction, preconcentration, and sample introduction, in a single step (Risticevic et al., 2009).

The availability of commercial SPME fibers contributed, to a high extent, to the current acceptance of the technique (Pawliszyn, 1997). Currently, a set of 10 SPME coatings are commercially available, PDMS being the most commonly employed in the literature. Besides, poly(acrylate) (PA), a highly polar polymeric liquid, and porous adsorbents immobilized onto the fiber by means of polyethylene glycol (PEG) or PDMS can be selected for the extraction and preconcentration of target compounds depending on their polarity and volatility. However, it is worth mentioning that the development of novel SPME coating materials with improved physicochemical properties is one of the current trends in analytical chemistry, which enables to solve current challenges in this area. In this sense, conducting polymers, molecularly imprinted polymers, sol–gel coatings, immunosorbents, ionic liquids, nanomaterials, and metal–organic frameworks are among the SPME coatings more commonly reported in the literature (Spietelun et al., 2010; Xu et al., 2013). Depending on the type of SPME coating, the preconcentration of target compounds can be performed by sorption or adsorption. This fact has an impact on the selectivity of the extraction process, as well as on the desorption conditions. Derivatization reactions can also be employed in SPME to increase the extractability of target analytes as well as to enable the compatibility of the miniaturized sample

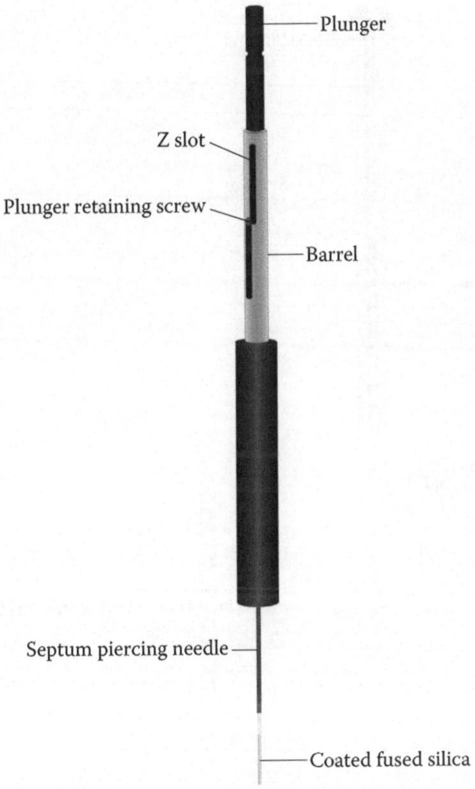

FIGURE 8.5 Schematic illustration of the SPME system.

pretreatment technique with analytical instrumentation. Relevant aspects on the derivatization reactions performed in SPME are described in Section 8.3.

Certain limitations can be mentioned for SPME, mainly related to the fiber coating durability, cost, fiber-to-fiber variability, and possible carryover. Further information on the theoretical and experimental aspects concerning SPME can be found in the literature (Dietz et al., 2006; Ouyang and Pawliszyn, 2006; Spietelun et al., 2010, 2013; Pawliszyn, 2011).

8.2.6 Inside Needle Capillary Adsorption Trap

Inside needle capillary adsorption trap (INCAT) is a solvent-free sample preparation technique that uses a hollow stainless steel needle filled with an internal coating as enrichment phase (McComb et al., 1997; Qin et al., 1997). This technique, based on both SPME and purge and trap methods, can be employed for the enrichment of target compounds from gas and liquid samples. The enrichment process is based on the introduction of a fluid sample into the needle and sorption of target compounds onto the inner surface of the device. Samples can be introduced passively to the trap by diffusion or actively by means of a syringe or a pump. A short length of GC column or an internal coating of sorbent material can act as the preconcentration medium. Desorption of the enriched target analytes can be finally carried out by rapid heating of the metallic needle at the injection port of a gas chromatograph. Figure 8.6 shows a schematic diagram of the INCAT system.

The configuration of the INCAT system shows certain advantages when compared with alternative sample pretreatment techniques. The preconcentration phase is protected by the metallic needle, so that it shows a high mechanical resistance, especially when compared with fragile silica-based SPME fibers. Furthermore, INCAT is a solvent-free approach that enables the integration of

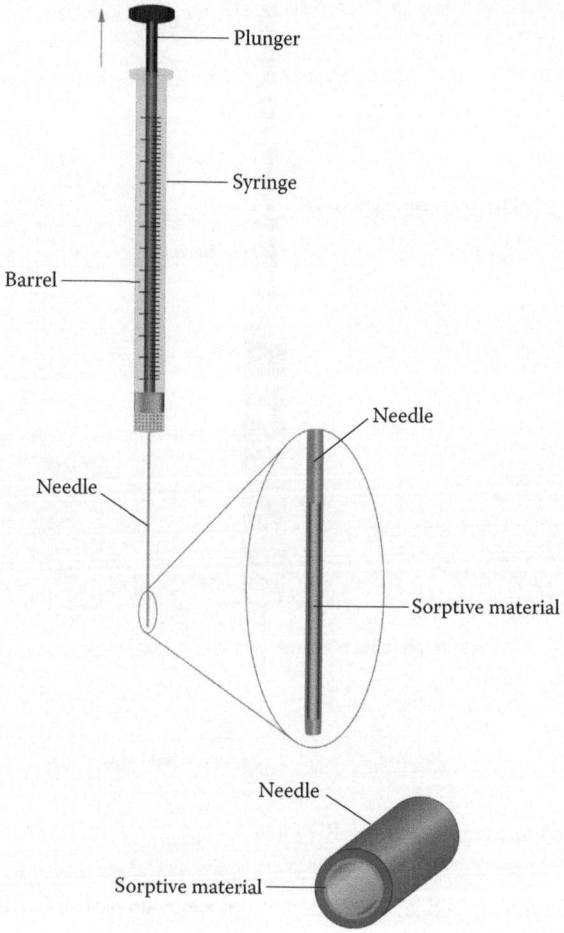

FIGURE 8.6 Schematic illustration of the INCAT device.

steps (sampling, extraction, enrichment, and sample introduction) to perform the analysis. However, the sample capacity is not particularly large due to the small internal diameter of the needle, and the elution zones of target compounds are slightly dispersed (Qin et al., 1997).

8.2.7 Needle Trap Device

As early as 1978, Raschdorf reported the first in-needle trap device for fragrance collection (Raschdorf, 1978). The sample preparation system consisted of a large needle packed with Tenax that required the modification of the standard inlet systems for introduction of desorbed compounds in the GC system (Lord et al., 2010). More recently, Pawliszyn and coworkers introduced the concept of a needle trap device (NTD) as an exhaustive extraction technique that combines the idea of active sampling and SPME (Koziel et al., 2001a; Pawliszyn, 2002a). The use of smaller-diameter needles (22 or 23 gauge) resulted in greater convenience for the combination of this sample preparation technique with analytical instrumentation (Lord et al., 2010).

NTD consists of blunt-type hypodermic stainless steel needles packed with an appropriate retaining material (sorbents or adsorbents) for the enrichment of target compounds in air. Gaseous samples are passed through the needle by a small side hole and analytes enriched by the retaining material. After preconcentration, analytes are mainly introduced in the gas chromatograph by

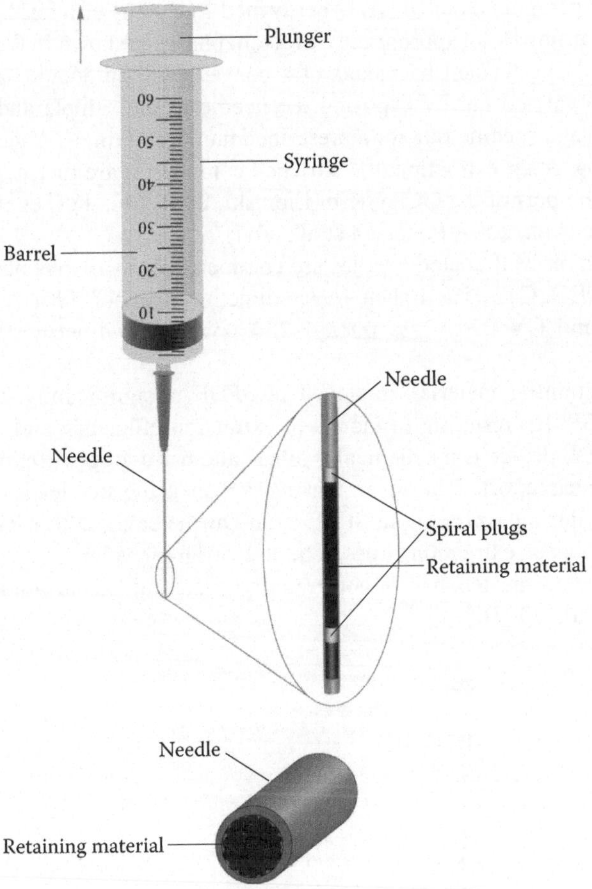

FIGURE 8.7 Schematic illustration of an NTD.

inserting the NTD into the injection port of the GC for desorption, even though other alternatives for analyte desorption have been reported (Lord et al., 2010). A schematic of a NTD is shown in Figure 8.7.

A variety of materials have been employed as retaining phase in NTDs, including PDMS, DVB, Tenax, or Carboxen, among others. In addition, the use of nanomaterials as stationary phase of NTDs has also been reported in the literature (Heidari et al., 2012).

NTD allows integrating sampling, preconcentration, and subsequent sample introduction for analysis. Furthermore, NTDs are robust and capable of performing exhaustive extraction. Interestingly, NTDs can be employed to extract free molecules and particles simultaneously. Specifically, this can be carried out by packing silanized glass wool in the front of a needle to trap particles and packing another sorbent afterward to retain free molecules (Lord et al., 2010). As for the limitations of the technique, one can mention the relatively low sampling capacity, rapid breakthrough of the NTD, poor reproducibility from needle to needle, and the possible dispersion of the elution zones of the analytes (Alonso et al., 2011).

8.2.8 Solid-Phase Dynamic Extraction

Solid-phase dynamic extraction (SPDE), also known as the "magic needle," is an inside-needle solventless extraction technique valid for liquid and gaseous samples (Lipinski, 2001). In SPDE, the stationary phase is protected inside the stainless steel needle of a 2.5 mL headspace syringe. The enrichment of target compounds in SPDE is performed by means of a headspace syringe with its needle coated on the inside by an appropriate extractant phase (as shown in Figure 8.8). Automation of SPDE is commonly performed by means of a conventional automatic sampler. The dynamic extraction process is carried out by exposing the needle to the sample and moving the syringe plunger up and down at a specific rate for a prescribed number of times. Analytes are thus retained in the internal coating. After extraction, the enriched compounds are thermally desorbed from the coating in the injection port of the GC system (Lipinski, 2001; Musshoff et al., 2002; Jochmann et al., 2007; Van Durme et al., 2007; Pokorska et al., 2011).

A number of different wall-coated needles are commercially available, namely, PDMS, PDMS/activated charcoal, PEG, CT5 (5% diphenyl/95% dimethylsiloxane), 1701 (14% cyanopropyl/86% dimethylsiloxane), and OV 225 (25% phenyl–25% cyanopropylmethylsilicone, 50% PDMS) (Chromtech, 2014).

The volume of retaining material employed in SPDE is significantly larger than that commonly employed in SPME, resulting in increased extraction efficiency and, therefore, sensitivity. Furthermore, the SPDE device is mechanically robust and has a longer lifetime when compared to SPME fibers. It has been reported, however, that every dispensing step leads to a significant loss of the retained analytes during aspiration in SPDE (Van Durme et al., 2007). This results in reduced extraction yield and longer extraction times. The use of larger volume aspiration, as well as the elimination of the dispensing step, has demonstrated to overcome this limitation of the SPDE technique (Van Durme et al., 2007).

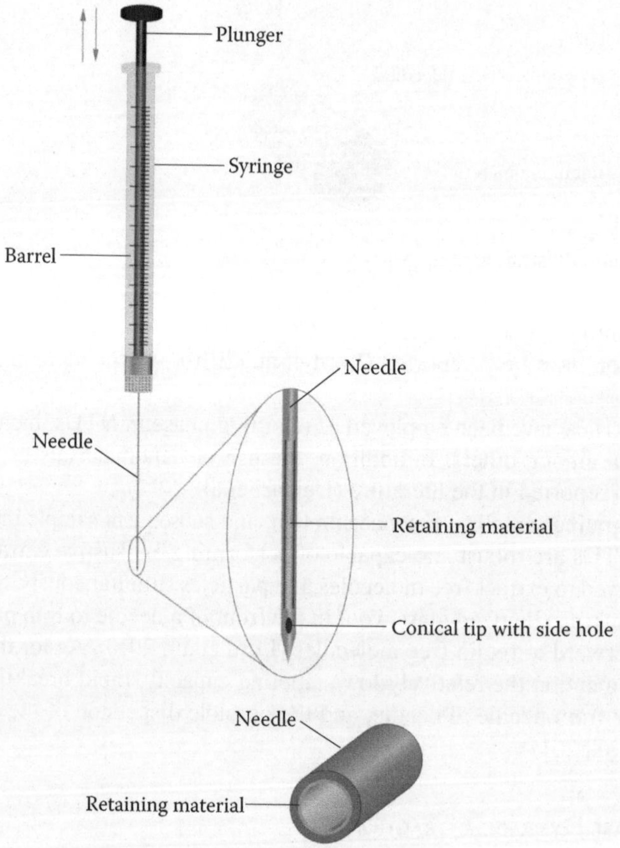

FIGURE 8.8 Schematic illustration of the SPDE system.

8.2.9 Gum-Phase Extraction

Gum-phase extraction (GPE) is a sample preparation technique that mainly employs polymeric sorbents for retention of target compounds from liquid and gaseous samples (Baltussen et al., 1997a, 1998, 1999a). In GPE, the sorptive phase material, mainly PDMS, is used in a packed bed configuration. Analytes are sorbed onto the polymeric trapping material and subsequently desorbed for analysis. Both liquid desorption and TD can be employed, the latter being the preferred option, as reported in Section 8.4.

Unlike sample preparation techniques that involve the use of adsorptive materials, analytes of interest are retained by solubility in the pure retaining material in GPE. Thus, the risk of permanent adsorption and/or decomposition of analytes is minimized by using sorptive enrichment on cartridges packed with PDMS particles instead of commonly used adsorbents (Baltussen et al., 1999a). The amount of sorptive phase employed in GPE is much larger than that used in alternative sample preparation techniques, such as SPME or OTT, resulting in greater sensitivity. Besides, GPE allows the use of high sampling flow rates, which results in increased sensitivity and reduced sampling times (Baltussen et al., 2002). The main drawback of GPE is the low breakthrough volumes obtained for weakly retained highly volatile compounds. This limitation can be overcome by using equilibrium gum-phase extraction (EGPE) (Baltussen et al., 1999b). In EGPE, sampling is continued until analytes are in equilibrium with the sorptive material. It should be mentioned that EGPE requires samples with a constant concentration over the sampling time. Under these conditions, accurate sampling of volatiles with maximum preconcentration are attained.

8.2.10 Single-Bed and Multibed Adsorption Systems

Adsorptive enrichment systems are based on the retention of target compounds onto the surface of adsorbent materials. Adsorption-based systems offer a wide range of possibilities for HAP monitoring in air samples. In fact, a variety of adsorbent materials can be used in single- and multibed systems, for active or passive sampling, with liquid desorption or TD. Remarkably, most of the methodologies recommended by the US EPA for the efficient retention of HAPs from ambient air are based on the use of both single- and multibed adsorbent systems (EPA, 1999).

Adsorbent materials should ideally fulfill a number of requirements, including complete enrichment of target compounds, low affinity to water, reduced adsorption capacity for air matrix constituents, surface homogeneity and inertness, and complete and fast desorption of target analytes (Dettmer and Engewald, 2002). A large number of materials can be used for adsorptive enrichment. In general, they can be classified into three general groups, namely, inorganic materials (mainly silica gel, zeolites, and alumina), carbon-based adsorbents (activated carbon, carbon molecular sieves, and graphitized carbon blacks), and porous organic polymers (cross-linked linear polymers, homogeneous cross-linked polymers, and heterogeneous cross-linked polymers) (Dettmer and Engewald, 2002).

With the aim of retaining a wide range of compounds from air samples, two or more adsorbents can be placed in series. When dealing with dynamic sampling, adsorbents are arranged in order of increasing adsorbent strength to avoid the irreversible retention of less volatile compounds in the strongest adsorbent (Ras et al., 2009). Thus, the least volatile compounds are retained on the weakest adsorbent placed at the front end of the tube, while more volatile analytes are trapped by stronger adsorbents. The most volatile compounds are retained by the strongest adsorbent placed at the far end of the tube (Harper, 2000). The retained target compounds are subsequently desorbed in the reverse direction (Figure 8.9). The appropriate selection of adsorbents should be made taking into account certain considerations, including the strength of the analyte–adsorbent interactions, artifacts, hydrophobicity, inertness, and mechanical strength (Woolfenden, 2010).

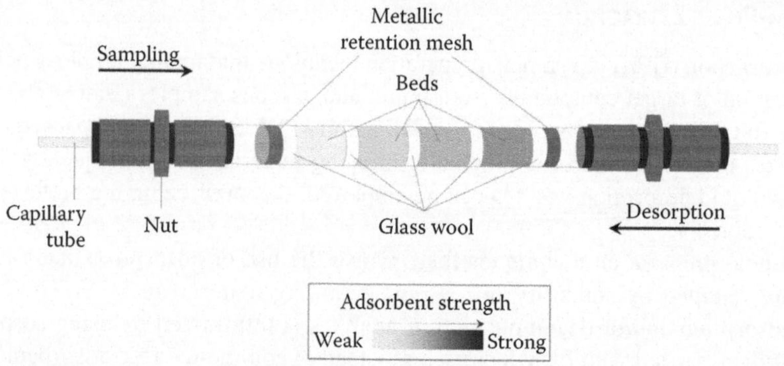

FIGURE 8.9 Multibed adsorption system for dynamic sampling.

In adsorption processes, analytes with lower affinity for the adsorbent can be displaced by compounds with a higher affinity. This is also true for water when dealing with adsorbents that show significant retention of water. Thus, these factors should be carefully considered when using adsorptive enrichment. Multibed adsorbents containing small amounts of Carboxen 1000 could be an attractive approach to avoid this limitation, as water vapor is not strongly retained by this adsorbent (Gawryś et al., 2001). Depending on the adsorbent used, ozone can also be an important interference, especially in the case of reactive unsaturated compounds (Dettmer and Engewald, 2003). A set of ozone scrubber materials can be found in the literature (Helmig, 1997).

8.2.11 Membrane-Based Techniques

In membrane-based techniques, membranes are used to enrich relevant air constituents while acting as a selective barrier between the extractant phase and the sample. Analytes are thus transferred from the sample (donor) into the acceptor phase through the membrane. Both porous and nonporous membranes can be employed for enriching target compounds (Demeestere et al., 2007). In the case of porous membranes, the extractant phase is immobilized within the pores of polymeric membranes by means of capillary forces. Semipermeable membrane devices (SPMDs), mainly based on a thin-walled polyethylene membrane containing triolein (Figure 8.10), have been extensively

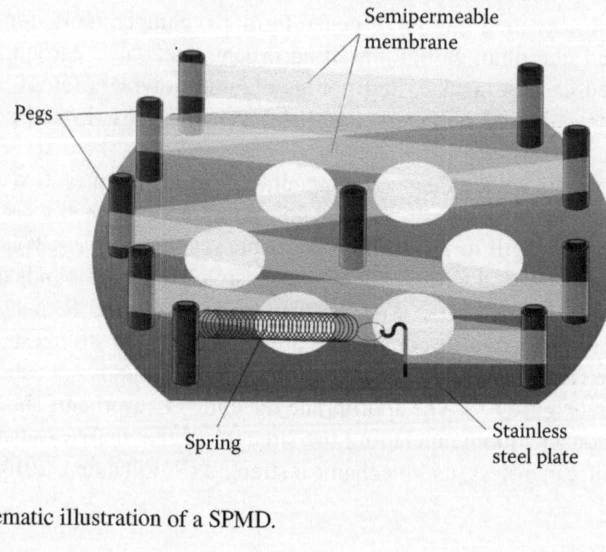

FIGURE 8.10 Schematic illustration of a SPMD.

employed as passive samplers for atmospheric compounds (Huckins et al., 1990; Petty et al., 1993; Ockenden et al., 1998a,b). Selective extraction can be achieved by using porous membranes due to their size exclusion characteristics. Thus, sample constituents may or may not pass the membrane depending on their molecular diameter. As for nonporous membranes, the enrichment of relevant compounds is mainly based on the solubility and diffusion rates of target compounds in the membrane material that can be composed of a polymeric material or a solvent (Pawliszyn, 2002b; Jakubowska et al., 2009).

Among membrane extraction techniques, the membrane extraction with a sorbent interface (MESI) is especially interesting for air analysis. A polymeric hollow fiber is used in MESI as the membrane, which permeates analytes from the surrounding air sample. Analytes inside the membrane are then transported by inert gas into a sorbent tube, commonly cooled, where target compounds are retained. TD finally enables their introduction into the GC for analysis (Jönsson and Mathiasson, 2001).

Membranes can be fabricated from a wide range of organic and inorganic materials. Both flat sheet membranes and tubular (hollow fiber) membranes can be employed for extraction and preconcentration. The tubular geometry provides faster mass transfer due to the higher surface area-to-volume ratio (Pinnau and Freeman, 2000).

Membrane-based techniques involve the use of reduced amounts of extractant phase and enable the achievement of relatively high enrichment factors as a consequence of the decreased extractant-to-sample volume ratio. Other advantages include the improved selectivity achieved (especially in the case of porous membranes) and the potential for automation. However, extended extraction times would be required to achieve very high enrichment factors needed for trace analysis. Furthermore, the enrichment of highly polar compounds is yet to be fully explored.

8.3 DERIVATIZATION STRATEGIES

Derivatization is commonly employed in analytical chemistry with the aim of improving the isolation, separation, and detection of target compounds. A wide variety of derivatization reactions have been reported in the literature aiming at enhancing the sensitivity, selectivity, extraction efficiency, stability of analytes, and overall quality of the data (Rosenfeld, 2003). It should be highlighted, however, that derivatization reactions can give rise to the formation of unexpected derivatives or by-products that can disturb the chromatogram (Little, 1999). In addition, they can involve the use of toxic and harmful reagents; contribute to the generation of wastes; and be laborious, tedious, and time consuming. For all these reasons, derivatization has been identified as an extra step of analytical procedures that should preferably be avoided. In fact, the eighth principle of green chemistry specifically recommends avoiding derivatization reactions whenever possible (Keith et al., 2007). Nevertheless, certain microreactions can offer significant opportunities for the application of a given sample preparation technique and can be valuable strategies to attain the abovementioned objectives (Lavilla et al., 2014). A highly recommended special issue, "Microreactions in Separation Science," has been recently published by the *Journal of Chromatography A* (Poole, 2013).

Analytical derivatization can be performed in combination with sample preparation techniques at different stages. When dealing with air samples, derivatization reactions can be performed during sample preparation or after the enrichment process.

8.3.1 Preconcentration with In Situ Derivatization

Simultaneous sampling and derivatization can be carried out to improve the enrichment and stability of analytes in the extractant phase. In general terms, in situ derivatization is performed by loading the extractant phase with a suitable derivatizing agent, so analytes are converted into appropriate analyte derivatives and effectively retained in the extractant phase. This fact can be exploited to integrate field sampling and sample preparation in a single step prior to air analysis.

Depending on the sample preparation technique considered and on the physicochemical properties of the derivatizing agent, the loading process can be performed by dissolution of the reagent in an appropriate extractant phase, by immersion into the reagent solution, or by static or dynamic exposure to the reagent vapors. Certain experimental parameters can be of paramount importance in the development of sample preparation methods with in situ derivatization, namely, reagent concentration, derivatization time, and temperature. Thus, these variables should be evaluated and controlled when implementing chemical reactions for simultaneous sampling and derivatization. The derivatizing agent should be nontoxic and insensitive to light, heat, and oxygen (Martos and Pawliszyn, 1998). Besides, it should react rapidly and quantitatively with the analyte, and the obtained derivatives should be compatible with the analytical instrumentation to avoid further derivatization. A number of in situ derivatization reactions can be performed depending on the functional groups of target analytes. For instance, the use of a carbonyl-specific derivatization agent such as *o*-2,3,4,5,6-(pentaflurobenzyl)hydroxylamine has been exploited for the in situ derivatization of aldehydes and ketones into their corresponding oximes (Koziel et al., 2001b; Ho and Yu, 2002; Bourdin and Desauziers, 2014). Besides, in situ conversion of ethylene oxide into 2-bromoethanol by means of hydrogen bromide has also been reported (Tsai and Wu, 2003).

8.3.2 Postextraction Derivatization

Postextraction derivatization is commonly performed to modify the chemical structure of retained analytes into analyte derivatives more suitable for the analysis. Thus, the enriched analytes are exposed to a suitable derivatizing agent for the prescribed time in order to ensure quantitative conversion. Postextraction derivatization can be a better option over in situ derivatization in certain cases, for instance, when the analyte derivatives show less favorable physicochemical properties for preconcentration than the original analytes, or the conditions needed for derivatization are far from the optimum conditions needed for preconcentration. In both cases, postextraction derivatization should be the preferred option. Furthermore, postextraction derivatization strategies are widely employed and can be implemented more easily than in situ derivatization, although the possibility to integrate sampling, enrichment, and derivatization is lost. Several examples of postextraction derivatization can be found in the literature. For instance, six derivatizing agents were evaluated for the postextraction conversion of phenol after its enrichment by SPME, based on alkylation, silylation, and acetylation reactions (Es-haghi et al., 2012). Postextraction derivatization of ammonia and volatile amines after their enrichment in an acid-coated annular denuder has also been carried out by formation of highly fluorescent sulfonatoindoles (Huang et al., 2009).

An interesting case of postextraction derivatization for GC analysis is named *in-port derivatization* (Bizkarguenaga et al., 2013; Wang et al., 2013). Specifically, in-port derivatization consists of the chemical reaction of extracted analytes at the high-temperature GC injection port, in such a way that the obtained derivatives present enhanced volatility, thermal stability, and detectability.

8.4 DESORPTION TECHNIQUES

After enrichment by the previously described sample preparation techniques, analytes need to be released from the retaining media and transferred to the corresponding analytical instrumentation for analysis. Sample preparation techniques can be combined with chromatographic systems by off-line or online approaches. Online systems are advantageous compared with off-line approaches, as they enable improved sensitivity, minimal sample contamination, and the possibility of totally automated analytical systems. However, its optimization can be more difficult to perform, and some compromises may be needed (Hyötyläinen, 2008). Both TD and liquid desorption serve the function of transferring the enriched analytes to the analytical column for separation and subsequent quantification.

8.4.1 Thermal Desorption

TD is currently considered as the technique of choice for sample introduction in GC. In TD, the extractant phase is heated at high temperatures (commonly in the range 200–300°C) in a flow of inert carrier gas and analytes are released and transferred to the chromatographic column by means of the carrier gas stream (Woolfenden, 2012).

The selection of appropriate TD conditions depends, to a high extent, on the physicochemical properties of both the analytes and the retention materials used in sample preparation. Thus, operation temperatures of retaining media should not be exceeded for optimal performance during TD to avoid detection of false-positive compounds.

TD provides significant advantages as a sample introduction technology for GC. Specifically, quantitative transfer of the retained analytes trapped on sorbent or adsorbent media can generally be achieved under optimal conditions due to the dynamic nature of this desorption process (Woolfenden, 2012). In addition, TD completely avoids the use of organic solvents, so maximum sensitivity can be attained with no dilution of the enriched compounds prior to the analysis, and interferences from solvents or solvent impurities are not an issue. Besides, the combination of solventless sample preparation techniques and TD is considered as the greener option when sample pretreatment is required.

In general terms, higher desorption temperatures are needed to efficiently transfer the enriched analytes into the chromatographic system when using adsorbents. This is especially true in the case of TD of strongly adsorbed analytes such as, for example, polar compounds retained in strong adsorbents such as activated carbon (Camel and Caude, 1995). Degradation of retaining materials at high temperatures can give rise to the appearance of degradation peaks in the chromatogram that can mask the ones of target compounds. Furthermore, the retaining media may react with the analytes during TD, resulting in artifact formation or permanent bond to the surface (Baltussen et al., 1999b).

8.4.2 Liquid Desorption

Liquid desorption generally involves solvent extraction of the analytes from the retaining media and subsequent injection of an aliquot of the extract into the chromatographic system for separation and detection. Liquid desorption is commonly carried out with passive sampling, being specially indicated for elution of analytes that strongly interact with the adsorbent, for example, polar compounds retained by activated carbon, as well as for thermally labile compound determination (Dettmer and Engewald, 2003). Liquid desorption can be performed either off-line or online, using a solvent compatible with the analytical instrumentation used. Thus, low-boiling-point organic solvents such as carbon disulfide or dichloromethane are commonly selected. Liquid desorption enables multiple analysis of a given sample. However, a significant dilution of the extract is produced prior to the analysis, and in turn, the sensitivity is reduced in a proportional manner. This fact is of special importance in trace analyte determination. In the latter case, partial evaporation of the solvent is a common practice to concentrate the analytes. However, this evaporation step can lead to additional problems, such as losses of volatile analytes or the introduction of artifacts (Camel and Caude, 1995).

The selection of the solvent is of paramount importance in liquid desorption, as nonquantitative desorption can occur depending on the partition coefficients of analytes between the involved phases. Furthermore, the solvent employed for desorption can interfere in the chromatographic analysis, by masking peaks of interest, by signal quenching, or by producing baseline disturbances (Woolfenden, 2012). Health and safety considerations should also be taken into account when selecting the solvent for liquid desorption.

8.5 SELECTED APPLICATIONS

8.5.1 Outdoor Air Analysis

Both natural and anthropogenic sources contribute significantly to the presence of air pollutants in the atmosphere. Vegetation is the main natural emission source, while chemical pollutants can be released by a number of anthropogenic sources, such as combustion sources; fuel storage and transport; solvent usage; and emissions from industrial activities, landfills, and hazardous waste facilities (Atkinson, 2000).

It has been reported that the presence of VOCs in ambient air shows a paramount role in the rise in ozone levels in the lower part of the atmosphere (troposphere), as well as in the stratospheric ozone depletion (WHO, 2000). While stratospheric ozone protects the earth from the sun's harmful UV irradiation, tropospheric ozone affects human health, vegetation, and materials (Jenkin and Clemitshaw, 2000). Furthermore, VOCs have been identified to contribute to the global greenhouse effect, acid rain, and photochemical smog. Last but not least, toxic and carcinogenic health effects on humans have been attributed to the presence of air pollutants.

HAPs are commonly determined in ambient air by chromatographic analysis after enrichment by using adsorptive-based systems followed by TD or liquid desorption, although several alternative sample preparation techniques described in Section 8.2 have been extensively reported. Selected applications of sample preparation techniques in the determination of different HAPs are summarized in Table 8.2.

8.5.2 Indoor Air Analysis

As early as 1985, Ott identified as an important area of research the one focused on evaluating the degree of exposure of the population to chemicals, as well as determining its sources, nature, extent, and severity (Ott, 1985). Assuming constant air concentration of an organic compound during the time that a person occupies a microenvironment, the integrated exposure can be calculated as the sum of the product of the concentration in each microenvironment and the time spent by a person in that microenvironment (Ott, 1985).

Inhalation of indoor air has been identified as the main exposure pathway to volatile contaminants in developed countries (Brown et al., 1994). In fact, it has been estimated that people spend, on average, more than 20 h per day indoors in industrial countries, considering time spent both at home and at the workplace or school (Hellweg et al., 2009). Importantly, the intake fractions in both occupational and household indoors have been reported to be of several orders of magnitude higher than those in the ambient environment for many pollutants, mainly due to the presence of significant indoor emission sources and lower dilution volumes (Hellweg et al., 2009). Building and construction materials, indoor equipment, outdoor air (via infiltration and ventilation), and human activities are among the potential sources of various HAPs (Salthammer and Bahadir, 2009; Marć et al., 2012). Several compounds detected in indoor air are known to cause many respiratory and circulatory diseases. Thus, the assessment of indoor emissions is needed for improving indoor air quality (Marć et al., 2012).

Several sample preparation techniques, including cryotrapping systems, impingers, OTTs, SPME, NTDs, accelerated SPDE, GPE, single- and multibed adsorption systems, or membrane-based extraction techniques, have been applied for the monitoring of HAPs in a large variety of indoor microenvironments. Selected applications of sample preparation techniques employed in the analysis of indoor air samples are collected in Table 8.3.

8.5.3 Breath Analysis

Breath air consists mainly of a mixture of nitrogen, oxygen, carbon dioxide, water vapor, and inert gases. A small fraction of the whole breath air (<100 ppm by volume) comprises a complex

TABLE 8.2
Selected Applications of Sample Preparation Methods for the Analysis of Outdoor Air Samples

Sample Preparation Technique	Retaining Media	Analytical Technique	Target Compounds	Sampling Site	Reference
Cryogenic trap	Glass wool trap (−186°C or −196°C)	GC-FPD	Reduced sulfur compounds	Tropical urban area	Campos et al. (2010)
Cryogenic trap	Glass bead trap (−170°C)	GC-FID and GC-MS	Alkanes, alkenes, alkynes, aromatic hydrocarbons, halogenated hydrocarbons, ethers, esters, ketones, furans, and VSCs	Urban area	Wang et al. (2012)
Cryogenic trap	Glass bead trap (−170°C)	GC-MS and GC-FID	MTBE, alkanes, alkenes, alkynes, aromatic hydrocarbons, and halogenated hydrocarbons	Urban areas	Chang et al. (2003)
Annular denuder	DNPH and phosphoric acid in acetonitrile	HPLC-UV	Formaldehyde and acetaldehyde	Urban area	Komazaki et al. (1999)
Annular denuder	Citric acid in water	HPLC-FLD	Ammonia and four alkylamines	Urban and semi-rural areas	Huang et al. (2009)
NTD	Tenax TA; Carbopack X	GC-MS	Aromatic hydrocarbons, aldehydes, ketones, and furans	Pedestrian area	Alonso et al. (2011)
GPE	PDMS	GC-MS and GC-ECD	PAHs and nitro-PAHs	Rural area	Baltussen et al. (1997b)
Adsorption system	Activated carbon; Carbotrap B	GC-MS and GC-FID	Aromatic hydrocarbons	Urban area	Czaplicka and Klejnowski (2002)
Adsorption system	Chromosorb 102	GC-FPD	Organophosphorus pesticides	Area nearby a pesticide-applied farm	Kawahara et al. (2005)
Adsorption system	Tenax GR and Carbosieve SIII	GC-MS	Alkanes, aromatic hydrocarbons, PAHs, alcohols, ketones, halogenated hydrocarbons, and terpenoids	Industrial, urban, and suburban areas	Jia et al. (2008)
Adsorption system	Activated carbon; Tenax TA and Carbograph 1TD	GC-MS	Alkanes, alkenes, aromatic hydrocarbons, PAHs, alcohols, ethers, furans, nitriles, and halogenated hydrocarbons	Area nearby a petrochemical complex	Ramírez et al. (2010)
Adsorption system	Carbograph 4	GC-MS	Benzene	Urban areas	Król et al. (2012)
SPMD	Triolein	GC-MS	PAHs and PCBs	Area nearby a wind tunnel	Söderström and Bergqvist (2004)

TABLE 8.3
Selected Applications of Sample Preparation Methods for the Analysis of Indoor Air Samples

Sample Preparation Technique	Retaining Media	Analytical Technique	Target Compounds	Sampling Sites	Reference
Cryogenic trap	Glass bead trap (−170°C)	GC-MS	Aromatic hydrocarbons, chlorofluorocarbons, and chlorinated hydrocarbons	Buildings in nuclear power plants	Hsieh et al. (2006)
Impinger	Water	HPLC-UV and IC	Thioglycolic acid, dithioglycolic acid, and ammonia	Beauty salon	Oikawa et al. (2012)
OTT	PDMS	GC-FID and GC-MS	Aromatic hydrocarbons and acetate esters	Woodworking shop	Dudek et al. (2002b)
SPME	PDMS	GC-MS	Alkanes, aromatics, and chlorinated hydrocarbons	Flat rooms and chemical laboratories	Gorlo et al. (1999)
SPME	PDMS; PEG	GC-MS	Phenol	Biologics production areas	Es-haghi et al. (2012)
SPME	PDMS	GC-NPD	Organophosphate triesters	Lecture room	Isetun et al. (2004)
NTD	MA/EGDMA; Tenax TA; activated carbon; DVB	GC-MS	Aromatic hydrocarbons	Rooms	Ueta et al. (2012)
NTD	Tenax TA; Carbopack X	GC-MS	Alkanes, aromatic hydrocarbons, acetate esters, aldehydes, ketones, furans, and chlorinated hydrocarbons	Chemical laboratories	Alonso et al. (2011)
SPDE	PDMS	GC-FID	Toluene	Room	Van Durme et al. (2007)
GPE	PDMS	GC-MS	Nicotine	Hospital rooms	Baltussen et al. (1999a)
Adsorption system	Chromosorb 102	GC-FPD	Organophosphorus pesticides	Homes and childcare facilities in an agricultural community	Kawahara et al. (2005)
Adsorption system	Tenax GR and Carbosieve SIII	GC-MS	Alkanes, aromatic hydrocarbons, PAHs, alcohols, ketones, halogenated hydrocarbons, and terpenoids	Residences in industrial, urban, and suburban areas	Jia et al. (2008)
Adsorption system	Carbograph 4	GC-MS	α-Pinene, 3-carene, and D-limonene	Residences in urban areas	Król et al. (2014)
Adsorption trap/SPME	Tenax TA/PA; PDMS; PDMS/DVB; CAR/PDMS; DVB/CAR/PDMS	GC-MS	Synthetic musks	Residential area	Regueiro et al. (2009)
MESI	Tenax	GC-FID	α-Pinene, eucalyptol, and γ-terpinene	Greenhouse	Liu et al. (2004)

TABLE 8.4
Selected Applications of Sample Preparation Methods for the Analysis of Breath Air Samples

Sample Preparation Technique	Retaining Media	Analytical Technique	Target Compounds	Application	Reference
Three-stage cryogenic trap	Glass bead (–150°C), Tenax (–30°C), and capillary focusing (–185°C) traps	GC-MS	Methanediol, dimethyl sulfide, and dimethyl disulfide	Breath air analysis	Ochiai et al. (2001)
SPME	PDMS; PA; CAR/PDMS; PEG/DVB	GC-MS	Ethanol, acetone, and isoprene	Breath air analysis	Grote and Pawliszyn (1997)
SPME	CAR/PDMS	GC-MS	Alkanes, alkenes, alkynes, aromatic hydrocarbons, carboxylic acids, alcohols, and ketones	Screen for chemical markers of colorectal cancer	Wang et al. (2014)
SPME	CAR/PDMS	GC-MS	Alkanes, alkenes, aromatic hydrocarbons, acetate esters, alcohols, aldehydes, ketones, amines, and VSCs	Screen for chemical markers of lung cancer	Bajtarevic et al. (2009)
SPME	DVB/CAR/PDMS	GC×GC-TOF-MS	Alkanes	Screen for chemical markers of allergic asthma	Caldeira et al. (2012)
NTD	Tenax, Carbopack X, and Carboxen 1000	GC-MS	Alkanes, aromatic hydrocarbons, aldehydes, ketones, and VSCs	Breath air analysis	Mieth et al. (2009)
NTD	Single bed: MA/EGDMA; triple bed: DVB, Carbopack X, and Carboxen 1000	GC-MS	Alkanes, alkenes, aromatic hydrocarbons, alcohols, aldehydes, ketones, and nitriles	Breath air analysis	Trefz et al. (2013)
NTD	Tenax TA; Carbopack X	GC-MS	Aromatic hydrocarbons, aldehydes, ketones, and furans	Breath air analysis of a smoker	Alonso et al. (2011)
Adsorption system	Carbopack Y, Carbopack B, Carbopack X, and Carboxen 1000	GC-FID	Alkanes, alkenes, aromatic hydrocarbons, alcohols, aldehydes, and ketones	Breath air analysis	Sanchez and Sacks (2003)

(Continued)

TABLE 8.4 (CONTINUED)
Selected Applications of Sample Preparation Methods for the Analysis of Breath Air Samples

Sample Preparation Technique	Retaining Media	Analytical Technique	Target Compounds	Application	Reference
Adsorption system	Carbopack B, Carbopack X, and Carboxen 1000	GC×GC-TOF-MS	Alkanes, alkenes, aromatic hydrocarbons, alcohols, aldehydes and ketones, and halogenated compounds	Breath air analysis, screen for biomarkers of active smoking, and occupational exposure	Sanchez and Sacks (2006)
MESI	XAD-2 resin	GC-FID and GC-IMS	Ethanol, acetone, chloroform, and BTEX	Breath air analysis after smoking and after swimming	Lord et al. (2002)
MESI	Carboxen	GC-FID	Methane and ethane	Breath air analysis	Yu and Pawliszyn (2004)

Note: BTEX, benzene, toluene, ethylbenzene, and xylene; CAR, Carboxen; CE, capillary electrophoresis; DNPH, 2,4-dinitrophenylhydrazine; DVB, divinylbenzene; ECD, electron capture detection; EGDMA, ethylene glycol dimethacrylate; EGPE, equilibrium gum-phase extraction; EPA, Environmental Protection Agency; FID, flame ionization detection; FLD, fluorescence detection; FPD, flame photometric detection; GC, gas chromatography; GPE, gum phase extraction; HAPs, hazardous air pollutants; HPLC, high-performance liquid chromatography; IC, ion chromatography; IMS, ion mobility spectrometry; INCAT, inside needle capillary absorption trap; MA, methacrylic acid; MESI, membrane extraction with a sorbent interface; MS, mass spectrometry; MTBE, methyl *tert*-butyl ether; NPD, nitrogen-phosphorus detection; NTD, needle trap device; NVINCs, nonvolatile inorganic compounds; NVOCs, nonvolatile organic compounds; OTT, open tubular trapping; PA, poly(acrylate); PAHs, polycyclic aromatic hydrocarbons; PCBs, polychlorinated biphenyls; PDMS, poly(dimethylsiloxane); PEG, polyethylene glycol; SPDE, solid-phase dynamic extraction; SPMDs, semipermeable membrane devices; SPME, solid-phase microextraction; SVINCs, semivolatile inorganic compounds; SVOCs, semivolatile organic compounds; TOF, time of flight mass analyzer; UV, ultraviolet detection; VINCs, volatile inorganic compounds; VOCs, volatile organic compounds; VSCs, volatile sulfur compounds; VVINCs, very volatile inorganic compounds; VVOCs, very volatile organic compounds.

combination of compounds present at different concentration levels (Alonso and Sanchez, 2013). From them, ammonia, acetone, isoprene, methanol, ethanol, propanol, and acetaldehyde are present at relatively high concentrations in exhaled breath (Buszewski et al., 2007).

The chemical composition of exhaled air can vary depending on the metabolic processes that take place in the human body. Breath constituents provide crucial information for the evaluation of human health risks, so that the analysis of breath air is becoming an important issue for early diagnosis of diseases, including lung cancer, inflammatory lung disease, diabetes, liver cirrhosis, or renal failure. In addition, the exposure to toxic compounds or environmental pollutants can also be evaluated by breath analysis (Alonso and Sanchez, 2013).

Only a fraction of exhaled breath is of importance for clinical purposes. Exhaled breath is composed of "dead space air," which comes from the mouth, nose, pharynx, trachea, and bronchi, and "alveolar air," which comes from the lungs. Alveolar sampling method is mandatory for biomarking recognition in breath analysis, as this fraction includes gaseous exchange between air and blood (Alonso and Sanchez, 2013). Commonly, alveolar sampling is performed by controlling the partial pressure of expired CO_2 during breath gas sampling by means of a capnograph.

Breath analysis provides a noninvasive, painless, and inexpensive way to identify potential volatile markers for early detection of diseases. A number of chemical substances have been identified in the literature as breath biomarkers. However, their significance is not always clear. This can be attributed to the limited knowledge on the metabolic pathways that give rise to the emission of these specific VOCs, the high levels of interferences, and the inter- and intraindividual variability (Di Francesco et al., 2005).

As VOCs are present in human breath air at low concentrations, preconcentration is commonly required for their determination with appropriate sensitivity and precision. Cryogenic enrichment, SPME, NTD, single- and multibed adsorption systems, and membrane-based extraction systems have been proposed in the literature for the enrichment of VOCs in exhaled air before their determination, mainly by GC-MS. Representative applications of these sample preparation techniques are detailed in Table 8.4.

8.6 CONCLUDING REMARKS

The determination of HAPs in air samples has become an important area of concern on both human health and environment. The applications of sample pretreatment techniques in air analysis are diverse, including detection of relevant HAPs in ambient and workspace air, measurement of chemical emission levels, monitoring personal exposure to HAPs, or identification of chemical biomarkers for diagnostic purposes.

The advances achieved in sample preparation have significantly contributed to the state of the art in air pollution monitoring. Nevertheless, further innovations and developments are yet expected on the simplification, automation, miniaturization, and portability of analytical systems for determining air constituents. Furthermore, the introduction of novel materials could potentially overcome certain limitations of conventionally used sorbents and adsorbents and importantly contribute to solve challenging issues of air monitoring. Much work is yet expected to be done in this challenging area, and sample preparation methods will presumably have a paramount role on future developments.

ACKNOWLEDGMENT

F. Pena-Pereira thanks Xunta de Galicia for financial support as a postdoctoral researcher of the I2C program.

REFERENCES

Alonso, M., Sanchez, J.M. 2013. Analytical challenges in breath analysis and its application to exposure monitoring. *TrAC Trends in Analytical Chemistry* 44: 78–89.

Alonso, M., Godayol, A., Antico, E., Sanchez, J.M. 2011. Needle microextraction trap for on-site analysis of airborne volatile compounds at ultra-trace levels in gaseous samples. *Journal of Separation Science* 34: 2705–2711.

Arthur, C.L., Pawliszyn, J. 1990. Solid phase microextraction with thermal desorption using fused silica optical fibers. *Analytical Chemistry* 62: 2145–2148.

Atkinson, R. 2000. Atmospheric chemistry of VOCs and NO. *Atmospheric Environment* 34: 2063–2101.

Bajtarevic, A., Ager, C., Pienz, M., Klieber, M., Schwarz, K. et al. 2009. Noninvasive detection of lung cancer by analysis of exhaled breath. *BMC Cancer* 9: 348.

Baltussen, E., Janssen, H.-G., Sandra, P., Cramers, C.A. 1997a. A novel type of liquid/liquid extraction for the preconcentration of organic micropollutants from aqueous samples: Application to the analysis of PAH's and OCP's in water. *HRC Journal of High Resolution Chromatography* 20: 395–399.

Baltussen, E., Janssen, H.-G., Sandra, P., Cramers, C.A. 1997b. A new method for sorptive enrichment of gaseous samples: Application in air analysis and natural gas characterization. *HRC Journal of High Resolution Chromatography* 20: 385–393.

Baltussen, E., David, F., Sandra, P., Janssen, H.-G., Cramers, C.A. 1998. Sorption tubes packed with polydimethylsiloxane: A new and promising technique for the preconcentration of volatiles and semi-volatiles from air and gaseous samples. *HRC Journal of High Resolution Chromatography* 21: 332–340.

Baltussen, E., Boer, A., Sandra, P., Janssen, H.-G., Cramers, C. 1999a. Monitoring of nicotine in air using sorptive enrichment on polydimethylsiloxane and TD-CGC-NPD. *Chromatographia* 49: 520–524.

Baltussen, E., David, F., Sandra, P., Janssen, H.-G., Cramers C. 1999b. Equilibrium sorptive enrichment on poly(dimethylsiloxane) particles for trace analysis of volatile compounds in gaseous samples. *Analytical Chemistry* 71: 5193–5198.

Baltussen, E., Cramers, C.A., Sandra, P.J.F. 2002. Sorptive sample preparation—A review. *Analytical and Bioanalytical Chemistry* 373: 3–22.

Bizkarguenaga, E., Iparragirre, A., Navarro, P., Olivares, M., Prieto, A. et al. 2013. In-port derivatization after sorptive extractions. *Journal of Chromatography A* 1296: 36–46.

Bonn, J., Redeby, J., Roeraade, J. 2009. Mixed sorbent phases for thick film open tubular traps. *Journal of Chromatographic Science* 47: 297–303.

Bourdin, D., Desauziers, V. 2014. Development of SPME on-fiber derivatization for the sampling of formaldehyde and other carbonyl compounds in indoor air. *Analytical and Bioanalytical Chemistry* 406: 317–328.

Brown, S.K., Sim, M.R., Abramson, M.J. 1994. Concentrations of volatile organic compounds in indoor air—A review. *Indoor Air* 4: 123–134.

Burger, B.V., Munro, Z. 1986. Headspace gas analysis: Quantitative trapping and thermal desorption of volatiles using fused-silica open tubular capillary traps. *Journal of Chromatography A* 370: 449–464.

Buszewski, B., Martyna, K., Ligor, T., Amann, A. 2007. Human exhaled air analytics: Biomarkers of diseases. *Biomedical Chromatography* 566: 553–566.

Caldeira, M., Perestrelo, R., Barros, A.S., Bilelo, M.J, Morête, A. et al. 2012. Allergic asthma exhaled breath metabolome: A challenge for comprehensive two-dimensional gas chromatography. *Journal of Chromatography A* 1254: 87–97.

Camel, V., Caude, M. 1995. Trace enrichment methods for the determination of organic pollutants in ambient air. *Journal of Chromatography A* 710: 3–19.

Campos, V.P., Oliveira, A.S., Cruz, L.P.S., Borges, J., Tavares, T.M. 2010. Optimization of parameters of sampling and determination of reduced sulfur compounds using cryogenic capture and gas chromatography in tropical urban atmosphere. *Microchemical Journal* 96: 283–289.

Chang, C.-C., Lo, S.-J., Lo, J.-G., Wang, J.-L. 2003. Analysis of methyl *tert*-butyl ether in the atmosphere and implications as an exclusive indicator of automobile exhaust. *Atmospheric Environment* 37: 4747–4755.

Chromtech. 2014. SPDE—The magic needle. http://www.chromtech.de/en/content/SPDE/~nm.18~nc.80/SPDE-.-The-Magic-Needle.html (accessed September 12, 2014).

Czaplicka, M., Klejnowski, K. 2002. Determination of volatile organic compounds in ambient air comparison of methods. *Journal of Chromatography A* 976: 369–376.

Demeestere, K., Dewulf, J., De Witte, B., Van Langenhove, H. 2007. Sample preparation for the analysis of volatile organic compounds in air and water matrices. *Journal of Chromatography A* 1153: 130–144.

Dettmer, K., Engewald, W. 2002. Adsorbent materials commonly used in air analysis for adsorptive enrichment and thermal desorption of volatile organic compounds. *Analytical and Bioanalytical Chemistry* 373: 490–500.

Dettmer, K., Engewald, W. 2003. Ambient air analysis of volatile organic compounds using adsorptive enrichment. *Chromatographia* 57: S339–S347.

Di Francesco, F., Fuoco, R., Trivella, M.G., Ceccarini, A. 2005. Breath analysis: Trends in techniques and clinical applications. *Microchemical Journal* 79: 405–410.

Dietz, C., Sanz, J., Cámara, C. 2006. Recent developments in solid-phase microextraction coatings and related techniques. *Journal of Chromatography A* 1103: 183–192.

Dudek, M., Kloskowski, A., Wolska, L., Pilarczyk, M., Namieśnik, J. 2002a. Using different types of capillary chromatographic columns as denudation traps: A comparison of sorption properties. *Journal of Chromatography A* 977: 115–123.

Dudek, M., Wolska, L., Pilarczyk, M., Zygmunt, B., Namieśnik, J. 2002b. The application of an open tubular trap in analysis of organic air pollutants. *Chemosphere* 48: 913–918.

EPA. 1999. *Compendium of Methods for the Determination of Toxic Organic Compounds in Ambient Air*. EPA, Cincinnati, OH.

EPA. 2013. The original list of hazardous air pollutants. http://www.epa.gov/ttn/atw/188polls.html (accessed September 12, 2014).

Es-haghi, A., Baghernejad, M., Bagheri, H. 2012. In situ solid-phase microextraction and post on-fiber derivatization combined with gas chromatography-mass spectrometry for determination of phenol in occupational air. *Analytica Chimica Acta* 742: 17–21.

Fish, B.R., Durham, J.L. 1971. Diffusion coefficient of SO_2 in air. *Environmental Letters* 2: 13–21.

Gawryś, M., Fastyn, P., Gawlowski, J., Gierczak, T., Niedzielski, J. 2001. Prevention of water vapour adsorption by carbon molecular sieves in sampling humid gases. *Journal of Chromatography A* 933: 107–116.

Górecki, T., Namieśnik, J. 2002. Passive sampling. *TrAC Trends in Analytical Chemistry* 21: 276–291.

Gorlo, D., Zygmunt, B., Dudek, M., Jaszek, A., Pilarczyk, M. et al. 1999. Application of solid-phase microextraction to monitoring indoor air quality. *Fresenius' Journal of Analytical Chemistry* 363: 696–699.

Grote, C., Pawliszyn, J. 1997. Solid-phase microextraction for the analysis of human breath. *Analytical Chemistry* 69: 587–596.

Gurjar, B.R., Molina, L.T., Ojha, C.S.P. 2010. *Air Pollution: Health and Environmental Impacts*. CRC Press, Boca Raton, FL.

Harper, M. 2000. Sorbent trapping of volatile organic compounds from air. *Journal of Chromatography A* 885: 129–151.

Heidari, M., Bahrami, A., Ghiasvand, A.R., Shahna, F.G., Soltanian, A.R. 2012. A novel needle trap device with single wall carbon nanotubes sol–gel sorbent packed for sampling and analysis of volatile organohalogen compounds in air. *Talanta* 101: 314–321.

Hellweg, S., Demou, E., Bruzzi, R., Meijer, A., Rosenbaum, R.K. et al. 2009. Critical review integrating human indoor air pollutant exposure within life cycle impact assessment. *Environmental Science & Technology* 43: 1670–1679.

Helmig, D. 1997. Ozone removal techniques in the sampling of atmospheric volatile organic trace gases. *Atmospheric Environment* 31: 3635–3651.

Ho, S.S.H., Yu, J.Z. 2002. Feasibility of collection and analysis of airborne carbonyls by on-sorbent derivatization and thermal desorption. *Analytical Chemistry* 74: 1232–1240.

Hsieh, L.-L., Chang, C.-C., Sree, U., Lo, J.-G. 2006. Determination of volatile organic compounds in indoor air of buildings in nuclear power plants, Taiwan. *Water, Air, and Soil Pollution* 170: 107–121.

Huang, G., Hou, J., Zhou, X. 2009. A measurement method for atmospheric ammonia and primary amines based on aqueous sampling, OPA derivatization and HPLC analysis. *Environmental Science & Technology* 43: 5851–5856.

Huckins, J.N., Tubergen, M.W., Manuweera, G.K. 1990. Semipermeable membrane devices containing model lipid: A new approach to monitoring the bioavailability of lipophilic contaminants and estimating their bioconcentration potential. *Chemosphere* 20: 533–552.

Hyötyläinen, T. 2008. On-line coupling of extraction with gas chromatography. *Journal of Chromatography A* 1186: 39–50.

Isetun, S., Nilsson, U., Colmsjö, A. 2004. Evaluation of solid-phase microextraction with PDMS for air sampling of gaseous organophosphate flame-retardants and plasticizers. *Analytical and Bioanalytical Chemistry* 380: 319–324.

Jakubowska, N., Zygmunt, B., Polkowska, Z., Zabiegała, B., Namieśnik, J. 2009. Sample preparation for gas chromatographic determination of halogenated volatile organic compounds in environmental and biological samples. *Journal of Chromatography A* 1216: 422–441.

Jenkin, M.E., Clemitshaw, K.C. 2000. Ozone and other secondary photochemical pollutants: Chemical processes governing their formation in the planetary boundary layer. *Atmospheric Environment* 34: 2499–2527.

Jia, C., Batterman, S., Godwin, C. 2008. VOCs in industrial, urban and suburban neighborhoods, part 1: Indoor and outdoor concentrations, variation, and risk drivers. *Atmospheric Environment* 42: 2083–2100.

Jochmann, M.A., Yuan, X., Schmidt, T.C. 2007. Determination of volatile organic hydrocarbons in water samples by solid-phase dynamic extraction. *Analytical and Bioanalytical Chemistry* 387: 2163–2174.

Jönsson, J.A., Mathiasson, L. 2001. Membrane extraction in analytical chemistry. *Journal of Separation Science* 24: 495–507.

Kataoka, H., Ishizaki, A., Nonaka, Y., Saito, K. 2009. Developments and applications of capillary microextraction techniques: A review. *Analytica Chimica Acta* 655: 8–29.

Kawahara, J., Horikoshi, R., Yamaguchi, T., Kumagai, K., Yanagisawa, Y. 2005. Air pollution and young children's inhalation exposure to organophosphorus pesticide in an agricultural community in Japan. *Environment International* 31: 1123–1132.

Keith, L.H., Gron, L.U., Young, J.L. 2007. Green analytical methodologies. *Chemical Reviews* 107: 2695–2708.

Kloskowski, A., Pilarczyk, M., Namieśnik, J. 2002. Denudation—A convenient method of isolation and enrichment of analytes. *Critical Reviews in Analytical Chemistry* 32: 301–335.

Kloskowski, A., Pettersson, J., Roeraade, J. 2004. Thick film traps with an irregular film. *Journal of Chromatography A* 1035: 159–165.

Kloskowski, A., Chrzanowski, W., Pilarczyk, M., Namiesnik, J. 2007. Modern techniques of sample preparation for determination of organic analytes by gas chromatography. *Critical Reviews in Analytical Chemistry* 37: 15–38.

Komazaki, Y., Hiratsuka, M., Narita, Y., Tanaka, S., Fujita, T. 1999. The development of an automated continuous measurement system for the monitoring of HCHO and CH_3CHO in the atmosphere by using an annular diffusion scrubber coupled to HPLC. *Fresenius' Journal of Analytical Chemistry* 363: 686–695.

Koziel, J.A., Odziemkowski, M., Pawliszyn, J. 2001a. Sampling and analysis of airborne particulate matter and aerosols using in-needle trap and SPME fiber devices. *Analytical Chemistry* 73: 47–54.

Koziel, J.A., Noah, J., Pawliszyn, J. 2001b. Field sampling and determination of formaldehyde in indoor air with solid-phase microextraction and on-fiber derivatization. *Environmental Science & Technology* 35: 1481–1486.

Król, S., Zabiegała, B., Namieśnik, J. 2010. Monitoring VOCs in atmospheric air: II. Sample collection and preparation. *TrAC Trends in Analytical Chemistry* 29: 1101–1112.

Król, S., Zabiegała, B., Namieśnik, J. 2011. Monitoring and analytics of semivolatile organic compounds (SVOCs) in indoor air. *Analytical and Bioanalytical Chemistry* 400: 1751–1769.

Król, S., Zabiegała, B., Namiésnik, J. 2012. Measurement of benzene concentration in urban air using passive sampling. *Analytical and Bioanalytical Chemistry* 403: 1067–1082.

Król, S., Namieśnik, J., Zabiegała, B. 2014. α-Pinene, 3-carene and D-limonene in indoor air of polish apartments: The impact on air quality and human exposure. *Science of the Total Environment* 468–469: 985–995.

Lavilla, I., Romero, V., Costas, I., Bendicho, C. 2014. Greener derivatization in analytical chemistry. *TrAC Trends in Analytical Chemistry* 61: 1–10.

Lipinski, J. 2001. Automated solid phase dynamic extraction—Extraction of organics using a wall coated syringe needle. *Fresenius' Journal of Analytical Chemistry* 369: 57–62.

Little, J.L. 1999. Artifacts in trimethylsilyl derivatization reactions and ways to avoid them. *Journal of Chromatography A* 844: 1–22.

Liu, X., Pawliszyn, R., Wang, L., Pawliszyn, J. 2004. On-site monitoring of biogenic emissions from *Eucalyptus dunnii* leaves using membrane extraction with sorbent interface combined with a portable gas chromatograph system. *The Analyst* 129: 55–62.

Lord, H., Yu, Y., Segal, A., Pawliszyn, J. 2002. Breath analysis and monitoring by membrane extraction with sorbent interface. *Analytical Chemistry* 74: 5650–5657.

Lord, H.L., Zhan, W., Pawliszyn, J. 2010. Fundamentals and applications of needle trap devices: A critical review. *Analytica Chimica Acta* 677: 3–18.

Marć, M., Zabiegała, B., Namieśnik, J. 2012. Testing and sampling devices for monitoring volatile and semivolatile organic compounds in indoor air. *TrAC Trends in Analytical Chemistry* 32: 76–86.

Marć, M., Zabiegała, B., Namieśnik, J. 2013. Miniaturized passive emission chambers for in situ measurement of emissions of volatile organic compounds. *Critical Reviews in Analytical Chemistry* 43: 55–61.

Martos, P.A., Pawliszyn, J. 1998. Sampling and determination of formaldehyde using solid-phase microextraction with on-fiber derivatization. *Analytical Chemistry* 70: 2311–2320.
McComb, M.E., Oleschuk, R.D., Giller, E., Gesser, H.D. 1997. Microextraction of volatile organic compounds using the inside needle capillary adsorption trap (INCAT) device. *Talanta* 44: 2137–2143.
Michulec, M., Wardenki, W., Partyka, M., Namieśnik, J. 2005. Analytical techniques used in monitoring of atmospheric air pollutants. *Critical Reviews in Analytical Chemistry* 35: 117–133.
Mieth, M., Kischkel, S., Schubert, J.K., Hein, D., Miekisch, W. 2009. Multibed needle trap devices for on site sampling and preconcentration of volatile breath biomarkers. *Analytical Chemistry* 81: 5851–5857.
Musshoff, F., Lachenmeier, D.W., Kroener, L., Madea, B. 2002. Automated headspace solid-phase dynamic extraction for the determination of amphetamines and synthetic designer drugs in hair samples. *Journal of Chromatography A* 958: 231–238.
Ochiai, N., Takino, M., Daishima, S., Cardin, D.B. 2001. Analysis of volatile sulphur compounds in breath by gas chromatography–mass spectrometry using a three-stage cryogenic trapping preconcentration system. *Journal of Chromatography B, Biomedical Sciences and Applications* 762: 67–75.
Ockenden, W.A., Prest, H.F., Thomas, G.O., Sweetman, A., Jones, K.C. 1998a. Passive air sampling of PCBs: Field calculation of atmospheric sampling rates by triolein-containing semipermeable membrane devices. *Environmental Science & Technology* 32: 1538–1543.
Ockenden, W.A., Sweetman, A.J., Prest, H.F., Steinnes, E., Jones, K.C. 1998b. Toward an understanding of the global atmospheric distribution of persistent organic pollutants: The use of semipermeable membrane devices as time-integrated passive samplers. *Environmental Science & Technology* 32: 2795–2803.
Oikawa, D., Takeuchi, W., Murata, S., Takahashi, K., Sekine, Y. 2012. Measurement of concentrations of thioglycolic acid, dithiodiglycolic acid and ammonia in indoor air of a beauty salon. *Journal of Occupational Health* 54: 370–375.
Ott, W.R. 1985. Total human exposure. *Environmental Science & Technology* 19: 880–886.
Ouyang, G., Pawliszyn, J. 2006. Recent developments in SPME for on-site analysis and monitoring. *TrAC Trends in Analytical Chemistry* 25: 692–703.
Partyka, M., Zabiegała, B., Namieśnik, J., Przyjazny, A. 2007. Application of passive samplers in monitoring of organic constituents of air. *Critical Reviews in Analytical Chemistry* 37: 51–78.
Pawliszyn, J.B. 1997. Method and Device for Solid Phase Microextraction and Desorption. US Patent 5691206.
Pawliszyn, J.B. 2002a. Needle Trap. US Patent 6481301 B2.
Pawliszyn, J. 2002b. Membrane extraction. In *Sampling and Sample Preparation for Field and Laboratory*, ed. Pawliszyn, J. Elsevier, Amsterdam, pp. 479–502.
Pawliszyn, J. 2011. *Handbook of Solid Phase Microextraction*. Elsevier, Amsterdam.
Pettersson, J., Aldaeus, F., Kloskowski, A., Roeraade, J. 2004. Ultra thick film open tubular traps with an increased inner diameter. *Journal of Chromatography A* 1047: 93–99.
Petty, J.D., Huckins, J.N., Zajicek, J.L. 1993. Application of semipermeable membrane devices (SPMDs) as passive air samplers. *Chemosphere* 27: 1609–1624.
Pinnau, I., Freeman, B.D. 2000. Formation and modification of polymeric membranes: Overview. In *Membrane Formation and Modification*, ed. Pinnau, I., Freeman, B.D. American Chemical Society, Washington, DC, pp. 1–22.
Pokorska, O., Dewulf, J., Van Langenhove, H. 2011. Accelerated solid-phase dynamic extraction for the analysis of biogenic volatile organic compounds in air. *International Journal of Environmental Analytical Chemistry* 91: 1206–1217.
Poole, C.F. 2013. Microreactions in separation science: Reagents and techniques; Foreword. *Journal of Chromatography A* 1296: 1.
Qin, T., Xu, X., Polák, T., Pacáková, V., Stulik, K. et al. 1997. A simple method for the trace determination of methanol, ethanol, acetone and pentane in human breath and in the ambient air by preconcentration on solid sorbents followed by gas chromatography. *Talanta* 44: 1683–1690.
Ramírez, N., Cuadras, A., Rovira, E., Borrull, F., Marcé, R.M. 2010. Comparative study of solvent extraction and thermal desorption methods for determining a wide range of volatile organic compounds in ambient air. *Talanta* 82: 719–727.
Ras, M.R., Borrull, F., Marcé, R.M. 2009. Sampling and preconcentration techniques for determination of volatile organic compounds in air samples. *TrAC Trends in Analytical Chemistry* 28: 347–361.
Raschdorf, F. 1978. Rapid measurements in the ppm and ppb region. *Chimia* 32: 478–483.
Regueiro, J., Garcia-Jares, C., Llompart, M., Lamas, J.P., Cela, R. 2009. Development of a method based on sorbent trapping followed by solid-phase microextraction for the determination of synthetic musks in indoor air. *Journal of Chromatography A* 1216: 2805–2815.

Risticevic, S., Niri, V.H., Vuckovic, D., Pawliszyn, J. 2009. Recent developments in solid-phase microextraction. *Analytical and Bioanalytical Chemistry* 393: 781–795.

Rosenfeld, J.M. 2003. Derivatization in the current practice of analytical chemistry. *Trends in Analytical Chemistry* 22: 785–798.

Salthammer, T., Bahadir, M. 2009. Occurrence, dynamics and reactions of organic pollutants in the indoor environment. *CLEAN—Soil, Air, Water* 37: 417–435.

Sanchez, J.M., Sacks, R.D. 2003. GC analysis of human breath with a series-coupled column ensemble and a multibed sorption trap. *Analytical Chemistry* 75: 2231–2336.

Sanchez, J.M., Sacks, R.D. 2006. Development of a multibed sorption trap, comprehensive two-dimensional gas chromatography, and time-of-flight mass spectrometry system for the analysis of volatile organic compounds in human breath. *Analytical Chemistry* 78: 3046–3054.

Seethapathy, S., Górecki, T., Li, X. 2008. Passive sampling in environmental analysis. *Journal of Chromatography A* 1184: 234–253.

Söderström, H.S., Bergqvist, P.-A. 2004. Passive air sampling using semipermeable membrane devices at different wind-speeds in situ calibrated by performance reference compounds. *Environmental Science & Technology* 38: 4828–4234.

Spietelun, A., Kloskowski, A., Chrzanowski, W., Namieśnik, J. 2013. Understanding solid-phase microextraction: Key factors influencing the extraction process and trends in improving the technique. *Chemical Reviews* 113: 1667–1685.

Spietelun, A., Pilarczyk, M., Kloskowski, A., Namieśnik, J. 2010. Current trends in solid-phase microextraction (SPME) fibre coatings. *Chemical Society Reviews* 39: 4524–4537.

Trefz, P., Rösner, L., Hein, D., Schubert, J.K., Miekisch, W. 2013. Evaluation of needle trap micro-extraction and automatic alveolar sampling for point-of-care breath analysis. *Analytical and Bioanalytical Chemistry* 405: 3105–3115.

Tsai, S.-W., Wu, K.-K. 2003. Determination of ethylene oxide by solid-phase microextraction device with on-fiber derivatization. *Journal of Chromatography A* 991: 1–11.

Ueta, I., Mizuguchi, A., Fujimura, K., Kawakubo, S., Saito, Y. 2012. Novel sample preparation technique with needle-type micro-extraction device for volatile organic compounds in indoor air samples. *Analytica Chimica Acta* 746: 77–83.

Van Durme, J., Demeestere, K., Dewulf, J., Ronsse, F., Braeckman, L. et al. 2007. Accelerated solid-phase dynamic extraction of toluene from air. *Journal of Chromatography A* 1175: 145–153.

Wang, D.K.W., Austin, C.C. 2006a. Determination of complex mixtures of volatile organic compounds in ambient air: An overview. *Analytical and Bioanalytical Chemistry* 386: 1089–1098.

Wang, D.K.W., Austin, C.C. 2006b. Determination of complex mixtures of volatile organic compounds in ambient air: Canister methodology. *Analytical and Bioanalytical Chemistry* 386: 1099–1120.

Wang, J.-L., Chang, C.-C., Lee, K.-Z. 2012. In-line sampling with gas chromatography-mass spectrometry to monitor ambient volatile organic compounds. *Journal of Chromatography A* 1248: 161–168.

Wang, Q., Ma, L., Yin, C.-R., Xu, L. 2013. Developments in injection port derivatization. *Journal of Chromatography A* 1296: 25–35.

Wang, C., Ke, C., Wang, X., Chi, C., Guo, L. et al. 2014. Noninvasive detection of colorectal cancer by analysis of exhaled breath. *Analytical and Bioanalytical Chemistry* 406: 4757–4763.

Wardencki, W., Namiésnik, J. 2003. Analytical techniques used in monitoring of atmospheric air and stack gases. In *New Horizon and Challenges in Environmental Analysis and Monitoring*, ed. Namieśnik, J., Chrzanowski, W., Zmijewska, P. Chem. Fac. Gdańsk University of Technology, Gdańsk, pp. 446–463.

Wardencki, W., Katulski, R.J., Stefański, J., Namieśnik, J. 2008. The state of the art in the field of non-stationary instruments for the determination and monitoring of atmospheric pollutants. *Critical Reviews in Analytical Chemistry* 38: 259–268.

WHO. 2000. *Air Quality Guidelines for Europe*. Regional Office for Europe, World Health Organization, Copenhagen.

Woolfenden, E. 2010. Sorbent-based sampling methods for volatile and semi-volatile organic compounds in air: Part 2. Sorbent selection and other aspects of optimizing air monitoring methods. *Journal of Chromatography A* 1217: 2685–2694.

Woolfenden, E. 2012. Thermal desorption for gas chromatography. In *Gas Chromatography*, ed. Poole, C.F. Elsevier, Amsterdam, pp. 235–289.

Xu, J., Zheng, J., Tian, J., Zhu, F., Zeng, F. et al. 2013. New materials in solid-phase microextraction. *TrAC Trends in Analytical Chemistry* 47: 68–83.

Yu, Y., Pawliszyn, J. 2004. On-line monitoring of breath by membrane extraction with sorbent interface coupled with CO_2 sensor. *Journal of Chromatography A* 1056: 35–41.

9 Passive Sampling Strategies for Environmental Monitoring in Air and Aquatic Environment

Anna-Akrivi Thomatou and Ioannis Konstantinou

CONTENTS

9.1 INTRODUCTION

Considerable amounts of chemicals such as pesticides, PPCPs, phenols etc. produced and used by humans are released into the environment through various point and nonpoint sources constituting an important class of environmental contaminants. After their emission on primary sources, chemicals are found on terrestrial and aquatic environments and emitted into the atmosphere, sometimes at long distances away from the area of initial use. The physicochemical properties of these chemicals are of broad spectrum and affect their environmental fate and adverse effects on human and ecosystem health. These chemicals are frequently persistent to environmental degradation; many of them can accumulate on animals and humans and are related to human and ecotoxicity. Therefore, monitoring of chemical pollutants is necessary although it has some difficulties, such as complex mixtures of chemicals, low concentrations, and different environmental matrices, such as indoor air, ambient air, surface and groundwaters, soil, and sediments (Seethapathy et al., 2008).

Until now, the presence of organic chemicals in the environment and their toxicological action have lead scientists to involve these chemicals in monitoring programs and campaigns in order to study their distribution and ecotoxicological effects. Monitoring programs are also conducted to satisfy the requirements of frameworks and directives that are incorporated in environmental legislation. The European Union has established the Water Framework Directive 2000/60/EC, a legislative framework to protect and improve water quality of surface waters such as lakes, rivers, transitional and coastal waters, and groundwater. Priority pollutants, such as pesticides, polychlorinated biphenyls, and PAHs, are examples of chemicals included in the Water Framework Directive (WFD). The WFD 2000/60/EC has established environmental quality standards (EQSs) and defined the annual average (AA-EQS) and maximum allowable concentration (MAC-EQS) for 33 priority substances and eight other pollutants with the aim of achieving good surface water chemical status (European Parliament and Council of the European Union, 2000).

In order to determine pollution levels and to check the compliance with the legislative standards, several monitoring campaigns should be conducted (Söderström et al., 2009). Monitoring studies are generally based on grab sampling. Nevertheless, this has a number of disadvantages. First of all, the analysis of samples provides only an instantaneous assessment of the contaminant's concentration at the time of sampling, so it is possible to miss peak inputs or the presence of contaminants at trace levels. The latter can be resolved by collecting large volumes of sample. However, this type of sampling is costly and time consuming. To increase the frequency of sampling, automatic samplers are used to take numerous samples on a daily or weekly basis. Unfortunately, this is costly and impractical since a safe position and specific pretreatment of samples are required. Likewise, grab sampling does not provide time-weighted average (TWA) concentrations since measurement of TWA concentrations requires continuous exposure (Vrana et al., 2005a).

Another approach used to monitor environmental contaminants is biomonitoring. Biomonitoring is an integrative technique, and a number of species can be used depending on the environment being investigated. Analysis of contaminants that accumulate in organisms' tissues can give an indication of the environmental concentration levels. However, the use of living organisms as samplers has some limitations. Firstly, there are several physiological, anatomical, and behavioral factors such as metabolism, depuration rates, excretion, stress, and viability that can affect the rate of bioaccumulation in organisms' tissues and, consequently, the determined levels. In addition, only species that do not migrate are useful in biomonitoring in order to reflect the contaminant concentrations at a specific location and their variation during the sampling period. Finally, the extraction procedure from animal tissues for instrumental analysis is particularly complex (Kingston et al., 2000; Kot et al., 2000; Vrana et al., 2005a; Söderström et al., 2009).

In recent years, passive sampling methods have been developed to supplement and replace spot sampling methods. Passive sampling enables the determination of TWA concentrations of contaminants for prolonged periods. The technique is based on the diffusion of environmental pollutants from a medium to a receiving phase due to different concentrations between the two phases. Thus, the analyte's concentration is measured as a weighted function of the sampling period. Passive sampling has many advantages. First of all, passive samplers are relatively small and easy to use, so it is practical to carry them to sampling sites located far from the laboratory. Another advantage is that passive sampling combines several methodological steps such as sampling, isolation, and preconcentration. In this case, analytical cost decreases since sample pretreatment is not necessary and decomposition of analytes declines since they are isolated in the passive sampler. Moreover, at every sampling site, only one device is needed as long as the sampling lasts, and therefore, there is no need for large number of samples (Kot et al., 2000; Górecki and Namieśnik, 2002; Kot-Wasik et al., 2007).

Table 9.1 represents the landmarks in the evolution of passive samplers, which can be used in environmental studies. Nowadays, there is a growing interest in passive sampling, which is still developing. In the past few years, there is a considerable evolution on passive sampler design, calibration, and usage. This can be confirmed from Figure 9.1, which presents the number of publications on passive sampling from 2000 to 2014, which keeps growing and proves the development of passive sampling.

Passive samplers have been widely applied in water and air monitoring studies in order to:

- Monitor the existence or not of the contaminants
- Observe the variation in concentrations of pollutants
- Modeling spatial distribution and sources of contaminants
- Study the speciation of contaminants
- Evaluate the fate and distribution among different environments
- Calculate TWA concentrations
- Estimate the toxicity of pollutants (Vrana et al., 2005a)

9.2 WATER PASSIVE SAMPLING

9.2.1 Modeling and Principles

The first passive sampling devices were designed for monitoring dissolved inorganic compounds such as heavy metals in water (Benes and Steinnes, 1974) and in sediment–water interface (Mayer, 1976; Hesslein, 1976). Since then, various methods of sampling metals from waters have been published (Srikameswaran et al., 1984; Morrison, 1987; Davison and Zhang, 1994).

In passive samplers, pollutants are trapped in a medium within the sampler i.e., the receiving phase which can be a solvent, chemical reagent, or a porous adsorbent. Figure 9.2 represents a general diagram of a passive sampler. The sampler is exposed to the examined medium and collects analytes by diffusion. Analytes pass through the membrane, and they are dissolved in the solvent or adsorbed onto the porous receiving phase (Stuer-Lauridsen, 2005).

Thus, passive sampling methods are based on either adsorption or absorption of pollutants from water. Adsorptive methods are based on the physical or chemical sorption of analytes on surfaces and depend on surface binding and surface area. Absorptive methods include analyte distribution on the receiving phase and depend mainly on the polarity of the pollutant and the solvent phase and the solubility of the pollutant in water and the solvent.

A passive sampler is designed to operate in two different accumulation regimes (Figure 9.3):

- The equilibrium phase (equilibrium passive samplers)
- The kinetic or time-integrative uptake phase (kinetic passive samplers)

TABLE 9.1
Milestones in the Development of Passive Techniques for Sampling of Analytes from Different Media

1973	First passive sampler for atmosphere
1976	Passive sampler for metals in sediments
1980	Organic vapor monitor (OVM) for VOCs in air
1984	First passive sampler for metals in water
1985	First passive sampler for organic micropollutants in soil
1987	First passive sampler filled with solvent for organic micropollutants in water
1988	Passive samplers filled with sorbent for the detection of contaminants in water and air
1990	SPMD
1990	PETREX tubes for sampling on soil
1992	Supported liquid membrane (SLM) devices
1993	PISCES
1993	Ogawa passive sampler for the determination of ozone, NO_2, SO_2
1994	DGT for metals and inorganic ions
1995	Detection of compounds in water at picogram per liter level
1995	Combination with bioassays
1995	GORE SORBER for soil
1996	Radiello
1997	Passive diffusion bag for sampling VOCs
1999	POCIS
1999	Ceramic dosimeter
1999	Twister-stir bar sorptive extraction
2000	Chemcatcher
2000	Analyst
2000	Tristearin-coated fiber glass sheets
2001	MESCO
2001	PUF disks
2001	POGs
2003	XAD 2 resin
2004	SPATT
2007	Ceramic toximeter for PAHs
2008	Needle trap device (NTD)
2009	Continuous flow integrative sampler (CFIS)
2011	Polypropylene glycol (PPG)-coated hollow fiber
2011	Five-layer passive organic dermal (POD) sampler
2011	Total gaseous Hg (TGM) passive sampler
2011	Multigas passive sampler (MGPS)
2012	Time-integrative sampler using in situ solvent extraction (TISIS)
2012	PFCs-POCIS for perfluorinated chemicals
2012	Passive samplers with Empore disks for monitoring carbamates and s-triazines
2014	Polymer inclusion membrane (PIM) for Zn(II)

Sorption of analytes in the receiving phase of a passive sampler follows usually first-order kinetics. It can be described by a primary linear phase, which is followed by an equilibrium phase. During the linear sorption, sampling rate and partition coefficients are constant and are not affected by analyte concentrations. With kinetic passive samplers, the TWA concentrations can be estimated and it takes a long time to reach equilibrium (Alvarez et al., 2007).

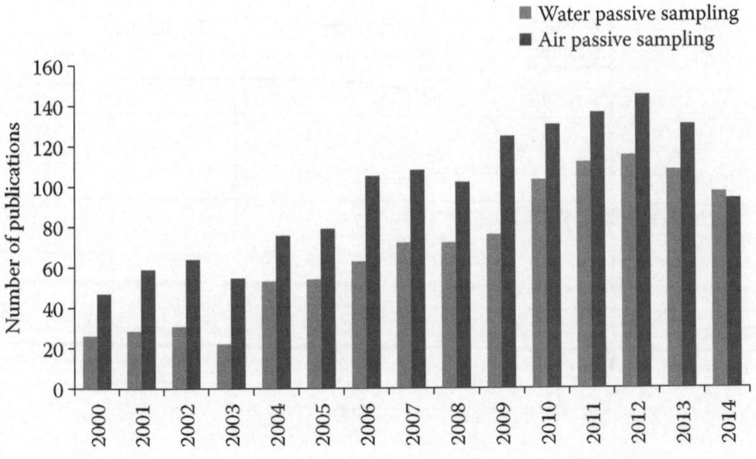

FIGURE 9.1 Number of publications on passive sampling published between 2000 and 2014.

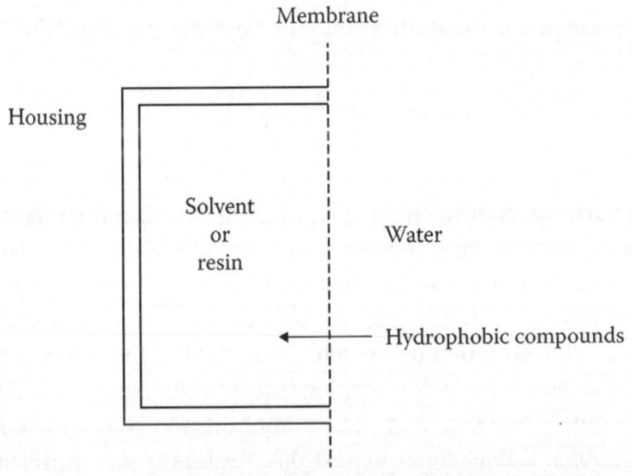

FIGURE 9.2 Schematic representation of a passive accumulation device. The membrane is typically a hydrophobic material, and the receiving solvent or resin acts as a sink for the compounds of interest. To increase the surface area, several of the samplers allow the membrane to completely enclose the receiving phase, thus forming a permeable housing. (Reprinted from *Environmental Pollution*, 136, Stuer-Lauridsen, F., Review of passive accumulation devices for monitoring organic micropollutants in the aquatic environment, 503–524, Copyright (2005), with permission from Elsevier.)

The exchange kinetics between the water phase and passive sampler is described by one-compartment mathematical model:

$$C_s(t) = C_w \frac{k_1}{k_2}(1 - e^{-k_2 t}), \tag{9.1}$$

where $C_s(t)$ is the concentration (μg g^{-1}) of the analyte in the sampler at exposure time t, C_w is the analyte concentration in the water (μg L^{-1}), and k_1 (L g^{-1} day^{-1}) and k_2 (day^{-1}) are the uptake and elimination rate constants, respectively.

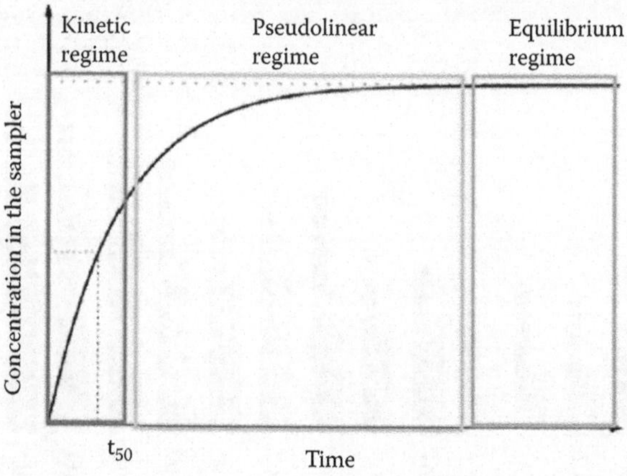

FIGURE 9.3 Passive sampling devices that operate in different accumulation regimes. (Reprinted with *Trends in Analytical Chemistry*, 36, Morin, N., C. Miège, J. Randon, and M. Coquery, Chemical calibration, performance, validation and applications of the polar organic chemical integrative sampler (POCIS) in aquatic environments, 144–175, Copyright (2012), with permission from Elsevier.)

Equilibrium passive samplers: In equilibrium passive sampling, Equation 9.1 reduces to

$$C_s = C_w \frac{k_1}{k_2} = C_w K, \tag{9.2}$$

where K (L g^{-1}) is the partition coefficient of the pollutant between the receiving phase and water. When the partition coefficient and the concentration in the receiving phase are known, it is feasible to estimate the analyte concentration in the water phase. The method requirements are that the response time has to be known in order to reach stable concentrations and must be shorter than the analyte's variations in the concentration being determined. Also, the environmental conditions cannot affect the analyte concentrations in the sampler. Equilibrium passive samplers are characterized by a fast attainment of equilibrium between analyte concentrations in sampled medium and passive sampler (Vrana et al., 2005a; Zabiegała et al., 2010a). Results from equilibrium passive sampling are comparable to those of spot sampling. For this reason, these samplers cannot estimate TWA concentrations (Kot-Wasik et al., 2007; Ouyang and Pawliszyn, 2007).

Kinetic passive samplers: Kinetic passive samplers never reach equilibrium during sampling period, and the analyte continues to be transferred constantly in the receiving phase; thus, the concentration between the sampled medium and the receiving phase follows a linear relationship. These samplers are characterized by high capacity so that analytes are constantly fortified during the exposure time. With kinetic passive samples, the TWA concentrations of analytes during the whole sampling period can be estimated (Zabiegała et al., 2010a,b,c).

As the sampler is exposed to the environment, there is insignificant desorption of analytes from the receiving phase to the environment and the sampler works in the linear phase (Vrana et al., 2005a). So, Equation 9.1 reduces to

$$C_s(t) = C_w k_1 t. \tag{9.3}$$

Equation 9.3 can be transformed in order to replace the uptake rate k_1 with the sampling rate R_s to the following equation:

$$C_s(t) = \frac{C_w R_s t}{M_s(t)}, \tag{9.4}$$

where $M_s(t)$ is the mass (mg) of the pollutant accumulated in the receiving phase after a deployment time period t and R_s is the sampling rate (L day^{-1}). In other words, R_s is the product of the first-order rate constant (k_1) for uptake of pollutant and the water volume, which gives the same chemical activity as the volume of receiving phase (Vrana et al., 2005a). The sampling rate R_s of the target analytes is usually calibrated during laboratory experiments (Kot-Wasik et al., 2007; Ouyang and Pawliszyn, 2007). Therefore, the TWA concentration C_w of the analyte in the aquatic environment can be calculated from the sampling rate (R_s) and the mass (M_s) accumulated in the receiving phase.

9.2.2 Sampler Design

All the kinetic passive samplers that have been designed have common features. The most important component is the presence of a membrane, the barrier between the environment, and the receiving phase. The membrane should enable the collection of analyte molecules into the receiving phase, providing large permeability and selectivity and limiting certain species from sampling. The membrane specifies also the proportion of molecules that are sampled at a given concentration and decreases the effects of the surrounding environment such as water turbulence, fouling, and temperature.

Based on the properties of the membrane, samplers come under two categories: diffusion based and permeation based. Diffusion-based devices collect analytes by diffusion through a static layer of water contained in well-defined openings in the sampler. Permeation-based devices collect analytes by permeation through a porous or nonporous membrane (Górecki and Namieśnik, 2002). In both cases, the transport of molecules is based on the different concentrations of analytes between water and the receiving phase enclosed in the membrane. Passive samplers are manufactured to maximize the quantity of analytes sampled so that detection limits will be lower (Vrana et al., 2005a).

The most commonly used devices can be separated in two general categories: (a) solvent-filled devices and (b) sorbent-filled devices. A review of the passive sampling devices that have been used for monitoring of pollutants in the water is presented in Table 9.2.

9.2.2.1 Passive Sampling Devices Filled with Solvent

The first approach of passive samplers filled with solvent was made with a hydrophilic dialysis bag constructed of regenerated cellulose, with a 1000 Da cutoff, and filled with 3 mL of hexane, which was developed by Södergren (1987). In this sampler, hydrophobic pollutants pass through the membrane and are absorbed in the receiving phase. The sampling method is similar to bioconcentration of organic pollutants into the fat tissues of aquatic organisms. The device is cheap and easy to use and does not need further processing or solvents. However, the polar membrane tends to prevent the adsorption of the hydrophobic compounds. This method is not in use a few years ago (Stuer-Lauridsen, 2005).

Bergqvist et al. (1993) introduced a semipermeable membrane for dialysis of lipophilic organic compounds (Bergqvist et al., 1993, 1999; Falandysz et al., 1996; Strandberg et al., 1998). Kot and Wasik (1999) developed a 2 cm diameter passive sampling device filled with solvent (commonly isooctane, cyclohexane). Analytes pass through the membrane and they are absorbed on organic solvents, which are then analyzed with chromatographic techniques with no further process needed.

These samplers were cheap and easy to use and have been used for monitoring dissolved organic compounds at various surface waters.

9.2.2.2 Semipermeable Membrane Devices

Huckins et al. (1990a) developed semipermeable membrane devices (SPMDs) to use them in accumulation of lipophilic substances. SPMDs are designed to sample chemicals dissolved in surface water, mimicking the bioconcentration of organic contaminants into the fatty tissues of organisms. The SPMD is an integrative sampler which accumulates organic pollutants over a deployment

TABLE 9.2
Overview of Published PAD Methods That Have Been Used In Situ

PAD Method	Content/Membrane	Target Compounds	Sampling Purpose[a]	Reference	Rank[b]
Solvent-filled dialysis membrane	Hexane or heptanes/dialysis membrane	Lipophilic	Screening	Södergren (1987)	M
Passive dosimetry	Activated carbon/perforated acrylic	p-xylene and atrazine	Quantitative, not tested in situ	DiGiano et al. (1989)	L
Permeation sampling	XAD-7, Tenax Ta/polycarbonate membrane	Phenolic compounds	Quantitative, not tested in situ	Zhang and Hardy (1989)	L
Semipermeable membrane devices (SPMD)	Triolein/polyethylene	Lipophilic	Quantitative, screening, toxicity tests	Huckins et al. (1990a)	H
Passive sampling device	C-18, XAD-4, 2,2,4-trimethylpentane/polyethylene	PCBs and pesticides	Screening in soil	Zabik et al. (1992)	L
Passive in situ concentration extraction sampler (PISCES)	Hexane/polyethylene membrane	Organochlorines	Screening	Litten et al. (1993)	L
Passive sampler (TMP)	Trimethylpentane/polyethylene	Chlordane and dieldrin	Quantitative	Peterson et al. (1995)	M
Empore disks	Filter with C18 resin/no membrane	Lipophilic	Equilibrium sampling	Verhaar et al. (1995)	L
Groundwater monitoring	Various adsorbents/Gore-Tex membrane	VOCs and SVOCs	Screening	Mehltretter and Sorge (1995)	L
In situ monitoring	Polyurethane/fiberglass	Aromatic compounds	Screening, toxicity tests	Madsen et al. (1996)	L
Groundwater monitoring	C18, XAD, Amberlite, and polyurethane/ceramic membrane	PAHs, PCBs and pesticides	Quantitative	Grathwohl and Schiedek (1997)	L
Passive sampler for volatile aromatics in water	Porapak Q, Tenax-TA/silicone polycarbonate	Monocyclic aromatics	Quantitative (via gas phase)	Lee and Hardy (1998)	L
Polar organic chemical integrated sampler (POCIS)	Various chromatographic materials/various membranes	Polar organic compounds	Quantitative	Alvarez et al. (1999)	L

(*Continued*)

TABLE 9.2 (CONTINUED)
Overview of Published PAD Methods That Have Been Used In Situ

PAD Method	Content/Membrane	Target Compounds	Sampling Purpose[a]	Reference	Rank[b]
Empore® disks in situ	Filter with C18 resin/polysulfone and polyethylene	Organic compounds	Quantitative	Kingston et al. (2000)	L[c]
LDPE and silicone passive sampling	No solvent or resin/LDPE or silicone	PAHs and PCBs	Booij et al. (2000)	Booij et al. (2000)	M
Groundwater monitoring	Dowex Optipore L-493/ceramic membrane	BTEX and naphthalenes	Quantitative	Martin et al. (2003)	L
Blue cotton	Phthalocyanine trisulfonate/no membrane	Phthalates and aromatics	Toxicity tests	Sayato et al. (1990)	L
TLC plate	Thin layer chromatography plate/no membrane	Organophosphates	Screening	LeBlanc et al. (2003)	L
Solid-phase micro extraction (SPME)	Various materials/no membrane	Various compounds	Screening, equilibrium sampling	Verbruggen et al. (2000)	L

Source: *Environmental Pollution*, 136, Stuer-Lauridsen, F., Review of passive accumulation devices for monitoring organic micropollutants in the aquatic environment, 503–524, Copyright (2005), with permission from Elsevier.

[a] Several sampling purposes are identified: determination of contaminant concentration in the water (quantitative), qualitative or semiquantitative determination of occurrence (screening), collection of material for toxicity tests (toxicity tests), and determination of concentration by equilibrium sampling (equilibrium sampling).

[b] Applicability ranking for monitoring based on annual average of studies published with the method since first published (as of August 2004). H (high) >1; M (medium) >0.5; L (low) ≤0.5.

[c] If combined with the studies involving Empore disk equilibrium sampling, a medium ranking is achieved.

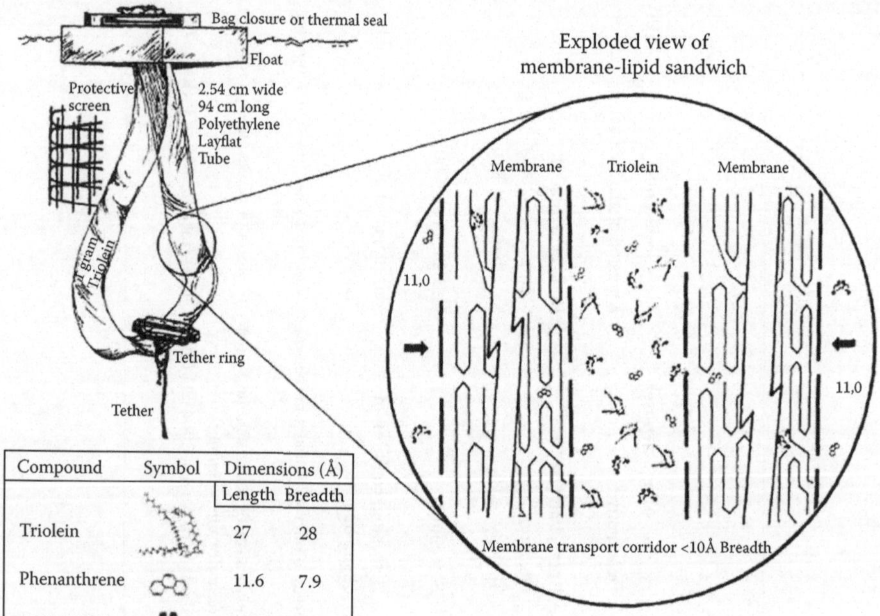

Compound	Symbol	Dimensions (Å)	
		Length	Breadth
Triolein		27	28
Phenanthrene		11.6	7.9
2,2′,5,5′-TCB		11.6	8.6

FIGURE 9.4 One of many possible configurations of lipid-containing SPMDs. (Reprinted with permission from Huckins et al. (1993), 2489–2496. Copyright 1993 American Chemical Society.)

period ranging from days to months. SPMD (Figures 9.4 and 9.5) is a low-density polyethylene (LDPE) membrane, which formed a tube film and it is filled with a high-molecular weight lipid, such as triolein (Huckins et al., 1990a, 1993; Bennett et al., 1996). Triolein was selected for use in SPMDs because it can be found in most organisms, results in low LDPE membrane permeability, and supplies a suitable reservoir for performance reference compounds (PRCs). LDPE was chosen because of its constancy in organic solvents, the low permeation rates of triolein, and its resistance to wearing and drilling (Huckins et al., 1990b; Meadows et al., 1993; Bergqvist et al., 1998).

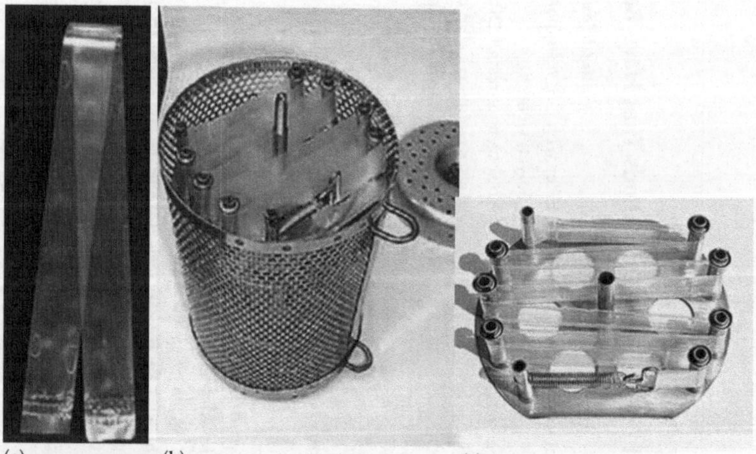

(a) (b) (c)

FIGURE 9.5 (a) Semipermeable membrane device (SPMD) and (b) SPMDs deployed inside a stainless steel canister and (c) mounted on deployment rack. (Reproduced from Bacaloni, A., S. Insogna, and I. Zoccolillo, *Air Quality Monitoring, Assessment and Management*, InTech, Rijeka, Croatia, pp. 223–240, 2011. With permission.)

TABLE 9.3
Classes or Specific Chemicals Known to Concentrate in SPMDs

PAHs	PAHs and Alkylated PAHs and Chlorinated Naphthalenes
Various hydrocarbons	Heterocyclic aromatics, cyclic hydrocarbons (e.g., decalin and alkylated decalins), and aliphatics
Halogenated hydrocarbons	PCB congeners, chlorinated dibenzofurans, chlorinated dibenzodioxins, polybrominated diphenyl ethers, chlorinated benzenes
Pesticides	Organochlorine pesticides such as HCHs, DDTs, cyclodienes, etc.
	Other pesticides: diazinon, endosulfans, pyrethroids, toxaphene, and trifluralin
Pharmaceuticals	Triclosan, triclocarban
Antifouling compounds	Tributyltin and its analogues
Phenols	Chlorinated anisoles and veratroles
	Alkyl phenols (nonylphenol)

LDPE has no fixed pores with size of about 1 nm, so that only truly dissolved compounds pass through the membrane and accumulate in the solvent. The characteristic of size limitation results to extremely low diffusion rates of triolein in LDPE. The hydrophobic membrane enables the hydrophobic compounds to pass into the sampler very easily, and compounds with log $K_{ow} > 3$ can be accumulated effectively (Huckins et al., 1993). SPMDs are not appropriate for sampling very large organic molecules, ionized organic compounds, and metals. Also, for compounds with log $K_{ow} < 3$, SPMDs may not work equally well as other sampling systems. Table 9.3 shows representative examples of chemicals that are accumulated in SPMDs.

Extraction of contaminants accumulated in SPMDs is performed by dialysis with organic solvents, hexane in most cases. Dialysis procedure has same disadvantages, such as long extraction times (24–72 h) and large volumes of solvent (100–900 mL). Therefore, it is proposed to use environmentally friendly procedures such as accelerated solvent extraction (ASE), microwave-assisted extraction, ultrasound assisted extraction, or headspace direct volatilization of the contaminants. After the extraction of contaminants, gel permeation chromatography is used to eliminate the coextractives and the extract can be finally analyzed with either GC (GC-FID, GC-ECD, GC-MS) or liquid chromatography techniques (HPLC-DAD, HPLC-FL, LC-MS). Figure 9.6 depicts a diagram for the selection of the appropriate extraction procedure for an analyte's determination in SPMDs (Esteve-Turrillas et al., 2008).

In general, SPMDs have been used for the determination of the TWA concentration of hydrophobic semivolatile organic pollutants in water and air, the determination of time-integrated flux of chemicals from dissolved and vapor phases, and the evaluation of bioconcentration factors (Huckins et al., 2006) (Table 9.4). SPMDs are easy to use, they can be calibrated, and they can be deployed for long periods. The concentrations of the analytes in the sampler depend on the concentration of the analytes in the water, temperature, sampling period, surface area, and octanol–water portioning coefficient (K_{ow}) (Kot et al., 2000).

9.2.2.3 Passive In Situ Concentration/Extraction Sampler

A passive in situ concentration/extraction sampler (PISCES) was developed by Litten et al. (1993). It consists of plumbing pieces and a 3.8 cm chrome-plated brass slip joint tee that are weld to a 10 cm slip joint arm. The end of the slip joint arm is sealed with Teflon® disk, and the ends of the tee are fitted with polyethylene membranes of 1 mm thickness. Viton® O-rings create a tight seal for the Teflon disk and the polyethylene membranes. The sampler is rinsed with acetone and then with hexane. Before use, it is rinsed again with hexane. The sampler is filled with 200 mL of hexane, sealed, and placed in the water column for a few days. The analytes are diffused through the membranes

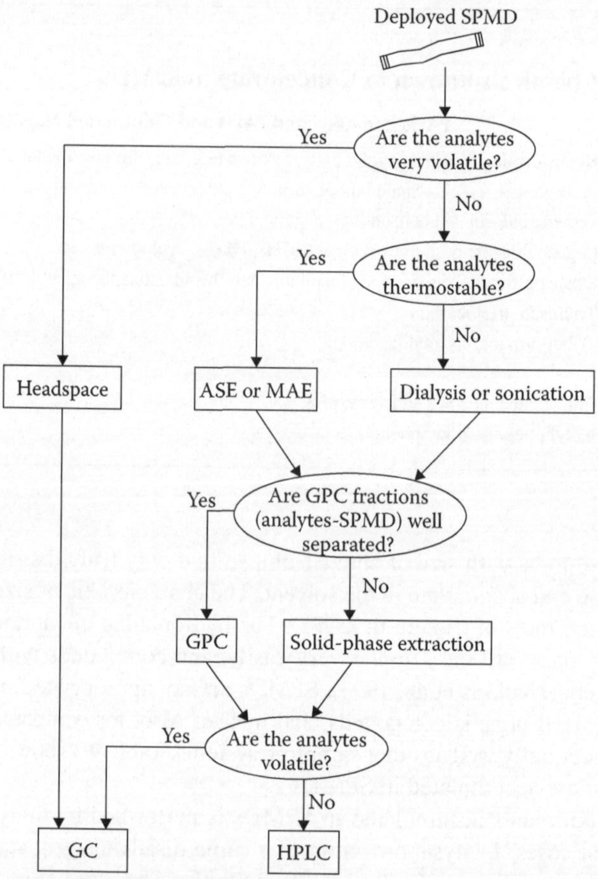

FIGURE 9.6 Flowchart to select an appropriate analytical procedure for analyte's determination in SPMDs. (Reprinted from *Talanta*, 74, Esteve-Turrillas, F.A., V. Yusà, A. Pastor, and M. de la Guardia, New perspectives in the use of semipermeable membrane devices as passive samplers, 443–457, Copyright (2008), with permission from Elsevier.)

from the water to the solvent. The solvent diffuses slowly out of the sampler, passes through the membrane, and impedes colonization organisms to form a biofilm and biofouling on membrane surface. This procedure does not influence the uptake of analytes in the passive sampler because the process of diffusion is guided by the activity gradient of the analyte dissolved in water and solvent (Litten et al., 1993). PISCES is an easy-to-use passive sampler and useful in the field for sampling low concentration levels of organic contaminants in the water (Kot et al., 2000) (Table 9.4).

9.2.2.4 Passive Samplers Filled with Sorbent

The first approach of such samplers is designed by DiGiano et al. (1988), consisting of two disks of acrylic plastic with 5 cm diameter which are closed together by an O-ring. The upper disk comprises 72 holes with a 1 mm diameter and a 1 cm length. The lower disk contained the receiving phase consisted of 1.5 g of 30–40-mesh Filtrasorb 400 granular activated carbon. Between the disks, there is a screen in order to prevent the receiving phase from coming out of the holes.

9.2.2.5 Chemcatcher

This passive sampler consists of a diffusion-limiting membrane and a solid receiving phase placed in a housing. Analytes pass through the membrane to the receiving phase, where they are accumulated. Different combinations of membrane and receiving phase are used in different designs of

TABLE 9.4
Application of Passive Samplers Filled with Solvent in Aquatic Environment

Passive Sampler	Type of Aquatic Environment	Type of Analytes	References
SPMD	Creek	PAHs	Lebo et al. (1996)
	Stream	Organochlorine pesticides, PCBs, PAHs	Vrana et al. (2001)
	Wetland	Organochlorine pesticides, organophosphate pesticides, pharmaceuticals, PAHs	Petty et al. (2004)
	Freshwater	Organochlorine pesticides, PCBs, PAHs	Verweij et al. (2004)
	River	PAHs	Shaw et al. (2004)
	Groundwater	Organic contaminants	Vrana et al. (2005b)
	Storm water	PAHs	Komorova et al. (2006)
	River, lake	PAHs, pesticides, industrial compounds	Rosen et al. (2006)
	River	PAHs, PCBs, hexachlorobenzene, *p,p'*-DDE	Allan et al. (2009)
	Marine water, river	Halogenated natural products	Vetter et al. (2009)
	Stream	Pesticides	Polidoro et al. (2009)
	River	EDCs	Grabic et al. (2010) Creusot et al. (2013)
	Coastal water	PAHs, PCBs, alkylphenols	David et al. (2010)
	Lake	PCBs	Schubauer-Berigan et al. (2012)
	Lake	Synthetic organic compounds	Alvarez et al. (2012)
	Estuary	PAHs, PCBs	Bustamante et al. (2012)
	River	PAHs	Bourgeault and Gourlay-Francé (2013)
	WWTP	Estrogen, androgen, aryl hydrocarbon	Jálová et al. (2013)
	Dam	Organochlorine pesticides, PCBs, PAHs	Amdany et al. (2014a,b,c)
	River	PCBs, PAHs	Vrana et al. (2014)
	Coastal water	Organochlorine pesticides, PCBs, PAHs	Vilmaz et al. (2014)
	Surface water	Organochlorine pesticides, PCBs, PAHs	Amdany et al. (2014b)
	River	Emerging contaminants	Alvarez et al. (2014a,b,c)
	Sewage lagoon	PPCPs	Hoque et al. (2014)
PISCES	River	PCBs	Litten et al. (1993)
	Lake	PAHs, PCBs	Polito and Hassett (2002)

Chemcatcher to monitor polar and nonpolar organic contaminants. Sampling rates and selectivity are adjusted by the selection of the diffusion-limiting membrane and the solid receiving phase. Table 9.5 includes a variety of sorbent diffusion-limiting membranes for all kinds of analytes, including polar and nonpolar organic analytes, organometallic compounds, and metals. Regarding the diffusion-limiting membrane, two kinds of polymeric membranes have been examined: (i) nonporous membranes including LDPE and (ii) microporous membranes including glass fiber, nylon, polycarbonate, polytetrafluoroethylene (PTFE), polyvinylidenedifluoride, cellulose acetate (CA), polysulfone (PS), polyethersulfone (PES), and regenerated cellulose. As mentioned earlier, membranes are barriers that protect the receiving phase from losses into water while permitting dissolved analytes to pass through to the sorbent and limit the permeation of microorganisms and particulates with size greater than the porous membrane (Greenwood et al., 2007). In most applications, Chemcatcher (Figure 9.7) consisted of a 47 mm C18 Empore™ disk as a receiving phase in both configurations and a diffusion-limiting membrane made by different materials, PS for the polar and polyethylene for the nonpolar contaminants, was used. The configuration is held from a

TABLE 9.5
Chemcatcher Configurations for Integrative Sampling of Various Pollutant Classes

Pollutant Class	Receiving Phase	Diffusion Membrane
Hydrophobic organic compounds (log $K_{ow} > 3$)	C_{18} Empore disk	Non-porous low density polyethylene (LDPE)
Hydrophilic organic compounds (log $K_{ow} < 3$)	C_{18} Empore disk	Microporous polysulfone (PS)
	SDB-RPS Empore disk	Microporous polyethersulfone (PES)
Metals	Chelating Empore disk	Microporous cellulose acetate (CA)
Mercury	Chelating Empore disk	Microporous polyethersulfone (PES)
Organotin compounds	C_{18} Empore disk	Microporous cellulose acetate (CA)

Source: *Comprehensive Analytical Chemistry 48: Passive Sampling Techniques in Environmental Monitoring*, 48, Greenwood, R., G. A. Mills, B. Vrana, I. J. Allan, R. Aguilar-Martinez et al., Monitoring of priority pollutants in water using Chemcatcher passive sampling devices, 199, Copyright (2007), with permission from Elsevier.

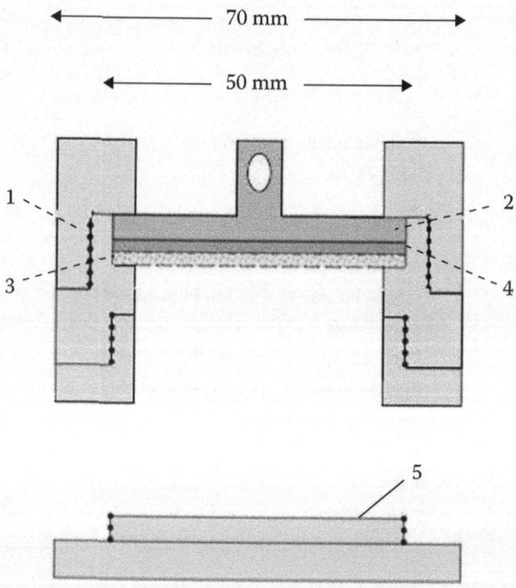

FIGURE 9.7 Schematic diagram of the Chemcatcher passive sampling device. PTFE body parts: components 1 and 4; receiving phase: component 2; diffusion-limiting membrane: component 3; screw cap: component 5. (Reprinted from *Environmental Pollution*, 142, Vrana, B., G. A. Mills, E. Dominiak, and R. Greenwood, Calibration of the Chemcatcher passive sampler for the monitoring of priority organic pollutants in water, 333–343, Copyright (2006), with permission from Elsevier.)

50 mm PTFE supporting layer. The active surface area of the Chemcatcher is 17.5 cm^2. The sampler is sealed up with rings that are screwed in order to keep the body parts of the sampler together and be filled with water. Sometimes, it is possible to use again the PTFE body, after a thorough washing procedure (Kingston et al., 2000).

Sampling rates for both polar and nonpolar compounds are affected by environmental conditions. Water temperature and turbulence, as well as biofouling on the surface of the diffusion-limiting membrane, must be taken into account when estimating TWA concentrations (Kingston et al., 2000). Several studies provide Chemcatcher sampling rates from calibration experiments for either polar (Gunold et al., 2008; Vermeirssen et al., 2013) or nonpolar compounds (Vrana et al., 2007; Aguilar-Martinez et al., 2008).

Analytes accumulated in Chemcatcher are extracted from the Empore disks in an ultrasonic bath for 5 min with 5 mL of acetone followed by 5 min with 5 mL of 50:50 v/v ethyl acetate/2,2,4-trimethylpentane mixture. Then, the extracts are combined and filtered through a drying cartridge that contains 1 g of sodium sulfate and concentrated under a stream of nitrogen. Subsequently, internal standard is added and followed by GC-MS analysis to determine the concentrations of the contaminants accumulated (Vrana et al., 2006).

Chemcatcher passive sampler has been also used extensively in aquatic monitoring of nonpolar organic contaminants such as PAHs, polychlorinated biphenyls, and organochlorine pesticides (Kot-Wasik et al., 2007).

9.2.2.6 Membrane-Enclosed Sorptive Coating

The membrane-enclosed sorptive coating (MESCO; Figure 9.8) consists of a twister bar used for stir bar sorptive extraction (component 1), surrounded by a dialysis membrane bag from regenerated cellulose (Spectra/Por 6, molecular weight cutoff 1000, 18 mm flat width, 30 mm length, component 2). In SBSE, the stir bar coated with PDMS is placed into the sample, stirred for a prearranged time, and thermally desorbed online into chromatographic system (Baltussen et al., 1998, 1999). The twister bar (15 mm length) consists of a magnetic core closed in a glass layer with 22 mg of PDMS (500 μm thickness, 24 μL volume). The dialysis membrane bag is filled with 3 mL of bidistilled water (component 3) and sealed with 35 mm Spectra Por 6 enclosures (component 4). After sampling, the MESCO sampler twister bar is rinsed with bidistilled water, dried with a paper cloth, and analyzed using chromatography–MS for the confirmation and quantification of analytes (Vrana et al., 2001).

9.2.2.7 Polar Organic Chemical Integrative Sampler

The polar organic chemical integrative sampler (POCIS) device was the first reported to sample polar organic compounds with log K_{ow} value less than 4. The POCIS (Figures 9.9 and 9.10) consists of a solid sequestration medium enclosed between two hydrophilic microporous PES membranes (Jones-Lepp et al., 2004; Petty et al., 2004; Alvarez et al., 2005, 2007). To impede sorbent loss, rings are used to form a compression of membranes and can be of metallic or polymeric material. PES membrane is the barrier between the receiving phase and the aquatic environment, so that molecules with cross-sectional diameters larger than 100 nm are specifically blocked. The surface area of the membranes exposed to water is 41 cm^2, and the sorbent mass is ≈228 mg. Therefore, a standard

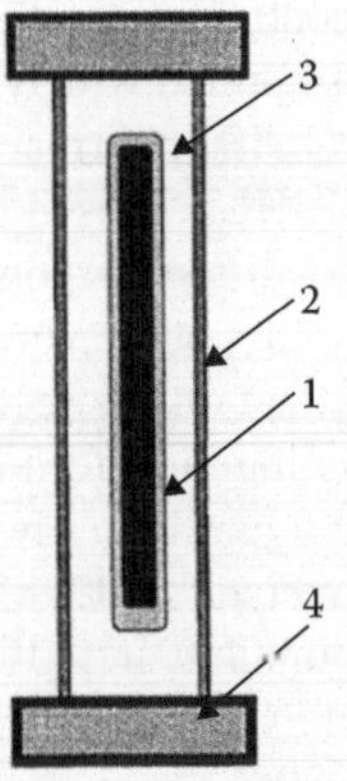

FIGURE 9.8 Schematic diagram of the MESCO passive sampling device. (Reprinted with permission from Vrana et al. (2001), 5191–5200. Copyright 2001 American Chemical Society.)

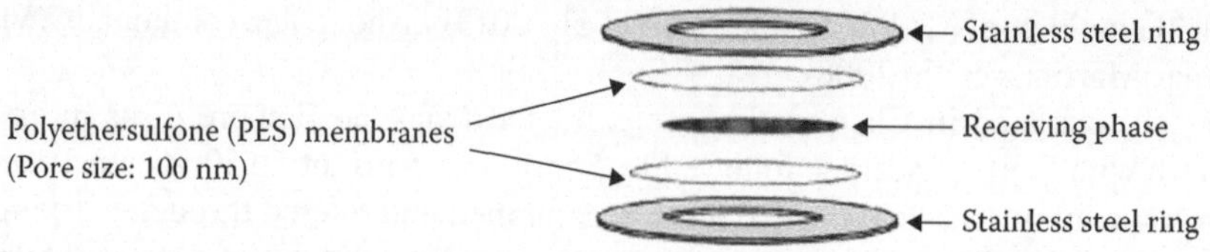

FIGURE 9.9 Disassembled view of the POCIS. (Reprinted from *TrAC Trends in Analytical Chemistry*, 36, Morin, N., C. Miège, J. Randon, and M. Coquery, Chemical calibration, performance, validation and applications of the polar organic chemical integrative sampler (POCIS) in aquatic environments, 144–175, Copyright (2012), with permission from Elsevier.)

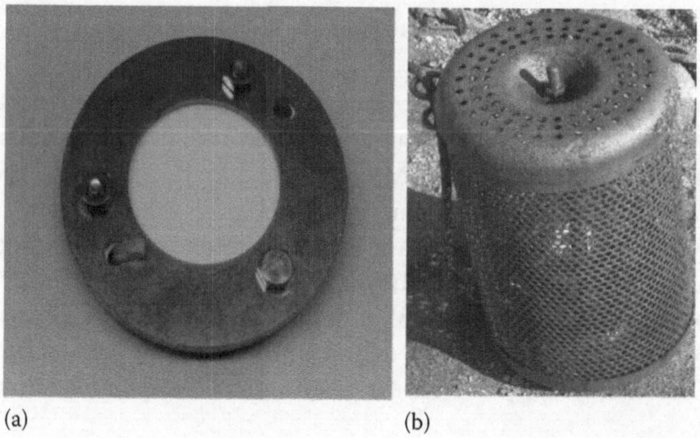

FIGURE 9.10 Polar organic chemical integrative sampler (a) and deployment canister (b).

POCIS has a surface area-to-sorbent mass ratio of ≈180 $cm^2\ g^{-1}$. Biofouling is possible when POCIS is exposed to surface waters for more than 2 weeks. Nevertheless, the microporous PES membrane seems to resist biofouling compared to other materials that are used in passive samplers.

POCIS is flexible, as the sorbent can change to target specific chemical classes. Many times, different configurations of POCIS are used together in the field to get more analytes sampled. Two configurations of POCIS are generally used. The generic configuration consists of the triphasic admixture of Isolute ENV + polystyrene divinylbenzene and Ambersorb 1500 carbon dispersed on S-X3 Biobeads. This mixture has excellent recoveries of many pesticides, natural and synthetic hormones, and other wastewater-related contaminants (Stuer-Lauridsen, 2005; Alvarez et al., 2005, 2007). Because the triphasic admixture is applied widely, it is assumed as the generic sorbent configuration. The other configuration of POCIS (pharmaceutical) uses the Oasis HLB sorbent for pharmaceutical compound isolation. Oasis HLB is used because many pharmaceuticals have multiple functional groups that strongly bind to the carbonaceous component of the sorbent. Oasis HLB is a sorbent used frequently for sampling of pharmaceuticals when using the SPE method, so the literature has a significant amount of data on the recoveries of analytes that accumulate at the sorbent (Alvarez et al., 2005, 2007). Sampling rates for both configurations have been published in previous studies, for either the generic configuration (Alvarez et al., 2007; Morin et al., 2012) or the pharmaceutical configuration (Söderström et al., 2009; Morin et al., 2012). Table 9.6 contains a list of chemicals that have been proved to accumulate in POCIS.

The extraction of the receiving phase of POCIS is performed by carrying the sorbent into an SPE cartridge or an ASE cell (Hernando et al., 2005) or by sonication (Kohoutek et al., 2008, 2010). The extraction of the pesticide POCIS (generic configuration) is accomplished with methanol/toluene/dichloromethane (1:1:8) and the pharmaceutical POCIS with methanol (Morin et al., 2012). In most

TABLE 9.6
Examples of Classes or Specific Chemicals Known to Concentrate in POCIS

Pollutant Class	Compounds
Pharmaceuticals	Acetaminophen
	Azithromycin
	Carbamazepine
	Propranolol
	Sulfa drugs
	Tetracycline antibiotics
Illicit drugs	Methamphetamine, MDMA
Several natural and synthetic hormones	17β-Estradiol
	17α-Ethynylestradiol
	Estrone
	Estriol
Triazine herbicides	Atrazine
	Cyanazine
	Hydroxyatrazine
	Terbuthylazine
Various polar pesticides	Alachlor
	Chlorpyrifos
	Diazinon
	Dichlorvos
	Diuron
	Isoproturon
	Metolachlor
Various household and industrial products and degradation products	Alkyl phenols (nonylphenol)
	Benzophenone
	Caffeine
	DEET
	Indole
	Triclosan
Urobilin (fecal contamination marker)	
Essentially any compound with log $K_{ow} \leq 3.0$	

Source: *Comprehensive Analytical Chemistry 48: Passive Sampling Techniques in Environmental Monitoring*, 48, Alvarez, D. A., J. N. Huckins, J. D. Petty, T. Jones-Lepp, F. Stuer-Lauridsen et al., Tool for monitoring hydrophilic contaminants in water: Polar organic chemical integrative sampler (POCIS), 171–197, Copyright (2007), with permission from Elsevier.

studies, LC-MS/MS or GC-MS systems, based on the analytes, followed the solvent extraction. Figure 9.11 represents the general processing for POCIS samples (Alvarez et al., 2007).

Although POCISs accumulate an extended number of contaminants, they are not appropriate for all analytes (Table 9.7). Specifically, contaminants with log K_{ow} > 3–4 may not accumulate in POCIS as in other samplers, such as SPMDs. Figure 9.12 represents the relationship of POCIS and SPMDs with compounds of different log K_{ow} values. The difference between the two samplers is attributed to the extra transfer resistance of the PES membrane (Alvarez et al., 2007). Consequently, an integrated sampling contains both samplers, POCIS and SPMDs, in order to get more information on compounds' concentrations.

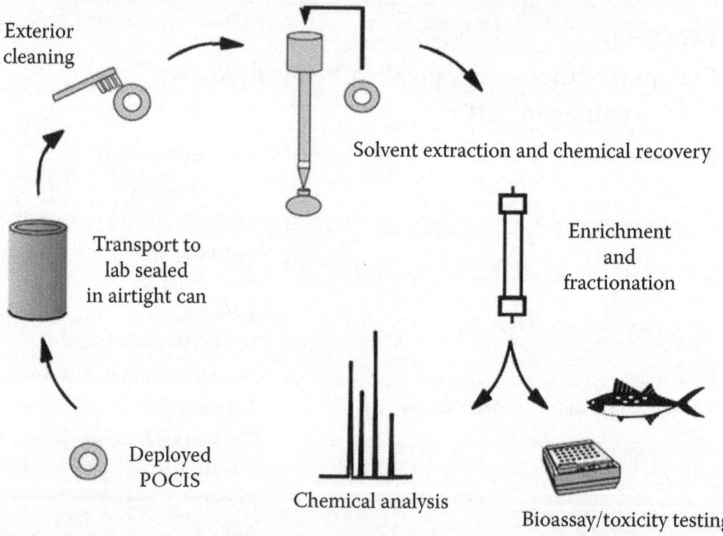

FIGURE 9.11 General POCIS processing scheme. (Reprinted from *Comprehensive Analytical Chemistry 48: Passive Sampling Techniques in Environmental Monitoring*, 48, Alvarez, D. A., J. N. Huckins, J. D. Petty, T. Jones-Lepp, F. Stuer-Lauridsen et al., Tool for monitoring hydrophilic contaminants in water: Polar organic chemical integrative sampler (POCIS), 171–197, Copyright (2007), with permission from Elsevier.)

9.2.3 Environmental Factors Affecting Passive Sampling

9.2.3.1 Water Flow

Water flow can change the width of the boundary layer, and this results in differences in the R_s of the target analytes because the uptake by passive samplers is regulated by the aqueous boundary layer at the membrane–water interface. Previous studies have shown that increases in water flow rates increase sampling rates generally by a factor of <2 (Harman et al., 2012). Li et al. (2010a) noticed that this effect depends on the compounds; thus, the increase in sampling rates is largest for more hydrophobic compounds. Alvarez et al. (2004) showed that increased water flow resulted in four to nine times higher R_s values. Likewise, other studies that have examined POCIS sampling rates on calibration experiments at different stirring rates showed a similar increase in sampling rates (Macleod et al., 2007; Kohoutek et al., 2010; Li et al., 2010a,b). On the other hand, Kingston et al. (2000) showed that uptake of diuron and atrazine in Chemcatcher is not affected by the boundary layer but by the membrane since it is relatively thicker than the boundary layer. However, while calibration experiments on water flow are easy to be conducted, they do not typically represent the environmental conditions and may inadequately present what is happening at the sampler in the field.

9.2.3.2 Temperature

Temperature seasonal and daily fluctuations can have significant effect on the uptake of the analytes. An increase in temperature can increase water solubility and desorption of organic contaminants from suspended particles, thus increasing the dissolved fraction of the analytes. Therefore, temperature can affect the uptake of the analytes, because passive samplers accumulate the dissolved fraction of the contaminants and those bound to small particles (Söderström et al., 2009). In previous studies, an increase in temperature from 15°C to 21°C increased the R_s of the studied pharmaceuticals with POCIS up to 100% (Togola and Budzinski, 2007), while a twofold increase in sampling rates from 5°C to 25°C for almost 30 PPCPs and endocrine-disrupting compounds (EDCs) was found elsewhere (Li et al., 2010a). A linear correlation of temperature and uptake of atrazine and diuron was also found in Chemcatcher (Kingston et al., 2000). Thus, it is necessary to calibrate the samplers at environmental relevant water temperature before field monitoring.

TABLE 9.7
Application of Passive Samplers Filled with Sorbent in Aquatic Environment

Passive Sampler	Type of Aquatic Environment	Type of Analytes	References
MESCO	River	α-Hexachlorocyclohexane, hexachlorobenzene	Paschke et al. (2006a,b)
	River	PAHs, PCBs, hexachlorobenzene, *p,p'*-DDE	Allan et al. (2009)
	Lake	Organic micropollutants	Neziri et al. (2011)
Chemcatcher	River	Hydrophobic organic compounds	Vrana et al. (2007)
	Stream	Pesticides	Schäfer et al. (2008a,b)
	Marine water	Organotin compounds	Aguilar-Martinez et al. (2008)
	River	PAHs, PCBs, hexachlorobenzene, *p,p'*-DDE	Allan et al. (2009)
	Marine water	Pesticides	Shaw et al. (2010)
	River	Estrogens	Kuster et al. (2010)
	Surface water	Organochlorine pesticides, PAHs	Vrana et al. (2010)
	Drain water	Amitrole	Sánchez-Bayo et al. (2013)
	River	Nonylphenol ethoxylates, nonylphenol	Ahkola et al. (2014)
POCIS	WWTP	Drugs	Jones-Lepp et al. (2004)
	River	Estrogens	Vermeirssen et al. (2005); Burki et al. (2006); Sellin et al. (2009a,b); Alvarez et al. (2013)
	Stream	Organic wastewater-related contaminants	Alvarez et al. (2005)
	River	Steroid hormones	Matthiessen et al. (2006)
	Estuary	Pharmaceuticals	Togola and Budzinski (2007)
	WWTP	PPCPs	MacLeod et al. (2007)
	River	Steroidogenic compounds	Kolok et al. (2007)
	River	EDCs, pharmaceuticals	Zhang et al. (2008)
	Surface water	Organochlorine pesticides, PCBs, PAHs	Alvarez et al. (2008)
	Marine water	EDCs	Arditsoglou and Voutsa (2008)
	River	EDCs	Rahman et al. (2008); Grabic et al. (2010); Jarosova et al. (2012); Creusot et al. (2013); Vallejo et al. (2013); Zhang et al. (2014)
	Marine water	Pharmaceuticals, pesticides, biocides	Martinez-Bueno et al. (2009)
	River	Pharmaceuticals	Bartelt-Hunt et al. (2009)
	WWTP	EDCs	Alvarez et al. (2009); Liscio et al. (2009)
	Oil platform	PAHs, alkylphenol	Harman et al. (2009a)
	River	Pesticides, hormones	Sellin et al. (2009b)
	Stream	Organic wastewater compounds	Bidwell et al. (2010)
	Lake	PPCPs, EDCs	Li et al. (2010a,b)
	River	Estrogenic UV filters	Fent et al. (2010)
	Stream	Fluoxetine	Bringolf et al. (2010)

(Continued)

TABLE 9.7 (CONTINUED)
Application of Passive Samplers Filled with Sorbent in Aquatic Environment

Passive Sampler	Type of Aquatic Environment	Type of Analytes	References
	Lake	Synthetic organic compounds	Rosen et al. (2010); Alvarez et al. (2012)
	Surface water, groundwater	PPCPs, herbicides	Dougherty et al. (2010)
	Oil production platform	PAHs, alkylphenols	Harman et al. (2010)
	River	Herbicides	Roubeix et al. (2010)
	Drinking water treatment plant	EDCs	Di Carro et al. (2010)
	WWTP	Pharmaceuticals, pesticides, biocides	Vermeirssen et al. (2010)
	Lakes	EDCs	Writer et al. (2010)
	Stream	Atrazine	Černoch et al. (2011)
	Offshore produced water recharges	PAHs, alkylphenols	Harman et al. (2011a,b,c)
	River, WWTP	Pharmaceuticals, alkylphenols, estrogenic hormones	Miège et al. (2011)
	River, WWTP	Steroidal estrogens	Rujiralai et al. (2011)
	WWTP	Pesticides, pharmaceuticals, steroid hormones	Bartelt-Hunt et al. (2011)
	Lake	Pesticides	Thomatou et al. (2011)
	WWTP	Estrogens	Kolok et al. (2012)
	WWTP, surface water	Sulfonamide antibiotics	Cernah et al. (2012)
	WWTP, surface water	Beta-blockers, hormones	Jacquet et al. (2012)
	River, wastewater	Pharmaceuticals, alkylphenols	Miège et al. (2012)
	Lake	Organic wastewater associated chemicals	Helm et al. (2012)
	Estuary	Perfluorinated alkyl carboxylates and sulfonates	Kaserzon et al. (2012)
	River	Pharmaceuticals, illicit drugs, personal care products	Jones-Lepp et al. (2012)
	Coastal water	Pharmaceuticals, alkylphenols, pesticides	Munaron et al. (2012)
	WWTP	Perfluorinated acids, sulfonates	Fedorova et al. (2013)
	Hospital wastewater	Pharmaceuticals	Bailly et al. (2013)
	River	Pesticides	Knight et al. (2013)
	Groundwater	Pesticides	Berho et al. (2013)
	WWTP	Estrogen, androgen, aryl hydrocarbon	Jálová et al. (2013)
	Surface water	Triclocarban, triclosan	Gautam et al. (2014)
	Estuary	Antibiotics, EDCs	Shi et al. (2014)
	Drinking water treatment plant	Pharmaceuticals, artificial sweetener	Metcalfe et al. (2014)
	WWTP	Naproxen, ibuprofen, triclosan	Amdany et al. (2014a,b,c)
	River	Atrazine	Dalton et al. (2014)
	Sewage collection system	Methamphetamine	Boles and Wells (2014)
	River	Synthetic steroids	Creusot et al. (2014)

(*Continued*)

TABLE 9.7 (CONTINUED)
Application of Passive Samplers Filled with Sorbent in Aquatic Environment

Passive Sampler	Type of Aquatic Environment	Type of Analytes	References
	Coastal water	Fragrances, phosphate flame retardants, pharmaceuticals, PAHs, PCBs, pesticides	Alvarez et al. (2014a,b,c)
	Tropical water	EDCs	Bayen et al. (2014)
	River	Anti-androgen mixture	Liscio et al. (2014)
	River	Emerging contaminants	Alvarez et al. (2014b)
	WWTP	Drugs of abuse	Rodayan et al. (2014)
	Sewage lagoon	PPCPs, artificial sweetener	Hoque et al. (2014)
	Surface water	Analgesics, psycholeptics, antidepressants, illicit drugs	Fedorova et al. (2014)
	Wastewater	Drugs of abuse	Yargeau et al. (2014)
	Offshore oil installation	Naphthenic acids	Harman et al. (2014)
	River	Pesticides, metabolites	Lissalde et al. (2014)

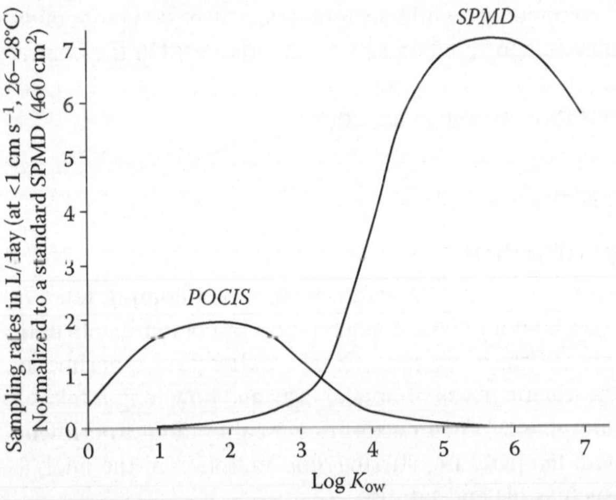

FIGURE 9.12 Comparison of POCIS and SPMD sampling rates for a wide range of organic chemical log K_{ow} values. (Reprinted from *Comprehensive Analytical Chemistry 48: Passive Sampling Techniques in Environmental Monitoring*, 48, Alvarez, D. A., J. N. Huckins, J. D. Petty, T. Jones-Lepp, F. Stuer-Lauridsen et al., Tool for monitoring hydrophilic contaminants in water: Polar organic chemical integrative sampler (POCIS), 171–197, Copyright (2007), with permission from Elsevier.)

9.2.3.3 Biofouling

Algal growth on the membrane's surface can prevent the transfer of the analytes into the receiving phase and affect the uptake to the sampler. The PES membrane, which is used in POCIS, was studied for biofouling and was proved to be less affected than SPMDs (Alvarez et al., 2004, 2007). Also, PES membrane was utilized in Empore disk-based samplers that were developed by Tran et al. (2007), assuming that biofouling will not affect uptake to these samplers too.

Fouling of POCIS affects differently the uptake of organic compounds depending on their polarity, with the less polar compounds being more affected. Therefore, it is complicated to correct uptakes for fouling using PRCs; thus, fouling should be regarded during deployment in the field.

Fouling has been also studied on other passive sampling devices for polar compounds, such as Chemcatcher and Empore disk-based receiving phases when exposed without membranes, and it was noticed that it might have reduced the uptake ability of the device because of the existence of a biofilm on the surface of the passive sampler (Tan et al., 2007; Schäfer et al., 2008a).

9.2.3.4 pH

pH can affect the R_s only if the neutral and ionized forms of selected contaminants are accumulated with different rates. According to Zhang et al. (2008), uptake in POCIS and R_s are not affected by pH when it varies from 4 to 10. They also supported that both neutral and ionized forms of the analytes are accumulated in the receiving phase. However, these data refer only to weak acids ($pK_a > 10$) while the accumulation of acid compounds ($pK_a < 7$) needs to be studied in order to conclude if pH affects their uptake to passive samplers or not. On the other hand, Li et al. (2011) studied the effect of pH on the sampling rates of some PPCPs in pharmaceutical POCIS configuration and they found that the sampling rates of acidic pharmaceuticals decreased and those of basic compounds increased when pH increased from 3 to 9, while the sampling rates of the neutral drug carbamazepine were not affected by pH change.

9.2.3.5 Salinity

Increasing salinity results to decreased water solubility and, therefore, to decreased uptake from passive samplers because, as previously mentioned, most passive samplers sampled the dissolved fraction of contaminants. Togola and Budzinski (2007) showed that salinity had low effect on the accumulation of acid compounds, while accumulation of basic compounds such as tricyclic antidepressant compounds was depended on salt concentrations. On the contrary, Zhang et al. (2008) studied the accumulation of basic compounds in POCIS and showed that salinity did not affect R_s of the target compounds (Söderström et al., 2009).

9.2.4 Calibration Methods

9.2.4.1 Laboratory Calibration

Calibration of passive samplers for the determination of sampling rates R_s includes exposure of samplers in water spiked with a known concentration of contaminants under regulated laboratory conditions (Vrana et al., 2005a; Booij et al., 2007). Calibration experiments are also performed to determine how long the kinetic phase of uptake lasts and how long it takes for contaminant uptake to reach the equilibrium phase. When performing a calibration experiment, the following parameters should be adjusted: temperature, stirring, concentration of the analytes, and other physical–chemical parameters such as pH and salinity.

Laboratory experiments have some advantages and disadvantages. The advantages are that R_s can be obtained from all the present molecules and are trustworthy because they are calculated under controlled conditions. On the other hand, it is necessary to use PRCs in the calibration method in order to adjust laboratory R_s as closer to real field R_s values, while calibration experiments are costly and time consuming.

The laboratory methodological approaches that have been used are (a) static renewals, (b) static depletion, and (c) flow-through systems.

9.2.4.1.1 Static Renewal

Static renewal calibration is widely used because of its simplicity (Alvarez, 1999; Alvarez et al., 2004, 2007; Jones-Lepp et al., 2004; MacLeod et al., 2007; Togola and Budzinski, 2007; Arditsoglou and Voutsa, 2008; Martinez-Bueno et al., 2009; Kohoutek et al., 2010). The advantages of this method are that it can prepare many replicates and test different environmental variables. Also, the concentrations of analytes remain constant because the experiments take place in dark-covered beakers, so the processes of degradation and evaporation are minimized. The experiments usually take place in small beakers (1–3 L) and are refreshed sporadically, usually daily (Alvarez et al., 2004). The frequency of water refreshing should be high in order to reduce the decrease in contaminant concentration provoked by uptake into the sampler. Daily refreshment is necessary when calibration experiments are conducted for about a month.

9.2.4.1.2 Static Depletion

Static depletion experiments are performed with a single spike of the contaminants in the water and monitoring the decrease of contaminant concentration in order to estimate R_s (Matthiessen et al., 2006; MacLeod et al., 2007; Bartelt-Hunt et al., 2011; Li et al., 2010b, 2011; Rujiralai et al., 2011). These experiments are proper when the compounds studied do not undergo degradation or adsorption in the container (Hernando et al., 2005; Matthiessen et al., 2006; Mazzella et al., 2007). High initial concentrations should be used because a decrease in contaminant's concentration follows (Bartelt-Hunt et al., 2011). Control compounds are required to estimate losses from degradation or hydrolysis. The membrane plays a major role because considerable portion of analytes may be absorbed on it, but the membrane is not analyzed or comprised in calculations (Alvarez, 1999; Vermeirssen et al., 2012).

9.2.4.1.3 Flow-Through Systems

Flow-through calibration experiments are conducted on the basis of constant supply of fortified water with known analyte concentrations to the exposure tank (Vrana et al., 2006). An advantage of this method is that passive samplers exhibit the same analyte concentration in water. Samplers are taken from the tank, and the receiving phase is analyzed to estimate the uptake of analytes. In flow-through systems, contaminant concentrations that are used are, in general, low. Therefore, the collection of large volumes of water samples will not influence the experiment as fortified water is flowing continuously. Also, mixtures prepared in acetone or methanol can be inserted directly into the tank or mixed in water and inserted into the calibration tank. Flow-through systems have been used for alkylphenols (Harman et al., 2008, 2009b) and EDCs (Zhang et al., 2008). The exposure system that was used by Harman et al. (2008) was composed of a 200 L stainless steel cylindrical tank that is protected from ambient light and contained within a temperature-controlled room. The mixture of analytes was injected at the bottom of the tank toward a stirring paddle, and samplers were placed around the stirrer. The whole system was used for 3 days before inserting the samplers to achieve steady-state water concentrations and minimize the effects of adsorption to the exposure system. Flow-through systems are particularly useful for the application of passive sampling in flowing waters or wastewaters such as rivers and the outflow of wastewater treatment plants.

9.2.4.2 In Situ Calibration

In situ calibration is a method where passive sampling devices are exposed in the field at the location where monitoring is about to take place. R_s is calculated using average concentration obtained from grab sampling (Harman et al., 2011a). The advantage of in situ calibration is that there is no need for PRCs, while R_s is based on realistic conditions. However, in situ calibration is costly and time consuming to conduct for each site studied. So, average in situ R_s can be determined in relation to different sites and different environmental conditions.

During in situ calibration, environmental conditions such as temperature, biofouling, turbulence, and physical–chemical parameters cannot be regulated. Until recently, there are no studies about the relationship of R_s with environmental parameters, so when performing in situ calibration, it is required to specify R_s for each field monitoring.

When a PRC method does not exist, in situ calibration provides a good approximation of TWA concentration because environmental factors, such as temperature, pH, flow, and salinity, are more comparable to field deployment than laboratory calibration conditions. This was demonstrated by Zhang et al. (2008). who calculated higher in situ R_s than laboratory calibration values when they performed calibration experiments for EDCs in a WWTP. Likewise, Mazzella et al. (2010) conducted an in situ calibration to compare laboratory R_s, corrected laboratory R_s, and in situ R_s. After the deployment, they concluded that in situ R_s was similar to the corrected R_s.

9.2.5 Performance Reference Compounds

A PRC is an analyte which is not present in the environment and is spiked in the receiving phase of the sampler prior to its exposure. PRCs are used to correct sampling rates which are usually different from in situ sampling rates. Therefore, PRCs are internal standards that regulate R_s according to corresponding environmental conditions. When using a PRC, it is necessary to take into account that some compounds are desorbed from the sorbent after spiking them (Mazzella et al., 2007). The desorption of the compounds has been observed in field and laboratory calibration studies (Harman et al., 2009a, 2011a), and it has to be measured during the deployment time.

To calculate the corrected sampling rates, it is inevitable to first calculate the elimination rate constant of the PRC:

$$\frac{C_s}{C_{s0}} = e^{-k_e t}, \tag{9.5}$$

where C_s is the concentration of the analyte in the receiving phase at time t (μg g^{-1}), C_{s0} is the concentration of the PRC in the receiving phase before exposure, and k_e is the elimination rate constant of the PRC.

Next, the following equation is used to calculate the corrected sampling rates:

$$R_{s(\text{corr})} = \left(\frac{k_{e\text{PRC(in situ)}}}{k_{e\text{PRC(lab)}}} \right) \times R_{s(\text{lab})}, \tag{9.6}$$

where $R_{s(\text{corr})}$ is the corrected sampling rate, $R_{s(\text{lab})}$ is the laboratory sampling rate, $k_{e\text{PRC(in situ)}}$ is the elimination constant of the PRC calculated in situ, and $k_{e\text{PRC(lab)}}$ is the elimination constant of the PRC measured in the laboratory (Morin et al., 2012).

Mazzella et al. (2010) proposed that deuterium-labeled desisopropyl atrazine could be used as a PRC in passive sampling, while Lissalde et al. (2011) used the same compound to calculate in situ sampling rates from laboratory calibrations. Deuteriorated standards could be an appropriate PRC for specific pollutant chemical groups sharing very similar chemical and physical properties with the target analytes.

Conclusively, due to the range of interactions between molecules and the sorbent, it is not possible for a PRC to be used for all the compounds (Harman et al., 2011b). Therefore, further research is required to determine which PRCs could be used for the correction of sampling rates of specific compounds or specific chemical families.

9.2.6 Use of Extracts in Bioassays

Passive samplers have also been used in biological testing in order to assess the overall toxicity of the sample (Shaw et al., 2009). The extracts obtained from the samplers are tested with a range of bioassays in order to evaluate the level and biological effects of water contaminants (Vrana et al., 2005a). Passive samplers have been used with a numerous bioassays such as the following:

- Estrogenicity (Vermeirssen et al., 2005; Burki et al., 2006; Liscio et al., 2009; Alvarez et al., 2013)
- Microtox (Sabaliūnas et al., 1999, 2000; Johnson et al., 2004; Muller et al., 2007; Alvarez et al., 2008; Roig et al., 2011)
- Mutatox (Sabaliūnas et al., 2000; Johnson et al., 2004)

- Ethoxyresorufin-*O*-deethylase activity (Johnson et al., 2004; Shaw et al., 2004; Harman et al., 2010)
- Sister chromatid exchange (Sabaliūnas et al., 2000)
- Enzyme-linked immunosorbent assay (Vermeirssen et al., 2005; Černoch et al., 2011)
- Yeast estrogen screen assay (Petty et al., 2004; Burki et al., 2006; Matthiessen et al., 2006; Alvarez et al., 2008; Greenwood et al., 2009)
- *Daphnia pulex* immobilization assay (Sabaliūnas et al., 2000)
- *D. magna* (Schäfer et al., 2007, 2008b)

Selecting the appropriate bioassay is a significant factor in the monitoring of contaminant toxicity. One important advantage of using bioassays in environmental monitoring is that they can supply a measure of all the compounds in an area, including unknown pollutants. However, the results from bioassay tests that include a number of contaminants, each one with its own toxicity and sampling rate, prevent the chance to estimate a precise TWA concentration that correlated to the calculated toxicity in the passive sampler. Another issue of combining passive sampling with bioassay tests is the representativeness of the sample, that is, if the mixture of compounds in the receiving phase corresponds to the mixture of compounds present in water during monitoring. The problem is when numerous compounds, with the same action, have different sampling rates. For example, if a compound is highly toxic but has low sampling rates, the toxicity will be underestimated. Therefore, variable sampling rates of contaminants lead to different mixtures in the receiving phase than those in the water, so the representativeness of the sample is low. Consequently, it must be examined if the sampling rates of known compounds are in a similar order (Harman et al., 2012).

9.3 AIR PASSIVE SAMPLING

9.3.1 Modeling and Principles

Mass transfer from outside into the sampler is a three-step process: transfer from ambient air to the air volume inside the chamber, from the air into the chamber to the sampler–air interface, and from the interface into the sampler (Figure 9.13). At the interface and into the sampler, molecular diffusion is the only transfer mechanism (Bartkow et al., 2005).

Molecular diffusion is described by Fick's first law from the equation

$$F = -DA\frac{dC}{dx}, \tag{9.7}$$

where F is the flux (mole time^{-1} [MT^{-1}]), D is the diffusion coefficient or diffusivity (length^2 time^{-1}, [L^{2}T^{-1}]), A is the cross-sectional area, and dC/dx is the concentration gradient (mole length^{-1}, [ML^{-1}]) perpendicular to A.

Diffusivity is the product of a velocity (LT^{-1}) and the distance that velocity applies, dx (L). The diffusive velocity can be expressed as a mass transfer coefficient k_i, which can be incorporated in Equation 9.7 to express flux in a more convenient way (Mackay, 1991):

$$F_i = k_i A \Delta C, \tag{9.8}$$

where ΔC is the difference in concentration among bulk phase i and interface. The flux from the air into the chamber to the sampler–air interface (F_V) can be given by the following equation:

$$F_V = k_V A_S (C_{VC} - C_{Vi}), \tag{9.9}$$

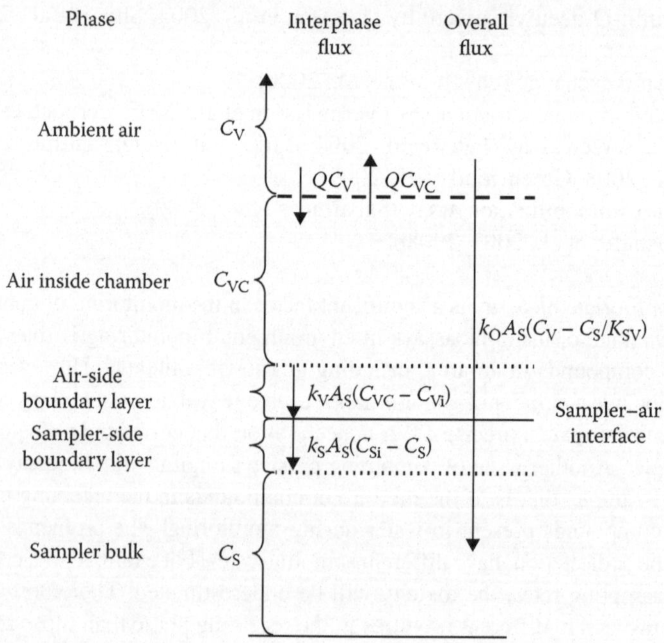

FIGURE 9.13 Schematic of mass fluxes between ambient air and passive samplers. (Reprinted from *Chemosphere*, 60, Bartkow, M. E., K. Booij, K. E. Kennedy, J. F. Müller, and D. W. Hawker, Passive air sampling theory for semivolatile organic compounds, 170–176, Copyright (2005), with permission from Elsevier.)

where k_V is the mass transfer coefficient for the air boundary layer, A_S is the area of the sampler, C_{VC} is the concentration inside the chamber, and C_{Vi} is the concentration at the air side of the interface. In the same way, the flux between the interface and the sampler bulk phase (F_S) is expressed from the equation

$$F_S = k_SA_S(C_{Si} - C_S), \quad (9.10)$$

where k_S is the mass transfer coefficient for the sampler bulk phase boundary layer, C_{Si} is the concentration at the sampler side of the interface, and C_S is the concentration in the bulk of the sampler (Bartkow et al. 2005).

Considering the equalization of the fluxes under the assumption of flux continuity (Equation 9.11) and the establishment of interface sorption equilibrium (Equation 9.12), the concentrations C_{Vi} and C_{Si} at the interface can be eliminated

$$F_V = F_S(\equiv F), \quad (9.11)$$

$$C_{Si} = K_{SV}C_{Vi}, \quad (9.12)$$

where K_{SV} is the sorption equilibrium constant at the interface.

The mass balance equation for the air contained in the chamber can be written as

$$QC_V - QC_{VC} - F_V = 0, \quad (9.13)$$

where Q (volume time^{-1}) is the rate of air supply to the chamber and C_V is the concentration in the ambient air. Replacing F_V from Equation 9.9, Equation 9.13 transforms to

$$F = k_O A_S \left(C_V - \frac{C_S}{K_{SV}} \right), \tag{9.14}$$

where k_O is the overall mass transfer coefficient and is equal to

$$\frac{1}{k_O} = \frac{1}{k_V} + \frac{1}{k_S K_{SV}} + \frac{1}{(Q/A_S)}. \tag{9.15}$$

Equation 9.15 expresses the all-out resistance ($1/k_O$), which is the sum of the resistance in the air boundary layer ($1/k_V$), the sampler boundary layer ($1/k_S K_{SV}$), and the protective chamber ($1/(Q/A_S)$). Equation 9.14 is called a "vapor-based" flux expression. Similarly, a "sampler-based" flux expression can be estimated from the following equation (Bartkow et al. 2005):

$$F = k_{OS} A_S (K_{SV} C_V - C_S), \tag{9.16}$$

where k_{OS} is defined as follows:

$$k_{OS} = \frac{k_O}{K_{SV}}. \tag{9.17}$$

The equation that regulates contaminant accumulation in passive samplers can be based either on the difference in sampler effective concentrations,

$$V_S = \frac{dC_S}{dt} = k_{OS} A_S (C_V K_{SV} - C_S), \tag{9.18}$$

or on the difference in vapor phase concentrations,

$$V_S = \frac{dC_S}{dt} = k_{OS} A_S \left(C_V - \frac{C_S}{K_{SV}} \right). \tag{9.19}$$

From Equation 9.19, the net flux, that is, the difference between uptake and elimination flux, can be expressed as

$$\frac{dC_S}{dt} = \left(k_O \frac{A_S}{V_S} \right) C_V - \left(k_O \frac{A_S}{K_{SV} V_S} \right) C_S, \tag{9.20}$$

which can be written in terms of uptake and elimination rate as

$$\frac{dC_S}{dt} = k_u C_V - k_e C_S, \tag{9.21}$$

where $k_u = k_O A_S / V_S$ and $k_e = k_O A_S / K_{SV} V_S$ are the first-order uptake and elimination rate constants, respectively.

In order to determine the first-order uptake and elimination rate constants, the overall mass transfer coefficient (k_O) can be simplified by supposing that the rate-limiting step is the mass transfer through either the air-side boundary layer ($k_O = k_V$) or the sampler-side boundary layer ($k_O = k_S K_{SV}$) or the air supply rate to the sampler chamber ($k_O = Q/A_S$) (Bartkow et al., 2005).

If mass transfer through the sampler-side boundary layer is the limiting step, uptake and elimination rate constants are determined by the following equations:

$$k_u = k_S K_{SV} \frac{A_S}{V_S}, \tag{9.22}$$

$$k_e = k_S K_S \frac{A_S}{V_S}. \tag{9.23}$$

If mass transfer through the air-side boundary layer is the limiting step, uptake and elimination constants are determined by the following equations:

$$k_u = k_V \frac{A_S}{V_S}, \tag{9.24}$$

$$k_e = k_V \frac{A_S}{V_S K_{SV}}. \tag{9.25}$$

Finally, if the flux to the sampler chamber is rate limiting, then uptake and elimination constants are determined by the following equations:

$$k_u = \frac{Q}{V_S}, \tag{9.26}$$

$$k_e = \frac{Q}{V_S K_{SV}}. \tag{9.27}$$

9.3.2 Uptake of Chemicals into Passive Samplers

The uptake of chemicals from the atmosphere into the passive samplers can be described in three stages (Figure 9.14): linear, curvilinear, and equilibrium stages.

9.3.2.1 Calculation of Sampling Rates

9.3.2.1.1 Linear Uptake

At the first stages of uptake, the concentration in the sampler (C_S) can be assumed zero; thus, the elimination rate is minor, so only uptake causes a significant change in concentration and Equation 9.21 can be simplified to

$$\frac{dC_S}{dt} \approx k_u C_V. \tag{9.28}$$

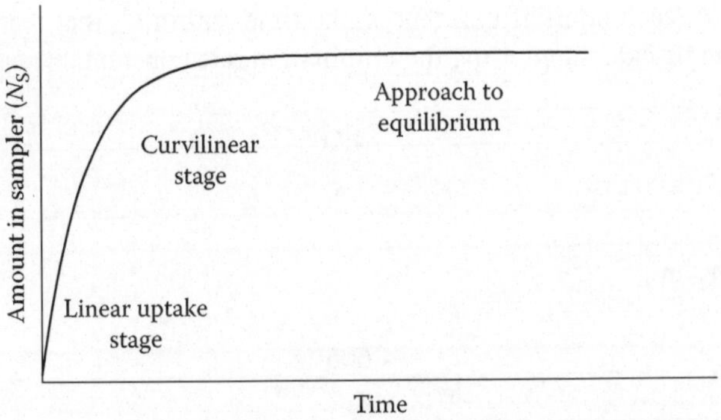

FIGURE 9.14 Theoretical uptake curve for passive air sampler showing the linear, curvilinear, and approach to equilibrium stages of uptake. (Reprinted from *Chemosphere*, 60, Bartkow, M. E., K. Booij, K. E. Kennedy, J. F. Müller, and D. W. Hawker, Passive air sampling theory for semivolatile organic compounds, 170–176, Copyright (2005), with permission from Elsevier.)

If the contaminant concentration in the ambient air remains constant, the equation integrates to

$$C_S = k_u C_V t, \tag{9.29}$$

which expresses the linear uptake of the pollutant into the sampler (Bartkow et al., 2005).

In order to calculate air concentrations (C_V), the amount of pollutant accumulated in passive samplers (M_S) can be determined by chemical analysis and Equation 9.29 can be transformed to

$$C_V = \frac{M_S}{k_u V_S t} = \frac{M_S}{R_S t}, \tag{9.30}$$

where the uptake is expressed in terms of the sampling rate, R_S (volume of air sampled per day, m^3 day^{-1}), and it is correlated to the physical characteristics of the sampler:

$$R_S = k_u V_S. \tag{9.31}$$

In the case of air-side limiting step and replacing k_u from Equation 9.24, then the sampling rate R_S is expressed as

$$R_S = k_V A_S, \tag{9.32}$$

i.e., the sampling rate is the product of air-side mass transfer coefficient and surface area of the sampler. Therefore, an increase in the surface area of the sampler increases the sampling rates, which, in practice, is very significant for compounds that exhibit very low concentrations in the atmosphere (Bartkow et al., 2005).

9.3.2.1.2 *Equilibrium Passive Sampling*

As the pollutant concentration in the sampler increases, the elimination rate becomes significant, the uptake follows a curvilinear stage, and integration of Equation 9.21 leads to

$$C_S = \frac{k_u}{k_e} C_V [1 - \exp(-k_e t)]. \tag{9.33}$$

The time needed to reach equilibrium, that is, the time within C_S that reaches 99% of the true equilibrium value, can be calculated using the elimination rate constant,

$$t_{eq} = \frac{4.605}{k_e}, \tag{9.34}$$

or the uptake rate constant,

$$t_{eq} = \frac{4.605}{k_u} K_{SV}. \tag{9.35}$$

The equilibrium time depends on the sampler and pollutant characteristics. Relationships between the physical characteristics of the sampler and the equilibrium time can be deduced by replacing the elimination or uptake rate with the expressions of Equations 9.22 through 9.25 depending on the limiting step considered. If the surface area-to-volume ratio of the sampler is increased, the equilibrium time decreases. This is followed when developing "equilibrium" samplers, such as polymer-coated glass (POG) sampler (polymer-coated glass) (Bartkow et al., 2005; Harner et al., 2003).

9.3.3 Passive Air Sampling Devices

Passive air samplers are designed in two types: tubes and badges. Tube-type samplers have a long length of diffusion path and low cross-sectional area, so they have low uptake rates. Badge-type samplers have short length of diffusion path and greater cross-sectional area, so these characteristics result in higher sampling rates.

9.3.3.1 Radiello Passive Sampler

The Radiello passive sampler is a device developed by the Fondazione Salvatore Maugeri (Padova, Italy). The Radiello system consists of three basic parts (Figure 9.15):

1. An adsorbing cartridge: a cylindrical cartridge placed into the diffusive body for the adsorption of pollutants. Different types of cartridges can be used to sample air pollutant compounds such as: benzene, toluene, ethylbenzene, and xylenes and VOCs, nitrogen dioxide (NO_2), sulfur dioxide (SO_2), and hydrogen fluoride (HF); aldehydes; ozone; hydrogen sulfide; ammonia; and HCl. The cartridge, with 5.8 mm diameter and 60 mm height, consists of a stainless steel grid filled with 530 ± 30 mg of activated carbon with 35–50 mesh.

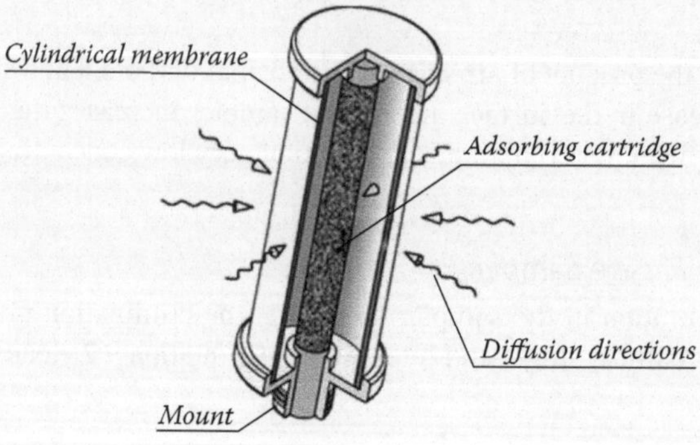

FIGURE 9.15 Design of the Radiello passive sampler (Cocheo et al. 1996). (Reproduced from Namieśnik, J. et al., *Analytical and Bioanalytical Chemistry* 381: 279–301, 2005. With permission.)

2. A diffusive body: it is cylindrical, too, with 16 mm diameter and 60 mm height, while the polyethylene wall has 1.7 mm thickness and 25 ± 5 μm porosity.
3. A base on which the diffusive body is screwed (Baechler et al., 2010).

Contaminants that are trapped in Radiello passive samplers are released in two ways, either with solvent extraction or with a two-step thermal desorption (TD). In solvent extraction, cartridges are eluted with an organic solvent and analyzed using either GC or HPLC analysis (Baechler et al., 2010; Evans and Stuart, 2011). In TD, at the first step, contaminants are transferred to a cold trap with carrier gas (helium) that was passed through the cartridge and cooled. The second step comprises the heating of the trap and transferring of contaminants with a carrier gas into the chromatographic system. GC-MS or gas chromatography coupled with selective (nitrogen phosphorus detector, ECD, etc.) or nonselective detectors (FID, thermal conductivity detector) is used to quantify the concentration of analytes. TD is preferred to solvent extraction because it is a more environmentally friendly technique and eliminates the use of organic solvents. Additionally, samples are injected directly into the chromatograph, so there is no need for further operation (Król et al., 2012; Marć et al., 2014a).

Radiello has high sampling rates for very volatile compounds, but Radiello's sensitivity declines as compound volatility declines too. Thus, Radiello's sensitivity depends on compound volatility. Until now, Radiello is proved effective in sampling organic and inorganic pollutants such as benzene–toluene–xylenes (BTX), VOCs, NO_2, SO_2, ozone (O_3), HF, and hydrogen chloride (HCl) (Namieśnik et al., 2005). Radiello passive sampler has been also used in occupational exposure experiments (Table 9.8).

9.3.3.2 Analyst Passive Sampler

Analyst passive sampler was developed in 2000 (Bertoni et al., 2000) to allow long-term sampling. Analyst (Figure 9.16) consists of a glass cylinder (20 mm i.d. × 20 mm diffusive path length) closed at the base and threaded near the opening to involve a plastic screw cap, which keeps a Teflon rubber septum drilled in the middle. The adsorbent is situated on the bottom within a stainless steel ring. The adsorbent is a granular mineral active charcoal with high specific surface area (1000 $m^2\ g^{-1}$). A finely woven stainless steel net, kept in place by a stainless steel ring, is placed in the open side of the device during sampling time to avoid accidental results of eddy diffusion (Underhill, 1993).

The denseness of the body and the inaction of the materials used reserve the absence of adverse actions, such as permeation through the walls during long-term sampling, inverse migrating through the walls during storage, and the absence of artifacts originating from the materials (Harper and Purnell, 1987).

For the sampling, a ring and a hook connected to the open side of the sampler are used in order to keep the sampler in the correct position during sampling. The sampler is constructed as an "extraction vial," which means that there is no need for further carriage and handling. When the sampling period ends, the cylinder is covered with the screw on top. The solvent and the internal standard are injected into the vial with a syringe that perforates the seal. Then, the portion of the solution that will be injected into the chromatograph is taken with a similar syringe (Bertoni et al., 2001a; Namieśnik et al., 2005). Analyst has been used for the sampling of volatile organic compounds, BTX, PAHs, nitrogen oxides, and sulfur dioxide (Table 9.8).

9.3.3.3 Polyurethane Foam Disks

Polyurethane foam (PUF) disks were developed in 2001 (Wilcockson and Gobas, 2001) to estimate chemical fugacity in fish tissues and other solid substrates. They are widely used (Shoeib and Harner, 2002; Pozo et al., 2004, 2006; Jaward et al., 2004; Harner et al., 2004, 2006a,b,c; Wilford et al., 2004; Klánová et al., 2006; Pozo et al., 2006; Gevao et al., 2006; Lee et al., 2007, Chaemfa et al., 2008) because they are not expensive and they are convenient in sample preparation, extraction, and handling. PUF disks have been used to monitor PAHs, PCBs, and organochlorine (OC) pesticides, PBDEs, and polychlorinated naphthalenes (PCNs) (Jaward et al., 2004) (Table 9.9). PUF disks have been also used in occupational exposure experiments (Bohlin et al., 2010).

TABLE 9.8
Application of Passive Samplers in Air Monitoring

Passive Sampler	Type of Monitored Air	Type of Analytes	References
Radiello	Indoor air	Benzene	Kouniali et al. (2003); Chatzis et al. (2005); Pérez Ballesta et al. (2006)
		Aldehydes	Clarisse et al. (2003); Meininghaus et al. (2003)
		Benzene, 1,3-butadiene	Strandberg et al. (2006a,b)
		BTEX, aldehydes	Allou et al. (2008)
		VOCs	Bruno et al. (2008); Missia et al. (2009); Gallego et al. (2011a,b); Kovacs et al. (2013)
	Ambient air	Ozone	Detimmerman et al. (2000); Ribas and Peñuelas (2006); Delgado Saborit and Esteve Cano (2007); Díaz-de-Quijano et al. (2009); Alejo et al. (2011)
		2-Butanone, BTX, phenol	Meybeck et al. (2000); Simon et al. (2004); Bruno et al. (2007)
		VOC	Vardoulakis et al. (2002); Bruno et al. (2008); Chiriac et al. (2009); Roukos et al. (2009); Amodio et al. (2013); Nunes et al. (2013)
		BTX	Angiuli et al. (2003); Bruno et al. (2005); Cocheo et al. (2009)
		Benzene	Murena and Vorraro (2003); Bhugwant and Siéja (2004); Pérez Ballesta et al. (2005); Król et al. (2012)
		1,3-Butadiene, benzene	Strandberg et al. (2005)
		Benzene, toluene	Bruno et al. (2006)
		Ozone, benzene	Burica et al. (2008)
		BTEX	Zabiegała et al. (2010a,b,c); Kerchich et al. (2011); Zabiegała et al. (2011); Kerchich and Kerbachi (2012); Lan and Binh (2012); Marć et al. (2014a,b)
		Ozone, NO_2	Martin et al. (2010)
		BTEX, acetone, ethanol, ozone	Roukos et al. (2011)
		VOCs, aldehydes, hydrogen sulfide	Mason et al. (2011)
		Formaldehyde, acetaldehyde	Evans and Stuart (2011)
		Ammonia	Day et al. (2012)
		H_2S	Pavilonis et al. (2013)
		NO_2, SO_2	Alejo et al. (2013)
	Personal exposure	Benzene	Crebelli et al. (2001); Skov et al. (2001); Manini et al. (2006); Lovreglio et al. (2007); Fracasso et al. (2010)
		Benzene, toluene, methyl tert-butyl ether	Sørensen et al. (2003)
		BTX	Fanou et al. (2006)
		BTEX	Manini et al. (2008)
		Benzene, toluene	Lovreglio et al. (2010)
		VOCs	Svecova et al. (2012)
		Perchloroethylene	Macca et al. (2012)
		1,3-Butadiene	Carrieri et al. (2014)

(*Continued*)

TABLE 9.8 (CONTINUED)
Application of Passive Samplers in Air Monitoring

Passive Sampler	Type of Monitored Air	Type of Analytes	References
	Workplace exposure	Formaldehyde	Bono et al. (2006)
		Benzene	Carrieri et al. (2006)
		BTX	Manini et al. (2006); Simon et al. (2006)
		1,3-Butadiene, benzene	Strandberg et al. (2014)
		BTEX, PAHs	Magagnotti et al. (2014)
Analyst	Indoor air	BTEX	Bertoni et al. (2001a,b)
		PAHs	Bertoni et al. (2004)
		Formaldehyde	Mabilia et al. (2010)
		VOCs	Moussaoui et al. (2012)
	Ambient air	Benzene	Allegrini et al. (1999)
		BTEX	Bertoni et al. (2000)
		Semivolatile PAHs	Bertoni et al. (2001b)
		VOCs	Bertoni et al. (2002a,b)
		BTX	Bertoni et al. (2002b); Wang et al. (2010a,b)
		Nitrogen oxides	De Santis et al. (2002)
		Ozone	Manes et al. (2003)
		Nitrogen oxides, sulfur dioxide, ammonia, BTX	De Santis et al. (2004)
		Nitrogen oxides, sulfur dioxide, BTX	Costabile et al. (2006)
		Formaldehyde, acetaldehyde	Villanueva et al. (2013)

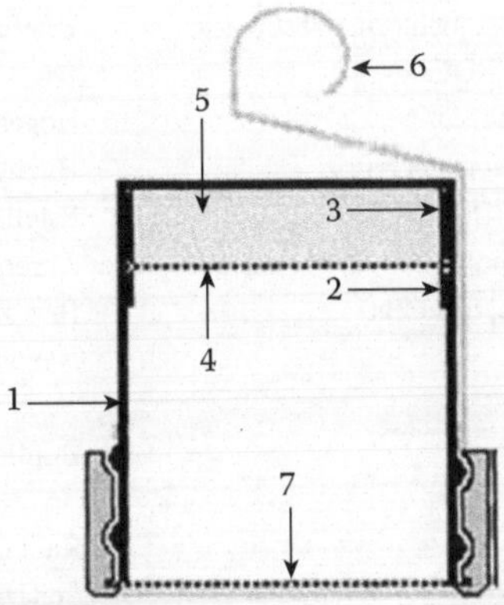

FIGURE 9.16 Structure of the Analyst sampler: 1 = glass cylinder (20 mm i.d. × 20 mm diffusive path length), 2 = retaining stainless steel ring, 3 = viewing stainless steel ring, 4 = stainless steel mesh, 5 = adsorbent bed, 6 = hook, and 7 = removable aluminum and stainless steel mesh shield (air barrier). (With kind permission from Springer Science+Business Media: *Chromatographia*, The internal consistency of the "analyst" diffusive sampler—A long-term field test, 54, 2001, 653–657, Bertoni, G., R. Tappa, and I. Allegrini.)

TABLE 9.9
Application of Passive Samplers in Air Monitoring

Passive Sampler	Type of Monitored Air	Type of Analytes	References
PUF disks	Indoor air	PAHs	Fiala et al. (2001)
		PBDEs	Wilford et al. (2004)
		PBDEs, PCBs	Harrad et al. (2006)
		PAHs, PCBs, PBDEs, organochlorine pesticides	Bohlin et al. (2008)
	Ambient air	PCBs, organochlorine pesticides	Harner et al. (2004); Chaemfa et al. (2008); Pozo et al. (2011); Tombesi et al. (2014); Bohlin et al. (2014)
		PAHs, PBDEs	Harrad and Hunter (2004); Gevao et al. (2006)
		PCNs	Herbert et al. (2005); Harner et al. (2006b); Hogarh et al. (2012a,b); Xu et al. (2014)
		PAHs, PCBs, organochlorine pesticides	Cupr et al. (2006); Klánová et al. (2006)
		PAHs	Harner et al. (2006ab,c); Santiago and Cayetano (2007); Lin et al. (2007); An et al. (2009); Cheng et al. (2013); Harner et al. (2013)
		PCBs, PBDEs	Hazrati and Harrad (2007); Li et al. (2012)
		PCDD/Fs, PCBs, PCNs	Mari et al. (2008)
		Pesticides	Gouin et al. (2008); Koblizkova et al. (2012); Meire et al. (2012)
		PCBs, PCNs, PBDEs	Baek et al. (2008)
		PBDEs	Chaemfa et al. (2009); Hearn et al. (2012)
		Organochlorine pesticides	Santiago and Cayetano (2011); Devi et al. (2011); Hogarh et al. (2014)
		PCDD/PCDFs	Ren et al. (2011)
		PCDD/Fs, PCBs, PBDEs	Li et al. (2011)
		PCBs, organochlorine pesticides, PAHs, PBDEs	Estellano et al. (2012); Pozo et al. (2012)
		PCNs, PCBs	Hogarh et al. (2012b); Hogarh et al. (2013)
		Total gaseous mercury	Huang et al. (2012)
		PAHs, PCBs, SVOCs	Estellano et al. (2013)
		Per- and polyfluoroalkyl substances	Ahrens et al. (2013)
		PCDD/Fs, di-PCBs	Heo and Lee (2014)
	Personal exposure	PBDEs	Hazrati et al. (2010)
	Workplace exposure	PAHs	Bohlin et al. (2010)
SPMD	Indoor air	PAHs	Rantalainen et al. (1999)
		PAHs, PCBs	Söderström and Bergqvist (2004); Strandberg et al. (2006a,b)
		BTEX	Esteve-Turrillas et al. (2007, 2009)
		PCBs	Kannan and Petrick (2009)

(*Continued*)

TABLE 9.9 (CONTINUED)
Application of Passive Samplers in Air Monitoring

Passive Sampler	Type of Monitored Air	Type of Analytes	References
	Ambient air	PCBs	Ockenden et al. (1998); Xu et al. (2013)
		PAHs	Bartkow et al. (2004); Piccardo et al. (2010); Zhu et al. (2013)
		PAHs, nitro-PAHs	Söderström et al. (2005)
		PCBs, organochlorine pesticides, PAHs, PBDEs	Gioia et al. (2006)
		Semi-volatile chlorinated organic compounds	Paschke et al. (2006a,b)
		Fluorotelomer alcohols	Fiedler et al. (2010)
	Workplace exposure	Organophosphorus pesticides	Esteve-Turrillas et al. (2008)
		PAHs	Bonetta et al. (2009)
		VOCs	Ly-Verdú et al. (2010)
XAD-2 resin	Ambient air	Persistent organic pollutants	Wania et al. (2003)
		Neutral perfluorinated alkyl substances	Loewen et al. (2008)
		PCBs, organochlorine pesticides	Liu et al. (2010); Baek et al. (2011)
		PCBs, organochlorine pesticides, PBDEs	Wang et al. (2010a,b)
		Pesticides	Schummer et al. (2012)
		PCBs, organochlorine pesticides, dechlorane plus	Baek et al. (2013)
		PCBs	Gueguen et al. (2013)
		Neutral per- and polyfluoroalkyl substances	Gawor et al. (2014)
		PAHs	Schummer et al. (2014)
	Workplace exposure	Sulfur mustard	Suryanarayana et al. (2001)
POGs	Indoor air	PCBs	Harner et al. (2003)
		PCBs, current use pesticides, perfluorinated compounds, PBDEs	Genualdi and Harner (2012)
	Ambient air	PCBs	Farrar et al. (2005a,b)
		PCBs, PAHs, organochlorine pesticides	Farrar et al. (2005b)
		PCBs, PAHs, PCNs, PBDEs	Farrar et al. (2006)
SPME	Indoor air	BTEX	Tumbiolo et al. (2006)
		Pesticides	Raeppel et al. (2011)
	Ambient air	Semivolatile compounds	Paschke et al. (2006a,b)
		Ozone	Lee and Tsai (2008)
		Pesticides	Wang et al. (2009)
	Personal exposure	Ethylene oxide	Tsai et al. (2004)

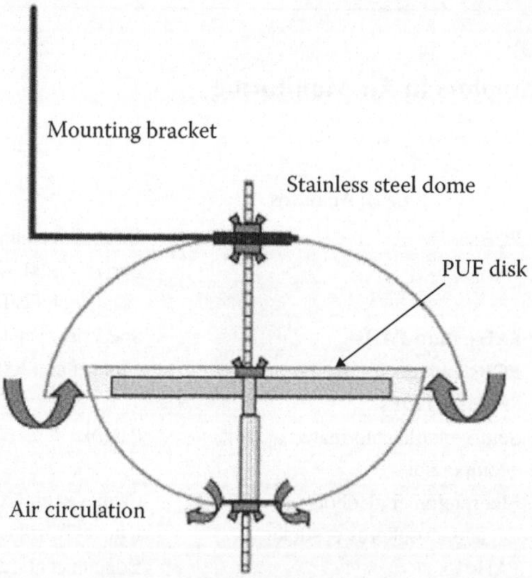

FIGURE 9.17 Design of the PUF disk passive air sampler. (Reprinted from *Environmental Pollution*, 144, Harner, T., K. Pozo, T. Gouin, A. M. Macdonald, H. Hung et al., Global pilot study for persistent organic pollutants (POPs) using PUF disk passive air samplers, 445–452, Copyright (2006), with permission from Elsevier.)

The sampler (Figure 9.17) consists of a foam disk (14 cm diameter and 1.35 cm thickness) placed in a stainless steel sampling chamber that comprises two domes. Air circulated through a 2.5 cm gap between the two domes. Two chambers exist, one to sample air indoors and the other to outdoors (Wilford et al., 2004). The uptake of pollutants has been studied and ranges between 3 and 5 m^3 day^{-1} (Shoeib and Harner, 2002; Pozo et al., 2004) have been reported.

In another approach, PUF was impregnated with XAD-4 resin in order to widen the range of compounds to be sampled to more volatile compounds (Shoeib et al., 2008).

After collection, all PUF disk samples are extracted using Soxhlet extraction for a period of 24 h. Then, the extracts are concentrated by rotary evaporation and analyzed by using gas chromatographic techniques (Tuduri et al., 2006).

9.3.3.4 Semipermeable Membrane Devices

As mentioned previously, Huckins et al. (1990a) developed an SPMD consisting of a polyethylene layflat tube that contains a thin film of large-molecular weight (≥600 Da) neutral lipid, such as triolein, for passive sampling of aquatic contaminants, but since then, it has been used to sample both indoor and outdoor atmospheres. Gas phase contaminants pass through the membrane and concentrate into the lipid phase.

SPMDs are used in occupational exposure experiments (Esteve-Turrillas et al., 2008; Bonetta et al., 2009; Ly-Verdú et al., 2010). The sampler is used for the monitoring of hydrophobic semivolatile organic compounds with log $K_{ow} > 3$ (Huckins et al., 1993), such as PAHs, polychlorinated biphenyls in indoor and outdoor air, and organochlorine pesticides (Namieśnik et al., 2005) (Table 9.9).

9.3.3.5 XAD Resin-Based Samplers

In this type of sampler the pollutants are sorpted onto the resin XAD-2, a styrene–divinyl benzene copolymer, as presented by Wania et al. (2003). The resin is put in a long cylinder made of fine stainless steel mesh kept in shape by two end caps, which is put in a protective shelter. The cylinder is filled with XAD-2 resin (60 mL of methanol slurry). The cylinder helps to protect the resin from environmental conditions, such as wind speed, precipitation, and temperature (Gioia et al., 2007).

The shelter is opened at both ends to allow air to circulate (Tuduri et al., 2012). The samplers have been used in long-range monitoring campaigns (Shen et al., 2005). The sampling rates are lower than those obtained with PUF disks or SPMDs; thus, they should be deployed for longer time periods for reaching the same sensitivity. The sampler is used for the monitoring of PCBs, OC pesticides, and PBDEs (Gioia et al., 2007) (Table 9.9).

The XAD-2 resin samples are extracted using Soxhlet extraction for 20–24 h with organic solvents. The extracts are concentrated, cleaned using silica columns, and eluted with an organic solvent. The eluates are concentrated and analyzed using GC-MS analysis (Baek et al., 2013; Jin et al., 2014).

9.3.3.6 Polymer-Coated Glass Samplers (POGs)

POGs consist of a thin film of ethylene vinyl acetate (EVA) that covers the inside and outside surfaces of a hollow glass cylinder. The sampler is prepared by dissolving 4–8 g of EVA pellets in 200–400 mL dichloromethane. Glass cylinders are immersed into the solution, and when they come out, dichloromethane evaporates and a coat of EVA is left on the glass cylinder. The surface area of POGs ranges between 280 and 300 cm^2. Uptake of pollutants in the sampler is based on the absorption of the analyte into the polymer (Gioia et al., 2007).

The sampler has been used in the monitoring of the distribution of PAHs, polychlorinated biphenyls, and organochlorine pesticides at an urban area (Farrar et al., 2005a,b) (Table 9.9).

9.3.3.7 Solid-Phase Microextraction

SPME (Figure 9.18) was first introduced in 1990 (Arthur and Pawlizsyn, 1990) and subsequently optimized and automated (Arthur et al., 1992). The sampler consists of a fused-silica fiber, coated with a thin layer of sorbent in a device that seems to be a syringe. It is a solvent-free sample preparation technique that blends sampling and preconcentration in one step and is directly inserted in a gas chromatograph. Its simplicity of use, relatively short sample processing time, and fiber reusability have made SPME an attractive choice for many analytical applications (Pawlizsyn, 1997, 1999; Zeng and Noblet, 2002).

SPME fibers can be employed in air sampling by using different configurations: (i) fully exposed SPME fiber, (ii) partially exposed SPME fiber, and (iii) retracted fiber in the needle at a given distance from the needle opening (Tuduri et al., 2012).

The device was subsequently applied for the TWA air sampling of formaldehyde, alkanes from C_5 to C_{15}, VOCs, volatile sulfur compounds, isocyanate, and *n*-valeraldehyde (Ouyang and Pawliszyn, 2007) and organic compounds in gas phase and in personal exposure assessment (Tsai et al., 2004) (Table 9.9).

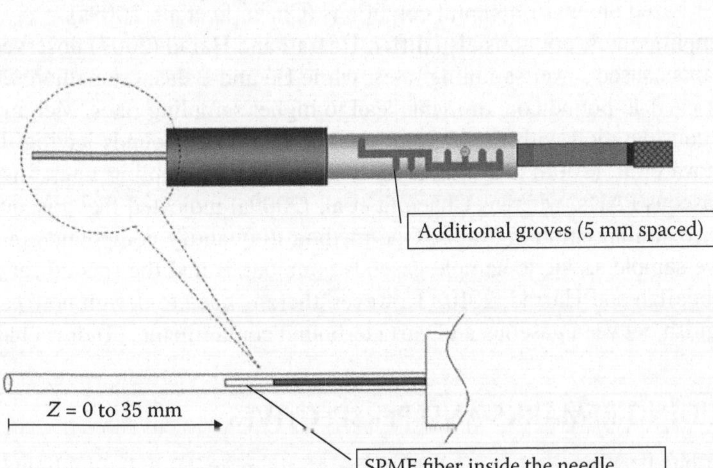

FIGURE 9.18 SPME TWA passive air sampling device. (Reprinted with permission from Chen and Pawliszyn (2003), 2004–2010. Copyright 2003 American Chemical Society.)

9.3.4 Factors Affecting Sampling Rates

The factors that have been proved to cause a variability in sampling rates are changes in air concentration, influence and change in temperature, wind speed, and the concentration and size of particles in the air (Hazrati and Harrad, 2006; Chaemfa et al., 2008, 2009b; Persoon and Hornbuckle, 2009).

9.3.4.1 Wind Speed and Temperature

Wind speed induces the size of sampling rates and along with wind direction is affected by the number of air masses sampled by the passive air sampler. A number of studies have indicated that increased wind speed reduces the thickness of the air-side boundary layer, which subsequently increases the sampling rates (Bartkow et al., 2005; Tuduri et al., 2006; Chaemfa et al., 2009a; Moeckel et al., 2009). Specifically, wind speed and sampling rates exhibit an exponential relationship (Moeckel et al., 2009), which is considerable when wind speed surpasses 3.5 m s^{-1} (Tuduri et al., 2006).

There are also other factors that affect wind speed and sampling rate relationship. These are seasonal variation, temperature and origin of the air mass. The effects of temperature are observed through changes in physical–chemical properties that control the uptake rate. Seasonal variation is also important, and increased sampling rates were observed during summer at warmer temperatures as particle phase compounds shifted toward the gas phase. This was also observed by Melymuk et al. (2011), who noticed an increase in sampling rates when temperature increased from 4°C to 16°C during deployment time.

The origin of the air mass is significant, too, because wind speed may be higher from one direction and lower from another, resulting in greater amount of air sampled from one direction due to higher sampling rates (Melymuk et al., 2011).

9.3.4.2 Particles

It is not clear yet how particle-bound contaminants affect sampling rates. First of all, according to previous studies (Chaemfa et al., 2009a; Bohlin et al., 2010; Abdallah and Harad, 2010; Melymuk et al., 2011), it is confirmed that PUF passive samplers accumulate an amount of air particulates. The quantity of particles sampled is still not clear. Chaemfa et al. (2009a) indicated that particles up to 1 μm are sampled, while Bohlin et al. (2010) assumed that even particles with size up to 2 μm can be sampled. Transfer of particles to the sampler follows four paths: sedimentation, impaction, interception, and diffusion (Bohlin et al., 2010). Small particles (<2 μm) diffuse into the sampler-like gas phase compounds (Chaemfa et al., 2009a; Bohlin et al., 2010), while sedimentation, impaction, and interception differ based on environmental conditions (Chaemfa et al., 2009a).

Regarding sampling rates, opinions also differ. Hazrati and Harad (2007) observed that particle-bound contaminants caused lower sampling rates, while He and Balasubramanian (2010) suggested the opposite, that particle-bound contaminants lead to higher sampling rates. Melymuk et al. (2011) support the first consideration with the high-molecular weight compounds having the lowest sampling rates. Klánová et al. (2008) also observed 10 times lower sampling rates for particle-bound PAHs than for gaseous PAHs, whereas Chaemfa et al. (2009a) indicated that gaseous and particle-bound PBDEs have similar sampling rates. One method to quantify compounds in both phases is to use two passive samplers, one to sample gaseous compounds and the second for particle-bound contaminants (Abdallah and Harrad, 2010). However, there is need to design new passive samplers that could distinguish between gaseous and particle-bound contaminants (Tuduri et al., 2012).

9.4 CONCLUDING REMARKS AND PERSPECTIVES

Passive sampling has been applied to a broad range of compounds, individual environments, and sampling projects. Although passive samplers have been used extensively for many years, passive sampling is still in progress. Passive samplers have been used for a long time in the short- and long-term monitoring of air quality in the local and global scales. On the contrary, the use of passive samplers for the monitoring of aquatic environments has been slower. It was not until recently that it was acknowledged that passive samplers are of great importance in water monitoring and water quality assessment within a legislative framework such as the European Union's WFD. It is expected that the aquatic monitoring will follow a development similar to that of air monitoring.

Many different devices have been developed, and among them, the most widely used passive samplers are POCIS, Chemcatcher, and SPMD for aquatic environment and Radiello, PUF disks, and SPMD for air monitoring. Passive sampling has some noteworthy advantages such as convenience in use, low cost, no need for costly and complex equipment, no need for operation, and estimation of TWA concentration. However, there are still some disadvantages that need to be overcome. The first limitation is the effect that environmental conditions have on sampling rates. Also, lack of calibration data is an issue that needs further research. Moreover, initial investigation about sorbents and extraction methods should be accomplished before laboratory calibration. Sampling rates can be very variable because of the different calibration methods, so it is needed to regulate calibrations to obtain reference sampling rates. Furthermore, laboratory calibrations need regulation considering also the frequency of sampling for water concentrations. Static calibrations are low cost, but flow-through calibrations are better approaches for the field. Also, in situ calibration could eliminate the effect of environmental conditions and the need of making routine calibrations. PRC approach needs further development to determine compounds to be used as internal standards during in situ calibration. In order to get more representative TWA concentrations, different points need more research:

- The use and performance of internal standards (PRC)
- The effect of environmental conditions and physicochemical characteristics
- Ideal duration of sampling period
- The ability of the sampler to detect low concentration levels

Finally, time-integrative and in situ collection of extracts for bioassays is an approach that gives an overall estimation of the compounds present in the environment and should be one of the first issues in the future. However, TWA concentration from toxicity levels of extracts is often underestimated, since extracts contain a mixture of compounds with different sampling rates. Therefore, an effort should be made to check if sampling rates of structurally related contaminants are in a similar range. Also, additional experiments are necessary to improve understanding of differences between the results of bioassays performed on passive sampler extracts and TWA concentrations.

Until now, the weakness of passive samplers (i.e., the semiquantitative results obtained) has been overlooked considering the important tasks that they undertook. However, from now on, the semiquantitative nature of passive sampling should be overcome and standard sampling protocols must be developed, along with new, more robust samplers, less sensitive to environmental conditions and adaptable to analytical and legislative needs of the monitoring studies.

ACKNOWLEDGMENTS

This research project has been cofinanced by the European Union (European Regional Development Fund [ERDF]) and Greek national funds through the Operational Program "THESSALY-MAINLAND GREECE AND EPIRUS-2007-2013" of the National Strategic Reference Framework (NSRF 2007–2013).

REFERENCES

Abdallah, M.A.E., Harrad, S. 2010. Modification and calibration of a passive air sampler for monitoring vapor and particulate phase brominated flame retardants in indoor air: Application to car interiors. *Environmental Science & Technology* 44: 3059–3065.

Aguilar-Martínez, R., Palacios-Corvillo, M.A., Greenwood, R., Mills, G.A., Vrana, B. et al. 2008. Calibration and use of the Chemcatcher® passive sampler for monitoring organotin compounds in water. *Analytica Chimica Acta* 618: 157–167.

Ahkola, H., Herve, S., Knuutinen, J. 2014. Study of different Chemcatcher configurations in the monitoring of nonylphenol ethoxylates and nonylphenol in aquatic environment. *Environmental Science and Pollution Research* 21: 9182–9192.

Ahrens, L., Harner, T., Shoeib, M., Koblizkova, M., Reiner, E.J. 2013. Characterization of two passive air samplers for per- and polyfluoroalkyl substances. *Environmental Science & Technology* 47: 14024–14033.

Alejo, D., Morales, M.C., Nuñez, V., Bencs, L., Van Grieken, R. et al. 2011. Monitoring of tropospheric ozone in the ambient air with passive samplers. *Microchemical Journal* 99: 383–387.

Alejo, D., Morales, M.C., De La Torre, J.B., Grau, R., Bencs, L. et al. 2013. Seasonal trends of atmospheric nitrogen dioxide and sulfur dioxide over North Santa Clara, Cuba. *Environmental Monitoring and Assessment* 185: 6023–6033.

Allan, I.J., Booij, K., Paschke, A., Vrana, B., Mills, G.A. et al. 2009. Field performance of seven passive sampling devices for monitoring of hydrophobic substances. *Environmental Science & Technology* 43: 5383–5390.

Allegrini, I., Bertoni, G., Tappa, R., Mabilia, R. 1999. Monitoring of benzene in urban and extra-urban areas by means of long-time exposure passive samplers. *Rivista dei Combustibili* 53: 225–232.

Allou, L., Marchand, C., Mirabel, Ph., Le Calvé, S. 2008. Aldehydes and BTEX measurements and exposures in university libraries in Strasbourg (France). *Indoor and Built Environment* 17: 138–145.

Alvarez, D.A. 1999. Development of an integrative sampling device for hydrophilic organic contaminants in aquatic environments. PhD thesis, University of Missouri-Columbia, Columbia, MO, p. 160.

Alvarez, D.A., Petty, J.D., Huckins, J.N., Jones-Lepp, T.L., Getting, D.T. et al. 2004. Development of a passive, in situ, integrative sampler for hydrophilic organic contaminants in aquatic environments. *Environmental Toxicology and Chemistry* 23: 1640–1648.

Alvarez, D.A., Stackelberg, P.E., Petty, J.D., Huckins, J.N., Furlong, E.T. et al. 2005. Comparison of a novel passive sampler to standard water-column sampling for organic contaminants associated with wastewater effluents entering a New Jersey stream. *Chemosphere* 61: 610–622.

Alvarez, D.A., Huckins, J.N., Petty, J.D., Jones-Lepp, T., Stuer-Lauridsen, F. et al. 2007. Tool for monitoring hydrophilic contaminants in water: Polar organic chemical integrative sampler (POCIS). In *Comprehensive Analytical Chemistry 48: Passive Sampling Techniques in Environmental Monitoring*, ed. Greenwood, R., Mills, G., Vrana, B. Elsevier, Amsterdam, pp. 171–197.

Alvarez, D.A., Cranor, W.L., Perkins, S.D., Clark, R.C., Smith, S.B. 2008. Chemical and toxicologic assessment of organic contaminants in surface water using passive samplers. *Journal of Environmental Quality* 37: 1024–1033.

Alvarez, D.A., Cranor, W.L., Perkins, S.D., Schroeder, V.L., Iwanowicz, L.R. et al. 2009. Reproductive health of bass in the Potomac, USA, drainage: Part 2. Seasonal occurrence of persistent and emerging organic contaminants. *Environmental Toxicology and Chemistry* 28: 1084–1095.

Alvarez, D.A., Rosen, M.R., Perkins, S.D., Cranor, W.L., Schroeder, V.L. et al. 2012. Bottom sediment as a source of organic contaminants in Lake Mead, Nevada, USA. *Chemosphere* 88: 605–611.

Alvarez, D.A., Shappell, N.W., Billey, L.O., Bermudez, D.S., Wilson, V.S. et al. 2013. Bioassay of estrogenicity and chemical analyses of estrogens in streams across the United States associated with livestock operations. *Water Research* 47: 3347–3363.

Alvarez, D.A., Maruya, K.A., Dodder, N.G., Lao, W., Furlong, E.T. et al. 2014a. Occurrence of contaminants of emerging concern along the California coast (2009–10) using passive sampling devices. *Marine Pollution Bulletin* 81: 347–354.

Alvarez, D., Perkins, S., Nilsen, E., Morace, J. 2014b. Spatial and temporal trends in occurrence of emerging and legacy contaminants in the Lower Columbia River 2008–2010. *Science of the Total Environment* 484: 322–330.

Amdany, R., Chimuka, L., Cukrowska, E., Kukučka, P., Kohoutek, J. et al. 2014a. Investigating the temporal trends in PAH, PCB and OCP concentrations in Hartbeespoort Dam, South Africa, using semipermeable membrane devices (SPMDs). *Water SA* 40: 425–436.

Amdany, R., Chimuka, L., Cukrowska, E., Kukučka, P., Kohoutek, J. et al. 2014b. Assessment of bioavailable fraction of POPS in surface water bodies in Johannesburg City, South Africa, using passive samplers: An initial assessment. *Environmental Monitoring and Assessment* 186: 5639–5653.

Amdany, R., Chimuka, L., Cukrowska, E. 2014c. Determination of naproxen, ibuprofen and triclosan in wastewater using the polar organic chemical integrative sampler (POCIS): A laboratory calibration and field application. *Water SA* 40: 407–414.

Amodio, M., De Gennaro, G., Marzocca, A., Trizio, L., Tutino, M. 2013. Assessment of impacts produced by anthropogenic sources in a little city near an important industrial area (Modugno, Southern Italy). *The Scientific World Journal* 2013: 150397.

An, J.G., Yim, U.H., Shim, W.J., Kim, G.B., Kim, S.K. et al. 2009. Seasonal variation of atmospheric polycyclic aromatic hydrocarbons (PAHs) on Anmyeon island. *Ocean and Polar Research* 31: 189–198.

Angiuli, L., Bruno, P., Caputi, M., Caselli, M., de Gennaro, G. et al. 2003. Radial Passive samplers for air quality monitoring in field comparison with a BTEX automatic analyser—Preliminary results. *Fresenius Environmental Bulletin* 12: 1167–1172.

Arditsoglou, A., Voutsa, D. 2008. Passive sampling of selected endocrine disrupting compounds using polar organic chemical integrative samplers. *Environmental Pollution* 156: 316–324.

Arthur, C.L., Pawliszyn, J. 1990. Solid phase microextraction with thermal desorption using fused silica optical fibers. *Analytical Chemistry* 62: 2145–2148.

Arthur, C.L., Killam, L.M., Buchholz, K.D., Pawliszyn, J. et al. 1992. Automation and optimization of solid-phase microextraction. *Analytical Chemistry* 64: 1960–1966.

Bacaloni, A., Insogna, S., Zoccolillo, L. 2011. Remote zones air quality. Persistent organic pollutants: Sources, sampling and analysis. In *Air Quality Monitoring, Assessment and Management*, ed. Mazzeo, N., pp. 223–240, InTech, Rijeka, Croatia.

Baechler, S., Comment, S., Delémont, O. 2010. Extraction and concentration of vapors from fire debris for forensic purposes: Evaluation of the use of Radiello passive air sampler. *Talanta* 82: 1247–1253.

Baek, S.Y., Choi, S.D., Lee, S.E.J., Chang, Y.S. 2008. Assessment of the spatial distribution of coplanar PCBs, PCNs, and PBDEs in a multi-industry region of South Korea using passive air samplers. *Environmental Science & Technology* 42: 7336–7340.

Baek, S.Y., Choi, S.D., Chang, Y.S. 2011. Three-year atmospheric monitoring of organochlorine pesticides and polychlorinated biphenyls in polar regions and the south pacific. *Environmental Science & Technology* 45: 4475–4482.

Baek, S.Y., Jurng, J., Chang, Y.S. 2013. Spatial distribution of polychlorinated biphenyls, organochlorine pesticides, and dechlorane plus in Northeast Asia. *Atmospheric Environment* 64: 40–46.

Bailly, E., Levi, Y., Karolak, S. 2013. Calibration and field evaluation of polar organic chemical integrative sampler (POCIS) for monitoring pharmaceuticals in hospital wastewater. *Environmental Pollution* 174: 100–105.

Baltussen, E., David, F., Sandra, P., Janssen, H.G., Cramers, C.A. 1998. Retention model for sorptive extraction–thermal desorption of aqueous samples: Application to the automated analysis of pesticides and polyaromatic hydrocarbons in water samples. *Journal of Chromatography A* 805: 237–247.

Baltussen, E., Sandra, P., David, F., Cramers, C. 1999. Stir bar sorptive extraction (SBSE), a novel extraction technique for aqueous samples: Theory and principles. *Journal of Microcolumn Separations* 11: 737–747.

Bartelt Hunt, S.L., Snow, D.D., Damon, T., Shockley, J., Hoagland, K. 2009. The occurrence of illicit and therapeutic pharmaceuticals in wastewater effluent and surface waters in Nebraska. *Environmental Pollution* 157: 786–791.

Bartelt-Hunt, S.L., Snow, D.D., Damon-Powell, T., Brown, D.L., Prasai, G. et al. 2011. Quantitative evaluation of laboratory uptake rates for pesticides, pharmaceuticals, and steroid hormones using POCIS. *Environmental Toxicology and Chemistry* 30: 1412–1420.

Bartkow, M.E., Huckins, J.N., Müller, J.F. 2004. Field-based evaluation of semipermeable membrane devices (SPMDs) as passive air samplers of polyaromatic hydrocarbons (PAHs). *Atmospheric Environment* 38: 5983–5990.

Bartkow, M.E., Booij, K., Kennedy, K.E., Müller, J.F., Hawker, D.W. 2005. Passive air sampling theory for semivolatile organic compounds. *Chemosphere* 60: 170–176.

Bayen, S., Segovia, E., Loh, L.L., Burger, D.F., Eikaas, H.S. et al. 2014. Application of polar organic chemical integrative sampler (POCIS) to monitor emerging contaminants in tropical waters. *Science of the Total Environment* 482–483: 15–22.

Benes, P., Steinnes, E. 1974. In situ dialysis for the determination of the state of trace elements in natural waters. *Water Research* 8: 947–953.

Bennett, E.R., Metcalfe, T.L., Metcalfe, C.D. 1996. Semi permeable membrane devices (SPMDS) for monitoring organic contaminants in the Otonabee River, Ontario. *Chemosphere* 33: 363–375.

Bergqvist, P.A., Strandberg, B., Rappe, C. 1993. Trace analysis—The future limits. *Analytical Proceedings* 30: 404–408.

Bergqvist, P.A., Strandberg, B., Rappe, C. 1998. Lipid removal using semipermeable membranes in PCDD and PCDF analysis of fat-rich environmental samples. *Chemosphere* 38: 933–943.

Bergqvist, P.A., Strandberg, B., Rappe, C. 1999. Lipid removal using semipermeable membranes (SPMs) in PCDD and PCDF analysis of fat-rich environmental samples. *Chemosphere* 38: 933–943.

Berho, C., Togola, A., Coureau, C., Ghestem, J.P., Amalric, L. 2013. Applicability of polar organic compound integrative samplers for monitoring pesticides in groundwater. *Environmental Science and Pollution Research* 20: 5220–5228.

Bertoni, G., Tappa, R., Allegrini, I. 2000. Assessment of a new device for the monitoring of benzene and other volatile aromatic compounds in the atmosphere. *Annali di Chimica* 90: 249–263.

Bertoni, G., Tappa, R., Allegrini, I. 2001a. The internal consistency of the "analyst" diffusive sampler—A long-term field test. *Chromatographia* 54: 653–657.

Bertoni, G., Tappa, R., Cecinato, A. 2001b. Environmental monitoring of semi-volatile polyciclic aromatic hydrocarbons by means of diffusive sampling devices and GC-MS analysis. *Chromatographia Supplies* 53: S312–S316.

Bertoni, G., Tappa, R., Bertuccio, L., Parmagnani, F. 2002a. Air monitoring of volatile aromatic compounds by means of long-term exposure diffusive samplers. *Annali di Chimica* 92: 353–361.

Bertoni, G., Ciuchini, C., Pasini, A., Tappa, R. 2002b. Monitoring of ambient BTX at Monterotondo (Rome) and indoor-outdoor evaluation in school and domestic sites. *Journal of Environmental Monitoring* 4: 903–909.

Bertoni, G., Ciuchini, C., Tappa, R. 2004. Long-term diffusive samplers for the indoor air quality evaluation. *Annali di Chimica* 94: 637–646.

Bhugwant, C., Siéja, B., Sacco, P. 2004. Distribution de la concentration de benzène autour des stations d'essences à La Réunion. *Pollution atmosphérique* 184: 455–467.

Bidwell, J.R., Becker, C., Hensley, S., Stark, R., Meyer, M.T. 2010. Occurrence of organic wastewater and other contaminants in cave streams in northeastern Oklahoma and northwestern Arkansas. *Archives of Environmental Contamination and Toxicology* 58: 286–298.

Bohlin, P., Jones, K.C., Tovalin, H., Strandberg, B. 2008. Observations on persistent organic pollutants in indoor and outdoor air using passive polyurethane foam samplers. *Atmospheric Environment* 42: 7234–7241.

Bohlin, P., Jones, K.C., Strandberg, B. 2010. Field evaluation of polyurethane foam passive air samplers to assess airborne PAHs in occupational environments. *Environmental Science & Technology* 44: 749–754.

Bohlin, P., Audy, O., Škrdlíková, L., Kukučka, P., Přibylová, P. et al. 2014. Outdoor passive air monitoring of semi volatile organic compounds (SVOCs): A critical evaluation of performance and limitations of polyurethane foam (PUF) disks. *Environmental Science: Processes & Impacts* 16: 433–444.

Boles, T.H., Wells, M.J.M. 2014. Pilot survey of methamphetamine in sewers using a Polar Organic Chemical Integrative Sampler. *Science of the Total Environment* 472: 9–12.

Bonetta, S., Carraro, E., Bonetta, S., Pignata, C., Pavan, I. et al. 2009. Application of semipermeable membrane device (SPMD) to assess air genotoxicity in an occupational environment. *Chemosphere* 75: 1446–1452.

Bono, R., Vincenti, M., Schiliro, T., Scursatone, E., Pignata, C. et al. 2006. *N*-Methylenvaline in a group of subjects occupationally exposed to formaldehyde. *Toxicology Letters* 161: 10–17.

Booij, K., Van Weerlee, E.M., Fischer, C.V., Hoedemaker, J. 2000. Passive sampling of organic contaminants in the water phase. Report from the Netherlands Institute for Sea Research: NIOZ. Report 2000-5, ISBN 0923-2310.

Booij, K., Vrana, B., Huckins, J.N. 2007. Theory, modeling, and calibration of passive samplers used in water monitoring. In *Passive Sampling Techniques in Environmental Monitoring, Vol 48—Comprehensive Analytical Chemistry*, ed. Greenwood. R., Mills, G., Vrana, B. Elsevier, Amsterdam, pp. 141–169.

Bourgeault, A., Gourlay-Francé, C. 2013. Monitoring PAH contamination in water: Comparison of biological and physico-chemical tools. *Science of the Total Environment* 454–455: 328–336.

Bringolf, R.B., Heltsley, R.M., Newton, T.J., Eads, C.B., Fraley, S.J. et al. 2010. Environmental occurrence and reproductive effects of the pharmaceutical fluoxetine in native freshwater mussels. *Environmental Toxicology and Chemistry* 29: 1311–1318.

Bruno, P., Caputi, M., Caselli, M., De Gennaro, G., De Rienzo, M. 2005. Reliability of a BTEX radial diffusive sampler for thermal desorption: Field measurements. *Atmospheric Environment* 39: 1347–1355.

Bruno, P., Caselli, M., De Gennaro, G., De Gennaro, L., Tutino, M. 2006. High spatial resolution monitoring of benzene and toluene in the Urban Area of Taranto (Italy). *Journal of Atmospheric Chemistry* 54: 177–187.

Bruno, P., Caselli, M., De Gennaro, G., Solito, M., Tutino, M. 2007. Monitoring of odor compounds produced by solid waste treatment plants with diffusive samplers. *Waste Management* 27: 539–544.

Bruno, P., Caselli, M., De Gennaro, G., Iacobellis, S., Tutino, M. 2008. Monitoring of volatile organic compounds in non-residential indoor environments. *Indoor Air* 18: 250–256.

Burki, R., Vermeirssen, E.L.M., Körner, O., Joris, C., Burkhardt-Holm, P. et al. 2006. Assessment of estrogenic exposure in brown trout (*Salmo trutta*) in a Swiss midland river: Integrated analysis of passive samplers, wild and caged fish, and vitellogenin mRNA and protein. *Environmental Toxicology and Chemistry* 25: 2077–2086.

Bustamante, J., Arana, G., de Diego, A., Madariaga, J.M. 2012. The use of SPMDs and implanted oysters for monitoring PAHs and PCBs in an aquatic environment in the estuary of Urdaibai (western Pyrenees). *Environmental Engineering and Management Journal* 11: 1707–1714.

Carrieri, M., Bonfiglio, E., Scapellato, M.L., Maccà, I., Tranfo, G. et al. 2006. Comparison of exposure assessment methods in occupational exposure to benzene in gasoline filling-station attendants. *Toxicology Letters* 162: 146–152.

Carrieri, M., Bartolucci, G.B., Paci, E., Sacco, P., Pigini, D. et al. 2014. Validation of a radial diffusive sampler for measuring occupational exposure to 1,3-butadiene. *Journal of Chromatography A* 1353: 114–120.

Černoch, I., Fránek, M., Diblíková, I., Hilscherová, K., Randák, T. et al. 2011. Determination of atrazine in surface waters by combination of POCIS passive sampling and ELISA detection. *Journal of Environmental Monitoring* 13: 2582–2587.

Chaemfa, C., Barber, J.L., Gocht, T., Harner, T., Holoubek, I. et al. 2008. Field calibration of polyurethane foam (PUF) disk passive air samplers for PCBs and OC pesticides. *Environmental Pollution* 156: 1290–1297.

Chaemfa, C., Wild, E., Davison, B., Barber, J.L., Jones, K.C. 2009a. A study of aerosol entrapment and the influence of wind speed, chamber design and foam density on polyurethane foam passive air samplers used for persistent organic pollutants. *Journal of Environmental Monitoring* 11: 1135–1139.

Chaemfa, C., Barber, J.L., Moeckel, C., Gocht, T., Harner, T. et al. 2009b. Field calibration of polyurethane foam disk passive air samplers for PBDEs. *Journal of Environmental Monitoring* 11: 1859–1865.

Chatzis, C., Alexopoulos, E.C., Linos, A. 2005. Indoor and outdoor personal exposure to benzene in Athens, Greece. *Science of the Total Environment* 349: 72–80.

Chen, Y., Pawliszyn, J. 2003. Time-weighted average passive sampling with a solid-phase microextraction device. *Analytical Chemistry* 75: 2004–2010.

Cheng, H., Deng, Z., Chakraborty, P., Liu, D., Zhang, R. et al. 2013. A comparison study of atmospheric polycyclic aromatic hydrocarbons in three Indian cities using PUF disk passive air samplers. *Atmospheric Environment* 73: 16–21.

Chiriac, R., Carré, J., Perrodin, Y., Vaillant, H., Gasso, S. et al. 2009. Study of the dispersion of VOCs emitted by a municipal solid waste landfill. *Atmospheric Environment* 43: 1926–1931.

Clarisse, B., Laurent, A.M., Seta, N., Le Moullec, Y., El Hasnaoui, A. et al. 2003. Indoor aldehydes: Measurement of contamination levels and identification of their determinants in Paris dwellings. *Environmental Research* 92: 245–253.

Cocheo, C., Boaretto, C., Pagani, D., Quaglio, F., Sacco, P. et al. 2009. Field evaluation of thermal and chemical desorption BTEX radial diffusive sampler Radiello® compared with active (pumped) samplers for ambient air measurements. *Journal of Environmental Monitoring* 11: 297–306.

Cocheo, V., Boaretto, C., Sacco, P. 1996. High uptake rate radial diffusive sampler suitable for both solvent and thermal desorption. *Journal of Occupational and Environmental Hygiene* 57: 897–904.

Costabile, F., Bertoni, G., Desantis, F., Wang, F., Weimin, H. et al. 2006. A preliminary assessment of major air pollutants in the city of Suzhou, China. *Atmospheric Environment* 40: 6380–6395.

Crebelli, R., Tomei, F., Zijno, A., Ghittori, S., Imbriani, M. et al. 2001. Exposure to benzene in urban workers: Environmental and biological monitoring of traffic police in Rome. *Occupational and Environmental Medicine* 58: 165–171.

Creusot, N., Tapie, N., Piccini, B., Balaguer, P., Porcher, J.M. et al. 2013. Distribution of steroid- and dioxin-like activities between sediments, POCIS and SPMD in a French river subject to mixed pressures. *Environmental Science and Pollution Research* 20: 2784–2794.

Creusot, N., Aït-Aïssa, S., Tapie, N., Pardon, P., Brion, F. et al. 2014. Identification of synthetic steroids in river water downstream from pharmaceutical manufacture discharges based on a bioanalytical approach and passive sampling. *Environmental Science & Technology* 48: 3649–3657.

Cupr, P., Klanova, J., Bartos, T., Flegrova, Z., Kohoutek, J. et al. 2006. Passive air sampler as a tool for long-term air pollution monitoring: Part 2. Air genotoxic potency screening assessment. *Environmental Pollution* 144: 406–413.

Dalton, R.L., Pick, F.R., Boutin, C., Saleem, A. 2014. Atrazine contamination at the watershed scale and environmental factors affecting sampling rates of the polar organic chemical integrative sampler (POCIS). *Environmental Pollution* 189: 134–142.

Day, D.E., Chen, X., Gebhart, K.A., Carrico, C.M., Schwandner, F.M. et al. 2012. Spatial and temporal variability of ammonia and other inorganic aerosol species. *Atmospheric Environment* 61: 490–498.

David, A., Gomez, E., Aït-Aïssa, S., Bachelot, M., Rosain, D. et al. 2010. Monitoring organic contaminants in small French coastal lagoons: Comparison of levels in mussel, passive sampler and sediment. *Journal of Environmental Monitoring* 12: 1471–1481.

Davison, W., Zhang, H. 1994. In situ speciation measurements of trace components in natural waters using thin-film gels. *Nature* 367: 546–548.

Delgado Saborit, J.M., Esteve Cano, V.J. 2007. Field comparison of passive samplers versus UV-photometric analyser to measure surface ozone in a Mediterranean area. *Journal of Environmental Monitoring* 9: 610–615.

De Santis, F., Dogeroglu, T., Fino, A., Menichelli, S., Vazzana, C. et al. 2002. Laboratory development and field evaluation of a new diffusive sampler to collect nitrogen oxides in the ambient air. *Analytical and Bioanalytical Chemistry* 373: 901–907.

De Santis, F., Fino, A., Menichelli, S., Vazzana, C., Allegrini, I. 2004. Monitoring the air quality around an oil refinery through the use of diffusive sampling. *Analytical and Bioanalytical Chemistry* 378: 782–788.

Detimmerman, F., Gerboles, M., Boudaa, S., Duchemann, C., De Saeger, E. 2000. Mapping of ozone levels during a photochemical episode using fast response diffusive sampler—An application: Varese monitoring campaign. *The Diffusive Monitor* 11: 9–14.

Devi, N.L., Qi, S., Chakraborty, P., Zhang, G., Yadav, I.C. 2011. Passive air sampling of organochlorine pesticides in a northeastern state of India, Manipur. *Journal of Environmental Sciences* 23: 808–815.

Díaz-de-Quijano, M., Peñuelas, J., Ribas, A. 2009. Increasing interannual and altitudinal ozone mixing ratios in the Catalan Pyrenees. *Atmospheric Environment* 43: 6049–6057.

Di Carro, M., Scapolla, C., Liscio, C., Magi, E. 2010. Development of a fast liquid chromatography-tandem mass spectrometry method for the determination of endocrine-disrupting compounds in waters. *Analytical and Bioanalytical Chemistry* 398: 1025–1034.

DiGiano, F.A., Elliott, D., Leith, D. 1989. Application of passive dosimetry to the detection of trace organic contaminants in water. *Environmental Science & Technology* 22: 1365–1367.

Dougherty, J.A., Swarzenski, P.W., Dinicola, R.S., Reinhard, M. 2010. Occurrence of herbicides and pharmaceutical and personal care products in surface water and groundwater around Liberty Bay, Puget Sound, Washington. *Journal of Environmental Quality* 39: 1173–1180.

Estellano, V.H., Pozo, K., Harner, T., Corsolini, S., Focardi, S. 2012. Using PUF disk passive samplers to simultaneously measure air concentrations of persistent organic pollutants (POPs) across the Tuscany region, Italy. *Atmospheric Pollution Research* 3: 88–94.

Estellano, V.H., Pozo, K., Silibello, C., Mulder, M.D., Efstathiou, C. et al. 2013. Characterization of urban pollution in two cities of the Puglia region in Southern Italy using field measurements and air quality (AQ) model approach. *Atmospheric Pollution Research* 5: 34–41.

Esteve-Turrillas, F.A., Pastor, A., de la Guardia, M. 2007. Assessing air quality inside vehicles and at filling stations by monitoring benzene, toluene, ethylbenzene and xylenes with the use of semipermeable devices. *Analytica Chimica Acta* 593: 108–116.

Esteve-Turrillas, F.A., Yusà, V., Pastor, A., de la Guardia, M. 2008. New perspectives in the use of semipermeable membrane devices as passive samplers. *Talanta* 74: 443–457.

Esteve-Turrillas, F.A., Ly-Verdú, S., Pastor, A., de la Guardia, M. 2009. Development of a versatile, easy and rapid atmospheric monitor for benzene, toluene, ethylbenzene and xylenes determination in air. *Journal of Chromatography A* 1216: 8549–8556.

European Parliament, Council of the European Union. 2000. Directive 2000/60/EC of the European Parliament and of the Council of 23 October 2000 establishing a framework for Community action in the field of water policy. *Official Journal of the European Communities* L 327, 22.12.2000, pp. 1–51.

Evans, A.M., Stuart, A.L. 2011. A passive sampling study of small-scale variations in ambient acetaldehyde and formaldehyde concentrations. *Air, Soil and Water Research* 4: 71–79.

Falandysz, J., Strandberg, L., Bergqvist, P.A., Kulp, S.E., Strandberg, B. et al. 1996. Polychlorinated naphthalenes in sediment and biota from the Gdansk basin, Baltic Sea. *Environmental Science & Technology* 30: 3266–3274.

Fanou, L.A., Mobio, T.A., Creppy, E.E., Fayomi, B., Fustoni, S. et al. 2006. Survey of air pollution in Cotonou, Benin—Air monitoring and biomarkers. *Science of the Total Environment* 358: 85–96.

Farrar, N.J., Harner, T.J., Sweetman, A.J., Jones, K.C. 2005a. Field calibration of rapidly equilibrating thin-film passive air samplers and their potential application for low-volume air sampling studies. *Environmental Science & Technology* 39: 261–267.

Farrar, N.J., Harner, T., Shoeib, M., Sweetman, A., Jones, K.C. 2005b. Field deployment of thin film passive air samplers for persistent organic pollutants: A study in the urban atmospheric boundary layer. *Environmental Science & Technology* 39: 42–48.

Farrar, N.J., Prevedouros, K., Harner, T., Sweetman, A.J., Jones, K.C. 2006. Continental scale passive air sampling of persistent organic pollutants using rapidly equilibrating thin films (POGs). *Environmental Pollution* 144: 423–433.

Fedorova, G., Golovko, O., Randak, T., Grabic, R. 2013. Passive sampling of perfluorinated acids and sulfonates using polar organic chemical integrative samplers. *Environmental Science and Pollution Research* 20: 1344–1351.

Fedorova, G., Randak, T., Golovko, O., Kodes, V., Grabicova, K., Grabic, R. 2014. A passive sampling method for detecting analgesics, psycholeptics, antidepressants and illicit drugs in aquatic environments in the Czech Republic. *Science of the Total Environment* 487: 681–687.

Fent, K., Zenker, A., Rapp, M. 2010. Widespread occurrence of estrogenic UV-filters in aquatic ecosystems in Switzerland. *Environmental Pollution* 158: 1817–1824.

Fiala, Z., Vyskocil, A., Krajak, V., Viau, C., Ettlerova, E. et al. 2001. Environmental exposure of small children to polycyclic aromatic hydrocarbons. *International Archives of Occupational and Environmental Health* 74: 411–420.

Fiedler, S., Pfister, G., Schramm, K.W. 2010. Partitioning of fluorotelomer alcohols (FTOH) to semipermeable membrane devices (SPMD). *Environmental Science and Pollution Research* 17: 420–428.

Fracasso, M.E., Doria, D., Bartolucci, G.B., Carrieri, M., Lovreglio, P. et al. 2010. Low air levels of benzene: Correlation between biomarkers of exposure and genotoxic effects. *Toxicology Letters* 192: 22–28.

Gallego, E., Roca, F.J., Perales, J.F., Guardino, X. 2011a. Evaluation of the effect of different sampling time periods and ambient air pollutant concentrations on the performance of the Radiello® diffusive sampler for the analysis of VOCs by TD-GC/MS. *Journal of Environmental Monitoring* 13: 2612–2622.

Gallego, E., Roca, F.J., Perales, J.F., Guardino, X. 2011b. Comparative study of the adsorption performance of an active multi-sorbent bed tube (Carbotrap, Carbopack X, Carboxen 569) and a Radiello® diffusive sampler for the analysis of VOCs. *Talanta* 85: 662–672.

Gautam, P., Carsella, J.S., Kinney, C.A. 2014. Presence and transport of the antimicrobials triclocarban and triclosan in a wastewater-dominated stream and freshwater environment. *Water Research* 48: 247–256.

Gawor, A., Shunthirasingham, C., Hayward, S.J., Lei, Y.D., Gouin, T. et al. 2014. Neutral polyfluoroalkyl substances in the global atmosphere. *Environmental Science: Processes & Impacts* 16: 404–413.

Genualdi, S., Harner, T. 2012. Rapidly equilibrating micrometer film sampler for priority pollutants in air. *Environmental Science & Technology* 46: 7661–7668.

Gevao, B., Al-Omair, A., Sweetman, A., Al-Ali, L., Al-Bahloul, M. et al. 2006. Passive sampler-derived air concentrations for polybrominated diphenyl ethers and polycyclic aromatic hydrocarbons in Kuwait. *Environmental Toxicology and Chemistry* 25: 1496–1502.

Gioia, R., Steinnes, E., Thomas, G.O., Mejier, S.N., Jones, K.C. 2006. Persistent organic pollutants in European background air: Derivation of temporal and latitudinal trends. *Journal of Environmental Monitoring* 8: 700–710.

Gioia, R., Jones, K.C., Harner, T. 2007. Tool for monitoring hydrophilic contaminants in water: Polar organic chemical integrative sampler (POCIS). In *Comprehensive Analytical Chemistry 48: Passive Sampling Techniques in Environmental Monitoring*, ed. Greenwood, R., Mills, G., Vrana, B. Elsevier, Amsterdam, pp. 33–56.

Górecki, T., Namieśnik, J. 2002. Passive sampling. *TrAC Trends in Analytical Chemistry* 21: 276–291.

Gouin, T., Wania, F., Ruepert, C., Castillo, L.E. 2008. Field testing passive air samplers for current use pesticides in a tropical environment. *Environmental Science & Technology* 42: 6625–6630.

Grabic, R., Jurcikova, J., Tomsejova, S., Ocelka, T., Halirova, J. et al. 2010. Passive sampling methods for monitoring endocrine disruptors in the Svratka and Svitava Rivers in the Czech Republic. *Environmental Toxicology and Chemistry* 29: 550–555.

Greenwood, R., Mills, G.A., Vrana, B., Allan, I.J., Aguilar-Martinez, R. et al. 2007. Monitoring of priority pollutants in water using Chemcatcher passive sampling devices. In *Comprehensive Analytical Chemistry 48: Passive Sampling Techniques in Environmental Monitoring*, ed. Greenwood, R., Mills, G., Vrana, B. Elsevier, Amsterdam, p. 199.

Grathwohl, P., Schiedek, T. 1997. Passive samplers as a long-term monitoring system for hydrophobic organic contaminants. In: Gottlieb, J. et al. (Ed.), Field Screening Conference, Karlsruhe '97. Kluwer Academic Publishers, 33–36.

Greenwood, R., Mills, G.A., Vrana, B. 2009. Potential applications of passive sampling for monitoring non-polar industrial pollutants in the aqueous environment in support of REACH. *Journal of Chromatography A* 1216: 631–639.

Gueguen, F., Stille, P., Millet, M. 2013. Optimisation and application of accelerated solvent extraction and flash chromatography for quantification of PCBs in tree barks and XAD-2 passive samplers using GC-ECD with dual columns. *Talanta* 111: 140–146.

Gunold, R., Schafer, R.B., Paschke, A., Schuurmann, G., Liess, M. 2008. Calibration of the Chemcatcher® passive sampler for monitoring selected polar and semi-polar pesticides in surface water. *Environmental Pollution* 155: 52–60.

Harman, C., Tollefsen, K.E., Bøyum, O., Thomas, K.V., Grung, M. 2008. Uptake rates of alkylphenols, PAHs and carbazoles in semipermeable membrane devices (SPMDs) and polar organic chemical integrative samplers (POCIS). *Chemosphere* 72: 1510–1516.

Harman, C., Thomas, K.V., Tollefsen, K.E., Meier, S., Boyum, O. et al. 2009a. Monitoring the freely dissolved concentrations of polycyclic aromatic hydrocarbons (PAH) and alkylphenols (AP) around a Norwegian oil platform by holistic passive sampling. *Marine Pollution Bulletin* 58: 1671–1679.

Harman, C., Bøyum, O., Thomas, K.V., Grung, M. 2009b. Small but different effect of fouling on the uptake rates of semipermeable membrane devices and polar organic chemical integrative samplers. *Environmental Toxicology and Chemistry* 28: 2324–2332.

Harman, C., Farmen, E., Tollefsen, K.E. 2010. Monitoring North Sea oil production discharges using passive sampling devices coupled with in vitro bioassay techniques. *Journal of Environmental Monitoring* 12: 1699–1708.

Harman, C., Reid, M., Thomas, K.V. 2011a. In situ calibration of a passive sampling device for selected illicit drugs and their metabolites in wastewater, and subsequent year-long assessment of community drug usage. *Environmental Science & Technology* 45: 5676–5682.

Harman, C., Allan, I.J., Bauerlein, P.S. 2011b. The challenge of exposure correction for polar passive samplers—The PRC and the POCIS. *Environmental Science & Technology* 45: 9120–9121.

Harman, C., Brooks, S., Sundt, R.C., Meier, S., Grung, M. 2011c. Field comparison of passive sampling and biological approaches for measuring exposure to PAH and alkylphenols from offshore produced water discharges. *Marine Pollution Bulletin* 63: 141–148.

Harman, C., Allan, I.J., Vermeirssen, E.L.M. 2012. Calibration and use of the polar organic chemical integrative sampler—A critical review. *Environmental Toxicology and Chemistry* 31: 2724–2738.

Harman, C., Langford, K., Sundt, R.C., Brooks, S. 2014. Measurement of naphthenic acids in the receiving waters around an offshore oil platform by passive sampling. *Environmental Toxicology and Chemistry* 33: 1946–1949.

Harner, T., Farrar, N.J., Shoeib, M., Jones, K.C., Gobas, F.A.P.C. 2003. Characterization of polymer-coated glass as a passive air sampler for persistent organic pollutants. *Environmental Science & Technology* 37: 2486–2493.

Harner, T., Shoeib, M., Diamond, M., Stern, G., Rosenberg, B. 2004. Using passive air samplers to assess urban—Rural trends for persistent organic pollutants: 1. Polychlorinated biphenyls and organochlorine pesticides. *Environmental Science & Technology* 38: 4474–4483.

Harner, T., Bartkow, M., Holoubek, I., Klanova, J., Wania, F. et al. 2006a. Passive air sampling for persistent organic pollutants: Introductory remarks to the special issue. *Environmental Pollution* 144: 361–364.

Harner, T., Shoeib, M., Gouin, T., Blanchard, P. 2006b. Polychlorinated naphthalenes in Great lakes air: Assessing spatial trends and combustion inputs using PUF disk passive air samplers. *Environmental Science & Technology* 40: 5333–5339.

Harner, T., Pozo, K., Gouin, T., Macdonald, A.M., Hung, H. et al. 2006c. Global pilot study for persistent organic pollutants (POPs) using PUF disk passive air samplers. *Environmental Pollution* 144: 445–452.

Harner, T., Su, K., Genualdi, S., Karpowicz, J., Ahrens, L. et al. 2013. Calibration and application of PUF disk passive air samplers for tracking polycyclic aromatic compounds (PACs). *Atmospheric Environment* 75: 123–128.

Harper, M., Purnell, C.J. 1987. Diffusive sampling—A review. *American Industrial Hygiene Association Journal* 48: 214–218.

Harrad, S., Hunter, S. 2004. Spatial variation in atmospheric levels of PBDEs in passive air samples on an urban–rural transect. *Organohalogen Compounds* 66: 3786–3792.

Harrad, S., Hazrati, S., Ibarra, C. 2006. Concentrations of polychlorinated biphenyls in indoor air and polybrominated diphenyl ethers in indoor air and dust in Birmingham, United Kingdom: Implications for human exposure. *Environmental Science & Technology* 40: 4633–4638.

Hazrati, S., Harrad, S. 2006. Causes of variability in concentrations of polychlorinated biphenyls and polybrominated diphenyl ethers in indoor air. *Environmental Science & Technology* 40: 7584–7589.
Hazrati, S., Harrad, S. 2007. Calibration of polyurethane foam (PUF) disk passive air samplers for quantitative measurement of polychlorinated biphenyls (PCBs) and polybrominated diphenyl ethers (PBDEs): Factors influencing sampling rates. *Chemosphere* 67: 448–455.
Hazrati, S., Harrad, S., Alighadri, M., Sadeghi, H., Mokhtari, A. et al. 2010. Passive air sampling survey of polybrominated diphenyl ether in private cars: Implications for sources and human exposure. *Iran Journal of Environmental Health Science & Engineering* 7: 157–164.
He, J., Balasubramanian, R. 2010. A comparative evaluation of passive and active samplers for measurements of gaseous semi-volatile organic compounds in the tropical atmosphere. *Atmospheric Environment* 44: 884–891.
Hearn, L.K., Kennedy, K., Hawker, D.W., Toms, L.M.L., Alberts, V. et al. 2012. Spatial mapping of city-wide PBDE levels using an exponential decay model. *Journal of Environmental Monitoring* 14: 643–650.
Helm, P.A., Howell, E.T., Li, H., Metcalfe, T.L., Chomicki, K.M. et al. 2012. Influence of nearshore dynamics on the distribution of organic wastewater-associated chemicals in Lake Ontario determined using passive samplers. *Journal of Great Lakes Research* 38: 105–115.
Heo, J., Lee, G. 2014. Field-measured uptake rates of PCDDs/Fs and dl-PCBs using PUF-disk passive air samplers in Gyeonggi-do, South Korea. *Science of the Total Environment* 491–492: 42–50.
Herbert, B.M.J., Halsall, C.J., Villa, S., Fitzpatrick, L., Jones, K.C. et al. 2005. Polychlorinated naphthalenes in air and snow in the Norwegian Arctic: A local source or an Eastern Arctic phenomenon? *Science of the Total Environment* 342: 145–160.
Hernando, M.D., Martínez-Bueno, M.J., Fernández-Alba, A.R. 2005. Seawater quality control of microcontaminants in fish farm cage systems: Application of passive sampling devices. *Boletin del Instituto Espanol de Oceanografia* 21: 37–46.
Hesslein, R.H. 1976. An in situ sampler for close interval pore water studies. *Limnology and Oceanography* 21: 912–914.
Hogarh, J.N., Seike, N., Kobara, Y., Masunaga, S. 2012a. Atmospheric polychlorinated naphthalenes in Ghana. *Environmental Science & Technology* 46: 2600–2606.
Hogarh, J.N., Seike, N., Kobara, Y., Habib, A., Nam, J.J. et al. 2012b. Passive air monitoring of PCBs and PCNs across East Asia: A comprehensive congener evaluation for source characterization. *Chemosphere* 86: 718–726.
Hogarh, J.N., Seike, N., Kobara, Y., Masunaga, S. 2013. Seasonal variation of atmospheric polychlorinated biphenyls and polychlorinated naphthalenes in Japan. *Atmospheric Environment* 80: 275–280.
Hogarh, J.N., Seike, N., Kobara, Y., Ofosu-Budu, G.K., Carboo, D. et al. 2014. Atmospheric burden of organochlorine pesticides in Ghana. *Chemosphere* 102: 1–5.
Hoque, M.E., Cloutier, F., Arcieri, C., McInnes, M., Sultana, T. et al. 2014. Removal of selected pharmaceuticals, personal care products and artificial sweetener in an aerated sewage lagoon. *Science of the Total Environment* 487: 801–812.
Huang, J., Choi, H.D., Landis, M.S., Holsen, T.M. 2012. An application of passive samplers to understand atmospheric mercury concentration and dry deposition spatial distributions. *Journal of Environmental Monitoring* 14: 2976–2982.
Huckins, J.N., Tubergen, M.W., Manuweera, G.K. 1990a. Semipermeable membrane devices containing model lipid: A new approach to monitoring the bioavailability of lipophilic contaminants and estimating their bioconcentration potential. *Chemosphere* 20: 533–552.
Huckins, J.N., Tubergen, M.W., Lebo, J.A., Gale, R.W., Schwartz, T.R. 1990b. Polymeric film dialysis in organic solvent media for cleanup of organic contaminants. *Journal of AOAC International* 73: 290–293.
Huckins, J.N., Manuweera, G.K., Petty, J.D., Mackay, D., Lebo, J.A. 1993. Lipid-containing semipermeable membrane devices for monitoring organic contaminants in water. *Environmental Science & Technology* 27: 2489–2496.
Huckins, J.N., Petty, J.D., Booij, K. 2006. *Monitors of Organic Chemicals in the Environment: Semipermeable Membrane Devices*. Springer, New York.
Jacquet, R., Miège, C., Bados, P., Schiavone, S., Coquery, M. 2012. Evaluating the polar organic chemical integrative sampler for the monitoring of beta-blockers and hormones in wastewater treatment plant effluents and receiving surface waters. *Environmental Toxicology and Chemistry* 31: 279–288.
Jálová, V., Jarošová, B., Bláha, L., Giesy, J.P., Ocelka, T. et al. 2013. Estrogen-, androgen- and aryl hydrocarbon receptor mediated activities in passive and composite samples from municipal waste and surface waters. *Environment International* 59: 372–383.

Jarosova, B., Blaha, L., Vrana, B., Randak, T., Grabic, R. et al. 2012. Changes in concentrations of hydrophilic organic contaminants and of endocrine-disrupting potential downstream of small communities located adjacent to headwaters. *Environment International* 45: 22–31.

Jaward, F.M., Farrar, N.J., Harner, T., Sweetman, A.J., Jones, K.C. 2004. Passive air sampling of PCBs, PBDEs, and organochlorine pesticides across Europe. *Environmental Science & Technology* 38: 34–41.

Jin, G., Cong, L., Wang, H., He, M., Li, J. et al. 2014. A simple and rapid analysis for gas-phase polycyclic aromatic hydrocarbons using an organic-solvent-based method. *Atmospheric Environment* 89: 367–372.

Johnson, B.T., Petty, J.D., Huckins, J.N., Lee, K., Gauthier, J. 2004. Hazard assessment of a simulated oil spill on intertidal areas of the St. Lawrence River with SPMD-TOX. *Environmental Toxicology* 19: 329–335.

Jones-Lepp, T.L., Alvarez, D.A., Petty, J.D., Huckins, J.N. 2004. Polar organic chemical integrative sampling and liquid chromatography–electrospray/ion-trap mass spectrometry for assessing selected prescription and illicit drugs in treated sewage effluents. *Archives of Environmental Contamination and Toxicology* 47: 427–439.

Jones-Lepp, T.L., Sanchez, C., Alvarez, D.A., Wilson, D.C., Taniguchi-Fu, R.L. 2012. Point sources of emerging contaminants along the Colorado River Basin: Source water for the arid Southwestern United States. *Science of the Total Environment* 430: 237–245.

Kannan, N., Petrick, G. 2009. Semi permeable membrane device reveals indoor and outdoor contamination of polychlorinated biphenyls (PCBs). *Toxicology and Environmental Health Sciences* 1: 91–96.

Kaserzon, S.L., Kennedy, K., Hawker, D.W., Thompson, J., Carter, S. et al. 2012. Development and calibration of a passive sampler for perfluorinated alkyl carboxylates and sulfonates in water. *Environmental Science & Technology* 46: 4985–4993.

Kerchich, Y., Kerbachi, R., Khatraoui, H. 2011. Ambient air levels of aromatic organic compounds in the urban area of Algiers, Algeria. *Asian Journal of Chemistry* 23: 323–330.

Kerchich, Y., Kerbachi, R. 2012. Measurement of BTEX (benzene, toluene, ethybenzene, and xylene) levels at urban and semirural areas of Algiers City using passive air samplers. *Journal of the Air & Waste Management Association* 62: 1370–1379.

Kingston, J.K., Greenwood, R., Mills, G.A., Morrison, G.M., Persson, L.B. 2000. Development of a novel passive sampling system for the time-averaged measurement of a range of organic pollutants in aquatic environments. *Journal of Environmental Monitoring* 2: 487–495.

Klánová, J., Kohoutek, J., Hamplova, L., Urbanova, P., Holoubek, I. 2006. Passive air sampler as a tool for long-term air pollution monitoring: Part 1. Performance assessment for seasonal and spatial variations. *Environmental Pollution* 144: 393–405.

Klánová, J., Eupr, P., Kohoutek, J., Harner, T. 2008. Assessing the influence of meteorological parameters on the performance of polyurethane foam-based passive air samplers. *Environmental Science & Technology* 42: 550–555.

Knight, L.A., Christenson, M.K., Trease, A.J., Davis, P.H., Kolok, A.S. 2013. The spring runoff in Nebraska's (USA) Elkhorn River watershed and its impact on two sentinel organisms. *Environmental Toxicology and Chemistry* 32: 1544–1551.

Koblizkova, M., Lee, S.C., Harner, T. 2012. Sorbent impregnated polyurethane foam disk passive air samplers for investigating current-use pesticides at the global scale. *Atmospheric Pollution Research* 3: 456–462.

Kohoutek, J., Babica, P., Bláha, L., Maršálek, B. 2008. A novel approach for monitoring of cyanobacterial toxins: Development and evaluation of the passive sampler for microcystins. *Analytical and Bioanalytical Chemistry* 390: 1167–1172.

Kohoutek, J., Maršálek, B., Bláha, L. 2010. Evaluation of the novel passive sampler for cyanobacterial toxins microcystins under various conditions including field sampling. *Analytical and Bioanalytical Chemistry* 397: 823–828.

Kolok, A.S., Snow, D.D., Kohno, S., Sellin, M.K., Guillette Jr., L.J. 2007. Occurrence and biological effect of exogenous steroids in the Elkhorn River, Nebraska, USA. *Science of the Total Environment* 388: 104–115.

Kolok, A.S., Miller, J.T., Schoenfuss, H.L. 2012. The mini mobile environmental monitoring unit: A novel bio-assessment tool. *Journal of Environmental Monitoring* 14: 202–208.

Kot, A., Wasik, A. 1999. Passive samplers as a long-term monitoring system for organic pollutants of natural waters. *V International Symposium Chemistry Forum '99*. April 19–21, 1999, Warsaw.

Kot, A., Zabiegała, B., Namieśnik, J. 2000. Passive sampling for long-term monitoring of organic pollutants in water. *TrAC Trends in Analytical Chemistry* 19: 446–459.

Kot-Wasik, A., Zabiegała, B., Urbanowicz, M., Dominiak, E., Wasik, A. et al. 2007. Advances in passive sampling in environmental studies. *Analytica Chimica Acta* 602: 141–163.

Kouniali, A., Cicolella, A., Gonzalez-Flesca, N., Dujardin, R., Gehanno, J.F. et al. 2003. Environmental benzene exposure assessment for parent-child pairs in Rouen, France. *Science of the Total Environment* 308: 73–82.

Kovacs, M.H., Roba, C., Roşu, C., Neamţiu, I., Gurzău, E. 2013. Assessment of air quality from primary school classrooms and the presence of volatile organic compounds (VOCs). *Carpathian Journal of Earth and Environmental Sciences* 8: 55–61.

Król, S., Zabiegała, B., Namieśnik, J. 2012. Measurement of benzene concentration in urban air using passive sampling. *Analytical and Bioanalytical Chemistry* 403: 1067–1082.

Kuster, M., De La Cal, A., Eljarrat, E., López De Alda, M.J., Barceló, D. 2010. Evaluation of two aquatic passive sampling configurations for their suitability in the analysis of estrogens in water. *Talanta* 83: 493–499.

Lan, T.T.N., Binh, N.T.T. 2012. Daily roadside BTEX concentrations in East Asia measured by the Lanwatsu, Radiello and Ultra I SKS passive samplers. *Science of the Total Environment* 441: 248–257.

LeBlanc, C.J., Stallard, W.M., Green, P.G., Schroeder, E.D. 2003. Passive sampling screening method using thin-layer chromatography plates. *Environmental Science & Technology* 37: 3966–3971.

Lebo, J.A., Zajicek, J.L., Orazio, C.E., Petty, J.D., Huckins, J.N. et al. 1996. Use of the semipermeable membrane device (SPMD) to sample polycyclic aromatic hydrocarbon pollution in a lotic system. *Polycyclic Aromatic Compounds* 8: 53–65.

Lee, H.L., Hardy, J.K. 1998. Passive sampling of monocyclic aromatic priority pollutants in water. *International Journal of Environmental Analytical Chemistry* 72: 83–97.

Lee, S.C., Harner, T., Pozo, K., Shoeib, M., Wania, F. et al. 2007. Polychlorinated naphthalenes in the global atmospheric passive sampling (GAPS) study. *Environmental Science & Technology* 41: 2680–2687.

Lee, I.S., Tsai SW. 2008. Passive sampling of ambient ozone by solid phase microextraction with on-fiber derivatization. *Analytica Chimica Acta* 610: 149–155.

Li, H., Vermeirssen, E.L.M., Helm, P.A., Metcalfe, C.D. 2010a. Controlled field evaluation of water flow rate effects on sampling polar organic compounds using polar organic chemical integrative samplers. *Environmental Toxicology and Chemistry* 29: 2461–2469.

Li, H., Vermeirssen, E.L.M., Helm, P.A., Metcalfe, C.D. 2010b. Sampling in the Great Lakes for pharmaceuticals, personal care products, and endocrine-disrupting substances using the passive polar organic chemical integrative sampler. *Environmental Toxicology and Chemistry* 29: 751–762.

Li, H., Helm, P.A., Paterson, G., Metcalfe, C.D. 2011. The effects of dissolved organic matter and pH on sampling rates for polar organic chemical integrative samplers (POCIS). *Chemosphere* 83: 271–280.

Li, Y., Geng, D., Liu, F., Wang, T., Wang, P. et al. 2012. Study of PCBs and PBDEs in King George Island, Antarctica, using PUF passive air sampling. *Atmospheric Environment* 51: 140–145.

Liscio, C., Magi, E., Di Carro, M., Suter, M.J.F., Vermeirssen, E.L.M. 2009. Combining passive samplers and biomonitors to evaluate endocrine disrupting compounds in a wastewater treatment plant by LC/MS/MS and bioassay analyses. *Environmental Pollution* 157: 2716–2721.

Liscio, C., Abdul-Sada, A., Al-Salhi, R., Ramsey, M.H., Hill, E.M. 2014. Methodology for profiling anti-androgen mixtures in river water using multiple passive samplers and bioassay-directed analyses. *Water Research* 57: 258–269.

Lissalde, S., Mazzella, N., Fauvelle, V., Delmas, F., Mazellier, P. et al. 2011. Liquid chromatography coupled with tandem mass spectrometry method for thirty-three pesticides in natural water and comparison of performance between classical solid phase extraction and passive sampling approaches. *Journal of Chromatography A* 1218: 1492–1502.

Lissalde, S., Mazzella, N., Mazellier, P. 2014. Polar organic chemical integrative samplers for pesticides monitoring: Impacts of field exposure conditions. *Science of the Total Environment* 488–489: 188–196.

Litten, S., Mead, B., Hassett, J. 1993. Application of passive samplers (PISCES) to locating a source of PCBs on the Black River, New York. *Environmental Toxicology and Chemistry* 12: 639–647.

Liu, W., Chen, D., Liu, X., Zheng, Z., Wen, Y. et al. 2010. Transport of semivolatile organic compounds to the Tibetan plateau: Spatial and temporal variation in air concentrations in mountainous western Sichuan, China. *Environmental Science & Technology* 44: 1559–1565.

Loewen, M., Wania, F., Wang, F., Tomy, G. 2008. Altitudinal transect of atmospheric and aqueous fluorinated organic compounds in Western Canada. *Environmental Science & Technology* 42: 2374–2379.

Lovreglio, P., Basso, A., Antelmi, A., Meliddo, G., Drago, I. et al. 2007. Influence of cigarette smoking on the excretion of urinary benzene in filling-station attendants. *Giornale italiano di medicina del lavoro ed ergonomia* 29: 291–294.

Lovreglio, P., Barbieri, A., Carrieri, M., Sabatini, L., Fracasso, M.E. et al. 2010. Validity of new biomarkers of internal dose for use in the biological monitoring of occupational and environmental exposure to low concentrations of benzene and toluene. *International Archives of Occupational and Environmental Health* 83: 341–356.

Ly-Verdú, S., Esteve-Turrillas, F.A., Pastor, A., de la Guardia, M. 2010. Determination of volatile organic compounds in contaminated air using semipermeable membrane devices. *Talanta* 80: 2041–2048.

Mabilia, R., Scipioni, C., Vegliò, F., Tomasi Scianò, M.C. 2010. Fractional factorial experiments using a test atmosphere to assess the accuracy and precision of a new passive sampler for the determination of formaldehyde in the atmosphere. *Atmospheric Environment* 44: 3942–3951.

Macca, I., Carrieri, M., Scapellato, M.L., Scopa, P., Trevisan, A. et al. 2012. Biological monitoring of exposure to perchloroethylene in dry cleaning workers. *Medicina del Lavoro* 103: 382–393.

Mackay, D. 1991. *Multimedia Environmental Models: The Fugacity Approach*. Lewis, Chelsea, MI.

MacLeod, S.L., McClure, E.L., Wong, C.S. 2007. Laboratory calibration and field deployment of the polar organic chemical integrative sampler for pharmaceuticals and personal care products in wastewater and surface water. *Environmental Toxicology and Chemistry* 26: 2517–2529.

Madsen, E.L., Thomas, C.T., Wilson, M.S., Sandoli, R.L., Bolotta, S.E. 1996. In situ dynamics of aromatic hydrocarbons and bacteria capable of AH metabolism in a coal tar waste contaminated field site. *Environmental Science & Technology* 30: 2414–2416.

Magagnotti, N., Picchi, G., Sciarra, G., Spinelli, R. 2014. Exposure of mobile chipper operators to diesel exhaust. *Annals of Occupational Hygiene* 58: 217–226.

Manes, F., De Santis, F., Giannini, M.A., Vazzana, C., Capogna, F. et al. 2003. Integrated ambient ozone evaluation by passive samplers and clover biomonitoring mini-stations. *Science of the Total Environment* 308: 133–141.

Manini, P., De Palma, G., Andreoli, R., Poli, D., Mozzoni, P. et al. 2006. Environmental and biological monitoring of benzene exposure in a cohort of Italian taxi drivers. *Toxicology Letters* 167: 142–151.

Manini, P., De Palma, G., Andreoli, R., Poli, D., Petyx, M. et al. 2008. Biological monitoring of low benzene exposure in Italian traffic policemen. *Toxicology Letters* 181: 25–30.

Marć, M., Namieśnik, J., Zabiegała, B. 2014a. BTEX concentration levels in urban air in the area of the Tri-City agglomeration (Gdansk, Gdynia, Sopot), Poland. *Air Quality, Atmosphere & Health* 7: 489–504.

Marć, M., Zabiegała, B., Namieśnik, J. 2014b. Application of passive sampling technique in monitoring research on quality of atmospheric air in the area of Tczew, Poland. *International Journal of Environmental Analytical Chemistry* 94: 151–167.

Mari, M., Schuhmacher, M., Feliubadaló, J., Domingo, J.L. 2008. Air concentrations of PCDD/Fs, PCBs and PCNs using active and passive air samplers. *Chemosphere* 70: 1637–1643.

Martin, H., Patterson, B.M., Davis, G.B., Grathwohl, P. 2003. Field trial of contaminant groundwater monitoring: Comparing time integrating ceramic dosimeters and conventional water sampling. *Environmental Science & Technology* 37: 1360–1364.

Martin, P., Cabañas, B., Villanueva, F., Gallego, M.P., Colmenar, I. et al. 2010. Ozone and nitrogen dioxide levels monitored in an urban area (Ciudad Real) in central southern Spain. *Water, Air, and Soil Pollution* 208: 305–316.

Martinez-Bueno, M.J., Hernando, M.D., Aguera, A., Fernandez-Alba, A.R. 2009. Application of passive sampling devices for screening of micro-pollutants in marine aquaculture using LC-MS/MS. *Talanta* 77: 1518–1527.

Mason, J.B., Fujita, E.M., Campbell, D.E., Zielinska, B. 2011. Evaluation of passive samplers for assessment of community exposure to toxic air contaminants and related pollutants. *Environmental Science & Technology* 45: 2243–2249.

Matthiessen, P., Arnold, D., Johnson, A.C., Pepper, T.J., Pottinger, T.G. et al. 2006. Contamination of headwater streams in the United Kingdom by oestrogenic hormones from livestock farms. *Science of the Total Environment* 367: 616–630.

Mayer, L.M. 1976. Chemical water sampling in lakes and sediments with dialysis bags. *Limnology and Oceanography* 21: 909–912.

Mazzella, N., Dubernet, J.F., Delmas, F. 2007. Determination of kinetic and equilibrium regimes in the operation of polar organic chemical integrative samplers: Application to the passive sampling of the polar herbicides in aquatic environments. *Journal of Chromatography A* 1154: 42–51.

Mazzella, N., Lissalde, S., Moreira, S., Delmas, F., Mazellier, P. et al. 2010. Evaluation of the use of performance reference compounds in an Oasis-HLB adsorbent based passive sampler for improving water concentration estimates of polar herbicides in freshwater. *Environmental Science & Technology* 44: 1713–1719.

Meadows, J., Tillitt, D., Huckins, J., Schroeder, D. 1993. Large-scale dialysis of sample lipids using a semipermeable membrane device. *Chemosphere* 26: 1993–2006.

Mehltretter, C., Sorge, H. 1995. Gore-Sorber screening surveys a passive sorptive method for detecting organic compounds in soil and groundwater. In: van den Brink, W.J., Bosman, R., Arendt, F. (Eds.), Contaminated Soil '95. Kluwer Academic Publishers, The Netherlands, 523–524.

Meininghaus, R., Kouniali, A., Mandin, C., Cicolella, A. 2003. Risk assessment of sensory irritants in indoor air-a case study in a French school. *Environment International* 28: 553–557.

Meire, R.O., Lee, S.C., Yao, Y., Targino, A.C., Torres, J.P.M. et al. 2012. Seasonal and altitudinal variations of legacy and current-use pesticides in the Brazilian tropical and subtropical mountains. *Atmospheric Environment* 59: 108–116.

Melymuk, L., Robson, M., Helm, P.A., Diamond, M.L. 2011. Evaluation of passive air sampler calibrations: Selection of sampling rates and implications for the measurement of persistent organic pollutants in air. *Atmospheric Environment* 45: 1867–1875.

Metcalfe, C., Hoque, M.E., Sultana, T., Murray, C., Helm, P. et al. 2014. Monitoring for contaminants of emerging concern in drinking water using POCIS passive samplers. *Environmental Science: Processes & Impacts* 16: 473–481.

Meybeck, M., Della Massa, J.P., Simon, V., Grasset, E., Torres, L. 2000. Etude de la distribution atmosphérique de Composés Organiques Volatils (COV) monoaromatiques: Benzène, Toluène, Xylènes (BTX) et du dioxyde d'azote sur l'agglomération toulousaine. *Pollution atmosphérique* 168: 569–582.

Miège, C., Budzinski, H., Jacquet, R., Soulier, C., Pelte, T. et al. 2011. Passive sampling using POCIS: Application for monitoring organic micropollutants in wastewater effluent and in surface water. *Techniques Sciences Methodes* 1–2: 80–94.

Miège, C., Budzinski, H., Jacquet, R., Soulier, C., Pelte, T. et al. 2012. Polar organic chemical integrative sampler (POCIS): Application for monitoring organic micropollutants in wastewater effluent and surface water. *Journal of Environmental Monitoring* 14: 626–635.

Missia, D.A., Tolis, E.I., Karanikolas, D., Kosmidis, C., Markou, A. et al. 2009. Volatile organic compounds measurements at schools in Kozani, Greece. In *Information Technologies in Environmental Engineering*, ed. Athanasiadis, I.N., Rizzoli, A.E., Mitkas, P.A., Marx Gómez, J. Springer, Berlin, pp. 203–214.

Moeckel, C., Harner, T., Nizzetto, L., Strandberg, B., Lindroth, A. et al. 2009. Use of depuration compounds in passive air samplers: Results from active sampling supported field deployment, potential uses, and recommendations. *Environmental Science & Technology* 43: 3227–3232.

Morin, N., Miège, C., Randon, J., Coquery, M. 2012. Chemical calibration, performance, validation and applications of the polar organic chemical integrative sampler (POCIS) in aquatic environments. *TrAC Trends in Analytical Chemistry* 36: 144–175.

Morrison, G.M.P. 1987. Bioavailable metal uptake rate determination in polluted waters by dialysis with receiving resins. *Environmental Technology Letters* 8: 393–402.

Moussaoui, Y., Cecinato, A., Assami, K., Meklati, B.Y. 2012. Indoor and outdoor voc emissions in urban areas of northern Algeria. *Fresenius Environmental Bulletin* 21: 190–198.

Muller, R., Tang, J.Y.M., Thier, R., Mueller, J.F. 2007. Combining passive sampling and toxicity testing for evaluation of mixtures of polar organic chemicals in sewage treatment plant effluent. *Journal of Environmental Monitoring* 9: 105–110.

Munaron, D., Tapie, N., Budzinski, H., Andral, B., Gonzalez, J.L. 2012. Pharmaceuticals, alkylphenols and pesticides in Mediterranean coastal waters: Results from a pilot survey using passive samplers. *Estuarine, Coastal and Shelf Science* 114: 82–92.

Murena, F., Vorraro, F. 2003. Vertical gradients of benzene concentration in a deep street canyon in the urban area of Naples. *Atmospheric Environment* 37: 4853–4859.

Namieśnik, J., Zabiegała, B., Kot-Wasik, A., Partyka, M., Wasik, A. 2005. Passive sampling and/or extraction techniques in environmental analysis: A review. *Analytical and Bioanalytical Chemistry* 381: 279–301.

Neziri, A., Lazo, P., Brummer, J., Paschke, A. 2011. New passive sampling systems for monitoring organic micropollutants and heavy metals in the Shkodra lake. *Journal of Environmental Protection and Ecology* 12: 2218–2224.

Nunes, T., Poceiro, C., Evtyugina, M., Duarte, M., Borrego, C. et al. 2013. Mapping anthropogenic and natural volatile organic compounds around Estarreja chemical industrial complex. *WIT Transactions on Ecology and the Environment* 174: 55–64.

Ockenden, W.A., Sweetman, A.J., Prest, H.F., Steinnes, E., Jones, K.C. 1998. Toward an understanding of the global atmospheric distribution of persistent organic pollutants: The use of semipermeable membrane devices as time integrated passive samplers. *Environmental Science & Technology* 32: 2795–2803.

Ouyang, G., Pawliszyn, J. 2007. Configurations and calibration methods for passive sampling techniques. *Journal of Chromatography A* 19: 226–235.

Paschke, A., Schwab, K., Brümmer, J., Schüürmann, G., Paschke, H. et al. 2006a. Rapid semi-continuous calibration and field test of membrane-enclosed silicone collector as passive water sampler. *Journal of Chromatography A* 1124: 187–195.

Paschke, A., Vrana, B., Popp, P., Schüürmann, G. 2006b. Comparative application of solid-phase microextraction fibre assemblies and semi-permeable membrane devices as passive air samplers for semivolatile chlorinated organic compounds: A case study on the landfill "Grube Antonie" in Bitterfeld, Germany. *Environmental Pollution* 144: 414–422.

Pavilonis, B.T., O'Shaughnessy, P.T., Altmaier, R., Metwali, N., Thorne, P.S. 2013. Passive monitors to measure hydrogen sulfide near concentrated animal feeding operations. *Environmental Science: Processes & Impacts* 15:1271–1278.

Pawliszyn, J. 1997. *Solid Phase Microextraction—Theory and Practice*. Wiley, New York.

Pawliszyn, J. (ed.). 1999. *Applications of Solid Phase Microextraction*. RSC, Cornwall.

Pérez Ballesta, P., Patier, F., Field, R., Galán, D., Baeza, A. et al. 2005. People campaign in Madrid: Assessment of outdoor, indoor and personal exposure to benzene. *Urban Air Quality*, Valencia, Spain, March 29–31, 2005.

Pérez Ballesta, P., Fild, R.A., Connolly, R., Cao, N. 2006. Population exposure to benzene: One day cross-sections in six European cities. *Atmospheric Environment* 40: 3355–3366.

Persoon, C., Hornbuckle, K.C. 2009. Calculation of passive sampling rates from both native PCBs and depuration compounds in indoor and outdoor environments. *Chemosphere* 74: 917–923.

Peterson, S.M., Apte, S.C., Batley, G.E., Coade, C. 1995. Passive sampling for chlorinated pesticides in estuarine waters. *Chemical Speciation & Bioavailability* 7: 83–88.

Petty, J.D., Huckins, J.N., Alvarez, D.A., Brumbaugh, W.G., Cranor, W.L. et al. 2004. A holistic passive integrative sampling approach for assessing the presence and potential impacts of waterborne environmental contaminants. *Chemosphere* 54: 695–705.

Piccardo, M.T., Stella, A., Pala, M., Balducci, D., Valerio, F. 2010. Field use of semipermeable membrane devices (SPMDs) for passive air sampling of polycyclic aromatic hydrocarbons: Opportunities and limitations. *Atmospheric Environment* 44: 1947–1951.

Polidoro, B.A., Morra, M.J., Ruepert, C., Castillo, L.E. 2009. Pesticide sequestration in passive samplers (SPMDs): Considerations for deployment time, biofouling, and stream flow in a tropical watershed. *Journal of Environmental Monitoring* 11: 1866–1874.

Polito, J.M., Hassett, J.P. 2002. Sampling for hydrophobic organic compounds using passive in-situ concentration extraction sampler (PISCES). *ACS Division of Environmental Chemistry, Preprints* 42: 714–719.

Pozo, K., Harner, T., Shoeib, M., Urrutia, R., Barra, R. et al. 2004. Passive-sampler derived air concentrations of persistent organic pollutants (POPs) on a north–south transect in Chile. *Environmental Science & Technology* 38: 6529–6537.

Pozo, K., Harner, T., Wania, F., Muir, D.C.G., Jones, K.C. et al. 2006. Toward a global network for persistent organic pollutants in air: Results from the GAPS study. *Environmental Science & Technology* 40: 4867–4873.

Pozo, K., Harner, T., Lee, S.C., Sinha, R.K., Sengupta, B. et al. 2011. Assessing seasonal and spatial trends of persistent organic pollutants (POPs) in Indian agricultural regions using PUF disk passive air samplers. *Environmental Pollution* 159: 646–653.

Pozo, K., Harner, T., Rudolph, A., Oyola, G., Estellano, V.H. et al. 2012. Survey of persistent organic pollutants (POPs) and polycyclic aromatic hydrocarbons (PAHs) in the atmosphere of rural, urban and industrial areas of Concepción, Chile, using passive air samplers. *Atmospheric Pollution Research* 3: 426–434.

Raeppel, C., Nief, M., Fabritius, M., Racault, L., Appenzeller, B.M. et al. 2011. Simultaneous analysis of pesticides from different chemical classes by using a derivatisation step and gas chromatography-mass spectrometry. *Journal of Chromatography A* 1218: 8123–8129.

Rahman, S., Moman, J., Angus, R.A., Peters, R.W. 2008. Treatment of endocrine disrupting compounds in water. *AIChE Annual Meeting, Conference Proceedings*.

Rantalainen, A.L., Hyötyläinen, T., Saramo, M., Niskanen, I. 1999. Passive sampling of PAHs in indoor air in Nepal. *Toxicological & Environmental Chemistry* 68: 335–348.

Ren, Z., Zhang, B., Lu, P., Li, C., Gao, L. et al. 2011. Characteristics of air pollution by polychlorinated dibenzo-*p*-dioxins and dibenzofurans in the typical industrial areas of Tangshan City, China. *Journal of Environmental Sciences* 23: 228–235.

Ribas, A., Peñuelas, J. 2006. Surface ozone mixing ratio increase with altitude in a transect in the Catalan Pyrenees. *Atmospheric Environment* 40: 7308–7315.

Rodayan, A., Majewsky, M., Yargeau, V. 2014. Impact of approach used to determine removal levels of drugs of abuse during wastewater treatment. *Science of the Total Environment* 487: 731–739.

Roig, N., Nadal, M., Sierra, J., Ginebreda, A., Schuhmacher, M. et al. 2011. Novel approach for assessing heavy metal pollution and ecotoxicological status of rivers by means of passive sampling methods. *Environment International* 37: 671–677.

Rosen, M.R., Rowe, T.G., Goodbred, S.L., Shipley, D.O., Arufe, J.A. 2006. Importance of land use, streamflow, and water quality on stream toxicity in the Lake Tahoe and Truckee river watersheds. *American Fisheries Society Symposium* 2006: 129–149.

Rosen, M.R., Alvarez, D.A., Goodbred, S.L., Leiker, T.J., Patiño, R. 2010. Sources and distribution of organic compounds using passive samplers in Lake Mead national recreation area, Nevada and Arizona, and their implications for potential effects on aquatic biota. *Journal of Environmental Quality* 39: 1161–1172.

Roubeix, V., Mazzella, N., Delmas, F., Coste, M. 2010. In situ evaluation of herbicide effects on the composition of river periphytic diatom communities in a region of intensive agriculture. *Vie et Milieu* 60: 233–241.

Roukos, J., Riffault, V., Locoge, N., Plaisance, H. 2009. VOC in an urban and industrial harbor on the French North Sea coast during two contrasted meteorological situations. *Environmental Pollution* 157: 3001–3009.

Roukos, J., Locoge, N., Sacco, P., Plaisance, H. 2011. Radial diffusive samplers for determination of 8-h concentration of BTEX, acetone, ethanol and ozone in ambient air during a sea breeze event. *Atmospheric Environment* 45: 755–763.

Rujiralai, T., Bull, I.D., Llewellyn, N., Evershed, R.P. 2011. In situ polar organic chemical integrative sampling (POCIS) of steroidal estrogens in sewage treatment works discharge and river water. *Journal of Environmental Monitoring* 13: 1427–1434.

Sabaliūnas, D., Ellington, J., Sabaliūniene, I. 1999. Screening bioavailable hydrophobic toxicants in surface waters with semipermeable membrane devices: Role of inherent oleic acid in toxicity evaluations. *Ecotoxicology and Environmental Safety* 44: 160–167.

Sabaliūnas, D., Lazutka, J.R., Sabaliūniene, I. 2000. Acute toxicity and genotoxicity of aquatic hydrophobic pollutants sampled with semipermeable membrane devices. *Environmental Pollution* 109: 251–265.

Sánchez-Bayo, F., Hyne, R.V., Kibria, G., Doble, P. 2013. Calibration and field application of Chemcatcher® passive samplers for detecting amitrole residues in agricultural drain waters. *Bulletin of Environmental Contamination and Toxicology* 90: 635–639.

Santiago, E.C., Cayetano, M.G. 2007. Polycyclic aromatic hydrocarbons in ambient air in the Philippines derived from passive sampler with polyurethane foam disk. *Atmospheric Environment* 41: 4138–4147.

Sayato, Y., Nakamuro, K., Ueno, H., Goto, R. 1990. Mutagenicity of adsorbates to a copper-phthalocyanine derivative recovered from municipal river water. *Mutation Research* 242: 313–317.

Santiago, E.C., Cayetano, M.G. 2011. Organochlorine pesticides in ambient air in selected urban and rural residential areas in the Philippines derived from passive samplers with polyurethane disks. *Bulletin of Environmental Contamination and Toxicology* 86: 50–55.

Schäfer, R.B., Caquet, T., Siimes, K., Mueller, R., Lagadic, L. et al. 2007. Effects of pesticides on community structure and ecosystem functions in agricultural streams of three biogeographical regions in Europe. *Science of the Total Environment* 382: 272–285.

Schäfer, R.B., Paschke, A., Liess, M. 2008a. Aquatic passive sampling of a short-term thiacloprid pulse with the Chemcatcher: Impact of biofouling and use of a diffusion-limiting membrane on the sampling rate. *Journal of Chromatography A* 1203: 1–6.

Schäfer, R.B., Paschke, A., Vrana, B., Mueller, R., Liess, M. 2008b. Performance of the Chemcatcher® passive sampler when used to monitor 10 polar and semi-polar pesticides in 16 Central European streams, and comparison with two other sampling methods. *Water Research* 42: 2707–2717.

Schubauer-Berigan, J.P., Foote, E.A., Magar, V.S. 2012. Using SPMDs to assess natural recovery of PCB-contaminated sediments in lake Hartwell, SC: I. A field test of new in-situ deployment methods. *Soil and Sediment Contamination* 21: 82–100.

Schummer, C., Tuduri, L., Briand, O., Appenzeller, B.M., Millet, M. 2012. Application of XAD-2 resin-based passive samplers and SPME-GC-MS/MS analysis for the monitoring of spatial and temporal variations of atmospheric pesticides in Luxembourg. *Environmental Pollution* 170: 88–94.

Schummer, C., Appenzeller, B.M., Millet, M. 2014. Monitoring of polycyclic aromatic hydrocarbons (PAHs) in the atmosphere of southern Luxembourg using XAD-2 resin-based passive samplers. *Environmental Science and Pollution Research* 21: 2098–2107.

Seethapathy, S., Górecki, T., Li, X. 2008. Passive sampling in environmental analysis. *Journal of Chromatography A* 1184: 234–253.

Sellin, M.K., Snow, D.D., Akerly, D.L., Kolok, A.S. 2009a. Estrogenic compounds downstream from three small cities in Eastern Nebraska: Occurrence and biological effect. *J Am Water Research Assoc* 45: 14–21.

Sellin, M.K., Snow, D.D., Schwarz, M., Carter, B.J., Kolok, A.S. 2009b. Agrichemicals in Nebraska, USA, watersheds: Occurrence and endocrine effects. *Environmental Toxicology and Chemistry* 28: 2443–2448.

Shaw, M., Tibbetts, I.R., Müller, J.F. 2004. Monitoring PAHs in the Brisbane River and Moreton Bay, Australia, using semipermeable membrane devices and EROD activity in yellowfin bream, Acanthopagrus australis. *Chemosphere* 56: 237–246.

Shaw, M., Eaglesham, G., Muller, J.F. 2009. Uptake and release of polar compounds in SDB-RPS Empore disks; implications for their use as passive samplers. *Chemosphere* 75: 1–7.

Shaw, M., Furnas, M.J., Fabricius, K., Haynes, D., Carter, S. et al. 2010. Monitoring pesticides in the Great Barrier Reef. *Marine Pollution Bulletin* 60: 113–122.

Shen, L., Wania, F., Lei, Y.D., Teixeira, C., Muir, D.C.G. et al. 2005. Atmospheric distribution and long-range transport of organochlorine pesticides in North America. *Environmental Science & Technology* 39: 409–420.

Shi, X., Zhou, J.L., Zhao, H., Hou, L., Yang, Y. 2014. Application of passive sampling in assessing the occurrence and risk of antibiotics and endocrine disrupting chemicals in the Yangtze Estuary, China. *Chemosphere* 111: 344–351.

Shoeib, M., Harner, T. 2002. Characterization and comparison of three passive air samplers for persistent organic pollutants. *Environmental Science & Technology* 36: 4142–4151.

Shoeib, M., Harner, T., Lee, S.C., Lane, D., Zhu, J. 2008. Sorbent-impregnated polyurethane foam disk for passive air sampling of volatile fluorinated chemicals. *Analytical Chemistry* 80: 675–682.

Simon, V., Baer, M., Torres, L., Olivier, S., Meybeck, M. et al. 2004. The impact of reduction in the benzene limit value in gasoline on airborne benzene, toluene and xylenes levels. *Science of the Total Environment* 334– 335: 177–183.

Simon, V., Borownjack, C., Touya, C. 2006. Benzene, toluene and xylene measurements in the vicinity of petrol stations (Toulouse—France). *WIT Transactions on the Built Environment* 89: 663–668.

Skov, H., Hansen, A.B., Lorenzen, G., Andersen, H.V., Løfstrøm, P. et al. 2001. Benzene exposure and the effect of traffic pollution in Copenhagen, Denmark. *Atmospheric Environment* 35: 2463–2471.

Södergren, A. 1987. Solvent-filled dialysis membranes simulate uptake of pollutants by aquatic organisms. *Environmental Science & Technology* 21: 855–859.

Söderström, H.S., Bergqvist, P.A. 2004. Passive air sampling using semipermeable membrane devices at different wind-speeds in situ calibrated by performance reference compounds. *Environmental Science & Technology* 38: 4828–4834.

Söderström, H., Hajšlová, J., Kocourek, V., Siegmund, B., Kocan, A. et al. 2005. PAHs and nitrated PAHs in air of five European countries determined using SPMDs as passive samplers. *Atmospheric Environment* 39: 1627–1640.

Söderström, H., Lindberg, R.H., Fick, J. 2009. Strategies for monitoring the emerging polar organic contaminants in water with emphasis on integrative passive sampling. *Journal of Chromatography A* 1216: 623–630.

Sørensen, M., Skov, H., Autrup, H., Hertel, O., Loft, S. 2003. Urban benzene exposure and oxidative DNA damage: Influence of genetic polymorphisms in metabolism genes. *Science of the Total Environment* 309: 69–80.

Srikameswaran, K., Gesser, H.D., Venkateswaran, M. 1984. Extraction of cadmium from natural waters by passive complexing agent loaded foam strips. *Journal of Environmental Sciences Health* A 19: 83–91.

Strandberg, B., Bergqvist, P.A., Rappe, C. 1998. Dialysis with semipermeable membranes as an efficient lipid removal method in the analysis of bioaccumulative chemicals. *Analytical Chemistry* 70: 526–533.

Strandberg, B., Sunesson, A.L., Olsson, K., Levin, J.O., Ljungqvist, G. et al. 2005. Evaluation of two types of diffusive samplers and adsorbents for measuring 1,3-butadiene and benzene in air. *Atmospheric Environment* 39: 4101–4110.

Strandberg, B., Sunesson, A.L., Sundgren, M., Levin, J.O., Sällsten, G. et al. 2006a. Field evaluation of two diffusive samplers and two adsorbent media to determine 1,3-butadiene and benzene levels in air. *Atmospheric Environment* 40: 7686–7695.

Strandberg, B., Gustafson, P., Söderström, H., Barregard, L., Bergqvist, P.A. et al. 2006b. The use of semipermeable membrane devices as passive samplers to determine persistent organic compounds in indoor air. *Journal of Environmental Monitoring* 8: 257–262.

Strandberg, B., Bergemalm-Rynell, K., Sallsten, G. 2014. Evaluation of three types of passive samplers for measuring 1,3-butadiene and benzene at workplaces. *Environmental Science: Processes & Impacts* 16: 1008–1014.

Stuer-Lauridsen, F. 2005. Review of passive accumulation devices for monitoring organic micropollutants in the aquatic environment. *Environmental Pollution* 136: 503–524.

Suryanarayana, M.V.S., Shrivastava, R.K., Pandey, D., Vaidyanathaswamy, R. et al. 2001. Simple time weighted average level air-monitoring method for sulfur mustard in work places. *Journal of Chromatography A* 907: 229–234.

Svecova, V., Topinka, J., Solansky, I., Sram, R.J. 2012. Personal exposure to volatile organic compounds in the Czech Republic. *Journal of Exposure Science and Environmental Epidemiology* 22: 455–460.

Tan, B.L.L., Hawker, D.W., Muller, J.F., Leusch, F.D.L., Tremblay, L.A. et al. 2007. Comprehensive study of endocrine disrupting compounds using grab and passive sampling at selected wastewater treatment plants in south east Queensland, Australia. *Environment International* 33: 654–669.

Thomatou, A.A., Zacharias, I., Hela, D., Konstantinou, I. 2011. Passive sampling of selected pesticides in aquatic environment using polar organic chemical integrative samplers. *Environmental Science and Pollution Research* 18: 1222–1233.

Togola, A., Budzinski, H. 2007. Development of polar organic integrative samplers for analysis of pharmaceuticals in aquatic systems. *Analytical Chemistry* 79: 6734–6741.

Tombesi, N., Pozo, K., Harner, T. 2014. Persistent organic pollutants (POPs) in the atmosphere of agricultural and urban areas in the Province of Buenos Aires in Argentina using PUF disk passive air samplers. *Atmospheric Pollution Research* 5: 170–178.

Tran, A.T.K., Hyne, R.V., Doble, P. 2007. Calibration of a passive sampling device for time-integrated sampling of hydrophilic herbicides in aquatic environments. *Environmental Toxicology and Chemistry* 26: 435–443.

Tsai, S.W., Tsai, S.T., Wang, V.S., Lai, J.S. *2004*. Laboratory and field validations of a solid-phase microextraction device for the determination of ethylene oxide. *Journal of Chromatography A* 1026: 25–30.

Tuduri, L., Harner, T., Hung, H. 2006. Polyurethane foam (PUF) disks passive air samplers: Wind effect on sampling rates. *Environmental Pollution* 144: 377–383.

Tuduri, L., Millet, M., Briand, O., Montury, M. 2012. Passive air sampling of semi-volatile organic compounds. *TrAC Trends in Analytical Chemistry* 31: 38–49.

Tumbiolo, S., Gal, J.F., Maria, P.C., Laborde, P., Teton, S. 2006. Rapid sampling of BTEX in air by SPME in the city of Nice and at the Nice-Côte d'Azur airport. *Pollution atmosphérique* 190: 185–195.

Underhill, D.W. 1993. Convective transport in diffusive samplers. *Talanta* 40: 785–90.

Vallejo, A., Prieto, A., Moeder, M., Usobiaga, A., Zuloaga, O. et al. 2013. Calibration and field test of the polar organic chemical integrative samplers for the determination of 15 endocrine disrupting compounds in wastewater and river water with special focus on performance reference compounds (PRC). *Water Research* 47: 2851–2862.

Verbruggen, E.M.J., van Loon, W.M.G.M., Tonkes, M., van Duijn, P., Seinen, W., Hermens, J.L.M. 1999. Biomimetic extraction as a tool to identify chemicals with high bioconcentration potential: An illustration by two fragrances in sewage treatment plant effluents and surface waters. *Environmental Science & Technology* 33: 801–806.

Vardoulakis, S., Gonzalez-Flesca, N., Fisher, B.E.A. 2002. Assessment of traffic-related air pollution in two street canyons in Paris: Implications for exposure studies. *Atmospheric Environment* 36: 1025–1039.

Verhaar, H.J.M., Busser, F.J.M., Hermens, J.L.M. 1995. Surrogate parameter for the baseline toxicity content of contaminated water: Simulating the bioconcentration of mixtures of pollutants and counting molecules. *Environmental Science & Technology* 29: 726–734.

Vermeirssen, E.L.M., Körner, O., Schönenberger, R., Suter, M.J.F., Burkhardt-Holm, P. 2005. Characterization of environmental estrogens in river water using a three pronged approach: Active and passive water sampling and the analysis of accumulated estrogens in the bile of caged fish. *Environmental Science & Technology* 39: 8191–8198.

Vermeirssen, E.L.M., Hollender, J., Bramaz, N., Van der Voet, J., Escher, B.I. 2010. Linking toxicity in algal and bacterial assays with chemical analysis in passive samplers deployed in 21 treated sewage effluents. *Environmental Toxicology and Chemistry* 29: 2575–2582.

Vermeirssen, E.L.M., Dietschweiler, C., Escher, B.I., van der Voet, J., Hollender, J. 2012. Transfer kinetics of polar organic compounds over polyethersulfone membranes in the passive samplers POCIS and Chemcatcher. *Environmental Science & Technology* 46: 6759–6766.

Vermeirssen, E.L.M., Dietschweiler, C., Escher, B.I., van der Voet, J., Hollender, J. 2013. Uptake and release kinetics of 22 polar organic chemicals in the Chemcatcher passive sampler. *Analytical and Bioanalytical Chemistry* 405: 5225–5236.

Verweij, F., Booij, K., Satumalay, K., Van Der Molen, N., Van Der Oost, R. 2004. Assessment of bioavailable PAH, PCB and OCP concentrations in water, using semipermeable membrane devices (SPMDs), sediments and caged carp. *Chemosphere* 54: 1675–1689.

Vetter, W., Haase-Aschoff, P., Rosenfelder, N., Komarova, T., Mueller, J.F. 2009. Determination of halogenated natural products in passive samplers deployed along the Great Barrier Reef, Queensland/Australia. *Environmental Science & Technology* 43: 6131–6137.

Villanueva, F., Colmenar, I., Mabilia, R., Scipioni, C., Cabañas, B. 2013. Field evaluation of the Analyst® passive sampler for the determination of formaldehyde and acetaldehyde in indoor and outdoor ambient air. *Analytical Methods* 5: 516–524.

Vrana, B., Popp, P., Paschke, A., Schűűrmann, G. 2001. Membrane-enclosed sorptive coating. An integrative passive sampler for monitoring organic contaminants in water. *Analytical Chemistry* 73: 5191–5200.

Vrana, B., Mills, G.A., Allan, I.J., Dominiak, E., Svensson, K. et al. 2005a. Passive sampling techniques for monitoring pollutants in water. *TrAC Trends in Analytical Chemistry* 24: 845–868.

Vrana, B., Paschke, H., Paschke, A., Popp, P., Schüürmann, G. 2005b. Performance of semipermeable membrane devices for sampling of organic contaminants in groundwater. *Journal of Environmental Monitoring* 7: 500–508.

Vrana, B., Mills, G.A., Dominiak, E., Greenwood, R. 2006. Calibration of the Chemcatcher passive sampler for the monitoring of priority organic pollutants in water. *Environmental Pollution* 142: 333–343.

Vrana, B., Mills, G.A., Kotterman, M., Leonards, P., Booij, K. et al. 2007. Modelling and field application of the Chemcatcher passive sampler calibration data for the monitoring of hydrophobic organic pollutants in water. *Environmental Pollution* 145: 895–904.

Vrana, B., Mills, G.A., Leonards, P.E.G., Kotterman, M., Weideborg, M. et al. 2010. Field performance of the Chemcatcher passive sampler for monitoring hydrophobic organic pollutants in surface water. *Journal of Environmental Monitoring* 12: 863–872.

Vrana, B., Klučárová, V., Benická, E., Abou-Mrad, N., Amdany, R. et al. 2014. Passive sampling: An effective method for monitoring seasonal and spatial variability of dissolved hydrophobic organic contaminants and metals in the Danube river. *Environmental Pollution* 184: 101–112.

Wang, J., Tuduri, L., Mercury, M., Millet, M., Briand, O. et al. 2009. Sampling atmospheric pesticides with SPME: Laboratory developments and field study. *Environmental Pollution* 157: 365–370.

Wang, F., Costabile, F., Liu, F., Hong, W., Fang, D. et al. 2010a. Ambient BTX measurements in Suzhou, China. *Environmental Monitoring and Assessment* 168: 21–31.

Wang, X.P., Gong, P., Yao, T.D., Jones, K.C. 2010b. Passive air sampling of organochlorine pesticides, polychlorinated biphenyls, and polybrominated diphenyl ethers across the Tibetan plateau. *Environmental Science & Technology* 44: 2988–2993.

Wania, F., Shen, L., Lei, Y.D., Teixeira, C., Muir, D.C.G. 2003. Development and calibration of a resin-based passive sampling system for monitoring persistent organic pollutants in the atmosphere. *Environmental Science & Technology* 37: 1352–1359.

Wilcockson, J.B., Gobas, F.A. 2001. Thin-film solid-phase extraction to measure fugacities of organic chemicals with low volatility in biological samples. *Environmental Science & Technology* 35: 1425–1431.

Wilford, B.H., Harner, T., Zhu, J., Shoeib, M., Jones, K.C. 2004. Passive sampling survey of polybrominated diphenyl ether flame retardants in indoor and outdoor air in Ottawa, Canada: Implications for sources and exposure. *Environmental Science & Technology* 38: 5312–5318.

Writer, J.H., Barber, L.B., Brown, G.K., Taylor, H.E., Kiesling, R.L. et al. 2010. Anthropogenic tracers, endocrine disrupting chemicals, and endocrine disruption in Minnesota lakes. *Science of the Total Environment* 409: 100–111.

Xu, Q., Zhu, X., Henkelmann, B., Schramm, K.W., Chen, J. et al. 2013. Simultaneous monitoring of PCB profiles in the urban air of Dalian, China with active and passive samplings. *Journal of Environmental Sciences (China)* 25: 133–143.

Xu, Y., Li, J., Chakraborty, P., Syed, J.H., Malik, R.N. et al. 2014. Atmospheric polychlorinated naphthalenes (PCNs) in India and Pakistan. *Science of the Total Environment* 466–467: 1030–1036.

Yargeau, V., Taylor, B., Li, H., Rodayan, A., Metcalfe, C.D. 2014. Analysis of drugs of abuse in wastewater from two Canadian cities. *Science of the Total Environment* 487: 722–730.

Yilmaz, A., Karacik, B., Henkelmann, B., Pfister, G., Schramm, K.W. et al. 2014. Use of passive samplers in pollution monitoring: A numerical approach for marinas. *Environment International* 73: 85–93.

Zabiegała, B., Kot-Wasik, A., Urbanowicz, M., Namieśnik, J. 2010a. Passive sampling as a tool for obtaining reliable analytical information in environmental quality monitoring. *Analytical and Bioanalytical Chemistry* 396: 273–296.

Zabiegała, B., Urbanowicz, M., Szymanska, K., Namiesnik, J. 2010b. Application of passive sampling technique for monitoring of BTEX concentration in urban air: Field comparison of different types of passive samplers. *Journal of Chromatographic Science* 48: 167–175.

Zabiegała, B., Urbanowicz, M., Namieśnik, J., Górecki, T. 2010c. Spatial and seasonal patterns of benzene, toluene, ethylbenzene, and xylenes in the Gdańsk, Poland and surrounding areas determined using Radiello passive samplers. *Journal of Environmental Quality* 39: 896–906.

Zabiegała, B., Sărbu, C., Urbanowicz, M., Namieśnik, J. 2011. A comparative study of the performance of passive samplers. *Journal of the Air & Waste Management Association* 61: 260–268.

Zabik, J.M., Aston, L.S., Seiber, J.N. 1992. Rapid characterization of pesticide residues in contaminated soils by passive sampling devices. *Environmental Toxicology & Chemistry* 11: 765–770.

Zeng, E.Y., Noblet, J.A. 2002. Theoretical considerations on the use of solid-phase microextraction with complex environmental samples. *Environmental Science & Technology* 36: 3385–3392.

Zhang, G.Z., Hardy, J.K. 1989. Determination of phenolic pollutants in water using permeation sampling. *Journal of Environmental Science & Health A* 24: 279–295.

Zhang, Z., Hibberd, A., Zhou, J.L. 2008. Analysis of emerging contaminants in sewage effluent and river water: Comparison between spot and passive sampling. *Analytica Chimica Acta* 607: 37–44.

Zhang, L.P., Wang, X.H., Li, Y.Y., Wu, Y.L., Zhang, Z.L. 2014. Determination of estrogenic compounds in water of Jiulong River using polar organic chemical integrative sampler. *Environmental Science* 35: 117–122.

Zhu, X., Zhou, C., Henkelmann, B., Wang, Z., Ma, X. et al. 2013. Monitoring of PAHs profiles in the urban air of Dalian, China with active high-volume sampler and semipermeable membrane devices. *Polycyclic Aromatic Compounds* 33: 265–288.

Section III

Quality Assurance

10 Quality Assurance and Validation

General Considerations and Trends

Roberta Galarini, Simone Moretti, and Giorgio Saluti

CONTENTS

10.1 INTRODUCTION

Validation and quality assurance (QA) are topics of increasing importance in analytical chemistry. The traditional approach, that is, prescribing standardized methods of analysis and describing them in official documents, is being used less and less. At present, the "performance criteria" approach is the most popular one: it involves the use of internally developed and validated methods that meet the required performance characteristics. However, the latter is per se a more risky approach for the quality of results since, generally, standardized methods are already externally validated (collaborative studies) and the laboratory should only verify its ability to meet the specified performance characteristics. On the other hand, this change (from standard to internal methods) has been necessary due to the better flexibility of the performance criteria strategy, fitting with the tremendous technological advances in instrumental techniques and the increased complexity of analytical demands with several matrix/analyte combinations. As a consequence, validation and QA are becoming more and more important, since they are fully responsible for guaranteeing the quality of results. Therefore, today not only must analytical chemists deeply know instrumental techniques and the chemistry of their test materials and analytes, but they are also expected to understand and apply the basic principles of good practices in analytical measurements and the necessary statistical tools. These good practices can be summarized in following principles:

- Analytical measurements should be made to satisfy an agreed requirement: the proposed analytical methodology needs to be delivered in the most cost-effective manner consistent with fitness for purpose.
- Analytical measurements should be made using methods and equipment which have been tested to ensure that they are fit for purpose: need for method validation.
- Analytical measurements made in one location should be consistent with those elsewhere: the traceability of results must be assured through calibration of equipments and the use of suitable reference materials.
- Organizations making analytical measurements should have well-defined quality control (QC) and QA procedures: the implementation of appropriate QA and QC procedures is typically backed up by formal accreditation through third-party assessment of the laboratory quality system.

There are numerous specialized articles and textbooks which deal with the topics of this chapter; therefore, for details and in-depth analysis, cross-references to these texts have been made. The goal here pursued consists of guiding readers through the multitude of concepts while selecting the most important for laboratory practices. Some significant cases are reported to immediately give practical examples, in the hope that the reader will be able to extract information for his/her own specific case and be able to apply it in daily routine work.

10.2 QUALITY SYSTEMS AND LABORATORY ACCREDITATION

The official definition of a quality system given by the International Organization for Standardization (ISO) is "the management system used to direct and control an organization with regard to quality." In other words, a quality system is a set of procedures and responsibilities which an organization adopts to make sure that it meets the requirements of its customers. QA and QC are both part of a quality management system. The Cooperation on International Traceability in Analytical Chemistry's and Eurachem's *Guide to Quality in Analytical Chemistry* (CITAC and Eurachem, 2002) gives the following definitions:

- QA: "QA describes the overall measures that a laboratory uses to ensure the quality of its operations." Typically, these measures include a quality system, trained staff, training procedures and records, suitable laboratory environment, equipment maintenance and calibration procedures, QC procedures, documented and validated methods, traceability and measurement uncertainty, preventive and corrective actions, internal audit and review procedures, complaints procedures, requirements for reagents, calibrants, measurement standards, reference materials, etc.
- QC: This includes "the operational techniques and activities that are used to fulfill requirements for quality." QC procedures are aimed at ensuring the quality of specific samples or batches of samples and include all the activities listed as internal quality control (IQC) and external quality control (EQC). The typical QC activities are analysis of reference materials/measurement standards, analysis of blank and spiked samples, analysis in duplicate, use of control charts, participation in proficiency testing (PT) schemes, etc.

There is some confusion as to what is meant by these terms, and many people think that QA and QC are simply interchangeable terms. In practice, QA is a planned system of review procedures and it is aimed at preventing nonconformities/defects. In analytical chemistry, the defects are unreliable results. QC, on the other hand, is the most basic level of quality. It is a system of routine technical activities whose purpose is to control the quality of products (the analytical results) by finding problems and defects. Figure 10.1 summarizes the four phases of QA (Funk et al., 2007).

Nowadays, the users of test results are particularly interested in certain characteristics of the methods. Firstly, they require that a measurement result be accompanied by a quantitative statement of its uncertainty in order to compare it with a specification limit or with other similar results. Moreover, they also require that analytical methods have a high sample throughput and are capable of quantifying very low concentrations. Since in practice it is impossible to guarantee these characteristics for every single test result, the role of quality systems is to ensure the quality of all the results obtained under certain conditions. In addition, in the current global scenario, an essential prerequisite of trade is that any product or service which is formally accepted in one economy must also be free to circulate in other economies without having to undergo extensive retesting. The World Trade Organization recognizes that nonacceptance of test or calibration results is a barrier to trade.

Accordingly, the accreditation process is the first essential step for facilitating mutual acceptance: it is a procedure by which an authoritative body gives formal recognition that a laboratory is competent at performing specific analyses or calibrations. ISO/IEC 17025:2005, *General Requirements for the Competence of Testing and Calibration Laboratories* (ISO, 2005), has gained worldwide acceptance as a common standard for accrediting testing and calibration laboratories, which use it to implement a quality system aimed at improving their ability to consistently produce valid results. ISO/IEC 17025 was first issued in 1999 by the ISO and the International Electrotechnical Commission (IEC) (ISO, 2000). At present it is the single most important standard for calibration and testing laboratories around the world.

In order to have formal recognition, laboratories need to demonstrate full compliance with ISO/IEC 17025 standard. ISO/IEC 17025 focuses greatly on documenting the process of any analysis performed by a laboratory. It includes two sections: the quality management system and the technical requirements of the accreditation process. As part of accreditation, a laboratory's quality management system is thoroughly evaluated on a regular basis to ensure continued technical competence and compliance with ISO/IEC 17025. This evaluation (external audit) can be granted only by a national accreditation body, which itself is scrutinized under the auspices of the International Laboratory Accreditation Cooperation. Laboratory accreditation is a voluntary, third party-reviewed process; however, in the global context, it has evolved from being a voluntary act at the beginning to becoming and remaining a competitive factor for some time, until finally reaching the stage of being

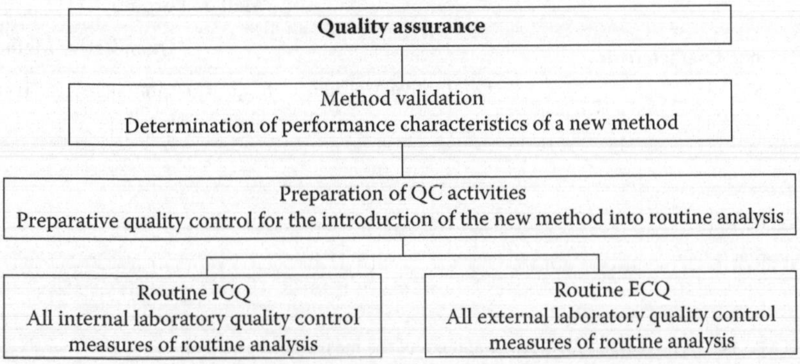

FIGURE 10.1 The four phases model of analytical QA.

considered a requirement for survival. Nowadays the legislation in several countries specifies that official laboratories must be accredited according to ISO/IEC 17025 (European Parliament, Council of the European Union, 2004; European Commission, 2009).

10.3 METHOD VALIDATION

A method is an exactly defined analytical procedure including (i) all stages of sample preparation, (ii) the instrument used, and (iii) the calibration equations applied to estimate the concentration of the measurand in the material under examination. In a quality system, this "defined analytical procedure" is described in the so-called standard operating procedure (SOP).

Method validation is the provision of objective evidence that a given item is adequate for an intended use. For methods internally developed by a laboratory, the (in-house) validation normally occurs immediately after the development phases. Since the QC activities are aimed at ascertaining if an analytical system is under statistical control, that is, if the determined performance characteristics are maintained during its operational lifetime, validation is also the starting point to plan the IQC during routine method use (Figure 10.1).

The experimental work to validate an analytical method should follow a documented process, such as a validation plan deciding in advance which performance characteristics have to be considered. The extent of validation mainly depends on method type, in other words, on method purpose. In Table 10.1, the parameters to be assessed for each method type are listed.

As shown, in methods for which the validation range does not include or reach the lower limits (major components), LOD and LOQ do not need to be part of the validation study. On the other hand, qualitative methods do not need necessarily to investigate precision and trueness.

Prior to defining the validation plan, the method origin and the validation data already available should be assessed, too. As to the method origin, the main distinction is between standardized (external) and internal methods, although in recent years, especially in trace analysis, there has been a shift from standardized to internal (performance-based) procedures. Obviously, analytical procedures published in scientific journals, prestigious as they may be, cannot be considered the same as a standardized method: only standardization bodies (e.g., European Committee for Standardization, CEN, International Organization for Standardization, ISO) and recognized technical organizations (e.g., Association of Official Analytical Chemists, AOAC, Nordic Committee on Food Analysis, NMKL) have discretion to issue standardized procedures. Even a fully validated standardized

TABLE 10.1
Method Type and Validation Requirements

Method Performance Characteristic (Parameter)[a]	Method Purpose		
		Quantitative Method	
	Qualitative Method	Major Components[b]	Trace Analysis[b]
Selectivity	x	x	x
Linearity		x	x
Accuracy (trueness and precision)		x	x
Limit of detection (LOD)	x		x
Limit of quantification (LOQ)			x
Ruggedness	x	x	x

[a] Uncertainty is a property of a measurement result and not of the measurement method.
[b] Major component: analyte concentration in the range 1–100%; trace analysis: analyte concentration less than 100 mg kg^{-1}.

method cannot be directly used without the laboratory first verifying its ability to meet the specified performance characteristics, implementing a partial validation. Partial validation is also carried out in the cases in which a fully validated internal/external method undergoes some changes as the extension of matrix types or of the measuring range (Irish National Accreditation Board, 2012).

The performance-based approach allows the laboratories to apply an internally developed method provided that the relevant validation study demonstrates that it meets the required characteristics. At present, from a legislative point of view, different scenarios coexist. In very few cases, the methods are defined (standardized). More frequently, the analytical procedures are explicitly specified; however, alternative methods can be used, if their results are at least as reliable as those produced by the specified methods. Directive 2009/90/EC (European Commission, 2009), which is on monitoring of water status, for example, takes another approach: no analytical methods are stipulated; only their performance criteria are. A similar trend is observable in food safety legislation (European Commission, 2002), since the shift from standardized (official) to internal methods (performance criteria) is a flexible approach allowing the laboratories to quickly incorporate advances in analytical techniques and to apply new methods to address new analytical problems in the globalized world. Clearly, the performance criteria approach involves more work for laboratories.

At the end of every validation study, the laboratory draws up a validation report containing a declaration concerning the scale of validation decided upon, the applied validation scheme (or a reference), the obtained results, and the final assessment. A declaration concerning the method's validity with, where appropriate, method release (approval) must be included. Finally, the report is signed by the individual responsible for the method and kept in the laboratory as part of the documentation for the method of analysis in question.

10.3.1 Performance Characteristics

10.3.1.1 Some Statistics

Validation studies, data comparisons, and QC involve the continuous application of some basic statistical tools. An inherent (natural) variation is associated with any real measurement: different values are generally observed also when replicate measurements of the same parameter are carried out on the same item. This variation follows the normal (Gaussian) distribution, and it is not the outcome of carelessness, but is simply due to the uncontrollable and inevitable variation in the activities necessary for the measurement, which is often a complex multistage process. This unavoidable variation is also called random error.

The arithmetic mean ($\overline{x}$) and standard deviation (s) of n replicate measurements are the best estimators of normally distributed samples. This is the starting point for data evaluation and, accordingly, for the estimation of the method performance characteristics such as trueness, precision, or robustness. Then the analyst needs to compare two experimental means or an experimental mean with an assigned value distinguishing between "natural" variability and "caused" variability. To do this, a t-test is applied. The analyst also needs to compare the precision (standard deviation or variance) of two analytical methods or of two operators (F-test). The basic statistical significance tests are reported in Table 10.2.

An important part of performing any test is the concept of the null hypothesis, H_0. This is the hypothesis that the value of the test parameter derived from the data is purely the result of random (natural) error in taking the sample measurements from the population of all possible values; the exact interpretation depends to some extent on the type of test being performed, but essentially, if the null hypothesis is true, then there is no significant difference between the sample and population values. To verify the hypothesis, the value obtained by applying the formula listed in Table 10.2 is compared with a critical value tabulated for the pertinent distribution: if this value is lower than the critical one (tabulated) at a certain level of confidence (e.g., 95%), the null hypothesis cannot be rejected.

The mean ($\overline{x}$) and standard deviation (s) are the estimators more frequently applied in analytical chemistry to summarize a set of normally distributed data. However, in some circumstances, the

TABLE 10.2
More Useful Statistical Tests in Analytical Chemistry

Comparison	Formula	Notes	Test Type
One sample mean ($\bar{x}$) to an accepted value (μ_0)	$t = \dfrac{\bar{x} - \mu_0}{s/\sqrt{n}}$	s is the standard deviation of the sample (not the population standard deviation), and n the number of observations carried out to estimate s.	t-test
Two sample means ($\bar{x}_1$ and $\bar{x}_2$)	$t = \dfrac{\bar{x}_1 - \bar{x}_2}{\sqrt{s_1^2/n_1 + s_2^2/n_2}}$	s_1 and s_2 are the standard deviations of the two samples, and n_1 and n_2 are the number of observations carried out to estimate s_1 and s_2.	t-test[a]
One sample variance (s^2) to a target variance (σ^2)	$T = (N-1)\dfrac{s^2}{\sigma^2}$	s^2 is the observed sample variance, and σ^2 is the target (population) variance. N is the number of observations.	Chi-square test
Two sample variances (s_1 and s_2)	$F = \dfrac{s_1^2}{s_2^2}$	The larger variance is always placed in the numerator ($s_1 > s_2$).	F-test[b]

[a] Where there are more than two groups or populations and the hypothesis is that all of the means are equal, an analysis of variance is the test required.

[b] Where there are more than two groups or populations and the hypothesis is that all of the variances are equal, other tests are applied (e.g., Hartley's and Bartlett's tests).

analyst can suspect the presence of outliers. Outliers are out-of-control data due to gross errors, that is, errors not attributable to either systematic or random errors associated with the sample, instrument, or procedure. If included in calculations, outliers will tend to strongly affect both mean and standard deviation, altering data evaluation. Therefore, in these cases the use of robust estimators is suggested: median instead of the arithmetical mean and median absolute deviation instead of standard deviation. These parameters are called "robust" because they are less affected by outliers.

10.3.1.2 Selectivity

Selectivity is usually the first validation parameter to be determined. When the analyst starts to develop the determination of a certain analyte based on the measurement of its signal, he or she should make sure that the acquired response is due only to the presence of the analyte in the investigated sample. According to the International Union of Pure and Applied Chemistry (IUPAC), the selectivity of a method is the extent to which it can determine particular analyte(s) in a complex mixture without interference from other components in the mixture (Vessman et al., 2001). In other terms, selectivity is the degree to which a method can identify and quantify the analyte accurately in the presence of interfering substances. Ideally, selectivity should be evaluated for any important interfering compound likely to be present, but to consider or test every potential matrix interference may be impracticable. It is therefore recommended that the cases likely to cause the worst interference be checked. European Decision 2002/657/EC (European Commission, 2002) dealing with the determination of residues of veterinary drugs in food recommends checking at least 20 different blank samples belonging to a certain matrix to prove method selectivity. It is particularly important to check interfering substances which, due to certain chemical principles or structures, are likely to respond to the test, for example, the capability of a measuring system with a mass spectrometer to discriminate between two compounds with similar structures (e.g., the analyte and its metabolite). As a general principle, the selectivity should be high enough for any interference to be ignored.

The terms *selectivity* and *specificity* are frequently used indifferently, although they differ in their essential meanings. Specificity is described by the IUPAC as the ultimate of selectivity: it is an absolute term and, thus, it cannot be graded. The term *specificity* suggests that no component other than the analyte contributes to the result. Hardly any method is that specific and, in general, the term *specificity* should be avoided.

10.3.1.2.1 Identity

The selectivity issue is related to analyte identification, that is, to the possibility of a false-positive identification (α error). To date, a systematic approach in determining the selectivity criteria is lacking. When using MS techniques coupled to chromatographic systems, a minimum number of ions are generally required to enable the definitive compound identification. An example of these selectivity criteria for analyte identification is reported in the Directorate-General for Health and Consumer Protection (SANCO) document for analytical QC and validation procedures for pesticide residues analysis in food and feed (Table 10.3) (European Commission, 2013). These criteria are mainly based on empirical studies and experts' judgments, but now a quantitative assessment of selectivity using probabilistic tools is possible (Berendsen et al., 2013; Stoev et al., 2013). The drawback of these more rigorous approaches lies in the difficulty of the necessary calculations. On the other hand, more stringent and clear rules could help to guarantee at least a more appropriate level of selectivity. For example, the SANCO document claims, "high *m/z* ions are more specific than low *m/z* ions (e.g., $m/z < 100$). However, high-*m/z* ions arising from loss of water or loss of common moieties may be of little use" (Stoev et al., 2013). No precise penalization for when less selective ions or transitions are chosen is indicated. Should transitions such as the loss of water be not considered or be clearly considered less in a hypothetical selectivity score system (European Commission, 2002)? In addition, in the last 10 years, tremendous advances in MS equipment along with the widespread diffusion of high-resolution mass analyzers based on TOF and Orbitrap technologies have made the

TABLE 10.3
Identification Criteria for Different MS Techniques (Document SANCO/12571/2013)

	MS Mode		
	Single-Stage MS (Unit Mass Resolution)	**Single-Stage MS (High Resolution/High Mass Accuracy)**	**MS/MS**
Typical systems (examples)	Quadrupole, ion trap, TOF	TOF, Orbitrap, Fourier transform MS, magnetic sector	QqQ, ion trap, hybrid MS (e.g., QTOF, Q-trap)
Acquisition mode	Full scan, limited *m/z* range, SIM	Full scan, limited *m/z* range, SIM	Selected/multiple reaction monitoring, full-scan product-ion spectra
Requirements for identification	≥3 diagnostic ions, preferably including the (quasi)molecular ion	≥2 diagnostic ions, preferably including the (quasi)molecular ion; mass accuracy <5 ppm; at least one fragment ion	≥2 product ions
		Ion Ratio(s): According to Table below[a]	
	Ion Ratio (Least/Most Intense Ion)	**Maximum Tolerance (Relative) for GC-EI-MS (%)**	**Maximum Tolerance (Relative) for LC-MS^n, LC-MS, GC-MS^n, GC-CI-MS[a] (%)**
	0.50–1.00	±10	±20
	0.20 0.50	±15	±25
	0.10–0.20	±20	±30
	<0.10	±50	±50

[a] These maximum tolerances are those reported in the previous version of the document (SANCO/12495/2011).

definition of criteria for each instrument type and configuration (single-stage or hybrid analyzers) more complicated (Berendsen et al., 2015).

Instrumental resources have been assigned an important role in reaching selectivity, but it must be kept in mind that selectivity can be ascribed only to a whole analytical process for a particular matrix/analyte combination. Also, the measuring range is fundamental: when concentration levels close to "zero" are required in the analysis of complex matrices, how much selectivity is needed? Strategies to improve selectivity cannot, therefore, ignore the other "side" of an analytical process, that is, the sample preparation steps. Valcarcel et al. (2001) proposed a scheme to exemplify the traps in reaching selectivity (Figure 10.2). The less information is available on the sample, the more

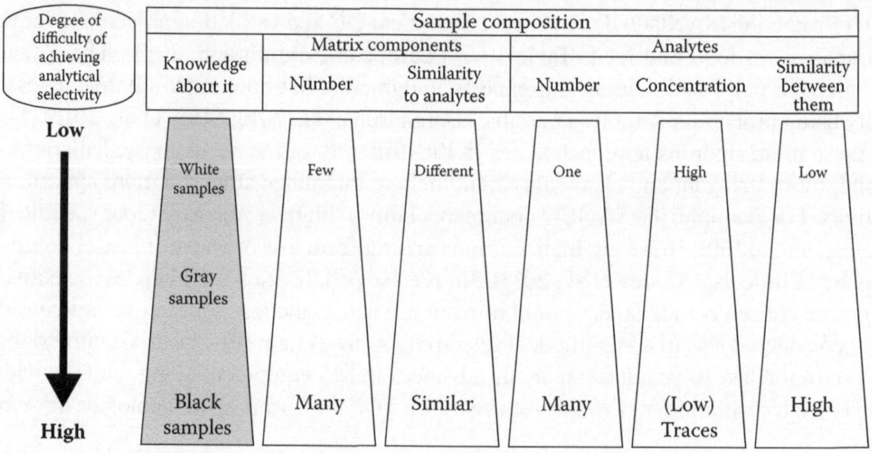

FIGURE 10.2 Factors influencing the selectivity of analytical processes.

TABLE 10.4
Possible Scoring System for Selectivity Required in Function of the Analytical Issue

	Parameter	Description/Examples	Score
Matrix	Complexity (number of matrix components)	Spring water, tap water	1
		Vegetables with a high water content	2
		Contaminated soils, biological fluids	3
	Similarity to analytes	Low	1
		Medium	2
		High	3
Analytes	Number	≤1 analyte peak/min[a]	1
		2–10 analyte peaks/min[a]	2
		>10 analyte peaks/min[a]	3
	Concentration/method LOQ ratio	>10	1
		>1	2
		1–0.3	3
	Similarity between them	Low	1
		Medium	2
		High	3
Total score			5–15

[a] Number of determined analytes divided by the length (min) of the chromatographic run. The description is only for example and should vary depending on the specific chromatographic technique (i.e., GC or LC).

TABLE 10.5
Possible Scoring System for Selectivity Delivered in Function of the Characteristics of the Candidate Method

Analytical Step	Description		Score
Sample preparation	SPE	Low-specificity sorbent (C18, C8, alumina, etc.)	2
		Medium-specificity sorbent (ionic exchange, etc.)	3
		High-specificity sorbent (immunosorbent/MIP, etc.)	4
	Liquid–liquid extraction		2
	Derivatization reaction		2
Instrumental determination	Relative retention time	±1%[a]	4
	Single-stage MS (unit mass resolution)	One point for each of n ions monitored	$1 \times n$
	MS/MS (unit mass resolution)	One transition	2.5
		More than one transition	Further 1.5 for each additional transition
	Single-stage MS (high resolution/high accuracy)	Two points for each of n ions monitored	$2 \times n$
	MS/MS (high resolution/high accuracy)	One transition	4.5
		More than one transition	Further 2.5 for each additional transition

[a] This tolerance is only for example and should vary depending on specific chromatographic technique (i.e., GC or LC).

difficult it is likely to be to reach selectivity. "White samples" have well-defined and constant properties, "gray samples" are those whose composition is known only in part, and, finally, "black samples" are of unknown composition and have an indefinite number and concentrations of interfering species.

Before developing an analytical method (candidate method), the analyst should take into account both the complexity of the problem and the resources available in his/her laboratory. The increasing diffusion of internally developed methods determining a high number of analytes (>100) in more and more complex matrices requires definite criteria to reach the "fit-for-purpose" level regarding selectivity. Tables 10.4 and 10.5 give a rough example of establishing the selectivity requirement with a scoring system to reach this in-trace analysis. Although it could be difficult to assign an exact score to a certain analytical problem and to a candidate method, the proposed scoring system has the undoubted advantage of forcing both legislative authorities and analysts to consider the analytical issue as a whole (Baldwin et al., 1997).

10.3.1.3 Calibration

The IUPAC guidelines for calibration in analytical chemistry (Danzer and Currie, 1998) defines calibration as the operation that determines the functional relationship between measured values (signal intensities), that is, the y variable, and analytical quantities characterizing types of analytes and their amount or concentration, that is, the x variable. Every instrument used in chemical analysis is characterized by a specific calibration function, that is, an equation relating the instrument output signal to the analyte content. This response function may or may not be linear: the exact form depends on the system being measured and the measurement process itself. The calibration curve is obtained by fitting an appropriate equation to a set of experimental data (calibration data) consisting of the measured responses to known concentrations of analyte.

The most common response function in instrumental analytical chemistry is linear, so the best-fit straight line through calibration data must be determined by setting the known concentrations in the x axis and the relevant acquired instrumental responses in the y axis. However, even when a linear instrument response function is expected, it should not be assumed that the calibration data are always linear. For example, in chromatographic methods, well-known causes of nonlinearities are the column or detector overloads. Linearity can be assessed initially by constructing a scatter plot response versus analyte amount (visual inspection). The data can then be evaluated by carrying out linear regression to establish the slope (b) and intercept (a) of the line and the correlation coefficient (r). Linear regression (or least-squares method) is the natural approach to analytical calibration: the best straight-line fit is mathematically obtained when the sum of the squares of the individual y axis residuals is at a minimum. Residuals are the differences between the experimental y values and the values predicted by the proposed line. There are numerous textbooks and technical papers dealing with the statistical approach to calibration (Danzer and Currie, 1998; Funk et al., 2007).

10.3.1.3.1 Types of Calibration

Irrespective of the functional relationship between the analyte concentration (amount) and signal, particular care must be taken in selecting the "calibrants," which are the measurement standards specifically used in calibrating (x variable). The standards (reference materials) can be essentially classified into two types: (i) The first are substance reference materials, prepared from single analytes. They include (a) pure substances characterized by chemical purity and/or trace purities and (b) standard solutions, prepared from pure substances (single or mixed) in the working solvents used. (ii) The second are matrix reference materials characterized by a composition of specified major, minor, and trace chemical constituents. They may be prepared from the matrices to be analyzed or by means of synthetic mixtures. Thus, these reference materials contain the analytes being determined plus the principal chemical compounds characterizing the matrix to be matched (Cuadros-Rodríguez et al., 2007). However, in practice, "matrix-matched" calibrants are frequently obtained by adding standard solutions of pure standards to freshly prepared real matrices free of the analyte.

Nowadays, due to wide diffusion of methods based on MS techniques, calibration is probably one of the most challenging topics in the setting up of new procedures. At first, it is fundamental to know whether a method is matrix-effect-free or not. The phenomenon of the matrix effect (ion suppression or enhancement) can occur, especially when MS analyzers are applied. It happens when molecules coeluting with the compound being analyzed alter the ionization efficiency of the ionization source. It probably originates from the competition between the analyte and the coeluting, undetected matrix components. Frequently, in trace analysis of complex samples, it is not possible to eliminate the matrix effects by sample preparation and/or chromatographic resolution. However, it may be possible to compensate for these effects by adopting specific calibration strategies.

Calibration approaches are mainly divided into external calibration (EC) and internal calibration (IC). In the former, a series of external standards is prepared separately (externally) from the sample. It involves a simple comparison of instrument responses from the sample to the responses from the target compounds in the calibrants. IC is achieved with an internal standard (IS), a chemical substance that is added in a constant amount to both test samples and calibrants. Calibration involves plotting the ratio of the analyte signal to the IS signal (y variable) as a function of analyte concentrations of the standards (x variable). This is done to correct for both random and systematic errors (losses during sample preparation, matrix effects, etc.). It is fundamental that the IS shows a behavior similar to that of the analyte.

The four main calibration types are listed in the next sections. To achieve correct calibration, the analyst must have in-depth knowledge of the behavior of the analytical process to identify the more appropriate choice in terms of cost and benefits, too. MS analyzers could hide traps, especially due to matrix effects, which may also vary within the same matrix (relative or sample-to-sample matrix effects). Figure 10.3 suggests a possible road map for calibration.

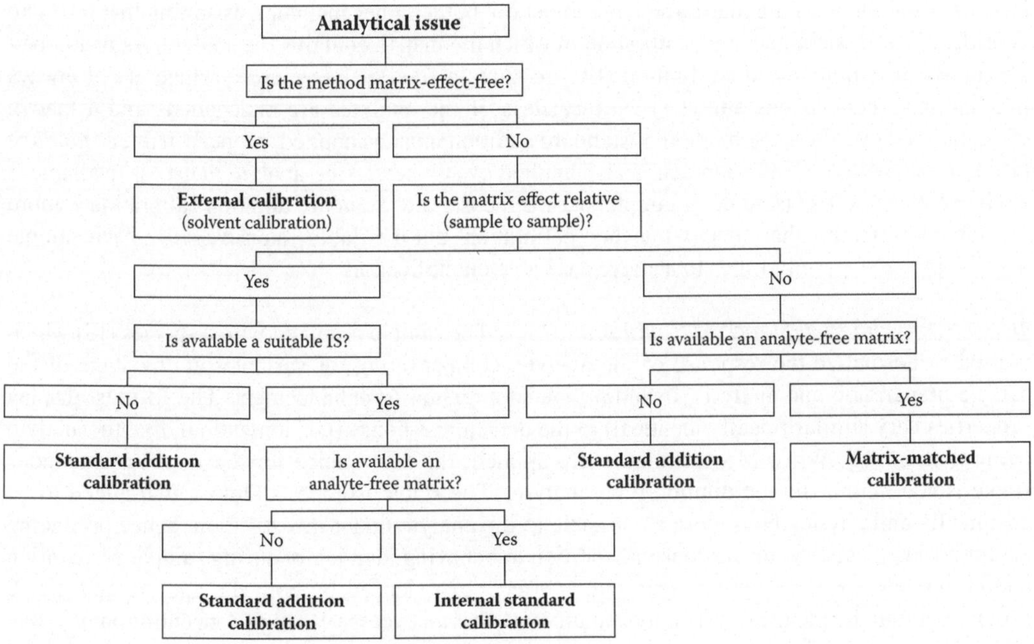

FIGURE 10.3 Road map to choosing the calibration approach.

10.3.1.3.1.1 External Solvent Calibration (EC) EC is the most common calibration methodology, and it is named so because the calibrants are not made up of the sample test portion but they are prepared and analyzed separately from the samples. In this sense, standards and samples form part of different analytical preparations which are measured in a sequential way. EC has also been termed as *solvent calibration* since the calibrants are prepared in the working solvent; accordingly, this calibration type is suitable only for matrix-effect-free methods, such as the traditional procedures using UV or fluorometric detection.

10.3.1.3.1.2 External Matrix-Matched Calibration (EC) Matrix-matched calibration is a particular type of EC in which the calibrants are prepared in a real sample that initially does not contain the analyte. There are two common ways to build matrix-matched curves: adding the analyte prior to or after the sample treatment. In the former case, both matrix effects and the systematic error due to method recovery are compensated for; in the latter, only matrix effects are corrected. The analyst must be aware of this during the validation study to assess the method's trueness correctly. Matrix effects affect the slope of the calibration line, and the comparison of the slopes obtained by preparing the calibrants in the working solvent or in the matrix allows studying these effects as described in the following example.

In this calibration type, it is important that there be little variation in the test sample composition since both the test sample and the calibrant must be affected in the same way by ion suppression or enhancement. Otherwise, a relative matrix effect exists, that is, the matrix effect changes not only, as obvious, among the different matrices but also within the same matrix (sample to sample). In this case, the standard addition or IS calibration should be applied.

10.3.1.3.1.3 Standard Addition Calibration (IC) Standard addition calibration is a particular type of IC, usually based on the analysis of several test portions, adding increasing amounts of analyte standard to many sample aliquots. From the data, a function can be obtained, which will have the same slope as the matrix-matched function, but a higher intercept, assuming that it can be considered as a matrix-matched calibration in which the matrix contains the analyte. As mentioned previously, matrix-matched calibration (EC) requires a free-analyte matrix, which is not always possible (e.g., endogenous analytes). Nevertheless, if the analytes are endogenous and a matrix effect has been verified, the method of standard addition must be applied, properly named *standard addition calibration*. This approach is also applied even when a free-analyte matrix is available if a relative matrix effect (sample to sample) has been observed. Standard addition calibration is more reliable and effective than matrix-matched calibration, but it is labor intensive since each sample must be prepared several times to achieve a reliable quantification.

10.3.1.3.1.4 Internal Standard Calibration (IC) The sample is spiked with a species (IS) which is used to normalize the response of the analyte, compensating for variables at any stage of the sample preparation and analysis, including ion suppression or enhancement. The IS must display properties very similar (ideally identical) to the detector response (i.e., ionization), like the analyte being determined. When MS techniques are applied, the best choice for the IS is an analogous stable isotope in an isotope dilution-type analysis. The stable isotope is almost guaranteed to be chemically and physically as close as possible to the analyte under investigation, hence producing an almost identical detector response in addition to behaving identically during sample preparation and chromatographic resolution. The matrix effects experienced by both the analyte and the IS should therefore be identical. It is important to note that an excessively high concentration of stable isotope IS may cause ion suppression itself, since it will coelute with the analyte of interest. Hence, the IS should be added at an appropriate concentration. Finally, although the use of isotopic dilution probably represents the best type of calibration, economic considerations are also involved in choosing this calibration approach (Figure 10.3). This is also in the light of the current trend toward multiresidue multiclass methods which may require tens of labeled compounds to adequately quantify all the analytes included in the method scope.

Example 10.1

A method for the quantitative determination of hexachlorobenzene (HCB) in tissues by GC-MS was developed. Three different curves were prepared to evaluate method linearity, recoveries, and matrix effects by using calibrants prepared, respectively, in neat solvent (SC), in tissue extracts spiked with pure standards after sample treatment (matrix-matched curve—MMC1), and in samples spiked with standards before extraction (matrix-fortified curve—MMC2). Six concentration points were prepared for each curve covering the expected working range of the real tissue samples. At first, no suitable IS (IS-labeled analogue) was commercially available. Later, after deeper research, a labeled analogue of the analyte was found (HCB-$^{13}C_6$), similarly performing IC. Figure 10.4 shows the curves obtained applying the linear calibration (least squares method). The comparisons among the three slopes (*b*) of SC, MMC1, and MMC2 curves without and then with the IS gave useful information about the analytical process.

In Figure 10.4a, the presence of the ion suppression phenomenon is immediately evident since the slope of MMC1 is lower than that of SC. The ratio between these slopes ($[b_{(MMC1)}/b_{(SC)}]_{EC} \times 100 = 84\%$) indicates the extent of this suppression. On the other hand, the ratio between the slopes of MMC2 and MMC1 ($[b_{(MMC2)}/b_{(MMC1)}]_{EC} \times 100 = 81\%$) represents the method's (real) recovery, that is, a measure of the losses of the analyte during the sample treatment steps (extraction, purification, derivatization, etc.). Since the absence of a relative matrix effect (sample to sample) was previously checked, MMC1 calibration was then able to compensate for the matrix effect, but, nonetheless, the results observed in real samples should be corrected for the bias (81%) due to the analyte losses (recovery). Otherwise, by applying the MMC2 calibration, the results were automatically corrected for both recovery and matrix effect.

The matrix effects as well as the losses during sample treatment were automatically corrected when a suitable IS was introduced (Figure 10.4b). In this case, the ratio $[b_{(MMC2)}/b_{(MMC1)}]_{IC}$ (× 100) equal to 100% represented an "apparent recovery," since the analyte and its labeled analogue

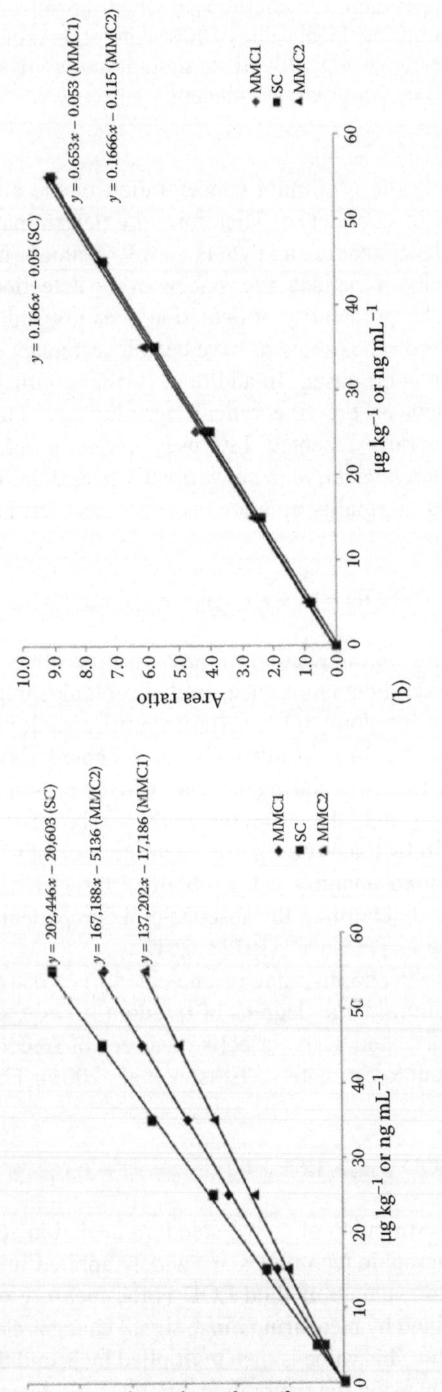

FIGURE 10.4 Calibration curves SC, MMC1, and MMC2 without (a) and then with IS (b).

have a similar behavior both during sample preparation and instrumental determination. That was also proved by the value of the ratio $[b_{(MMC1)}/b_{(SC)}]_{IC} \times 100 = 102\%$, revealing that the IS efficiently compensated for the ion suppression. Obviously, also when the selected IS is adequate, slight deviations of the slope ratio from the ideal value of 100% are observable due to the unavoidable presence of random errors. A range (e.g., 90–110%) should then be fixed, and only outside that range the matrix effect should be considered significant.

10.3.1.4 Limit of Detection

The LOD (or detection limit) is the minimum concentration of the analyte that can reliably be detected with a specified level of confidence. Therefore, the determination of the detection limit requires some degree of confidence that the analyte is above a blank or background level.

Despite the apparent simplicity of the idea, the concept of the detection limit hides some pitfalls. This arises from the fact that the probability of detection does not suddenly change from zero to unity as some threshold is crossed. The problems have been investigated statistically in some detail, and a range of decision criteria is proposed. In addition, further confusion arises because there is currently no universal agreement on the "theoretical" terminology. The terms *limit of detection* and *detection limit* are not generally accepted. ISO uses *minimum detectable net concentration*, whereas IUPAC prefers *minimum detectable (true) value* (Currie, 1995; ISO, 1997).

Under certain circumstances (normality and homoscedasticity), the general form for estimating LOD from experimental data is

$$\mathrm{LOD} = x_0 + t_{(1-\alpha,\nu)}s_0 + t_{(1-\beta,\nu)}s_0,$$

where x_0 is a correction term for taking account of any baseline offset or reading obtained from the blank sample, s_0 is the standard deviation measured in a blank sample, and t is the one-tailed Student's t value at ν degrees of freedom and at certain significance levels α and β. It is now clear that LOD estimation requires deciding in advance the accepted risk of false-positive results (α error, i.e., the probability of wrongly declaring the analyte to be present) and false-negative results (β error, i.e., the probability of wrongly declaring the analyte to be absent). The value assigned to t determines this accepted risk. In fact, a more rigorous definition of LOD (international vocabulary of metrology [VIM]) is "measured quantity value, obtained by a given measurement procedure, for which the probability of falsely claiming the absence of a component in a material is β, given a probability α of falsely claiming its presence" (BIPM, 2012).

Generally, $\alpha = \beta = 0.05$ and the chosen value of one-sided t is 1.65. Although 1.65 is the tabulated value for one-sided t assuming large degrees of freedom ($\nu = \infty$), some authors prefer to use that instead of the (greater) t tabulated at the effective degrees of freedom, that is, the number of experiments (minus 1) carried out to estimate s_0 (Ellison et al., 2009). Therefore, the value of LOD is given by

$$\mathrm{LOD} = x_0 + 1.65s_0 + 1.65s_0 = x_0 + 3.3s_0.$$

The key point is the reliable estimation of s_0, the standard deviation at the zero, level which may be no simple matter. Chromatographic techniques are widely applied in trace analysis, and one of the most reported approaches to evaluate method LOD is the measure of the signal-to-noise ratio (*S*/*N*). The noise level is determined by measuring range signal changes close to the retention time of the analyte on the chromatogram. This value is then multiplied by 3, and the signal value obtained is converted into a concentration by using the calibration curve. This, however, makes the determined value of *S*/*N* rather subjective, as it depends on the zone selected and on the way the interfering peaks are integrated. Moreover, technical evolution of MS equipment has resulted in very low-noise systems that have made the *S*/*N* measurement increasingly difficult and in some modes of operation impossible. This is especially true when using ultralow-noise modes such as HRMS or MS/MS, where the noise is essentially zero. On the other hand, the *S*/*N* approach provides a good estimate when static measurements such as those determined by spectrophotometers (UV–visible, inductively coupled plasma (ICP)–MS, fluorometers, atomic absorption spectrometry (AAS)) are involved.

The difficulty with the determination of blank variance in chromatographic methods is avoided by the EPA determining variance at zero concentration by fortifying samples at low level. The method detection limit (MDL), which substantially coincides with LOD, is then defined as follows:

$$\mathrm{MDL} = t_{(1-\alpha,\nu)} s_0,$$

where t is one-sided Student's t, s_0 is the standard deviation of replicate spikes at low level, α is the accepted level for false-positive error probability ($\alpha = 0.01$), and ν $(n - 1)$ is the number of degrees of freedom. This relationship is applied under certain simplifications (normality and homoscedasticity, i.e., the variances at low level and at zero level are not different). For chromatographic techniques, $x_0 = 0$. The MDL estimation requires replicate sample preparation (at least 7–10 replicates are recommended), and the level of spiking should be one to five times the estimated MDL. Otherwise, the chosen spiking level could be too high to assure a reliable estimation of s_0.

The MDL definition considers only the false-positive error (α) contrasting with the VIM definition. This is the consequence of the confusion introduced by the earlier terminology about the detection, which explicitly recognized only the false-positive error, making in effect the probability of false-negative equal to 50% ($\beta = 0.5$) (Currie, 1995). Later the so-called hypothesis testing school took into account both error types (α and β). Another example of this discrepancy is given by Commission Decision 2002/657/EC (European Commission, 2002), which rightly defines two parameters as a function of the two different errors: CCα (decision limit) and CCβ (detection capability). Unfortunately, the Commission Decision substantially considers CCα as an LOD, although it takes into account only the false-positive error ($\beta = 0.5$). Figure 10.5 summarizes this concept in both the signal and concentration domains. Following the hypothesis testing school (Currie, 1995), CCβ should be assumed as the LOD, instead of CCα (Antignac et al., 2003).

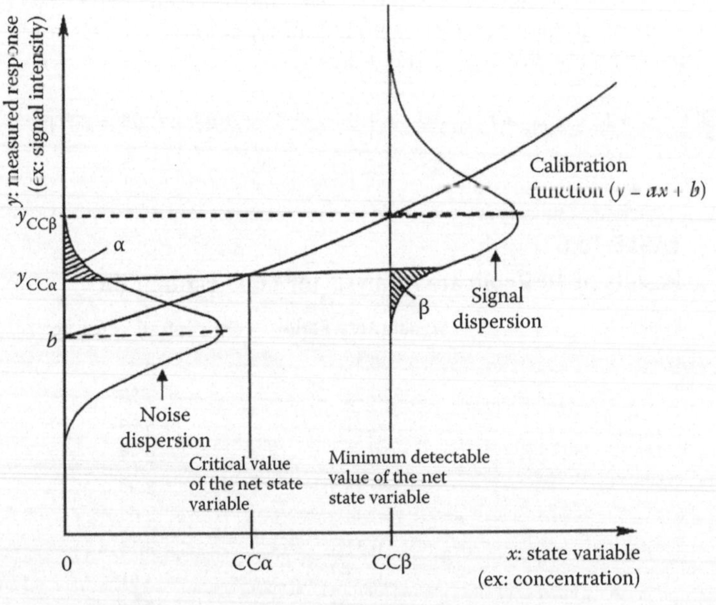

FIGURE 10.5 Definitions of CCα and CCβ according to ISO 11843-1 document. (Reprinted from *Analytica Chimica Acta*, 483, Antignac, J., B. Le Bizec, F. Monteau, and F. Andre, Validation of analytical methods based on mass spectrometric detection according to the "2002/657/EC" European decision: Guideline and application, 325–334, Copyright (2003), with permission from Elsevier.)

The confusion has been further incited by the use of similar multiplicative factors for s_0 and also when the two approaches (only the probability of α error or both) are applied. That happens because there is no general agreement about the choice of the degrees of freedom ν (experimental or supposing a large number of experiments, i.e., $\nu = \infty$?) and of the level of confidence of α (0.05 or 0.01?). For example, in the MDL equation, the one-sided Student's t at six degrees of freedom ($\alpha = 0.01$) is 3.1, which is very close to the multiplicative value of 3.3 (1.65 × 2) obtained applying the z distribution values ($\alpha = \beta = 0.05$) for a large number of experiments (ideally $\nu = \infty$). The confusion is increased by the frequent use of the generic multiplicative factor of 3 without any statistical consideration. Finally, it must be considered that if the SOP of method involves blank correction, some modification in the earlier mentioned equations should be carried out (Ellison et al., 2009; Eurachem, 2014). A practical example of the traps in the estimation of the detection limit is reported in the following.

Example 10.2

During the validation study of the previously introduced method for the determination of HCB in fish by GC-MS, the performance characteristics LOD/LOQ were also evaluated. Taking into account the fit-for-purpose criterion, the statutory requirements involved an LOQ of method lower than, or equal to, 3 μg kg^{-1} (European Commission, 2009). Based on experience, the analyst estimated a detection limit of about 1 μg kg^{-1}; therefore, 10 replicate measurements were carried out spiking blank muscle sample at 3 μg kg^{-1}. The measured area ratios (area of native HCB divided by the area of the corresponding IS, HCB-$^{13}C_6$) are listed in Table 10.6. The concentrations were obtained by dividing these area ratios by the slope (sensitivity) of the calibration curve.

Applying the MDL formula ($s_0 = 0.55$ μg kg^{-1}),

$$\text{MDL} = t_{(1-0.01,9)} s_0 = 2.82 \times 0.55 = 1.5 \ \mu\text{g kg}^{-1}.$$

Therefore, the chosen spiking level (3 μg kg^{-1}) was suitable, being one to five times the estimated MDL. On the other hand considering both error types ($\alpha = \beta = 0.05$), for nine degrees of freedom, the tabulated one-tailed t value was 1.83:

$$\text{LOD} = x_0 + t_{(1-0.05,9)} s_0 + t_{(1-0.05,9)} s_0 = 1.83 \times 0.55 + 1.83 \times 0.55 = 2.0 \ \mu\text{g kg}^{-1}.$$

TABLE 10.6
Results of 10 Replicate Analysis for LOD Estimation

Replicate	Signal (Area Ratio)	Concentration (μg kg^{-1})
1	0.32	3.37
2	0.21	2.22
3	0.31	3.23
4	0.26	2.77
5	0.28	2.99
6	0.39	4.09
7	0.31	3.26
8	0.34	3.61
9	0.26	2.76
10	0.24	2.52
Mean	0.29	3.08
Standard deviation (s_0)	0.05	0.55

The latter equation was the correct way to calculate the LOD. If the level of confidence for false-positive error probability was stated at 0.01 instead of 0.05, the LOD was given by

$$\text{LOD} = x_0 + t_{(1-0.01,9)}s_0 + t_{(1-0.05,9)}s_0 = 2.82 \times 0.55 + 1.83 \times 0.55 = 2.6\ \mu\text{g kg}^{-1}.$$

LOQ is often fixed at about three times the LOD (see Section 10.3.1.5), concluding that the developed method was not fit for purpose (LOQ ≈ 6 μg kg^{-1} > 3 μg kg^{-1}). Therefore, the analyst shall modify the candidate method to improve its LOQ/LOD.

To summarize, (i) the LOD estimation should consider both α and β errors and their confidence levels and (ii) in the function of the number of experimental replicates the relevant one-sided t values shall be applied. It must be underlined that the LOD determined during the validation study is established with poor accuracy because 10 measurements are not able to give a good estimate of σ_0, that is, the standard deviation of the blank population. The interval estimate of a certain σ in the function of the number of observations (N) carried out to evaluate s is possible by applying the chi-square distribution:

$$\sqrt{\frac{(N-1)s^2}{\chi^2_{1-\alpha/2,N-1}}} < \sigma < \sqrt{\frac{(N-1)s^2}{\chi^2_{\alpha/2,N-1}}}.$$

For example, if $N = 10$, σ is included between about 70% and about 180% of the observed s ($\alpha = 0.05$: $\chi^2_{0.025,9} = 19.023$ and $\chi^2_{0.975,9} = 2.700$). In the case of MDL = 1.5 μg kg^{-1} (see Example 10.2), the MDL "true" value can therefore vary from 1.0 to 2.7 μg kg^{-1}. Accordingly, when in routine analysis the expected concentrations are close to the method limits (trace analysis), the fit-for-purpose LOD (or LOQ) should be continuously monitored during the IQC activities, adding a specific control material such as a blank test sample spiked at the required LOD or LOQ level. In the case of the HCB example, the analyst could insert a blank tissue freshly spiked at 3 μg kg^{-1} to verify daily LOQ

The LOD is frequently confused with the method's sensitivity. The sensitivity of an analytical method is its capability to discriminate small differences in the concentration or mass of the test analyte. In practical terms, sensitivity is the slope of the calibration curve that is obtained by plotting the response against the analyte concentration or mass. This is not a particularly important performance characteristic (Eurachem, 2014).

10.3.1.5 Limit of Quantification

The limit of determination or LOQ is the lowest level of analyte that can be determined with acceptable performance. "Acceptable performance" is variously considered by different guidelines to include precision, precision and trueness, or measurement uncertainty. When using the measurement uncertainty to establish the fitness for purpose of LOQ, for example, LOQ may be the lowest method concentration at which the relative uncertainty is still lower than 20%.

To focus on this topic, it is important to have some ideas about the dependence between uncertainty ($u(x)$) and analyte concentration (x). Generally, in chemical measurement, over a large range of analyte concentrations, dominant contributions to the overall uncertainty vary proportionately to the level of analyte, that is $u(x) \propto x$. In such cases, the relative standard uncertainty ($u(x)/x$) is about constant. Reaching low concentrations (close to zero), the relative uncertainty increases to the point where it is not pertinent to refer to a quantitative result (LOQ) and, at yet lower concentrations, to a point where the signal is not recognizable against the noise (LOD). This is the range in which the constant contribution (the standard deviation of blank s_0) becomes dominant. According to its definition (about $3s_0$), at LOD the relative uncertainty reaches about 33%.

The general trend of measurement uncertainty $u(x)$ as a function of the analyte concentration x is shown in Figure 10.6. The dotted line parallel to the x axis represents the constant component s_0,

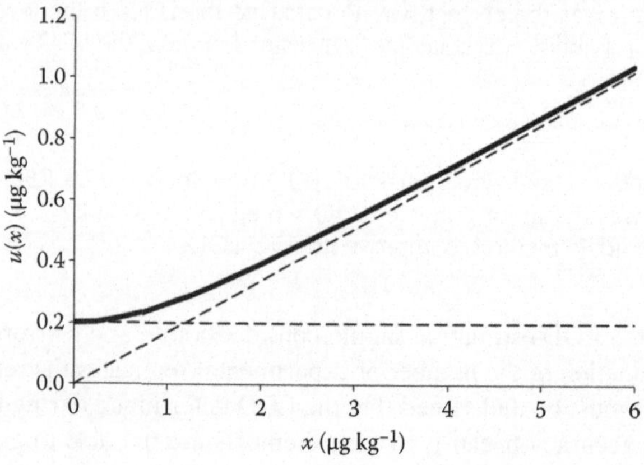

FIGURE 10.6 Variation of uncertainty $u(x)$ with the analyte concentration x at levels close to zero.

whereas the dotted line starting from the axis origin is the proportional one, s_1x. The thick curve is the overall (combined) uncertainty function $u(x)$ (CITAC and Eurachem, 2012).

Mathematically, $u(x)$ is given by (nonlinear regression)

$$u(x) = \sqrt{s_0^2 + (xs_1)^2}.$$

When $xs_1 \gg s_0$, the proportional term prevails and $u(x)/x$ is about constant ($u(x)/x \approx s_1$). On the other hand, at levels close to zero, $xs_1 \ll s_0$ and then it is the absolute standard uncertainty which is about constant ($u(x) \approx s_0$). In the intermediate zone ($xs_1 \approx s_0$), both terms contribute significantly to the overall uncertainty and some curvature is visible (Figure 10.6). This intermediate zone approximately corresponds to the concentration range from LOD to LOQ.

For example, state arbitrarily for a certain method that $s_0 = 0.2$ µg kg^{-1} and $s_1 = 0.17$. Solving the $u(x)$ function, the relative measurement uncertainty reaches 33% and 20% when $x = 0.7$ and 2 µg kg^{-1}, respectively. These two concentrations are estimates of the method's LOD and LOQ, respectively. The LOQ also must be fixed about three times higher than the LOD.

From another point of view, applying the formula introduced in the previous section ($\alpha = \beta = 0.05$; $\nu = \infty$),

$$\text{LOD} = x_0 + t_{(1-0.05,\infty)}s_0 + t_{(1-0.05,\infty)}s_0 = 1.64 \times 0.2 + 1.64 \times 0.2 = 0.7 \text{ µg kg}^{-1}.$$

The obtained LOD concentration agrees with that calculated using the uncertainty function: $u(x) = \sqrt{s_0^2 + (xs_1)^2}$.

Detection and quantification limits are fundamental in trace analysis when one has to decide whether a contaminant is present below or above a specification limit. Accordingly, some regulations state that the LOD and/or the LOQ of the applied analytical method must be lower than a certain fraction of the relevant specification limit. In the performance criteria approach, this condition is able to assure the fitness for purpose of the used analytical method.

10.3.1.6 Accuracy

Accuracy is defined as the "closeness of agreement between a measured quantity value and a true quantity value of a measurand" (BIPM, 2012). *Accuracy* is a qualitative term and a property of a single measurement result. It is expressed by two components of the analytical method, trueness (bias) and precision, and does not require extra validation work. Since in analytical chemistry the total error is composed of systematic (determinate) and random (indeterminate) errors, accuracy covers a combination of both. Systematic and random errors are linked to the method's trueness and precision, respectively. The VIM definitions are listed as follows:

- Random measurement error: component of measurement error that in replicate measurements varies in an unpredictable manner. Random errors are unavoidable due to the fact that every physical measurement is affected by some uncertainty (BIPM, 2012). Precision depends on the distribution of the random error and is unrelated to a true value (accepted reference value or conventional true value).
- Systematic measurement error: component of measurement error that in replicate measurements remains constant or varies in a predictable manner (BIPM, 2012). Measurement trueness is inversely related to systematic measurement error, but is not related to random measurement error. The term *bias* is used as estimate of a systematic measurement error. Systematic measurement error, and its causes, can be known or unknown. However, it should always be removed or, if it is not possible, a correction should be applied to compensate for it. A typical case of unavoidable systematic error is the recovery in trace analysis due to the losses of analyte during the sample treatment steps.

A frequently used analogy to explain the link between precision, trueness, and accuracy is the of the "dartboard" metaphor.

10.3.1.6.1 Precision

Precision is the closeness of agreement between indications or measured quantity values obtained by replicate measurements on the same or similar objects under specified conditions (BIPM, 2012). Precision is estimated with test samples that have been taken through the whole analytical procedure. These test samples may be reference materials, control materials, or real test materials (samples). Precision is quantitatively expressed in terms of imprecision and computed as either the standard deviation (s), variance (s^2), or coefficient of variation (CV, or also relative standard deviation [RSD]) of the test results. The larger the standard deviation, the lesser the precision. The "specified conditions" are divided into three main categories, giving rise to three precision types (Table 10.7).

The two extreme precision measures are reproducibility and repeatability: for this reason, within-laboratory reproducibility is also called intermediate precision. Reproducibility gives the largest expected precision because it is certainly obtained by varying all the factors that may affect the results; that is, results are obtained in different laboratories, with different operators using different equipment. This precision can be estimated only by interlaboratory studies, and for this reason, it is also called interlaboratory reproducibility to distinguish it from within-laboratory (intralaboratory) reproducibility. Collaborative tests are expensive, and, generally, they are carried out to "standardize" a candidate method. However, the widespread diffusion of the performance criteria approach involves more and more the application of internally developed methods. For a single laboratory setting an internal procedure, therefore, the most useful precision measure is the intermediate precision (s_{wR}) obtained by varying all possible factors, because it gives an idea of the sort of variability that the laboratory can expect of its results. As it will be cleared later, a suitable estimation of the intermediate precision is essential to obtain the measurement uncertainty and to set control limits for the IQC, too.

If the analytical method is intended for use over a wide concentration interval, a relationship between precision and concentration should be estimated by testing several concentration levels across the working range. As introduced in Section 10.3.1.5, generally, a linear relation exists between the analyte concentration and the observed standard deviations (uncertainty) and, therefore, the RSD can be considered approximately constant. This is true at concentrations higher than the method's LOQ and over a limited interval (about two orders of magnitude) (CITAC and Eurachem, 2012).

10.3.1.6.1.1 Precision Study When an internal quantitative method is under validation, repeatability and the within-laboratory reproducibility are studied. For this purpose, it is advisable to fix

TABLE 10.7
Conditions for Assessing Precision and the Utility of the Resulting Estimates

Notation	Conditions of Replication	Notes
Repeatability—within-run (or intrabatch) precision (s_r)	Replication on separate test portions of the same material in the same run. Same method, laboratory, operator, equipment and within short interval of time.	"Short interval of time" is the length of an analytical "run." It has limited applicability, but it can be used in IQC to have a first estimate of s_{wR} (see Section 10.4).
Within-laboratory reproducibility or intermediate precision (s_{wR})	Replication in separate runs. Same material, method and laboratory, but at least two of the factors affecting the variability are changed (i.e., operator, instrument, calibration, batches of reagents). Thus, different types of s_{wR} estimates are obtained depending on the factors changed from one measurement to the next. For instance, if the results are observed on different days and by different operators, we obtain the (time + operator)-different intermediate precision.	This type of precision is addressed in IQC (Section 10.4).
Reproducibility (s_R) (1)	Replication by the same method in different laboratories. Same material.	This is the estimate obtained by the collaborative (interlaboratory) test.
Reproducibility (s_R) (2)	Replication by various methods in different laboratories. Same material.	This is the estimate obtained from the results of a single round of a proficiency tests. In this case, s_R is generally greater than that obtained from collaborative studies.

in advance an experimental design with n replicates of the same test sample analyzed in p runs. In each of the p runs, a factor is varied (operator, instrument, calibration, etc.). The experimental design corresponds to a two-factor fully nested design as reported in Figure 10.7.

Analyzing the results obtained by the analysis of variance (ANOVA), the repeatability and the within-laboratory precision are estimated. The repeatability variance $\left(s_r^2\right)$ reflects the within-run precision, and the within-laboratory variance $\left(s_{wR}^2\right)$ reflects both the within- and the between-run precision $\left(s_{run}^2\right)$:

$$s_{wR}^2 = s_r^2 + s_{run}^2.$$

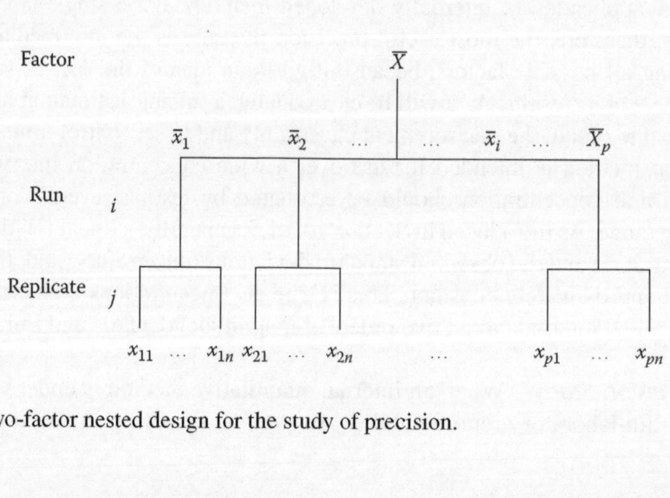

FIGURE 10.7 Two-factor nested design for the study of precision.

The number of degrees of freedom (ν) for repeatability is given by $\nu_{sr} = (n - 1)p$, whereas for within-laboratory reproducibility, $\nu_{swR} = n \times p - 1$ (Irish National Accreditation Board, 2012).

As repeatability and within-laboratory reproducibility could vary depending on analyte concentration, the experimental design must be replicated at different concentration levels (minimum of three) investigating the method precision in the measuring range. One of these levels should be close to the lower value of the measuring range.

Example 10.3

In order to evaluate the repeatability and intermediate precision of the previously introduced method (HCB in tissue), 3-day experiments (runs: $p = 3$) were planned in the validation, varying operator, instrument, and calibration status of the instrument (operator + calibration + instrument). On each day, six independent replicates ($n = 6$) were performed, spiking a blank tissue at environmental quality standard level (10 µg kg^{-1}). The results of single experiments together with mean s_r and s_{wR} are shown in Table 10.8. For calculation details, see, for example, Ellison et al. (2009).

10.3.1.6.1.2 The Horwitz Equation Regulations based on the performance criteria approach often establish the acceptability of the method precision based on the determined concentration of the analyte. These values of acceptability are generally based on the Horwitz equation (Horwitz et al., 1980). Examining the results of over 50 interlaboratory collaborative studies, Horwitz and collaborators noticed a striking pattern in the RSDs. As the concentration of the analyte decreased over two orders of magnitude, the RSD of reproducibility (RSD_R) increased by a factor of 2. Therefore, at 100% concentration of the analyte, the RSD_R was about 2%, at 1% the RSD_R was about 4%, and at 0.01% (100 ppm) the RSD_R was about 8%. This pattern persisted at least down to sub-ppm levels, but is not valid below the method's LOQ. The Horwitz relationship is the following:

$$RSD_R = 2^{(1-0.5 \log C)},$$

TABLE 10.8
Results (µg kg^{-1}) of the Validation Study for Precision (Spiked Tissue with 10 µg kg^{-1} of HCB)

	Run		
Replicate	**1**	**2**	**3**
1	10.6	10.5	9.9
2	11.8	9.3	10.7
3	11.2	10.0	10.8
4	10.0	10.5	10.0
5	10.3	9.0	11.8
6	10.0	10.2	10.7
Mean (µg kg^{-1})	11.1	9.92	10.6
s (µg kg^{-1})	0.73	0.63	0.68
RSD (%)	6.59	6.36	6.42
Overall mean (µg kg^{-1})		10.5	
s_r (µg kg^{-1})[a]		0.68	
s_{wR} (µg kg^{-1})[a]		0.75	

[a] Calculated by applying ANOVA. Degrees of freedom were $\nu_{sr} = 15$ and $\nu_{swR} = 17$.

where C is the mass fraction expressed as a power (exponent) of 10 (e.g., 1 mg g^{-1} = 10^{-3}) and RSD_R is the RSD observed in reproducibility (interlaboratory) conditions. The Horwitz equation is also directly applicable to predict within-laboratory reproducibility.

Further investigations found that at the extreme concentration ranges (<120 μg kg^{-1}, >13.8%), precision was overestimated by Horwitz equation. As a result, some adjustments were carried out as shown in Table 10.9.

It must be noted that in this adjusted version at trace concentrations (<120 μg kg^{-1}), the RSD_R remains constant (22%).

For example, to calculate the σ_R at 420 μg kg^{-1} (ppb) for a certain analyte, convert the concentration into a mass fraction ($C = 0.00000042 = 4.2 \times 10^{-7}$) and insert this value in the relevant Horwitz equation:

$$\sigma_R = 0.02 \times (4.20 \times 10^{-7})^{0.85} = 7.66 \times 10^{-8} = 76.6\ \mu\text{g kg}^{-1}.$$

The RSD is given by

$$RSD_R\ (\%) = \frac{7.66 \times 10^{-8}}{4.20 \times 10^{-7}} \times 100 = 18.2\%.$$

Alternatively, the RSD_R (%) is directly obtained by applying the following equation (Table 10.9):

$$RSD_R\ (\%) = 2 \times (4.2 \times 10^{-7})^{-0.15} = 18.2\%.$$

The analyst would also know if the intermediate (experimental) precision (s_{wR} = 65.2 μg kg^{-1}) is significantly better (lower) than the predicted value (σ_R). This experimental s_{wR} has been estimated via ANOVA by carrying out 18 experiments ($n = 6$, $p = 3$: six replicates in three different runs). Applying the chi-square test,

$$T = \frac{(N-1)s^2}{\sigma^2} = \frac{17 \times 65.2^2}{76.6^2} = 12.3.$$

Since T is higher than the critical value of the χ distribution ($\chi^2_{0.975,17} = 7.56$), the null hypothesis cannot be rejected, that is, there is insufficient evidence at the 95% level of confidence to conclude that the method provides better precision than the predicted Horwitz value (Ellison et al., 2009). To reject the null hypothesis, the T value must be lower than the critical tabulated value of $\chi^2_{0.975,N-1}$ or higher than $\chi^2_{0.025,N-1}$. Otherwise, to compare two experimental variances, an F-test is applicable (Table 10.2).

The use of the Horwitz equation or its modification is limited to analytical methods that express measurands as a concentration of mass; it does not apply to empirical analytes (moisture, ash, etc.) or physical properties (color, viscosity, etc.).

TABLE 10.9
Reproducibility as a Function of Analyte Concentration (Modified Horwitz Equation)

Concentration Range (Mass Fraction C)	Equation for Reproducibility Values (s_R)	Equation for Relative Reproducibility Values (RSD_R, %)	Reproducibility Values (RSD_R, %)
<120 μg kg^{-1} (<1.2 × 10^{-7})	$\sigma_R = 0.22C$	RSD = 22	22
120 μg kg^{-1}: 13.8% (1.2 × 10^{-7}–1.38 × 10^{-1})	$\sigma_R = 0.02C^{0.85}$	RSD = $2C^{-0.15}$	22–2.7
>13.8% (>1.38 × 10^{-1})	$\sigma_R = 0.01C^{0.5}$	RSD = $C^{-0.5}$	≤2.7

The Horwitz function is widely used as a benchmark for the performance of analytical methods, via a measure called HorRat (Horwitz ratio). HorRat is a normalized performance parameter indicating the acceptability of methods of analysis with regard to interlaboratory and within-laboratory reproducibility. The HorRat is calculated as follows:

$$\text{HorRat} = \frac{\text{RSD}_{\text{R/wR}}\ (\%)}{\text{PRSD}_{\text{R}}\ (\%)},$$

where $\text{RSD}_{\text{R/wR}}$ (%) is the experimental RSD (from s_{R} or s_{wR}) and PRSD_{R} (%) is the corresponding predicted RSD calculated using the relevant equations in Table 10.9. The concentration C (found or added) must always be expressed as a mass fraction. Many regulations give the HorRat index as the criterion for judging the suitability of method precision instead of the Horwitz equation: values falling within the range of 0.5–2.0 are generally considered acceptable. Finally, HorRat can be applied not only to inter- or intralaboratory reproducibility evaluation, but also to judge method repeatability. When precision is obtained in repeatability conditions, RSD_{r} must range between 0.5 (1/2) and 0.66 (2/3) of the RSD_{R} calculated by means of the Horwitz equation or its modifications. Therefore, if RSD_{r} is at maximum two-thirds of RSD_{R}, under repeatability conditions the accepted HorRat (r) values are between 0.3 and 1.3:

$$\text{HorRat}(r) = \frac{\text{RSD}_{\text{r}}\ (\%)}{\text{PRSD}_{\text{R}}\ (\%)}.$$

10.3.1.6.2 Trueness

Precision describes the variation in results obtained from replicate analyses of a sample, but it gives no information about the closeness of these results to the true concentration of the analyte in the sample. Trueness describes the "closeness of agreement between the average of an infinite number of replicate measured quantity values and a reference quantity value" (BIPM, 2012). In the past, *trueness* and *accuracy* were often used synonymously. However, in current usage, these terms are different: accuracy regards how closely a single measurement agrees with the reference value, whereas trueness regards the extent to which an average of a series of measurements agrees with the reference value. In the first case, imprecision contributes to the lack of agreement, whereas in the second, imprecision is minimized (ideally removed entirely) by use of an average.

Trueness is quantitatively expressed in terms of "bias," which is the difference between an accepted reference value (x_{ref}) and the expectation of test results ($\bar{x}_{\text{lab}}$):

$$\text{bias} = \bar{x}_{\text{lab}} - x_{\text{ref}}$$

The bias of an analytical method is usually determined by applying it to a relevant "reference" object, such as a (certified) reference material, spiked materials, and the reference method. It is important to specify the conditions under which the test results are obtained: different conditions do not necessarily lead to the same expectation, that is, the same mean value. Thus, the reference object will be measured at least $n = 6$ times (better $n > 10$) under within-laboratory reproducibility conditions, which correspond to those applied in routine method use. The bias may be a function of analyte content. Many methods show a proportional bias, usually called recovery:

$$\text{recovery} = \frac{\bar{x}_{\text{lab}}}{x_{\text{ref}}}.$$

Recovery information is especially important in ultratrace analysis, including such difficult fields as the determination of pesticide and veterinary drug residues.

10.3.1.6.2.1 Trueness Study As already mentioned, trueness is assessed comparing the results obtained using the procedure under validation (x_{lab}) with a reference value (x_{ref}). One of the problems when planning a trueness study is the suitable choice of this reference. There are a number of options, but not all references have the same level of traceability. Traceability concerns the direct or indirect linking of analytical results to the international metric system of units (International System of Units [SI]); thereby, it is the way of achieving comparability in chemical analysis. The most common references are listed in Table 10.10 in the order of their level of traceability (from maximum to minimum). Reference materials shall also be similar to real-life samples as closely as possible and be available with different analyte concentrations.

The result evaluation of trueness study is performed at first by applying a *t*-test (Table 10.2) to investigate whether the measured bias is significant or not, that is, whether the bias is really different from zero. The general formula is given by

$$t_{sper} = \frac{\text{bias}}{u(\text{bias})} = \frac{\text{bias}}{\sqrt{\left[u(\bar{x}_{lab})\right]^2 + \left[u(x_{ref})\right]^2}}.$$

TABLE 10.10
Approaches to Assessing a Method's Trueness

Approach	Definition
Certified reference materials (CRMs)	Reference materials accompanied by documentation issued by an authoritative body and providing one or more specified property values with associated uncertainties and traceabilities, using valid procedures (BIPM, 2012). The CRM should be similar to the routine samples.
Reference materials	Materials sufficiently homogeneous and stable with reference to specified properties, which have been established to be fit for its intended use in measurement or in examination of nominal properties (BIPM, 2012). Examples of reference materials include materials whose characteristics have been defined by a reference material producer, but whose values are not accompanied by an uncertainty statement; materials analyzed in the laboratory for use as reference materials; and materials distributed in a proficiency test.
Reference measurement procedure (method)	Procedure accepted as providing measurement results fit for their intended use in assessing measurement trueness of measured quantity values obtained from other measurement procedures for quantities of the same kind, in calibration, or in characterizing reference materials (BIPM, 2012). This is a useful option when checking an alternative to, or modification of, an established standard method already validated and in use in the laboratory. Both methods are used to analyze a number of typical test materials, preferably covering a useful range of concentrations fairly evenly. Comparison of the results over the whole range by a suitable statistical method would demonstrate any bias between the methods.
PT	The reference value corresponds to the value assigned by the provider of the PT. This approach is applicable only if a certain number of PT rounds are available for the same analyte(s).
Spiked samples	These references have the lowest level of traceability, and the analyst should resort to spiked samples when other references are not available. A typical test material is analyzed by the method under validation both in its original state and after the addition of a known mass of the analyte to the test portion (spiking). The difference between the two results is called the surrogate recovery (or marginal recovery). Recoveries significantly different from unity indicate that a bias is affecting the method. Strictly speaking, recovery studied with spiked samples only assess bias due to effects operating on the added analyte and the same effects do not necessarily apply to the same extent to the native analyte, which can also be influenced by additional factors. Accordingly, good surrogate recovery is not a guarantee of trueness, while poor surrogate recovery certainly is an indication of lack of trueness.

TABLE 10.11
Approaches and Equations for Estimating the Uncertainty Bias

Experimental Approach	*u*(bias)	Symbols
Replicate determinations using a CRM or other reference materials	$u(\text{bias}) = \sqrt{\frac{s_{\text{lab}}^2}{n} + u(x_{\text{ref}})^2}$	s_{lab}: standard deviation of the *n* measurements on the (certified) reference material; $u(x_{\text{ref}})$: standard (combined) uncertainty reported in the certificate
Comparison with a reference (standardized) method	$u(\text{bias}) = \sqrt{\frac{s_{\text{lab}}^2}{n} + u(x_{\text{ref}})^2}$	s_{lab}: relative standard deviation of the *n* measurements with the method under validation; $u(x_{\text{ref}})$: standard (combined) uncertainty of the reference method
Evaluation of results in participation to PT	$u(\text{bias}) = \sqrt{\frac{\sum (b_i - \bar{b})^2}{n-1}}$	b_i: observed bias in each round of the PT scheme; $\bar{b}$: average bias from all PT rounds; *n*: number of PT rounds
Replicate determinations using a spiked material	$u(\text{bias}) = \sqrt{\frac{s_{\text{lab}}^2}{n} + u(x_{\text{spiked}})^2}$	s_{lab}: relative standard deviation from the spiking experiment; *n*: number of replicates from the spiking experiment; $u(x_{\text{spiked}})$: uncertainty associated to the nominal concentration of the spike

The bias uncertainty *u*(bias) is introduced in the denominator of the t_{sper} equation, and it is always made up of two components: the uncertainty associated with knowledge of the reference value (x_{ref}) and the uncertainty associated with the experimental measurements to check the method bias ($\bar{x}_{\text{lab}}$). Examples of *u*(bias) estimation in the most important cases are detailed in Table 10.11 (Linsinger, 2008).

As will be seen later, the estimation of bias uncertainty is fundamental not only to determining the method's trueness but also, along with the uncertainty associated with random error (precision), to estimating the overall measurement uncertainty.

The concentration ($\bar{x}_{\text{lab}}$) measured in trueness assessment to calculate the bias is the final result obtained when applying the method under validation after the removal of any previously known systematic error. If the *t*-test reveals a significant method bias, the analyst should remove it or, if it is not possible, should correct the result to compensate for the known (unavoidable) systematic measurement error. A typical case of an unavoidable systematic error is the recovery in trace analysis due to the losses of the analyte during the sample treatment steps (see Section 10.3.1.6). The bias removal can be performed in different ways, that is, with the improvement of the procedure steps, with a proper calibration (internal standardization, standard addition, etc.), or, at the end, with recovery correction of the result for the estimated bias. This is an important topic in analytical chemistry and the analyst must be aware too of the difference between "recovery" and "apparent recovery" (Thorburn Burns et al., 2002). In any case, there is unanimous agreement that recovery correction is necessary with respect to the ultimate goal of achieving comparability of measurement results worldwide. To conclude the comparison between x_{ref} and $\bar{x}_{\text{lab}}$ should indicate the presence of any significant residual (unknown) bias also after the removal or correction for all recognized significant bias.

Example 10.4

The assessment of trueness of method for the determination of HCB in tissues was carried out by analyzing a CRM (National Institute of Standards and Technology [NIST] Standard Reference Material [SRM] 1947—Lake Michigan fish tissue). The HCB certified value and its uncertainty were 7.48 ± 0.66 µg kg^{-1}. Six independent analyses were carried out in two different days by two different operators for a total of 12 replicates. The measured average value ($\bar{x}_{\text{lab}}$) and its standard deviation s_{lab} were 8.07 and 0.42 µg kg^{-1}, respectively. The application of IC (HCB-$^{13}C_6$) permitted to have results already corrected for recovery. The comparison between the two concentrations (certified and experimental) was done with a *t*-test as follows:

$$t_{sper} = \frac{\text{bias}}{u(\text{bias})} = \frac{\left|\bar{x}_{lab} - x_{ref}\right|}{\sqrt{s_{lab}^2/n + u(x_{ref})^2}} = \frac{\left|8.07 - 7.48\right|}{\sqrt{0.42^2/12 + 0.33^2}} = 1.7,$$

where $u(x_{ref})$ is the standard uncertainty associated with x_{ref}. It must be noted that the uncertainty reported in CRM certificates is generally expressed as an expanded uncertainty $U(x_{ref})$ and, therefore, the standard uncertainty $u(x_{ref})$, is obtained dividing $U(x_{ref})$ by 2 (see Section 10.3.1.8). To verify the null hypothesis, t_{crit} is generally set equal to 2. Since $t_{sper} < t_{crit}$, there was no evidence that the observed bias was significantly different from zero (or recovery different from 1).

When reference materials are not available, the trueness assessment can be performed by using the data of the spiking study already performed for precision assessment (Table 10.10). The way to evaluate trueness was exactly the same as the case of the CRM, although the level of traceability was lower. The reference concentration x_{ref} was the spiked concentration, and the equation for t_{sper} was given by

$$t_{sper} = \frac{\text{bias}}{u(\text{bias})} = \frac{\left|\bar{x}_{lab} - x_{ref}\right|}{\sqrt{s_{lab}^2/n + u(x_{spike})^2}} = \frac{\left|10.5 - 10\right|}{\sqrt{0.85^2/18 + 0.2^2}} = 1.8.$$

Based on the information on the purity of the material used (standard solutions) to prepare the spiked samples and the accuracy of the volumetric glassware and analytical balances used, the uncertainty $u(C_{spike})$ associated with the knowledge of the nominal concentration of HCB in the spiked tissues was about 0.2 µg kg–^{1} (CITAC and Eurachem, 2012). Also, with this approach, the observed bias was not significant (t_{sper} <2) and the method can be considered unbiased.

10.3.1.7 Robustness

Robustness is a measure of the capacity of an analytical procedure to remain unaffected by small but deliberate variations in procedural parameters listed in the documentation; it provides an indication of the suitability and reliability of the method or procedure during normal use (Vander Heyden et al., 2001). A robustness test is an experimental setup for evaluating the robustness of methods by means of an intralaboratory study. *Robustness* and *ruggedness* are closely related terms, and they are often used as synonyms. However, ruggedness indicates the "robustness" studied in interlaboratory condition. A discussion about ruggedness is beyond the scope of this chapter.

Methods are made up of a certain number of separate steps which are carefully described in the SOP, namely, the detailed, written instructions reporting the procedure operations to achieve uniformity of the performance while it is being carried out. However, each analyst is expected to use their own judgment to a certain extent when following the prescribed instructions. Robustness tests examine the effect that operational parameters have on the analysis results. Seven main steps can be identified during a robustness study: (i) choice of the factors/variables to be tested, (ii) definition of the different levels of *what* for the factors, (iii) selection of the experimental design and definition of the experimental protocol (complete experimental setup), (iv) definition of the responses to be determined, (v) experiments and measurement of the chosen responses, (vi) calculation of effects and their statistical and/or graphical analysis, and (vii) the drawing of chemically relevant conclusions from the statistical analysis and, if necessary, taking of measures to improve the performance of the method.

During the first step, a number of method factors (variables) are selected. These factors may regard sample treatment (composition of extraction mixture, evaporation temperature, evaporation time, etc.) or instrumental determination. For instance, when the LC technique is applied, the pH, flow rate, column temperature, injection volume, or mobile phase composition can be tested. On the other hand, for GC methods, the injector temperature, the column temperature, and the slope of the temperature are possible variables. Not only quantitative factors can be investigated, but qualitative ones such as reagent source or column manufacturer can be too.

In the second step, the levels of factors are selected. The choice of the levels (variation range) is important: a wrong choice (unrealistic) can make the robustness experiment useless. Generally, the quantitative factors range between two values (one with a small letter and one with a capital letter). These are chosen as the smallest and the greatest values of the factor expected during normal use of the procedure, or the nominal value specified in the procedure and a changed value.

In the third step, the choice of the experimental study is made. Although the univariate approach is feasible, where only one variable at a time is changed, experimental designs allow robustness studies to be carried out using the minimum number of proofs. The best-known experimental plan is the Plackett–Burman scheme, which requires $2n$ experiments to test up to $2^n - 1$ variables. Plackett and Burman developed designs which are suitable only when the interactions are negligible, since they cannot deal with factor interactions. In analytical chemistry, the Plackett–Burman scheme is also called the Youden scheme (European Commission, 2002). The basic idea is not to study one alteration at a time but to introduce several variations at once. Table 10.12 reports a typical scheme with eight experiments for a maximum of seven variables. A, B, C, D, E, F, and G (capital letters) denote the highest or the nominal values for seven possible different factors that could influence the results if their nominal values are changed slightly. Their alternative values are denoted by the corresponding small letters: a, b, c, d, e, f, and g. When the variables are less than seven, dummy variables are introduced. A dummy factor is an imaginary factor for which the change from one level to the other has no physical meaning.

From the experiments performed, a number of responses can be determined. For chromatographic methods, responses describing a quantity such as the content of analyte and/or peak areas are used the most (step iv).

Robustness testing should be performed at the end of method development or at the beginning of its validation (step v). When the method robustness is found to be acceptable, the method can be further validated, and when the validation results are regarded as satisfactory, it can be routinely applied. Otherwise, performing the robustness study late in the validation procedure involves the risk that when a method is found to not be robust, it needs to be redeveloped and reoptimized, and at this stage, much effort and resources have already been spent on optimization and validation.

During step vi, the quantitative influence of the variables is determined, verifying if this influence falls within a previously specified tolerance range, that is, the intermediate precision of the

TABLE 10.12
Example of a Typical Plackett–Burman (Youden) Experimental Design for a Maximum of Seven Factors

	Experiment Number								
Factor	**1**	**2**	**3**	**4**	**5**	**6**	**7**	**8**	D_i
A/a	A	A	A	A	a	a	a	a	0.17
B/b	B	B	b	b	B	B	b	b	2.43
C/c	C	C	C	c	C	c	C	c	−0.13
D/d	D	D	d	d	d	d	D	D	0.08
E/e	E	E	E	e	e	E	e	E	−0.08
F/f	F	F	f	F	F	f	f	F	−0.32
G/g	G	G	g	G	g	G	G	g	0.23
Result	s	t	u	v	w	x	y	z	
Concentration (μg kg^{-1})	10.2	10.5	7.8	7.9	9.8	10.4	8	7.5	

method performed in nominal conditions: if this happens, the method is robust for that factor in the chosen range. The problem is that the estimate of intermediate precision (s_{wR}) is often not available when the robustness test is performed (i.e., at the start of the validation study). Other approaches are possible to evaluate this tolerance as shown in the following example (Vander Heyden et al., 2001). In the last step (step vii), if a factor is judged as critical, the analyst can decide to review the method or to indicate clearly in the SOP the critical factor.

Example 10.5

A robustness study for HCB determination in tissue was carried out by applying the Plackett–Burman design with eight experiments for seven factors (variables). The HCB concentrations were used as responses (μg kg^{-1}), spiking eight portions of a blank tissue at 10 μg kg^{-1}. The experimental plan with the observed responses (s, t, u, …) and the calculated effects (D_i) is reported in Table 10.12.

The effect of each factor (D_i) was estimated by subtracting the mean of the results obtained with the factor at the highest level (capital letter) from the mean of the results obtained with it at the lowest level (small letter). For example, for factor A/a, the difference $D_{A/a}$ was calculated as follows:

$$D_{A/a} = \frac{(s+t+u+v)}{4} - \frac{(w+x+y+z)}{4},$$

$$D_{A/a} = \frac{(10.2+10.5+7.8+7.9)}{4} - \frac{(9.8+10.4+8.0+7.5)}{4} = 0.17\ \mu g\ kg^{-1}.$$

The differences (D_i) for each factor were calculated in the same way (Table 10.12). If variations in one or two factors are affecting the analysis results, their differences will be significantly larger than those for the other factors. From a glance at Table 10.12, the difference $D_{B/b}$ seems the most suspect. In order to evaluate whether the differences were statistically significant or not, an experimental t (t_{sper}) was calculated for each factor and then compared with the relevant t_{crit} (two-tailed at 95% confidence):

$$t_{sper} = \frac{|D_i|}{SE}.$$

SE represents an estimate of the standard error obtained in nominal (not altered) conditions, that is, the "natural" variability of the method. The basic idea is to compare the natural variability with the variability observed in altered conditions. For this purpose, some authors use the F-test, too (European Commission, 2002). When the intermediate experimental precision is not available as in Case 10.1, SE can be estimated by using the two approaches described Cases 10.2 and 10.3 (Vander Heyden et al., 2001).

The experimental values for t of each factor obtained from applying the three approaches are summarized in Table 10.13. The relevant two-tailed critical t values at 95% confidence are reported, too. Irrespective of the approach, factor B was always critical ($t_{sper} > t_{crit}$). The analyst can either develop the method again or underline the need to carefully control this factor in the SOP.

10.3.1.8 Measurement Uncertainty

The measurement uncertainty is a "non-negative parameter characterizing the dispersion of the quantity values being attributed to a measurand, based on the information used" (BIPM, 2012). Due to further inevitable "errors" with each procedural step, repeating the entire analysis of a sample can lead to analytical results that are not identical, but that deviate from one another to

CASE 10.1

The standard deviation (s_{wR}) evaluated in within-laboratory reproducibility conditions was equal to 0.75 µg kg^{-1} ($\nu = 17$; see Table 10.8). The standard error SE was given as

$$SE = \frac{2s_{wR}}{\sqrt{N}} = \frac{2 \times 0.75}{\sqrt{8}} = 0.53\ \mu g\ kg^{-1},$$

where N was the number of experiments carried out within the experimental plan ($N = 8$). Therefore, for each factor, the D_i was obtained as follows:

$$t_{sper} = \frac{|D_i|}{SE} = \frac{|D_i|}{0.53}.$$

CASE 10.2

Dummy variables approach: To apply this approach, at least three dummy variables are required; otherwise, the power of the t-test to detect any significance is low and the use of the Dong algorithm is to be preferred. A possibility is to perform an experimental design with more experiments than those strictly required by the number of (real) factors. In this example, for instance, a 12-experiment Plackett–Burman plan (instead of an 8-experiment one) should be implemented to apply the dummy variables approach.

Anyway, we suppose C, E, and G were three dummy factors. The standard error SE of the nominal method was estimated by using the formula

$$SE = \sqrt{\frac{\sum D_i^2}{n}} = \sqrt{\frac{D_{C/c}^2 + D_{E/e}^2 + D_{G/g}^2}{3}} = \sqrt{\frac{0.016 + 0.006 + 0.051}{3}} = 0.155\ \mu g\ kg^{-1}.$$

Therefore, for each factor, D_i was calculated:

$$t_{sper} = \frac{|D_i|}{SE} = \frac{|D_i|}{0.155}.$$

a greater or lesser extent. Consequently, only a result interval, or confidence interval, in which the true analytical result lies with a definite probability of error (usually 5%) may be indicated. The half width of this interval, that is, the distance of the lower or upper interval limit from the mean value, is the measurement uncertainty. This parameter became popular among the analytical chemists when the first version of ISO/IEC 17025 in 1999 (ISO, 2000) forced the accredited laboratories to include it in the method validation study. Until the 1990s, legislative compliance was generally judged without including measurement uncertainty: a maximum specification limit was exceeded simply when the measurement result exceeded its value (noncompliant result). However, only with its uncertainty a result becomes really comparable with specification limits or with measurements carried out in other laboratories. At present, the uncertainty declaration is required

CASE 10.3

Dong algorithm approach: An initial estimate of the error of an effect was obtained in the following way:

$$s_0 = 1.5 \times \text{median } |D_i| = 1.5 \times 0.17 = 0.26 \ \mu\text{g kg}^{-1},$$

where median $|D_i|$ was the median value of the seven $|D_i|$ (2.43, 0.32, 0.23, 0.17, 0.13, 0.08, 0.08) listed in Table 10.12. From s_0, a final estimation of the standard error SE was derived as follows:

$$\text{SE} = \sqrt{\frac{\sum D_i^2}{m}} = \sqrt{\frac{0.08^2 + 0.08^2 + 0.13^2 + 0.17^2 + 0.23^2 + 0.32^2}{6}} = 0.189 \ \mu\text{g kg}^{-1},$$

where m was the number of absolute effects smaller than $2.5s_0 = 0.66$ µg kg^{-1}. Since only $|D_{B/b}|$ was higher than 0.66 µg kg^{-1}, in the SE calculation this effect has not been included ($m = 6$). Therefore, the differences D_i of each factor were calculated as follows:

$$t_{\text{sper}} = \frac{|D_i|}{\text{SE}} = \frac{|D_i|}{0.189}.$$

The use of the median (robust estimator) must be noted. By its nature, the robustness study could introduce outliers, and so a robust estimator also permits to evaluate the SE, a measure of the natural variability of the method, in the presence of outliers.

when the uncertainty is relevant to the validity or application of the test results, when a customer's instructions require it, or when the uncertainty affects compliance to a specification limit (ISO, 2000). The other important aspect in the compliance assessment is the choice of the decision rules. For example, in the presence of an upper limit there are four kinds of results as shown in Figure 10.8. Cases i and iv are not questionable: the maximum specification limit (upper limit) is exceeded in the former and not exceeded in the latter. On the other hand, the interpretation of cases ii and

TABLE 10.13
Interpretation of the Robustness Study Using the Three Proposed Approaches

Approach Factor	Intermediate Precision t_{sper}	Dummy Variables t_{sper}	Dong Algorithm t_{sper}
A/a	0.33	1.13	0.93
B/b	4.57[a]	15.7[a]	12.9[a]
C/c	0.24	–	0.66
D/d	0.14	0.48	0.40
E/e	0.14	–	0.40
F/f	0.61	2.10	1.72
G/g	0.42	–	1.19
t_{crit}	$t_{0.05,17}$ 2.11	$t_{0.05,3}$ 3.18	$t_{0.05,6}$ 2.45

[a] Not robust.

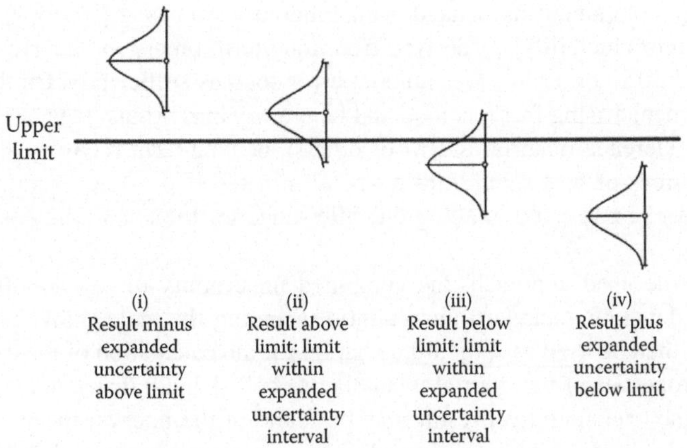

FIGURE 10.8 Assessment of compliance with an upper limit.

iii depends on the decision rules chosen. Decision rules give a prescription for the acceptance or rejection of a product based on the measurement result, its uncertainty, and the specification limit or limits, taking into account the acceptable level of probability of making a wrong decision. On this basis, an "acceptance zone" and a "rejection zone" are determined, such that if the measurement result lies in the acceptance zone the product is declared compliant and if the measurement is in the rejection zone the product is declared noncompliant (CITAC and Eurachem, 2007). The decision rule which is widely applied is that a result is not compliant if the measured value exceeds the limit by the expanded uncertainty, namely. the limit is exceeded only if the measured value minus the expanded uncertainty is above the limit. Applying this in both cases ii and iii (Figure 10.8), the product is judged as compliant.

There are two main approaches to estimating the measurement uncertainty: the bottom–up and the top–down.

- The bottom–up (CITAC and Eurachem, 2012) or component-by-component approach suggests proceeding as follows: First, make a list of all possible sources of uncertainty and calculate the uncertainty related to each of them. Then, express as standard deviations the different contributions that were identified. Finally, combine these standard deviations by using the error propagation law so as to obtain the combined (standard) type uncertainty (u_{comb}). This strict approach makes it possible to clearly identify and quantify the uncertainties related to each distinct source or process step. Knowing the value that is specific to each uncertainty source may be useful later for optimizing the analytical procedure. Despite that, the application of the bottom–up approach is progressively declined due to its complexity and the unavoidable underestimation of some uncertainty sources.
- The top–down approach is probably the most common method of estimating measurement uncertainty by using the existing information from validation studies, data from IQC, and participation in PT schemes. An example of this pragmatic calculation is provided by Technical Report 537 of Nordtest (Magnusson et al., 2012), written for environmental testing laboratories in the Nordic countries in order to support the implementation of the concept of measurement uncertainty in their routine measurements. The aim is to provide a practical, comprehensible, and common way to calculate measurement uncertainty without necessarily knowing its sources. The combined standard uncertainty is estimated combining random (u(wR)) and systematic (u(bias)) uncertainty components:

$$u_{comb} = \sqrt{u(\mathrm{wR})^2 + u(\mathrm{bias})^2}\,.$$

The uncertainty component associated with random errors is generally estimated from the within-laboratory reproducibility s_{wR} derived from long-term precision data (IQC, CRM analysis, and validation studies) in order to cover random error sources sufficiently. On the other hand, the uncertainty component arising from method and laboratory bias u(bias) may be estimated from PT data, QC using a reference material (CRM or others), or long-term recovery studies. As already mentioned, the sources of bias should always be eliminated if possible. If not, the measurement result should always be corrected, making due allowance for the uncertainty associated with this correction.

Irrespective of the used approach, the combined uncertainty (u_{comb}) is nothing other than a "global" standard deviation (standard uncertainty) including the main individual standard uncertainties, and it can then be used as standard deviation for the calculation of the confidence interval. Considering the properties of the normal (Gaussian or bell-shaped) distribution, the probability of error with which the true analytical result may be found in the interval mean value ± 1 standard deviation (mean value ± u_{comb}) is approximately 32%. For a maximum probability of error of 5%, the interval must be doubled. For this reason, the measurement uncertainty is generally reported as an expanded uncertainty (U) obtained by multiplying the combined (standard) uncertainty by an appropriate coverage factor $k = 2$. Applying a coverage factor k equal to 3, the error is reduced to about 0.3%.

Example 10.6

The IQC of the validated method for the determination of HCB in tissue was carried out by analyzing two CRMs for a period of about 6 months: European Reference Material (ERM®-BB430 (pork fat) and SRM 1947 (fish tissue). The data from the certificate and from the experimental determinations carried out by the laboratory are summarized in Table 10.14. Applying the top–down approach, the analyst estimated the measurement uncertainty as follows:

The quantification of the relative method and laboratory bias (%) was obtained by using the formula

$$u'(\text{bias}) = \sqrt{u'(C_{ref})^2 + (\text{RMS}'_{bias})^2},$$

where $u'(C_{ref})$ was the relative (certified) uncertainty of two CRMs,

$$u'(C_{ref}) = \sqrt{\frac{\left[u'(C_{ref\_CRM\text{-}1})^2 + u'(C_{ref\_CRM\text{-}2})^2\right]}{2}} = 4.41\%,$$

and RMS'_{bias} (root-mean-square) was the square root of the sum of squares of relative bias values divided by the number of bias estimates ($n = 2$):

$$\text{RMS}'_{bias} = \sqrt{\frac{8.74^2 + 9.36^2}{2}} = 9.10\%.$$

The u' (bias) was therefore

$$u'(\text{bias}) = \sqrt{u'(C_{ref})^2 + \text{RMS}_{bias}^2} = \sqrt{4.41^2 + 9.10^2} = 10.07\%.$$

The quantification of relative uncertainty due to random effects was directly obtained using the experimental RSD (RSD_{wR}) observed during the measurements performed on CRMs. The RMS of

TABLE 10.14
Example of Estimation of Measurement Uncertainty (HCB in Tissues)

Error Type	Parameter[a]	Applied Formula or Description	ERM-BB430	SRM 1947
Systematic error	C_{ref} (μg kg^{-1})	Certified reference value of CRM	193	7.48
	$U(C_{ref})$ (μg kg^{-1})	Certified expanded uncertainty	17	0.66
	$u'(C_{ref,i})$ (%)	$\frac{U}{2 \times C_{ref}} \times 100$	4.40	4.41
	$u'(C_{ref,i})$ (%)	$\sqrt{\frac{\sum [u(C_{ref,i})]^2}{n}}$	4.41	
	$C_{lab,i}$ (μg kg^{-1})	Mean of laboratory results	209.9	8.18
	bias'_i (%)	$\frac{C_{lab,i} - C_{ref,i}}{C_{ref,i}} \times 100$	4.41	8.47
	RMS'_{bias} (%)	$\sqrt{\frac{\sum (\text{bias}'_i)^2}{n}}$	9.05	
	$u'(\text{bias})$ (%)	$u(\text{bias}) = \sqrt{u'(C_{ref})^2 + (\text{RMS}'_{bias})^2}$	10.07	
Random error	s_{wR} (μg kg^{-1})	Standard deviation of laboratory results (within-lab reproducibility)	19.09	0.79
	$\text{RSD}_{wR,i}$ (%)	$\frac{s_{wR,i}}{C_{lab,i}} \times 100$	9.10	9.71
	$\text{RMS}_{RSD,wR}$ (%)	$\sqrt{\frac{\sum (\text{RSD}_{wR,i})^2}{n}}$	9.41	

[a] Parameters marked with prime (′) and RSD are relative amounts.

the two RSD_{wR} values ($\text{RMS}_{RSD,wR}$ = 9.41%) was inserted to estimate the relative combined (standard) uncertainty as follows:

$$u'_{comb} = \sqrt{u'(\text{RMS}_{RSD,wR})^2 + u'(\text{bias})^2} = \sqrt{10.07^2 + 9.41^2} = 13.8\%.$$

Multiplying the relative combined uncertainty for the coverage factor 2 (level of confidence of approximately 95%), the relative expanded uncertainty (U) was estimated (27.6%). This relative uncertainty is valid approximately starting from the method's LOQ to the upper concentration range for about two orders of magnitude (Section 10.3.1.5) (CITAC and Eurachem, 2012).

When the available data are not from CRM analysis, but from other sources (proficiency test, internal control quality, validation data, etc.), the way to estimate the measurement uncertainty is similar (Magnusson et al., 2012; European Commission, 2013).

10.3.1.8.1 Reporting Results and Significant Figures

A quantitative test yields a value expressed in SI units followed (if required) by its measurement uncertainty. Once the expanded uncertainty (U) has been calculated for a specified level of confidence (typically 95%), the test result x and the expanded uncertainty U should be reported as $x \pm U$

and accompanied by a statement of confidence. For HCB, considering a relative expanded uncertainty of 27.7%, the result should be expressed as follows:

$$HCB = (13.0 \pm 3.6)\ \mu g\ kg^{-1}.$$

Comparing this result with a maximum specification limit equal to 10.0 μg kg^{-1} and applying the most common decision rule, 13.0 μg kg^{-1} does not exceed the specification limit (13.0 − 3.6 = 9.4 μg kg^{-1} < 10.0 μg kg^{-1}).

The number of significant figures used to express the result depends on the measurement uncertainty, which is generally reported with one or two significant figures (CITAC and Eurachem, 2012). In this example, the relative expanded uncertainty for HCB was about 28% giving an "absolute" expanded uncertainty equal to 3.6 μg kg^{-1}: the use of more than three significant figures is useless. In the (unrealistic) case that the estimated relative expanded uncertainty associated with HCB measurement was 4%, the result would be expressed with four significant figures as follows:

$$HCB = (13.00 \pm 0.52)\ \mu g\ kg^{-1}.$$

It must be noted that when the confidence interval was narrowed, the interpretation changed and the specification limit was considered exceeded (13.00 − 0.52 = 12.48 μg kg^{-1} > 10.0 μg kg^{-1}).

Finally, the number of significant figures used to express the specification limit by enforcement authorities has to be taken into account, too. For instance, if the maximum limit is expressed as 10 μg kg^{-1}, a compliant result (result minus its expanded uncertainty) will lie from 0 to 15.4 μg kg^{-1}. On the other hand, if the limit is expressed as 10.0 μg kg^{-1}, a compliant result will lie from 0 to 15.04 μg kg^{-1}.

10.4 QUALITY CONTROL

QC is the combination of the operational techniques and activities that are used to fulfill requirements for quality. It encompasses all activities that bring an analysis under statistical control. Statistical control implies that factors affecting the uncertainty of measurement results have not deteriorated since method validation. In other terms, QC ensures that factors determining the magnitude of uncertainty do not change during the routine use of an analytical method over long periods. Together with validation, QC forms the mainstay of quality practice in chemical analysis. QC procedures relate to ensuring the quality of specific samples or batches of samples and include IQC and EQC.

10.4.1 Internal Quality Control

10.4.1.1 Choice of Control Materials

When a QC program is established, it is essential to have in mind the requirements for the analytical results and for what purposes the analytical results are produced (fitness for purpose). First, the analyst sets up the control program stating the type of QC materials, the QC charts, the control limits, and the control frequency. When running batches of samples with calibration solutions and unknowns (test samples or materials), there are a number of extra samples that are analyzed that cover different aspects of QC (QC materials). The QC materials are subjected to the same or part of the same measurement procedure as that used for test materials and, therefore, although standard solutions or solvent blanks undoubtedly contribute to ICQ, they are not considered QC materials.

QC materials can be classified in four types: The first two are type I, CRM—matrix CRM, and type II, in-house or synthetic materials. This latter type is usually prepared by the laboratory. They can be either stable, homogeneous routine (test) samples, or synthetic in-house materials prepared by spiking suitable materials with standard solution. The next two are type III, blank samples, and

TABLE 10.15

No.	Determination	QC Material Type	QC	Chart
1	Ni in low-alloy steel with X-ray fluorescence	Routine sample (steel sample)	IV	X
2	N–NH_4 in water with indophenol blue method	In-house sample; synthetic (standard) solution of $(NH_4)_2SO_4$—duplicate determination	II	X and R
3	Pb in water by ICP-MS	In-house sample; routine sample (lake water) stabilized with HNO_3	II	X
4	β-HCH in biological material with GC	Cod liver oil BCR598 with β-HCH (16 μg kg^{-1})	I	X
		An analyte-free fatty material	III	X
5	Pesticides in surface water with LC-MS/MS	Routine sample spiked randomly in duplicate (water sample)	II	X and R
6	Total phosphorus in water by using spectrophotometric method	Routine sample—duplicate determination (water sample)	IV	R%

Note: HCH, hexachlorocyclohexane.

type IV, routine samples. Unlike the routine samples of type II, these samples are not handled by the laboratory (Hovind et al., 2011).

In Table 10.15, some examples of all four QC types are given to help understand the strategies of choice and the implemented control charts (Hovind et al., 2011). To choose the appropriate QC materials, there are some preliminary questions: (1) Are the analytes stable? (2) Are suitable CRMs available? (3) Are the test (routine) samples expected to contain target analytes? For example, if the routine samples are not expected to contain the target analyte(s), they are unable to directly control method precision with duplicate analysis; therefore, in this case routine samples must be spiked randomly with a known amount of the analyte(s) by using a standard solution of pure substances prior to carrying out duplicate analysis (case 5, Table 10.15). In case 6 (total phosphorus in water), on the other hand, the analyte is always present in routine samples and precision control is possible without any spiking (duplicate analysis). In the latter case, therefore, an R%-chart is used to monitor the QC results (see Section 10.4.3.2). To control the method's trueness, a QC material with an expected (reference) value is necessary (CRM or other material with an expected concentration). QC materials should be representative of the test (routine) samples analyzed and should reproduce similar uncertainty. As well as surrogate recovery (Table 10.10), results obtained by analyzing spiked QC material (surrogate) could be affected by a lower uncertainty than that of real test samples since the speciation, binding, and physical form of the added analyte are not necessarily the same as those of the native one. However, it must be underlined that generally some organic contaminants such as pesticides or veterinary drugs are not stable, precluding the availability of CRMs or other reference materials. To control these methods, the QC materials are necessarily in-house materials (e.g., blank samples) freshly spiked with standard solutions of pure materials immediately prior to analyze each batch. In this context, it should be stressed that calibrators and standard solutions used to spike in-house materials must have a different origin. Otherwise, systematic errors could be not detected.

For each matrix included in the analytical run the relevant control sample must be provided, however in some cases, if matrix equivalence has been previously established, a QC material belonging to a certain matrix can also control the other ones.

10.4.1.2 Frequency of Using Quality Control Materials

The proportion of control materials to be inserted into a run should be the minimum that will provide adequate control. This is difficult to specify in detail, but some broad principles are applicable. As a general rule, between 5% and 15% insertions are often used. Obviously short runs will need at least one control material and multiple insertions of a material in a run may benefit from the use of mean and range control charts. The frequency of inclusion generally depends on cost and time constraints, counterbalanced by the risk entailed in subsequent analytical errors. In many cases, the range of concentrations of the analyte encountered in real samples is relatively small, and, therefore, a single control material is appropriate, with the analyte in midrange. If, on the other hand, the analyte concentration varies widely in the real samples, we should consider using two different control materials, with concentrations in the upper and lower zones of the usual range. If the analysis is for checking whether the concentration of a certain constituent in a material exceeds its legal or contractual limit or not, the analyte concentration should be near or equal to this limit. Where the limit is close to the detection limit or is zero, as for banned substances, a control material assuring the maintenance of the method LOQ or LOD estimated during the validation study must be provided.

In practice, different QC materials can be inserted in an analytical run. Generally, the SOP also features a section reporting IQC activities that are performed within each run and the criteria for accepting the results. The technician responsible for the procedure execution must consider these criteria in order to decide whether the results are reliable enough to be released.

10.4.2 External Quality Control

EQC provides evidence of both the quality of the laboratory performances and the proficiency of the individual analyst. Furthermore, like IQC, EQC can assist in identifying needs for training and changes in laboratory procedures. It is generally achieved with (i) PT and (ii) analysis of CRMs.

10.4.2.1 Proficiency Testing

The aim of PT is to allow laboratories to monitor and improve the accuracy of their analytical measurements. The standard that governs the accreditation of testing laboratories, ISO/IEC 17025 (ISO, 2000), makes statements about participation in PT in point 5.9.1: "The laboratory shall have QC procedures for monitoring the validity of tests and calibrations undertaken.... This monitoring ... may include ... b) participation in interlaboratory comparison or PT programmes." Unlike other interlaboratory tests (e.g., collaborative trials), in a proficiency test the laboratories are free to select the test method. PT is generally applied only to methods that are performed routinely and, accordingly, the materials and analytes should be typical of the participants' routine work. In its usual form, the provider of the PT scheme distributes portions of a homogeneous and sufficiently stable material to each participant, who analyzes the material under typical conditions and reports the result to the provider. Each distribution of PT materials is referred to as a "round." The provider compiles the results and informs the participants of the outcome, usually in the form of a score relating to the accuracy of the result. The score by the PT provider adds value to the raw analytical data produced by the participant. The purpose of providing a normalized performance evaluation is to make all PT results comparable, so that participants can immediately evaluate the quality of their performance. The most common scoring system is the z-score:

$$z = \frac{x - X}{\sigma_p},$$

where x is the result submitted by the participant, X is the assigned value, and σ_p is the standard deviation for the proficiency (the target value). Although the data evaluation is carried out by the provider of the PT, the participants must be aware of the statistical procedures used to determine z-scores. The interpretation of z-scores is simple, and it is based on the properties of the normal (Gaussian) distribution: $|z| \leq 2$, satisfactory performance; $2 < |z| \leq 3$, questionable performance; and $|z| > 3$, unsatisfactory performance.

In addition, there are other scoring systems such as z′-score, zeta-score, and E_n-number. These scores differ from the most common z-score only for the "uncertainty" inserted at the denominator.

According to the Eurachem (2011), there are essentially five approaches to obtaining the assigned value x_a, an estimate of the true value, and five approaches to determining the standard deviation σ_p, that is, the acceptable range of participant results, too. Table 10.16 summarizes the main ones.

When an unsatisfactory z-score (>3) is given in a proficiency test, action should be taken. This is mandatory for accredited laboratories following the ISO/IEC 17025 standard. However, when the assigned score is calculated on the basis of the participating consensus, there are a number of issues to consider

TABLE 10.16
Main Ways to Setting Assigned Value (x_a) and Standard Deviation (σ_p) to Calculate z-Score in PT

Approach Description	Advantages	Disadvantages
	Assigned Value (x_a)	
Formulation: the addition of a known amount or concentration of analyte to a base material containing none.	Providing a traceable assigned value.	Being a spiked material, it may not simulate the difficulty of real sample preparation procedures, which include extraction and speciation. Recovery problems may well arise. It may be difficult to achieve the necessary homogeneity of the bulk material to distribute to participants.
Certified reference values: when the PT material is a CRM, its certified reference value is used as the assigned value.	Providing a traceable assigned value.	CRMs are generally expensive. Appropriate CRMs are often not available.
Consensus value from participants: it is based on the results obtained by the participants. The consensus value is usually estimated using robust statistical techniques.	It is the most straightforward and the most common. In some cases, for example, when using natural matrix samples, it is often the only way to establish an estimate of the true value.	There may be no real consensus in the results reported by participants. The consensus may be biased by the general use of faulty methodology.
	Standard Deviation (σ_p)	
Prescribed value: σ_p is set at a value required for a specific task or by legislation. The σ_p becomes equivalent to a fitness-for-purpose statement for the measurement method.	σ_p is directly related to fitness for purpose criteria for a specific application.	The fitness for purpose criteria might not be known.
From a general model: σ_p is derived from a general model for the reproducibility of the measurement method.	Easy to calculate. Some analytical sectors use models as criteria for fitness for purpose (e.g., Horwitz equation).	The true reproducibility of a certain method may differ substantially from the value given by the model. Reproducibility could depend not only on the concentration level of the parameter but also on the parameter itself or on the measurement procedure.
From data obtained in a round of a PT scheme: the σ_p is the robust standard deviation of the results reported by all the participants.	Obtained directly from data from the PT round.	The value may vary substantially from round to round, making it difficult for a laboratory to assess the z-score trend over the course of time.

such as the number of participants and the percentage of unsatisfactory results. In the final report, the provider should declare how both the assigned values and the standard deviation were calculated and the criteria by which the data were considered too poor to assign z-scores. Finally, in some cases a multimodal distribution is evident. In this circumstance, it is not possible to identify a consensus value.

10.4.2.2 Certified Reference Materials

CRMs are cornerstones of modern analytical QA because they allow for calibration of instruments, validation, and QC of methods and laboratories based on the traceability and comparability of measurement results. Therefore, when available, CRMs can be regarded as the ultimate standards of trueness, thanks to their high degree of traceability. Target and range-certified values may be generated from data produced by several laboratories by using different analytical methods or calculated from data obtained by the use of one analytical method (usually a reference method).

CRMs are often recommended for use as control materials in ICQ of analysis in that the CRM is a direct route to an acceptable traceability. However, it is better to keep the concepts of QC and traceability distinct. The use of a CRM (where available) on a scale suitable for ICQ would be expensive, but on a lesser scale, it is useless. On the other hand, the analysis of CRMs can provide an occasional check of an ongoing analytical system, as well as assess the method's trueness during the validation study. In this instance, it is better to regard this action as a kind of one-laboratory proficiency test (Thompson and Lowthian, 2011).

10.4.3 Control Charts

Control charts involve plotting data in a time-ordered sequence. These data are, generally, the results obtained from successive analyses of a QC material. Thus, they allow the monitoring of the performance of the measurement system over time, detecting unusual variations in measurement results or the occurrence of a bias. The principle is that IQC data are graphically plotted so that that they can easily be compared and interpreted. Consequently, a control chart must be simple to use, understand, and act upon. Two types of control charts are commonly used in laboratories: accuracy or X-charts for QC samples, including reagent blanks, incurred reference materials, or spiked samples, and precision or range charts for replicate or duplicate analyses. These charts are essential tools for QC. Computer-generated and maintained lists or databases with values, limits, and trending may be used as an alternative to control charts.

10.4.3.1 X-Charts

The X-chart or Shewhart chart is probably the most widely used control chart. It is a plot, in time order, of the average of the n measurements or, frequently, of a single measurement taken on a certain QC material. The centerline (CL), warning lines (WLs), and action lines (ALs) are marked parallel to the x axis. Where reference materials are in use for control purposes and the method is known to be unbiased (or the results corrected for the known bias), the CL is set at the assigned value. Where there is no assigned value, the overall average of the progressively collected data is used as CL. WLs are placed at two standard deviations to either side of the target line. In the absence of evidence to the contrary, it is reasonable to assume that the distribution of random errors in the process under observation follows the normal distribution. Accordingly, 95% of results from assays under statistical control will fall between the two WLs. ALs are normally placed at three standard deviations to either side of the target line, and 99.7% of normally distributed results should be between the ALs. If the point to be plotted is the average of a number of measurements, say n, the upper action limit (UAL) and the lower action limit (LAL) are set as follows:

$$\text{UAL} = \text{CL} + 3\frac{\sigma}{\sqrt{n}} \quad \text{and} \quad \text{LAL} = \text{CL} - 3\frac{\sigma}{\sqrt{n}},$$

where σ is the standard deviation of individual measurements. The upper warning limit (UWL) and the lower warning limit (LWL) are thus obtained:

$$\text{UWL} = \text{CL} + 2\frac{\sigma}{\sqrt{n}} \quad \text{and} \quad \text{LWL} = \text{CL} - 2\frac{\sigma}{\sqrt{n}}.$$

Frequently, especially in trace analysis, the costs and time constraints or the paucity of a certain control material may render it necessary to do a single measurement of the control material. In this circumstance $n = 1$; therefore,

$$\text{UAL} = \text{CL} + 3\sigma \quad \text{and} \quad \text{LAL} = \text{CL} - 3\sigma,$$

where σ is the standard deviation of individual measurements. Similarly, UWL and LWL are obtained by multiplying σ by 2 instead of 3. Normally, no more than 1 out of 20 consecutive results should fall outside the WLs. If this frequency is exceeded, or if a result falls outside the ALs, the method is out of control. The scatter of the assay results for the QC material around the target line provides an indication of the precision of the method, while the mean of the assay results relative to the target value indicates whether there is any bias (consistent deviation) in the results. There is no general agreement on how control charts are to be used, and many different sets of rules are recommended. The following general rules give guidance about what are to be considered "abnormal" (not Gaussian) patterns: (i) seven consecutive values on one side of the central line; (ii) seven consecutive values showing an ascending or a descending trend; and (iii) two out of three consecutive values outside of the warning limits. In the regular day-to-day use of an X-chart, an aliquot from an appropriate QC control material is analyzed with every batch of samples and the measured concentration of the variable in the aliquot is plotted on the chart.

10.4.3.2 R-Charts

While the X-chart shows how well the mean values of subgroups or single values correlate with the grand average (process mean), it does not provide any information about the distribution of individual results within and between the subgroups. In contrast, the R-chart (R = range) serves above all the purpose of precision control (Howarth, 1995; Mullins, 2003). The range R is defined as the difference between the largest (x_{max}) and the smallest (x_{min}) single result among the n measurements carried out in each subgroup. In order to construct an R-chart, the R_i (range of each subgroup i) must be calculated:

$$R_i = x_{max} - x_{min}.$$

The average $\bar{R}$ of the within-run ranges forms the CL of the R-chart:

$$\bar{R} = \frac{\sum R_i}{N},$$

where N is the number of subgroups.

The R-chart has the same general features and is used in the same way as the X-chart: it has a CL and upper and lower control limits, and similar rules for detecting out-of-control conditions are applied to the plotted points. The basis for calculating the control limits is different, since sample ranges do not follow a normal distribution. This technical difference presents no difficulties in practice as tables of constants (D_3 and D_4) are reported in several specialized textbooks (Montgomery, 2009). The LAL and the UAL are therefore obtained from the mean range $\bar{R}$ by multiplying it by the appropriate D_3 and D_4 constants, taking into the account the number of multiple determinations n and the significance level:

$$\mathrm{LAL} = D_3\bar{R}, \quad \mathrm{UAL} = D_4\bar{R}.$$

Constant factors are also available to calculate LWL and UWL (Funk et al., 2007). Following the convention of the three standard deviations, for a sample size less than seven ($n < 7$), the LAL would be negative and the value of D_3, therefore, would be equal to zero (Mullins, 2003). When the R-chart is in its simplest and common form, that is, when performing a duplicate determination per analysis batch ($n = 2$), the values of D_3 and D_4 are 0 and 3.267, respectively, at a probability level of 99.7%. In addition, to insert the UWL too, for duplicate results, $\bar{R}$ is multiplied by 2.809 ($P = 95\%$). These values (3.267 and 2.809) have a level of confidence equivalent to the values of 3 and 2 in the calculation of action and warning limits in the X-charts.

The duplicate determinations should be performed on test (routine) samples as often as possible to reproduce as much as possible the real process uncertainty. For instance, at the beginning of each analysis batch, one of the routine samples is identified as the "control sample" and it is analyzed again at the end of the series by performing a second determination. A complication of this duplication strategy of routine samples (incurred analyte) is that the concentrations found (x) can vary considerably within a run and the dispersion of the differences (R) will vary accordingly. A frequent expedient is to postulate a constant value of RSD at concentration higher than the method's LOQ (Thompson and Lowthian, 2011). Similarly to uncertainty $u(x)$ (see Section 10.3.1.5), a nonlinear regression function can be postulated between s_r (repeatability) and x (analyte concentration). Thus, the building of an R%-chart in which R% is the relative range is feasible. From an empirical point of view, the Horwitz equation also states that the RSD varies very slowly when the analyte concentration increases.

Finally, if the analyte is not naturally present, the R-chart is constructed by means of spiked control materials performed in duplicate by using a test sample free of the analyte.

10.4.3.3 Setting Control Limits

10.4.3.3.1 Statistical Control Limits

Statistical control limits are thus set according to the observed performance of the analytical system, irrespective of the demands of fitness for purpose. For the X-chart, the control limits should be based on the mean μ and the standard deviation σ of the process (i.e., the analytical system) under test. Under the normal assumption, a result falling outside the action limits ($\mu \pm 3\sigma$) implies that the analyst can reasonably consider the system to be "out of control." We infer that factors affecting the process have deteriorated and need to be restored to their original conditions. The implication of an out-of-control condition is that the uncertainty established at validation is no longer credible. However, the parameters μ and σ are unknown and must be replaced with their estimated values $\bar{x}$ and s, respectively. This is not a simple task. For example, it should not be assumed that μ will be exactly equal to the certified value of a CRM (QC material type I) or the theoretical value of an in-house QC material (type II): although the assigned value μ of these QC materials is known, the true value around which the actual determination will fluctuate is not. In addition, many observations are needed to have a reliable estimate of the standard deviation σ and they can be made only during routine use of the method. IQC is based on between-run precision (s_{wR}), that is, closeness of agreement between results obtained in separate runs of analysis.

These problems can be avoided by the initial use of an interim chart based on validation statistics: if within-laboratory reproducibility (s_{wR}) is available from validation data, this value is used to set the warning ($2s_{wR}$) and action limits ($3s_{wR}$). If until then only the repeatability has been determined, the repeatability value (s_r) is multiplied by a factor between 1.5 and 2.0 times to avoid false alarms: this is generally the observed interval for the ratio s_{wR}/s_r (s_r range between 0.5 [1/2] and

0.66 [2/3] of the s_{wR} or s_R as introduced in Section 10.3.1.6.1). When enough routine data have been accumulated for reasonably stable run-to-run control limits to be estimated, the interim chart is replaced by a permanent one (Thompson and Lowthian, 2011).

10.4.3.3.2 Target Control Limits

Statistical control limits are calculated from the historical results obtained during the analysis of the QC materials or validation data. Another possibility is to set control limits on the basis of empirically derived or statutorily required exclusion limits, that is, as so-called target control limits. This strategy can be applied to the different control charts such as X- and R-charts, and it can be particularly appropriate in certain cases, for example, if no normal distribution of the control values is present (e.g., blank values), no sufficient database is available for the determination of statistical characteristics (e.g., rarely conducted analyses), or a particular quality of the analytical results needs to be guaranteed in order to meet the specifications of an exclusion limit. With target control limits, there are no warning or action control limits, but only an upper and a lower exclusion limit. These exclusion limits can be derived from legal specifications, parameter-related standards, and analytical QC instruction cards, the (lowest) guaranteed laboratory-internal precision and trueness specifications. The target value chart is constructed with the aid of the exclusion limits; thus, a preliminary period for the determination of control chart limits may be dropped.

Example 10.7

The validated GC-MS method for the determination of HCB was applied to analyze fish tissues within an official monitoring plan. The QC activities were planned (Figure 10.1), and QC materials were chosen on the following basis (Section 10.4.1.1): the analyte was stable, suitable CRMs were available, and the test (routine) samples were not generally free from target analytes. The frequency of QC was established mainly on the basis of the analysis cost (financial cost and analysis time) and of the risk associated with a wrong result. The cost was high, allowing the analysis of only 12 samples per batch; on the other hand, the risk was evaluated as medium because the method was applied in an environmental monitoring plan as fact-finding survey. Accordingly, the number of IQC materials included in each analytical batch was reduced to a minimum of three. Nevertheless, these controls represent 25% of the batch (3/12). Long and complicated analyses and analyses carried out rarely are the most onerous from the point of view of QC, whose principles are more easily applicable to high-throughput analytical systems.

The acceptance criteria and relevant corrective actions of QC, together with the materials used, were thoroughly described in the method's SOP.

The planned IQC is reported as follows:

- X-chart using a procedural (reagent) blank (type III) to monitor laboratory contamination.
- X-chart using an in-house QC material (type II) to monitor the systematic effects: this material was a routine sample analyzed in a previous batch, homogenized, and stored in several test portions. Provisional CL (HCB concentration) was progressively calculated from the first 20 run data, and at first it was estimated to be about 48 μg kg^{-1}. The between-run standard deviation was fixed using the s_{wR} observed during the validation study at 10 μg kg^{-1} (Table 10.11, Section 10.3.1.6.1), considering the RSD_{wR} to be about constant in the involved concentration range. Accordingly, UAL and LAL were 58 and 38 μg kg^{-1}, respectively (s_{wR} = 3.4 μg kg^{-1}). After the first 20 runs, the data scrutiny was carried out (Figure 10.9a). The 13th run was suspected and eliminated from the data set used to establish the definitive limits: this result could excessively enlarge the control limits. This observation was replaced with the 21st run. Definitive CL and control limits were set using the experimental mean and the standard deviation: CL = 47 μg kg^{-1} and s_{wR} = 6.3 μg kg^{-1}, with UAL and LAL equal to 60 and 28 μg kg^{-1}, respectively (Figure 10.9b). The s_{wR} obtained from Table 10.8 underestimated the run-to-run process variability, probably because it was obtained in a limited time (3 weeks). On the other hand, the s_{wR} (6.3 μg kg^{-1}) obtained during 4 months on QCs was surely more representative of the real process variability.

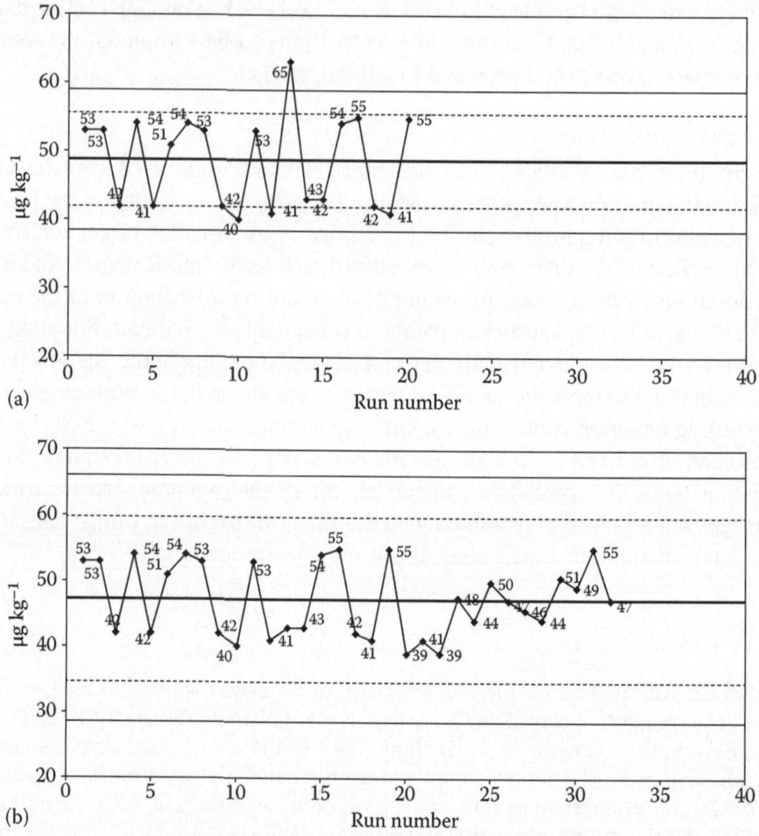

FIGURE 10.9 (a) Provisional and (b) definitive X-charts for HCB determination in tissue.

- R%-chart with routine sample (type IV) to monitor the precision: In this example, the QC material used to build the X-chart should be also used to carry out duplicate analysis controlling precision with an R-chart. In this way, only three samples are saved for IQC, improving method throughput. However, to deal with the use of an R%-chart, in each run it is supposed to choose a test sample analyzed in duplicate (x_1 and x_2). The results observed during the first 20 runs are listed in Table 10.17. The incurred levels and, accordingly, the range values R_i were variable, but as expected the relative range values (R%) were quite constant (from 5.4% to 7.9%). The R%-chart was obtained by plotting R% versus the run number. The UAL and the UWL were calculated by multiplying the mean R% (6.7%) by 3.267 and 2.809, respectively.

The planned EQC is reported as follows:

- Since CRMs were available (e.g., NIST SRM 1947), they should certainly occasionally be analyzed along with test materials (one analysis for each 10 batches?), regarding the outcome as a kind of one-participant proficiency test (EQC) (Thompson and Lowthian, 2011). A "pseudo-z-score," z', is then calculated as follows:

$$z' = \frac{x_{\text{lab}} - x_{\text{ref}}}{\sqrt{\sigma_p^2 + u(x_{\text{ref}})^2}},$$

where x_{lab} is the result of laboratory, x_{ref} is the certified value, $u(x_{\text{ref}})$ is the uncertainty of the certified value ($U(x_{\text{ref}})/2$), and σ_p represents the fitness-for-purpose uncertainty (such

TABLE 10.17
IQC Data of Duplicate Determinations of HCB in Routine Samples (Tissue) ($i = 20, n = 2$)

Run	x_1 (μg kg^{-1})	x_2 (μg kg^{-1})	$\bar{x}$ (μg kg^{-1})	R_i (μg kg^{-1})	$R\% = \frac{x_1 - x_2}{\bar{x}} \times 100$
1	11.8	10.9	11.4	0.9	7.9
2	17.0	16.1	16.6	0.9	5.4
3	42.0	39.1	40.6	2.9	7.2
4	31.4	29.2	30.3	2.2	7.3
5	23.0	21.5	22.3	1.5	6.7
6	17.2	16.2	16.7	1.0	6.0
7	35.4	33.2	34.3	2.2	6.4
8	25.7	24.0	24.9	1.7	6.8
9	37.6	35.5	36.6	2.1	5.7
10	36.8	34.0	35.4	2.8	7.9
11	13.2	12.3	12.8	0.9	7.1
12	28.0	26.3	27.2	1.7	6.3
13	21.7	20.4	21.1	1.3	6.2
14	9.4	8.8	9.1	0.6	6.6
15	7.0	6.5	6.8	0.5	7.4
16	33.0	31.0	32.0	2.0	6.3
17	14.8	14.0	14.4	0.8	5.6
18	43.8	41.0	42.4	2.8	6.6
19	10.5	9.9	10.2	0.6	5.9
20	54.0	50.3	52.2	3.7	7.1
Mean			24.8	1.7	6.6
UAL					21.6
UWL					18.6

a test would be pointless, however, unless $u(x_{ref}) < \sigma_p/3$). This use of CRMs provides an ongoing external reference (EQC) and can therefore be considered comparable to participation in a proficiency test.

- Participation in PT (yearly). This participation was useful for checking the method's performances in different animal tissues at different concentrations, completing the EQC together with CRM data.

REFERENCES

Antignac, J., Le Bizec, B., Monteau, F., Andre, F. 2003. Validation of analytical methods based on mass spectrometric detection according to the "2002/657/EC" European decision: Guideline and application. *Analytica Chimica Acta* 483: 325–334.

Baldwin, R., Bethem, R.A., Boyd, R.K., Budde, W.L., Cairns, T., Gibbons, R.D., Henion, J.D., Kaiser, M.A., Lewis, D.L., Matusik, J.E., Sphon, J.A., Stephany, R., Trubey, R.K. 1997. ASMS Fall Workshop: Limits to confirmation, quantitation, and detection. *Journal of the American Society for Mass Spectrometry* 8: 1180–1190.

Berendsen, B.J.A., Stolker, L.A.M., Nielen, M.W.F. 2013. The (un)certainty of selectivity in liquid chromatography tandem mass spectrometry. *Journal of the American Society for Mass Spectrometry* 24: 154–163.

Berendsen, B.J.A., Wegh, R.S., Meijer, T., Nielen, M.W.F. 2015. The assessment of selectivity in different Quadrupole-Orbitrap mass spectrometry acquisition modes. *Journal of the American Society for Mass Spectrometry* 26: 337–346.

BIPM. 2012. *JCGM 200:2012: International Vocabulary of Metrology—Basic and General Concepts and Associated Terms*, 3rd ed. International Bureau of Weights and Measures, Sèvres, France.

CITAC, Eurachem. 2002. *Guide to Quality in Analytical Chemistry.* http://www.eurachem.org (accessed December 12, 2014).

CITAC, Eurachem. 2007. *Use of Uncertainty Information in Compliance Assessment*, 1st ed. http://www.eurachem.org (accessed December 12, 2014).

CITAC, Eurachem. 2012. *Quantifying Uncertainty in Analytical Measurement*, 3rd ed. http://www.eurachem.org (accessed December 12, 2014).

Cuadros-Rodríguez, L., Bagur-González, M.G., Sánchez-Viñas, M., González-Casado, A., Gómez-Sáez, A.M. 2007. Principles of analytical calibration/quantification for the separation sciences. *Journal of Chromatography A* 1158: 33–46.

Currie, L.A. 1995. Nomenclature in evaluation of analytical methods, including detection and quantification capabilities. *Pure and Applied Chemistry* 67: 1699–1723.

Danzer, K., Currie, L.A. 1998. Guidelines for calibration in analytical chemistry—Part 1: Fundamentals and single component calibration. *Pure and Applied Chemistry* 70: 993–1014.

Ellison S.L.R., Barwick, V.J., Duguid Farrant, T.J. 2009. *Practical Statistics for the Analytical Scientist: A Bench Guide*, 2nd ed. RSC, Cambridge.

Eurachem. 2011. *Selection, Use and Interpretation of Proficiency Testing (PT) Schemes*, 2nd ed. http://www.eurachem.org (accessed December 12, 2014).

Eurachem. 2014. *The Fitness for Purpose of the Analytical Methods*, 2nd ed. http://www.eurachem.org (accessed January 8, 2015).

European Commission. 2002. Commission Decision 2002/657/EC of 12 August 2002 implementing Council Directive 96/23/EC concerning the performance of analytical methods and the interpretation of results, *Official Journal of the European Communities* L221, pp. 8–36.

European Commission. 2009. Commission Directive 2009/90/EC of 31 July 2009 laying down, pursuant to Directive 2000/60/EC of the European Parliament and of the Council, technical specifications for chemical analysis and monitoring of water status. *Official Journal of the European Communities* L201, pp. 36–38.

European Commission. 2013. SANCO/12571/2013: Guidance document on analytical quality control and validation procedures for pesticide residues analysis in food and feed, 2013. http://www.eurl-pesticides.eu (accessed December 12, 2014).

European Parliament, Council of the European Union. 2004. Regulation No 882/2004 of the European Parliament and of the Council of 29 April 2004 on official controls performed to ensure the verification of compliance with feed and food law, animal health and animal welfare rules. *Official Journal of the European Communities* L165, pp. 1–141.

Funk, W., Dammann, V., Donnevert, G. 2007. *Quality Assurance in Analytical Chemistry,* Wiley, Weinheim.

Horwitz, W., Kamps, L.R., Boyer, K.W. 1980. Quality assurance in the analysis of foods and trace constituents. *Journal of the Association of Official Analytical Chemists* 63: 1344–1354.

Hovind, H., Magnusson, B., Krysell, M., Lund, U., Mäkinen, I. 2011. *Internal Quality Control—Handbook for Chemical Laboratories*, NT TR 569, ed. 4. http://www.nordtest.info (accessed December 12, 2014).

Howarth, R. J. 1995. Quality control charting for the analytical laboratory: Part 1. Univariate methods—A review. *Analyst* 120: 1851–1873.

Irish National Accreditation Board. 2012. Guide to Method Validation for Quantitative Analysis in Chemical Testing Laboratories. http://www.inab.ie (accessed December 12, 2014).

ISO. 1997. International Standard ISO 11843-1: Capability of Detection—Part 1: Terms and Definitions. ISO, Geneva.

ISO. 2000. International Standard ISO/IEC 17025:1999: General Requirements for the Competence of Testing and Calibration Laboratories. ISO, Geneva.

ISO. 2005. International Standard ISO/IEC 17025:2005: General Requirements for the Competence of Testing and Calibration Laboratories. ISO, Geneva.

Linsinger, T.P.J. 2008. Use of recovery and bias information in analytical chemistry and estimation of its uncertainty contribution. *TrAC Trends in Analytical Chemistry* 27: 916–923.

Magnusson, B., Näikki, T., Hovind, H., Krysell, M. 2012. *Handbook for Calculation of Measurement Uncertainty in Environmental Laboratories*, NT TR 537, ed. 3.1. http://www.nordtest.info (accessed December 12, 2014).

Montgomery, D.C. 2009. *Introduction to Statistical Quality Control*, 6th ed. Wiley, Jefferson City, MO.

Mullins, E. 2003. *Statistics for the Quality Control Chemistry Laboratory.* Royal Society of Chemistry, Cambridge.

Stoev, G., Xuan, Y., Scigelova, M. 2013. Quantitative assessment of the contribution of high resolution mass spectrometry analysis to the reliability of compound confirmation. *Talanta* 98: 19–27.

Thompson, M., Lowthian, P.J. 2011. *Notes on the Statistics and the Data Quality for Analytical Chemists.* Imperial College Press, London.

Thorburn Burns, D., Danzer, K., Townshend, A. 2002. Use of the terms "recovery" and "apparent recovery" in analytical procedures. *Pure and Applied Chemistry* 74: 2201–2205.

Valcarcel, M., Goèmez-Hens, A., Rubio, S. 2001. Selectivity in chemistry revisited. *TrAC Trends in Analytical Chemistry* 20: 386–393.

Vander Heyden, Y., Nijhuis, A., Smeyers-Verbeke, J., Vandeginste, B.G.M., Massart, D.L. 2001. Guidance for robustness/ruggedness tests in method validation. *Journal of Pharmaceutical and Biomedical Analysis* 24: 723–753.

Vessman, J., Raluca, I.S., Van Staden, J.F., Danzer, K., Lindner, W. et al. 2001. Selectivity in analytical chemistry. *Pure and Applied Chemistry* 73: 1381–1386.

11 Proficiency Testing in Environmental Analysis

Achievements and Challenges

Charalampos Alexopoulos, Elias Kakoulidis, and Eugenia Lampi

CONTENTS

11.1 INTRODUCTION

11.1.1 Environmental Issues/Concerns

At present, a large number of environmental concerns have been raised (Cao and Orrù, 2014) that are related to anthropogenic effects on the environment, which can cause a gradual degradation of its quality and inflict devastating results (e.g., as a result of water and surface pollution) (Kemp, 2004).

An indicative and comprehensive list of environmental issues and concerns (Alma, 1993; Thornes and Wainwright, 2004) as well as of the harmful aspects of human activity on the biophysical environmental effects is given in Table 11.1.

TABLE 11.1
Causes and Effects in Environmental Concerns

Causes	Effects
• Industrialization (Kemp, 2004) • Overpopulation • Intensive exploitation (land and water) (Kemp, 2004; Zhang et al., 2015) • Intensive (animal and crop) farming (Zhang et al., 2015) • Intensive irrigation • Use of pesticides/herbicides • Nuclear and radiation accidents (radioactive leaks) (Buesseler et al., 2012; Steinhauser et al., 2014) • Nuclear weapons • Environmental accidents (chemical industry accidents, pipeline bursts, tanker spills) (Landrigan and Fuller, 2014) • Bad practices in waste management (poor disposal techniques, mishandling of hazardous waste materials) (Mehrotra and Singhal, 2000; Melnyk et al., 2014)	• Resource depletion (overexploitation of natural resources, deforestation) • Climate changes (greenhouse gases, global warming) (Jones, 2008; Mendelsohn, 2001) • Environmental degradation, bioaccumulation • Land/water pollution (wastewater, soil contamination, superfund sites, urban runoff, groundwater contamination, algal bloom, marine pollution, radioactive contamination) (Ritter et al., 2002; Fuerhacker et al., 2009; Ganoulis, 2009; Karaouzas et al., 2011; Melnyk et al., 2014) • Air pollution (smog, ozone depletion, VOCs, chlorofluorocarbons [CFCs], particulate matter) (Jacob, 1999; Schwela, 2000; Jones, 2008; Romieu et al., 2012) • Water scarcity • Nuclear fallout, nuclear waste disposal (Buesseler et al., 2012) • Environmental health and public health issues (air quality, lead poisoning, sick building syndrome, environmental impact of coal industry, electromagnetic fields/radiation, health: respiratory diseases, leukemia, etc.) (Schwela, 2000; Lewtas, 2007; Liu et al., 2008; Landrigan and Fuller, 2014; Steinhauser et al., 2014) • Waste concerns (hazardous waste that could contain carcinogenic or teratogenic compounds, toxic waste, leachate, waste disposal) (Boffetta and Nyberg, 2003; Hashim and Boffetta, 2014; Melnyk et al., 2014)

This situation leads to an increasing demand not only for tests and measurements in the field of environmental protection but also for verification of product specifications in order to avoid environmental contamination. Additionally, taking into account, on the one hand, that the earth, the sea, and the atmosphere are universal and, on the other hand, the globalized economy, the comparability and international acceptability of these tests and measurements are necessary.

11.1.2 Traceability and Comparability of Measurement Results

Measurement results play a major role in the areas of environmental quality monitoring (air, water, and soil), environmental protection, environmental pollution (polluted aqueous environment, untreated wastewater, hazardous sludge/sediment, soil degradation, harmful air-emissions, contaminated foodstuff), exploitation of water, mineral, and energy resources, and consumers' health protection.

In all these fields, the quality and the acceptability of the measurement results, as part of a decision-making process for consumer and environmental protection and release of products, need to be ensured and established and demonstrated based on a documented quality control (QC) system (Zhang, 2007).

It is reasonable that quality and acceptability should be expected of real and measured quantities, and this can be achieved through the comparability of test results and traceability (Figure 11.1). Traceability or metrological traceability according to the international vocabulary of metrology (third edition, 2012) is a "property of a *measurement result* whereby the result can be related to a reference through a documented unbroken chain of *calibrations*, each contributing to the *measurement uncertainty*." A "reference" can be considered a *measurement unit* through its practical realization; a *measurement procedure*, including the measurement unit for a *nonordinal quantity*; or a *measurement standard*.

However, in the more critical fields dealing with environmental protection, such as chemical, microbiological, and biological measurements and examinations, traceability to a measurement standard, even to a unique measurement procedure, is not always achievable.

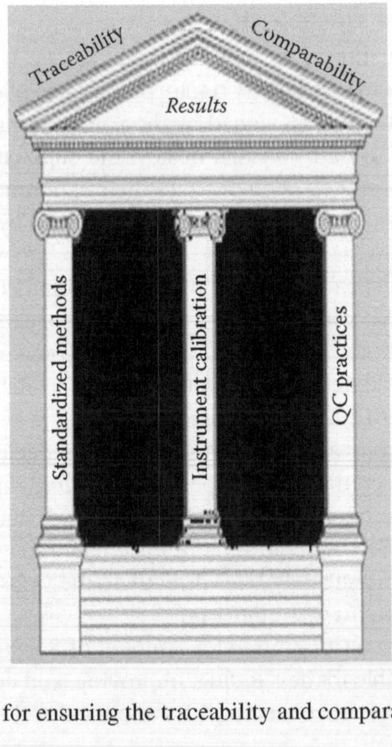

FIGURE 11.1 The construction for ensuring the traceability and comparability of results.

In order to promote traceability and comparability in the reported environmental measurement results (Skurikhin, 1993; Quevauviller, 2005; Tanimoto et al., 2007), laboratories as part of the national metrological infrastructure shall implement an appropriate quality management system (including quality assurance [QA] and QC measurement) which is based on the standards developed by international standardization–harmonization activities and which are assessed and approved by accreditation and certification bodies. This international system is based on three major pillars: standardization, metrology, and conformity assessment.

The basic tools used by environmental laboratories to assure and demonstrate the quality of their measurement results are the following:

- The use of measurement methods, fit for purpose (implementation of official/standardized methods and/or validated analytical procedures) for the production of reliable and comparable analytical results. To this end, standardized methods of analysis for the determination of numerous environmental contaminants such as disinfection byproducts, trace metals in air/water/soil/sediment, PCBs, PAHs, and pesticide residues in feed and food have been developed and validated (Ashley et al., 2009, 2012). Before use, laboratories should validate or verify methods.
- The use of appropriate measurement standards, certified reference materials (CRMs) and reference materials (RMs) (Inagaki et al., 2004; Ulberth, 2006; Duewer et al., 2009; Jakopič et al., 2013).
- The proper calibration of measuring instruments (Prohaska et al., 2000; Turk et al., 2001; Bercaru et al., 2003; Inagaki et al., 2004; Ulberth et al., 2006; Wise et al., 2006; Duewer et al., 2009).
- The implementation of QC practices by applying an internal QC system and an external quality control (EQC) by participation in interlaboratory comparisons (ILCs) and proficiency testing schemes (PTSs) on a regular basis (Harvey et al., 2008; Kong et al., 2008; Bao et al., 2009; Harper et al., 2009; Sanyal and Rani, 2009; Caurant et al., 2010).

11.1.2.1 International Measurement System

The leading role on the dissemination of measurement standards is held by the Comité international des poids et mesures, while regional metrology organizations also play a vital role to this end. Apart from the mutual recognition arrangement signed by the participating countries, the equivalence of measurement standards is controlled by means of key and supplementary comparisons between national metrology institutes in order to demonstrate their calibration and measurement capabilities (CMCs). International equivalence of RMs is also determined by special comparisons between metrology institutes.

It is widely recognized that the calibration of analytical equipment relies on the development of suitable, internationally accepted calibrants, while a method's trueness is also assessed by means of appropriate RMs. Expert laboratories participate in key comparison by using primary methods of analysis with minimum stated uncertainties in order to determine the degree of equivalence of their determinations with an agreed consensus value, the key comparison reference value. Successful participation in key comparisons is a prerequisite for declaration of their CMCs. In 2014, the International Laboratory Accreditation Cooperation requires that all measurement results for accredited laboratories should be traced to metrological institutes with declared CMCs on pertinent measurement areas. Should the reader require more information on the agreement and dissemination of measurement standards, he/she may refer directly to the website of the Bureau international des poids et mesures (BIPM) (http://www.bipm.org/).

In this respect, the Consultative Committee for Amount of Substance: Metrology in Chemistry and Biology (CCQM)—responsible for developing, improving, and documenting the equivalence of national standards (CRMs and reference methods) for chemical and biological measurements—has planned and organized an ongoing key comparison program. Three of its working groups have already been involved in key comparisons in order to support the metrological infrastructure concerning environmental analysis. During the past 15 years, the Working Groups on Gas Analysis, Inorganic Analysis, Electrochemical Analysis, and Organic Analysis have organized several key comparisons that aim to address the measurement of environmental factors (see the key comparison database, maintained by BIPM on its website [http://kcdb.bipm.org/default.asp]). Comprehensive lists of key comparisons are presented in Tables 11.2 through 11.4.

TABLE 11.2
CCQM Gas Analysis Working Group Key Comparisons

Reference No.	Description	Coordinating Laboratory	Start Date
CCQM-K7	Benzene–toluene–xylene in N_2/air	NIST	1999
CCQM-K15	SF_6, CFCs—emission levels	KRISS	2003
CCQM-K22	VOCs in air	NMIJ	2003
CCQM-K26.a	Reactive gases—ambient levels—NO in N_2	NPL	2003
EUROMET.QM-K26	Reactive gases—ambient levels—NO in N_2	NPL	2012
CCQM-K26.b	Reactive gases—ambient levels—SO_2 in air	NPL	2003
CCQM-K52	CO_2 in air (360–400 ppm)	VSL	2006
CCQM-K68	N_2O at ambient levels	KRISS	2008
CCQM-K71	Multicomponent gas stack emissions	VSL	2008
CCQM-K82	Methane in air	BIPM	2009
CCQM-K83	Halocarbons in air	NIST	2011

Note: KRISS, Korea Research Institute of Standards and Science; NMIJ, National Metrology Institute of Japan; NPL, National Physical Laboratory; VSL, Van Swinden Laboratory.

TABLE 11.3
CCQM Inorganic Analysis Working Group Key Comparisons

Reference No.	Description	Coordinating Laboratory	Start Date
CCQM-K31	Arsenic in fish or shellfish	NIST	2002
CCQM-K43	Methylmercury in salmon	IRMM	2004
CCQM-K43.1	As, Hg, Se, and methylmercury content in marine fish	NMIJ	2005
CCQM-K44	Trace metals in sewage sludge	IRMM	2004
CCQM-K45	Toxic metals in food (tin in tomato paste)	LGC	2005
CCQM-K49	Toxic and essential elements in bovine liver	NIST	2006
CCQM-K56	Trace elements in soybean powder	NIM	2006
CCQM-K59	Determination of nitrite and nitrate in calibration solutions and natural water	SMU/NRC	2006–2007
CCQM-K70	Determination of low levels of Hg in natural water	PTB/BAM/LNE	2008
CCQM-K75	Toxic metals in algae	IAEA	2009
CCQM-K89	Trace and essential elements in Herba Ecliptae	GLHK	2010
CCQM-K97	Arsenobetaine in solution and fish	NMIJ/NIM	2011
CCQM-K108	Determination of arsenic species, total arsenic, and cadmium in brown rice flour	NMIJ	2013

Note: BAM, Bundesanstalt für Materialforschung und-prüfung; GLHK, Government Laboratory of Hong Kong; IAEA, International Atomic Energy Agency; IRMM, Institute for Reference Materials and Measurements; LNE, Laboratoire national de métrologie et d'essais; NIM, National Institute of Metrology; NRC, National Research Council of Canada; PTB, Physikalisch-Technische Bundesanstalt; SMU, Slovak Institute of Metrology.

TABLE 11.4
CCQM Organic Analysis Working Group Key Comparisons

Reference No.	Description	Coordinating Laboratory	Start Date
CCQM-K5	*p,p′*-Dichlorodiphenyldichloroethylene in fish oil	LGC	1999
CCQM-K25	PCBs in sediments (PCBs 28, 101,153,170)	NIST/NRC	2001
CCQM-K50	PAHs in soils/sediments/particulates (a, soil; b, particulate)	CENAM/BAM	2006
CCQM-K85	Antifungals in food: malachite green in fish tissue (fresh-frozen salmon)	LGC	2010
CCQM-K95	Organochlorine pesticides in tea	GLHK/NIM	2011

Note: CENAM, Centro Nacional de Metrología.

11.1.3 Environmental Protection and Regulatory Framework

Environmental pollution is defined as the release of contaminants into the natural environment that could cause adverse changes.

The major categories of the environmental pollution are listed as follows:

- Air pollution is caused by the release of chemicals and particulates into the atmosphere (e.g., gaseous pollutants: carbon monoxide, sulfur dioxide, CFCs, nitrogen oxides, VOCs, and particulate matter: PM_{10} to $PM_{2.5}$). In the 2014 report issued by the World Health Organization (WHO, 2014), it is confirmed that air pollution constitutes one of the world's largest environmental health risk.
- Water pollution is caused by the discharge of untreated domestic sewage and discharge of wastewater from commercial and industrial waste into surface waters. Furthermore, the release of waste into surface runoff, waste disposal, and the introduction of leaching compounds from agricultural activities into the groundwater play also major roles.
- Soil contamination occurs when chemical compounds (e.g., heavy meals, PCBs, PAHs, methyl *tert*-butyl ether (MTBE), pesticide residues, and chlorinated hydrocarbons) are released into the soil. Bioaccumulation refers to the process whereby certain chemical compounds such as pesticide residues, mycotoxins, PCBs, PAHs, or heavy metals either through passive absorption or via the food chain gain access to and accumulate in tissues or internal organs (biomagnification) of organisms (Suedel et al., 1994; Gray, 2002).
- Radioactive contamination could result from nuclear fallout, disposal of radioactive waste, or a nuclear accident.

The initiative of the Kyoto Protocol, adopted in Kyoto and enforced in 2005, is to implement the United Nations Framework Convention on Climate Change to reduce greenhouse gas concentrations/emissions in order to prevent dangerous anthropogenic interference to the climate system. A commitment period was proposed in 2012 (Doha Amendment), and recent negotiations were held in Paris, 2014, in order to agree on a post-Kyoto legal framework.

The international regulatory framework for the air quality standards lies on (a) the European Union's (EU) Air Quality Directive and the US National Ambient Air Quality Standards, setting maximum permitted concentrations for specific atmospheric pollutants and (b) the North American Air Quality Index, representing the corresponding risk in the form of a scale. In the EU, the Air Framework Directive 2008/50/EC was adopted on May 21, 2008, where in the United States, a large number of hazardous air pollutants are regulated under the Clean Air Act (42 US Code 7401).

Since pollutants may be released in the environment from various sources, a strategy in the EU for dealing with pollution of water from chemicals was set out by the Water Framework Directive 2000/60/EC and its amendments. In order to improve the quality of the monitoring data obtained under the WFD, the European Commission adopted Directive 2009/90/EC laying down technical specifications for chemical analysis and monitoring of water status. A list of priority substances, reviewed and updated every 4 years, was adopted in Decision 2455/2001/EC. In order to fulfill monitoring requirements in the EU, two types of environmental quality standards (EQSs) for the priority substances have been established for surface waters (Directive 2008/105/EC): (a) the annual average concentration for protection against long-term exposure and chronic effects and (b) the maximum allowable concentration for avoiding irreversible consequences to the ecosystem due to short-term acute exposure.

In order to effectively protect the marine environment across Europe, the EU's Marine Strategy Framework Directive (2008/56/EC) has been adopted.

In the United States, the environmental laws and regulations constitute the driving forces behind almost all the environmental sampling and analytical work (Title 40 "Protection of Environment" of the Code of Federal Regulations, 40 CFR).

The Federal Occupational Safety and Health Administration probably constitutes the most important guide for industrial hygienists and safety professionals in environmental health and safety (OSHA, 1998).

In the United Kingdom, the occupational exposure to lead is regulated by the Control of Lead at Work Regulations (Stationery Office Ltd., 2002) and the exposure to cadmium and chromium by the Control of Substances Hazardous to Health Regulations 2004.

In 2007, a new EU regulatory framework for chemicals called REACh (Registration, Evaluation, Authorization and Restriction of Chemicals) was enforced, aiming at improving the protection of human health and the environment by the identification of the intrinsic properties of chemical compounds, registered to the Chemical Abstracts Service.

11.1.4 Interlaboratory Comparisons: Proficiency Testing Schemes

11.1.4.1 Introduction: Terminology

The main terms and definitions concerning PTSs are provided in the ISO/IEC 17043:2010, *Conformity Assessment: General Requirements for Proficiency Testing*:

- ILC is the organization, performance, and evaluation of measurements or tests on the same or similar items by two or more laboratories in accordance with predetermined conditions.
- PT constitutes the evaluation of participant performance against preestablished criteria by means of ILCs.
- A PT scheme is PT designed and operated in one or more rounds for a particular area of testing, measurement, calibration, or inspection.
- A test item of a PTS is a sample, product, RM, piece of equipment, measurement standard, data set, or other information used for PT.
- The PT provider is responsible for all tasks in the development and operation of a PTS.
- The participant could be a laboratory, an organization, or an individual that receives proficiency test items and submits results for review by the PT provider.

11.1.4.2 Brief Description of a Proficiency Testing Scheme

Proficiency testing (PT) is a type of EQC applied to evaluate laboratory accuracy. The PT provider submits samples for examination to the participating laboratory according to a predetermined distribution program, which is announced to the participants.

The samples are dispatched to all subscribers at the same time, and they have a predefined period in order to analyze them as part of their normal routine and submit their results to the PTS provider.

After the deadline, the organizer evaluates the submitted results and issues a final report, showing how closely the results of each participant agree with the assigned value. The assessment and evaluation of statistical data in the final report not only characterizes current laboratory performance but provide laboratory management with information to improve long-term laboratory performance.

11.1.4.3 Participants' Benefits from Regular Participation in Environmental Proficiency Testing Schemes

Laboratories performing determination of high-concern substances in environmental samples, feed, and foodstuff (under national and international legislation limits on the concentration levels) is fundamentally important to participate in PTSs.

Participation in PTSs provides the following (Figure 11.2):

1. *Independent evaluation of the laboratory's performance*: Participation in PTSs constitutes a key role in the evaluation of laboratory competence for the performance of specific tests and measurements in the environmental sector; the successful participation in PTSs could serve as an instant confirmation of laboratory competency. By participating in PTSs, the laboratory receives regular benchmarking to other laboratories on the grounds of technical proficiency, measurement bias, repeatability, and measurement uncertainty. Therefore, mutual trust among laboratories performing environmental analysis could be established.
2. *Third-party demonstration of technical competence evidence to accreditation bodies and regulatory-competent authorities*: Participation in PTSs and use of environmental CRMs are considered as essential QA actions for laboratory accreditation according to ISO/IEC 17025:2005. ISO/IEC 17011:2004 requires accreditation bodies to take account of the laboratory performance in PTSs. Environmental legislation foresees accreditation of laboratories performing environmental tests, in accordance with ISO/IEC 17025; therefore, participation in PTSs is necessary tool to prove the laboratory's competency and is a fundamental constituent of the assessment of the laboratory by accreditation bodies. Satisfactory results of a laboratory in proficiency tests play an important role in the laboratory's approval by the relevant competent authorities in order to be involved in official environmental surveillance. The results may also provide additional value and trust toward customers and any external users of laboratory services.
3. *Feedback in case of measurement bias*: In case of an unsatisfactory or unacceptable laboratory's performance in environmental PTSs, the implementation of corrective actions should be initiated. By participating in PTSs regularly, systematic errors may be detected,

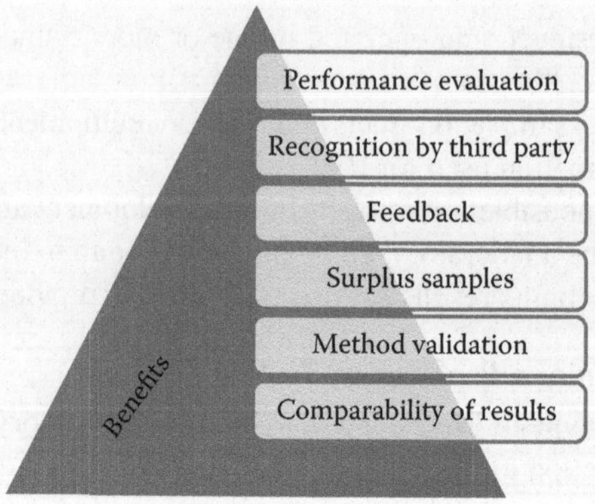

FIGURE 11.2 Benefits of environmental laboratories from regular participation in PTSs.

deviations in measurement results could be monitored, and risk management through an early warning of potential problems could be established.

4. *Surplus samples of environmental PTSs as RMs or additional QC samples*: Any remaining amount of the environmental PT items may serve as a QC sample. Moreover, surplus environmental PTS samples are available from the majority of PTSs providers and may serve not only as QC samples but also as a tool for method validation or verification, training and evaluation, etc.
5. *Data for method validation (Raposo et al., 2009; Chen et al., 2014)*: A PTS could provide information for validation or verification of method applied in the laboratory by using the statistical information on the assigned value, standard deviation, and performance of the other participants (Senseman et al., 2003; Julshamn et al., 2007; Senyuva et al., 2010; Wong et al., 2010; Medina-Pastor et al., 2011; Ramsey et al., 2011; Sack et al., 2011; Solfrizzo et al., 2011; Julshamn et al., 2013; Nanita et al., 2013; Yoshinari et al., 2013). Furthermore, participation in environmental PTSs provides useful information in method validation, as the distributed test material and the final report may be used to determine or verify crucial parameters: repeatability, reproducibility of the method, measurement uncertainty, etc. (Horwitz et al., 2001; Brera et al., 2007; Garí and Grimalt, 2010; Galea et al., 2012; McClure and Lee, 2012; Uhlig et al., 2013; McClure and Lee, 2014). Validation of a new developed method through an ILC is more reliable.
6. *Comparability of measurement results using different methods-techniques-operators*: The final report of each PTS could serve as a valuable tool for the comparative evaluation of different methods, techniques, and operators (Gratz et al., 2000; Violante et al., 2009; Matissek et al., 2010).

The participation of laboratories in environmental PTS activities is important for the participants in order to enhance the analytical skills of their specialists, demonstrate technical competence in measuring environmental parameters, and enhance the worldwide comparability of the environmental measurement results.

Significant information on comparisons of different validated methods, new equipment (Gratz et al., 2000; Violante et al., 2009; Matissek et al., 2010), and available analytical capabilities and expertise is provided, leading to methodological harmonization and mutual recognition of measurement results at both the national and international levels and to identification of competent laboratories in specific areas of interest (e.g., network), particularly in situations of environmental emergency.

PTSs could be designed in order to monitor and demonstrate the performance and testing capabilities of the network members in the environmental field and to identify gaps and problem areas that need further consideration.

11.1.4.4 Accreditation of Proficiency Testing Scheme Providers: ISO/IEC 17043:2010

The standard ISO/IEC 17043:2010 (ISO, 2010) specifies the requirements (a) for the competence of providers of PTSs and (b) for the development and operation of PTSs. As stated in the standard, these requirements are intended to be general for all types of PTSs and they can be used as a basis for specific technical requirements for particular fields of application.

Accreditation of PTS providers according to this standard is a tool to ensure the implementation of internationally acceptable procedures for the establishment of a quality management system and application of reliable and statistically valid processes for sample handling, homogeneity, and stability and estimation of the participants' evaluation parameters.

Laboratories operating a quality management system according to ISO 17025:2005 can use the accreditation of PTS providers according to ISO/IEC 17043:2010 as a verification of the technical competence of their services instead of applying other suppliers' assessment procedures.

11.2 PROFICIENCY TESTING SCHEMES IN ENVIRONMENTAL SECTION

11.2.1 Introduction

Exposure to nonpathogenic, toxic contaminants creates a significant risk to human health and ecosystem integrity. Nowadays, a discussion is raised about the following:

- The primary sources of toxic contaminants which are found in environmental matrices and media (Post et al., 2012; Zareitalabad et al., 2013; González-Castanedo et al., 2015; Vanderford et al., 2014)
- The pathways through which they may be transported (foodstuff, drinking water, air) (D'Hollander et al., 2010; Zareitalabad et al., 2013; González-Castanedo et al., 2015)
- The factors affecting these multi/many-transport flow paths (D'Hollander et al., 2010; Zareitalabad et al., 2013)
- The risks that these contaminants pose to human and environmental health (D'Hollander et al., 2010; Klecka et al., 2010)

11.2.2 Parameters/Analytes

Many chemical compounds, including contaminants from large-scale use (Sun et al., 2013) (e.g., industrial [effluents], municipal [sewage/wastewater discharge], agro-field sector [agricultural runoff], atmospheric deposition, etc.) that have been banned or restricted have been identified in the food chain in recent studies (Smith, 2009). Food chain chemical contamination by many persistent and potentially bioaccumulative compounds (Suedel et al., 1994; Gray, 2002; Guéguen et al., 2011), such as dioxins, PCBs, chlorinated paraffins, brominated diphenyl ethers, and perfluorinated compounds may occur in several pathways (Ritter et al., 2002; Brack et al., 2007; Weber et al., 2008).

The contaminants of interest included in this review are heavy metals (Devier et al., 2005), PCBs/dioxins (Devier et al., 2005; Weber et al., 2008), phthalate esters (Chai et al., 2014), mycotoxins, radioactive compounds (Zheng et al., 2013), trihalomethanes, PAHs (Chang et al., 2006; Albinet et al., 2007; Lin et al., 2014), disinfection by-products, pesticide residues (Hela et al., 2005; Konstantinou et al., 2006; Chopra et al., 2011), total petroleum hydrocarbons (TPHs)/diesel range organics, benzene–toluene–ethylbenzene–xylenes (BTEX)/MTBE, VOCs, halogenated hydrocarbons, drug residues (Castiglioni et al., 2008; Kosma et al., 2014; Evgenidou et al., 2015), brominated flame retardants (BFRs), cyanotoxins/microcystins, asbestos, nitrogen oxides, particulate matter (Wu et al., 2007), contaminants of emerging concern (Vanderford et al., 2014), endocrine-disrupting compounds (Kuch and Ballschmiter, 2001; Arditsoglou and Voutsa, 2008; Annamalai and Namasivayam, 2015), personal care products (Bu et al., 2013), and pharmaceuticals (Stackelberg et al., 2004; Baquero et al.. 2008; Martinez, 2009; Rizzo et al., 2013; Rodriguez-Mozaz et al., 2014).

The microbiological pollution is beyond the scope of this chapter.

11.2.3 Substrate in Environmental Proficiency Testing Schemes

The substrates that are used in environmental PTSs are classified as follows:

- Water: drinking/potable water, surface water, groundwater, seawater, reprocessed, effluent
- Waste: sludge, sediment, wastewater, influent/effluent, hazardous waste
- Soil: agricultural soil, contaminated land (contaminated soil and leachates), soil hazardous waste
- Foodstuff: fishery products, honey, dairy products, oil/fats, meat samples, fruits, vegetables
- Air and emissions: environmental air samples, dust samples, air filters, fly ash, gaseous emissions, stack emissions
- Others: plant samples, fertilizers, treated timber, cosmetics, paint samples

11.2.4 Selecting the Appropriate Proficiency Testing Scheme

The selection of the appropriate PTS is of primary importance. Laboratories performing tests in the environmental field have to develop an adequate PT participation strategy. In order to design the participation plan, they have to define the following important key points, which play major roles:

1. The scope of their testing activities in comparison to the PTS's provided tests
2. The adequate and appropriate frequency of their participation and the frequency of the PTS runs
3. The suitability of the procedures applied by the PTS provider in comparison to the laboratory's needs, such as location, time, sample stability, dissemination procedures, level of analytes under determination, appropriate substrate test items, quantity, homogeneity, and traceability to national or international standards when relevant
4. The accreditation of PTS provider according to ISO/IEC 17043
5. The availability of the participants' evaluation criteria (e.g., for the estimation of the participants' performance in the trial)
6. The participation cost
7. The provider's policy on confidentiality and reliability
8. The timetable of the final report

It should be noted that accreditation bodies, in agreement with accredited or under-accreditation laboratories, should have access to the laboratories' results and to details of PTS design, statistics, procedures for estimating the assigned value and the target value for standard deviation, guidelines for the participants, and the final report.

11.2.4.1 Test Samples for Distribution: Analytes to Be Determined

Sample or items matrices which are distributed in a PTS should be of the same nature when tested in laboratories (Freville et al., 2010). In general, naturally contaminated samples (real samples) or preparations made of natural samples are preferable for distribution, so that the participants should not be obliged to follow preparations different from those required for natural samples, which could involve additional errors.

The sample material which is prepared for the PTS should be adequately homogenous, and the provider should clearly declare the procedure and the statistical analysis which were used to prove homogeneity. It is generally accepted that among samples, the homogeneity standard deviation is 0.3 times the target value for the standard deviation.

The PTS organizer should provide adequate guidelines for sample conservation, environmental effects (e.g., temperature), substrate, and analyte stability.

There are PTS providers that insist on the distribution of totally synthetic test samples to the participants in cases where naturally contaminated samples are easily available. Several PTS providers do not perform stability studies on the distributed samples or fail to predict the metabolites of the analyte (degradation route), which is required to be determined.

11.2.4.2 Parameters for Evaluation of Participants' Performance

Participating laboratories are evaluated based on the difference of their results from the assigned value for the measurand or analyte. Afterward, a performance score is granted to the participants. Statistical analysis is performed in order to estimate the parameters for evaluating the participants, including estimation of outliers and strugglers (Miyashita et al., 2013; Sykes et al., 2013; Yarita et al., 2015). The statistical tools which are used are described in ISO 13528, the ISO 5725 series, and Association of Analytical Communities protocol. This analysis should provide simple, clear, and comprehensible results.

11.2.4.2.1 Assigned Value $\hat{X}$

The assigned value is estimated by various procedures as a reference value traceable either to a national or an international standard or to a CRM, the robust mean or median of the values provided by expert laboratories, as a formulation value, or in the majority of cases as a consensus value expressed as the median of the results provided by the participants.

11.2.4.2.2 Target Value of Standard Deviation for Proficiency Assessment σ_p

The target value of standard deviation is the criterion which is used to evaluate participants, and, consequently, participants shall be informed of the way of its estimation. Several formulas are used to calculate its value. In the case of chemical contaminants, the target value of the standard deviation is expressed as a function of the concentration of the analyte. One popular way is the use of the modified Horwitz equation.

11.2.4.2.3 Performance Score

For evaluation of the participating laboratories, a performance score is granted. The most used is the z-score, which is calculated as

$$z = \frac{x - \hat{X}}{\sigma_p},$$

where x is the participant's reported result, $\hat{X}$ is the assigned value, and σ_p is the target value of the standard deviation for proficiency assessment.

- Results with $|z| \leq 2$ are designated as acceptable or satisfactory.
- A result with $2 < |z| < 3$ shall be considered to give a "warning signal" according to ISO 13528:2005 and is designated as questionable according to the International Union of Pure and Applied Chemistry's (IUPAC) International Harmonized Protocol (IHP).
- A result with $|z| > 3$ shall be considered to give an "action signal" according to ISO 13528:2005 and is designated as unacceptable or unsatisfactory according to the IUPAC's IHP.

11.2.4.2.4 Performance Evaluation in Multianalyte Determinations

Performance scores are calculated on the basis of the numerical results which are provided by the participants (Otake et al., 2014). However, in multiparametric/multianalyte determinations such as determination of pesticide residues where a target list is provided to the participants, in case of false-negative results (i.e., residues were present in the distributed test sample, but these were not identified even though the sample was specifically analyzed), the PTS providers could evaluate the participants' performance in different ways.

In such a case, the provider/organizer may decide to classify the laboratories into two groups (named A and B). According to this, participating laboratories which detect successfully a high percentage of the analytes (e.g., pesticides) present in the distributed test sample (e.g., at least 90%) and reported no false-positive results are considered to demonstrate "sufficient scope" and therefore could be classified into category A. Other PTS providers evaluate the z-score of false-negative results as if the reported value was zero. Some providers flag the false-negative result as an outlying value in order to emphasize that the residue was not identified, while others decide not to give the participant a warning signal and do not evaluate such performance. However, in Eurachem's 8th International Workshop on Proficiency Testing, held in Berlin in October 2014, it was pointed out that the situation should be specifically examined and further details should be provided in the revision of ISO 13528:2005, which is in progress.

In case of the determination of multiparameters/analytes (e.g., pesticide residues by LC-MS/MS, trace metals by inductively coupled plasma mass spectrometry [ICP-MS]) (Škrbić et al., 2013) in a test sample, apart from the z-score evaluation (Vander Heyden and Smeyers-Verbeke, 2007), a combined z-score (e.g., rescaled sum of z-scores, sum of squared z-scores, average of the squared z-scores) is also proposed by several providers as a tool for the evaluation of the overall performance of the participating laboratories (Medina-Pastor et al., 2010). However, its application should be considered carefully.

11.3 PROFICIENCY TESTING SCHEME PROVIDERS IN THE ENVIRONMENTAL SECTOR

11.3.1 List of Environmental Proficiency Testing Scheme Providers Worldwide

Table 11.5 contains the PTS providers in the environmental sector worldwide. A more detailed database of PTSs providers is hosted in the European Proficiency Testing Information System website (http://www.eptis.bam.de). Apart from that, cooperations of PTS providers focusing on specific areas of concern (e.g., occupational exposure, chemical weapons convention) have been established. Brief descriptions are given in the following sections.

11.3.2 Proficiency Testing for Water Framework Directive

The mission of the Proficiency Testing for Water Framework Directive (PT-WFD) network is to provide all interested parties with harmonized, from the performance evaluation viewpoint, PTSs in order to support the implementation of the WFD.

The EU PT-WFD network consists of the following members:

1. Association Générale des Laboratoires d'Analyse de l'Environnement, Hallennes lez Haubourdin, France
2. AQS Baden-Württemberg at the Institute for Sanitary Engineering, Water Quality and Solid Waste Management, Universität Stuttgart, Germany
3. Bureau Interprofessionnel des Etudes Analytiques, Gennevilliers, France
4. ielab Calidad, S.L., Alicante, Spain
5. Institut für Hygiene und Umwelt, Hamburg, Germany
6. IWW Water Centre, Mülheim an der Ruhr, Germany
7. Kemijski Inštitut, Ljubljana, Slovenia
8. LGC Standards Proficiency Testing, Bury, United Kingdom
9. Proftest SYKE, Helsinki, Finland
10. Quality Consult, Rome, Italy
11. Scheme for Chemical Measurement Assessment (SCHEMA), Athens, Greece
12. Wessling International Research and Education Center, Budapest, Hungary

AQUAREF, L'Institut national de l'environnement industriel et des risques, and Joint Research Centre (JRC)–Institute for Reference Materials and Measurements (IRMM) are used as external advisors in order to support the PT-WFD network.

Unfortunately, at the end of 2013, some members withdrew from the PT-WFD network, and its future is currently not clear.

11.3.3 European Union Reference Laboratories

The scope of the European Union Reference Laboratories (EURLs), hosted by the JRC, is to ensure the reliability of measurement results in the EU, to coordinate the network of the national reference

TABLE 11.5
List of Available PTS Providers in the Environmental Field

	PTS Provider	Organization/ Country	Accreditation	Analytes/Parameters—Substrate
1	ERA/Waters Corporation (http://www.eraqc.com)	Waters Company, United States	ISO/IEC 17043:2010	• Radiochemistry (Radchem, MRaD) • Soil (metals, nutrients, BTEX/MTBE, PCBs, PAHs, TPHs, pesticide residues) • Water supply (pesticide residues, trace metals, dioxin, haloacetic acids, semivolatiles, PCBs) • Water pollution (trace metals, TPHs, BTEX/MTBE, pesticide residues, PAHs, PCBs) • Air and emissions (inorganics, metals, semivolatiles, volatiles) • Underground storage tank (BTEX/MTBE, diesel range organics, TPHs)
2	Global Proficiency (http://www.global-proficiency.com)	Australia, New Zealand	ISO/IEC 17043:2010	• WaterChek (trace metals, trihalomethanes, disinfection by-products, PAHs, pesticide residues in potable water, effluent) • SoilChek (metals and extractable metals in soils) • PlantChek (metals in plant samples) • FertChek (metals in synthetic fertilizers) • TimberChek (metals, light organic solvent preservative in treated timber) • Foodstuff (parameters in dairy, fats/oils, beverages, meat samples)
3	Association Générale des Laboratoires d'Analyse de l'Environnement (http://www.association-aglae.fr)	France	ISO/IEC 17043:2010	• Water, soil, sediments, waste • Disinfection by-products • Heavy metals • PAHs • Chlorophenols, alkylphenols • Organotin, cyanotoxins, bisphenol • Pesticide residues • Diphenyl ether compounds • Bromodiphenyl ethers • BTX, volatile organohalogens, chlorobenzenes
4	Chinese Academy of Inspection and Quarantine, CTC account (http://www.lab.caiqtest.com)	China	ISO/IEC 17043:2010	Heavy metals, drug residues, toxins, genetically modified, cosmetics, animal feed, and water chemistry analysis

(*Continued*)

TABLE 11.5 (CONTINUED)
List of Available PTS Providers in the Environmental Field

	PTS Provider	Organization/ Country	Accreditation	Analytes/Parameters—Substrate
5	Bureau Interprofessionnel des Etudes Analytiques (International Bureau for Analytical Studies) (http://www.bipea.org)	France	ISO/IEC 17043:2010	• Water: trihalomethanes, PBDEs, alkylphenols, anilines, organic residues, trace metals, disinfection by-products, organotins, chlorophenols, hydrocarbons • Soil: heavy metals • Activated sludges and sediments: PAHs, PCBs, organohalogenated insecticides, heavy metals • Organic fertilizers: heavy metals • Foodstuff: mycotoxins, trace elements, PCBs, dioxins, PAHs, pesticide residues.
6	LGC (http://www.lgcpt.com)	LGC Standards Proficiency Testing, United Kingdom	ISO/IEC 17043:2010	Water and environmental PTSs: • Air PT/STACKS: air and stack emissions (heavy metals, VOCs, BTEX) • AQUACHECK: water, agricultural soils, and sludges (inorganic, organic, and metal analytes in a range of water types; determination of radiochemical and ecotoxicological parameters—trace metals, halogenated solvents, disinfection by-products, PAHs, PCBs, pesticide residues, adsorbable organic halogens) • CONTEST: contaminated land (leachates and solutions from contaminated land, PAHs, PCBs, TPHs, BTEX, metals) • Food and feed PTSs (trace elements, mycotoxins, antibiotics, pesticide residues) • AIMS: asbestos in materials science
7	Environment Canada (http://www.ec.gc.ca)	Canada	ISO/IEC 17043:2010	Trace metals in water, soil, sediment

(*Continued*)

TABLE 11.5 (CONTINUED)
List of Available PTS Providers in the Environmental Field

	PTS Provider	Organization/ Country	Accreditation	Analytes/Parameters—Substrate
8	RT Corporation, RTC Environmental Proficiency Testing (http://www.rt-corp.com)	RT Corporation, Sigma-Aldrich RTC, United States	ISO/IEC 17043:2010	• Air PT: trace metals, PCBs, semivolatile compounds, PAHs, nitrogen oxides • Solids/soils hazardous waste PTS: pesticide residues, BTEX/MTBE, PAHs, PCBs, metals • Water pollution/waste water: trace metals, PAHs, TPHs, VOCs, pesticide residues, PCBs • Water supply/drinking water: pesticide residues, PCBs, VOCs, trihalomethanes, trace metals, disinfection by-products
9	Proftest SYKE (http://www.syke.fi/proftest/en)	SYKE, Finland	ISO/IEC 17043:2010	• Trace metals in water/soil/wastewater/sludge • Radon comparison (^{222}Rn) in groundwater • Organics (C_{10}–C_{40}, VOCs, PCBs) in water/soil/sediment
10	Proficiency Testing Australia (http://www.pta.asn.au)	Australia	ISO/IEC 17043:2010	• Metals, PCBs, pesticide residues in water • Metals in foodstuff • Asbestos (slide fiber count, sample identification) • Metals on filter paper (air and emissions)
11	Institute for Environmental Reference Materials (http://www.ierm.com.cn)	Ministry of Environmental Protection, China	ISO/IEC 17043:2010	PT programs on water, air and soil • Water: heavy metals, BTEX, phthalate esters, pesticide residues, PAHs, PCBs • Gas workplace ambient materials: BTEX, inorganics, BTEX • Soil: heavy metals
12	Associazione per l'unificazione nel Settore dell'Industria Chimica (http://www.unichim.it)	Italy	ISO/IEC 17043:2010	• Dioxins (DIOX) • PCBs (PCBS) • Hydrocarbons (IDRO) • PAHs (IPAS) • Heavy metals (META) • Pesticides • Halogenated solvents • Soils, sediments, fly ash, water
13	National Measurement Institute (http://www.measurement.gov.au)	Department of Industry, Australia	ISO/IEC 17043:2010	• Hydrocarbons in soil and water • Metals in water, soil, and food • Pesticides in water, soil, fruits, and vegetables

(Continued)

TABLE 11.5 (CONTINUED)
List of Available PTS Providers in the Environmental Field

	PTS Provider	Organization/ Country	Accreditation	Analytes/Parameters—Substrate
14	KWR (http://www .kwrwater.nl)	KWR Watercycle Research Institute, the Netherlands	ISO/IEC 17043:2010	• PAHs PCBs, pesticide residues, volatile halogenated hydrocarbons, trace metals, pharmaceuticals, and mineral oil in drinking water, surface water, seawater, and wastewater • Water (drinking water, surface water, groundwater, wastewater, cooling water): halogenated hydrocarbons, PAHs, heavy metals, pesticides, pharmaceuticals, PCBs
15	NSI Lab Solutions (http://www.nsi-es .com/; http://www .nsilabsolutions.com)	NC, United States	ISO/IEC 17043:2010	• Drinking water fields of PT: trace metals, disinfection by-products, VOCs, asbestos, pesticides, carbamates, and Vydate, chlorinated acid herbicides, adipate/phthalates, PCBs/dioxins, PAHs • Nonpotable water: trace metals, BTEX, volatile halocarbons, volatile petroleum hydrocarbons, PCBs, pesticides, PAHs, phthalates • Solid and chemical materials: heavy metals, volatile aromatics, PCBs, pesticides, petroleum hydrocarbons • Soil/hazardous waste: metals, oil, and grease, PCBs, pesticide residues, VOCs, PAHs
16	Canadian Association for Laboratory Accreditation Inc. (CALA) (http://www .cala.ca)	CALA, Canada	ISO/IEC 17043:2010	• Water: pesticide residues, PAHs, PCBs, BTEX, VOCs, trace metals, trihalomethanes • Air filter: metals, asbestos • Soil/sediments: metals, PAHs, BTEX, PCBs, VOCs, pesticide residues
17	Rede Metrologica (http://www.rede metrologica.com.br)	Brazil	NBR ISO/IEC 17043:2011	• Heavy metals, BTEX, trihalomethanes, VOCs in water, effluents • Veterinary drugs in foodstuff

(*Continued*)

TABLE 11.5 (CONTINUED)
List of Available PTS Providers in the Environmental Field

	PTS Provider	Organization/ Country	Accreditation	Analytes/Parameters—Substrate
18	Laboratory Environmental Analysis Proficiency Scheme (http://fapas.com/proficiency-testing-schemes/leap/)	Food and Environment Research Agency, United Kingdom.	ISO/IEC 17043:2010	• Potable water chemistry: for analysis of inorganics, metals, and organics • Effluent, wastewater and contaminated land, and hazardous waste: for analysis of inorganics, metals and organics in water and metals in soil • Potable water chemical contamination emergency: unknown chemically contaminated potable water sample on an emergency short term is dispatched and analysis for gross α and gross β radioactivity only is also required
19	INTER2000 QM (http://www.inter2000.cat)	Ministry of Agriculture, Livestock, Fisheries, Food and Natural Environment; Laboratori Agroalimentari of Generalitat (Government of Catalonia), Spain	All PT tests are organized taking into consideration ISO/IEC 17043	• Agro-food sector laboratories • Water, wastewater, soil, foodstuff • Trace metals and VOCs in drinking water • Trace metals in foodstuff
20	AQS Baden-Württemberg (http://www.iswa.uni-stuttgart.de/ch/aqs/)	Department of Hydrochemistry and Analytical Quality Assurance, Institute for Sanitary Engineering, Water Quality and Solid Waste Management, Germany	All PTSs are in accordance with ISO/IEC 17043:2010 and DIN 38402-45:2003	PAHs, alkylphenols, pesticide residues, VOC/BTEX, PBDEs, pharmaceuticals in drinking water/groundwater/raw water/wastewater/surface water
21	Department for Agrobiotechnology (IFA-Tulln), University of Natural Resources and Life Sciences (BOKU), in cooperation with Umweltbundesamt GmbH, Environment Agency, Austria (http://www.ifatest.at)	Center for Analytical Chemistry, IFA-Tulln, BOKU, in cooperation with Umweltbundesamt GmbH, Austria	All PTSs are performed according to ISO/IEC 17043	• Organotin compounds, herbicides/pesticides, PAHs, BTEX and MTBE, pharmaceuticals, industrial chemicals in water • Chlorinated hydrocarbons, BTEX in air samples
22	Quality Reliable LabServices S.L/qRL Services (http://www.qrlservices.com)	Spain	ISO 9001:2008	Metals and total organic carbon in surface water, seawater, drinking water. soil, and sludge

(Continued)

TABLE 11.5 (CONTINUED)
List of Available PTS Providers in the Environmental Field

	PTS Provider	Organization/ Country	Accreditation	Analytes/Parameters—Substrate
23	Quality Consult interlaboratory PT (http://www.aqc.it)	Quality Consult, Italy	ISO/IEC 17043:2010	• Drinking water (metals, organochlorine pesticides, PAHs, VOCs) • Urban wastes (sewage sludge and wastewater: metal and leaching test) • Contaminated soil (PAHs, BTEX, pesticides) • Groundwater (total hydrocarbons)
24	ILT Interlaboratory Test (http://www.ptinter laboratory.com)	ILT Interlaboratory Test S.A., Argentina	The regulation governing PTSs is ISO/ IEC 17043	Metals, phthalates, pesticides, aflatoxins, and antibiotic residues in water, soil, and foodstuff
25	Instituto Nacional de Controle de Qualidade em Saúde (http://www .incqs.fiocruz.br)	Brazil	NBR ISO/IEC 17043:2011	Pesticide residues and mycotoxins in foodstuff
26	MOL Labs Ltda (http:// www.mollabs.com)	Colombia	ISO/IEC 17043:2010	Trace metals in residual water, pharmaceuticals, and foodstuff
27	Deutsches Referenzbüro für Lebensmittel-Ringversuche (http:// www.drrr.de)	Germany	ISO/IEC 17043:2010	• Water/environment (drinking water/surface/reprocessed water, effluent): pharmaceuticals, PAHs, heavy metals, pesticide residues • Foodstuff: trace metals
28	Inspection and Quarantine Technical Center of Liaoning Entry-Exit Inspection and Quarantine Bureau/Liaoning Academy of Inspection and Quarantine Bureau (LCNIQ) (http://www .lnciq.gov.cn/en)	General Administration of Quality Supervision, Inspection and Quarantine (AQSIQ), China	ISO/IEC 17043:2010	Pesticide residues, heavy metals, and mycotoxins in foodstuff
29	LCNIQ (http://www .iqtc.cn)	AQSIQ, China	ISO/IEC 17043:2010	Vegetable oil: mycotoxins, consumers products: phthalate esters, heavy metals, PBDEs
30	Collaborative Testing Services, Inc. (http:// www.collaborative testing.com/)	United States	ISO/IEC 17043:2010	• Agricultural Laboratory Proficiency program: soil analyses, botanicals analyses, water analyses • Fasteners and Metals program
31	North American Proficiency Testing Program (http://www .naptprogram.org)	Soil Science Society of America, United States		• Specific soil analyses • Specific plant analyses • Specific water analyses • Soil environmental analyses

(Continued)

TABLE 11.5 (CONTINUED)
List of Available PTS Providers in the Environmental Field

	PTS Provider	Organization/ Country	Accreditation	Analytes/Parameters—Substrate
32	NIL (http://www.nil .org.cn)	China NIL Research Center for Proficiency Testing, China	ISO/IEC 17043:2010	Chemical analysis: metals or alloys, ore or mineral, pottery, water, etc.
33	ieLab (http://www.ielab .es)	Spain	ISO/IEC 17043:2010	• Metals, petroleum hydrocarbons, PAHs, pesticide residues, microcystins, acrylamide in potable and raw water • Metals in wastewater, soil, and sludge • Metals in stack emissions
34	Dalian CFAPA Testing Technology Company Limited (http://www .cfapa.com)	China	ISO/IEC 17043:2010	Foodstuff: heavy metals, pesticide residues, pharmaceuticals
35	Phenova (http://www .phenova.com)	Phenomenex Company, United states	ISO/IEC 17043:2010	• Water supply: trace metals, pesticide residues, PCBs, trihalomethanes, organic disinfection by-products, semivolatiles • Water pollution: oil and grease, trace metals, BTEX, PAHs, PCBs, pesticide residues • Soils/hazardous Waste: TPHs, trace metals, BTEX, PCBs
36	Qualinova (http://www .labnova.com)	Labnova Distibuciones S.L., Spain	The organizational system complies with the requirements of ISO/IEC 17043	Heavy metals in wastewater, drinking water, soil
37	Relacre (http://www .relacre.pt)	Portugal	ISO/IEC 17043	• Water, sludge, gaseous emissions/ air quality • Heavy metals, total hydrocarbons, VOCs
38	National Laboratory Association (http:// www.home.nla.org.za)	South Africa	In accordance with ISO/IEC 17043:2010	Pesticide formulation
39	Gabinete de Servicios para la Calidad (http:// www.intercomparativos .com)	Spain	ISO/IEC 17043:2010	• Metals, pesticide residues, PAHs, trihalomethanes in water (drinking, residual, seawater) • Metals in foodstuff
40	Jabatan Kimia Malaysia PTS (http://www .kimia.gov.my)	Department of Chemistry Malaysia, Ministry of Science, Technology and Innovation, Malaysia		Pesticide residues in foodstuff, trace metals in water, lead in air

(Continued)

TABLE 11.5 (CONTINUED)
List of Available PTS Providers in the Environmental Field

	PTS Provider	Organization/ Country	Accreditation	Analytes/Parameters—Substrate
41	RTI International (http://www.rti.org)	United States	ISO/IEC 17043:2010	• Determination of asbestos-in-bulk and asbestos-in-air test samples • The ELPAT is part of AIHA's Environmental Lead Laboratory Accreditation Program (lead-in-paint, lead-in-soil and lead-in-dust PT materials)
42	Shandong Technical Center for Inspection and Quarantine of Shandong Entry-Exit Inspection and Quarantine Bureau (SDCIQ), Inspection and Quarantine Technology Center of SDCIQ (http://www.ptplab.net)	AQSIQ, China	ISO/IEC 17043:2010	Food and agricultural products: heavy metals, pesticide residues, mycotoxins, phthalate esters
43	Quasimeme Laboratory Performance Studies, Quality Assurance of Information for Marine Environmental Monitoring in Europe (http://www.quasimeme.org)	Wageningen Evaluating Programmes for Analytical Laboratories, Wageningen University, the Netherlands	ISO/IEC 17043:2010	• Metals, halogenated organics, VOCs, organotins, PAHs, pesticide residues in seawater • Trace metals, PAHs, organotins, BFRs, chlorinated paraffins, PFASs in biota • Trace metals, PAHs, organotins, PFASs, BFRs, PCBs in sediment • Trace metals, PCBs, PAHs, chlorinated organics in fish/shellfish
44	National Atomic Energy Agency and Institute of Nuclear Chemistry and Technology (http://www.ichtj.waw.pl)	Poland	In accordance with ISO/IEC 17043	Radionuclides in water, milk powder, wheat flour and soil
45	SCHEMA (http://www.gcsl.gr/index.asp?a_id=372)	General Chemical State Laboratory, Chemical Metrology Service, Greece	ISO/IEC 17043:2010	• Heavy metals, pesticide residues, PAHs in potable water • Phthalate esters, heavy metals, pesticide residues, mycotoxins, PAHs in foodstuff • Pesticide residues, PAHs, heavy metals in surface water, wastewater

(*Continued*)

TABLE 11.5 (CONTINUED)
List of Available PTS Providers in the Environmental Field

	PTS Provider	Organization/ Country	Accreditation	Analytes/Parameters—Substrate
		WFD Network		
A	AIHA Proficiency Analytical Testing Programs (http://www.aihapat.org)	United Sates	ISO/IEC 17043:2010	• Industrial Hygiene Proficiency Analytical Testing: metals, silica, asbestos, organic solvents • ELPAT: lead determination in air samples, paint, soil and dust samples • Bulk Asbestos Proficiency Analytical Testing • Beryllium Proficiency Analytical Testing • Asbestos Analyst Testing
B	WASP/Health and Safety Laboratory + Network	United Kingdom		• Analysis of occupational hygiene and environmental air samples
C	IAEA			For the members of the ALMERA network: • Soil sampling • Gamma-emitting radionuclides in water and soil • ^{210}Po/water • ^{226}Ra/soil • Anthropogenic radionuclides in water, seaweed, and sediment sample Collaborated with the UNEP Mediterranean Action Plan on the determination of organic contaminants and trace metals in marine samples

Note: AIHA, American Industrial Hygiene Association; ALMERA, Analytical Laboratories for the Measurement of Environmental Radioactivity; BTX, benzene–toluene–xylenes; DIN, Deutsches Institut für Normung e.V.; DIOX, Dioxins; ELPAT, Environmental Lead Proficiency Analytical Testing Program, IAEA, International Atomic Energy Agency; META, heavy metals; NBR, Norma Brasileira Regulamentadora; PCBs, polychlorinated biphenyls; PFAS, perfluorooctanesulfonic acid; SCHEMA, Scheme for Chemical Measurement Assessment; UNEP, United Nations Environment Programme; WASP, Workplace Analysis Scheme for Proficiency.

laboratories (NRLs) of the member states, and to provide support to EU policies and to chemical measurement system in the EU (Jakopič et al., 2013; Jobbágy et al., 2014).

The EURLs and the JRC organize proficiency tests through the operation of the following:

- Regular European Interlaboratory Measurement Evaluation Program in the field of nuclear measurements: nuclear fuel cycle materials
- Interlaboratory Measurement Evaluation Program for Nuclear Signatures in the Environment (Truyens et al., 2013): measurements of the isotopic abundances of characteristic elements—implementation of the Treaty on the Non-Proliferation of Nuclear Weapons
- International Measurement Evaluation Programme (Cordeiro et al., 2014, 2015; Fiamegos et al., 2014)

In the field of air quality monitoring, the European Reference Laboratory for Air Pollution organizes ILCs for air pollutants which are included in the European Air Quality Directives. Regular tests take place for gaseous organic and inorganic air pollutants as well as for pollutants in the particulate phase. These ILCs are organized to support the external control needs of the National Air Quality Reference Laboratories which participate in the AQUILA Network.

The EURL in the Netherlands, hosted by RIKILT, in the Wageningen University and Research Centre, the Netherlands, is a center of expertise for analysis of biological residues accumulated in food production animals, for the presence of hormonal growth-promoting compounds, sedatives, and mycotoxins. This EURL organizes PTSs and workshops and supplies analytical standards and RMs.

The four EURLs for pesticide residues in food and feed (EURL for Fruits and Vegetables, EURL for Feed and Cereals, EURL for Food of Animal Origin and Commodities with High Fat Content, EURL for Single Residue Methods) regularly organize European Union Proficiency Tests provided to the NRLs and Official Laboratories involved in the official control of pesticide residues within the EU Member States.

11.3.4 International Atomic Energy Agency

For nearly 40 years, the International Atomic Energy Agency (IAEA) conducts PTSs and ILCs in different environmental matrices on a regular basis (IAEA, 2011, 2012, 2014). The IAEA's Environmental Laboratories are the coordinators of the Analytical Laboratories for the Measurement of Environmental Radioactivity (ALMERA) worldwide network, currently (November 2014) consisted of 149 laboratories from 84 member states.

IAEA has developed collaborations with the United Nations Environment Programme Mediterranean Action Plan for evaluating the performance of the laboratories from the Mediterranean countries on the determination of organic contaminants and trace elements in marine samples. Upon a Japanese request, the analytical performance of 45 Japanese laboratories was also evaluated under the IAEA umbrella.

One of the objectives of IAEA is to also provide advice and assistance on marine radioactivity to member states and intergovernmental bodies, such as the Oslo–Paris Commission for Protection of the Marine Environment of the North-Eastern Atlantic, the Joint Group of Experts on the Scientific Aspects of Marine Environmental Protection, and the Barcelona Convention.

Since 2004, IAEA has included within its ALMERA activities the organization of PT and ILC schemes on an annual basis (Table 11.6) (Shakhashiro, 2008; Belli, 2009), the development and validation of analytical procedures for the determination of radionuclides in environmental samples (including soil, water, air aerosols, and unprocessed food products), and the organization of training courses and workshops. To this end and taking into account that milk products constitute the major targets for radiation contamination in a radiation emergency, a training course was organized in Daejeon, South Korea, in November 2014, where a recently developed method on the rapid determination of "radioactive strontium" in milk has been presented in selected experts from the ALMERA network laboratories (Aurelien Pitois, 2014).

Through the ALMERA PTSs (Table 11.6), the participating laboratories provide demonstration of their technical competence and evidence of reliable and timely determination of radionuclides in distributed samples for both routine and emergency environmental monitoring, leading to enhanced worldwide comparability of environmental radioactivity measurement results.

11.3.5 Proficiency Testing Scheme Providers in the Field of Occupational Exposure, Safety, and Health

11.3.5.1 Health and Safety Laboratory, United Kingdom–LGC

Since occupational exposure to toxic compounds is regulated, the employers are required to monitor the exposure and assess the health exposure risk in order to provide respiratory protective equipment

TABLE 11.6
ALMERA PTSs for the Last 5 Years

Year	PTS Code	Distributed Samples	Analytes to Be Determined
2014	IAEA-TEL-2014-04	Three water samples, one seaweed sample, one sediment sample, and one water sample from oil field	Anthropogenic and natural gamma-emitting radionuclides ^{90}Sr, ^{226}Ra, ^{238}U, ^{235}U, and ^{234}U in the water samples; gamma-emitting radionuclides in the seaweed and sediment samples; and ^{226}Ra in the water sample from oil field
	IAEA-TEL-2014-03	Three water samples, one seaweed sample, and one lake sediment sample	Gamma-, beta-, and alpha-emitting radionuclides in water, biota, and sediment samples
2013		Three water samples and one flour sample	Artificial and natural gamma-emitting radionuclides in water samples and ^{134}Cs and ^{137}Cs in the flour sample
2012	IAEA-TEL-2012-03	Three water samples, one hay sample, and one soil sample	Anthropogenic gamma emitters in water in the presence of natural background, of radiocesium (^{134}Cs and ^{137}Cs) in hay, and of ^{90}Sr, ^{137}Cs, transuranium, and natural radioisotopes in soil sample
2011	IAEA-MEL-2011-06	Organic pollutants in biota sample	Chlorinated pesticides, petroleum hydrocarbons, and PCBs
	IAEA-MEL-2011-05	Sediment sample	Trace elements and methyl mercury in sediment sample: Al, As, Cd, Co, Cr, Cu, Fe, Hg (total and methyl Hg), Li, Mn, Ni, Pb, Se, Sb, Sn, V, and Zn
	IAEA-TEL-2011-04	Soil and water samples	Natural and artificial radionuclides in soil and water: ^{3}H, ^{60}Co, ^{133}Ba, ^{134}Cs, ^{137}Cs, ^{152}Eu, and ^{241}Am in water; ^{40}K, ^{90}Sr, ^{137}Cs, radioisotopes from the uranium and thorium decay series, ^{238}Pu, ^{239}Pu, ^{240}Pu, and ^{241}Am in soil
	IAEA-TEL-2011-03	One soil sample and three water samples	In soil: ^{40}K, ^{90}Sr, ^{137}Cs, ^{210}Pb, ^{226}Ra, ^{234}U, ^{238}U, ^{238}Pu, ^{239}Pu, ^{240}Pu, and ^{241}Am In water: ^{3}H, ^{60}Co, ^{134}Cs, ^{137}Cs, and ^{152}Eu
	IAEA-MEL-2011-02	Sediment Sample	Al, As, Cd, Co, Cr, Cu, Fe, Hg (total and methyl Hg), Li, Mn, Ni, Pb, Se, Sb, Sn, V and Zn
2010	IAEA-CU-2010-04	One soil sample and five water samples	^{226}Ra, ^{234}U, ^{238}U, gross alpha/beta in water, ^{226}Ra in soil
	IAEA-CU-2010-03	One soil sample and five water samples	Natural radionuclides in water and ^{226}Ra in soil
	IAEA-CU-2010-02	One sterilized sewage sludge sample	As, Ba, Cd, Co, Cr, Cu, Fe, Hg, Ni, Pb, Se, Sr, and Zn
2009	IAEA-CU-2009-04 IAEA/AQ/16	Four air-filter samples	^{54}Mn, ^{57}Co, ^{59}Fe, ^{60}Co, ^{65}Zn, ^{109}Cd, ^{133}Ba, ^{134}Cs, ^{137}Cs, ^{152}Eu, and ^{241}Am in filters 01, 02, and 03; the participating laboratories were provided with a control filter 04 containing only ^{60}Co and ^{133}Ba

to employees. The UK Health and Safety Laboratory (HSL), one of the leading health and safety research bodies and consultancy providers, has been organizing the Workplace Analysis Scheme for Proficiency (WASP) concerning the determination of multianalytes in air samples (Stacey et al., 2003; Grzebyk et al., 2005; Stacey, 2006). Concerning the organization of PTSs in the determination of heavy metals in air, the WASP PTSs provider dispatched to the participating laboratories mixed cellulose ester filters fortified with known amounts of heavy metals for performing their analysis/determinations (Stacey and Wright, 2001; Stacey, 2006; Stacey and Butler, 2008).

Recently, a new PTS for laboratories performing chemical analysis of air monitoring samples (e.g., filters, sorbent tubes, stack/impinger solutions), under the name AIR-PT, was created from the collaboration between HSL and LGC (a merger of the HSL's WASP and LGC's STACKS PTS has been performed) (HSL, 2014). The AIR-PTSs will be run by LGC and will be supported by HSL technical expertise.

In France, apart from the External Quality Control Campaign (organized upon request of the French labor ministry) in the determination of lead in whole blood (F-EQCPbB), other PTSs are organized in order to improve the participating laboratories' performance in the determination of occupational exposure to toxic compounds (Pineau et al., 2014; ALASCA PTSs: Langlois et al., 2008).

11.3.5.2 National Institute for Occupational Safety and Health

In the United States, the National Institute for Occupational Safety and Health (NIOSH), which constitutes the US federal agency that conducts research activity, provides recommendations how to predict/address emerging problems in the area of occupational health and safety and provides assistance for assessing occupational health exposure and health hazard evaluations.

NIOSH, established by the Occupational Safety and Health Act of 1970, is part of the US Centers for Disease Control and Prevention, in the US Department of Health and Human Services.

Proficiency Analytical Testing (PAT) programs (Schlecht et al., 1986; Shulman et al., 1992; Schlecht and Shulman, 1995; Ashley et al., 2009) are developed and organized by NIOSH in collaboration with the American Industrial Hygiene Association. The reported data/submitted results in these PAT programs could be useful in supporting the NIOSH research in sampling and analytical methodologies (Schlecht et al., 1997, 2011; Eller et al., 1999; Ashley et al., 2003, 2010, 2012; Oatts et al., 2012).

11.3.6 Organization for the Prohibition of Chemical Weapons

The Organization for the Prohibition of Chemical Weapons (OPCW) organizes PTSs on a regular basis in the field of environmental analyses for the presence of scheduled compounds and their metabolites (Konopski et al., 2014).

The accredited (by an internationally recognized accreditation body) laboratories, which have successful performance in at least three consecutive OPCW PTSs are considered capable to be certified as OPCW designated laboratories. The designated laboratories should participate regularly in the OPCW PTSs, where three environmental fortified samples, generally soil, water, wipes, or other nonsoil solid samples, are usually prepared.

11.4 SCHEMA PTS PROVIDER: PROTOCOL OF SCHEMA PTSs

11.4.1 General Schedule: Planning and Annual Program Announcement

The annual SCHEMA PT scheme program is announced once a year at the end of June. The choice of new test samples to be distributed is based on the identification of the market needs, the monitoring of the Rapid Alert System for Food and Feed alert system, deviations in characteristics/composition, and feasibility studies. Targeted PTSs (tailor-made), upon identification of potential participants, should also be developed in order to satisfy particular customers' needs.

The annual SCHEMA's PTS program, which is announced on the website of General Chemical State Laboratory (GCSL) (http://www.gcsl.gr) and delivered by e-mail to the participants, includes details for each PTS, such as the substrate to be dispatched, the analytes to be determined, and the dispatch date.

The interested laboratories are requested to make a registration in the schemes they plan to participate in, so as to facilitate design details (bulk material to be used, number of containers, etc.). The laboratories may enroll in SCHEMA PTSs at any time provided the registration deadline is still open. Otherwise, the laboratory should contact SCHEMA PTS provider to check the possibility of participation in case that the samples have not yet been distributed.

In each SCHEMA PTS, 3 weeks before sample dispatch, the participants receive a participation form containing all the crucial details (description of the test material/quantity/container, date of sample dispatch, deadline for submission of results, issue date of final PTS report, etc.).

The samples are dispatched on the specified dates according to the SCHEMA PTS schedule. In case of changes in the distribution program, the updated one will be communicated by e-mail to all participants.

The period for performing the analysis by the participating laboratories may vary, depending on the nature of the samples and analytes. In general, it does not exceed the period of 4 working weeks.

SCHEMA organizes environmental PTSs on the determination of the following:

- Heavy metals in water, wastewater, and foodstuff
- Pesticide residues in water and foodstuff
- PAHs in water and foodstuff
- Mycotoxins in foodstuff

11.4.2 Organization of SCHEMA Proficiency Testing Schemes

11.4.2.1 Documentation

The main documents associated with the initial phase of SCHEMA PTSs are the following.

- *Participation form*: This form provides information on the type of samples (substrate, etc.) and analytes to be determined, crucial dates for the SCHEMA PTS, and the participation fees.
- *Sample receipt acknowledgement form*: The participants are kindly requested to inform the provider about the condition of the delivered test samples by completing and returning the sample receipt form.
- *Cover letter*: Important details about sample handling, treatment, storage, etc., are mentioned. The unique participating laboratory code for this scheme and the deadline for result submission are also provided.
- *Results reporting form*: A preformatted results' sheet is supplied to enable consistency in the statistical treatment of the results. The deadline for result submission is also included.
- *Questionnaire*: The participants are requested to briefly describe their experimental procedures/applied methods on a preformatted questionnaire, provided with sample dispatch. This information is presented in tabular format in the appendix of each PT report.

11.4.2.2 Application of ISO/IEC 17043:2010 Requirements

Among the PTS provider requirements included in ISO/IEC 17043:2010 (ISO, 2010), the design and planning of the PTS and the evaluation of participant performance play a critical role. In line with these requirements, SCHEMA does not subcontract the following:

- Planning of the PTS
- Evaluation of performance
- Issuance of the final reports

Other activities may be subcontracted, such as preparation of PT items/test samples and homogeneity studies. The criteria for sufficient homogeneity and stability of PT materials/test samples are determined in relation to their impact on the evaluation of participant performance.

11.4.2.3 Preparation of Test Samples: Parameters/Analytes

Test samples should meet the requirements set by the SCHEMA PTS provider. Naturally contaminated samples (real samples) or preparations made of natural samples are preferable for distribution, so that the participants should not be obliged to perform preparations different from those required for natural samples, which could involve additional errors. The evaluation of laboratory performance through the participation in PTSs could provide an effective insight into the accuracy of the routine measurements, only in the case that PT test samples are treated in the same way as routine analysis (e.g., no further replicates and no different process should be performed). In case naturally contaminated samples are not available, analytes could be added or synthetic samples could be prepared.

Test sample preparation (production, packaging, and labeling) is controlled and performed either at the premises of the Chemical Metrology Service, GCSL, Greece, or is performed by subcontractors following SCHEMA's prescribed requirements (analytes and concentration range, storage conditions, containers, etc.). To this end, SCHEMA PTS provider maintains an up-to-date current list of reviewed subcontractors.

Prior to the distribution of the test materials to the participants, the PT items are subjected to homogeneity assessment (when necessary) according to international requirements (e.g., ISO 13528:2005 [ISO, 2005], the IUPAC's IHP [Thompson et al., 2006]). In both cases, 10 samples are randomly selected and analyzed in duplicate in order to evaluate the homogeneity of the batch. Duplicate analyses are also performed close to the deadline for the submission of results in order to evaluate test sample stability (e.g., according to ISO 13528:2005 [ISO, 2005]).

The technical details for sample preparation are recorded by the SCHEMA PTS provider and are included in the final report of the SCHEMA PTS, together with the statistical analysis of sample homogeneity data.

11.4.2.4 Dispatch of Test Samples

Packaging and sample distribution procedures are considered carefully to ensure the stability and characteristics of the distributed test samples. The distribution of the SCHEMA test samples is contracted to a delivery service and operated according to a predefined schedule. The test materials are traceably dispatched to the participating laboratories by a unique courier tracking number.

11.4.2.5 Submission of Results

Participating laboratories may determine the parameters they are interested in by an analytical method of their choice (however, participants are strongly encouraged to apply official and/or validated analytical methods). They are requested to submit their results by a particular deadline by using the provided results form. Results submitted after the aforementioned date are not accepted.

11.4.3 Statistical Evaluation

In order to evaluate the submitted results, statistical analysis based on well-established statistical procedures/protocols according to internationally accepted standards (e.g., ISO 13528:2005 [ISO, 2005], the IUPAC's IHP [Thompson et al., 2006]) is performed.

Several criteria may be employed to evaluate participant performance. In most cases, the deviation of the participant's result from the assigned value is compared to the standard deviation for the proficiency assessment and a z-score for each submitted result is calculated.

11.4.3.1 Identification of Outliers

By using appropriate statistical methodology (Grubbs's [Grubbs, 1969], Dixon's [Rorabacher, 1991], and Hampel's tests), possible outliers within the submitted results are identified. The outliers are flagged in the final report in each PTS. The normality of the distribution of the submitted results is evaluated by the Kolmogorov–Smirnov test. Kernel density plots are constructed and qualitatively evaluated.

11.4.3.2 Determination of the Assigned Value $\hat{X}$

A number of possible approaches for the determination of the assigned value in a test material are available. The choice of the best approach is determined by taking into account: (a) the analyte being examined, (b) its level, and (c) the substrate in which the determination will be performed. The assigned value is clearly specified in each scheme, and its choice is explained in the final PTS report. Assigned value can be estimated as follows.

11.4.3.2.1 Consensus Value from Expert Laboratories

Expert laboratories (or reference laboratories) should have to demonstrate a satisfactory performance in the determination of analytes in the environmental PTS, using their validated methods known to be highly accurate and comparable to methods in general use. Therefore, a consensus value from expert laboratories could be used as the assigned value in the PTS.

11.4.3.2.2 Formulation Value

In case that the environmental test samples to be distributed are prepared by spiking an analyte-free matrix with an analyte of known purity, the formulated concentration could be used as the assigned value in the PTS.

11.4.3.2.3 Reference Value

By using a primary method of chemical measurement (e.g., isotope dilution MS [Yarita et al., 2015]), the reference value determined in the environmental PTS item is traceable to a national or international standard. This value may be ascribed as the assigned value (Hon et al., 2013).

11.4.3.2.4 Certified Reference Value

In case where CRMs are distributed as proficiency test items, the certified reference value may be used as the assigned value for the PTS.

11.4.3.2.5 Consensus of Participants' Results

This is the common practice, appearing in the majority of the PTSs. The use of the consensus relies on estimating the most suitable measure of central tendency from the participants' submitted results.

However, for the best assignment of the assigned value consultation with experts members of the Advisory Committee, previous experience and data from scientific literature are applied, specifically when the number of participants is limited or when the reported results are widely dispersed.

In addition, in SCHEMA's PTSs, the uncertainty of the assigned value (u_x) is estimated and compared to the target value for the standard deviation. In case that the uncertainty of the assigned value (u_x) is significantly different from the target value of the standard deviation, the likelihood that participating laboratories receive a false message for a warning or an action is possible. In such cases, "provisional" z-scores are issued.

11.4.3.3 Target Standard Deviation for the Proficiency Assessment

The target standard deviation for the proficiency assessment used to calculate z-scores is assigned by the PTS coordinator and is not the standard deviation calculated from results submitted by the participating laboratories.

The target standard deviation for the proficiency assessment should be an accurate estimate of the between-laboratory coefficient of variation and should meet the purpose of the PTS.

According to the SCHEMA protocol, the value of the target standard deviation may be calculated from the following:

- Reproducibility or precision of international standards
- Prescribed value (arising from legal or regulatory sources)
- Generalized models (e.g., modified Horwitz equation)
- Participant's results (robust standard deviation is preferred)
- Experience from previous PTSs

11.4.3.4 Evaluation of Laboratory Performance

In SCHEMA PTSs, laboratory performance is evaluated by the calculated z-score for each submitted result.

Therefore, z-scores (a) are used as a fixed reference point for the assessment of laboratory performance at a given time, (b) could serve as a valuable tool in order to monitor trends for an individual laboratory or group of laboratories, and (c) may provide a benchmark for progressive improvement.

11.4.4 Final Report

The final report is issued on the specified date, according to the schedule announced in the participation form, usually within 4 weeks of the deadline for results submission.

The final report contains the following:

- Summary (statistical summary data are also included)
- Introduction with general description and scope of the PTS
- Description of distributed test items—test sample selection, preparation, homogeneity and stability testing, sample distribution data
- Statistical evaluation of the submitted results (definition of terms and statistics used
- Discussion of results/conclusions—(interpretation of graphical data, individual analyte, summaries, participants' comments)
- References (bibliographic references, regulations, etc.)
- Plots of z-scores for each analyte (e.g., Figure 11.3) and kernel density plots (e.g., Figure 11.4), Youden plots, scatter plots

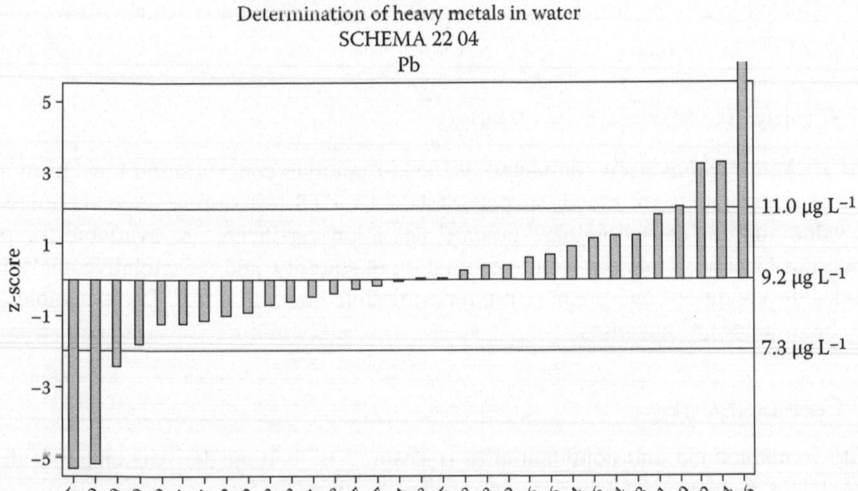

FIGURE 11.3 SCHEMA 22 04 PTS: z-scores of the submitted results of Pb determination in drinking water.

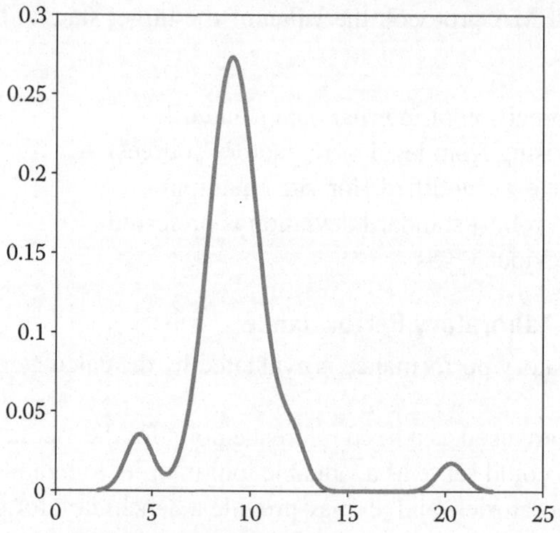

FIGURE 11.4 SCHEMA 22 04 PTS: kernel density plot of the submitted results for Pb determination in drinking water.

- Result presentation—result tables including laboratory codes, submitted results, and corresponding z-scores
- Annexes (details on participants' sample preparation and analytical methods are included)

11.4.5 Feedback

The final report is accompanied by an assessment questionnaire, which participants are kindly requested to complete and return to SCHEMA within a week. In this way, participants have the opportunity to give their feedback to the organization and make any suggestions for future improvements (e.g., expansion of parameters/analytes, different levels, different substrates, new PTSs).

Annual feedback is also provided. Participants are requested to review all the SCHEMA PTSs they participated in and evaluate SCHEMA's annual performance. All these replies are considered and evaluated following weighted analysis, providing information which is used to improve SCHEMA's services.

11.4.6 Surplus Test Materials and Reports

After a SCHEMA PTS has been completed, values for analyte concentration have been assigned, and the final report has been issued, surplus SCHEMA PTS test samples accompanied with an assigned value and an interlaboratory standard deviation certificate are available for purchase. These materials have been retested for the estimation of stability, and their relative analysis certificate includes the conditions and the maximum permissible interval of use. These samples are very reliable to be used for QC purposes.

11.4.7 Confidentiality

Anonymity is maintained and confidentiality is ensured by coding the participants' identity. In SCHEMA PTSs, participating laboratories receive the final report, providing all the submitted results and their evaluation for the individual round but without identifying individual laboratories. To ensure participant confidentiality, a unique laboratory code number is assigned to each participant in the scheme (in the cover letter accompanying the PT item).

Furthermore, data and submitted results reported in the SCHEMA's reports may not be reproduced or published elsewhere without the explicit permission of SCHEMA.

11.5 CONCLUSIONS

Thousands of testing measurements, which affect a significant number of financial, social, and safety decisions, are taken worldwide daily. Consequently, the need for reliable and comparable measurements, under the light of metrological requirements, is self-evident and inevitable. Then principal target of metrology is to ensure that measurements are what they are intended to be.

The crucial role and impact of environmental measurements on decisions regarding consumer and environmental protection and sustainable development have been accepted. For example, decisions characterizing water as potable or food as safe are based on measurements.

It is evident that environmental measurements should be reliable and comparable among different involved parties, and this lies on an international metrological system. Within this infrastructure, ILCs and PTSs constitute a fundamental instrument.

REFERENCES

Albinet, A., Leoz-Garziandia, E., Budzinski, H., Viilenave, E. 2007. Polycyclic aromatic hydrocarbons (PAHs), nitrated PAHs and oxygenated PAHs in ambient air of the Marseilles area (South of France): Concentrations and sources. *Science of the Total Environment* 384: 280–292.

Alma, P. 1993. *Environmental Concerns*. Cambridge University Press, Cambridge.

Annamalai, J., Namasivayam, V. 2015. Endocrine disrupting chemicals in the atmosphere: Their effects on humans and wildlife. *Environment International* 76C: 78–97.

Arditsoglou, A., Voutsa, D. 2008. Determination of phenolic and steroid endocrine disrupting compounds in environmental matrices. *Environmental Science and Pollution Research* 15: 228–236.

Ashley, K., Howe, A.M., Demange, M., Nygren, O. 2003. Sampling and analysis considerations for the determination of hexavalent chromium in workplace air. *Journal of Environmental Monitoring* 5: 707–716.

Ashley, K., Brisson, M.J., Howe, A.M., Bartley, D.L. 2009. Interlaboratory evaluation of a standardized ICP-MS method for the determination of trace beryllium in air filter samples. *Journal of Occupational and Environmental Hygiene* 6(12): 745–750.

Ashley, K., Brisson, M.J., Howe, A.M. 2010. Interlaboratory evaluation of a standardized ICP-MS method for the determination of trace elements in air-filter samples: Preliminary results. *Analytical Methods* 2: 1823–1826.

Ashley, K., Sculman, S.A., Brisson, M.J., Howe, A.M. 2012. Interlaboratory evaluation of trace element determination in workplace air filter samples by ICP-MS. *Journal of Environmental Monitoring* 14: 360–367.

Aurelien Pitois, A. 2014. IAEA environment laboratories: Checking for radionuclides in dairy food products, ALMERA training course demonstrates state-of-the-art method for detecting radiostrontium in milk. http://www.iaea.org/newscenter/news/checking-radionuclides-dairy-food-products (accessed December 19, 2014).

Bao, L., Bao, Z., Zhang, Y., Liang, C., Lu, N. et al. 2009. Aflatoxin testing in peanuts: A proficiency assessment scheme for Chinese analytic laboratories. *Journal of AOAC International* 92: 481–486.

Baquero, F., Martínez, J.L., Cantón, R. 2008. Antibiotics and antibiotic resistance in water environments. *Current Opinion in Biotechnology* 19: 260–265.

Belli, M., de Zorzi, P., Sansone, U., Shakhashiro, A., Gondin da Fonseca, A. et al. 2009. A soil sampling intercomparison exercise for the ALMERA network. *Journal of Environmental Radioactivity* 100: 982–987.

Bercaru, O., Gawlik, B.M., Ulberth, F., Vandecasteele, C. 2003. Reference materials for the monitoring of the aquatic environment—A review with special emphasis on organic priority pollutants. *Journal of Environmental Monitoring* 5: 697–705.

Boffetta, P., Nyberg, F. 2003. Contribution of environmental factors to cancer risk. *British Medical Bulletin* 68: 71–94.

Brack, W., Klamer, H.J., López de Alda, M., Barceló, D. 2007. Effect-directed analysis of key toxicants in European river basins: A review. *Environmental Science and Pollution Research* 14: 30–38.

Brera, C., Debegnach, F., Minardi, V., Pannunzi, E., De Santis, B. et al. 2007. Immunoaffinity column cleanup with liquid chromatography for determination of aflatoxin B1 in corn samples: Interlaboratory study. *Journal of AOAC International* 90: 765–772.

Bu, Q., Wang, B., Huang, J., Deng, S., Yu, G. 2013. Pharmaceuticals and personal care products in the aquatic environment in China: A review. *Journal of Hazardous Materials* 262: 189–211.

Buesseler, K.O., Jayne, S.R., Fisher, N.S., Rypina, I.I., Baumann, H. et al. 2012. Fukushima-derived radionuclides in the ocean and biota off Japan. *Proceedings of the National Academy of Sciences of the United States of America* 109: 5984–5988.

Cao, G., Orrù, R. 2014. *Current Environmental Issues and Challenges*. Springer, New York.

Castiglioni, S., Zuccato, E., Chiabrando, C., Fanelli, R., Bagnati, R. 2008. Mass spectrometric analysis of illicit drugs in wastewater and surface water. *Mass Spectrometry Reviews* 27: 378–394.

Caurant, A., Lalère, B., Schbath, M.C., Stumpf, C., Sutour, C. et al. 2010. A new certified reference material for benzene measurement in air on a sorbent tube: Development and proficiency testing. *Analytical and Bioanalytical Chemistry* 398: 2265–2277.

Chai, C., Cheng, H., Ge, W., Ma, D., Shi, Y. 2014. Phthalic acid esters in soils from vegetable greenhouses in Shandong Peninsula, East China. *PLoS One* 9: e95701.

Chang, K.F., Fang, G.C., Chen, J.C., Wu, Y.S. 2006. Atmospheric polycyclic aromatic hydrocarbons (PAHs) in Asia: A review from 1999 to 2004. *Environmental Pollution* 142: 388–396.

Chen, Y., Reddy, R.M., McLaughlin, M.A., Al-Taher, F., Yettella, R.R. et al. 2014. Method validation and proficiency testing for determination of total arsenic in apple juice by inductively coupled plasma/mass spectrometry. *Journal of AOAC International* 97: 1143–1150.

Chopra, A.K., Sharma, M.K., Chamoli, S. 2011. Bioaccumulation of organochlorine pesticides in aquatic system—An overview. *Environmental Monitoring and Assessment* 173: 905–916.

Cordeiro, F., Calderón, J., Gonçalves, S., Lourenço, M.H., Robouch, P. et al. 2014. IMEP-115: Determination of methylmercury in seafood by elemental mercury analysis: Collaborative study. *Journal of AOAC International* 97: 593–597.

Cordeiro, F., Llorente-Mirandes, T., López-Sánchez, J.F., Rubio, R., Sánchez Agullo, A. et al. 2015. Determination of total cadmium, lead, arsenic, mercury and inorganic arsenic in mushrooms: Outcome of IMEP-116 and IMEP-39. *Food Additives & Contaminants: Part A: Chemistry, Analysis, Control, Exposure & Risk Assessment* 32: 54–67.

Devier, M.H., Augagneur, S., Budzinski, H., Le Menach, K., Mora, P. et al. 2005. One-year monitoring survey of organic compounds (PAHs, PCBs, TBT), heavy metals and biomarkers in blue mussels from the Arcachon Bay, France. *Journal of Environmental Monitoring* 7: 224–240.

D'Hollander, W., de Voogt, P., De Coen, W., Bervoets, L. 2010. Perfluorinated substances in human food and other sources of human exposure. *Reviews of Environmental Contamination and Toxicology* 208: 179–215.

Duewer, D.L., Lippa, K.A., Long, S.E., Murphy, K.E., Sharpless, K.E. et al. 2009. Demonstrating the comparability of certified reference materials. *Analytical and Bioanalytical Chemistry* 395: 155–169.

Eller, P.M., Feng, H.A., Song, R.S., Key-Schwartz, R.J., Esche, C.A., Groff, J.H. 1999. Proficiency Analytical Testing (PAT) silica variability 1990–1998. *American Industrial Hygiene Association Journal* 60: 533–539.

Evgenidou, E.N., Konstantinou, I.K., Lambropoulou, D.A. 2015. Occurrence and removal of transformation products of PPCPs and illicit drugs in wastewaters: A review. *Science of the Total Environment* 505C: 905–926.

Fiamegos, Y., Cordeiro, F., Robouch, P., Emteborg, H., Charoud-Got, J. et al. 2014. Feed premix: A difficult matrix for the accurate determination of trace elements—The outcome of IMEP-114 and IMEP-36. *Food Additives & Contaminants: Part A: Chemistry, Analysis, Control, Exposure & Risk Assessment* 31: 63–72.

Freville, L., Moulut, J.C., Grzebyk, M., Kauffer, E. 2010. Producing samples for the organization of proficiency tests: Study of the homogeneity of replicas produced from two atmosphere generation systems. *Annals of Occupational Hygiene* 54: 659–670.

Fuerhacker, M. 2009. EU Water Framework Directive and Stockholm Convention: Can we reach the targets for priority substances and persistent organic pollutants? *Environmental Science and Pollution Research* 1: S92–S97.

Galea, K.S., Searl, A., Sánchez-Jiménez, A., Woldbæk, T., Halgard, K. et al. 2012. Oil mist and vapour concentrations from drilling fluids: Inter- and intra-laboratory comparison of chemical analyses. *Annals of Occupational Hygiene* 56: 61–69.

Ganoulis, J. 2009. *Risk Analysis of Water Pollution*, 2nd ed. Wiley, Weinheim.

Garí, M., Grimalt, J.O. 2010. Use of proficiency testing materials for the calculation of detection and quantification limits in the analysis of organochlorine compounds in human serum. *Analytical and Bioanalytical Chemistry* 397: 1383–1387.

González-Castanedo, Y., Sanchez-Rodas, D., Sánchez de la Campa, A.M., Pandolfi, M., Alastuey, A. et al. 2015. Arsenic species in atmospheric particulate matter as tracer of the air quality of Doñana Natural Park (SW Spain). *Chemosphere* 119: 1296–1303.

Gratz, L.D., Bagley, S.T., Leddy, D.G., Johnson, J.H., Chiu, C. et al. 2000. Interlaboratory comparison of HPLC-fluorescence detection and GC/MS: Analysis of PAH compounds present in diesel exhaust. *Journal of Hazardous Materials* 74: 37–46.

Gray, J.S. 2002. Biomagnification in marine systems: The perspective of an ecologist. *Marine Pollution Bulletin* 45: 46–52.

Grubbs, F. 1969. Procedures for detecting outlying observations in samples. *Technometrics* 11: 1–21.

Grzebyk, M., Kauffer, E., Fréville, L. 2005. Proposal to adapt the Workplace Analysis Scheme for Proficiency (WASP) programme to fibre counting tests. *Annals of Occupational Hygiene* 49: 325–334.

Guéguen, M., Amiard, J.C., Arnich, N., Badot, P.M., Claisse, D. et al. 2011. Shellfish and residual chemical contaminants: Hazards, monitoring, and health risk assessment along French coasts. *Reviews of Environmental Contamination and Toxicology* 213: 55–111.

Harper, M., Slaven, J.E., Pang, T.W. 2009. Continued participation in an asbestos fiber-counting proficiency test with relocatable grid slides. *Journal of Environmental Monitoring* 11: 434–438.

Harvey, B.W., Winstead, W.G. 2008. Analysis of the analytical performance of laboratories participating in two major U.S. bulk asbestos proficiency testing programs. *Journal of Occupational and Environmental Hygiene* 5: 455–462.

Hashim, D., Boffetta, P. 2014. Occupational and environmental exposures and cancers in developing countries. *Annals of Global Health* 80: 393–411.

Hela, D.G., Lambropoulou, D.A., Konstantinou, I.K., Albanis, T.A. 2005. Environmental monitoring and ecological risk assessment for pesticide contamination and effects in Lake Pamvotis, northwestern Greece. *Environmental Toxicology and Chemistry* 24: 1548–1556.

Hon, P.Y., Wan, M.Y., Sin, D.W., Wong, Y.C. 2013. Performance evaluation of polycyclic aromatic hydrocarbons analysis in sediment from proficiency testing with metrological reference values. *Analytical and Bioanalytical Chemistry* 405: 7241–7250.

Horwitz, W., Jackson, T., Chirtel, S.J. 2001. Examination of proficiency and control recovery data from analyses for pesticide residues in food: Sources of variability. *Journal of AOAC International* 84: 919–935.

HSL. 2014. AIR PT Scheme (Formerly WASP & STACKS). http://www.hsl.gov.uk/centres-of-excellence/proficiency-testing-schemes/wasp.aspx (accessed November 4, 2014).

IAEA. 2011. ALMERA Proficiency Test: Determination of natural and artificial radionuclides in soil and water (IAEA-TEL-2011-04). http://www-pub.iaea.org/MTCD/Publications/PDF/IAEA-AQ-32_web.pdf (accessed December 19, 2014).

IAEA. 2012. World-wide proficiency test, IAEA-TEL-2012-03, on the determination of anthropogenic gamma-emitters in water in the presence of natural background, of radio-caesium in hay and of both natural and anthropogenic radionuclides in soil samples. http://nucleus.iaea.org/rpst/ReferenceProducts/Proficiency_Tests/IAEA-TEL-2012-03/index.htm (accessed December 19 2014).

IAEA. 2014. World Wide Open Proficiency Test, IAEA-TEL-2014-03, on the determination of gamma, beta and alpha emitting radionuclides in water, biota and sediment samples. http://nucleus.iaea.org/rpst/ReferenceProducts/Proficiency_Tests/IAEA-TEL-2014-03/index.htm (accessed December 19, 2014).

Inagaki, K., Takatsu, A., Kuroiwa, T., Nakama, A., Eyama, S. et al. 2004. Certified sediment reference materials for trace element analysis from the National Metrology Institute of Japan (NMIJ). *Analytical and Bioanalytical Chemistry* 378: 1271–1276.

ISO. 2005. ISO 13528:2005: Statistical Methods for Use in Proficiency Testing by Interlaboratory Comparisons. ISO, Geneva.

ISO. 2010. ISO/IEC 17043:2010: Conformity Assessment—General Requirements for Proficiency Testing. ISO, Geneva.

Jacob, D.J. 1999. *Introduction to Atmospheric Chemistry*. Princeton University Press, Princeton.

Jakopič, R., Sturm, M., Kraiem, M., Richter, S., Aregbe, Y. 2013. Certified reference materials and reference methods for nuclear safeguards and security. *Journal of Environmental Radioactivity* 125: 17–22.

Jobbágy, V., Merešová, J., Wätjen, U. 2014. Critical remarks on gross alpha/beta activity analysis in drinking waters: Conclusions from a European interlaboratory comparison. *Applied Radiation and Isotopes* 87: 429–434.

Jones, J.C. 2008. *Atmospheric Pollution*. J.C. Jones & Ventus. ISBN 978-87-7681-416-8.

Julshamn, K., Maage, A., Norli, H.S., Grobecker, K.H., Jorhem, L. et al. 2007. Determination of arsenic, cadmium, mercury, and lead by inductively coupled plasma/mass spectrometry in foods after pressure digestion: NMKL interlaboratory study. *Journal of AOAC International* 90: 844–856.

Julshamn, K., Maage, A., Norli, H.S., Grobecker, K.H., Jorhem, L. et al. 2013. Determination of arsenic, cadmium, mercury, and lead in foods by pressure digestion and inductively coupled plasma/mass spectrometry: First action 2013.06. *Journal of AOAC International* 96: 1101–1112.

Karaouzas, I., Lambropoulou, D.A., Skoulikidis, N.T., Albanis, T.A. 2011. Levels, sources and spatiotemporal variation of nutrients and micropollutants in small streams of a Mediterranean River basin. *Journal of Environmental Monitoring* 13: 3064–3074.

Kemp, D.D. 2004. *Exploring Environmental Issues: An Integrated Approach*. Psychology Press, Oxford.

Klecka, G., Persoon, C., Currie, R. 2010. Chemicals of emerging concern in the Great Lakes Basin: An analysis of environmental exposures. *Reviews of Environmental Contamination and Toxicology* 207: 1–93.

Kong, M.F., Chan, S., Wong, Y.C. 2008. Proficiency testing program for the determination of total arsenic, cadmium, and lead in seawater shrimp. *Journal of AOAC International* 91: 858–864.

Konopski, L., Liu, P., Wuryani, W., Sliwakowski, M. 2014. OPCW Proficiency Test: A practical approach also for interlaboratory test on the detection and identification of pesticides in environmental matrices. *Scientific World Journal* 2014: 542357.

Konstantinou, I.K., Hela, D.G., Albanis, T.A. 2006. The status of pesticide pollution in surface waters (rivers and lakes) of Greece: Part I—Review on occurrence and levels. *Environmental Pollution* 141: 555–570.

Kosma, C.I., Lambropoulou, D.A., Albanis, T.A. 2014. Investigation of PPCPs in wastewater treatment plants in Greece: Occurrence, removal and environmental risk assessment. *Science of the Total Environment* 466–467: 4214–38.

Kuch, H.M., Ballschmiter, K. 2001. Determination of endocrine-disrupting phenolic compounds and estrogens in surface and drinking water by HRGC-(NCI)-MS in the picogram per liter range. *Environmental Science & Technology* 35: 3201–3206.

Landrigan, P.J., Fuller, R. 2014. Environmental pollution and occupational health in a changing world. *Annals of Global Health* 80: 245–246.

Langlois, E., Boulet, A., Kauffer, E. 2008. ALASCA proficiency testing scheme for occupational hygiene laboratories. *Journal of Environmental Monitoring* 10: 1460–1466.

Lewtas, J. 2007. Air pollution combustion emissions: Characterization of causative agents and mechanisms associated with cancer, reproductive, and cardiovascular effects. *Mutation Research* 636: 95–133.

Lin, Y., Qiu, X., Ma, Y, Ma, J., Zheng, M. et al. 2014. Concentrations and spatial distribution of polycyclic aromatic hydrocarbons (PAHs) and nitrated PAHs (NPAHs) in the atmosphere of North China, and the transformation from PAHs to NPAHs. *Environmental Pollution* 196C: 164–170.

Liu, G., Niu, Z., Van Niekerk, D., Xue, J., Zheng, L. 2008. Polycyclic aromatic hydrocarbons (PAHs) from coal combustion: Emissions, analysis, and toxicology. *Reviews of Environmental Contamination and Toxicology* 192: 1–28.

Martinez, J.L. 2009. Environmental pollution by antibiotics and by antibiotic resistance determinants. *Environmental Pollution* 157: 2893–2902.

Matissek, R., Raters, M., van Haren, W., Fledderus, K. 2010. Determination of ochratoxin A in liquorice products using HPLC based analytical methods—Part I: Proficiency test of methods commonly used by the confectionary industry. *Mycotoxin Research* 26: 93–99.

McClure, F.D., Lee, J.K. 2012. Variances and uncertainties of the sample laboratory-to-laboratory variance (S(L)2) and standard deviation (S(L)) associated with an interlaboratory study. *Journal of AOAC International* 95: 244–251.

McClure, F.D., Lee, J.K. 2014. Use of prediction methods to assess laboratory bias and mean values associated with an interlaboratory study for method validation and/or proficiency testing. *Journal of AOAC International* 97: 624–629.

Medina-Pastor, P., Mezcua, M., Rodríguez-Torreblanca, C., Fernández-Alba, A.R. 2010. Laboratory assessment by combined z score values in proficiency tests: Experience gained through the European Union proficiency tests for pesticide residues in fruits and vegetables. *Analytical and Bioanalytical Chemistry* 397: 3061–3070.

Medina-Pastor, P., Valverde, A., Pihlstrom, T., Masselter, S., Gamon, M. et al. 2011. Comparative study of the main top–down approaches for the estimation of measurement uncertainty in multiresidue analysis of pesticides in fruits and vegetables. *Journal of Agricultural and Food Chemistry* 59: 7609–7619.

Mehrotra, A.K., Singhal, R.K. 2000. Environmental issues and waste management in energy and mineral production. In *Proceedings of the Sixth International Symposium, Calgary, Alberta, Canada*. CRC Pres, Boca Raton, FL 49–55, 69–73.

Melnyk, A., Kuklińska, K., Wolska, L., Namieśnik, J. 2014. Chemical pollution and toxicity of water samples from stream receiving leachate from controlled municipal solid waste (MSW) landfill. *Environmental Research* 135: 253–261.

Mendelsohn, R. 2001. *Global Warming and the American Economy—A Regional Assessment of Climate Change Impacts*. Edward Elgar, Cheltenham, UK.

Miyashita, S., Inagaki, K., Naito, S., Shindoh, K., Narukawa, T. et al. 2013. Identification of possible technical problems in determination of the major inorganic constituents of brown-rice flour by evaluating proficiency test results. *Analytical and Bioanalytical Chemistry* 405: 8347–8362.

Nanita, S.C., Stewart, D.J., Rodgers, C.A., Tesfai, E., Schwarz, T. et al. 2013. Analytical method and interlaboratory study for the quantitation of aminocyclopyrachlor residues in vegetation by liquid chromatography/tandem mass spectrometry. *Journal of AOAC International* 96: 1473–1481.

Oatts, T.J., Hicks, C.E., Adams, A.R., Brisson, M.J., Youmans-McDonald, L.D. et al. 2012. Preparation, certification and interlaboratory analysis of workplace air filters spikes with high-fired beryllium oxide. *Journal of Environmental Monitoring* 14: 391–401.

OSHA. 1998. The Federal Occupational Safety and Health Administration, Respiratory Protection Standard (29 CFR Part 1910.134) ("Respiratory Protection," Code of Federal Regulations Title 29, Part 1910.134, 1998. Appendix A).

Otake, T., Yarita, T., Aoyagi, Y., Numata, M., Takatsu, A. 2014. Evaluation of the performance of 57 Japanese participating laboratories by two types of z-scores in proficiency test for the quantification of pesticide residues in brown rice. *Analytical and Bioanalytical Chemistry* 406: 7337–7344.

Pineau, A., Otz, J., Guillard, O., Fauconneau, B., Dumont, G. et al. 2014. Interlaboratory comparison study on lead in blood, in external quality assessment schemes since 1996: A progress report. *Toxicology Mechanisms and Methods*. 24: 396–403.

Post, G.B., Cohn, P.D., Cooper, K.R. 2012. Perfluorooctanoic acid (PFOA), an emerging drinking water contaminant: A critical review of recent literature. *Environmental Research* 116: 93–117.

Prohaska, T., Quétel, C.R., Hennessy, C., Liesegang, D., Papadakis, I. et al. 2000. SI-traceable certification of Cu, Cr, Cd and Pb in sediment and fly ash candidate reference materials. *Journal of Environmental Monitoring* 2: 613–620.

Quevauviller, P. 2005. Traceability of environmental chemical analyses: Can theory match practice? *Analytical and Bioanalytical Chemistry* 382: 1800–1807.

Ramsey, M.H., Geelhoed, B., Wood, R., Damant, A.P. 2011. Improved evaluation of measurement uncertainty from sampling by inclusion of between-sampler bias using sampling proficiency testing. *Analyst* 136: 1313–1321.

Raposo, F., de la Rubia, M.A., Borja, R., Alaiz, M., Beltrán, J. et al. 2009. An interlaboratory study as useful tool for proficiency testing of chemical oxygen demand measurements using solid substrates and liquid samples with high suspended solid content. *Talanta* 80. 329–337.

Ritter, L., Solomon, K., Sibley, P., Hall, K., Keen, P. et al. 2002. Sources, pathways, and relative risks of contaminants in surface water and groundwater: A perspective prepared for the Walkerton inquiry. *Journal of Toxicology and Environmental Health A* 65: 1–142.

Rizzo, L., Manaia, C., Merlin, C., Schwartz, T., Dagot, C. et al. 2013. Urban wastewater treatment plants as hotspots for antibiotic resistant bacteria and genes spread into the environment: A review. *Science of the Total Environment* 447: 345–360.

Rodriguez-Mozaz, S., Chamorro, S., Marti, E., Huerta, B., Gros, M. et al. 2014. Occurrence of antibiotics and antibiotic resistance genes in hospital and urban wastewaters and their impact on the receiving river. *Water Research* 69C: 234–242.

Romieu, I., Gouveia, N., Cifuentes, L.A., de Leon, A.P., Junger, W. et al. 2012. HEI Health Review Committee: Multicity study of air pollution and mortality in Latin America (the ESCALA study). *Research Report (Health Effects Institute)* 171: 5–86.

Rorabacher, D.B. 1991. Statistical treatment for rejection of deviant values: Critical values of Dixon's Q parameter and related subrange ratios at the 95% confidence level. *Analytical Chemistry* 63: 139–146.

Sack, C., Smoker, M., Chamkasem, N., Thompson, R., Satterfield, G. et al. 2011. Collaborative validation of the QuEChERS procedure for the determination of pesticides in food by LC-MS/MS. *Journal of Agricultural and Food Chemistry* 59: 6383–6411.

Sanyal, D., Rani, A. 2009. Proficiency test for chemical laboratories for the analysis of a pesticide in a formulated product: Interlaboratory study. *Journal of AOAC International* 92: 271–278.

Schlecht, P.C., Shulman, S.A. 1986. Performance of asbestos fiber counting laboratories in the NIOSH proficiency analytical testing (PAT) program. *American Industrial Hygiene Association Journal* 47: 259–269.

Schlecht, P.C., Shulman, S.A. 1995. Phase contrast microscopy asbestos fiber counting performance in the Proficiency Analytical Testing program. *American Industrial Hygiene Association Journal* 56: 480–489.

Schlecht, P.C., Song, R., Groff, J.H., Feng, H.A., Esche, C.A. 1997. Interlaboratory and intralaboratory variabilities in the Environmental Lead Proficiency Analytical Testing (ELPAT) Program. *American Industrial Hygiene Association Journal* 58: 779–786.

Schlecht, P., O'Connor, P.F., Key-Schwartz, R., Lunsford, A., Gagnon, Y. 2011. *NIOSH Manual of Analytical Methods 5th Edition*—New resources and direction. *Journal of Occupational and Environmental Hygiene* 8: D59–D62.

Schwela, D. 2000. Air pollution and health in urban areas. *Reviews on Environmental Health* 15: 13–42.

Senseman, S.A., Mueller, T.C., Riley, M.B., Wauchope, R.D., Clegg, C. et al. 2003. Interlaboratory comparison of extraction efficiency of pesticides from surface and laboratory water using solid-phase extraction disks. *Journal of Agricultural and Food Chemistry* 51: 3748–3752.

Senyuva, H.Z., Gilbert, J., Stroka, J., Biselli, S., De Girolamo, A. et al. 2010. Determination of fumonisins B1 and B2 in corn by LC/MS with immunoaffinity column cleanup: Interlaboratory study. *Journal of AOAC International* 93: 611–721.

Shakhashiro, A. 2008. The IAEA's ALMERA Network proficiency test on the determination of gamma-emitting radionuclides: A test of results comparability. *Applied Radiation and Isotopes* 66: 1722–1725.

Shulman, S.A., Groff, J.H., Abell, M.T. 1992. Performance of laboratories measuring silica in the Proficiency Analytical Testing program. *American Industrial Hygiene Association Journal* 53: 49–56.

Škrbić, B., Héberger, K., Durišić-Mladenović, N. 2013. Comparison of multianalyte proficiency test results by sum of ranking differences, principal component analysis, and hierarchical cluster analysis. *Analytical and Bioanalytical Chemistry* 405: 8363–8375.

Skurikhin, I.M. 1993. New metrological characteristics of analytical methods of analysis used for safety control of food and environmental materials. *Journal of AOAC International* 76: 262–267.

Smith, S.R. 2009. A critical review of the bioavailability and impacts of heavy metals in municipal solid waste composts compared to sewage sludge. *Environment International* 35: 142–156.

Solfrizzo, M., De Girolamo, A., Gambacorta, L., Visconti, A., Stroka, J. et al. 2011. Determination of fumonisins B1 and B2 in corn-based foods for infants and young children by LC with immunoaffinity column cleanup: Interlaboratory validation study. *Journal of AOAC International* 94: 900–908.

Stacey, P.R. 2006. The Performance of laboratories analyzing heavy metals in the Workplace Analysis Scheme for Proficiency (WASP). Annals of Occupational Hygiene 50: 417–425.

Stacey, P.R., Wright, M.D. 2001. A proficiency testing scheme for aromatic hydrocarbons in air by the manual thermal desorption-GC method: A comparison of laboratory performance with the uncertainty requirements of the European Union Ambient Air Directive. *Journal of Environmental Monitoring* 3: 425–431.

Stacey, P.R., Tylee, B., Bard, D., Atkinson, R. 2003. The performance of laboratories analysing alpha-quartz in the Workplace Analysis Scheme for Proficiency (WASP). *Annals of Occupational Hygiene* 47: 269–277.

Stacey, P.R., Butler, O. 2008. Performance of laboratories analysing welding fume on filter samples: Results from the WASP proficiency testing scheme. *Annals of Occupational Hygiene* 52: 287–295.

Stackelberg, P.E., Furlong, E.T., Meyer, M.T., Zaugg, S.D., Henderson, A.K. et al. 2004. Persistence of pharmaceutical compounds and other organic wastewater contaminants in a conventional drinking-water-treatment plant. *Science of the Total Environment* 329: 99–113.

Stationery Office Ltd., Health and Safety, The Control of Lead Regulations, 2002, No. 2676.

Steinhauser, G., Brandl, A., Johnson, T.E. 2014. Comparison of the Chernobyl and Fukushima nuclear accidents: A review of the environmental impacts. *Science of the Total Environment* 470–471: 800–817.

Suedel, B.C., Boraczek, J.A., Peddicord, R.K., Clifford, P.A., Dillon, T.M. 1994. Trophic transfer and biomagnification potential of contaminants in aquatic ecosystems. *Reviews of Environmental Contamination and Toxicology* 136: 21–89.

Sun, Y., Huang, H., Sun, Y., Wang, C., Shi, X.L. et al. 2013. Ecological risk of estrogenic endocrine disrupting chemicals in sewage plant effluent and reclaimed water. *Environmental Pollution* 180: 339–344.

Sykes, M., Thompson, M., Reynolds, S. 2013. Pesticide residues in food-based proficiency test materials, spiking values versus consensus assigned values. *Journal of Agricultural and Food Chemistry* 61: 4205–4209.

Tanimoto, H., Mukai, H., Sawa, Y., Matsueda, H., Yonemura, S. et al. 2007. Direct assessment of international consistency of standards for ground-level ozone: Strategy and implementation toward metrological traceability network in Asia. *Journal of Environmental Monitoring* 9: 1183–1193.

Thompson, M., Ellison, S.L.R., Wood, R. 2006. The International Harmonized Protocol for the Proficiency Testing of analytical chemistry laboratories (IUPAC Technical Report). *Pure and Applied Chemistry* 78: 145–196.

Thornes, J.B., Wainwright, J. 2004. *Environmental Issues in the Mediterranean: Processes and Perspectives from the Past and Present*. Routledge, London.

Truyens, J., Stefaniak, E.A., Aregbe, Y. 2013. NUSIMEP-7: Uranium isotope amount ratios in uranium particles. *Journal of Environmental Radioactivity* 125: 50–55.

Turk, G.C., Yu, L.L., Salit, M.L., Guthrie, W.F. 2001. Using inductively coupled plasma-mass spectrometry for calibration transfer between environmental CRMs. *Fresenius Journal of Analytical Chemistry* 370: 259–263.

Uhlig, S., Eichler, S., Gowik, P. 2013. Reliability of precision data obtained from interlaboratory studies. *Journal of AOAC International* 96: 466–470.

Ulberth, F. 2006. Certified reference materials for inorganic and organic contaminants in environmental matrices. *Analytical and Bioanalytical Chemistry* 386: 1121–1136.

Vander Heyden, Y., Smeyers-Verbeke, J. 2007. Set-up and evaluation of interlaboratory studies. *Journal of Chromatography A* 1158: 158–167.

Vanderford, B.J., Drewes, J.E., Eaton, A., Guo, Y.C., Haghani, A. et al. 2014. Results of an interlaboratory comparison of analytical methods for contaminants of emerging concern in water. *Analytical Chemistry* 86: 774–782.

Violante, F.G., Bastos, L.H., Cardoso, M.H., Rodrigues, J.M., Gouvêa, A.V. et al. 2009. Proficiency testing for the determination of pesticides in mango pulp: A view of the employed chromatographic techniques and the evaluation of laboratories' performance. *Journal of Chromatographic Science* 47: 833–839.

Weber, R., Gaus, C., Tysklind, M., Johnston, P., Forter, M. et al. 2008. Dioxin- and POP-contaminated sites—Contemporary and future relevance and challenges: Overview on background, aims and scope of the series. *Environmental Science and Pollution Research* 15: 363–393.

WHO. 2014. 7 million premature deaths annually linked to air pollution. http://www.who.int/mediacentre/news/releases/2014/air-pollution/en/ (accessed December 12, 2014).

Wise, S.A., Poster, D.L., Kucklick, J.R., Keller, J.M., Vanderpol, S.S. et al. 2006. Standard reference materials (SRMs) for determination of organic contaminants in environmental samples. *Analytical and Bioanalytical Chemistry* 386: 1153–1190.

Wong, J., Hao, C., Zhang, K., Yang, P., Banerjee, K. et al. 2010. Development and interlaboratory validation of a QuEChERS-based liquid chromatography-tandem mass spectrometry method for multiresidue pesticide analysis. *Journal of Agricultural and Food Chemistry* 58: 5897–5903.

Wu, Y.S., Fang, G.C., Lee, W.J., Lee, J.F., Chang, C.C. et al. 2007. A review of atmospheric fine particulate matter and its associated trace metal pollutants in Asian countries during the period 1995–2005. *Journal of Hazardous Materials* 143: 511–515.

Yarita, T., Otake, T., Aoyagi, Y., Kuroiwa, T., Numata, M. et al. 2015. Proficiency testing for determination of pesticide residues in soybean: Comparison of assigned values from participants' results and isotope-dilution mass spectrometric determination. *Talanta* 132: 269–277.

Yoshinari, T., Tanaka, T., Ishikuro, E., Horie, M., Nagayama, T. et al. 2013. Inter-laboratory study of an LC-MS/MS method for simultaneous determination of fumonisin B1, B2 and B3 in corn. *Shokuhin Eiseigaku Zasshi* 54: 266–276.

Zareitalabad, P., Siemens, J., Hamer, M., Amelung, W. 2013. Perfluorooctanoic acid (PFOA) and perfluorooctanesulfonic acid (PFOS) in surface waters, sediments, soils and wastewater—A review on concentrations and distribution coefficients. *Chemosphere* 91: 725–732.

Zhang, Ch. 2007. *Fundamentals of Environmental Sampling and Analysis*. Wiley, Hoboken, NJ.

Zhang, Y., Wang, P., Wang, L., Sun, G., Zhao, J. et al. 2015. The influence of facility agriculture production on phthalate esters distribution in black soils of northeast China. *Science of the Total Environment* 506–507: 118–125.

Zheng, J., Tagami, K., Uchida, S. 2013. Release of plutonium isotopes into the environment from the Fukushima Daiichi Nuclear Power Plant accident: What is known and what needs to be known. *Environmental Science & Technology* 47: 9584–9595.

12 Reference Methods for the Measurement of Pollutants in Environmental Matrices

Leo M. L. Nollet

CONTENTS

12.1 UNITED STATES

12.1.1 Air

The EPA (US EPA, n.d(a)) is continuously working on the development of reference and equivalent methods for measuring key air pollutants.

The EPA operates a nationwide air monitoring network to measure six primary air pollutants, including carbon monoxide, lead, sulfur dioxide, ozone, nitrogen dioxide, and PM (PM_{10} and $PM_{2.5}$). One mission of the EPA is to protect human health and the environment.

The Clean Air Act (US EPA, n.d.(b)) requires the EPA to set the National Ambient Air Quality Standards (NAAQSs) (US EPA, n.d.(d)) (Table 12.1) for these six pollutants.

TABLE 12.1
National Ambient Air Quality Standards (NAAQS)

Pollutant (Final Rule Cite)		Primary/Secondary	Averaging Time	Level	Form
Carbon monoxide (76 FR 54294, August 31, 2011) (Federal Register, n.d.(a))		Primary	8 h	9 ppm	Not to be exceeded more than once per year
			1 h	35 ppm	
Lead (73 FR 66964, November 12, 2008) (Federal Register, n.d.(c))		Primary and secondary	Rolling 3-month average	0.15 μg m^{-3}[a]	Not to be exceeded
Nitrogen dioxide (75 FR 6474, February 9, 2010) (US EPA, n.d.(e)); (61 FR 52852, October 8, 1996) (Federal Register, n.d.(d))		Primary	1 h	100 ppb	98th percentile of 1 h daily maximum concentrations, averaged over 3 years
		Primary and secondary	Annual	53 ppb[b]	Annual mean
Ozone (73 FR 16436, March 27, 2008) (Federal Register, n.d.(e))		Primary and secondary	8 h	0.075 ppm[c]	Annual fourth highest daily maximum 8 h concentration, averaged over 3 years
Particle pollution (December 14, 2012) (US EPA, n.d.(k))	$PM_{2.5}$	Primary	Annual	12 μg m^{-3}	Annual mean, averaged over 3 years
		Secondary	Annual	15 μg m^{-3}	Annual mean, averaged over 3 years
		Primary and secondary	24 h	35 μg m^{-3}	98th percentile, averaged over 3 years
	PM_{10}	Primary and secondary	24 h	150 μg m^{-3}	Not to be exceeded more than once per year on average over 3 years
Sulfur dioxide (75 FR 35520, June 22, 2010); (38 FR 25678, September 14, 1973) (Federal Register, n.d.(c))		Primary	1 h	75 ppb[d]	99th percentile of 1 h daily maximum concentrations, averaged over 3 years
		Secondary	3 h	0.5 ppm	Not to be exceeded more than once per year

Source: US EPA, NAAQS table, https://www.epa.gov/criteria-air-pollutants/naaqs-table, EPA, Washington, DC, (n.d.(d)).
Note: Data as of October 2011.

[a] Final rule signed October 15, 2008: The 1978 lead standard (1.5 μg m−3 as a quarterly average) remains in effect until 1 year after an area is designated for the 2008 standard, except that in areas designated nonattainment for the 1978 standard, the 1978 standard remains in effect until implementation plans to attain or maintain the 2008 standard are approved.

[b] The official level of the annual NO_2 standard is 0.053 ppm, equal to 53 ppb, which is shown here for the purpose of clearer comparison to the 1 h standard.

[c] Final rule signed March 12, 2008: The 1997 ozone standard (0.08 ppm, annual fourth highest daily maximum 8 h concentration, averaged over 3 years) and related implementation rules remain in place. In 1997, EPA revoked the 1 h ozone standard (0.12 ppm, not to be exceeded more than once per year) in all areas, although some areas have continued obligations under that standard (anti-backsliding). The 1 h ozone standard is attained when the expected number of days per calendar year with maximum hourly average concentrations above 0.12 ppm is less than or equal to 1.

[d] Final rule signed June 2, 2010: The 1971 annual and 24 h SO_2 standards were revoked in that same rulemaking. However, these standards remain in effect until 1 year after an area is designated for the 2010 standard, except in areas designated for nonattainment for the 1971 standards, where the 1971 standards remain in effect until implementation plans to attain or maintain the 2010 standard are approved.

EPA scientists also conduct methods development research to develop ways of accurately and reliably measuring these six criteria pollutants in ambient air. These methods, called federal reference methods (FRMs), are used by states and other monitoring organizations to assess implementation actions needed to attain NAAQSs. FRMs are the gold standard of air pollution monitoring systems.

To allow innovation and advance new technologies, EPA also reviews, tests, and approves other methods, called federal equivalent methods (FEMs), which are based on different sampling and/or analyzing technologies than FRMs but must be as accurate as FRMs.

EPA scientists review and establish NAAQS on a 5-year cycle based on new health and environmental research data. Scientists in the EPA's Office of Research and Development review and ensure that methods are based on modern technology and have adequate instrument sensitivity to accurately measure pollutants.

Recently, EPA scientists reviewed the NAAQS for PM and finalized the revised standards and measurement methodologies for $PM_{2.5}$ (US EPA, n.d.(c); US EPA, n.d.(f)). Other recent NAAQS reviews include revised methods and performance specifications for carbon monoxide (76 FR 54294) (Federal Register, n.d.(a)) and sulfur dioxide (73 FR 66964) (Federal Register, n.d.(c)).

EPA researchers are continually evaluating new potential FRMs and FEMs to foster innovation and improved measurement of atmospheric pollutants. Adopting new technologies improves the EPA's ability to measure air pollution in new ways and locations.

New approved methods are formally announced through publication in the Federal Register, while approved modifications to existing FRMs and/or FEMs are announced through the EPA's public websites. EPA scientists provide air monitoring technical guidance to states and tribal organizations.

12.1.2 Water

On the website of the EPA (US EPA, n.d.(h)) the reader finds following links:

- Analytical methods and laboratories (US EPA, n.d.(e)): Here one can learn about the EPA-approved analytical methods and laboratories that water systems must use when analyzing samples to meet federal monitoring requirements or to demonstrate compliance with federal water regulations.
- Drinking water laboratory certification: The EPA offers guidance for laboratories seeking its certification to conduct drinking water analysis.
- Safe Drinking Water Act (SDWA) methods (US EPA, n.d.(c); US EPA., n.d.(i)): Listed here approved analytical methods for analyzing water samples according to federal monitoring requirements and drinking water regulations under SDWA. See Section 12.1.2.1.
- Research and risk assessment: Via this link, people learn about the National Water Program Research Strategy, the semiannual newsletter, and ongoing or completed research on drinking water quality and ambient water quality.
- Surface water standards and guidance: Information on water quality criteria and water quality standards, fish and shellfish advisories, fish tissue studies, and requirements for cooling water intake structures are provided here.
- Wastewater technology: The link Wastewater Technology provides information on wastewater technology including a host of publications and reports and links dedicated to biosolids and industry-effluent guidelines.
- Water laboratory alliance: The EPA provides the water sector with a nationwide network of laboratories capable of analyzing water samples in the event of chemical, biological, or radiochemical contamination.
- Clean Water Act (CWA) methods (US EPA, n.d.(j); Federal Register, n.d.(f); US EPA; n.d.(g)): In this section, the EPA provides information regarding laboratory methods for analyzing the chemical, physical, and biological components of wastewater and other environmental samples that are required under CWA authority.

12.1.2.1 EPA-Approved Methods

Water systems must use EPA-approved analytical methods when analyzing samples to meet federal monitoring requirements or to demonstrate compliance with drinking water regulations. Approved methods are listed in the Code of Federal Regulations (CFR) after publication in a final rule or as part of an expedited approval. Approved methods are developed by the EPA, other government agencies, universities, consensus methods organizations, water laboratories, and instrument manufacturers. On the page SDWA methods (see Section 12.1.2.2), the reader finds detailed information on the following:

- Analytical methods approved for drinking water compliance monitoring
- Sources of drinking water methods
- Technical notes on drinking water methods
- Related links

12.1.2.2 Clean Water Act Analytical Methods

The EPA publishes laboratory analytical methods or test procedures that are used by industries and municipalities to analyze the chemical, physical and biological components of wastewater and other environmental samples that are required by regulations under the authority of the CWA. Most of these methods are published as regulations in the CFR.

The EPA's Office of Science and Technology manages the CWA methods program, working with EPA regions, states, and stakeholders.

The reader can explore links such as the following:

- CWA methods not contained in EPA regulations
- Other agency methods
- Analytic methods for the oil and gas extraction point source category
- Oil and grease measurements
- Cyanide measurements
- Mercury measurements
- Microbiological measurements
- Detection and quantitation federal advisory committee
- Effluent guidelines
- Toxic and priority pollutants
- Whole effluent toxicity

12.2 EUROPEAN UNION

12.2.1 Air

In the study by the European Commission (n.d.(c)), the reader finds information on the implementation of ambient air quality legislation.

12.2.1.1 Understanding the Problem

In position papers, the understanding of the impact of individual pollutants and their sources and evolution of concentrations helps develop the policy and supports effective abatement.

A number of pollutants are covered by the First Daughter Directive (1999/30/EC):

- Position paper on NO_2 (European Commission, n.d.(w))
- Position paper on Pb (European Commission, n.d.(ff))
- Position paper on SO_2 (European Commission, n.d.(aa))
- Second position paper on PM (December 2004) (European Commission, n.d.(z)) and the first position paper on PM_{10} (European Commission, n.d.(d))

Two pollutants are covered by the Second Daughter Directive (2000/69/EC):

- Position paper on CO (European Commission, n.d.(g))
- Position paper on benzene (European Commission, n.d.(e))

Ozone is covered by the Third Daughter Directive (2002/3/EC):

- Ozone position paper (European Commission, n.d.(v))

A number of pollutants are covered by the Fourth Daughter Directive (2004/107/EC):

- Position paper on As, Cd, and Ni (European Commission, n.d.(f)) and annexes (European Commission, n.d.(h))
- Position paper on PAH (European Commission, n.d.(e)) and annexes (European Commission, n.d.(f))
- Position paper on mercury (European Commission, n.d.(c))

12.2.1.2 Methods

12.2.1.2.1 Methods and Criteria

The Framework Directive, as well as its daughter directives, requires the assessment of the ambient air quality existing in member states on the basis of common methods and criteria. In the directives (European Commission, n.d.(a)), the minimum assessment requirements are written. They are linked to the specific concentration thresholds as well as the population within each air quality zone or agglomeration. While continuous monitoring is mandated in specific cases, modeling is always encouraged in order to provide better information on the spatial distribution of concentrations.

The commission with the extensive support of the national experts in Working Group on Implementation has prepared several guidances to facilitate implementation of these provisions:

- Guidance report on preliminary assessment under EC air quality directives (European Commission, n.d.(n))
- Guidance on assessment under the European Union (EU) air quality directives (European Commission, n.d.(o))

In order to ensure comparability between measurements of PM_{10} from the First Daughter Directive, the commission prepared the following guidance:

- Guidance to member states on PM_{10} monitoring and intercomparisons with the reference method (European Commission, n.d.(o))

Regarding the Third Daughter Directive relating to ozone, the commission has prepared a guidance for implementing this directive. Annex III includes guidance on the strategy for the measurement of ozone precursors in ambient air as requested in the Third Daughter Directive:

- Decision 2004/279/EC: Guidance for implementing the ozone directive (European Commission, 2004)

Another guidance, which is based on Note 2003/3 of the former Clean Air For Europe Working Group on Implementation, gives recommendations on how to assess air quality around point sources:

- Guidance on assessment around point sources (European Commission, n.d.(m)).

When the monitoring is being used for assessment, it is extremely important to appropriately select the siting of the monitoring station. Siting has a bearing on the ability to use the monitoring information to assess compliance in the specific area and to obtain further information in regard to exposure, source apportionment that supports further development of air policy, as well as air pollution management in the area.

The EU Air Quality Directive 2008/50/EC (European Commission, n.d.(i)) requires that a minimum of one rural background station is installed every 100,000 km^2 for measuring $PM_{2.5}$ in order to get data on the total mass concentration and the chemical speciation. This minimum is larger than the total surface area of several member states. The directive further states that member states shall set up at least one measuring station or may by agreement with adjoining member states set up together one or several common measuring stations. This document provides guidance to member states regarding the selection and the implementation of these stations (commission staff working paper establishing guidelines for the agreements on setting up common measuring stations for $PM_{2.5}$ under Directive 2008/50/EC on ambient air quality and cleaner air for Europe [European Commission, n.d.(i)]).

Directive 2008/50/EC provides member states with the possibility to subtract the contribution of natural sources under certain conditions before comparing the ambient air pollutant concentrations to the limit values. This document aims to give guidance on which sources can be regarded as natural in this context and on methods to quantify and subtract the contribution of these sources:

- Commission staff working paper establishing guidelines for demonstration and subtraction of exceedances attributable to natural sources under Directive 2008/50/EC on ambient air quality and cleaner air for Europe (European Commission, n.d.(j))

Member states may under Directive 2008/50/EC indicate that for designated zones or agglomerations within which limit values for PM_{10} are exceeded in ambient air, such exceedance is due to the resuspension of particulates following winter sanding or salting of roads. This document recommends methods for the determination of the contributions of these processes to the ambient concentrations of PM:

- Commission staff working paper establishing guidelines for determination of contributions from the resuspension of particulates following winter sanding or salting of roads under Directive 2008/50/EC on ambient air quality and cleaner air for Europe (European Commission, n.d.(k))

12.2.1.2.2 Work in Progress about Monitoring Stations

In 2007, a study was conducted for the commission that would facilitate more harmonized approach to monitoring stations (interim report on representativeness and classification of air quality monitoring stations [European Commission, n.d.(u)]).

The commission is reviewing and updating the available guidance documents, where it was considered necessary in order to include more recent implementation experience and potential new elements brought forward through the new revised legislation (European Commission, n.d.(y)).

12.2.1.2.3 Reference Measurement Methods

For each regulated pollutant, a reference measurement method has been prescribed. At the time of the adoption of the directives, standardized methods have not been developed for all pollutants. The situation has since developed, and an updated list of reference measurement method is now available, which is explicitly referenced in the new proposal for ambient air quality directive:

- SO_2: EN 14212:2005
- NO_2 and NO_x: EN 14211:2005
- Pb, Cd, As, Ni: EN 14902: 2005
- PM_{10}: EN 12341:1999

- $PM_{2.5}$: EN 14907:2005
- Benzene: EN 14662:2005
- CO: EN 14626:2005

The standards can be obtained through the national standardization organization. See the study by the European Committee for Standardization (n.d.).

12.2.1.2.4 Work in Progress about Standards

The development of the new standards for pollutants under the Fourth Daughter Directive, as well as periodic review and potential revision of the developed standards continues under European Committee for Standardisation (CEN) Technical Committee 264. For the organization and the work of CEN, see http://www.cen.eu (European Committee for Standardization, n.d.).

In addition, the certified reference materials (CRMs) for the calibration of measurements of heavy metals and PAH is (currently being) developed at Joint Research Centre Institute for Reference Materials and Measurements (JRC-IRMM) (European Commission, n.d.(s)).

12.2.1.2.5 Equivalence

Nonreference measurement methods can also be used provided they respect provisions for equivalence set out in the directives (see, for example, Directive 2008/50/EC, Annex VI). A commission working group on equivalence has prepared a document describing principles and methodologies to be used for the demonstration of the equivalence of alternative (nonreference) measurement methods to the reference methods described by the European Norm (EN) standard methods. This document has been discussed at a workshop in May 2007, in Ispra (European Commission, n.d.(t)), and subsequently updated. The air quality committee established under Directive 2008/50/EC has endorsed the new guidance for the implementation of Directive 2008/50/EC:

- Guidance for the demonstration of equivalence of ambient air monitoring methods from January 2010 (European Commission, n.d.(p))

The corresponding tool to facilitate the use of the guidance (in particular for checking the equivalence of nonreference methods for PM monitoring) has been developed in MS Excel.

The template provides the formulas and the macros with the statistical calculations needed.

12.2.1.2.6 Modeling

In the directive, modeling is considered to provide supplemental information on air quality monitoring and is to be used where modeling is not mandatory. Modeling is becoming a principal assessment tool that is validated by monitoring and provides much more comprehensive information in regard to public exposure and supports the identification of sources and future projections based on different measures scenarios.

The use of modeling has also been developed under specific initiatives such as HARMO (n.d.). Under the sixth research framework program, an Air4EU project (n.d.) has recently been concluded that facilitates the combined use of monitoring and modeling for air quality assessment through the developed guidance and information technology tools.

The Forum for Air Quality Modelling (FAIRMODE) (n.d.) of modelers and users has been established to support the widespread and harmonized use of models through model validation and intercomparison exercises and through the management of the modeling network. The FAIRMODE webpage contains links to its current activities.

12.2.1.2.7 Objective Estimation

Objective estimation is reserved for the air quality zones with very good air quality and no large conurbations. It is usually combined with modeling. By the identification of local pollution sources and information of regional air quality, an estimation of concentration of a regulated pollutant is made.

12.2.1.3 Ensuring Quality of Assessment Information

Ensuring the quality of assessment information either generated through monitoring, modeling, or objective estimation is one of the paramount provisions of the directives. Data quality objectives (DQOs) are prescribed that define maximum allowed uncertainty, time coverage, and data coverage.

While member states are responsible for ensuring appropriate quality assurance of the assessment as well as the appropriate quality control of the information provided to the public and through the assessment reports, the commission set up a community-wide process, managed by the JRC. The JRC organizes intercomparison exercises for the national reference laboratories and manages AQUILA (European Commission, n.d.(t))—a network of national reference laboratories that follows the implementation of assessment by monitoring, serves as exchange forum, and provides expert advice to the commission.

AQUILA has recently prepared an extensive document that summarizes the roles and the responsibilities of the national reference laboratories and of the network itself and describes the quality assurance procedures and EU-wide comparisons. The document also includes the interpretation of obligations for the national reference laboratories and the monitoring networks under the new directive (European Commission, n.d.(u)). The document received positive opinion when presented at the air quality committee under the directive on July 1, 2009 (AQUILA—Roles and requirements for quality of measurements [European Parliament and Council of the European Union. 2013c]).

The European Environment Agency together with its Topic Centre is supporting the efforts through further quality control of the reported assessment information and by facilitating the exchange of best practices by the data providers within the European Environment Information and Observation Network.

The new directive (European Commission, n.d.(t)) does not significantly change provisions for the assessment but does introduce some streamlining and further quality requirements and assessment requirements for fine PM, $PM_{2.5}$.

12.2.1.4 Water Protection and Management (Water Framework Directive)

The EU has established a common framework for water protection and management (WFD) (European Commission, n.d.(q)), which aims to protect and restore aquatic ecosystems and to guarantee long-term, sustainable water usage for individuals, businesses, and the natural world. Directive 2000/60/EC of the European Parliament and of the Council of October 23, 2000 establishes a framework for community action in the field of water policy (European Parliament and Council of the European Union, 2013b).

This framework directive has several objectives such as preventing and reducing pollution, promoting sustainable water usage, protecting the environment, improving the state of aquatic ecosystems, and reducing the effects of floods and droughts.

It has established a framework for the protection of the following:

- Inland surface waters
- Groundwater
- Transitional Waters
- Coastal waters

Its ultimate objective is/was to achieve a good status for all EU waters by 2015.

12.2.2 River Basins and Districts

Member states have to identify all the river basins lying within their national territory and to assign them to individual river basin districts. River basins covering the territory of more than one member state are assigned to an international river basin district.

Member states are to designate a competent authority for the application of the rules provided for in this framework directive within each river basin district.

Each member state shall produce the following:

- An analysis of the characteristics of each river basin district
- A review of the impact of human activity on water
- An economic analysis of water use
- A register of protected areas, that is, areas identified as requiring special protection (areas identified for drinking water abstraction and other areas listed in Annex IV of the directive)
- A survey of all bodies of water used for abstracting water for human consumption and producing more than 10 m^3 per day or serving more than 50 persons

This analysis should be revised every 6 years thereafter.

Member states shall encourage participation by all stakeholders in the implementation of this framework directive, specifically with regard to the management plans for river basin districts. Projects from the management plans must be submitted for a public consultation for a minimum period of 6 months.

Member states shall introduce arrangements to ensure that effective, proportionate, and dissuasive penalties are imposed in the event of breaches of the provisions of this framework directive. A temporary deterioration in bodies of water does not constitute a breach of the directive, if it is the result of exceptional and unforeseeable circumstances related to an accident, natural causes, or a case of force majeure.

In Table 12.2, the reader finds amendments and corrections to Directive 2000/60/EC.

The successive amendments and corrections to Directive 2000/60/EC have been incorporated in the original text.

TABLE 12.2
Ammendments and Corrections to Directive 2000/60/EC

Act	Entry into Force—Date of Expiry	Deadline for Transposition in the Member States	Official Journal (OJ)
Directive 2000/60/EC (European Commission, n.d.(t))	December 22, 2000	December 22, 2003	OJ L 327 of December 22, 2000
Amending Act(s)	Entry into Force	Deadline for Transposition in the Member States	Official Journal (OJ)
Decision 2455/2001/EC (European Parliament and Council of the European Union, 2001)	December 16, 2001	–	OJ L 331 of December 15, 2001
Directive 2008/32/EC (European Parliament and Council of the European Union, 2008b)	March 21, 2008	–	OJ L 81 of March 20, 2008
Directive 2009/31/EC (European Parliament and Council of the European Union, 2009)	June 25, 2009	June 25, 2011	OJ L 140 of June 5, 2009
Directive 2013/39/EU (European Parliament and Council of the European Union, 2013a)	September 13, 2013	September 14, 2015	OJL 226 of August 24, 2013
Directive 2013/64/EU (Council of the European Union, 2013)	January 1, 2014	Different according to the articles	OJ L 352 of December 28, 2013

Directive 2013/39/EU (European Commission, 2007) is about the communication from the commission to the European Parliament and the Council—toward sustainable water management in the EU—first stage in the implementation of the Water Framework Directive 2000/60/EC.

In Annex I of Directive 2008/105/EC (European Parliament and Council of the European Union. 2008a), the reader finds a list of priority substances in the field of water policy, that is, those which pose a significant risk to the environment.

12.2.2.1 Reference Materials Environment

European reference materials (ERMs) are a tool for improving the confidence in, and the mutual recognition of, test results and certificates in the European market. The certified reference materials (CRMs) comply with high metrological requirements, ensuring the traceability of measurements results, and are the end point of the traceability chain, thus being primary standards in chemistry.

The ERM concept (BAM, n.d.(a)) is a joint collaboration of three major ERM producers who guarantee to

- Use the most advanced principles currently available described in International Organization for Standardization guides 34 and 35 for the production of CRMs
- Rigorously demonstrate homogeneity and stability for all materials and guarantee the certified value for every single unit over the complete shelf life of the materials
- Be transparent in their approaches for the production of CRMs

ERM® is the registered trademark for ERMs.

The ERM concept is a collaboration between the following:

- The Institute for Reference Materials and Measurements (European Commission, n.d.(s); European Commission, n.d.(x)) of the European Commission's JRC, Geel, Belgium
- The Bundesanstalt für Materialforschung und -prüfung (BAM, n.d.(a)) (Federal Institute for Materials Research and Testing), Germany
- LGC, United Kingdom (LGC, n.d.)
- Other institutes involved in reference materials: the ERA (n.d.) and the International Atomic Energy Agency (n.d.)

12.3 OTHER REFERENCES RELATED TO REFERENCE METHODS OF POLLUTANTS

12.3.1 Standard Methods for the Examination of Water and Wastewater

Since 1905, the *Standard Methods for the Examination of Water and Wastewater* has represented "the best current practice of American water analysts." This comprehensive reference covers all aspects of water and wastewater analysis techniques. *Standard Methods* is a joint publication of the American Public Health Association, the American Water Works Association, and the Water Environment Federation (Standard Methods, n.d.; Rice et al., 2012).

12.3.2 Bundesanstaltfür Materialforschung und -prüfung

Bundesanstaltfür Materialforschung und -prüfung (BAM) (n.d.(f)) has worked out a number of reference materials, reference procedures, and reference organisms.

12.3.3 European Metrology Research Programme

The Joint Research Project (JRP) ENV08, Traceable Measurements for Monitoring Critical Pollutants under the European Water Framework Directive (WFD) 2000/60/EC, is a multipartner transnational project within the targeted call Environment of the European Metrology Research Programme (EMRP) (European Metrology Research Programme, n.d.(f); European Metrology Research Programme, n.d.(a)). The EMRP represents a metrology-focused European program of coordinated research and development that facilitates closer integration of the research programs of the National Metrology Institutes.

The WFD specifies a list of 33 priority water pollutants. For these substances, environmental quality standards (EQS), equivalent to the maximum allowable concentrations, had been defined at very low levels. EQS values refer to the whole water body, that is, include contaminants that are associated to suspended solids or colloids present in the natural water. For several priority pollutants, primary analytical methods capable of measuring the target analytes at the EQS level in natural water are still missing. The JRP will develop and validate such methods for three selected compounds, namely, tributyltin, PBDEs, and PAHs.

Furthermore, the JRP aims to develop concepts for the preparation of reference materials that contain the target analytes at the EQS level in a matrix as close as possible to natural water. Such materials serve as a quality assurance tool for monitoring laboratories to assess their measurement capabilities. The JRP therefore helps to implement the WFD and to improve the quality of measurements carried out by test laboratories in the field of water analysis and helps to achieve the final objective of the WFD, namely, good quality of surface, ground, and coastal water in the EU by 2015.

ENV08 is coordinated by BAM and involves 13 European institutes.

12.3.4 Reference Methods for Marine Pollution Studies

The Regional Seas Programme, a part of the United Nations Environmental Programme, lists reference methods for marine pollution studies (United Nations Environment Programme, n.d.).

12.3.5 United Kingdom

The European Commission, acting through the CEN has produced a series of standard methods for monitoring air pollutants. These documents outline minimum performance requirements for analyzers, in order to ensure that measurement methods comply with the DQOs set down in the Ambient Air Quality Directive (2008/50/EC) (Department for Environment, Food and Rural Affairs, n.d.(a)).

In 2005, after extensive discussion and research across the EU, a series of standard methods was published:

- EN 14211:2005: for nitrogen oxides
- EN 14212:2005: for sulfur dioxide
- EN 14625:2005: for ozone
- EN 14626:2005: for carbon monoxide
- EN 14907:2005: for $PM_{2.5}$
- EN 12341:1998: for PM_{10}; already published as a standard method

The standard methods are used as a template for instrument performance testing. The standards specify a series of tests and requirements for analyzers to achieve based on both laboratory- and field-based studies. The results from these tests can then be used to calculate an uncertainty budget in order to show whether the instrument performance (in its tested configuration) fulfills the DQO. This information can then be used as a stamp of approval for the use of specific analyzers (in their tested configuration) in national monitoring networks. This process is known as *type approval*.

12.3.5.1 UK-Type Approval of Gaseous Analyzers: MCERTS

For gaseous analyzers, the type approval testing process is managed in the United Kingdom by the Environment Agency under its Environment Agency's Monitoring Certification Scheme (MCERTS)

TABLE 12.3
AURN Measurement Techniques

O_3	UV absorption
NO/NO_2	Chemiluminescence
SO_2	UV fluorescence
CO	Infrared absorption
PM_{10} and $PM_{2.5}$	Tapered element oscillating microbalance
	Beta attenuation monitor
	Gravimetric monitor
	Filter dynamics measurement system

scheme with certification provided by Scientific Instrument Research Association (SIRA), the appointed certification body. The MCERTS performance standards mirror the requirements of the CEN standard methods, and decisions are made by a certification committee. The relevant performance standard for gases is continuous ambient air monitoring systems. UK testing against the standards began in 2005, and a number of instrument suppliers have successfully had their gaseous analyzers type-approved under these schemes. Further details can be found in the study by SIRA (2015).

For measurements in the UK national network, the specific conditions and facilities at each site have been used to calculate measurement uncertainties for individual analyzers. These results are published every six months and are available in the study by the Department for Environment, Food and Rural Affairs (2015).

12.3.5.2 UK-Deployed Monitoring Methods

The techniques used for monitoring within the United Kingdom's national compliance monitoring network, the Automatic Urban and Rural Network (AURN) are summarized in Table 12.3. With the exception of the automatic PM_{10} analyzers, the reference methods of measurement are defined in the relevant EU directives.

For PM, the AURN uses methods that have demonstrated equivalence to the reference method but which (unlike the reference method) allow continuous online monitoring.

12.3.6 Australia

In the study by the Department of Environment and Conservation (n.d.(a)), the reader finds information on approved methods for the sampling and analysis of water pollutants in New South Wales. The approved methods for the sampling and air pollutants in New South Wales may be found in the study by the Department of Environment and Conservation (n.d.(b)).

FURTHER READING

Schneider, T., de Koning, H.W., Brasser, L.J. Air pollution reference measurement methods and systems. *Proceedings of the International Workshop*, Bilthoven, December 12–16, 1977. Organized by the National Institute of Public Health, Bilthoven, The Netherlands, and cosponsored by the World Health Organization.

International Council of Scientific Unions. 1975. *Environmental Pollutants—Selected Analytical Methods: Scope 6*. Butterworths, London.

International Council of Scientific Unions. 2013. *Scientific Committee on Problems of the Environment: Working Group on Methodology of Determination of Toxic Substances in the Environment*. International Council of Scientific Unions, Paris.

Spiegel, J., Maystre, L.Y. 1998. Environmental pollution control. In *Encyclopaedia of Occupational Health and Safety*, 4th ed., ed. Stellman, J.M. International Labor Office, Geneva. http://www.ilocis.org/documents/chpt55e.htm (accessed November 11, 2015).

Vesilind, P.A., Peirce, J.J., Weiner, R.F. 2013. *Environmental Pollution and Control*, 4th ed. Elsevier, Amsterdam.

REFERENCES

Air4EU. n.d. Air4EU. http://www.air4eu.nl/. Air4EU, Utrecht (accessed November 11, 2015).

BAM. n.d.(a) Bundesanstaltfür Materialforschung und -prüfung. BAM, Bundesanstaltfür Materialforschung Germany. http://www.bam.de/de/index.htm (accessed November 11, 2015).

BAM. n.d.(b). What is the concept? BAM, Bundesanstaltfür Materialforschung Germany. http://www.erm-crm.org/about_ERM/the_concept/Pages/index.aspx (accessed November 11, 2015).

Council of the European Union. 2013. Council Directive 2013/64/EU of 17 December 2013 amending Council Directives 91/271/EEC and 1999/74/EC, and Directives 2000/60/EC, 2006/7/EC, 2006/25/EC and 2011/24/EU of the European Parliament and of the Council, following the amendment of the status of Mayotte with regard to the European Union. *Official Journal of the European Communities* L353. http://eur-lex.europa.eu/legal-content/EN/TXT/?uri=celex:32013L0064 (accessed November 11, 2015).

Department for Environment, Food and Rural Affairs. n.d.(a) Air quality monitoring methods. Department for Environment, Food and Rural Affairs. http://uk-air.defra.gov.uk/networks/monitoring-methods?view=eu-standards (accessed November 11, 2015).

Department for Environment, Food and Rural Affairs. n.d.(b). Monitoring network reports. Department for Environment, Food and Rural Affairs. http://uk-air.defra.gov.uk/library/reports?section_id=13 (accessed November 11, 2015).

Department of Environment and Conservation. n.d.(a). Approved methods for the sampling and analysis of water pollutants in New South Wales. Department of Environment and Conservation, New South Wales. http://www.environment.nsw.gov.au/resources/water/approvedmethods-water.pdf (accessed November 11, 2015).

Department of Environment and Conservation. n.d.(b). Approved methods for the sampling and analysis of air pollutants in New South Wales. Department of Environment and Conservation, New South Wales. http://www.epa.nsw.gov.au/resources/legislation/07001amsaap.pdf (accessed November 11, 2015).

ERA. n.d. ERA. ERA, Golden, CO. http://www.eraqc.com/ (accessed November 11, 2015).

European Commission. 2004. 2004/279/EC: Commission Decision of 19 March 2004 concerning guidance for implementation of Directive 2002/3/EC of the European Parliament and of the Council relating to ozone in ambient air. *Official Journal of the European Communities* L087. http://eur-lex.europa.eu/legal-content/EN/TXT/?uri=CELEX:32004D0279 (accessed November 11, 2015).

European Commission. 2007. Communication from the Commission to the European Parliament and the Council—Towards sustainable water management in the European Union—First stage in the implementation of the Water Framework Directive 2000/60/EC. http://eur-lex.europa.eu/legal-content/EN/TXT/?uri=celex:52007DC0128 (accessed November 11, 2015).

European Commission. n.d.(a). Air quality—Existing legislation. European Commission, Brussels. http://ec.europa.eu/environment/air/quality/legislation/existing_leg.htm (accessed November 11, 2015).

European Commission. n.d.(b). Ambient air pollution by AS, CD and NI compounds: Position paper. European Commission, Brussels. http://ec.europa.eu/environment/archives/air/pdf/pp_as_cd_ni.pdf (accessed November 11, 2015).

European Commission. n.d.(c). Ambient air pollution by mercury (Hg) position paper. European Commission, Brussels. http://ec.europa.eu/environment/archives/air/pdf/pp_mercury.pdf (accessed November 11, 2015).

European Commission. n.d.(d). Ambient air pollution by particulate matter position paper. European Commission, Brussels. http://ec.europa.eu/environment/archives/air/pdf/pp_pm.pdf (accessed November 11, 2015).

European Commission. n.d.(e). Ambient air pollution by polycyclic aromatic hydrocarbons. European Commission, Brussels. http://ec.europa.eu/environment/archives/air/pdf/pp_pah.pdf (accessed November 11, 2015).

European Commission. n.d.(f). Ambient air pollution by polycyclic aromatic hydrocarbons: Position paper annexes. European Commission, Brussels. http://ec.europa.eu/environment/archives/air/pdf/annex_pah.pdf (accessed November 11, 2015).

European Commission. n.d.(g). Ambient air pollution: Carbon monoxide position paper. European Commission, Brussels. http://ec.europa.eu/environment/archives/air/pdf/pp_co.pdf (accessed November 11, 2015).

European Commission. n.d.(h). Annex A: Members of the working group. European Commission, Brussels. http://ec.europa.eu/environment/archives/air/pdf/annex_as_cd_ni.pdf (accessed November 11, 2015).

European Commission. n.d.(i). Commission staff working paper establishing guidelines for the agreements on setting up common measuring stations for PM2.5 under Directive 2008/50/EC on ambient air quality and cleaner air for Europe. European Commission, Brussels. http://ec.europa.eu/environment/air/quality/legislation/pdf/sec_2011_77.pdf (accessed November 11, 2015).

European Commission. n.d.(j). Commission staff working paper establishing guidelines for demonstration and subtraction of exceedances attributable to natural sources under the Directive 2008/50/EC on ambient air quality and cleaner air for Europe. European Commission, Brussels. http://ec.europa.eu/environment/air/quality/legislation/pdf/sec_2011_0208.pdf (accessed November 11, 2015).

European Commission. n.d.(k). Commission staff working paper establishing guidelines for demonstration and subtraction of exceedances attributable to natural sources under the Directive 2008/50/EC on ambient air quality and cleaner air for Europe. European Commission, Brussels. http://ec.europa.eu/environment/air/quality/legislation/pdf/sec_2011_0207.pdf (accessed November 11, 2015).

European Commission. n.d.(l). Council directive on ambient air quality assessment and management working group on benzene. European Commission, Brussels.http://ec.europa.eu/environment/archives/air/pdf/ppbenzene.pdf (accessed November 11, 2015).

European Commission. n.d.(m). Guidance on air quality assessment around point sources under the EU Air Quality Directive 2008/50/EC. European Commission, Brussels. http://ec.europa.eu/environment/air/quality/legislation/pdf/Guidance%20on%20assessment%20around%20point%20sources%20AQCincluded_final.pdf (accessed November 11, 2015).

European Commission. n.d.(n). Guidance on assessment under the EU air quality directives. European Commission, Brussels. http://ec.europa.eu/environment/air/pdf/guidanceunderairquality.pdf (accessed November 11, 2015).

European Commission. n.d.(o). Guidance on assessment under the EU air quality directives. European Commission, Brussels. http://ec.europa.eu/environment/archives/air/pdf/guidanceunderairquality.pdf (accessed November 11, 2015).

European Commission. n.d.(p). Guide to the demonstration of equivalence of ambient air monitoring methods. European Commission, Brussels. http://ec.europa.eu/environment/air/quality/legislation/pdf/equivalence.pdf (accessed November 11, 2015).

European Commission. n.d.(q). National Air Quality Reference Laboratories and the European Network—AQUILA: Roles and requirements for measurement traceability, accreditation, quality assurance/quality control, and measurement comparisons, at national and European Levels. European Commission, Brussels. http://ec.europa.eu/environment/air/quality/legislation/pdf/aquila.pdf (accessed November 11, 2015).

European Commission. n.d.(r). Implementation of ambient air quality legislation. European Commission, Brussels. http://ec.europa.eu/environment/air/quality/legislation/assessment.htm (accessed November 11, 2015).

European Commission. n.d.(s). Joint Research Centre (JRC). European Commission, Brussels. https://ec.europa.eu/jrc/en/institutes/irmm (accessed November 11, 2015).

European Commission. n.d.(t). JRC in brief. European Commission, Brussels. https://ec.europa.eu/jrc/en/about/ies/ (accessed November 11, 2015).

European Commission. n.d.(u). New air directive. European Commission, Brussels. http://ec.europa.eu/environment/air/quality/legislation/directive.htm (accessed November 11, 2015).

European Commission. n.d.(v). Ozone position paper. European Commission, Brussels. http://ec.europa.eu/environment/archives/air/pdf/documents/pos_paper.pdf (accessed November 11, 2015).

European Commission. n.d.(w). Position paper on air quality: Nitrogen oxide. European Commission, Brussels. http://ec.europa.eu/environment/archives/air/pdf/pp_no2.pdf (accessed November 11, 2015).

European Commission. n.d.(x). Reference materials for environmental analysis. European Commission, Brussels. https://ec.europa.eu/jrc/en/research-topic/reference-materials-environmental-analysis (accessed November 11, 2015).

European Commission. n.d.(y). Representativeness and classification of air quality monitoring stations. European Commission, Brussels. http://ec.europa.eu/environment/air/quality/legislation/pdf/report_uba.pdf (accessed November 11, 2015).

European Commission. n.d.(z). Second position paper on particulate matter. European Commission, Brussels. http://ec.europa.eu/environment/archives/cafe/pdf/working_groups/2nd_position_paper_pm.pdf (accessed November 11, 2015).

European Commission. n.d.(aa). SO_2 position paper. European Commission, Brussels. http://ec.europa.eu/environment/archives/air/pdf/pp_so2.pdf (accessed November 11, 2015).

European Commission. n.d.(bb).Air quality daughter directives: Position paper on lead. European Commission, Brussels. http://ec.europa.eu/environment/archives/air/pdf/pp_pb.pdf (accessed November 11, 2015).

European Committee for Standardization. n.d. European Committee for Standardization. European Commission, Brussels. http://www.cen.eu/Pages/default.aspx (accessed November 11, 2015).

European Metrology Research Programme. n.d. About the EMRP. European Metrology Research Programme, Berlin. http://www.emrponline.eu/ (accessed November 11, 2015).

European Metrology Research Programme. n.d. Traceable measurements for monitoring critical pollutants under the European Water Framework Directive (WFD 2000/60/EC). European Metrology Research Programme, Berlin. http://www.emrp-waterframeworkdirective.bam.de/en/home/index.htm (accessed November 11, 2015).

European Parliament, Council of the European Union. 2001. Decision No 2455/2001/EC of the European Parliament and of the Council of 20 November 2001 establishing the list of priority substances in the field of water policy and amending Directive 2000/60/EC. *Official Journal of the European Communities* L331. http://eur-lex.europa.eu/legal-content/EN/TXT/?uri=celex:32001D2455 (accessed November 11, 2015).

European Parliament, Council of the European Union. 2008a. Directive 2008/105/EC of the European Parliament and of the Council of 16 December 2008 on environmental quality standards in the field of water policy, amending and subsequently repealing Council Directives 82/176/EEC, 83/513/EEC, 84/156/EEC, 84/491/EEC, 86/280/EEC and amending Directive 2000/60/EC of the European Parliament and of the Council. *Official Journal of the European Communities* L348. http://eur-lex.europa.eu/legal-content/EN/TXT/?uri=celex:32008L0105 (accessed November 11, 2015).

European Parliament, Council of the European Union. 2008b. Directive 2008/32/EC of the European Parliament and of the Council of 11 March 2008 amending Directive 2000/60/EC establishing a framework for Community action in the field of water policy, as regards the implementing powers conferred on the Commission. *Official Journal of the European Communities* L81. http://eur-lex.europa.eu/legal-content/EN/TXT/?uri=celex:32008L0032 (accessed November 11, 2015).

European Parliament, Council of the European Union. 2009. Directive 2009/31/EC of the European Parliament and of the Council of 23 April 2009 on the geological storage of carbon dioxide and amending Council Directive 85/337/EEC, European Parliament and Council Directives 2000/60/EC, 2001/80/EC, 2004/35/EC, 2006/12/EC, 2008/1/EC and Regulation (EC) No 1013/2006. *Official Journal of the European Communities* L140. http://eur-lex.europa.eu/legal-content/EN/TXT/?uri=celex:32009L0031 (accessed November 11, 2015).

European Parliament, Council of the European Union. 2013a. Directive 2013/39/EU of the European Parliament and of the Council of 12 August 2013 amending Directives 2000/60/EC and 2008/105/EC as regards priority substances in the field of water policy. *Official Journal of the European Communities* L226. http://eur-lex.europa.eu/legal-content/EN/TXT/?uri=celex:32013L0039 (accessed November 11, 2015).

European Parliament, Council of the European Union. 2013b. Good-quality water in Europe (EU Water Directive). http://eur-lex.europa.eu/legal-content/EN/TXT/HTML/?uri=URISERV:l28002b&rid=3#amendingact (accessed November 11, 2015).

European Parliament, Council of the European Union. 2013c. Directive 2000/60/EC of the European Parliament and of the Council of 23 October 2000 establishing a framework for Community action in the field of water policy. http://eur-lex.europa.eu/legal-content/EN/TXT/?uri=celex:02000L0060-20130913 (accessed November 11, 2015).

FAIRMODE. n.d. Forum for air quality modelling in Europe. FAIRMODE, European Union. http://fairmode.jrc.ec.europa.eu/(accessed November 11, 2015).

Federal Register. n.d.(a). Review of National Ambient Air Quality Standards for carbon monoxide; Final rule. Office of the Federal Register, College Park, MD. http://www.gpo.gov/fdsys/pkg/FR-2011-08-31/html/2011-21359.htm (accessed November 11, 2015).

Federal Register. n.d.(b). National Ambient Air Quality Standards for particulate matter; Final rule. Office of the Federal Register, College Park, MD. http://www.gpo.gov/fdsys/pkg/FR-2013-01-15/pdf/2012-30946.pdf (accessed November 11, 2015).

Federal Register. n.d.(c). National Ambient Air Quality Standards for lead; Final rule. Office of the Federal Register, College Park, MD. http://www.gpo.gov/fdsys/pkg/FR-2008-11-12/html/E8-25654.htm (accessed November 11, 2015).

Federal Register. n.d.(d). National Ambient Air Quality Standards for nitrogen dioxide; Final rule. Office of the Federal Register, College Park, MD. http://www.gpo.gov/fdsys/pkg/FR-1996-10-08/html/96-25786.htm (accessed November 11, 2015).

Federal Register. n.d.(e). National Ambient Air Quality Standards for Ozone; Final Rule. Office of the Federal Register, College Park, MD. http://www.gpo.gov/fdsys/pkg/FR-2008-03-27/html/E8-5645.htm (accessed November 11, 2015).

Federal Register. n.d.(f). Part 136—Guidelines Establishing Test Procedures for the Analysis of Pollutants. Office of the Federal Register, College Park, MD. http://www.gpo.gov/fdsys/pkg/CFR-2014-title40-vol23/xml/CFR-2014-title40-vol23-part136.xml (accessed November 11, 2015).

HARMO. n.d. Initiative on "Harmonisation within Atmospheric Dispersion Modelling for Regulatory Purposes." HARMO, European initiative. http://www.harmo.org/default.asp (accessed November 11, 2015).

International Atomic Energy Agency. n.d. IAEA reference materials. International Atomic Energy Agency, Vienna. https://nucleus.iaea.org/rpst/referenceproducts/About/index.htm (accessed November 11, 2015).

LGC. n.d. LGC standards. LGC, UK. http://www.lgcstandards.com/GB/en/?gclid=CPz7t4nDiMkCFUPnwgodmjUHKw (accessed November 11, 2015).

Rice, E.W., Baird, R.B., Eaton, A.D., Clesceri, L.S., eds. 2012. *Standard Methods for the Examination of Water and Wastewater*, 22nd ed. American Water Works Association, Northwest, Washington, DC.

SIRA. n.d. Leading standards in compliance and safety. SIRA, UK. http://www.siraenvironmental.com/UserDocs/mcerts/MCERTSCertifiedProductsCAMS.pdf (accessed November 11, 2015).

Standard Methods. n.d. Standard methods for the examination of water and wastewater. Standard Methods, USA. https://www.standardmethods.org/ (accessed November 11, 2015).

United Nations Environment Programme. n.d. United Nations Environment Programme, Nairobi. http://www.unep.ch/regionalseas/pubs/rsrm.htm (accessed November 11, 2015).

US EPA. n.d.(a). Air science. EPA, Washington, DC. http://www2.epa.gov/science-and-technology/air-science (accessed November 11, 2015).

US EPA. n.d.(b). List of designated reference and equivalent methods. EPA, Washington, DC.http://www.epa.gov/ttn/amtic/files/ambient/criteria/reference-equivalent-methods-list.pdf (accessed November 11, 2015).

US EPA. n.d.(c). Methods approved to analyze drinking water samples to ensure compliance with regulations. EPA, Washington, DC. http://www2.epa.gov/dwanalyticalmethods (accessed November 11, 2015).

US EPA. n.d.(d). NAAQS table. EPA, Washington, DC. https://www.epa.gov/criteria-air-pollutants/naaqs-table (accessed November 11, 2015).

US EPA. n.d.(e). Nitrogen dioxide. EPA, Washington, DC. http://www3.epa.gov/airquality/nitrogenoxides/ (accessed November 11, 2015).

US EPA. n.d.(f). Overview of the Clean Air Act and air pollution. EPA, Washington, DC. http://www.epa.gov/air/caa/ (accessed November 11, 2015).

US EPA. n.d.(g). Summary of the Clean Water Act. EPA, Washington, DC. http://www2.epa.gov/laws-regulations/summary-clean-water-act (accessed November 11, 2015).

US EPA. n.d.(h). Water science. EPA, Washington, DC. http://water.epa.gov/scitech/ (accessed November 11, 2015).

US EPA. n.d.(i). Approved drinking water analytical methods. EPA, Washington, DC. http://www2.epa.gov/dwanalyticalmethods/approved-drinking-water-analytical-methods (accessed November 11, 2015).

US EPA. n.d.(j). Clean Water Act analytical methods. EPA, Washington, DC. http://www2.epa.gov/cwa-methods (accessed November 11, 2015).

US EPA. n.d.(k).Particulate matter (PM) pollution. EPA, Washington, DC. http://www3.epa.gov/airquality/particlepollution/ (accessed November 11, 2015).

Section IV

Air Pollutants

13 Air Pollutants in the Outdoor Environment (NO_x, SO_2, VOCs, HAPs [CO, O_3])

Carolina Santamaría, David Elustondo, Esther Lasheras, and Jesús Miguel Santamaría

CONTENTS

13.1 VOLATILE ORGANIC COMPOUNDS

13.1.1 Volatile Organic Compounds Outdoors

According to the EPA (2015b), a VOC is any compound of carbon (excluding CO, CO_2, etc.) that participates in atmospheric photochemical reactions. Other definitions (Council of the European Union, 1999) take into account some of their physicochemical characteristics and consider VOCs as organic compounds with a vapor pressure greater than 10 Pa at 25°C, a boiling point of up to 260°C at atmospheric pressure, and 15 or less carbon atoms, emphasizing the volatile characteristics of this group of compounds.

VOCs are present at extremely low concentrations (ppb or ppt [Williams, 2004; Parra et al., 2006]) and have deep effects on the atmosphere, because they keep the oxidative atmospheric photochemistry running and act as ozone and fine particulate matter (PM) precursors (Williams and Kopmann, 2007). Almost any anthropogenic activity produces a release of VOCs to the atmosphere, so they are present in rural, industrial, and urban areas. They may contribute to climate change and are also considered as toxic (WHO, 2000; EPA, 2015a), and recent regulations have approved the need for VOC assessment at urban and suburban sites (European Parliament and Council of the European Union, 2008).

Sometimes it is very difficult to distinguish between biogenic volatile organic compounds (BVOCs) and anthropogenic VOCs, because many of these compounds are common to both sources (Durana et al., 2006; Reimann and Lewis, 2007). The main natural sources for VOCs are bacterial degradation processes, oceans, and vegetation, which is the most important of all of them. Some authors (Guenther, 2002) estimated the contribution of BVOCs as 10 times larger than the sum of all anthropogenic emissions, and almost half of the natural emissions correspond to isoprene, a compound released by deciduous trees (Kesselmeier and Staudt, 1999). Monoterpenes are also an abundant family of organic compounds, α-pinene being responsible for half of the emissions of this group, mainly released by conifers (Hakola et al., 2003).

Emissions of BVOCs are usually estimated from mathematical models developed from qualitative data of emissions recorded with portable analysis units equipped with photoionization detectors (PIDs) (Guenther et al., 1996; Klinger et al., 1998). However, these models need to be validated with quantitative data recorded for the different plant families, genera, and environmental conditions. Quantification becomes of critical relevance when effective strategies to reduce concentrations of pollutants have to be formulated (Karlik et al., 2002; Bouvier-Brown et al., 2012). Also, evaluation of ambient levels of BVOCs is an essential work to understand the processes undergone during transport, diurnal variation, or the importance of these pollutants at remote sites (Detournay et al., 2013; Greenberg et al., 2014).

Many recent research works have focused on identification and quantification of BVOCs emitted from different ecosystems such as remote forests (Bouvier-Brown et al., 2012; Kivimäenpää et al., 2012; Detournay et al., 2013; Llusia et al., 2013), urban forests (Hellén et al., 2012; Kim et al., 2013; Préndez et al., 2013; Wang et al., 2013; Curtis et al., 2014), urban sites (Filella and Peñuelas, 2006), volcanic areas and waste landfills (Tassi et al., 2013), boreal peat land (Faubert et al., 2010), and infected (Joó et al., 2010) and infested trees (Amin et al., 2012).

Sampling of BVOCs was done, pumping air through solid traps containing different combinations of adsorbents, using silica cartridges (Sep-Pak cartridges) (Faubert et al., 2010; Joó et al., 2010; Amin et al., 2012; Bouvier-Brown et al., 2012; Kivimäenpää et al., 2012; Detournay et al., 2013; Kim et al., 2013; Llusia et al., 2013; Préndez et al., 2013; Tassi et al., 2013) or collecting whole air samples in stainless steel canisters (Amin et al., 2012; Wang et al., 2013) or directly collected into the cold trap of a thermal desorber (Hellén et al., 2012).

Separation and analysis of BVOCs is usually carried out by GC, using conventional columns and identification and quantification is typically performed using mass-selective detectors (MSDs). The analysis of samples through using such configurations allows to discriminate between bark beetle-infested (higher concentrations of β-pinene, 3-carene, β-phellandrene, estragole, and *p*-cymene) or aphid-infected trees (higher concentrations of linalool, α-farnesene, and (*E*)-β-ocimene) from healthy trees from the identification of the individual BVOCs emitted by the different studied trees (Joó et al., 2010; Amin et al., 2012); to study the effect of global warming on the decrease in the water table of a boreal peat land from the increase in the concentration of monoterpene emissions under such conditions (Faubert et al., 2010); to identify the changes in BVOC emissions from different tree species with the time of the year (Hellén et al., 2012; Kim et al., 2013; Llusia et al., 2013); or to use BVOCs as tracers to distinguish between natural and anthropogenic sources in a volcanic area (Tassi et al., 2013).

More sophisticated GC detector configurations were described by Detournay et al. (2013) and Wang et al. (2013), consisting of bidimensional systems equipped with two different columns connected to two FIDs in the first case and to a FID and an MSD in the second one. Bidimensional chromatographs allowed the separation, identification, and quantification of extremely volatile and heavier BVOCs.

Amin et al. (2012) reported the use of a five-channel gas chromatograph equipped with two FIDs and an electron capture detector in conjunction with a mass spectrometer and four different separation columns for the analysis of whole air samples collected in stainless steel canisters. They also used a GC-MS analyzer for the analysis of BVOCs sampled on sorbent traps. Although most of the

BVOCs found in the solid traps were also found in the canister samples, limonene was not detected in the sorbent-trapped samples. The VF-5ms column used for BVOC separation does not resolve limonene from β-phellandrene, one of the BVOCs present in both samples but showing much higher concentrations near infested pine trunks. The difference between the electron impact mass spectra of β-phellandrene and limonene was used to confirm a higher concentration of the first one with respect to limonene in both types of samples.

In most of the mentioned papers, identification and quantification of BVOCs were done according to the retention time of the chromatographic peaks and the mass spectra (Amin et al., 2012; Hellén et al., 2012; Detournay et al., 2013; Llusia et al., 2013; Préndez et al., 2013; Tassi et al., 2013; Wang et al., 2013; Curtis et al., 2014), using a database to compare spectral data and an external calibration standard to determine BVOCs concentrations.

Assessment of the atmospheric budget of anthropogenic VOCs is usually performed from samples collected near the main emission sources. Rural sites (Parra et al., 2006; Tsai et al., 2008; Olszowski et al., 2012), remote places (deserts) (Yassaa et al., 2011), or coast sites (Liu et al., 2012), urban areas (Monteiro Martins et al., 2007; Parra et al., 2009; Park et al., 2010; Kawashima and Murakami, 2014; Kota et al., 2014), industrial areas (Turner et al., 2006; Ras-Mallorquí et al., 2007; Jia et al., 2008; Choi et al., 2009), urban areas affected by industrial emissions (Badol et al., 2008; Geng et al., 2008; Roukos et al., 2009; Ramírez et al., 2012; Louie et al., 2013) or major roads (Wang and Zhao, 2008), solid waste municipal and industrial landfill sites and treatment plants (Dincer et al., 2006; Rodríguez-Navas et al., 2012; Li et al., 2013; Gallego et al., 2014), wildfires (Evtyugina et al., 2013; Garcia-Hurtado et al., 2014), and petroleum refineries (Wei et al., 2014) are the usually studied sites.

As in the case of BVOCs, sampling of VOCs is typically done by either collecting whole air samples (Görgényi et al., 2000; Dincer et al., 2006; Kim et al., 2008a; Wang and Zhao, 2008; Choi et al., 2009; Park et al., 2010; Evtyugina et al., 2013; Li et al., 2013; Louie et al., 2013; Kawashima and Murakami, 2014; Kota et al., 2014) or specifically adsorbing VOCs on sorbent materials (Turner et al., 2006; Monteiro Martins et al., 2007; Ras-Mallorquí et al., 2007; Parra et al., 2009; Yassaa et al., 2011; Rodríguez-Navas et al., 2012; Ramírez et al., 2012; Gallego et al., 2014). Using Sep Pak cartridges (Parra et al., 2009; Louie et al., 2013) and pumping air samples through adsorbent tubes (Geng et al., 2008; Choi et al., 2009; Li et al., 2013) are also common ways of sampling.

VOCs can also be passively sampled through the exposure of samplers containing the most suitable adsorbent (Parra et al., 2009; Roukos et al., 2009; Juan-Peiró et al., 2012; Olszowski et al., 2012; Gallego et al., 2014; Garcia-Hurtado et al., 2104) or on a thermodesorption unit (Hellén et al., 2012) or first collected in a canister and then preconcentrated on the desorption unit (Badol et al., 2008; Wang and Zhao, 2008; Kawashima and Murakami, 2014).

Finally, VOC sampling could also be done in place and continuously by an online high-performance gas chromatographs equipped with a commonly used detector such as an FID and/or an electron capture detector (ECD) (Tsai et al., 2008; Liu et al., 2012). Tsai et al. (2008) and Kuo et al. (2014) monitored continuously VOCs using a commercial auto-two-dimensional (auto-2D) GC system.

Separation, identification, and quantitation of VOCs are usually done by performing GC for separation and three different detectors, MSD, FID, and ECD, used individually or combined. Target compounds were identified according to their retention times and mass spectra and quantified by multipoint external calibration method (Li et al., 2013).

GC–thermal conversion–isotope ratio mass spectrometry (IRMS) was used when isotopic analysis of atmospheric organic compounds and in particular of their stable carbon isotope ratio ($\delta^{13}C$) is used to identify the sources of VOCs (Turner et al., 2006; Kawashima and Murakami, 2014).

The use of combined detector systems has been reported for the analysis of different type of samples. Roukos et al. (2009), who studied the characteristics of urban and industrial harbor air, used GC equipped with dual-detection FID for VOC quantification and MS for qualitative analysis.

VOCs were analyzed, after preconcentration, by GC with a multidetector system consisting of an MS mass analyzer for sulfur compounds, two FIDs to measure hydrocarbons, and two ECDs for halocarbons (Louie et al., 2013).

The use of 2D chromatography permits a good separation of both heavy and light compounds, resolves the coelution problem, and prevents peak widening of heavier compounds. Finally, the development of a completely new and automated smart multichannel 2D micro-GC system which integrates a preconcentrator, microfabricated columns, an on-column gas detector (based on detection of changes in the refractive index or thickness of a PDMS film), thermal injectors, and gas pumps, as well as an algorithm for the control/operation of the system, was reported by Liu et al. (2013).

13.1.2 Volatile Organic Compounds in Natural Waters

The main anthropogenic sources of VOCs to aquatic environments are the effluents of urban and industrial activities, including wastewater, atmospheric deposition, urban and rural runoff, transport and/or transformation of fossil fuels, and natural sources (petrogenic and biogenic) (Dewulf and Van Langenhove, 1997). Anthropogenic sources of these compounds are well-known but natural sources, as marine sources are also important: oceans may act as a sink or reservoir of these compounds, and there is little information about the mechanisms inducing the production of these compounds by seawater organisms, such as macro- and microalgae.

In this sense, Huybrechts et al. (2005) studied the presence of selected VOCs in a specific area of the southern North Sea (chlorinated short-chain hydrocarbons, monocyclic aromatic hydrocarbons, and chlorinated monocyclic aromatic hydrocarbons). Surface seawater samples from the Irish Sea and the Liverpool Bay, areas of intense human activities, were studied by Bravo-Linares et al. (2007) to quantify and established the distribution of halocarbons and other selected VOCs which may have both natural and anthropogenic sources. Similar studies were reported by Cavalcante et al. (2010) in Fortaleza (Brazil), He et al. (2013) in the Yellow Sea and East China Sea, and Moliner-Martínez et al. (2013) in the Mediterranean Sea and estuaries, where they studied the distribution, emission sources, and influence of seasonal temperature fluctuations and precipitations on the concentration of VOCs in oceanic and estuarine environments.

Rainfall characteristics were analyzed by Mahbub et al. (2012) from the development of a prediction framework for semivolatile organic compounds (SOCs) and nonvolatile organic compounds under the dynamic influence of climate change.

Some recent experimental studies have dealt with the optimization of procedures for removing benzene–toluene–ethylbenzene–xylenes (BTEX) from water (Carvalho et al., 2012) and with the improvement of microextraction methods for the determination of these compounds in water samples (Hashemi et al., 2012).

The collection of water samples is done using large-volume containers (Huybrechts et al., 2005; Bravo-Linares et al., 2007; Mahbub et al., 2012; He et al., 2013), the samples being transferred to smaller containers sometimes spiked with acid mixtures added to prevent microbial degradation and surrogate or internal standard solutions (Cavalcante et al., 2010; Huybrechts et al., 2005; Bravo-Linares et al., 2007). VOCs could be extracted from water samples by liquid–liquid extraction with hexane as the exchange solvent and then cleaned up with different materials (Carvalho et al., 2012; Hashemi et al., 2012; Moliner-Martínez et al., 2013) until a homogeneous solution was achieved (Hashemi et al., 2012). VOCs in seawater samples were retained in a multisorbent trap containing Tenax TA, Carboxen 1000, and Carboxen 1001 (Huybrechts et al., 2005; He et al., 2013) or on SPME fibers (Bravo-Linares et al., 2007; Moliner-Martínez et al., 2013).

Analysis of VOCs could be performed with GC-MS (Huybrechts et al., 2005; Bravo-Linares et al., 2007; Mahbub et al., 2012), GC-FID (Hashemi et al., 2012; Moliner-Martínez et al., 2013), GC-ECD (He et al., 2013), GC-PID-FID connected in series (Cavalcante et al., 2010), or HPLC-UV (Carvalho et al., 2012) systems, and injection of samples was done in many different ways.

13.1.3 Volatile Organic Compounds in Soils

Soil is a complex and heterogeneous matrix with a porous structure that contains both inorganic and natural organic components. Currently, soils are subjected to intensive pollution with chemical compounds and they play an important role in the environment by preventing the contamination of adjacent ecosystems. When contaminating agents reach the soil, they become involved in different types of physicochemical interactions with mineral and organic components. VOCs are the most prevalent but less studied mobile contaminants of soil and sediments and are especially hazardous as potential sources of long-term contamination of groundwater, the atmosphere, and crops. The influence of the soil nature on the sorption and retention of VOCs (Serrano and Gallego, 2006) and the use of surfactants for remediation of subsoil and aquifers contaminated with VOCs (Long et al., 2013) were studied recently.

However, most of the research studies published for the past few years were devoted to the development and improvement of experimental procedures for the extraction, analysis, and quantification of VOCs in soil samples. A protocol based on purge-and-trap GC-MS method was developed for the simultaneous determination of different gasoline additives and their degradation products in soil samples (Rosell et al., 2006); the determination of BTEX in soils was improved with a fast method based on a multivariate calibration method to extract information from signals obtained from headspace mass spectrometry (HS-MS) equipment (Esteve-Turrillas et al., 2007), coupling a headspace autosampler with a PTV-GC-MS (Pérez Pavón et al., 2009) or developing a method based on simplified QuEChERS extraction followed by large-injection volume-fast GC-MS, which meets the requirements of green chemistry (García Pinto et al., 2011). Chlorinated pollutants present in soil samples were determined after liquid extraction with ethyl acetate followed by the addition of water to moisten the samples and salting-out partitioning of water with anhydrous $MgSO_4$ (García Pinto et al., 2010; Herrero Martín et al., 2010).

Soil samples were collected at contaminated (petrol station, high-traffic area, and abandoned petrol station) and noncontaminated sites, directly from the surface or at depths deeper than 0.5 m (petrol station) (Rosell et al., 2006; Esteve-Turrillas et al., 2007). Also, soils of different natures (agricultural, beach sand, bentonite clay, peat soil) (Serrano and Gallego, 2006; Pérez Pavón et al., 2009; García Pinto et al., 2010, 2011; Herrero Martín et al., 2010; Long et al., 2013) were sampled to study the influence of soil characteristics on the processes of sorption and retention of VOCs.

Analysis of VOCs present in soil samples was performed, in most of the cases, using a combination of GC-MS detector (Rosell et al., 2006; Serrano and Gallego, 2006; Esteve-Turrillas et al., 2007; Pérez Pavón et al., 2009; García Pinto et al., 2011). Alternatively, a micro-ECD (Herrero Martín et al., 2010; García Pinto et al., 2011) and an FID (Long et al., 2013) were also used as detectors after the chromatographic separation. Esteve-Turrillas et al. (2007) developed a fast method for BTEX analysis in soils, using an HS-MS equipment where the chromatographic column had been replaced by a silica tube transfer line.

13.1.4 Volatile Organic Compounds in Biological Samples

Recent research papers have evaluated the relationship between VOCs and biological samples from very different points of view: (a) screening of VOCs emitted for marine species to evaluate their contribution to overall VOCs emission (Colomb et al., 2008; Fratini et al., 2012); (b) characterization of odor profiles of different fungal species by analysis of several VOCs and use of profiles as patterns for a chemotyping tool (Müller et al., 2013); (c) evaluation of VOC profile exhaled for healthy and infected pets, which could be used to diagnose several diseases (De Oliveira et al., 2008); and (d) evaluation of VOC blood levels in runners and motorbike taxi drivers exposed to heavy traffic conditions (Avogbe et al., 2011; Blair et al., 2011).

VOCs emitted from biological samples were collected in the headspace of glass containers (algae suspensions, fungi growth, hair of healthy and infected dogs, shellfish, venous blood samples)

(Colomb et al., 2008; De Oliveira et al., 2008; Blair et al., 2011; Fratini et al., 2012; Müller et al., 2013), and extraction was done using SPME techniques (De Oliveira et al., 2008; Blair et al., 2011; Fratini et al., 2012; Müller et al., 2013). Personal exposure was evaluated from VOC sampling by using Radiello passive air samplers (Blair et al., 2011) or gas adsorbent badges containing an activated charcoal cartridge (Avogbe et al., 2011).

Assessment of phytoplankton production of isoprene and halogenated compounds from two different phytoplankton species was studied by HS-GC-MS system (Colomb et al., 2008). The results confirm the emission of isoprene, halogenated compounds, and other VOCs, but not all phytoplankton species emitted the same VOCs. The isolation of phytoplankton key species for local oceanic or coastal region will be useful to define the local sources of specific VOCs. The same analytical system, HS-GC-MS combined with SPME and chemometric tools (soft independent modeling of class analogy and principal component analysis) was used by De Oliveira et al. (2008) for the development of a complete methodology able to identify and evaluate specific VOCs profiles in the hair of dogs infected with visceral leishmaniasis. Successful identification of several biomarkers of the disease was possible when following the designed strategy. A similar procedure based on HS-SPME-GC-MS (Fratini et al., 2012) was employed to study VOCs in Atlantic shellfish species. The goal of the research work was the identification of volatile compounds as markers of the natural flavor and freshness of seafood. The results show that different shellfish species present different VOC profiles, allowing the identification of specific volatile compound contributors to flavor.

HS-SPME-GC-MS was used by Müller et al. (2013) to analyze VOCs released by different fungal species, trying to find characteristic biomarkers for the different fungal strains.

Exposure to BTEX of nonelite athletes was evaluated by SPME-GC-MS (Blair et al., 2011) from VOC analysis of ambient air samples (near a busy road) and from venous blood samples collected after the exercise phase. The results confirm the hypothesis that short-duration high-intensity roadside running increases BTEX concentration in blood.

13.2 POLYCYCLIC AROMATIC HYDROCARBONS

13.2.1 Polycyclic Aromatic Hydrocarbons in Outdoor Air

PAHs are a group of lipophilic anthropogenic chemicals that are ubiquitously distributed in the environment. PAHs consist of two or more fused benzene rings in various arrangements, and they form an important class of environmental contaminants because some of them exhibit carcinogenic or mutagenic potential (McGowin, 2006). PAHs are formed from the incomplete combustion of organic material such as oil, petroleum gas, coal, and wood, which are commonly used in energy production.

Once formed, PAHs are mainly transported and deposited through atmospheric pathways, this being the main controlling mechanism for the fate of airborne toxics (Crimmins and Baker, 2006; Fang et al., 2008). Although atmospheric PAHs partitioned between PM and the gas phase, they associate predominantly with fine and coarse PM. Breathing particulate-phase PAHs is considered a remarkable hazard to human health, and monitoring the levels of these compounds in urban areas has become a matter of primary importance. The European Union requires continuous air-quality monitoring with long-term assessment of benzo[*a*]pyrene (BaP), used as marker for carcinogenic risk of PAHs in ambient air. Directive 2004/107/EC (European Parliament and Council of the European Union, 2004) establishes a target value for BaP of 1 ng m^{-3} and a long-term objective of 0.1 ng m^{-3}.

PAHs, derivatives (mainly nitro-substituted PAHs), and other persistent organic pollutants are extracted from PM samples collected from urban and industrial atmospheres from cities of different sizes throughout the world (Karthikeyan et al., 2006; Yusà et al., 2006; Liu et al., 2007; Crimmins and Baker, 2006; Menichini et al., 2007a,b; Bohlin et al., 2008; Callén et al., 2008a,b; Fang et al., 2008; Halsall et al., 2008; Saarnio et al., 2008; Aldabe et al., 2012; Wu et al., 2014) and also at rural or remote areas considered as background sites (Menichini et al., 2007a; Yusà et al., 2006; Sofowote et al., 2010; Aldabe et al., 2012; Barrado et al., 2013) or pollutant sites because of the presence of fires (Souza et al., 2014).

Atmospheric fine and coarse PM are most commonly collected using large-volume samplers, on quartz filters of different dimensions (Crimmins and Baker, 2006; Karthikeyan et al., 2006; Yusà et al., 2006; Fang et al., 2008; Saarnio et al., 2008; Sofowote et al., 2010; Souza et al., 2014; Wu et al., 2014). The use of polyurethane foam (PUF) sheets to capture PAHs present in the gas phase (mainly benzo[*a*]anthracene [BaA] and chrysene [CHR]) has been also reported (Menichini et al., 2007a; Callén et al., 2008a,b; Saarnio et al., 2008; Sofowote et al., 2010). Liu et al. (2007) collected directly road dust from the surfaces of the pavement to evaluate the contribution of petrogenic and pyrogenic sources on the PAH content.

Before analysis, extraction of the organic compounds from the quartz filters or the PUFs is necessary. Most analytical methods have been developed based on US EPA method TO-13 (EPA, 1999). Soxhlet extraction (Fang et al., 2008; Bohlin et al., 2008; Callén et al., 2008a,b; Halsall et al., 2008; Aldabe et al., 2012; Souza et al., 2014), closed-vessel microwave-assisted extraction (MAE) (Karthikeyan et al., 2006; Aldabe et al., 2012), ASE (Yusà et al., 2006; Menichini et al., 2007b), or ultrasonic extraction (Menichini et al., 2007b ; Wu et al., 2014) are the most commonly described procedures for filter processing and PAH extraction from airborne PM.

GC-MS is the more commonly used technique for PAH analysis due to its greater separation efficiency compared to those of HPLC methods. Most of the samples outlined earlier were analyzed using GC-MS systems equipped with standard PTV split–splitless inlet and capillary column for separation of organic compounds.

Identification of individual PAHs was performed from the comparison of the retention times in the chromatographic column and the mass spectra of the samples with the data obtained from PAH standards. Quantification of individual compounds was based on comparison of peak areas with those of the standards, and recoveries were calculated from addition of an internal standard to samples before the extraction process. The use of statistical tools allowed the main pollution sources and associations of PAH concentrations with atmospheric or seasonal characteristics. Optimization of extraction and injection procedures was studied by several authors in order to improve the recovery and sensitivity of analyzed PAHs. The use of less hazardous mixtures of organic solvents and a lower-extraction temperature PM, improving the recovery of PAHs, was reported by Karthikeyan et al. (2006). Crimmins et al. (2006) and Yusà et al. (2006) made use of GC-MS systems to optimize the use of the PTV inlet with larger volumes of the extracted samples and obtain higher analytical sensitivities and improved temporal resolution. Fang et al. (2006) found that total individual concentrations of PAHs extracted from $PM_{2.5}$ were higher than those found in $PM_{2.5-10}$ and traffic was the main source for PAHs. Similar results were published by Liu et al. (2008), Wu et al. (2014), and Saarnio et al. (2008); not only traffic but also coal, fuel oil, and biomass combustion were the most important sources of PAHs whose concentrations followed a seasonal trend and were found to be higher in winter than in any other season of the year. Different diurnal patterns were reported by Souza et al. (2014) for the quantification of PAHs depending on the family of compounds studied. Sofowote et al. (2010) reported that local sources were more important than long-range source regions and an important seasonal influence in the values of gas- and particle-phase PAHs. Saarnio et al. (2008) found that benzo[*k*]fluoroanthene is a more suitable marker for carcinogenic PAHs than BaP. Callén et al. (2008b) found than stationary sources related to industry were the dominant pollution sources in Zaragoza, although vehicular emissions and natural gas home heating were also significant contributing sources. The importance of the height at which sampling of PAHs is done was studied in Rome (Menichini et al., 2007b) at the road level and at sixth- to ninth-floor buildings. The authors found that the relevant vertical gradient for PAH concentrations at 20–30 m height were 20–70% lower than values found at the road level, after the samples were analyzed by GC/low-resolution MS. When background levels were monitored at remote sites (Menichini et al., 2007a; Aldabe et al., 2012), concentrations for particulate-phase PAHs were found to be two orders of magnitude lower than in urban sites. Again, a seasonal influence in PAH concentrations showing higher values in winter than in summer was found.

Halsall et al. (2008) reported the analysis of PAHs combining the use of GC-MS and HPLC equipped with a fluorescence detector. Also, characterization of magnetic properties of aerosol was carried out, enabling high-resolution spatial and temporal sampling since this property can be used as specific marker for traffic-related pollution. The study concluded that there is little contribution of outdoor air in indoor spaces, even when sampling was done at the city center of Lancaster.

13.2.2 Polycyclic Aromatic Hydrocarbons in Natural Waters

PAHs released to the atmosphere through incomplete combustion processes or originated after oil spills or some industrial processes are deposited into the aquatic environment and associated with suspended PM because of their high partition coefficient and low solubility in water. They are present in trace concentrations, and, consequently a preconcentration step will be necessary before analysis is carried out. The hazardous properties of these organic compounds make monitoring of PAHs in different aquatic environments a priority for many researchers and public institutions. In this sense, the WFD (Directive 2000/60/EC [European Parliament and Council of the European Union, 2000]) and daughter directives listed the priority pollutants to be monitored and defined the monitoring periods, motivating the development of new or improved methods for sampling and analysis of PAHs in aquatic environments.

The fact that these compounds are present in trace concentrations drives most of the published research to the development of methods that allowed the simultaneous sampling and preconcentration of PAHs in different aqueous environments (natural water, rainwater, and seawater). Devices based on SPME (Hsieh et al., 2006; Bagheri et al., 2007), ASE (Olivella, 2006a,b), SBSE (Pérez-Carrera et al., 2007; Prieto et al., 2007), and SPMDs (Wang et al., 2009a; Marrucci et al., 2013) have been successfully used for sensitive, robust, and reliable sampling.

Separation and identification of PAHs in water samples were generally performed by using GC-MS systems, equipped with capillary columns. A mass spectrometer was operated in electron impact ionization (EI) mode (Hsieh et al., 2006), and the mass spectra of the target compounds was acquired and quantified in the SIM mode (Hsieh et al., 2006). Quantitation was done by calibration with internal standards at several concentration levels. Recoveries were calculated by spiking standards into the sample solutions.

The results obtained after using the combination of ionic liquid SPME with GC-MS confirmed the versatility of the whole system since the ionic liquid coating could be specifically modified according to the analyte to be studied (Hsieh et al., 2006). In a similar way, Bagheri et al. (2007) modified an SPME device, growing electrochemically a film of a conducting polymer such as polyaniline, claiming the use of this polymer as the fiber in future SPME devices. Olivella et al. (2006a,b) developed and optimized a procedure for the efficient sampling and fast elution of PAHs from large-volume water samples, in both rainwater and surface (lake) water, using ASE followed by GC-MS.

Devices based on SBSE and combined with a TD-GC-MS system allowed the analysis of water samples with low concentrations of PAHs and low limits of detection and are described as reproducible, showing good repeatability and sensitivity after being used in the analysis of seawater and interstitial water (Pérez-Carrera et al., 2007; Prieto et al., 2007).

Marrucci et al. (2013) made use of a GC-MS system whose detector was operated in EI mode. The GC-MS was equipped with an ion trap that allowed the system to be used in MS/MS mode. The MS/MS process was carried out sometimes via collision-induced dissociation to observe the resonant excitation of some of the compounds analyzed. Better results were obtained in single ion storage mode for other analytes. The use of SPMDs for sampling seawater and the analysis using the earlier mentioned analytical technique allowed authors to detect very low concentrations of PAHs in seawater and find significant differences between the compositions of individual PAHs found at the two sampling sites. A similar sampling device was used by Wang et al. (2009a) to analyze the water column at different points upstream the Three Gorges Reservoir in China, using a conventional GC-MS system. Higher concentrations of PAHs were found near the dam, although the authors found a significant relationship between sampling rate and water flow velocity, meaning that the slowdown of the water close to the dam might have increased the concentrations of organic pollutants.

13.2.3 Polycyclic Aromatic Hydrocarbons in Soils

Soil is a complex environmental matrix that shelters a great amount of biological activities whose integrity could be altered by the presence of anthropogenic pollutants. The European Soil Portal (European Soil Data Centre, 2015) defines soil contamination as the presence of anthropogenic alteration in the natural soil environment. Soils could be contaminated by PAHs when pollutants are spilled or contaminated surface waters percolate, so contaminated soil can be found in the proximity of major chemical industries, former industrial sites, coke production facilities, or places where ashes from waste incinerators are deposited, among others.

For the last few years, research in contaminated soils is mainly centered on three different areas: optimization of PAH extraction procedures (Richter et al., 2006; Pena et al., 2007; Wang et al., 2007), quantification of the amount of PAHs removed from soils or released to soils (Viglianti et al., 2006; Moret et al., 2007), characterization of PAHs in polluted soil samples (Bergknut et al., 2006; Yang and Hofmann, 2009), and description of spatial distribution of contaminants (Bengtsson and Törneman, 2009; Bengtsson et al., 2010).

Extraction of PAHs from contaminated soils is the most time-consuming step, and high amounts of solvents need to be used. Pressurized solvent extraction (PSE) (Richter et al., 2006), ultrasound-assisted pressurized solvent extraction (US-PSE) (Richter et al., 2006), MSPD technique (Pena et al., 2007), Soxhlet extraction (Wang et al., 2007), MAE (Wang et al., 2007), and ASE (Wang et al., 2007) are the usual extraction techniques. Results indicate that optimization of extraction processes managed to improve both aspects, reducing the time and the amount of solvents necessary to complete the process, when MSPD, US-PSE, and ASE are used with real contaminated samples: soil samples collected in the vicinity of a petroleum refinery (Richter et al., 2006) and contaminated agricultural soil (Pena et al., 2007). Analysis of extracts was performed using two different chromatographic techniques, HPLC with fluorescence detection (Pena et al., 2007) and conventional GC-MS (Richter et al., 2006; Wang et al., 2007).

When it comes to quantification of PAHs accumulated in soil samples and the study of their spatial distribution, conventional GC-MS is the chosen analytical technique in most of the cases.

Different types of contaminated samples were studied by Bergknut et al. (2006) and Yang and Hofman (2009): soils from industrial sites, antiskid sand, urban dust, ash samples from municipal solid waste incinerators, and river floodplain soils with different compositions of carbonaceous materials. GC-MS (Bergknut et al., 2006; Yang and Hofmann, 2009) and GC–time-of-flight mass spectrometry (TOFMS) (Bergknut et al., 2006) were the analytical techniques used by the authors. Contaminated soils are samples with a very complex chemical composition, and the presence of coeluting peaks in the chromatograms is quite common. The use of GC-TOFMS allows the resolution of those coeluting peaks, making possible the identification of 962 peaks in a pooled sample and between 123 and 527 peaks in individual samples. Yang and Hofman (2009) analyzed PAHs previously extracted by ASE by using Milli-Q water as extraction solvent. They made use of the high affinity of PAHs to natural organic content to use water as the solvent for the extraction procedure since the presence of dissolved organic carbon enhanced their solubility in aqueous media.

This same affinity between PAHs and carbonaceous materials was used by Bengtsson and Törneman (2009) and Bengtsson et al. (2010) to describe the spatial extent and the degree of environmental risk to mixtures of PAHs, the spatial heterogeneity and the correlation of PAH biodegradation and soil respiration, and possible associations with soil properties (pH, organic carbon content,

PAH concentrations, soil respiration, water holding capacity, etc.). Quantification of extracted PAHs was done using a GC-FID, and peak identification was confirmed by GC-MS.

HPLC equipped with a programmable spectrofluorometer (Moret et al., 2007) or UV–visible detector (Viglianti et al., 2006) were used as an analytical technique to quantify PAH extracts obtained from samples where the organic pollutants have been either removed from soils (Viglianti et al., 2006) or released to soils (Moret et al., 2007). Aged-contaminated soils were studied in the presence of an efficient removal agent as cyclodextrin: addition of cyclodextrin to aqueous washing solutions improved the removal efficiency several times compared to that for just water (Viglianti et al., 2006), highly promising results which could be extremely useful for subsurface remediation of aged soils. Moret et al. (2007) studied the amount of PAHs released into soil from old railway ties stored in different places and the possible transfer to vegetation growing in the vicinity of the storage place. Analysis of the extracts was done by HPLC equipped with a programmable spectrofluorometer set at optimized excitation and emission wavelengths to obtain the best sensitivity and selectivity. The results show that PAH amounts decrease rapidly with increasing distance from the source of contamination.

In all cases, PAH concentrations were calculated using external standard curves and recoveries were obtained after the samples were spiked with an internal standard.

13.2.4 Polycyclic Aromatic Hydrocarbons in Sediments

Sediments are potential reservoirs for contaminants when they accumulate in highly impacted urban and industrial environments, mainly harbors and estuaries. Coastal, oceanic, and riverine sediments contaminated with PAHs are a source of exposure of these organic pollutants to benthic and pelagic organisms since they tend to accumulate in bottom sediments with time because of their hydrophobic and persistent nature.

The level of pollution and the discrimination between pyrogenic (PAHs generated through incomplete combustion of organic matter in industrial operations, power plants using fossil fuels, waste incinerators, and exhaust from vehicles) and petrogenic (soot or gas brought to marine environments through direct atmospheric fallout and surface runoff and from crude oil and petroleum products) sources of PAH, are the main goals of most of the research in this area. Coastal and riverine environments (Boonyatumanond et al., 2006; Christensen et al., 2010), estuaries (Vane et al., 2007), urban lakes, harbors, and shipping waterway (Kim et al., 2008b; Slater et al., 2008) are some of the selected sites for PAH pollution monitoring. Extraction of PAHs from polluted samples was performed using all the available procedures, PSE (Boonyatumanond et al., 2006), sonication (Vane et al., 2007), ASE (Kim et al., 2008b), microwave extraction (Slater et al., 2008), and Soxhlet extraction (Christensen et al., 2010), always followed by a cleanup step using the corresponding chromatography column.

The preferred analytical technique for PAH extraction was conventional GC-MS, although some authors also made use of HPLC coupled to a fluorescence detector (Vane et al., 2007). Once identified and quantified, concentrations of individual PAHs are used to calculate different ratios in order to identify the petrogenic or pyrogenic origin of studied PAHs. Boonyatumanond et al. (2006) not only found low to moderate contamination by PAHs in monitored sites but also detected pyrogenic PAHs at several remote locations from MP/P ratios, defined as the sum of the different methylphenanthrene derivatives (MP) to phenanthrene (P), which indicates pyrogenic origin of PAHs when results are lower than 1. Phenanthrene to anthracene ratio was used by Vane et al. (2007) to classify the origin of studied PAHs as petrogenic, when ratios were greater than 10. Ratios lower than 10 corresponded to pyrogenic PAHs, but, as there are exceptions to this behavior, combination of these ratios with others obtained from different pairs of isomers gave a more reliable estimation of the origin of the PAH source. BaA/CHR and fluorine/pyrene are examples of additional PAHs ratios which serve the origin identification purpose. Kim et al. (2008b) made use of similar indexes, C_2-dibenzothiophene/C_2-phenanthrenes and C_3-dibenzothiophenes/C_3-phenanthrenes, to identify different sources of oil having different sulfur contents or C_2-CHRs/C_2-phenantrenes to differentiate harbor and shipping waterway PAHs from the urban lake sediments PAHs. Carbon isotope ratios were measured by GC–/IRMS and were also used as tracers of contaminant sources since they are quite resistant to chemical and biological alterations. Isotope ratios confirm the unique pyrogenic origin of the complex mixture of PAHs found in the urban lake sediments.

Methane concentration in porewater sediments was characterized by GC/IRMS together with PAHs and bacterial phospholipid fatty acids (PLFAs), the last two analyzed by GC-MS, to assess the impact of high concentrations of PAHs in living organisms as the sedimentary microbial communities (Slater et al., 2008). The isotopic results for methane illustrate the relevance of the technique for the analysis of geochemical parameters in the characterization of microbial activities in the environment. Results show the negative impact of PAH deposition on microbial communities because elevated levels of PAHs were found associated to low concentrations of PLFAs, an indicator of the occurrence of certain microbial community members.

Search for new indicators which facilitate the discrimination of pollution sources was developed by Christensen et al. (2010), who analyzed sections of selected ion chromatograms to remove variations unrelated to chemical composition (baselines, retention time shifts, or changes in detector sensitivity), retaining only variations in PAH composition that can be used to distinguish samples based on their source input. The studied method allowed the discrimination of the pollution sources in the studied bay and also the identification of new potential indicator ratios involving coelution peaks: ratios of C_1-BaA to C_1-CHRs and C_1-fluoroanthenes to C_1-pyrenes were found to be good indicators for distinguishing between pyrogenic and petrogenic sources. The obtained results indicate that the main sources of PAH were pyrogenic mixed with a fraction of high-molecular weight petrogenic PAHs.

13.2.5 Polycyclic Aromatic Hydrocarbons in Biological Matrices

Biological matrices are frequently used as pollution bioindicators or bioaccumulators because their treatment and analysis in the laboratory is easier than that of air samples. When an organism can be selected as bioindicator, it has to be able to accumulate a certain amount of pollutant or change its life cycle as a response to the presence of the pollutant. Lichens (Domeño et al., 2006), tree leaves (Ratola et al., 2006a; De Souza Pereira et al., 2007; Orecchio, 2007), marine samples (fish, mammals, polychaete) (Liguori et al., 2006; Pena et al., 2006; Cheung et al., 2007; Liang et al., 2007; Myers et al., 2008; Tairova et al., 2009) are some of the biological samples typically used as PAH bioindicators. Plant species are ideal biomarkers because they present a lipid-rich cuticle, whose nature allows the accumulation of organic pollutants. Marine organisms are exposed to pollutants previously concentrated in marine sediments, especially in coastal waters.

Most efforts are devoted to the development of efficient and selective methods for the extraction of PAHs in all kind of samples, always followed by a cleanup step. ASE (Liguori et al., 2006; Ratola et al., 2006b), MAE (Pena et al., 2006), Soxhlet extraction (Cheung et al., 2007; Liang et al., 2007), ultrasonic extraction (Ratola et al., 2006b; Leroye et al., 2010), and dynamic sonication-assisted solvent extraction method (Domeño et al., 2006) have been selected as the most efficient techniques, depending on studied samples. When dealing with marine samples, the cleanup step is of great relevance to eliminate the fat content of samples. Gel permeation chromatography, use of silica or alumina cartridges, and use of Florisil columns are the most used cleanup procedures.

Combination of optimized extraction methods with accurate analytical procedures, using conventional GC-MS as an analytical technique in most of the cases, is an extremely useful tool for the evaluation of food contamination and safety of freshwater and marine fish.

Cheung et al. (2007) studied the bioaccumulation of PAHs in two different parts of the bodies of freshwater and marine fish, the axial and ventral muscles, suggesting the use of the ventral muscle as indicator tissue due to its high fat content, associated with high PAH content. However, as aqueous and dietary routes are both important for PAH uptake, fish living in mangrove swamps and intertidal mudflats accumulate greater amounts of PAHs in the viscera than in muscles, pointing to this tissue as a promising indicator of PAH levels because it contains much higher concentrations (Liang et al., 2007).

Tairova et al. (2009) evaluated the use of 1-hydroxypyrene, a pyrene metabolite, as a biomarker of PAH exposure in the marine polychaete *Nereis diversicolor.* The identification and quantification of the biomarker was carried out by using three different analytical techniques: HPLC–fluorescence detection, synchronous fluorescence spectroscopy, and GC-MS, finding significant and similar correlations between total PAH and 1-hydroxypyrene by using both HPLC-fluorescence and GC-MS.

Marine mammals are also exposed to PAHs through the ingestion of contaminated prey species, particularly benthic invertebrates, so the livers of adult female sea otters were analyzed to determine PAH content (Kannan and Perrotta, 2008). Analysis was done by Soxhlet extraction followed by GC-MS, and the results show that concentrations of PAHs in sea otters have declined during the time decade monitoring was carried out, suggesting a decrease in the emission of PAHs in the area of study, the PAH concentrations being significantly correlated with lipid content in the otter livers.

Orecchio (2007) analyzed leaves of *Quercus ilex* L., bioaccumulators of PAHs, to assess the degree of contamination of the studied area and to confirm the use of these leaves as biomonitors of PAHs. Optimization of PAH extraction in an ultrasonic bath and an analysis with a GC-MS system was performed, and the author found that *Quercus* leaves could effectively be used as biomonitors of PAH contamination even for retrospective analysis. A wide range of leaf PAHs concentrations was detected in the studied area showing strong pollution gradients related to different emission sources and also to different degrees of air dispersion.

De Souza Pereira et al. (2007) chose an epiphytic bromeliad, *Tillandsia usneoides* (L.), as a possible biomonitor for the immission of pollutants in highly contaminated areas. During this study, monitoring of total deposition rates of PAHs and other pollutants was done in a very industrialized area in Rio de Janeiro State (Brazil) by using an endemic bromeliad species. PAHs were analyzed by HRGC-MS). The chosen bromeliad species was found to be a sensitive, reliable, and low-cost biomonitor for organic pollutants in ambient air in tropical climates.

Wang et al. (2009b) made use of the physicochemical properties of PAHs to characterize their distribution in different types of samples. Accumulation trends and distribution of 16 PAHs in soil, moss, and reindeer dung was the main goal of this study. Soil is considered as one of the major sinks of PAHs, and vegetation plays an important role in removing PAHs form the atmosphere. For poorly metabolizable SOCs, one of the primary elimination mechanisms from body is via feces. Extraction of PAHs from the different samples was done using an ultrasonic bath, and extracts were analyzed by conventional GC-MS. Differences found in the PAH distribution in soil, moss, and reindeer dung might be related to the physicochemical properties of individual PAHs and the different accumulation routes of PAHs in the different samples. Higher proportions of low-molecular weight PAHs (two- and three-ringed PAHs) and lower proportions of high-molecular weight PAH (five- and six-ringed PAHs) were found when the data are compared with results from other locations. Physicochemical properties are suitable parameters for characterizing the distribution of PAHs in the different studied samples.

13.3 NITROGEN OXIDES

Nitrogen oxides (NO_x: NO + NO_2) are important air pollutants, playing an important role in the chemistry of the atmosphere (Manahan, 2010). NO_x are released into the air from combustion processes and take part in a variety of environmental problems (Davidson et al., 2012): formation of tropospheric ozone, production of acid rain, global warming, and depletion of stratospheric ozone.

Gas-phase chemiluminescence through the reaction of NO with ozone is the favored method for NO_x analysis (Hodgeson, 1974), which allows detecting trace concentrations of NO and NO_2 (previously converted to NO):

$$NO + O_3 \leftrightarrow NO_2 + h\nu \text{ (light 0.6–3 μm)}.$$

Nitrogen dioxide can also be analyzed by other chemiluminiscent methods. Wendel et al. (1983) detected NO_2 through the chemiluminiscent reaction between this species and luminol in alkaline solution. Gaffney et al. (1998) measured both NO_2 and peroxyacyl nitrate (PAN) at sub-ppb concentrations by using a modified luminol detection system coupled to a capillary gas chromatographic column (3 m in length, DB-1, 0.53 mm, 3.00 μm) with helium as carrier gas. The system was capable of analyzing NO_2 in less than 1 min, applying an instrumentation that is simpler than that used by other researchers (Blanchard et al., 1993). The luminol detector has detection sensitivity comparable to that of an ECD system for PANs, and it could be improved by thermoelectric cooling to lower the dark current on the photomultiplier tube. Potential disadvantages of the system include temperature dependence and nonlinearity at very low NO_2 concentrations.

More recently, Milligan et al. (2002) used a relatively new technique, selected ion flow tube MS for the quantification of multiple trace gases above soils, including NO and NO_2. A major advantage of this method is that real-time monitoring of NO_x emissions is possible without the need for preconcentration, trapping, or chromatographic separation.

Nitrogen dioxide can also be reliably measured in the atmospheric air by use of passive samplers. Since the first passive sampler described by Palmes and Gunnison (1973), a variety of samplers have been developed (Hafkenscheid et al., 2009). All of them rely on diffusion of contaminant molecules across concentration gradients defined by Fick's law. Diffused contaminants are trapped on filters coated with different chemical reagents. NO_2 is normally chemiadsorbed onto triethanolamine (TEA) as nitrite, being quantified by visible spectrophotometry or ion chromatography (IC). This reaction can also be achieved actively by drawing a known volume of air through a sampling tube containing TEA. After sampling, samples are desorbed using an aqueous TEA solution and analyzed by nitrite by IC provided with a conductivity detector. The limit of quantification for NO_2 is about 1 ppb, which is suitable for the study of natural and urban areas.

Nitrous oxide, also known as laughing gas, is another important nitrogen oxide. The largest anthropogenic source of N_2O emissions is agriculture (fertilized agricultural soils and livestock manure) (Hénault et al., 2012). It is also a by-product of fuel and biomass burning. On the other hand, the largest natural source of nitrous oxide is soil under natural vegetation, reaching up to 60% of natural emissions.

The most widely used analytical method for measuring N_2O is GC with an ECD (Rapson and Dacres, 2014). ECD techniques rely on using large gas headspace injections (0.5–5 mL) through packed columns, such as a Porapak Q or HayeSep Q. In general, Ar or N_2 with 5% CH_4 is the preferred carrier gas (Wang et al., 2010a). If GC is coupled with an FID, then it is possible to measure other gases of interest at the same time, such as CO_2, CH_4, and CO. CO_2 and CO must first be catalytically converted to CH_4, prior to analysis with an FID (Laan et al., 2009; Thompson et al., 2009).

Using an ECD, LODs of 30 ppb with a precision of 0.18–0.4 ppb have been obtained. These LODs can be improved to 18 ppb through the use of an extractive technique, such as SPME (Drescher and Brown, 2006).

13.4 OZONE

Ozone is a naturally occurring chemical that can be found in both the stratosphere (the so-called ozone layer, 10–40 km above the earth) and the troposphere (the ground-level layer, 0–10 km above the earth). Tropospheric ozone is not emitted directly into the air but is created by chemical reactions between nitrogen oxides (NO_x) and VOCs in the presence of sunlight (Finlayson-Pitts and Pitts, 1993). While in the stratosphere ozone protects us from the damaging UV radiation, in the troposphere, due to its extremely high oxidizing power, ozone can promote a variety of health problems (Madronich et al., 2015), also having harmful effects on sensitive vegetation and ecosystems (Cape, 2008).

A variety of techniques have been developed to measure O_3, the most common being the absorption of UV light by the ozone molecule at a wavelength of 254 nm. The UV analyzer operates by comparing absorption measurements of the sample air with measurements of the same sample air after removal of only the ozone by an ozone scrubber (Jiménez et al., 2007).

Besides automatic monitoring techniques, in recent years passive sampling of gaseous pollutants has received considerable attention. Ozone samplers have been widely used for ambient monitoring all over the world, and among them, the Ogawa & Company USA, Inc. sampler (Koutrakis et al., 1993) has probably been the most commonly used.

The ozone passive sampler uses a filter coated with a nitrite-based solution that is oxidized to nitrate by ozone. After exposure, the filter is extracted with ultrapure (Milli-Q) water and the filter extract is analyzed by IC to determine the nitrate ion concentration (Salem et al., 2009), which is used to calculate the total amount of ozone collected. This technique allows determining ozone concentrations as low as 0.2 ppb.

13.5 CARBON MONOXIDE

Carbon monoxide is an important component of the troposphere, and it is considered as one of the most toxic gases in urban air. For this reason, its concentration is regulated by the most important international environmental agencies and the routine and accurate measurement of its level has become very important. Several methods have been developed for the determination of CO, infrared spectroscopy being the most commonly used the, as it is suitable for continuous measurements. However, this method can be interfered by other gaseous compounds; moreover, this technique is not suitable to determine small amounts of gas samples.

GC methods have been used for specific determination of CO in complex gaseous matrices (Kaminski et al., 2003) or in those where a low detection limit is necessary. Thermal conductivity detectors (TCDs) have been widely used for the determination of CO and CO_2 (Miller et al., 1980; Lodge, 1989; Wardas et al., 2000) in environments with high concentration since a TCD is not able to detect contents below a few hundred parts per million. Lodge (1989) established a chromatographic method to separate and determine O_2, N_2, CO, CO_2, and CH_4 in gas samples, using a dual column/dual TCD system.

Carbon monoxide must be usually trapped in a sorbent sampler prior to being injected to the GC, although some authors such as Wardas et al. (2000) have developed chromatographic systems that can introduce the sample directly by using the adsorbent Carbosieve II as the stationary phase.

In contrast to the aforesaid GC-TCD systems, FID is the favorite detector for sensitive determinations (Capasso and Inguaggiato, 1998; Juntarawijit et al., 2000; Ueta et al., 2013). When this detector is coupled to GC, carbon monoxide can first be reduced to methane by a methanizer. These methods are used in very specific determinations that usually require detection limits below 1 ppm. Kaminski et al. (2003) developed a method for determining trace amounts of CO, CH_4, and CO_2 in refinery hydrogen gases or in air. They modified GC-FID by adding a short molecular sieve 5A packed column to a Porapak Q column prior to reach the methanizer. This method reached detection limits of 0.15, 0.15, and 0.20 μg L^{-1} for CO, CH_4, and CO_2, respectively. Ueta et al. (2013) created a GC-FID system based on the introduction of a packed-capillary column and a methanizer to assess the photocatalytic properties of TiO_2 for gaseous VOCs.

Although GC-TCD and GC-FID are the most common chromatographic methods for determining CO, other methods have been developed to solve specific analytical problems. Thus, Cárdenas et al. (1998) used a reduction gas analyzer to measure carbon monoxide. The instrument consisted of a GC coupled to a UV absorption detector with detection based on the reaction of CO on a heated bed of HgO to produce $Hg_{(g)}$, which was determined by the UV detector.

Finally, some authors have designed specific chromatographic systems combining different analytical techniques to respond to new analytical demands. Thus, Zeng et al. (1994) combined GC with IRMS to measure the stable carbon isotope ratios of methane and carbon monoxide in environmental samples. The isotopic data of these compounds could help to understand their mechanism of production. The system consisted of a gas chromatograph, an interface with a combustion furnace, and an IRMS. This GC-combustion-IRMS system allowed them to determine the isotope ratio for VOCs without pretreatment.

13.6 SULFUR DIOXIDE

Sulfur dioxide is one of the major pollutants of air, and it is considered a primary criteria pollutant. SO_2 is released into the air from fuel combustion and is responsible for local air pollution and can exacerbate smog formation and acid rain. Although the emission of SO_2 has gradually decreased during the past few decades because of the control measures carried out by international environmental agencies, sulfur dioxide is still routinely measured in both air quality monitoring surveys and indoor environments. The most employed method to determine this pollutant is UV fluorescence, which allows for monitoring gas concentration in a continuous way. One example of this UV fluorescence continuous system was used by Mallik et al. (2012) in an air quality survey carried out in India. The SO_2 is excited using a 214 nm wavelength and a band-pass filter centered around 350 nm to measure the fluorescence. Other continuous gas analyzers, designed to direct determinations of multiple gases in stack or exhaust pipes, use different systems such as electrochemical nondispersive infrared sensor or infrared absorption gas filter correlation.

More recently, Sulzer et al. (2012) modified a conventional proton-transfer reaction (PTR)–mass spectrometry instrument to transform it into a universal trace gas analyzer. To that end, they slightly modified a PTR-MS instrument to be capable not only of ionizing trace compounds via PTRs but also of ionizing via charge-transfer-reactions, obtaining higher ionization energy. These minor changes allows using the PTR-MS to determine not only trace VOCs but also very important substances as CO, CO_2, CH_4, NO_x, and SO_2.

From an environmental point of view, GC has been widely used to determine reduced sulfur compounds (H_2S, dimethylsulfide, mercaptans, alkyl sulfides) due to their volatile nature (Yassaa et al., 2006; Bachu et al., 2009; Gras et al., 2009; Swan et al., 2015). In contrast, there is barely evidence of GC methods for SO_2 analysis. One example is the method developed by Wang et al. (2010b), who integrated a gas chromatograph with an in-house-built absorption–desorption device capable of measuring SO_2 at near-ppb level in order to detect SO_2 impurities in the coal gas-derived H_2 fuel. In this technique, the sample gas is collected by an SO_2 absorbent, which is later desorbed at a high temperature prior to gas analysis in a gas chromatograph equipped with a flame photometric detector.

When continuous measures are not necessary, a simple method to determine SO_2 lies in trapping the gas by using specific chemical reagents that transform SO_2 to SO_4^{2-}. Afterward, SO_4^{2-} is extracted and analyzed by IC.

In some of these methods, gas samples are passed through diffusion scrubbers or wet denuders with specific solutions that react with SO_2. Thus, Polkowska-Motrenko et al. (1998) developed a system to simultaneously determine SO_2 and SO_3 in combustion flue gases from a coal-fired electric plant. Air sample was passed through a scrubber with an alkaline solution (NaOH and formaldehyde), where both gases were absorbed. In the case of Komazaki et al. (1999), they developed an automated and continuous system by using a diffusion scrubber coupled to IC. Deionized water was used as a scrubbing solution for collecting HCl, HNO_3, SO_2, and NH_3, which were dissociated to ion species (Cl^-, NO_3^-, SO_4^{2-}, and NH_4^+, respectively). The scrubber was connected to concentrator columns, where samples were concentrated and introduced to the ion chromatograph, where all the analyte ions were separated and determined.

In other cases, gas samples passively diffuse through impregnated filters (Bralic et al., 2012) or passive samplers (Ayers et al., 1998, Yamada et al., 1999, Campos et al., 2010). Absorbents more frequently used for the passive sampling of SO_2 are sodium tetrachloromercurate (Na_2HgCl_4), potassium carbonate (K_2CO_3), TEA, sodium hydroxide (NaOH), and sodium carbonate (Na_2CO_3) (Cox, 2003; Cruz et al., 2004). Finally, filters are leached at the laboratory by using different solutions such as deionized water or H_2O_2 and analyzed by IC.

REFERENCES

Aldabe, J., Santamaría, C., Elustondo, D., Parra, A., Foan, L. et al. 2012. Polycyclic aromatic hydrocarbons (PAHs) sampled in aerosol phase at different sites of the western Pyrenees in Navarra (Spain). *Environmental Engineering and Management Journal* 11: 1049–1058.

Amin, H., Atkins, P.T., Russo, R.S., Brown, A.W., Sive, B. et al. 2012. Effect of bark beetle infestation on secondary organic aerosol precursor emissions. *Environmental Science & Technology* 46: 5696–5703.

Avogbe, P.H., Ayi-Fanou, L., Cachon, B., Chabi, N., Debende, A. et al. 2011. Hematological changes among Beninese motor-bike taxi drivers exposed to benzene by urban air pollution. *African Journal of Environmental Science & Technology* 5: 467–472.

Ayers, G.P., Keywwood, M.D., Gillet, R., Manins, P.C., Malfroy, H. et al. 1998. Validation of passive diffusion samplers for SO_2 and NO_2. *Atmospheric Environment* 32: 3587–3592.

Bachu, S., Pooladi-Darvish, M., Hong, H.F. 2009. Chromatographic partitioning of impurities (H_2S) contained in a CO_2 stream injected into a deep saline aquifer—Part 2: Effects of flow conditions. *International Journal of Greenhouse Gas Control* 3: 468–473.

Badol, C., Locoge, N., Léonardis, T., Galloo J.-C. 2008. Using a source-receptor approach to characterize VOC behavior in a French urban area influenced by industrial emissions—Part I: Study area description, data set acquisition and qualitative data analysis of the data set. *Science of the Total Environment* 389: 441–452.

Bagheri, H., Babanezhad, E., Es-haghi, A. 2007. An aniline-based fiber coating for solid phase microextraction of polycyclic aromatic hydrocarbons from water followed by gas chromatography–mass spectrometry. *Journal of Chromatography A* 1152: 168–174.

Barrado, A.I., García, S., Castrillejo, Y., Barrado, E. 2013. Exploratory data analysis of PAH, nitro-PAH and hydroxy-PAH concentrations in atmospheric PM_{10}-bound aerosol particles: Correlations with physical and chemical factors. *Atmospheric Environment* 67: 385–393.

Bengtsson, G., Törneman, N. 2009. A spatial approach to environmental risk assessment of PAH contamination. *Risk Analysis* 29: 48–61.

Bengtsson, G., Törneman, N., Yang, X. 2010. Spatial uncoupling of biodegradation, soil respiration, and PAH concentration in a creosote contaminated soil. *Environmental Pollution* 158: 2865–2871.

Bergknut, M., Frech, K., Andersson, P.L., Haglund, P., Tysklind, M. 2006. Characterization and classification of complex PAH samples using GC-pMS and GC-TOFMS. *Chemosphere* 65: 2208–2215.

Blair, C., Walls, J., Davies, N.W., Jacobson, G.A. 2011. Volatile organic compounds in runners near a roadway: Increased blood levels after short-duration exercise. *British Journal of Sports Medicine* 44: 731–735.

Blanchard, P., Shepson, P.B., Schiff, H.I., Drummond, J.W. 1993. Development of a gas chromatograph for trace level measurement of peroxyacetyl nitrate by using chemical amplification. *Analytical Chemistry* 65: 2472–2477.

Bohlin, P., Jones, K.C., Tovalin, H., Strandberg, B. 2008. Observations on persistent organic pollutants in indoor and outdoor air using passive polyurethane foam samplers. *Atmospheric Environment* 42: 7234–7241.

Boonyatumanond, R., Wattayakorn, G., Togo, A., Takada, H. 2006. Distribution and origins of polycyclic aromatic hydrocarbons (PAHs) in riverine, estuarine and marine sediments in Thailand. *Marine Pollution Bulletin* 52: 942–956.

Bouvier-Brown, N.C., Schade, G.W., Misson L., Lee, A., MacKay, M. et al. 2012. Contributions of biogenic volatile organic compounds to net ecosystem carbon flux in a ponderosa pine plantation. *Atmospheric Environment* 60: 527–533.

Bralic, M., Buljac, M., Peris, N., Buzuk, M., Dabic, P. et al. 2012. Monthly and seasonal variations of NO_2, SO_2 and black-smoke located within the sport district in urban area, City of Split, Croatia. *Croatica Chemica Acta* 85: 139–145.

Bravo-Linares, C.M., Mudge, S.M., Loyola-Sepulveda, R.H. 2007. Occurrence of volatile organic compounds (VOCs) in Liverpool Bay, Irish Sea. *Marine Pollution Bulletin* 54: 1742–1753.

Callén, M.S., de la Cruz, M.T., López, J.M., Murillo, R., Navarro, M.V. et al. 2008a. Some inferences on the mechanism of atmospheric gas/particle partitioning of polycyclic aromatic hydrocarbons (PAHs) at Zaragoza (Spain). *Chemosphere* 73: 1357–1365.

Callén, M.S., de la Cruz, M.T., López, J.M., Murillo, R., Navarro, M.V. et al. 2008b. Long-range atmospheric transport and local pollution sources on PAH concentrations in a South European urban area: Fulfilling of the European Directive. *Water, Air, & Soil Pollution* 190: 271–285.

Campos, V.P., Cruz, L.P.S., Godoi, R.H.M., Godoi, A.F.L., Tavares, T.M. 2010. Development and validation of passive samplers for atmospheric monitoring of SO_2, NO_2, O_3 and H_2S in tropical areas. *Microchemical Journal* 96: 132–138.

Capasso, G., Inguaggiato, S. 1998. A simple method for the determination of dissolved gases in natural water: An application to thermal water from Vulcano Island. *Applied Geochemistry* 13: 631–642.

Cape, J.N. 2008. Surface ozone concentrations and ecosystem health: Past trends and a guide to future projections. *Science of the Total Environment* 40: 257–269.

Cárdenas, L.M., Austin, J.F., Burgess, R.A., Clemitshaw, K.C., Dorling, S. et al. 1998. Correlations between CO, NO_y, O_3 and non-methane hydrocarbons and their relationships with meteorology during winter 1993 on the North Norfolk coast, U.K. *Atmospheric Environment* 32: 3339–3351.

Carvalho, M.N., da Motta, M., Benachour, M., Sales, D.C.S., Abreu, C.A.M. 2012. Evaluation of BTEX and phenol removal from aqueous solution by multi-solute adsorption onto smectite organoclay. *Journal of Hazardous Materials* 239–240: 95–101.

Cavalcante, R.M., de Andrade, M.V.F., Marins, R.V., Oliveira L.D.M. 2010. Development of a head-space chromatography (HS-GC-PID-FID) method for the determination of VOCs in environmental aqueous matrices: Optimization, verification and elimination of matrix effect and VOC distribution on the Fortaleza Coast, Brazil. *Microchemical Journal* 96: 337–343.

Cheung, K.C., Leung, H.M., Kong, K.Y., Wong, M.H. 2007. Residual levels of DDTs and PAHs in freshwater and marine fish from Hong Kong markets and their health risk assessment. *Chemosphere* 66: 460–468.

Choi, S.-W., Park, S.-W., Lee, C.-S., Kim, H.-J., Bae, S. et al. 2009. Patterns of VOC and BTEX concentration in ambient air around industrial sources in Daegu, Korea. *Journal of Environmental Science and Health, Part A* 44: 99–107.

Christensen, J.H., Tomasi, G., de Lemos Scotfield, A., Guadalupe Meniconi, M.F. 2010. A novel approach for characterization of polycyclic aromatic hydrocarbon (PAH) pollution patterns in sediments from Guanabara Bay, Rio de Janeiro, Brazil. *Environmental Pollution* 158: 3290–3297.

Colomb, A., Yassaa, N., Williams, J., Peeken, I., Lochte, K. 2008. Screening volatile organic compounds (VOCs) emissions from five marine phytoplankton species by head space gas chromatography/mass spectrometry (HS-GC/MS). *Journal of Environmental Monitoring* 10: 325–330.

Council of the European Union. 1999. Council Directive 1999/13/CE of 11 March 1999 on the limitation of emissions of volatile organic compounds due to the use of organic solvents in certain activities and installations. *Official Journal of the European Communities* L085.

Cox, R.M. 2003. The use of passive sampling monitor forest exposure to O_3, NO_2 and SO_2: A review and some case studies. *Environmental Pollution* 126: 301–311.

Crimmins, B.S., Baker, J.E. 2006. Improved GC/MS methods for measuring hourly PAH and nitro-PAH concentrations in urban particulate matter. *Atmospheric Environment* 40: 6764–6779.

Cruz, L.P.S., Campos, V.P., Silva, A.M.C., Tavares, T.M. 2004. A field evaluation of a SO_2 passive sampler in tropical industrial and urban air. *Atmospheric Environment* 38: 6425–6429.

Curtis, A.J., Helmig, D., Baroch, C., Daly, R. and Davis, S. 2014. Biogenic volatile organic compound emissions from nine tree species used in an urban tree-plantations program. *Atmospheric Environment* 95: 634–643.

Davidson, E.A., David, M.B., Galloway, J.N., Goodale, C.L., Haeuber, R. et al. 2012. Excess nitrogen in the U.S. environment: Trends, risks, and solutions. *Issues in Ecology* 15: 1–16.

De Oliveira, L.S., Rodrigues, F.d.M., De Oliveira, F.S., Mesquita, P.R.R., Leal, D.C. et al. 2008. Headspace solid phase microextraction/gas chromatography–mass spectrometry combined to chemometric analysis for volatile organic compounds determination in canine hair: A new tool to detect dog contamination by visceral leishmaniasis. *Journal of Chromatography B* 875: 392–398.

De Souza Pereira, M., Heitmann, D., Reifenhäuser, W., Ornellas Meire, R., Silva Santos, L. et al. 2007. Persistent organic pollutants in atmospheric deposition and biomonitoring with *Tillandsia usneoides* (L.) in an industrialized area in Rio de Janeiro state, southeast Brazil—Part II: PCB and PAH. *Chemosphere* 67: 1736–1745.

Detournay, A., Sauvage, S., Riffault, V., Wroblewski, A., Locoge, N. 2013. Source and behavior of isoprenoid compounds at a southern France remote site. *Atmospheric Environment* 77: 272–282.

Dewulf, J., Van Langenhove, H. 1997. Chlorinated C_1- and C_2-hydrocarbons and monocyclic aromatic hydrocarbons in marine waters: An overview on fate processes, sampling, analysis and measurements. *Water Research* 31: 1825–2125.

Dincer, F., Odabasi, M., Muezzinoglu, A. 2006. Chemical characterization of odorous gases at a landfill site by gas-chromatography-mass spectrometry. *Journal of Chromatography A* 1122: 222–229.

Domeño, C., Blasco, M., Sánchez, C., Nerín, C. 2006. A fast extraction technique for extracting polycyclic aromatic hydrocarbons (PAHs) from lichens samples used as biomonitors of air pollution: Dynamic sonication versus other methods. *Analytica Chimica Acta* 569: 103–112.

Drescher, S.R., Brown, S.D. 2006. Solid phase microextraction–gas chromatographic mass spectrometric determination of nitrous oxide evolution to measure denitrification in estuarine soils and sediments. *Journal of Chromatography A* 1133: 300–304.

Durana, N., Navazo, M., Gómez, M.C., Alonso, L., García, J.A. et al. 2006. Long term hourly measurements of 6 non-methane hydrocarbons in an urban area: Main results and contribution of non-traffic sources. *Atmospheric Environment* 40: 2860–2872.

EPA. 1999. Compendium Method TO-13A: Determination of Polycyclic Aromatic Hydrocarbons (PAHs) in Ambient Air Using Gas Chromatography/Mass Spectrometry (GC/MS). EPA, Washington, DC, pp. 1–42.

EPA. 2015a. Air toxics website: The original list of hazardous air pollutants. http://www.epa.gov/ttn/atw/allabout.html (accessed September 22, 2016).

EPA. 2015b. Definition of volatile organic compounds (VOC). http://www.epa.gov/ttn/naaqs/ozone/ozonetech/def_voc.htm (accessed April 29, 2015).

Esteve-Turrillas, F.A., Armenta, S., Garrigues, S., Pastor, A., De la Guardia, M. 2007. Headspace-mass spectrometry determination of benzene, toluene and the mixture of ethylbenzene and xylene isomers in soil samples using chemometrics. *Analytica Chimica Acta* 587: 89–96.

European Parliament, Council of the European Union. 2000. Directive 2000/60/EC of the European Parliament and of the Council of 23 October 2000 establishing a framework for Community action in the field of water policy. *Official Journal of the European Communities* L327.

European Parliament, Council of the European Union. 2004. Directive 2004/107/EC of the European Parliament and of the Council of 15 December 2004 relating to arsenic, cadmium, mercury, nickel and polycyclic aromatic hydrocarbons in ambient air. *Official Journal of the European Communities* L23.

European Parliament, Council of the European Union. 2008. Directive 2008/50/EC of the European Parliament and of the Council of 21 May 2008 on ambient air quality and cleaner air for Europe. *Official Journal of the European Communities* L152.

European Soil Data Centre. 2015. Soil contamination. http://eusoils.jrc.ec.europa.eu/library/themes/contamination/ (accessed February 3, 2015).

Evtyugina, M., Calvo, A.I., Nunes, T., Alves, C., Fernandes, A.P. et al. 2013. VOC emission of smouldering combustion from Mediterranean wildfires in central Portugal. *Atmospheric Environment* 64: 339–348.

Fang, G.-C., Wu, Y.-S., Chen, J.-C., Chang, C.-N., Ho, T.-T. 2006. Characteristic of polycyclic aromatic hydrocarbon concentrations and source identification for fine and coarse particulate at Taichung Harbor near Taiwan Strait during 2004–2005. *Science of the Total Environment* 366: 729–738.

Faubert, P., Tiiva, P., Rinnan, Å., Räty, S., Holopainen, K.J. et al. 2010. Effect of vegetation removal and water table drawdown on the non-methane biogenic volatile organic compound emission in boreal peatland microcosms. *Atmospheric Environment* 44: 4432–4439.

Filella, I., Peñuelas, J. 2006. Daily, weekly and seasonal time courses of VOCs concentrations in a semi-urban area near Barcelona. *Atmospheric Environment* 40: 7752–7769.

Finlayson-Pitts, B.J., Pitts, J.N. 1993. Atmospheric chemistry of tropospheric ozone formation: Scientific and regulatory implications. *Air and Waste* 43: 1091–1100.

Fratini, G., Lois, S., Pazos, M., Parisi, G., Medina, I. 2012. Volatile profile of Atlantic shellfish species by HS-SPME GC/MS. *Food Research International* 48: 856–865.

Gaffney, J.S., Bornick, R.M., Chen Y.-H., Marley, N.A. 1998. Capillary gas chromatographic analysis of nitrogen dioxide and PANs with luminol chemiluminescent detection. *Atmospheric Environment* 32: 1445–1454.

Gallego, E., Perales, J.F., Roca, F.J., Guardino X. 2014. Surface emission determination of volatile organic compounds (VOC) from a closed industrial waste landfill using a self-designed static flux chamber. *Science of the Total Environment* 470–471: 587–599.

García Pinto, C., Fernández Laespada, M.E., Herrero Martín, S., Casas Ferreira, A.M., Pérez Pavón, J.L. et al. 2010. Simplified QuEChERS approach for the extraction of chlorinated compounds from soil samples. *Talanta* 81: 385–391.

García Pinto, C., Herrero Martín, S., Pérez Pavón, J.L., Moreno Cordero, B. 2011. A simplified Quick, Easy, Cheap, Effective, Rugged and Safe approach for the determination of trihalomethane and benzene, toluene, ethylbenzene and xylenes in soil matrices by fast gas chromatography with mass spectrometry detection. *Analytica Chimica Acta* 689: 129–1361.

Garcia-Hurtado, E., Pey, J., Borrás, E., Sánchez, P., Vera, T. et al. 2014. Atmospheric PM and volatile organic compounds released from Mediterranean shrubland wildfires. *Atmospheric Environment* 89: 85–92.

Geng, F., Tie, X., Xu, J., Zhou, G., Peng, L. et al. 2008. Characterizations of ozone, NOx and VOCs measured in Shanghai, China. *Atmospheric Environment* 42: 6873–6883.

Görgényi, M., Dewulf, J., Van Langenhove, H. 2000. Matrix effect in the cryotrapping of gas samples in the ppbv range. *Chromatographia* 51: 461–466.

Gras, R., Luong, J., Carter, V., Sieben, L., Cortes, H. 2009. Practical method for the measurement of alkyl mercaptans in natural gas by multi-dimensional gas chromatography, capillary flow technology and flame ionization detection. *Journal of Chromatography A* 1216: 2776–2782.

Greenberg, J.P., Guenther, A., Turnipseed, A., Jiang, X., Seco, R. et al. 2014. A tethered-balloon PTRMS sampling approach for rapid surveying of landscape-scale biogenic VOC fluxes. *Atmospheric Measurement Techniques* 7: 979–999.

Guenther, A. 2002. The contribution of reactive carbon emissions from vegetation to the carbon balance of terrestrial ecosystems. *Chemosphere* 49: 837–844.

Guenther, A., Otter, L., Zimmerman, P., Greenberg, J., Scholes, R. et al. 1996. Biogenic hydrocarbon emissions from southern African savannas. *Journal of Geophysical Research* 101: 25859–25865.

Hafkenscheid, T., Fromage-Mariette, A., Goelen, E., Hangartner, M., Pfeffer, U. et al. 2009. *Review of the Application of Diffusive samplers in the European Union for the Monitoring of Nitrogen Dioxide in Ambient Air*. Office for Official Publications of the European Communities, Luxembourg.

Hakola, H., Tarvainen, V., Laurila, T., Hiltunen, V., Hellén, H. et al. 2003. Seasonal variation of VOC concentrations above a boreal coniferous forest. *Atmospheric Environment* 37: 1623–1634.

Halsall, C.J., Maher, B.A., Karloukovski, V.V., Shah, P., Watkins, S.J. 2008. A novel approach to investigating indoor/outdoor pollution links: Combined magnetic and PAH measurements. *Atmospheric Environment* 42: 8902–8909.

Hashemi, M., Jahanshahi, N., Habibi, A. 2012. Application of ultrasound-assisted emulsification microextraction for determination of benzene, toluene, ethylbenzene and *o*-xylene in water samples by gas chromatography. *Desalination* 288: 93–97.

He, Z., Yang, G.-P., Lu, X-L., Zhang, H.-H. 2013. Distributions and sea-to-air fluxes of chloroform, trichloroethylene, tetrachloroethylene, chlorodibromomethane and bromoform in the Yellow Sea and the East China Sea during spring. *Environmental Pollution* 177: 28–37.

Hellén, H., Tykkä, T., Hakola, H. 2012. Importance of monoterpenes and isoprene in urban air in northern Europe. *Atmospheric Environment* 59: 59–66.

Hénault, C., Grossel, A., Mary, B., Roussel, M., Léonard, J. 2012. Nitrous oxide emission by agricultural soils: A review of spatial and temporal variability for mitigation. *Pedosphere* 22: 426–433.

Herrero Martín, S., García Pinto, C., Pérez Pavón, J.L., Moreno Cordero, B. 2010. Determination of trihalomethanes in soil matrices by simplified quick, easy, cheap, effective, rugged and safe extraction and fast gas chromatography with electron capture detector. *Journal of Chromatography A* 1217: 4883–4889.

Hodgeson, J.A. 1974. A review of chemiluminescent techniques for air pollution monitoring. *Toxicological & Environmental Chemistry Reviews* 2: 81–97.

Hsieh, Y.-N., Huang, P.-C., Sun, I-W., Whang, T.-J., Hsu, C.-Y. et al. 2006. Nafion membrane-supported ionic liquid-solid phase microextraction for analyzing ultra trace PAHs in water samples. *Analytica Chimica Acta* 557: 321–328.

Huybrechts, T., Dewulf, J., Van Langenhove, H. 2005. Priority volatile organic compounds in surface waters of the southern North Sea. *Environmental Pollution* 133: 255–264.

Jia, C., Batterman, S., Godwin, C. 2008. VOCs in industrial, urban and suburban neighborhoods—Part I: Indoor and outdoor concentrations, variation and risk drivers. *Atmospheric Environment* 42: 2083–2100.

Jiménez, A.M., Navas, M.J., Galán, G. 1997. Air analysis: Determination of ozone by chemiluminescence. *Applied Spectroscopy Reviews* 32: 141–149.

Joó, É., Van Langenhove, H., Šimpraga, M., Steppe, K., Amelynck, C. et al. 2010. Variation of biogenic volatile organic compound emission pattern of *Fagus sylvatica* L. due to aphid infection. *Atmospheric Environment* 44: 227–234.

Juan-Peiró, L., Bernhammer, A., Pastor, A., De la Guardia, M. 2012. The use of Amberlite adsorbents for green chromatography determination of volatile organic compounds in air. *Journal of Analytical Methods in Chemistry* 2012: 1–6.

Juntarawijit, C., Poovey, H.G., Rando, R.J. 2000. Determination of carbon monoxide with a modified zeolite sorbent and methanization–gas chromatography. *American Industrial Hygiene Association Journal* 61: 410–414.

Kaminski, M., Kartanowicz, R., Jastrzebski, D., Kaminski, M.M. 2003. Determination of carbon monoxide, methane and carbon dioxide in refinery hydrogen gases and air by gas chromatography. *Journal of Chromatography A* 989: 277–283.

Kannan, K., Perrotta, E. 2008. Polycyclic aromatic hydrocarbons (PAHs) in livers of California sea otters. *Chemosphere* 71: 649–655.

Karlik, J.F., McKay, A.H., Welch, J.M., Winer, A.M. 2002. A survey of California plant species with a portable VOC analyzer for biogenic emission inventory development. *Atmospheric Environment* 36: 5221–5233.

Karthikeyan, S., Balasubramanian, R., Wei See, S. 2006. Optimization and validation of a low temperature microwave-assisted extraction method for analysis of polycyclic aromatic hydrocarbons in airborne particulate matter. *Talanta* 69: 79–86.

Kawashima, H., Murakami, M. 2014. Measurements of the stable carbon isotope ratio of atmospheric volatile organic compounds using chromatography, combustion and isotope ratio mass spectrometry coupled with thermal desorption. *Atmospheric Environment* 89: 140–147.

Kesselmeier, J., Staudt, M. 1999. Biogenic volatile organic compounds (VOC): An overview on emission, physiology and ecology. *Journal of Atmospheric Chemistry* 33: 23–88.

Kim, K.-H., Shon, Z.-H., Kim, M.-Y., Sunwoo, Y., Jeon, E.-C. et al. 2008a. Major aromatic VOC in the ambient air in the proximity of an urban landfill facility. *Journal of Hazardous Materials* 150: 754–764.

Kim, M. Kennicutt II, M.C., Qian, Y. 2008b. Source characterization using compound composition and stable carbon isotope ratio of PAHs in sediments from lakes, harbor and shipping waterway. *Science of the Total Environment* 389: 367–377.

Kim, S.-Y., Jiang, X., Lee, M., Turnipseed, A., Guenther, A. et al. 2013. Impact of biogenic volatile organic compounds on ozone production at the Taehwa Research Forest near Seoul, South Korea. *Atmospheric Environment* 70: 447–453.

Kivimäenpää, M., Magsarjav, N., Ghimire, R., Markkanen, J.-M., Heijari, J. et al. 2012. Influence of tree provenance on biogenic VOC emissions of Scots pine (*Pinus sylvestris*) stumps. *Atmospheric Environment* 60: 477–485.

Klinger, L.F., Greenberg, J., Guenther, A., Tyndall, G., Zimmerman, P. et al. 1998. Patterns of volatile organic compound emissions along a savanna-rainforest gradient in central Africa. *Journal of Geophysical Research* 103: 1443–1454.

Komazaki, Y., Hamada, Y., Hashimoto, S., Fujita, T., Tanaka, S. 1999. Development of an automated, simultaneous and continuous measurement system by using diffusion scrubber coupled to ion chromatography for monitoring trace acidic and basic gases (HCl, HNO_3, SO_2 and NH_3) in the atmosphere. *Analyst* 124: 1151–1157.

Kota, S.H., Park, C., Hale, M.C., Werner, N.D., Schade, G.W. et al. 2014. Estimation of VOC emission factors from flux measurements using a receptor model and footprint analysis. *Atmospheric Environment* 82: 24–35.

Koutrakis, P., Wolfson, J.M., Bunyaviroch, A., Froelich, S.E., Hirano, K. et al. 1993. Measurement of ambient ozone using a nitrite-coated filter. *Analytical Chemistry* 65: 209–214.

Kuo, C.-P., Liao, H.-T., Chou, C.C.-K., Wu, C.-F. 2014. Source apportionment of particulate matter and selected volatile organic compounds with multiple time resolution data. *Science of the Total Environment* 472: 880–887.

Laan, S.V.D., Neubert, R.E.M., Meijer, H.A.J. 2009. A single gas chromatograph for accurate atmospheric mixing ratio measurements of CO_2, CH_4, N_2O, SF_6 and CO. *Atmospheric Measurement Techniques* 2: 549–559.

Leroyer, A., Jeandel, F., Maitre, A., Howsam, M., Deplanque, D. et al. 2010. 1-Hydroxypyrene and 3-hydroxybenzo[*a*]pyrene as biomarkers of exposure to PAH in various environmental exposure situations. *Science of the Total Environment* 408: 1166–1173.

Li, G., Zhang, Z., Sun, H., Chen, J., An, T. et al. 2013. Pollution profiles, health risk of VOCs and biohazards emitted from municipal solid waste transfer station and elimination by an integrated biological-photocatalytic flow system: A pilot-scale investigation. *Journal of Hazardous Materials* 250–251: 147–154.

Liang, Y., Tse, M.F., Young, L., Wong, M.H. 2007. Distribution patterns of polycyclic aromatic hydrocarbons (PAHs) in the sediments and fish at Mai Po Marshes Nature Reserve, Hong Kong. *Water Research* 41: 1303–1311.

Liguori, L., Heggstad, K., Hove, H.T., Julshamn, K. 2006. An automated extraction approach for isolation of 24 polyaromatic hydrocarbons (PAHs) from various marine matrixes. *Analytica Chimica Acta* 573–574: 181–188.

Liu, M., Cheng, S.B., Ou, D.N., Hou, D.N., Hou, L.J. et al. 2007. Characterization, identification of road dust PAHs in central Shanghai areas, China. *Atmospheric Environment* 41: 8785–8795.

Liu, W.-T., Hsieh, H.-C., Chen, S.-P., Chang, J.S., Lin, N.-H. et al. 2012. Diagnosis of air quality through observation and modeling of volatile organic compounds (VOCs) as pollution tracers. *Atmospheric Environment* 55: 56–63.

Liu, J., Seo, J.H., Li, Y., Chen, D., Kurabayashi, K. et al. 2013. Smart multi-channel two dimensional micro-gas chromatography for rapid workplace hazardous volatile organic compounds measurement. *Lab on a Chip* 13: 818–825.

Llusia, J., Peñuelas, J., Guenther, A., Rapparini, F. 2013. Seasonal variations in terpene emission factors of dominant species in four ecosystems in NE Spain. *Atmospheric Environment* 70: 149–158.

Lodge, J.P. 1989. *Methods of Air Sampling and Analysis*, 3rd ed. Lewis, Chelsea, MI.

Long, A., Zhang, H., Lei, Y. 2013. Surfactant flushing remediation of toluene contaminated soil: Optimization with response surface methodology and surfactant recovery by selective oxidation with sulfate radicals. *Separation and Purification Technology* 118: 612–619.

Louie, P.K.K., Ho, J.W.K., Tsang, R.C.W., Blake, D.R., Lau, A.K.H. et al. 2013. VOCs and OVOCs distribution and control policy implications in Pearl River Delta region, China. *Atmospheric Environment* 76: 125–135.

Madronich, S., Shao, M., Wilson, S.R., Solomon, K.R., Longstreth, J.D. et al. 2015. Changes in air quality and tropospheric composition due to depletion of stratospheric ozone and interactions with changing climate: Implications for human and environmental health. *Photochemical and Photobiological Sciences* 14: 149–169.

Mahbub, P., Goonetilleke, A., Ayoko, G.A. 2012. Prediction of the wash-off traffic related semi- and non-volatile organic compounds from urban roads under climate change influenced rainfall characteristics. *Journal of Hazardous Materials* 213–214: 83–92.

Mallik, C., Venkataramani, S., Lal, S. 2012. Study of a high SO_2 event observed over and urban site in Western India. *Asia-Pacific Journal of Atmospheric Sciences* 48: 171–180.

Manahan, E. 2010. *Environmental Chemistry*, 9th ed. CRC Press, Boca Raton, FL.

Marrucci, A., Marras, B., Campisi, S.S., Schintu, M. 2013. Using SPMDs to monitor seawater concentrations of PAHs and PCBs in marine protected areas (Western Mediterranean). *Marine Pollution Bulletin* 75: 69–75.

McGowin, A.E. 2006. Polycyclic aromatic hydrocarbons. In *Chromatographic Analysis of the Environment*, ed. Nollet, L.M.L. Taylor & Francis, Boca Raton, FL, pp. 555–616.

Menichini, E., Iacovella, N., Monfredini, F., Turrio-Baldassarri, L. 2007a. Atmospheric pollution by PAHs, PCDD/Fs and PCBs simultaneously collected at a regional background site in central Italy and at an urban site in Rome. *Chemosphere* 69: 422–434.

Menichini, E., Iacovella, N., Monfredini, F., Turrio-Baldassarri, L. 2007b. Relationship between indoor and outdoor air pollution by carcinogenic PAHs and PCBs. *Atmospheric Environment* 41: 9518–9529.

Miller, D.L., Woods, J.S., Grubaugh, K.W., Jordon, L.M. 1980. Analysis of stack gases using a portable gas chromatographic sampling and analyzing system. *Environmental Science & Technology* 14: 97–100.

Milligan, D.B., Wilson, P.F., Mautner, M.N., Freeman, C.G., McEwan, M.J. et al. 2002. Real-time, high-resolution quantitative measurement of multiple soil gas emissions: Selected ion flow tube mass spectrometry. *Journal of Environmental Quality* 31: 515–524.

Moliner-Martínez, Y., Herraez-Hernandez, R., Verdú-Andres, J., Campíns-Falcó, P., Garrido-Palanca, C. et al. 2013. Study of the influence of temperature and precipitations on the levels of BTEX in natural waters. *Journal of Hazardous Materials* 263P: 131–138.

Monteiro Martins, E., Arbilla, G., Favilla Bauerfeldt, G., de Paula Murilo. 2007. Atmospheric levels of aldehydes and BTEX and their relationship with vehicular fleet changes in Rio de Janeiro urban area. *Chemosphere* 67: 2096–2103.

Moret, S., Purcaro, G., Conte, L.S. 2007. Polycyclic aromatic hydrocarbon (PAH) content of soil and olives collected in areas contaminated with creosote released from old railway ties. *Science of the Total Environment* 386: 1–8.

Müller, A., Faubert, P., Hagen, M., zu Castell, W., Polle, A., Schnitzler, J.-P., Rosenkranz, M. 2013. Volatile profiles of fungi—Chemotyping of species and ecological functions. *Fungal Genetics and Biology* 54: 25–33.

Myers, M.S., Anulacion, B.F., French, B.L., Reichert, W.L., Laetz, C.A. et al. 2008. Improved flatfish health following remediation of a PAH-contaminated site in Eagle Harbor, Washington. *Aquatic Toxicology* 88: 277–288.

Niki, T., Fujinaga, T., Watanabe, M.F., Kinoshita, J. 2004. Simple determination of dimethylsulfide (DMS) and dimethylsulfoniopropionate (DMSP) using solid-phase microextraction and gas chromatography-mass spectrometry. *Journal of Oceanography* 60: 913–917.

Olivella, M.A. 2006a. Isolation and analysis of polycyclic aromatic hydrocarbons from natural water using accelerated solvent extraction followed by gas chromatography-mass spectrometry. *Talanta* 69: 267–275.

Olivella, M.A. 2006b. Polycyclic aromatic hydrocarbons in rainwater and surface waters of Lake Maggiore, a subalpine lake in Northern Italy. *Chemosphere* 63: 116–131.

Olszowski, T., Tomaszewska, B., Góralna-Wlodarczyk, K. 2012. Air quality in non-industrialised area in the typical Polish countryside based on measurements of selected pollutants in immission and deposition phase. *Atmospheric Environment* 50: 139–147.

Orecchio, S. 2007. PAHs associated with the leaves of *Quercus ilex* L.: Extraction, GC-MS analysis, distribution and sources; Assessment of air quality in the Palermo (Italy) area. *Atmospheric Environment* 41: 8669–8680.

Palmes, E.D., Gunnison, A.F. 1973. Personal monitoring device for gaseous contaminants. *American Industrial Hygienists Association Journal* 34: 78–81.

Park, C., Schade, G.W., Boedeker, I. 2010. Flux measurements of volatile organic compounds by the relaxed eddy accumulation method combined with a GC-FID system in urban Houston, Texas. *Atmospheric Environment* 44: 2605–2614.

Parra, M.A., González, L., Elustondo, D., Garrigó, J., Bermejo, R. et al. 2006. Spatial and temporal trends of volatile organic compounds (VOS) in a rural area of northern Spain. *Science of the Total Environment* 370: 157–167.

Parra, M.A., Elustondo, D., Bermejo, R., Santamaría, J.M. 2009. Ambient levels of volatile organic compounds (VOC) and nitrogen dioxide (NO_2) in a medium size city in Northern Spain. *Science of the Total Environment* 407: 999–1009.

Pena, T., Pensado, L., Casais, C., Mejuto, C. Phan-Tan-Luu, R. et al. 2006. Optimization of a microwave-assisted extraction method for the analysis of polycyclic aromatic hydrocarbons from fish samples. *Journal of Chromatography A* 1121: 163–169.

Pena, M.T., Casais, M.C., Mejuto, M.C., Cela, R. 2007. Optimization of the matrix solid-phase dispersion sample preparation procedure for analysis of polycyclic aromatic hydrocarbons in soils: Comparison with microwave-assisted extraction. *Journal of Chromatography A* 1165: 32–38.

Pérez-Carrera, E., León León, V.M., Gómez Parra, A., González-Mazo, E. 2007. Simultaneous determination of pesticides, polycyclic aromatic hydrocarbons and polychlorinated biphenyls in seawater and interstitial marine water samples, using stir bar sorptive extraction–thermal desorption–gas chromatography–mass spectrometry. *Journal of Chromatography A* 1170: 82–90.

Pérez Pavón, J.L., Herrero Martín, S., García Pinto, C., Moreno Cordero, B. 2009. Programmed temperature vaporizer based method for the sensitive determination of trihalomethanes and benzene, toluene, ethylbenzene and xylenes in soils. *Journal of Chromatography A* 1216: 6063–6070.

Polkowska-Motrenko, H., Kulisa, K., Dybczynski, R. 1998. Ion chromatographic determination of SO_2 and SO_3 in flue gases in pilot plant for removal of SO_2 and NO_x by electron beam treatment. *Chemia Analityczna* 43: 365–373.

Pooladi-Darvish, M., Hong, H., Stocker, R.K., Bennion, B., Theys, S. et al. 2009. Chromatographic partitioning of H_2S and CO_2 in acid gas disposal. *Journal of Canadian Petroleum Technology* 48: 52–57.

Préndez, M., Carvajal, V., Corada, K., Morales, J., Alarcón, F. et al. 2013. Biogenic volatile organic compounds from the urban forest of the Metropolitan Region, Chile. *Environmental Pollution* 183: 143–150.

Prieto, A., Zuloaga, O., Usobiaga, A., Etxebarria, N., Fernández, L.A. 2007. Development of a stir bar sorptive extraction and thermal desorption–gas chromatography–mass spectrometry method for the simultaneous determination of several persistent organic pollutants in water samples. *Journal of Chromatography A* 1174: 40–49.

Ramírez, N., Cuadras, A., Rovira, E., Borrull, F., Marcé, R.M. 2012. Chronic risk assessment of exposure to volatile organic compounds in the atmosphere near the largest Mediterranean industrial site. *Environment International* 39: 200–209.

Rapson, T.D., Dacres, H. 2014. Analytical techniques for measuring nitrous oxide. *Trends in Analytical Chemistry* 54: 65–74.

Ras-Mallorquí, M.R., Marcé-Recasens, R.M., Borrull-Ballarín, F. 2007. Determination of volatile organic compounds in urban and industrial air from Tarragona by thermal desorption and gas chromatography–mass spectrometry. *Talanta* 72: 941–950.

Ratola, N., Lacorte, S., Alves, A., Barceló, D. 2006a. Analysis of polycyclic aromatic hydrocarbons in pine needles by gas chromatography–mass spectrometry: Comparison of different extraction and clean-up procedures. *Journal of Chromatography A* 1114: 198–204.

Ratola, N., Lacorte, S., Alves, A., Barceló, D. 2006b. Analysis of polycyclic aromatic hydrocarbons in pine needles by gas chromatography–mass spectrometry: Comparison of different extraction and clean-up procedures. *Journal of Chromatography A* 1114: 198–204.

Reimann, S., Lewis, A.C. 2007. Anthropogenic VOCs in volatile organic compounds in the atmosphere. In *Volatile Organic Compounds in the Atmosphere*, ed. Kopmann, R. Blackwell, Hoboken, NJ, pp. 33–81.

Richter, P., Jiménez, M., Salazar, R., Maricán, A. 2006. Ultrasound-assisted pressurized solvent extraction for aliphatic and polycyclic aromatic hydrocarbons from soil. *Journal of Chromatography A* 1132: 15–20.

Rodríguez-Navas, C., Fortaleza, R., Cerdà, V. 2012. Use of thermal desorption-gas chromatography–mass spectrometry (TD-GC-MS) on identification of odorant emission focus by volatile organic compounds characterisation. *Chemosphere* 89: 1426–1436.

Rosell, M., Lacorte, S., Barceló, D. 2006. Simultaneous determination of methyl *tert*-butyl ether, its degradation products and other gasoline additives in soil samples by closed-system purge-and-trap gas chromatography–mass spectrometry. *Journal of Chromatography A* 1132: 28–38.

Roukos, J., Riffault, V., Locoge, N., Plaisance, H. 2009. VOC in an urban and industrial harbor on the French North Sea coast during two contrasted meteorological situations. *Environmental Pollution* 157: 3001–3009.

Saarnio, K., Sillanpää, M., Hillano, R., Sandell, E., Pennanen, A.S. et al. 2008. Polycyclic aromatic hydrocarbons in size-segregated particulate matter from six urban sites in Europe. *Atmospheric Environment* 42: 9087–9097.

Salem, A.A., Soliman, A.A., El-Haty, I.A. 2009. Determination of nitrogen dioxide, sulfur dioxide, ozone and ammonia in ambient air using the passive sampling method associated with ion chromatographic and potentiometric analyses. *Air Quality, Atmosphere and Health* 2: 133–145.

Serrano, A., Gallego, M. 2006. Sorption study of 25 volatile organic compounds in several Mediterranean soils using headspace–gas chromatography–mass spectrometry. *Journal of Chromatography A* 1118: 261–270.

Slater, G.F., Cowie, B.R., Harper, N., Droppo, I.G. 2008. Variation in PAH inputs and microbial community in surface sediments of Hamilton Harbour: Implications to remediation and monitoring. *Environmental Pollution* 153: 60–70.

Sofowote, U.M., Hung, H., Rastogi, A.K., Westgate, J.N., Su, Y. et al. 2010. The gas/particle partitioning of polycyclic aromatic hydrocarbons collected at a sub-Arctic site in Canada. *Atmospheric Environment* 44: 4919–4926.

Souza, K.F., Carvalho, L.R.F., Allen, A.G., Cardoso, A.A. 2014. Diurnal and nocturnal measurements of PAHs, nitro-PAHs and oxy-PAH compounds in atmospheric particulate matter of a sugar cane burning region. *Atmospheric Environment* 83: 193–201.

Sulzer, P., Edtbauer, A., Hartungen, E., Jürschik, S., Jordan, A. et al. 2012. From conventional proton-transfer-reaction mass spectrometry (PTR-MS) to universal trace gas analysis. *International Journal of Mass Spectrometry* 321–322: 66–70.

Swan, H.B., Ivey, J.P., Jones, G.B., Eyre, B.D. 2015. The validation and measurement uncertainty of an automated gas chromatograph for marine studies of atmospheric dimethylsulfide. *Analytical Methods* 7: 3893–3902.

Tairova, Z.M., Giessing, A.M.B., Hansen, R., Andersen, O. 2009. 1-Hydroxypyrene as a biomarker of PAH exposure in the marine polychaete *Nereis diversicolor*. *Marine Environmental Research* 67: 38–46.

Tassi, F., Capecchiacci, F., Giannini, L., Vougioukalakis, G.E., Vaselli, O. 2013. Volatile organic compounds (VOCs) in air from Nisyros Island (Dodecanese Archipelago, Greece): Natural versus anthropogenic sources. *Environmental Pollution* 180: 111–121.

Thompson, R.L., Manning, A.C., Gloor, E. Schultz U., Seifert, T. et al. 2009. In-situ measurements of oxygen, carbon monoxide and greenhouse gases from Ochsenkopf tall tower in Germany. *Atmospheric Measurement Techniques* 2: 573–591.

Tsai, D.-H., Wang, J.-L., Wang, H.-H., Chan, C.-C. 2008. A study of ground-level ozone pollution, ozone precursors and subtropical meteorological conditions in central Taiwan. *Journal of Environmental Monitoring* 10: 109–118.

Turner, N., Jones, M., Grice, K., Dawson, D., Ioppolo-Armanios, M. et al. 2006. δ13C of volatile organic compounds (VOCS) in airborne samples by thermal desorption–gas chromatography–isotope ratio–mass spectrometry (TD-GC-IR-MS). *Atmospheric Environment* 40: 3381–3388.

Ueta, I., Mizuguchi, A., Tani, K., Kawakubo, S., Saito, Y. 2013. Rapid temperature-programmed separation of carbon monoxide and carbon dioxide on a packed capillary column in gas chromatography: Application to the evaluation of photocatalytic activity of TiO_2. *Analytical Sciences* 29: 673–676.

Vane, C.H., Harrison, I., Kim, A.W. 2007. Assessment of polyaromatic hydrocarbons (PAHs) and polychlorinated biphenyls (PCBs) in surface sediments on the Inner Clyde Estuary, UK. *Marine Pollution Bulletin* 54: 1287–1306.

Viglianti, C., Hanna, K., de Brauer, C., Germain, P. 2006. Removal of polycyclic aromatic hydrocarbons from aged-contaminated soil using cyclodextrins: Experimental study. *Environmental Pollution* 140: 427–435.

Wang, P., Zhao, W. 2008. Assessment of ambient volatile organic compounds (VOCs) near major roads in urban Nanjing, China. *Atmospheric Research* 89: 289–297.

Wang, W., Meng, B., Lu, X., Liu, Y., Tao, S. 2007. Extraction of polycyclic aromatic hydrocarbons and organochlorine pesticides from soils: A comparison between Soxhlet extraction, microwave-assisted extraction and accelerated solvent extraction techniques. *Analytica Chimica Acta* 602: 211–222.

Wang, J., Bi, Y., Pfister, G., Henkelmann, B., Zhu, K. et al. 2009a. Determination of PAH, PCB and OCP in water from the Three Gorges Reservoir accumulated by semipermeable membrane devices (SPMD). *Chemosphere* 75: 1119–1127.

Wang, Z., Ma, X., Na, G., Lin, Z., Ding, Q. et al. 2009b. Correlations between physicochemical properties of PAHs and their distribution in soil, moss and reindeer dong at Ny-Ålesund of the Arctic. *Environmental Pollution* 157: 3132–3136.

Wang, Y., Wang, H., Ling, J. 2010a. A new carrier gas type for accurate measurement of N_2O by GC-ECD. *Advances in Atmospheric Sciences* 27: 1322–1330.

Wang, H., Zhou, L., Liu, G., Li, W., Shi, S. et al. 2010b. Detection of ppb level SO_2 in H_2 by and adsorption–desorption gas chromatography method. *International Journal of Hydrogen Energy* 35: 2994–2996.

Wang, J.-L., Chew, C., Chang, C.-Y., Liao, W.-C., Lung, S.-C.C. et al. 2013. Biogenic isoprene in subtropical urban settings and implications for air quality. *Atmospheric Environment* 79: 369–379.

Wardas, W., Makowski, A., Czogala, J., Nowak, J. 2000. Determination of carbon monoxide in main and side streams of cigarette smoke by gas chromatography with Carbosieve stationary phase. *Chemia Analityczna* 45: 529–535.

Wei, W., Cheng, S., Li, G., Wang, G., Wang, H. 2014. Characteristics of volatile organic compounds (VOCs) emitted from a petroleum refinery in Beijing, China. *Atmospheric Environment* 89: 358–366.

Wendel, G.J., Stedman, D.H., Cantrell, C.A., Damrauer, L.D. 1983. Luminol-based nitrogen dioxide detector. *Analytical Chemistry* 55: 937–940.

WHO. 2000. *Air Quality Guidelines for Europe*, 2nd ed. WHO Regional Office for Europe, København, Denmark. http://www.euro.who.int/document/e71922.pdf (accessed September 22, 2016).

Williams, J. 2004. Organic trace gases in the atmosphere: An overview. *Environmental Chemistry* 1: 125–136.

Williams, J., Kopmann, R. 2007. Volatile organic compounds in the atmosphere: An overview. In *Volatile Organic Compounds in the Atmosphere*, ed. Kopmann, R. Blackwell, Hoboken, NJ, pp. 1–32.

Wu, Y., Zheng, X., Zhang, S., Song, S., Li, J. et al. 2014. Characterization and source apportionment of particulate PAHs in the roadside environment in Beijing. *Science of the Total Environment* 470–471: 76–83.

Yamada, E., Kimura, M., Tomozawa, K., Fuse, Y. 1999. Simple analysis of atmospheric NO_2, SO_2 and O_3 in mountains by using passive samplers. *Environmental Science & Technology* 33: 4141–4145.

Yang, Y., Hofmann, T. 2009. Aqueous accelerated solvent extraction of native polycyclic aromatic hydrocarbons (PAHs) from carbonaceous river floodplain soils. *Environmental Pollution* 157: 2604–2609.

Yassaa, N., Colomb, A., Lochte, K., Peeken, I., Williams, J. 2006. Development and application of a headspace solid-phase microextraction and gas chromatography/mass spectrometry method for the determination of dimethylsulfide emitted by eight marine phytoplankton species. *Limnology and Oceanography—Methods* 4: 374–381.

Yassaa, N., Ciccioli, P., Brancaleoni, E., Frattoni, M., Meklati, B.Y. 2011. Ambient measurement of selected VOCs in populated and remote sites of the Sahara desert. *Atmospheric Research* 100: 141–146.

Yusà, V., Quintas, G., Pardo, O., Pastor, A., de la Guardia, M. 2006. Determination of PAHs in airborne particles by accelerated solvent extraction and large-volume injection–gas chromatography–mass spectrometry. *Talanta* 69: 807–815.

Zeng, Y., Mukai, H., Bandow, H., Nojiri, Y. 1994. Application of gas chromatography–combustion–isotope ratio mass spectrometry to carbon isotopic analysis of methane and carbon monoxide in environmental samples. *Analytica Chimica Acta* 289: 195–204.

Section V

Residues in Different Matrices

14 Pharmaceuticals and Personal Care Products

Sergiane Souza Caldas, Ana Laura Venquiaruti Escarrone, and Ednei Gilberto Primel

CONTENTS

14.1 PHARMACEUTICALS AND PERSONAL CARE PRODUCTS

14.1.1 Definition and Introductions of PPCPs

PPCPs are defined by the Environmental Protection Agency (EPA) as any product used by individuals for personal health or cosmetic reasons or used by agribusiness to enhance growth or livestock health. PPCPs comprise a diverse collection of thousands of chemical substances, including prescription and over-the-counter therapeutic drugs, veterinary drugs, fragrances, lotions, and cosmetics (EPA, 2013).

In the last decade, PPCPs have been of increasing concern to researchers due to their pseudopersistent properties and potential harm to human beings and the environment (Hu and Zhou, 2011). There are many unanswered questions with regard to the biological effect of the presence of pharmaceuticals to the flora, fauna, and humans. Very little is known about the long-term effect of the pharmaceutical residues, and it is unknown if the combination of drugs that share a common mechanism of action exhibits toxic synergistic effect (Petrović et al., 2014).

Thus, the interest in measuring emerging contaminants, especially those which are not currently covered by existing regulations and may have potential effects on the environment and human health, even at trace concentrations, has been quickly increasing. Consequently, the general knowledge about PPCP fate has gradually improved (Koo et al., 2010; Kosma et al., 2010). Since the mid-1990s, studies related to the occurrence, effects, and risks of PPCPs in the environment have more than doubled or even tripled (Bu et al., 2013).

It is now recognized that these compounds are released into the environment through different pathways and the introduction of PPCPs in aquatic ecosystems comes from various sources (Sanchez et al., 2011). The major sources of PPCPs in the environment are human activities (e.g., bathing, shaving, swimming), illicit drugs, veterinary drug use, agribusiness, residues from pharmaceutical manufacturing, and residues from hospitals (EPA, 2013).

The importance of individuals adding chemicals to the environment has been largely overlooked, and the discovery of PPCPs in water and soil shows that even simple activities such as shaving, using lotion, or taking medication affect the environment in which we live. PPCPs are not typically persistent but are constantly released to receiving environments (Karnjanapiboonwong et al., 2011).

While PPCPs are continuously discharged to the environment without any restrictions, they are generally only slightly transformed or even unchanged into the WWTPs. Removal of PPCPs during wastewater treatment is rarely complete (McClellan and Halden, 2010; Subedi et al., 2011), and the efficiency depends on the nature of the target pharmaceuticals, the type of wastewater treatment technology implemented, age of the activated sludge, hydraulic retention time, environmental conditions (e.g., dilution of wastewater effluent, rainfall, temperature, and level of sunlight), and physical properties including the adsorption capacity of compound on sludge (Kosma et al., 2010). This inefficient elimination in WWTPs, in many cases, is related to the high polarity combined with the low biodegradability exhibited by some pharmaceutical compounds. An important consequence of this incomplete removal is the fact that PPCPs remaining in treated wastewater can leach to groundwater supplies and surface water downstream from wastewater effluent discharges that can be the source water used for drinking (Karnjanapiboonwong et al., 2011). This problem is worse in developing countries, where, in many places, there is no wastewater treatment (Spongberg et al., 2011).

Besides the concern about the presence of PPCPs in the environment, most of the compounds detected in the environmental samples are not included in world legislations (Petrović et al., 2014).

Since PPCPs present a large variety of chemical structures and very low concentrations in the environment, techniques to quantify and assess their occurrence have been optimized and studied (Suárez et al., 2008). The analytical methods are based mainly on GC or LC coupled to MS, and an increasing number of PPCPs are being detected in both liquid and solid environmental compartments (Petrovic et al., 2010; Primel et al., 2012; Aydin and Talinli, 2013; Blair et al., 2013).

14.1.2 Sample Preparation Techniques

The presence of PPCPs in different compartments of the environment has become a well-known issue during the last decade. As a result, many methods to determine these contaminants in environmental samples have been studied (Wille et al., 2011a).

Despite the advancement of analytical chemistry, the direct injection of environmental samples in the chromatographic system is almost always impossible, due to the complexity of the sample and the low levels in which these compounds are present in the samples. Therefore, an analyte preconcentration procedure is almost always necessary in order to achieve high enrichment factors. Although, in some studies, the direct injection of water samples was done successfully (Berset et al., 2010; Speksnijderet al., 2010), usually, a sample preparation technique is employed.

As can be observed in published reviews about the determination of PPCPs and organic compounds in environmental samples (Buchberger 2011; Primel et al., 2012), solid phase extraction (SPE) is the preferred technique for sample preparation of water samples, since it provides a convenient, cost-effective alternative to liquid–liquid extraction (LLE) and it has been shown to be valuable for extracting, concentrating, and cleaning up molecules with a wide range of physicochemical properties.

SPE of PPCP residues usually relies on sorbent materials such as octadecyl-bonded silica (C18) (Caldas et al., 2013), Oasis HLB (polystyrene–DVB–*N*-vinypyrrolidone terpolymer) (Crouse et al., 2012; Aydin and Talinli, 2013; Chen et al., 2013), Oasis MCX (a strong cation-exchange mixed-mode polymer) (Shraim et al., 2012), Strata-X (Cardoso et al., 2011; Lajeunesse et al., 2012; Matamoros et al., 2012), and some alternative sorbents, for example, coconut charcoal as sorbent (Reyes-Contreras et al., 2012) and molecularly imprinted polymers (Kubo et al., 2014).

Although SPE is the technique with more environmental applications, especially due to the variety of commercially available sorbents which cover the extraction of compounds with a wide range of physicochemical properties (Bratkowska et al., 2011), since 2010, other techniques such as dispersive liquid–liquid extraction (DLLME) (Zhang and Lee, 2012), hollow fiber liquid-phase microextraction (HF-LPME) (Zhang et al., 2013), passive samplers (Wille et al., 2011a), and sit bar solid extraction (SBSE) (Bratkowska et al., 2011; Gilart et al., 2013) have also been used for the extraction of PPCPs from water samples coupled to liquid chromatography–mass spectrometry (LC-MS) and gas chromatography–mass spectrometry (GC-MS), since they are faster and environmentally friendlier than SPE.

The combination of extraction techniques is also found in the literature review. A method using ionic liquid-based ultrasound-assisted dispersive liquid–liquid microextraction (IL-USA-DLLME) with determination by LC was employed for the extraction and preconcentration of four UV filters from three different water matrices. The procedure employed the IL extraction solvent 1-hexyl-3-methylimidazolium tris(pentafluoroethyl)trifluorophosphate in a small amount of methanol, and an ultrasound-assisted process was applied to accelerate the formation of the fine cloudy solution. The method showed acceptable recoveries and RSDs below 6.3% (Zhang and Lee, 2012).

Related mainly to water contamination, literature shows the potential of passive samplers or equilibrium-based sampling techniques for evaluation of the fate of PPCPs, especially in the marine environment in recent years. An analytical method for the quantification of a high number of pharmaceuticals and pesticides in passive sampling devices was developed by Wille et al. (2011a). The analytical procedure consisted of extraction by using methanol/acetonitrile (1:1 v/v) followed by detection with ultra-high performance liquid chromatography (UHPLC) coupled to high-resolution and high-mass accuracy Orbitrap MS. Propanolol, carbamazepine, and seven pesticides were found to be very abundant in the passive samplers (Wille et al., 2011a).

For solid sample extraction, there are a more restricted number of available methods. The analytical procedures include extraction of the contaminants from the solid using ultrasound-assisted extraction (UAE) (Yu and Wu, 2012), pressurized liquid extraction (PLE) (Chen et al., 2013; Wille et al., 2011b), microwave assisted extraction (MAE) (Hu and Zhou, 2011), solid–liquid extraction (SLE), solid phase microextraction (SPME) (Wang et al., 2011), and modified QuEChERS (Martínez Bueno et al., 2013).

The choice of the most appropriate technique will depend on many factors, and in some studies, a comparison among different extraction techniques is carried out. A comparison among MAE, simultaneous distillation-solvent extraction (SDSE), Soxhlet extraction (SE), and ultrasound probe (UP) for the extraction of polycyclic musks from sediment samples was done. MAE and UP proved to be more effective pretreatment techniques than SE and SDSE, with high recovery, repeatability, accuracy, and efficiency; little solvent consumption; and acceptable matrix effects (Hu and Zhou, 2011).

The number of analytical methodologies reported in the past few years for the extraction of several therapeutic families in biota samples is more restricted. The extraction techniques most reported in the literature used solvent extraction (Shanmugam et al., 2010; Gelsleichter and Szabo, 2013), SPE (Azzouz et al., 2011), PLE (Huerta et al., 2013), and SPME (Wang et al., 2011). This kind of sample usually demands extensive extraction and cleanup procedures to obtain extracts prior to analysis since these matrices are complex and PPCP residues are quantified in extremely low levels in this matrix.

Techniques such as MAE and PLE have emerged as efficient ways of increasing automation, shortening extraction time, and reducing the amount of organic solvents (Nieto et al., 2010). A method for measuring PPCPs in fish tissue using PLE combined with silica gel cleanup, gel permeation chromatography, and GC-MS/MS with ion trap was developed. This analytical improvement served to reduce the incurred cost, time, and loss of potential target analytes associated with independent cleanup steps (Huerta et al., 2013). The main limitations of PLE are that the selectivity toward the analytes during extraction is not as high as might be desired and that many interferents may be coextracted, and the analytes are sometimes diluted, especially when a high number of cycles are used (Nieto et al., 2010).

Another type of sample preparation that is a trend in the publications is nonlethal in vivo determination. In a study, published in 2012, the authors optimized an SPME of nonlethal in vivo determination of pharmaceuticals in fish muscle by using LC-MS. A new thin film based on C18 was optimized to improve SPME sensitivity and extraction kinetics. Rainbow trouts (*Oncorhynchus mykiss*) were exposed for 4 days to effluents from a municipal wastewater plant. Thin-film microextraction (TFME) and regular C18 fibers were inserted into the dorsal–epaxial muscle of the fishes. Sample extracts obtained from fibers were desorbed and analyzed. The C18 TFME phase successfully quantified fluoxetine, venlafaxine, sertraline, paroxetine, and carbamazepine in the muscle of living fish. This study demonstrates improved in vivo SPME sensitivity (recovery) and extraction rates during preequilibrium sampling of target pharmaceuticals in fish tissue due to the improved C18 extraction phase geometry (Togunde et al., 2012).

14.2 CHROMATOGRAPHY–MASS SPECTROMETRY-BASED APPROACHES

The simultaneous determination of different analytes requires sufficiently sensitive chromatographic techniques and analytical methods to detect and quantify PPCPs at trace levels. Although there has been an increase in the available methods for the quantification and determinations of PPCPs, the determination at trace concentration levels in environmental samples remains an important challenge (Reif et al., 2011).

As previously said PPCPs have been separated and quantified using GC and LC techniques usually using a mass spectrometer. The determination using liquid chromatography with ultraviolet detection (LC-UV) techniques is becoming increasingly rare, due to the lack of sensitivity and selectivity (especially for the analysis of complex matrices), since these compounds are generally present at microgram and nanogram levels (Gago-Ferrero et al., 2013).

Choosing between GC and LC is generally based on the physicochemical qualities of the target analyte. Thermally labile compounds that do not volatilize at temperatures of up to 250°C and that are strongly polar are more appropriate to be separated by LC (Dean, 2003). LC is usually used to determine more polar and less volatile compounds, while GC-MS is used to identify and quantify volatile or volatizable compounds and metabolites (Pietrogrande and Basaglia, 2007).

In relation to the mass spectrometer employed, although single-quadrupole instruments were successfully used when LC-MS procedures were first developed to determine residues of PPCPs, more complex mass analyzers such as the triple quadrupole (QqQ), time of flight (TOF), and ion trap are the most widely used instrumental techniques for drug analysis (Nieto et al., 2010).

14.2.1 Gas Chromatography–Mass Spectrometry

GC determinations have been carried out directly or after a derivatization step.

Due to the fact that most of PPCPs (i.e., parabens and some UV filters) are highly polar, thermally fragile compounds and lack sufficient volatility, a derivatization step is required to transform them into more volatile compounds and to make them suitable for GC analysis (Pietrogrande and Basaglia, 2007; Nieto et al., 2010). The ideal condition for a derivatization reaction is to permit detection of the compounds containing polar function groups with adequate signal-to-noise (S/N) ratio; this encourages complete derivatization (>90%) and is time efficient (Pietrogrande and Basaglia, 2007).

Derivatization is usually applied in polar analytes prior to analysis in order to reduce their polarity. Silylation is the most widely used derivatization technique, and normally, it does not require a purification step, and the derivatives can be injected directly into the GC. The introduction of a silyl group(s) can serve to enhance mass spectrometric properties of the derivatives, by producing more favorable diagnostic fragmentation patterns for use in structure investigations and for use in trace analyses in GC-MS under selected ion monitoring mode (Shanmugam et al., 2010).

Many variables are involved in the success of derivatization, and one of them is the derivatizing agent. Derivatizing agents such as the *N*-methyl-*N*-(*tert*-butyldimethysilyl)trifluoroacetamide (MTBSTFA) have been used for the derivatization of bisphenol A, estrone, nonylphenol, octylphenol, acetylsalicylic acid, carbamazepine, clofibric acid, diclofenac, gemfibrozil, ibuprofen, ketoprofen, naproxen, paracetamol, and triclosan in sewage sludge samples (Yu and Wu, 2012); triclosan and methyltriclosan in fish and suspended particulate matter (Rüdel et al., 2013); *N*-methyl-*N*-(trimethylsilyl)trifluoroacetamide (MSTFA) for parabens in human breast cancerous tissue

(Shanmugam et al., 2010); pentafluorobenzoyl chloride (PFBOCl) for the derivatization of 4-*t*-octylphenol, 4-nonylphenol, bisphenol A, estrone, 17β-estradiol, and triclosan in surface water and sediment samples (Wang et al., 2012); trimethylsilyl diazomethane for ibuprofen, naproxen, warfarin, cotinine, caffeine, and salicylic acid in effluents and receiving waters (Crouse et al., 2012); and *N,O*-bis(trimethylsilyl)trifluoroacetamide (BSTFA) with 1% of trimethylchlorosilane (TMCS) for derivatization of 20 pharmaceuticals in edible animal tissues (Azzouz et al., 2011).

Rice and Mitra (2007) evaluated the derivatizing agents MSTFA + 1% TMCS, trimethylsilyl iodide (TMSI), Tri-Sil TBT (3:3:2 TMSI/BSA [*N,O*-bis(trimethylsilyl) acetamide]/TMCS), and BSTFA + 1% TMCS for the derivatization of caffeine, diphenhydramine HCl, 17β-estradiol, epicoprostanol, ibuprofen, ketoprofen, musk ketone, naproxen, and triclosan. The authors conclude that the reaction with MSTFA + 1% TMCS permitted the GC-MS detection of all target compounds. None of the polar compounds were observed after reaction with TMSI, with the exception of ibuprofen. Thus, derivatization with BSTFA + 1% TMCS was chosen, as it proved to be the most effective, allowing all polar compounds to be readily detected (Rice and Mitra, 2007).

However, in the context of routine work, the usual but extensive and time-consuming derivatization step is one of the major drawbacks of GC-MS, especially when working with complex matrices (Kosma et al., 2010).

Some compounds such as synthetic musk fragrances (SMFs) (Chase et al., 2012) can be directly determined without a derivatization step. Polar drugs were also directly analyzed by GC-MS without the derivatization step. The usual derivatization step was avoided, and the compounds salicylic acid, ibuprofen, paracetamol, caffeine, phenazone, gemfibrozil, naproxen, triclosan, fenofibrate, diclofenac, and carbamazepine were successfully analyzed (Kosma et al., 2010).

Separation in GC is based on the vapor pressures of volatilized compounds and their affinities for the liquid stationary phase, as they pass down the column in a carrier gas (Dean, 2003). The practice of GC is usually carried out in capillary columns, typically 15–30 m long, with an internal diameter of 0.1–0.3 mm. The separation of PPCPs during GC analysis has been carried out mainly using nonpolar stationary phases such as the one containing 5% phenyl and 95% dimethylpolysiloxane (Kosma et al., 2010; Crouse et al., 2012; Yu et al., 2013). Less frequently, other stationary phases are found, such as the ones containing 100% dimethylpolysiloxane (Shanmugam et al., 2010) (Table 14.1).

Most of the GC-MS published methods for PPCP analysis in the past few years report identification based on electron impact ionization (EI) and single-quadrupole MS as the preferred GC detection system (Chase et al., 2012; Crouse et al., 2012; Yu and Wu, 2012; Yu et al., 2013).

EI is most popular. In EI, gas analyte molecules are bombarded by energetic electrons (70 eV), leading to the generation of a molecular radical ion that can subsequently generate ionized fragments. One important feature of electron ionization spectra is that they are highly reproducible, which means that mass spectral libraries can be used for identification of unknowns (Santos and Galceran, 2003).

GC-MS/MS analysis has also been used (Hu and Zhou, 2011; Reyes-Contreras et al., 2012). MS/MS techniques are a good choice once they improve the selectivity and sensitivity, decreasing the matrix background in the chromatograms. This is very advantageous mainly in the case of PPCPs, which are quantified in low levels and in matrices with high levels of background interference (Watson and Sparkman, 2007).

14.2.2 Liquid Chromatography–Mass Spectrometry

Contrary to GC-MS analysis, a derivatization step for PPCP determination by LC-MS is hardly ever observed. Usually, the methods include simultaneous determination of several PPCPs, with a wide range of polarities and acid–base properties, and in this case, LC is the technique of choice.

In particular, it is the most appropriate when analyzing metabolites and other transformation products, which usually have higher polarity than the parent compounds (Gago-Ferrero et al., 2013).

PPCP separation has been carried out using reversed phase. The stationary phase C18 is, by far, the most used in the analysis of PPCPs by LC-MS/MS (Table 14.2).

TABLE 14.1
Analytical Characteristics of Some GC-MS Methods Published since 2010 for the Determination of PPCPs in Environmental Samples

Sample	PPCPs	Sample Preparation	*R* (%)	RSDs (%)	Derivatizing Agent	Capillary Column	Ion Source	Mass Analyzer	Reference
Edible animal tissues	20 pharmaceuticals	SPE	92–101	<7	BSTFA + 1% TMCS	DB-5	EI	Quadrupole	Azzouz et al. (2011)
Water and sediment	SMFs	SPE SBSE Solvent extraction	>50		–	DB-5		Quadrupole	Chase et al. (2012)
Surface and effluent	6 pharmaceuticals	SPE	71–114			ZB-5MS		Quadrupole	Crouse et al. (2012)
sediments	2 polycyclic musks	MAE SDSE Soxhlet ultrasound	81–102	2–6		DB5-MS	EI	Quadrupole	Hu and Zhou (2011)
Surface water and restored wetland	17 PPCPs	SPE	>90	<20	TMSH[a]	TRB5-MS	EI	Quadrupole	Matamoros et al. (2012)
Fish muscle tissue	Triclosan and methyl-triclosan	PLE	100.6–101.0	<1.9	MSTFA	VF-5MS	CI	Quadrupole	Rüdel et al. (2013)
Fish tissue	26 PPCPs	PLE	64–131	<28	MSTFA	HP-5MS	EI	Ion trap	Subedi et al. (2011)
Human breast tissue	Parabens	Solvent extraction	96–113	<13.6	MSTFA	DB-1	EI	Quadrupole	Shanmugam et al. (2010)
Surface water and sediment	6 estrogenic compounds	SPE	76–119		PFBOCl	DB-35	CI	Quadrupole	Wang et al. (2012)
Sewage sludge	EDCs and PPCPs	UAE and cleanup with SPE	57.9–103.1	1.3–9.5	MTBSTFA	HP5-MS	EI	Quadrupole	Yu and Wu (2012)

[a] Trimethylsulfonium hydroxide.

TABLE 14.2
Analytical Characteristics of Some LC-MS Methods Published since 2010 for the Determination of PPCPs in Environmental Samples

Sample	PPCPs	Sample Preparation	*R* (%)	RSDs (%)	Column	Mobile Phase	Interface	Mass Analyzer	Reference
Wastewater and river water	25 pharmaceuticals	SPE	56–98	<20	C18	Methanol, water, formic acid	ESI	Quadrupole	Sim et al. (2010)
Soils and sediments	17 pharmaceuticals	SPE PLE UAE	Water: 71.2–99.3 Soil: 35.4–105.3 Sediment 42.1–97.8	Water<21.4, soil/sediment <19.1	C18	Methanol, water, formic acid (0.1%), acetonitrile, ammonium acetate (10 mM)	ESI	QqQ	Vazquez-Roig et al. (2010)
Wastewater	20 PPCPs	SPE	96–126	2–16	C18	Acetonitrile, methanol, formic acid (0.3%), ammonium formate (0.1%)	ESI	QqQ	Behera et al. (2011)
River waste water	14 PPCPs	SBSE	80–100 (most)	≤6	C18	Acetonitrile, water, formic acid (pH 3)	ESI	QqQ	Bratkowska et al. (2011)
River waters	4 estrogens 3 anti-inflammatories	SPE	82–102	<20	C18	Methanol, water, ammonia (0.2%)	ESI	QTRAP	Cao et al. (2010)
Surface water	10 PPCPs	SPE	62.1–125.4	≤14.6	C18 C8	Acetonitrile, ammonium acetate (20 mM), ammonium formate (20 mM), formic acid (0.3%)	ESI	QqQ	Koo et al. (2010)
Sewage	13 pharmaceuticals	SPE	93.2–106.8		C18	Methanol, water, acetonitrile, formic acid (0.2%), ammonium hydroxide (pH 3.2)	ESI	QqQ	Yuan et al. (2013)
Surface water	14 pharmaceuticals	SPE	61–98	7.9–14.9	Hypersil Gold column	Acetonitrile, water, formic acid (1%), NH_4OH (50 mM)	ESI	QqQ	Aydin and Talinli (2013)

(Continued)

TABLE 14.2 (CONTINUED)
Analytical Characteristics of Some LC-MS Methods Published since 2010 for the Determination of PPCPs in Environmental Samples

Sample	PPCPs	Sample Preparation	*R* (%)	RSDs (%)	Column	Mobile Phase	Interface	Mass Analyzer	Reference
Sewage sludge	30 pharmaceuticals	SPE—wastewater PLE—sludge	63.9–709	<133	C18	Water, methanol, formic acid, ammonium acetate	ESI	QqQ	Chen et al. (2013)
Bull shark plasma	Pharmaceuticals	Solvent extraction			C18	Acetonitrile, water, ammonium acetate (10 mM)	ESI		Gelsleichter and Szabo (2013)
Environmental waters	PPCPs	SBSE	<1–80	<17	C18	Water and acetonitrile pH 3 with formic acid	ESI	QqQ	Gilart et al. 2013)
Fish tissues	Pharmaceuticals	PLE		20	HSS T3, BEH C18	Methanol, acetonitrile, formic acid (10 mM), ammonium formate (pH 3.2), ammonium acetate/ammonia (5 mM)	ESI	Ion trap	Huerta et al. (2013)
Water, sediment, mussel	104 PPCPs	SPE	Water 33.3–420 sediment 20.9–5099 mussel 20.9–509	Water <172 sediment <45.9 mussel <26.7	C18	Acetonitrile, methanol, formic acid (0.1%), ammonium formate (0.1%), ammonium acetate (0.1%), acetic acid (01%)	ESI	QqQ	Klosterhaus et al. (2013)
Wastewater and biosolids	14 pharmaceuticals	SPE solvent extraction USA	Wastewater 72–102 biosolids 63–97	Wastewater <5, biosolids <10	C18	Acetonitrile, ammonium bicarbonate (5 mM) pH 7.8	ESI	QqQ	Lajeunesse et al. (2012)
Wastewater	156 polar organic contaminants	SPE and LLE	>80		C18	Water, acetonitrile, acetic acid (0.1%)	ESI	QqQ	Loos et al. (2013)
Wastewater influent and effluent	PPCPs and pesticides	SPE	>67 for most		C18	Water, acetonitrile, formic acid (0.1%)	ESI	Ion trap	Bueno et al. (2012)
Salt and fresh surface water	34 PPCPs	SPE	43–115	<5	C18	Water, acetonitrile, formic acid (0.1%)	ESI	QqQ	Spongberg et al. (2011)

(*Continued*)

TABLE 14.2 (CONTINUED)
Analytical Characteristics of Some LC-MS Methods Published since 2010 for the Determination of PPCPs in Environmental Samples

Sample	PPCPs	Sample Preparation	*R* (%)	RSDs (%)	Column	Mobile Phase	Interface	Mass Analyzer	Reference
Water	24 PPCPs	SPE	>70		C18	Water, methanol, ammonium acetate (2 mM)	ESI	QqQ	Tran et al. (2013)
Water and food	4 pharmaceuticals	HF-LPME	34–115	<12	C18	Methanol and water	ESI	QqQ	Zhang et al. (2013)
River waters	12 PPCPs	SPE	≥74	≤9	C18	Not cited	ESI	QqQ	Murata et al. (2011)
Fish muscle	8 PPCPs	SPE, SLE SPME	45.3–94.7	Not cited	C18	Methanol, water, ammonium acetate (5 mM)	ESI	QTRAP	Wang et al. (2011)
Marine mussels	2 pharmaceuticals 6 transformation products	Modified QuEChERS	67–110	18	C18	Acetonitrile, water, formic acid (0.1%)	ESI	Orbitrap	Martínez Bueno et al. (2013)
Marine organisms	11 pharmaceuticals	PLE	90–106	10–26	C18	Acetonitrile, water, isopropanol, formic acid (0.08%)	ESI	QqQ	Wille et al. (2011b)
Water	16 pharmaceuticals	Passive sampling devices	37.9–105.2	20	C18	Acetonitrile, water, formic acid	ESI	Orbitrap	Wille et al. (2011a)
Sewage sludge, sediment	19 PPCPs	UAE	63–119	<15	C18	Acetonitrile, water, formic acid (0.2%)	ESI	QqQ	Yu et al. (2011)
Fish homogenate, liver muscle	20 PPCPs	PLE	40–140	≤20	HSS T3 C18	Methanol, formic acid (10 mM), ammonium formate (pH 3.2)	ESI	QqQ-ion trap	Huerta et al. (2013)
Vegetables	PPCPs	UAE	87.1–123.5	0.2–18.2	C18	Methanol, water formic acid (0.001%)	ESI	QqQ	Wu et al. (2012)
Water sediment Sewage	12 PPCPs	SPE PFE	71.5–155.6	1.0–15.7	C18	Methanol, formic acid (0.3%), ammonium acetate (5 mM)	ESI		Zhu et al. (2013)

The most used mobile phase in the published studies is mixtures involving the solvents methanol and/or acetonitrile and water (Vazquez-Roig et al., 2010; Aydin and Talinli, 2013; Huerta et al., 2013; Loos et al., 2013). In order to obtain satisfactory ionization and improved sensitivity, ammonium acetate (Chen et al., 2013), ammonium formate (Huerta et al., 2013), ammonium hydroxide (Yuan et al., 2013), ammonium bicarbonate (Lajeunesse et al., 2012), formic acid (Bueno et al., 2012), and acetic acid (Loos et al., 2013) are often included as mobile phase additives. In positive mode, formic acid is the most used, at concentrations typically between 0.001% and 1.0% (Wu et al., 2012; Aydin and Talinli, 2013), and in the negative mode, ammonium acetate and ammonium formate are the preferred ones, at concentrations ranging from 2 to 20 mM (Koo et al., 2010; Tran et al., 2013).

The use of mobile phase additives usually increases the signals since these additives cause a wide variation in the analyte response (Fernandez-Alba, 2004).

In LC-MS determinations, the interface plays an important role. The electrospray ionization (ESI) interface is the most used interface. In ESI, a liquid that is coming from the analytical column is passed through a capillary, at atmospheric pressure, maintained at high voltage. The liquid stream breaks up with the formation of highly charged droplets, which are desolvated as they pass through the atmospheric pressure region of the source. Desolvation is assisted by a stream of nitrogen. Analyte ions are obtained from these droplets (Ardrey, 2003).

When working with LC-MS–based techniques, an important parameter to take into consideration is the matrix effect (ME).

The matrix effect in LC-MS (signal suppression or enhancement) is not fully understood, but it is believed to result from the competition between the analyte ions and the matrix components for access from the droplet surface to the gas-phase emission (Aga, 2010). This phenomenon is considered as the Achilles heel of the LC-MS (Taylor, 2005).

Many different mechanisms of ion suppression have been proposed; most are specific to the ionization technique used. In fact, although this phenomenon can have potential deleterious effects on ESI and APCI, evidence indicates that the ESI interface is more likely to be affected (Van Eeckhaut et al., 2009). However, some studies on environmental samples have demonstrated that APCI is not completely free of the ME (Lissalde et al., 2011).

If the ME cannot be eliminated by improving the cleanup, the chromatography separation, or the mass spectrometer conditions, an appropriate calibration technique to compensate for signal alterations must be used. In this case, the options are the standard addition, matrix-matched calibration, and internal standard addition. The most used method is the internal standard calibration, using isotopically labeled standards that are structurally similar to the target analytes to compare the instrument response during the analysis (Wang et al., 2011; Wille et al., 2011b).

14.3 OCCURRENCE IN ENVIRONMENTAL SAMPLES

PPCPs have been investigated in many environmental matrices, such as water bodies, sediment, soil, and biological samples. In contrast to other organic contaminants, PPCP determination in atmospheric environment has not attracted attention; probably due to their high polarity and low vapor pressure, they have low potential to escape to the atmosphere.

Water bodies can be considered susceptible to contamination by PPCPs since these compounds can reach the water sources after their ingestion and excretion, external application, or disposal.

PPCPs have been found in different water bodies, such as surface water (Blair et al., 2013), wastewater (Kosma et al., 2014), rivers (Bratkowska et al., 2011), seawater (Tashiro and Kameda, 2013), groundwater (Schaider et al., 2014), and drinking water (Silveira et al., 2013).

Among the PPCPs that have been detected in water bodies are the antibiotics (Koo et al., 2010; Murata et al., 2011; Aydin and Talinli, 2013; Blair et al., 2013), fibrates (Kosma et al., 2010; Spongberg et al., 2011), caffeine (Aydin and Talinli, 2013; Blair et al., 2013; Klosterhaus et al., 2013; Tran et al., 2013; Zhu et al., 2013; Kosma et al., 2014), triclosan (Blair et al., 2013; Gilart et al., 2013), SMFs (Chase et al., 2012), nonsteroidal anti-inflammatory drugs (NSAIDs) (Spongberg et al.,

2011; Crouse et al., 2012; Klosterhaus et al., 2013; Yu et al., 2013; Zhu et al., 2013), parabens (Gilart et al., 2013), UV filters (Gilart et al., 2013), illicit drugs (Klosterhaus et al., 2013), and analgesics (Behera et al., 2011; Tran et al., 2013).

As a result of the distribution of the PPCPs in water sources, sediment and soil can be contaminated. In sediment samples, PPCPs such as triclosan and triclocarban (Shanmugam et al., 2010; Yu et al., 2011; Blair et al., 2013), antibiotics (Blair et al., 2013), SMFs (Chase et al., 2012), caffeine (Klosterhaus et al., 2013), antihistamine (Long et al., 2013), NSAIDs (Stewart et al., 2014), and analgesics (Stewart et al., 2014) have been detected.

Other solid samples have also been studied, and ibuprofen was detected in soil samples (Karnjanapiboonwong et al., 2011), while antibiotics (Chen et al., 2013), analgesics (Yu and Wu, 2012), NSAIDs (Yu and Wu, 2012), and triclosan (Karnjanapiboonwong et al., 2011; Yu et al., 2011, 2013; Yu and Wu, 2012) have been detected in sludge samples.

The presence of PPCPs in biological samples has attracted recent attention as well as monitoring of water samples. Parabens had been found in cancerous tissues (Shanmugam et al., 2010); antibacterials in edible animal tissues (Azzouz et al., 2011); polycyclic musk fragrance (Subedi et al., 2011), triclosan (Rüdel et al., 2013), and anti-inflammatories (Huerta et al., 2013) in fish tissues; and caffeine, metoprolol, cotinine, and ketoprofen in marine mussels (Martínez Bueno et al., 2013).

14.4 CONCLUSIONS

The concern about the possible contamination of PPCPs and the lack of knowledge about the risk for the environment and human and aquatic life have aroused interest in developing methods for the investigation of these contaminants in different matrices. Since the methods are aimed at obtaining better precision and sensitivity and accurately quantifying contaminants present at trace concentrations in the environment, the MS-based chromatographic methods are the preferred ones.

For the extraction of PPCPs from environmental samples, different sample preparation techniques have been employed, with highlight on the ones that enable multiresidue analysis.

The use of LC-MS for qualitative and quantitative analyses of PPCPs is preferred instead that of GC-MS, since these compounds can be determined without a derivatization step in LC, and most of them are polar and thermally labile. Besides, the developments in the field of MS have transformed this technique into an important key for the analysis of PPCPs as environmental contaminants, due to the high selectivity and sensitivity that this technique provides.

With the progress in analytical science and the advances in MS techniques, information about the environmental contamination is becoming available, enabling more understanding and discussion about the presence of these contaminants in the environment. These developments suggest that new measurement and monitoring approaches need to be used more widely in future studies.

REFERENCES

Aga, D.S. 2010. *Fate of Pharmaceuticals in the Environment and in Water Treatment Systems*. CRC Press, Boca Raton, FL.

Ardrey, R.E. 2003. *Liquid Chromatography–Mass Spectrometry: An Introduction*. Wiley, New York.

Aydin, E., Talinli, I. 2013. Analysis, occurrence and fate of commonly used pharmaceuticals and hormones in the Buyukcekmece Watershed, Turkey. *Chemosphere* 90: 2004–2012.

Azzouz, A., Badredine, S., Evaristo, B. 2011. Determination of residual pharmaceuticals in edible animal tissues by continuous solid-phase extraction and gas chromatography–mass spectrometry. *Talanta* 84: 820–828.

Behera, S.K., Kim, H.W., Oh, J.E., Park, H.S. 2011. Occurrence and removal of antibiotics, hormones and several other pharmaceuticals in wastewater treatment plants of the largest industrial city of Korea. *Science of the Total Environment* 409: 4351–4360.

Berset, J.D., Brenneisen, R., Mathieu, C. 2010. Analysis of llicit and illicit drugs in waste, surface and lake water samples using large volume direct injection high performance liquid chromatography–electrospray tandem mass spectrometry (HPLC–MS/MS). *Chemosphere* 81: 859–866.

Blair, B.D., Crago, J.P., Hedman, C.J., Klaper, R.D. 2013. Pharmaceuticals and personal care products found in the Great Lakes above concentrations of environmental concern. *Chemosphere* 93: 2116–2123.

Bratkowska, D., Marcé, R.M., Cormack, P.A.G., Borrull, F., Fontanals, N. 2011. Development and application of a polar coating for stir bar sorptive extraction of emerging pollutants from environmental water samples. *Analytica Chimica Acta* 706: 135–142.

Bu, Q., Wang, B., Huang, J., Deng, S., Yu, G. 2013. Pharmaceuticals and personal care products in the aquatic environment in China: A review. *Journal of Hazardous Materials* 262: 189–211.

Buchberger, W.W. 2011. Current approaches to trace analysis of pharmaceuticals and personal care products in the environment. *Journal of Chromatography A* 1218: 603–618.

Bueno, M.J., Gomez, M.J., Herrera, S., Hernando, M.D., Agüera, A. et al. 2012. Occurrence and persistence of organic emerging contaminants and priority pollutants in five sewage treatment plants of Spain: Two years pilot survey monitoring. *Environmental Pollution* 164: 267–273.

Caldas, S.S., Bolzan, C.M., Guilherme, J.R., Silveira, M.A., Escarrone, A.L. et al. 2013. Determination of pharmaceuticals, personal care products, and pesticides in surface and treated waters: Method development and survey. *Environmental Science and Pollution Research* 20: 5855–5863.

Cao, J.L., Shi, J.H., Han, R., Li, Y.X., Yang, Z.F. 2010. Seasonal variations in the occurrence and distribution of estrogens and pharmaceuticals in the Zhangweinanyun River System. *Chinese Science Bulletin* 55: 3138–3144.

Cardoso, L.V., Tomasini, D., Sampaio, M.R.F., Caldas, S.S., Kleemann, N. et al. 2011. Optimization and validation of a method using SPE and LC-APCI-MS/MS for determination of pharmaceuticals in surface and public supply water. *Journal of the Brazilian Chemical Society* 22: 1944–1952.

Chase, D.A., Chase, D.A., Karnjanapiboonwong, A., Fang, Y., Cobb, G.P. et al. 2012. Occurrence of synthetic musk fragrances in effluent and non-effluent impacted environments. *Science of the Total Environment* 416: 253–260.

Chen, Y., Yu, G., Cao, Q., Zhang, H., Lin, Q. et al. 2013. Occurrence and environmental implications of pharmaceuticals in Chinese municipal sewage sludge. *Chemosphere* 93: 1765–1772.

Crouse, B.A., Ghoshdastidar, A.J., Tong, Z.A. 2012. The presence of acidic and neutral drugs in treated sewage effluents and receiving waters in the Cornwallis and Annapolis River watersheds and the Mill CoveSewage Treatment Plant in Nova Scotia, Canada. *Environmental Research* 112: 92–99.

Dean, J.R. 2003. *Methods for Environmental Trace Analysis*. Wiley, New York, Vol. 12.

EPA. 2013. Research. http://www.epa.gov/ppcp/ (accessed October 15, 2013).

Fernandez-Alba, A.R. 2004. *Chromatographic-Mass Spectrometric Food Analysis for Trace Determination of Pesticide Residues*. Elsevier, Amsterdam.

Gago-Ferrero, P., Silvia Diaz-Cruz, M., Barceló, D. 2013. Liquid chromatography tandem mass spectrometry for the multi-residue analysis of organic UV filters and their transformation products in the aquatic environment. *Analytical Methods* 5: 355–366.

Gelsleichter, J., Szabo, N.J. 2013. Uptake of human pharmaceuticals in bull sharks (*Carcharhinus leucas*) inhabiting a wastewater-impacted river. *Science of the Total Environment* 456–457: 196–201.

Gilart, N., Miralles, N., Marcé, R.M., Borrull, F., Fontanals, N. 2013. Novel coatings for stir bar sorptive extraction to determine pharmaceuticals and personal care products in environmental waters by liquid chromatography and tandem mass spectrometry. *Analytica Chimica Acta* 774: 51–60.

Hu, X., Zhou, Q. 2011. Comparisons of microwave-assisted extraction, simultaneous distillation-solvent extraction, Soxhlet extraction and ultrasound probe for polycyclic musks in sediments: Recovery, repeatability, matrix effects and bioavailability. *Chromatographia* 74: 489–495.

Huerta, B., Jakimska, A., Gros, M., Rodríguez-Mozaz, S., Barceló, D. 2013. Analysis of multi-class pharmaceuticals in fish tissues by ultra-high-performance liquid chromatography tandem mass spectrometry. *Journal of Chromatography A* 1288: 63–72.

Karnjanapiboonwong, A., Suski, J.G., Shah, A.A., Cai, Q., Morse, A.N. et al. 2011. Occurrence of PPCPs at a wastewater treatment plant and in soil and groundwater at a land application site. *Water, Air, & Soil Pollution* 216: 257–273.

Klosterhaus, S.L., Grace, R., Hamilton, M.C., Yee, D. 2013. Method validation and reconnaissance of pharmaceuticals, personal care products, and alkylphenols in surface waters, sediments, and mussels in an urban estuary. *Environment International* 54: 92–99.

Koo, S.H., Jo, C.H., Shin, S.K., Myung, S.W. 2010. Simultaneous determination and occurrences of pharmaceuticals by solid-phase extraction and liquid chromatography–tandem mass spectrometry (LC-MS/MS) in environmental aqueous samples. *Bulletin of the Korean Chemical Society* 31: 1193.

Kosma, C.I., Lambropoulou, D.A., Albanis, T.A. 2010. Occurrence and removal of PPCPs in municipal and hospital wastewaters in Greece. *Journal of Hazardous Materials* 179: 804–817.

Kosma, C.I., Lambropoulou, D.A., Albanis, T.A. 2014. Investigation of PPCPs in wastewater treatment plants in Greece: Occurrence, removal and environmental risk assessment. *Science of the Total Environment* 466: 421–438.

Kubo, T., Kuroda, K., Tominaga, Y., Naito, T., Sueyoshi, K. et al. 2014. Effective determination of a pharmaceutical, sulpiride, in river water by online SPE-LC–MS using a molecularly imprinted polymer as a preconcentration medium. *Journal of Pharmaceutical and Biomedical Analysis* 89: 111–117.

Lajeunesse, A., Smyth, S.A., Barclay, K., Sauvé, S., Gagnon, C. 2012. Distribution of antidepressant residues in wastewater and biosolids following different treatment processes by municipal wastewater treatment plants in Canada. *Water Research* 46: 5600–5612.

Lissalde, S., Mazzella, N., Fauvelle, V., Delmas, F., Mazellier, P. et al. 2011. Liquid chromatography coupled with tandem mass spectrometry method for thirty-three pesticides in natural water and comparison of performance between classical solid phase extraction and passive sampling approaches. *Journal of Chromatography A* 1218: 1492–1502.

Long, E.R., Dutch, M., Weakland, S., Chandramouli, B., Benskin, J.P. 2013. Quantification of pharmaceuticals, personal care products, and perfluoroalkyl substances in the marine sediments of Puget Sound, Washington, USA. *Environmental Toxicology and Chemistry* 32: 1701–1710.

Loos, R., Carvalho, R., António, D.C., Comero, S., Locoro, G. et al. 2013. EU-wide monitoring survey on emerging polar organic contaminants in wastewater treatment plant effluents. *Water Research* 47: 6475–6487.

Martínez Bueno, M.J., Boillot, C., Fenet, H., Chiron, S., Casellas, C. et al. 2013. Fast and easy extraction combined with high resolution-mass spectrometry for residue analysis of two anticonvulsants and their transformation products in marine mussels. *Journal of Chromatography A* 1305: 27–34.

Matamoros, V., Arias, C.A., Nguyen, L.X., Salvadó, V., Brix, H. 2012. Occurrence and behavior of emerging contaminants in surface water and a restored wetland. *Chemosphere* 88: 1083–1089.

McClellan, K., Halden, R.U. 2010. Pharmaceuticals and personal care products in archived US biosolids from the 2001 EPA national sewage sludge survey. *Water Research* 44: 658–668.

Murata, A., Takada, H., Mutoh, K., Hosoda, H., Harada, A. et al. 2011. Nationwide monitoring of selected antibiotics: Distribution and sources of sulfonamides, trimethoprim, and macrolides in Japanese rivers. *Science of the Total Environment* 409: 5305–5312.

Nieto, A., Borrull, F., Pocurull, E., Marcé, R.M. 2010. Pressurized liquid extraction: A useful technique to extract pharmaceuticals and personal-care products from sewage sludge. *Trends in Analytical Chemistry* 29: 752–764.

Petrovic, M., Farré, M., de Alda, M.L., Perez, S., Postigo, C. et al. 2010. Recent trends in the liquid chromatography–mass spectrometry analysis of organic contaminants in environmental samples. *Journal of Chromatography A* 1217: 4004–4017.

Petrović, M., Škrbić, B., Živančev, J., Ferrando-Climent, L., Barcelo, D. 2014. Determination of 81 pharmaceutical drugs by high performance liquid chromatography coupled to mass spectrometry with hybrid triple quadrupole–linear ion trap in different types of water in Serbia. *Science of the Total Environment* 468–469: 415–428.

Pietrogrande, M.C., Basaglia, G. 2007. GC-MS analytical methods for the determination of personal-care products in water matrices. *Trends in Analytical Chemistry* 26: 1086–1094.

Primel, E.G., Caldas, S.S., Escarrone, A.L.V. 2012. Multi-residue analytical methods for the determination of pesticides and PPCPs in water by LC-MS/MS: A review. *Central European Journal of Chemistry* 10: 876–899.

Reif, R., Santos, A., Judd, S.J., Lema, J.M., Omil, F. 2011. Occurrence and fate of pharmaceutical and personal care products in a sewage treatment works. *Journal of Environmental Monitoring* 13: 137–144.

Reyes-Contreras, C., Domínguez, C., Bayona, J.M. 2012. Determination of nitrosamines and caffeine metabolites in wastewaters using gas chromatography mass spectrometry and ionic liquid stationary phases. *Journal of Chromatography A* 1261: 164–170.

Rice, S.L., Mitra, S. 2007. Microwave-assisted solvent extraction of solid matrices and subsequent detection of pharmaceuticals and personal care products (PPCPs) using gas chromatography–mass spectrometry. *Analytica Chimica Acta* 589: 125–132.

Rüdel, H., Böhmer, W., Müller, M., Fliedner, A., Ricking, M. et al. 2013. Retrospective study of triclosan and methyl triclosan residues in fish and suspended particulate matter: Results from the German Environmental Specimen Bank. *Chemosphere* 91: 1517–1524.

Sanchez, W., Sremski, W., Piccini, B., Palluel, O., Maillot-Maréchal, E. et al. 2011. Adverse effects in wild fish living downstream from pharmaceutical manufacture discharges. *Environment International* 37: 1342–1348.

Santos, F.J., Galceran, M.T. 2003. Modern developments in gas chromatography–mass spectrometry-based environmental analysis. *Journal of Chromatography A* 1000: 125–151.

Schaider, L.A., Rudel, R.A., Ackerman, J.M., Dunagan, S.C., Brody, J.G. 2014. Pharmaceuticals, perfluorosurfactants, and other organic wastewater compounds in public drinking water wells in a shallow sand and gravel aquifer. *Science of the Total Environment* 468–469: 384–393.

Shanmugam, G., Ramaswamy, B.R., Radhakrishnan, V., Tao, H. 2010. GC-MS method for the determination of paraben preservatives in the human breast cancerous tissue. *Microchemical Journal* 96: 391–396.

Shraim, A., Diab, A., Alsuhaimi, A., Niazy, E., Metwally, M. et al. 2012. Analysis of some pharmaceuticals in municipal wastewater of Almadinah Almunawarah. *Arabian Journal of Chemistry*. http://dx.doi.org/10.1016/j.arabjc.2012.11.014

Silveira, M.A.K., Caldas, S.S., Guilherme, J.R., Costa, F.P., Guimarães, B.S. et al. 2013. Quantification of pharmaceuticals and personal care product residues in surface and drinking water samples by SPE and LC-ESI-MS/MS. *Journal of the Brazilian Chemical Society* 24: 1385–1395.

Sim, W.-J., Lee, J.-W., Oh, J.-E. 2010. Occurrence and fate of pharmaceuticals in wastewater treatment plants and rivers in Korea. *Environmental Pollution* 158: 1938–1947.

Speksnijder, P., Van Ravestijn, J., De Voogt, P. 2010. Trace analysis of isothiazolinones in water samples by large-volume direct injection liquid chromatography tandem mass spectrometry. *Journal of Chromatography A* 1217: 5184–5189.

Spongberg, A.L., Witter, J.D., Acuña, J., Vargas, J., Murillo, M. et al. 2011. Reconnaissance of selected PPCP compounds in Costa Rican surface waters. *Water Research* 45: 6709–6717.

Stewart, M., Olsen, G., Hickey, C.W., Ferreira, B., Jelić, A. et al. 2014. A survey of emerging contaminants in the estuarine receiving environment around Auckland, New Zealand. *Science of the Total Environment* 468–469: 202–210.

Suárez, S., Carballa, M., Omil, F., Lema, J.M. 2008. How are pharmaceutical and personal care products (PPCPs) removed from urban wastewaters? *Reviews in Environmental Science and Biotechnology* 7: 125–138.

Subedi, B., Mottaleb, M.A., Chambliss, C.K., Usenko, S. 2011. Simultaneous analysis of select pharmaceuticals and personal care products in fish tissue using pressurized liquid extraction combined with silica gel cleanup. *Journal of Chromatography A* 1218: 6278–6284.

Tashiro, Y., Kameda, Y. 2013. Concentration of organic sun-blocking agents in seawater of beaches and coral reefs of Okinawa Island, Japan. *Marine Pollution Bulletin*.

Taylor, P.J. 2005. Matrix effects: The Achilles heel of quantitative high-performance liquid chromatography–electrospray–tandem mass spectrometry. *Clinical Biochemistry* 38: 328–334.

Togunde, O.P., Oakes, K.D., Servos, M.R., Pawliszyn, J. 2012. Optimization of solid phase microextraction for non-lethal in vivo determination of selected pharmaceuticals in fish muscle using liquid chromatography–mass spectrometry. *Journal of Chromatography A* 1261: 99–106.

Tran, N.H., Hu, J., Ong, S.L. 2013. Simultaneous determination of PPCPs, EDCs, and artificial sweeteners in environmental water samples using a single-step SPE coupled with HPLC–MS/MS and isotope dilution. *Talanta* 113: 82–92.

Van Eeckhaut, A., Lanckmans, K., Sarre, S., Smolders, I., Michotte, Y. 2009. Validation of bioanalytical LC–MS/MS assays: Evaluation of matrix effects. *Journal of Chromatography B* 877: 2198–2207.

Vazquez-Roig, P., Segarra, R., Blasco, C., Andreu, V., Picó, Y. 2010. Determination of pharmaceuticals in soils and sediments by pressurized liquid extraction and liquid chromatography tandem mass spectrometry. *Journal of Chromatography A* 1217: 2471–2483.

Wang, S., Oakes, K.D., Bragg, L.M., Pawliszyn, J., Dixon, G. et al. 2011. Validation and use of in vivo solid phase micro-extraction (SPME) for the detection of emerging contaminants in fish. *Chemosphere* 85: 1472–1480.

Wang, L., Ying, G.-G., Chen, F., Zhang, L.-J., Zhao, J.-L. et al. 2012. Monitoring of selected estrogenic compounds and estrogenic activity in surface water and sediment of the Yellow River in China using combined chemical and biological tools. *Environmental Pollution* 165: 241–249.

Watson, J.T., Sparkman, O.D. 2007. *Introduction to Mass Spectrometry: Instrumentation, Applications, and Strategies for Data Interpretation*. Wiley, New York.

Wille, K., Claessens, M., Rappé, K., Monteyne, E., Janssen, C.R. et al. 2011a. Rapid quantification of pharmaceuticals and pesticides in passive samplers using ultra high performance liquid chromatography coupled to high resolution mass spectrometry. *Journal of Chromatography A* 1218: 9162–9173.

Wille, K., Kiebooms, J.A.L., Claessens, M., Rappé, K., Vanden Bussche, J. et al. 2011b. Development of analytical strategies using U-HPLC-MS/MS and LC-ToF-MS for the quantification of micropollutants in marine organisms. *Analytical and Bioanalytical Chemistry* 400: 1459–1472.

Wu, X., Conkle, J.L., Gan, J. 2012. Multi-residue determination of pharmaceutical and personal care products in vegetables. *Journal of Chromatography A* 1254: 78–86.

Yu, Y., Huang, Q., Cui, J., Zhang, K., Tang, C. et al. 2011. Determination of pharmaceuticals, steroid hormones, and endocrine-disrupting personal care products in sewage sludge by ultra-high-performance liquid chromatography–tandem mass spectrometry. *Analytical and Bioanalytical Chemistry* 399: 891–902.

Yu, Y., Wu, L. 2012. Analysis of endocrine disrupting compounds, pharmaceuticals and personal care products in sewage sludge by gas chromatography–mass spectrometry. *Talanta* 89: 258–263.

Yu, Y., Wu, L., Chang, A.C. 2013. Seasonal variation of endocrine disrupting compounds, pharmaceuticals and personal care products in wastewater treatment plants. *Science of the Total Environment* 442: 310–316.

Yuan, S.-L., Li, X.-F., Jiang, X.-M., Zhang, H.-X., Zheng, S.-K. 2013. Simultaneous determination of 13 psychiatric pharmaceuticals in sewage by automated solid phase extraction and liquid chromatography–mass spectrometry. *Chinese Journal of Analytical Chemistry* 41: 49–56.

Zhang, Y., Lee, H.K. 2012. Ionic liquid-based ultrasound-assisted dispersive liquid–liquid microextraction followed high-performance liquid chromatography for the determination of ultraviolet filters in environmental water samples. *Analytica Chimica Acta* 750: 120–126.

Zhang, H., Du, Z., Ji, Y, Mei, M. 2013. Simultaneous trace determination of acidic non-steroidal anti-inflammatory drugs in purified water, tap water, juice, soda and energy drink by hollow fiber-based liquid-phase microextraction and ultra-high pressure liquid chromatography coupled to tandem mass spectrometry. *Talanta* 109: 177–184.

Zhu, S., Chen, H., Li, J. 2013. Sources, distribution and potential risks of pharmaceuticals and personal care products in Qingshan Lake basin, Eastern China. *Ecotoxicology and Environmental Safety* 96: 154–159.

15 Endocrine-Disrupting Chemicals

Presence in Environmental Matrices

*Ramiro Vallejo-Rodríguez, Mario Murillo-Tovar, Leonel Hernández-Mena, and Alberto López-López**

CONTENTS

15.1 INTRODUCTION

The US EPA has defined *endocrine-disrupting chemicals* (EDCs) as exogenous agents that interfere with the synthesis, secretion, transport, binding, action, or elimination of natural hormones responsible for maintaining homeostasis and reproduction in live beings (US EPA, 1997).

* Corresponding author: alopez103@yahoo.com.

Considering the above definition, EDCs are synthetic compounds that were designed as a pesticide (organophosphates, methoxychlor, dichlorodiphenyltrichloroethane), plasticizer (phthalates), flame retardant (polybrominated biphenyl ether), dioxins, surfactant, biocide, pharmaceutical agents (diethylstilbestrol), or solvent but now have been found to have a side effect that, when adsorbed into the body, causes them to either mimic or block hormones and disrupt the body's normal functions (Schug et al., 2011; Rogers et al., 2013).

Actually, "new" contaminants or "emerging contaminants" (ECs) also exist which represent a risk to humans and are not regulated by competent sanitary and environmental authorities. Today, ECs are an object of study (US EPA, 2009, 2008). This group of contaminants includes natural estrogens such as 17β-estradiol, estrone, and estriol and synthetic estrogens such as 17α-ethinylestradiol and mestranol, in addition to some pharmaceuticals, veterinary drugs, antiseptics, and personal care products. ECs are not necessarily endocrine disruptors, but their low biodegradability contributes to the presence of the disruption of the endocrine systems of living organisms, as in the case of carbamazepine, diclofenac, diazepam, carbamazepine, and gemfibrozil, to name a few (Ibarluzea et al., 2004; Petrovic et al., 2004; Daughton, 2005; Esplugas et al., 2007; USGS, 2008; Eckstein and Sherk, 2011).

The EDCs are considered as ECs, because some of them are in the process of regularization by the environmental authorities (Barceló, 2003; Esplugas et al., 2007) and are compounds that, in small amounts, can cause disruption of the endocrine system and hormonal control in living aquatic ecosystems and wildlife (Esplugas et al., 2007; Caliman and Gavrilescu, 2009; Waye and Trudeau, 2011).

Anthropogenic activities, particularly industrial activities, generate and utilize a large range of EDCs, such as alkylphenols, dioxins, bisphenol A, polycyclic aromatic hydrocarbons (PAHs), styrene, and phthalates.

15.2 EFFECTS ON WILDLIFE AND HUMAN HEALTH

The endocrine system consists of many interacting tissues that talk to each other and the rest of the body using signaling mediated by molecules called hormones. A hormone is a molecule produced by an endocrine gland that travels through the blood to produce effects on distant cells and tissues via integrated complex interacting signaling pathways usually involving hormone receptors (Bergman et al., 2012). Hormones interface with cells that contain matching receptors in or on their surfaces. The hormone binds with the receptor, much like a key would fit into a lock. The hormones, or keys, need to find compatible receptors, or locks, to work properly. Once a receptor and a hormone bind, the receptor carries out the hormone's instructions by either altering the cell's existing proteins or turning on genes that will build a new protein. Both of these actions create reactions throughout the body (US EPA, 2013). Endocrine disruptors are chemicals that interfere in some way with hormone action and, in so doing, can alter endocrine function such that it leads to adverse effects on human and wildlife health. Thus, EDCs act like hormones (Bergman et al., 2012).

15.2.1 Environmental Health Effects

Numerous investigations have reported the effects of EDCs on the endocrine system of human beings such as sexual differentiation, ovarian function, sperm production and fertilization, cryptorchidism, and hypospadias, including changes in thyroid hormones (Ibarluzea et al., 2004; Mitra et al., 2004; Acosta et al., 2005; Daughton, 2005; Anway and Skinner, 2006; Guzmán and Zambrano, 2007; Bourguignon and Parent, 2010; Morales-Suárez-Varela et al., 2011).

A large number of studies on a laboratory level have also reported the effects of EDCs and ECs on the endocrine systems of fish, reptiles, birds, and mammals (disruption of reproductive functions) such as sexual differentiation, ovarian function, production of sperm, and fertilization (Acosta et al., 2005; Anway and Skinner, 2006; Guzmán and Zambrano, 2007; Caliman and Gavrilescu, 2009; Waye and Trudeau, 2011).

15.2.2 Effects on Human Health

Most of the environmental EDCs that mimic or block hormones are those categorized as estrogenic and are from natural dietary sources or are human-made pollutants; only a few have antiandrogenic or antiestrogenic activity (McLachlan et al., 2006; Cederroth et al., 2012; Vom Saal et al., 2012). This disrupting can occur by altering normal hormone levels, inhibiting or stimulating the production and metabolism of hormones, or changing the way that hormones travel through the body, thus affecting the functions that these hormones control (Schug et al., 2011).

Regarding male health, recent reports suggest the potential risk of environmental EDCs to men's reproductive health (Meeker, 2010), between them, downward trends in semen quality (Swan et al., 2003; Diamanti-Kandarakis et al., 2009) and testosterone levels (Andersson et al., 2007; Travison et al., 2007), along with increased rates of testicular (Huyghe et al., 2003; Fowler et al., 2012) and thyroid cancers (Enewold et al., 2009) and gynecomastia (Haynes and Mookadam, 2009) and increased rates of hypospadias (malformation of the urogenital tract) and cryptorchidism (maldescent of the testis) (Fowler et al., 2012).

In the case of female reproductive health, the effects are also numerous. The great sensitivity of both breast and uterine tissues to estrogens makes the female reproductive system especially vulnerable to compounds in the environment that mimic estrogen (McLachlan et al., 2006). In the past decades, trends of earlier age at onset of puberty have been reported in girls in both the United States and Europe (Mul et al., 2001; Parent et al., 2003).

The major human exposure to EDCs occurs in the consumption of drink water, personal care products, and food, particularly vegetables, fruits, and nuts, which contain relatively high levels of compounds that make nuclear hormone receptors such as estrogen and androgen receptors active or inactive (Safe, 2005; Bonefeld-Jorgense et al., 2007; Fowler et al., 2012; Rogers et al., 2013). Actual dates suggest that early life exposure to toxic chemicals (such as EDCs) probably is associated with increases in the rates of many of the most common human diseases at later stages of life, including asthma, learning and behavioral problems, early puberty, infertility, breast and prostate cancers, Parkinson's disease, obesity, and other diseases (Novikova et al., 2008; Yaoi et al., 2008; Ma, 2009; Perera et al., 2009). It is probable that human exposure to environmental EDCs (estrogens) has greatly increased over the past 40 years with increased agriculture and industry (including plastic manufacturing) production, popularity of herbal supplements and soy products, and use of pharmacological estrogens (McLachlan et al., 2006).

15.3 PRESENCE IN ENVIRONMENTAL MATRICES

15.3.1 Surface Water and Groundwater

EDCs and ECs are dumped directly into the environment by various anthropogenic activities, and even though some industrial and domestic effluents are previously treated, these compounds persist and are found in effluents from WWTPs (Miao et al., 2005; Gatidou et al., 2007; Benotti et al., 2009; Al-Odaini et al., 2011; Vallejo-Rodríguez et al., 2011a). As a consequence, these compounds have been identified in surface water (lakes, rivers, lagoon systems, etc.) (Bila et al., 2005; Suárez et al., 2007; Benotti et al., 2009; Matamoros et al., 2010; Moring, 2012); through leaching, these compounds reach groundwater (Gibson et al., 2007; Hsin-Chang et al., 2009).

15.3.2 Industrial Wastewater

A common characteristic of the EDCs is the recalcitrance or persistence against natural or controlled biological degradation (Esplugas et al., 2007; Koh et al., 2007; Beltrán et al., 2008; Mao et al., 2010; Xu et al., 2012). In this context, water pollution from EDCs also represents a technical problem for treatment and purification of water, given that the conventional biological processes (aerobic, anaerobic, coagulation–flocculation, filtration) and disinfection with chlorine treatments are not capable of removing or degrading the compounds (Kim et al., 2007; Benotti et al., 2009; Basile et al., 2011).

EDCs and ECs are commonly found in domestic wastewaters and in effluents from municipal wastewater treatment plants (Nakada et al., 2006; Kim et al., 2007; Xu et al., 2012) but mostly persist and are found in surface water in concentrations of microgram per liter to nanogram per liter. Furthermore, surface waters are the source of potable water supplies, thereby putting the populations that consume this water at risk (Gibson et al., 2007; Kim et al., 2007; Sharma, 2008; Benotti et al., 2009; Basile et al., 2011).

15.3.3 Presence in Solid Matrices (Soil)

The increasing amounts of liquid and solid wastes discharged in urbanized regions and areas with intensive agricultural practices contribute to raising EDCs' presence in both terrestrial and aquatic environments. In these areas, EDCs can be detected at concentrations ranging from few nanograms per liter to some milligrams per liter (Lee et al., 2004; Sharma et al., 2009).

Most EDCs have polar characteristics and also are highly hydrophilic with low partition coefficient (K_{ow}) and are partially soluble in aqueous media. Their presence and bioavailability in the aquatic environment depend on the interactions between solids and water in sludge and sediment. In many cases, some EDC concentration in water is higher. This is due to transformation processes involving photolysis, biodegradation, and dilution in water environment, whereas in other cases, persistent organic pollutants bioaccummulate and are adsorbed by the sludge, which suggest that the hydrophobicity of a chemical substance will determine whether EDCs will bioaccumulate in the solid phase or not (Caliman and Gavrilescu, 2009).

For this reason, one of the most important parameters to be considered for chemical loss during adsorption onto solid matrix is the organic-carbon-normalized sorption coefficient (K_{OC}). The chemical partitioning behavior of EDCs in aqueous media also depends on the partition coefficient (K_{ow}). Hence, the greater the values are for both parameters, the more those compounds sorb upon organic carbons such as suspended solids, nonpolar fats, lipids, and mineral oils, whereas leaching of ECDs from the soil or sediment may be attributed to low level of K_{OC} and K_{ow}.

Sorption of EDCs by soils and sediments has a critical influence on the fate of chemical pollutants in the environment, including a direct impact on bioavailability, transport by waters, and rates of transformation. Sorption of highly hydrophobic compounds such as PAHs, PCBs, and pesticides has been reported in literature (Spark and Swift, 2002; Hur and Schlautman, 2004).

15.3.4 Presence in Air

That many EDCs such as polychlorinated dibenzo-*p*-dioxins (PCDDs), polychlorinated dibenzofurans (PCDFs), and PAHs, among others, are anthropogenic compounds that are formed during various combustion and industrial processes is well known. Due to their persistence and semivolatility, these groups of compounds can be transported over long distances. Therefore, atmospheric transport is a primary pathway for its transference to terrestrial and aquatic ecosystems via deposition (Chao et al., 2004).

PCDDs and PCDFs present in the atmosphere come mainly from anthropogenic activities, particularly combustion or other thermal processes involving organic matters and chlorine (Everaert and Baeyens, 2002; Ni et al., 2009). There are several subcategories of waste incineration, which include municipal solid waste incineration, hazardous waste incineration, medical waste incineration, light fraction shredder waste incineration, sewage sludge waste incineration, waste wood and waste biomass incineration, and destruction of animal carcasses (US EPA, 2003).

Meanwhile, PAHs come mainly from incomplete combustion formed during the burning or pyrolysis of organic matter. Some of them are among the strongest known carcinogens (IARC, 1984; Finlayson-Pitts and Pitts, 2000). Their widespread presence is due to their emissions from a wide range of combustion sources including gasoline and diesel engines (Westerholm et al., 1991), biomass burning of agricultural fields and forest (Jenkins et al., 1996), and wood combustion (Finlayson-Pitts and Pitts, 2000).

15.4 ANALYTICAL METHODOLOGIES USED FOR DETERMINATION

15.4.1 Environmental and Analytical Issues

In general, trace analysis of EDCs, in environmental samples as either solid or liquid, often requires laborious sample pretreatment such as extraction, sample cleanup, and preconcentration. These previous steps must be not avoided, because direct determination is usually not possible due to interferences from complex mixture of organic and inorganic chemical constituents that exist in large amounts in matrix. However, the sample preparation conditions not only should be efficient for reducing chemicals interferences but also should afford recoveries of analytes enough for reaching the detection limits in analytical instruments because the EDCs in environmental samples are found at concentrations as low as parts per million or parts per trillion.

There are several techniques for sample pretreatment that have been used with this propose; however, the most widely employed are described.

15.4.2 Sample Preparation Techniques

15.4.2.1 Liquid–Liquid Extraction

Due to the low concentrations of the EDCs (μg L^{-1} and ng L^{-1}) in water, the extraction procedures are generally applied for concentrations of compounds of interest in an aqueous matrix. A liquid–liquid extraction (LLE) is one of the most frequently used separation techniques in environmental, clinical, and industrial laboratories. For example, drinking water supplies are routinely monitored for trihalomethanes ($CHCl_3$, $CHBrCl_2$, $CHBr_2Cl$, and $CHBr_3$) for their known carcinogenicity. Before the analysis by GC, trihalomethanes are separated from their aqueous matrix by LLE using pentane.

LLE is usually accomplished with a separatory funnel. The two liquids are placed in the separatory funnel and shaken to increase the surface area between the phases. When the extraction is complete, the liquids are allowed to separate, with the denser phase settling to the bottom of the separatory funnel. LLE may also be carried out in the sample container by adding the extracting solvent when the sample is collected (Dean, 1998; Mitra, 2003).

In LLE, the sample is distributed or partitioned between two immiscible solvents in which the analyte and matrix have different solubilities. The main advantage of this approach is the wide availability of pure solvents and the use of low-cost apparatus. However, this procedure generates a large amount of waste; for this reason, in recent years, it has been replaced by other alternative methods such as SPE that offered similar efficiencies.

Despite the use of high amounts of organic solvents in LLE, this procedure offers advantages such as the following:

- It lends itself to rapid and very selective separations that are usually highly efficient.
- Partition coefficients are often approximately independent of concentration down to tracer levels and, therefore, can be applied to a wide range of concentrations.
- It can usually be followed by back extraction into aqueous solvents or, in some cases, the solution can be used directly in subsequent procedures. This also provides significant pre-analysis concentration of the analyte.

Therefore, so far, many studies have carried out the extraction of compounds using LLE; for example, Deok-Hie Park et al. (2011) determined PCDDs and dibenzofurans in wastewater; specifically, they compared the efficiency of recovery with SPE method and concluded no significant differences in recoveries between the two methods. Also, Hsin-Chang et al. (2009) analyzed estrogenic compounds in wastewater using liquid chromatography (LC)–MS/MS with electrospray and atmospheric pressure photoionization following desalting extraction; in this study, the relevance of LLE was the elimination of matrix interference in wastewater.

15.4.2.2 Solid-Phase Extraction

SPE is a technique that allows concentrating and purifying an aqueous sample by absorption of a cartridge in solid phase, followed by elution of the analyte with a suitable solvent and subsequent instrumental analysis (Thurman and Mills, 1998). There are a variety of cartridges packed with different sorbents that allow the extraction and purification of the compounds according to the polarity and dissociation constant of the analyte and the pH of the aqueous sample.

However, the chemical substituent bonded to the silica gel determines the mode of action, while the capacity of the sorbent in milligrams per gram (mg g^{-1}), which may be sorbed, is related to both the chemistry of the phase and the loading weight of the carbon (Thurman and Mills, 1998). For normal-phase SPE, cyanopropyl (CN), aminopropyl (NH_2), and diol functional groups are chemically bonded to the silica gel and they may sometimes be used for reversed-phase applications, especially for the removal of hydrophobic solutes from water or other polar solvents (Thurman and Mills, 1998).

The widespread use of the C18 for reversed-phase sorption is a result of the popular use of C18 columns in HPLC. For this reason, it gained immediate popularity for SPE applications (Thurman and Mills, 1998; Jeannot et al., 2002; Koutsouba et al., 2003; Gatidou et al., 2007; Tian et al., 2010; Huerta-Fontela et al., 2011; Kusk et al., 2011; Samaras et al., 2011; Vallejo-Rodríguez et al., 2011b). The C18 sorbents are available in both trifunctional and monofunctional forms. If the pH is adjusted to acid values, it is critical to use the trifunctional silica in order to prevent hydrolysis of the hydrocarbon group from the surfaces of the sorbent. However, if polar molecules are to be sorbed and they contain basic sites, the monofunctional sorbents will have greater capacity because of the free silanol sites that are available for secondary interactions. Thus, the sources of the C18 sorbent may be selected to enhance the isolation that is required (Thurman and Mills, 1998; Koh et al., 2007). Polymeric sorbents, such as styrene divinylbenzene and carbon, also are used for reversed-phase SPE. These sorbents were some of the classical reversed-phase sorbents introduced in the 1960s (Thurman and Mills, 1998; Belfroid et al., 1999; Boyd et al., 2003).

The SPE technique using C18 cartridges has been widely used in the extraction of EDCs in natural waters and wastewaters. Another application was the identification of 12 pesticides in water sample with an SPE procedure on C18 cartridges (Martínez-Vidal et al., 2000). On the other hand, the simultaneous extraction and quantification of different estrogens, pesticides, and bisphenol A in natural waters have been developed on LiChrolut RP-18 cartridges (Rodriguez-Mozaz et al., 2004).

Ion-exchange sorbents usually contain both weak and strong cation and anion functional groups bonded to the silica gel. The typical ion-exchange sorbent consists of a resin of a cross-linked styrene–divinylbenzene polymer. Recently, several manufacturers are producing styrene–divinylbenzene-based ion-exchange SPE sorbents that are more rugged than silica-bonded sorbents because the styrene–divinylbenzene sorbents are stable from pH 1 to 13. Actually, the hydrophilic–lipophilic balance (HLB) cartridges are often used because of their wide applicability under different conditions of pH and polarity of the analytes to be extracted. Due to their advantages, they have been used by various researchers in the determination of drugs, steroids, and various compounds in surface waters, WWTP effluent, or spiked waters, by either GC-MS (Öllers et al., 2001; Jeannot et al., 2002; Rodríguez et al., 2003; Lin et al., 2005) or MS/MS (Hoai et al., 2003; Quintana et al., 2004; Trenholm et al., 2006).

The polarity of the compounds and pH are very important in the selection process of cartridge for sample extraction. EDC analysis included estrogens, 4-nonylphenol, bisphenol A, and 4-*tert*-octylphenol as the target compounds. HLB copolymer cartridges showed the best recoveries overall (57–118%) and were therefore used for further testing. Later, the researchers analyzed the effect of pH on the extraction efficiency in the cartridge. The results show that the extraction recovery for all target compounds remains high and relatively similar at pH 4.15 and 7.96 (Liu et al., 2004). It can be inferred from these studies that the advantage of using HLB cartridges in the SPE technique is the simultaneous extraction of nonpolar and polar compounds in a wide range of pH. These features have been implemented in the simultaneous determination of neutral and acidic pharmaceuticals and pesticides in natural waters (Öllers et al., 2001); in the extraction for the determination of five acidic nonsteroidal anti-inflammatory pharmaceuticals in sewage water (Rodríguez et al., 2003); in an analytical method for the simultaneous determination of five anti-inflammatory drugs and other EDCs in wastewater influents and effluents from WWTPs (Santos et al., 2005); in the analysis of acidic pharmaceuticals (potential EDCs) and two pesticides in untreated wastewater and spring waters, using a simple extraction procedure (Gibson et al., 2007); and in the extraction, before chromatographic analysis, of influent and effluent of WWTPs containing three natural estrogenic compounds and two synthetic estrogenic compounds (Hsin-Chang et al., 2009).

15.4.2.3 Solid-Phase Microextraction

A more recent extraction technique, SPME, was introduced by Pawliszyn (1997). SPME consists of an absorption and a desorption step. In the absorption step, a coated fused silica fiber extracts the analytes from the sample matrix. In the desorption step, the analytes are desorbed from the fiber and introduced into the analytical column for separation (Peñalver et al., 2002). Usually, SPME is combined with GC, placing the fiber in the hot injector of the gas chromatograph, where the analytes are thermally desorbed. The SPME method preserves all the advantages, such as simplicity and low cost, of site sampling and does not require solvents (Helaleh et al., 2001). For example, an SPME procedure was compared with a SPE-GC-MS method (C18 cartridge) for the analysis of NPEO and bromononylphenol ethoxylates in water samples. There was good agreement between the results from both methods, but the SPME procedure showed some advantages such as lower detection limits, a shorter analysis time, and the avoidance of organic solvents (Díaz et al., 2002).

The SPME method has been applied in the analysis of water and wastewater. An example is the extraction of phenols in different water samples by using poly(acrylate) coating fibers (Helaleh et al., 2001). An SPME method in the trace analysis of technical nonylphenol, bisphenol A, and 17α-ethinylestradiol was developed (Braun et al., 2003). The analysis of nine pharmaceutical drugs in wastewater by using the SPME technique for the sample preparation was carried out, using PDMS-DVB as the SPME fiber coating (Feitosa de Lima-Gomes et al. 2011).

15.4.2.4 Solid Matrices

EDCs have been reported in several types of solid samples such as soil (Xu et al., 2008; Llorca et al., 2009; Gibson et al., 2010), sludge (Jeannot et al., 2002; Maurício et al., 2006; Nie et al., 2009; Vega et al., 2011; Yu and Wu, 2012), sediments (Eljarrat et al., 2001; Liu et al., 2004; Hibberd et al., 2009; Gong et al., 2011), and airborne particulate matter (Sidhu et al., 2005; Xie et al., 2006; Blackwell et al., 2011; Salapasidou et al., 2011). In general, the analytical procedure includes four steps, namely, extraction, concentration, purification, and analysis. The extraction is carried out by either Soxhlet extraction (Jeannot et al., 2002; Xie et al., 2006), ultrasonic extraction (Maurício et al., 2006; Xu et al., 2008; Llorca et al., 2009; Yu and Wu, 2012), supercritical fluid extraction (Wu et al., 2008; Vega et al., 2010), accelerated solvent extraction (Shen, 2005; Reid et al., 2009), or microwave-assisted extraction (Liu et al., 2004; Hibberd et al., 2009; Vega et al., 2011) with volatile organic solvents. Subsequently, the organic extracted is concentrated using a rotatory evaporator or a Kuderna-Danish concentrator, and then, it is dried under a stream of nitrogen.

15.5 INSTRUMENTAL

The chromatographic techniques, either GC or LC, coupled with MS or MS/MS, have become the most commonly used instrumentation for the determination of several of these EDCs due to their capacity of identification by structural information, which is like the fingerprint of compounds, instead of using only retention time.

The LC-MS/MS technique has shown to be a versatile technique that is applicable for the majority of polar or thermolabile EDCs (e.g., antibiotics). However, the LC-MS/MS method is relatively expensive. In comparison with LC, the analysis of EDCs by GC-MS, although more limited in scope, offers a useful and sensitive method for the determination of EDCs and is more affordable for the majority of laboratories. However, often, the samples require additional derivatization steps following extraction to obtain the less polar and/or more volatile target compounds. The general steps in the analysis of EDCs by GC-MS are collecting the sample, extraction, derivatization, and, finally, the identification and quantification of the compounds (Kolpin et al., 2002; Flemming and Bent, 2003).

15.5.1 Gas Chromatography Coupled with Mass Spectrometry

GC-MS is the most commonly used analytical technique for the determination of EDCs in either liquid (Sarmah et al., 2006; Jackson and Sutton, 2008; Stasinakis et al., 2008; Migowska et al., 2012), solid (Tan et al., 2008; Jia et al., 2012; Langdon et al., 2012; Bao et al., 2013; Rhind et al., 2013), or air (Correa et al., 2004; Fu and Kawamura, 2010) environmental samples due to its simplicity, robustness, and resolution power. However, often, compounds with polar functional groups such as hydroxyl and amines (Mol et al., 2000; Farajzadeh et al., 2014) require additional derivatization steps following extraction and cleanup to obtain the less polar and/or more volatile target compounds what improve the sensitivity (Asimakopoulos et al., 2012). Despite the advantage offered by derivatization, that procedure can be laborious (Helaleh et al., 2001); it can have risk of contamination and is time consuming. These drawbacks could be dealt with by the development of in situ derivatization methods (Helaleh et al., 2001; Brossa et al., 2002; Basheer and Lee, 2004, Yang et al., 2006).

The fluctuations during instrumental analysis and recoveries from the procedure are two important aspects to consider before using GC-MS to identify and quantify EDCs; both are corrected by using compounds with similar physicochemical properties as the target compounds. However, it is difficult to find chemical substances that fulfill those requirements. Generally, deuterated compounds are used (Planas et al., 2006; Prieto et al., 2010, 2011); however, less uncertainty could be reached by use of compounds isotopically labeled with ^{13}C (Gibson et al., 2005).

In order to improve the reliability of identification of EDCs in low concentrations and complex matrices and/or with much interference, most recently, sequential MS, also called triple-quadrupole or tandem GC-MS, has been used (Padilla-Sánchez et al., 2010). The analysis of environmental complex mixtures by tandem GC-MS provides rather structural information compared with conventional GC-MS, allowing the identification of primary and secondary molecular fragments from the breakdown of target compounds, which reduces the probability of spectral interferences (Trenholm et al., 2006; Mansilha et al., 2010).

Nowadays, the capacity of MS instruments to identify EDCs in complex environmental mixtures has been coupled with new developments in GC. Recently, two-dimensional gas chromatography (GC×GC) has been applied to the determination of EDC in environmental samples such as water (Lima et al., 2013), soil (Casas et al., 2011), and particulate matter (Alliot et al., 2014). The main advantage of this combination is the resolution power and the use of three identification criteria, conventional retention time, structural information provided by MS, and separation of compounds in classes by polarity and vapor pressure in GC×GC system, which diminished the probability of bias during identification.

15.5.2 Liquid Chromatography Coupled with Mass Spectrometry

The presence of EDCs in the environment represents a serious problem of public health; however, their determination is an analytical challenge, because EDCs are found in low environmental concentrations in samples with a high-complexity chemical, since EDCs are constituted from a mixture of several classes of compounds with different chemical properties. Hence, the use of analytical tools that are highly efficient and fast that allow estimate reliability of the real concentrations in diverse environmental matrix is necessary.

In this sense, the technique LC-MS/MS has shown to be a versatile technique that is applicable for the fast and sensitive quantitation of small molecules, peptides, and proteins, among others in complex matrices such biological tissues and environmental samples. Such coupling is more sensitive and significantly more specific than other LC detectors (UV–vis, fluorescence, refractive index), routinely achieving nanogram to picogram per milliliter detection. LC-MS/MS can analyze compounds that lack a suitable chromophore and can identify components in unresolved chromatographic peaks, reducing the need for ideal chromatographic conditions.

The instrumentation comprises an HPLC attached, via a suitable interface, to a mass spectrometer. Compounds that are thermally labile, exhibit high polarity, or have a high molecular mass may all be analyzed using HPLC-MS, for example, proteins.

Solutions derived from samples of interest are injected onto an HPLC column; compounds are separated on the basis of their relative interaction with the chemical coating of these particles (stationary phase) and the solvent eluting through the column (mobile phase). Components eluting from the chromatographic column are then introduced to the mass spectrometer via an interface. The two most common interfaces used for HPLC-MS are the electrospray ionization and the atmospheric pressure CI interfaces (Gates, 2013).

15.5.2.1 Electrospray Ionization

In electrospray ionization mode, the analyte is introduced to the source at flow rates about 1 μL min^{-1}. The solution with the analyte flow passes through the electrospray needle that has a high potential difference (with respect to the counter electrode) applied to it (in the range from 2.5 to 4 kV). This helps the spraying of charged droplets from the needle with a surface charge of the same polarity to the charge on the needle. The droplets are repelled from the needle toward the source sampling cone on the counter electrode. As the droplets traverse the space between the needle tip and the cone, solvent evaporation occurs (Figure 15.1).

Once the solvent evaporation occurs, the droplet shrinks until it reaches the point that the surface tension can no longer sustain the charge at which point, a "coulombic explosion" occurs and the droplet is partitioned, what produces smaller droplets that can repeat the process until charged analyte molecules are obtained. The charged analyte molecules can be singly or multiply charged.

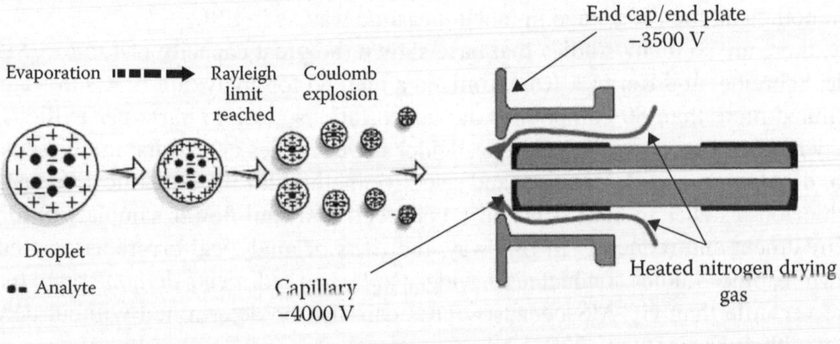

FIGURE 15.1 Electrospray ionization process. (Courtesy of Iowa State University, Ames, IO, http://www.cif.iastate.edu.)

This is a very soft method of ionization as very little residual energy is retained by the analyte upon ionization. It is the generation of multiply charged molecules that enables high-molecular weight components such as proteins. The major disadvantage of the technique is that very little (usually no) fragmentation is produced although this may be overcome through the use of tandem mass spectrometric techniques such as MS/MS or MS.

15.5.2.2 Atmospheric Pressure Chemical Ionization

Atmospheric pressure chemical ionization (APCI) is an ionization method analogous to CI. The main difference is that APCI is carried out at atmospheric pressure and has its primary applications in the areas of ionization of low-mass compounds (APCI is not suitable for the analysis of thermally labile compounds). In APCI, the analyte solution is introduced into a pneumatic nebulizer and desolvated in a heated quartz tube before interacting with the corona discharge, creating ions.

The primary ionization of the surrounding atmosphere is very fast, on the order of microseconds, and it is possible to generate positive and negative ions, depending on the selected ionization mode. The primary reagent ions generally react further, and the most thermodynamically favored (most stable) reagent ions will be formed. In a normal atmosphere and using positive APCI, these will often be protonated water clusters. These reagent ions will subsequently be used to ionize the analytes of interest. If the generated reagent ions are more stable than the protonated analyte, the desired reaction will not be favored and no ionization of the analyte will occur.

A typical proton transfer reaction at atmospheric pressure is shown in the following, where MH^+ is the reagent ion and A denotes an analyte:

$$MH^+ + A \leftrightarrow AH^+ + M.$$

The protonation of the analyte will proceed only if the gas phase basicity of A is higher than that of the reagent ion MH^+. The gas phase basicity is defined in the following reaction:

$$A + H^+ \leftrightarrow AH^+.$$

The protonated analyte will easily cluster with solvent molecules, resulting in the following reaction when water is used as the solvent:

$$H_3O^+ (H_2O)n + A \Leftrightarrow AH^+ (H_2O)b + (1 + n - b)H_2O.$$

These primary ions collide with the vaporized solvent molecules to form secondary reactant gas ions (see Figure 15.2). These reactant gas ions then undergo repeated collisions with the analyte, resulting in the formation of analyte ions. The high frequency of collisions results in a high ionization efficiency and thermalization of the analyte ions. This results in spectra of predominant molecular species and adduct ions with very little fragmentation. Once the ions are formed, they enter the pumping and focusing stage in much the same way as in ESI.

Actually, there are so many studies that have shown the great capacity of LC-MS/MS analysis; for example, Schreiber and Kern (2010) optimized a method for analyzing EDCs in water samples; they determined more than 80 compounds at concentrations close to parts per trillion. Also, this methodology has been used for analyzing pesticides in vegetables and fruits; in this sense, Grimalt et al. (2006) developed a rapid, sensitive, and selective method for the quantification and confirmation of trichlorfon residues in kaki fruit (flesh and peel) and cauliflower samples, minimizing the sample pretreatment and reducing, in this way, the risks of analytical errors associated with this step. Meanwhile, other studies conducted in sludge, soils, and sediments demonstrated that LC-MS/MS is more versatile than GC-MS because EDCs can also be determined without derivatization (Benijts et al., 2004; Kuster et al., 2004a,b).

In summary, LC-MS/MS has been constituted as one of the most powerful analytical tools for determining EDC complex matrices, such as wastewater, biological tissues, and sludge, among others.

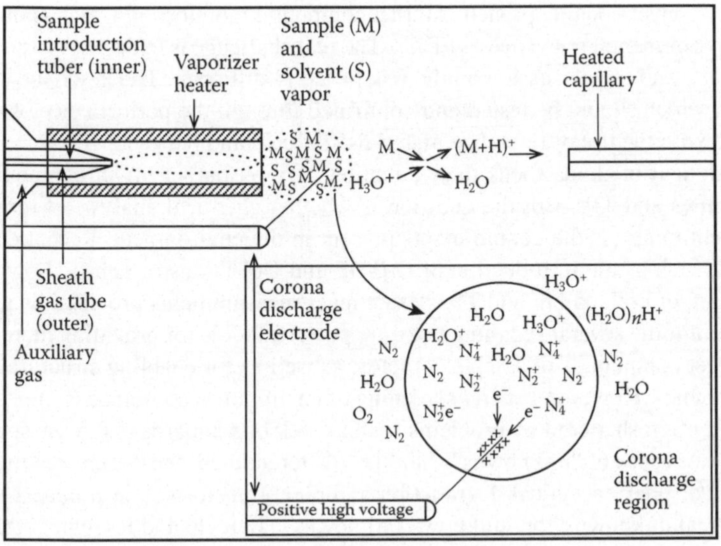

FIGURE 15.2 Ionization process occurring in the APCI source. (Adapted from Byrdwell, W.C., *Modern Method for Lipid Analysis by Liquid Chromatography/Mass Spectrometry and Related Techniques*, AOCS Press, Urbana, IL, 2005. With permission.)

15.6 PERSPECTIVES OF ENDOCRINE-DISRUPTING COMPOUNDS IN THE ENVIRONMENT AND FUTURE REGULATIONS OF ENDOCRINE-DISRUPTING COMPOUNDS

15.6.1 Analytical Challenges in Identifying and Quantifying Endocrine-Disrupting Compounds

EDCs are generally present at trace concentrations and in complex matrices, requiring highly selective and sensitive analytical methods for their measurement and treatment technologies for their removal in water, soil, and air. The wide range of different compound classes requires a variety of analytical approaches and techniques, making it challenging to understand all of the different chemicals in the environment and in human and wildlife tissues.

In general, complicated and time-consuming extraction and purification steps are required, followed by chromatographic techniques often coupled to MS. The majority of established analytical methods focus on dioxins and dioxin-like PCBs, indicator PCBs, organochlorine pesticides, BFRs, PFCs, phthalates, bisphenol A, and a large number of halogenated phenolic and nonphenolic compound classes known to exert ED effects (TEDX, 2011). Indeed, large numbers of halogenated phenols have been identified in living organisms in the environment. Analytical standards are required for routine quantitative analysis. They are available for most pesticides, pharmaceuticals, PAHs, metals, and many halogenated organics with persistent characteristics. However, they are typically not available for transformation products (TPs) nor for the vast array of organic chemicals in commerce. Thus, targeted analyses based on analytical standards provide the necessary sensitivity and selectivity but lead to a fragmented picture of the occurrence of EDCs in the environment, wildlife, and humans. Consequently, there is a high demand for developing screening analytical methods that will accommodate a wide variety of analytical functional groups at low detection levels (Bergman et al., 2012).

The introduction of new chromatographic techniques, such as fast LC, fast GC, GC×GC, and advanced tandem and hybrid MS detection systems (i.e., QqTOF, quadrupole linear ion trap [QqLIT], and Orbitrap), has improved analysis of complex matrices and TPs. However, the main drawback of the conventional approach is target compound monitoring, which is often insufficient to assess the environmental relevance of EDCs. The real challenge is to decide on the significance of the chemical data, and the discussion on the risk posed is still open. The key issue is the quality of analytical data, which should be tested and confirmed through the performance of interlaboratory tests, combined with the use of reference materials (Barceló and Petrovic, 2008).

Current analytical methods focus their attention only on parent target compounds and rarely include metabolites and TPs. But the question is whether chemical analysis of target compounds alone is sufficient to assess the contaminants present in the environment. Recent developments in the field of MS, such as the introduction of QqTOF and QqLIT instruments, allow the simultaneous determination of both parent and TPs. Exact mass measurements provided by QqTOF and the capability of combining several scan functions are powerful tools for providing more accurate identification of target compounds in sample complex, as well as for enabling structural elucidation of unknown compounds. However, general screening of an unknown substance is time consuming and expensive and is often shattered by problems such as lack of standards and mass spectral libraries. In this respect, two types of "unknowns" should be differentiated: the "suspected unknowns," such as pesticide and/or pharmaceutical degradation and phototransformation products in the environment, and the "real unknowns" or "unknown unknowns" (Barceló and Petrovic, 2008).

A very powerful tool is the combined approach of TOF-MS and IT-MS instruments and hybrids such as QqTOF-MS and QqLIT-MS, which enable either accurate mass determination or fragmentation patterning based on MS experiments, thus providing complementary structural information. Combination of ultraperformance liquid chromatography (UPLC)-QqTOF-MS and HPLC-IT-MS^n or LC-QqLIT-MS in the same basic format can be applied as a universal generic strategy to find, characterize, and confirm unknown compounds in environmental and food analyses. The availability of the precursor or product full mass spectra throughout each HPLC or UPLC chromatogram and either accurate mass measurements or multiple stages of MS will provide qualitative information that can be used to ascertain whether metabolites or any other compounds are present in the sample. The knowledge of exact analyte masses allows the determination of the possible molecular formula and chemical structure of the suspected metabolites. The sequential fragmentations will provide a better understanding of the fragmentation patterns. It seems feasible that this approach could be useful in providing information on metabolites while simultaneously achieving their quantification (Barceló and Petrovic, 2008).

Effect-directed analysis (EDA) and quantitative structure–activity relationship (QSAR)-directed nontarget analysis are two emerging techniques that are helping to address this challenge (Hecker and Hollert, 2009; Schymanski et al., 2009; Helbling et al., 2010). EDA involves bioassay-directed fractionation techniques to decrease the complexity of the sample matrix before identification of the endocrine-active compounds. Typical of these are estrogenic and androgenic in vitro assays (Weiss et al., 2009, 2011; Hecker and Holler, 2009). QSAR-directed analyses involve identification of TPs from structural information and identification from full-scan high-resolution mass spectra using postacquisition data processing. Nontarget analyses of candidate chemicals for which no standards are available can sometimes be accomplished by very high-resolution MS. However, improved prediction systems for theoretical fragmentation patterns, retention times, and ionization behavior are needed to widely apply this technology (Bergman et al., 2012).

15.6.2 Regulatory Perspectives and Needs

National and international governments are in the process of establishing testing programs and strategies to assess the safety of currently used chemicals with regard to their potential to interact with the endocrine system of humans and wildlife, resulting in potential impacts on reproduction, growth, and/or development. Specifically, the United States, Japan, the European Union, and the Organization for Economic Cooperation and Development have established testing approaches and regulatory frameworks with the aim to assess the risks associated with chemicals that have endocrine-disrupting properties. While there has been a large amount of efforts over the past two decades in context of assessment of chemical safety, no comparable attempts to harmonize and mutually accept testing strategies and decision-making criteria for environmental monitoring and assessment exist to date for EDCs. In fact, many of the current environmental programs such as the European WFD or the US Clean Water Act do not explicitly test for EDCs, and considering the unique requirements and end points required to assess the endocrine potential of a sample, these programs are unlikely to appropriately address exposure to these chemicals. This is of great concern since EDCs are ubiquitous in the environment, especially in aquatic ecosystems. One of the most important sources of EDCs in the environment is the effluent from sewage treatment plants. Many EDCs such as the natural and synthetic estrogens 17β-estradiol and 17α-ethinylestradiol, respectively, are not completely removed with conventional wastewater treatment systems. In recognition of these concerns, in Europe, there is increasing pressure to further develop advanced wastewater treatment methods, such as advanced oxidation process, membrane bioreactors, and adsorption processes for a broad application in municipal wastewater treatment. Another issue is the continuing lack of understanding of the environmental relevance of the phenomenon of ED. A great number of studies have been conducted to describe potential ED in wild and laboratory animals. Most of these studies relied on biomarkers of estrogenicity such as vitellogenin induction in males and mild histological alterations (i.e., occurrence of testicular oocytes), and to date—with few exceptions—no convincing evidence of population-relevant impacts of exposure to EDC in the wild exists. In conclusion, while there has been a great deal of research and efforts in context with the hazard assessment and regulation of EDCs, there are still a large number of remaining uncertainties and issues. These range from animal right concerns due to significant increases in the use of animals to fulfill testing requirements; associated needs for alternative testing concepts such as in vitro, in silico, and modeling approaches; lack of understanding of the relevance of the exposure of people and wildlife to EDCs; and the need for inclusion of EDCs in current environmental programs such as the WFD (Hecker and Hollert, 2011).

REFERENCES

Acosta, K., Cuvilier, L., Pedersen, B. 2005. Chemicals that affect reproduction: From Rachel to REACH. PhD thesis, European Union Regulation of Endocrine Disrupting—Institut for Miljo, Teknologi og Samfund, Roskilde University, Roskilde, Denmark.

Alliot, F., Moreau-Guigon, E., Bourges, C., Desportes, A., Teil, J.M. et al. 2014. A multi-residue method for characterization of endocrine disruptors in gaseous and particulate phases of ambient air. *Atmospheric Environment* 92: 1–8.

Al-Odaini, N.A., Zakaria, M.P., Yaziz, M.I., Surif, S., Abdulghani, M. 2011. The occurrence of human pharmaceuticals in wastewater effluents and surface water of Langat River and its tributaries, Malaysia. *International Journal of Environmental Analytical Chemistry* 1: 1–20.

Andersson, A.M., Jensen, T.K., Juul, A., Petersen, J.H., Jorgensen, T. et al. 2007. Secular decline in male testosterone and sex hormone binding globulin serum levels in Danish population surveys. *Journal of Clinical Endocrinology & Metabolism* 92: 4696–4705.

Anway, M.D., Skinner, M.K. 2006. Epigenetic transgenerational actions of endocrine disrupters. *Endocrinology* 147: S43–S49.

Asimakopoulos, A.G., Thomaidis, N.S., Koupparis, M. 2012. Recent trends in biomonitoring of bisphenol A, 4-*t*-octylphenol, and 4-nonylphenol. *Toxicology Lettters* 210: 141–154.

Bao, L.J., Jia, F., Crago, J., Zeng, E.Y., Schlenk, D. et al. 2013. Assessing bioavailability of DDT and metabolites in marine sediments using solid-phase microextraction with performance reference compounds. *Environmental Toxicology and Chemistry/SETAC* 32: 1946–1953.

Barceló, D. 2003. Emerging pollutants in water analysis. *TrAC Trends in Analytical Chemistry* 22: 10.

Barceló, D., Petrovic, M. 2008. *Emerging Contaminants from Industrial and Municipal Waste: Removal Technologies*, Springer, Berlin.

Basheer, C., Lee, H.K. 2004. Analysis of endocrine disrupting alkylphenols, chlorophenols and bisphenol-A using hollow fiber-protected liquid-phase microextraction coupled with injection port-derivatization gas chromatography–mass spectrometry. *Journal of Chromatography A* 1057: 163–169.

Basile, T., Petrella, A., Petrella, M., Boghetich, G., Petruzzelli, V. et al. 2011. Review of endocrine-disrupting-compound removal technologies in water and wastewater treatment plants: An EU perspective. *Industrial & Engineering Chemistry Research* 50: 8389–8401.

Belfroid, A.C., Van der Horst, A., van der Horst, B., Vethaak, D. 1999. Analysis and occurrence of estrogenic hormones and their glucuronides in surface water and waste water in The Netherlands. *Science of the Total Environment* 225: 101–108.

Beltrán, F.J., Aguinaco, A., García-Aray, F., Oropesab, A. 2008. Ozone and photocatalytic processes to remove the antibiotic sulfamethoxazole from water. *Water Research* 42: 3799–3808.

Benijts, T., Dams, R., Lambert, W., De Leenheer, A. 2004. Countering matrix effects in environmental liquid chromatography-electrospray ionization tandem mass spectrometry water analysis for endocrine disrupting chemicals. *Journal of Chromatography A* 1029: 153–159.

Benotti, M.J., Trenholm, R.A., Vanderford, B.J., Holady, J.C., Stanford, B.D. et al. 2009. Pharmaceuticals and endocrine disrupting compounds in U.S. drinking water. *Environmental Science & Technology* 43: 597–603.

Bergman, Å., Heindel, J.J., Jobling, S., Kidd, K.A., Zoeller, R.T. 2012. *State of the Science of Endocrine Disrupting Chemicals—2012*. WHO, Geneva. http://www.who.int/iris/bitstream/10665/78102/1/WHO_HSE_PHE_IHE_2013.1_eng.pdf (accessed November 21, 2013).

Bila, D.M., Montalvão, A.F., Dezotti, M. 2005. Ozonation of the endocrine disrupter 17β-estradiol and estrogenic activity evaluation. *IOA 17th World Ozone Congress*. Strasbourg, France. pp. VI.2.2.1–VI.2.2.7.

Blackwell, B.R., Cai, Q., Smith, P.N., Cobb, G.P. 2011. Liquid chromatography–tandem mass spectrometry analysis of 17α-trenbolone, 17β-trenbolone and trendione in airborne particulate matter. *Talanta* 85: 1317–1323.

Bonefeld-Jorgensen, E.C., Long, M., Hofmeister, M.V., Vinggaard, A.M. 2007. Endocrine-disrupting potential of bisphenol A, bisphenol A dimethacrylate, 4-*n*-nonylphenol, and 4-*n*-octylphenol in vitro: New data and a brief review. *Environmental Health Perspective* 115: 69–76.

Bourguignon, J.P., Parent, A.S. 2010. Early homeostatic disturbances of human growth and maturation by endocrine disrupters. *Current Opinion in Pediatrics* 22: 470–477.

Boyd, G.R., Reemtsma, H., Grimm, D.A., Mitra, S. 2003. Pharmaceuticals and personal care products (PPCPs) in surface and treated waters of Louisiana, USA and Ontario, Canada. *Science of the Total Environment* 311: 135–149.

Braun, P., Moeder, M., Schrader, S., Popp, P., Kuschk, P. et al. 2003. Trace analysis of technical nonylphenol, bisphenol A and 17-ethinylestradiol in wastewater using solid-phase microextraction and gas chromatography–mass spectrometry. *Journal of Chromatography A* 988: 41–51.

Brossa, L., Marcé, R.M., Borrull, F., Pocurull, E. 2002. Application of on-line solid-phase extraction-gas chromatography–mass spectrometry to the determination of endocrine disruptors in water samples. *Journal of Chromatography A* 963: 287–294.

Byrdwell, W.C. 2005. Qualitative and quantitative analysis of triacylglycerols by atmospheric pressure ionization (APCI and ESI) mass spectrometry techniques. In *Modern Method for Lipid Analysis by Liquid Chromatography/Mass Spectrometry and Related Techniques*, ed. Byrdwell, W.C. AOCS, Champaign, IL, pp. 298–412.

Caliman, F.A., Gavrilescu, M. 2009. Pharmaceuticals, personal care products and endocrine disrupting agents in the environment—A review. *CLEAN—Soil, Air, Water* 37: 277–303.

Casas-Ferreira, A.M., Möder, M., Fernández-Laespada, M.E. 2011. Stir bar sortive extraction of parabens, triclosan and methyl triclosan from soil, sediment and sludge with *in situ* derivatization and determination by gas chromatography–mass spectrometry. *Journal of Chromatography A* 1218: 3837–3844.

Cederroth, Ch.R., Zimmermann, C., Nef, S. 2012. Review: Soy, phytoestrogens and their impact on reproductive health. *Molecular and Cellular Endocrinology* 355: 192–200.

Chao, M.R., Hu, C.W., Chen, Y.L., Chang-Chien, G.P., Lee, W.J. et al. 2004. Approaching gas particle partitioning equilibrium of atmospheric PCDD/Fs with increasing distance from an incinerator: Measurements and observations on modeling. *Atmospheric Environment* 38: 1501–1510.

Correa, O., Rifai, H., Raun, L., Suarez, M., Koenig, L. 2004. Concentrations and vapor-particle partitioning of polychlorinated dibenzo-*p*-dioxins and dibenzofurans in ambient air of Houston, TX. *Atmospheric Environment* 38: 6687–6699.

Daughton, C.G. 2005. Emerging chemicals as pollutants in the environment: A 21st century perspective. *Renewable Resources Journal* 23: 7–8.

Dean, J.R. 1998. *Extraction Methods for Environmental Analysis*. Wiley, New York.

Deok-Hie, P., Young-Youn, Y., Choi, J., Young-Hee, L., Hye-Sung, C. 2011. A solid-phase extraction method for analyzing trace amounts of polychlorinated dibenzo-*p*-dioxins and dibenzofurans in waste water. *Mass Spectrometry Letters* 2: 12–15.

Diamanti-Kandarakis, E., Bourguignon, J.P., Giudice, L.C., Hauser, R., Prins, G.S. et al. 2009. Endocrine disrupting chemicals: An Endocrine Society scientific statement. *Endocrine Reviews* 30: 293–342.

Díaz, A., Ventura, F., Galceran, M.T. 2002. Development of a solid-phase microextraction method for the determination of short-ethoxy-chain nonylphenols and their brominated analogs in raw and treated water. *Journal of Chromatography A* 963: 159–167.

Eckstein, G., Sherk, G.W. 2011. Alternative strategies for managing pharmaceutical and personal care products in water resources. Texas Tech University Center for Water Law and Policy, Lubbock, TX. http://ssrn.com/abstract=1885091 (accessed June 12, 2013).

Eljarrat, E., Lo, M.J., Petrovic, M. 2001. Analysis and environmental levels of endocrine-disrupting compounds in freshwater sediments. *TrAC Trends in Analytical Chemistry* 20: 637–648.

Enewold, L., Zhu, K., Ron, E., Marrogi, A.J., Stojadinovic, A. et al. 2009. Rising thyroid cancer incidence in the United States by demographic and tumor characteristics, 1980–2005. *Cancer Epidemiology Biomarkers & Prevention* 18: 784–791.

Esplugas, S., Bila, D.M., Krause, L.G.T., Dezotti, M. 2007. Ozonation and advanced oxidation technologies to remove endocrine disrupting chemicals (EDCs) and pharmaceuticals and personal care products (PPCPs) in water effluents. *Journal of Hazardous Materials* 149: 631–642.

Everaert, K., Baeyens, J. 2002. The formation and emission of dioxins in large scale thermal processes. *Chemosphere* 46: 439–448.

Farajzadeh, M.A., Nouri, N., Khorram, P. 2014. Derivatization and microextraction methods for determination of organic compounds by gas chromatography. *TrAC Trends in Analytical Chemistry* 55:14–23.

Feitosa de Lima-Gomes, P.C., Yamashita, J.B., Domingues-Nazario, C.E., Santos-Neto, A.J., Von Wolff, M.A. et al. 2011. Optimization of in situ derivatization SPME by experimental design for GC-MS multiresidue analysis of pharmaceutical drugs in wastewater. *Journal of Separation Science* 34: 436–445.

Finlayson-Pitts, B.J., Pitts, J.N. 2000. *Chemistry of the Upper and Lower Atmosphere Theory, Experiments, and Applications*. Academic, San Diego, CA.

Flemming, I., Bent, H.S. 2003. Evaluation of Analytical Chemical Methods for Detection of Estrogens in the Environment. Working Report No. 44, Danish Environmental Protection Agency, Vol. 4, pp. 1–69. http://www.2.mst.dk/udgiv/publications/2003/87-7972-968-1/pdf/87-7972-970-3.pdf (accessed July 14, 2013).

Fowler, P.A., Bellingham, M., Sinclair, K.D., Evans, P.N., Pocar, P. et al. 2012. Review: Impact of endocrine-disrupting compounds (EDCs) on female reproductive health. *Molecular and Cellular Endocrinology* 355: 231–239.

Fu, P., Kawamura, K. 2010. Ubiquity of bisphenol A in the atmosphere. *Environmental Pollution* 158: 3138–3143.

Gates, P. 2013. High performance liquid chromatography mass spectrometry (HPLC/MS). University of Bristol, Bristol, UK. http://www.bris.ac.uk/nerclsmsf/techniques/hplcms.html (accessed November 18, 2013).

Gatidou, G., Thomaidis, N.S., Stasinakis, A.S., Lekkas, T.D. 2007. Simultaneous determination of the endocrine disrupting compounds nonylphenol, nonylphenol ethoxylates, triclosan and bisphenol A in wastewater and sewage sludge by gas chromatography–mass spectrometry. *Journal of Chromatography A* 1138: 32–41.

Gibson, R., Wang, M.J., Padgett, E., Beck, A.J. 2005. Analysis of 4-nonylphenols, phthalates, and polychlorinated biphenyls in soils and biosolids. *Chemosphere* 61: 1336–1344.

Gibson, R., Becerril-Bravo, E., Silva-Castro, V., Jiménez, B. 2007. Determination of acidic pharmaceuticals and potential endocrine disrupting compounds in wastewaters and spring waters by selective elution and analysis by gas chromatography–mass spectrometry. *Journal of Chromatography A* 1169: 31–39.

Gibson, R., Durán-Álvarez, J.C., Estrada, K.L., Chávez, A., Jiménez-Cisneros, B. 2010. Accumulation and leaching potential of some pharmaceuticals and potential endocrine disruptors in soils irrigated with wastewater in the Tula Valley, Mexico. *Chemosphere* 81: 1437–1445.

Gong, J., Ran, Y., Chen, D.Y., Yang, Y. 2011. Occurrence of endocrine-disrupting chemicals in riverine sediments from the Pearl River Delta, China. *Marine pollution Bulletin* 63: 556–563.

Grimalt, S., Sancho, J.V., Pozo, O.J., García-Baudin, J.M., Fernández-Cruz, M.L. et al. 2006. Analytical study of trichlorfon residues in kaki fruit and cauliflower samples by liquid chromatography–electrospray tandem mass spectrometry. *Journal of Agricultural and Food Chemistry* 54: 1188–1195.

Guzmán, C., Zambrano, E. 2007. Compuestos disruptores endocrinos y su participación en la programación del eje reproductive. *Revista de Investigación Clínica* 59: 73–81.

Haynes, B.A., Mookadam, F. 2009. Male gynecomastia. *Mayo Clinic Proceedings* 84: 672.

Hecker, M., Hollert, H. 2011. Endocrine disruptor screening: Regulatory perspectives and needs. *Environmental Sciences Europe* 23: 15.

Helaleh, M.I.H., Fujii, S., Korenaga, T. 2001. Column silylation method for determining endocrine disruptors from environmental water samples by solid phase micro-extraction. *Talanta* 54: 1039–1047.

Helbling, D.E., Hollender, J., Kohler, H.P.E., Singer, H., Fenner, K. 2010. High-throughput identification of microbial transformation products of organic micropollutants. *Environmental Science & Technology* 44: 6621–6627.

Hibberd, A., Maskaoui, K., Zhang, Z., Zhou, J.L. 2009. An improved method for the simultaneous analysis of phenolic and steroidal estrogens in water and sediment. *Talanta* 77: 1315–1321.

Hoai, P.M., Tsunoi, S., Ike, M., Kuratani, Y., Kudou, K. et al. 2003. Simultaneous determination of degradation products of nonylphenol polyethoxylates and their halogenated derivatives by solid-phase extraction and gas chromatography-tandem mass spectrometry after trimethylsilylation. *Journal of Chromatography A* 1020: 161–171.

Hsin-Chang, C., Han-Wen, K., Wang-Hsien, D. 2009. Determination of estrogenic compounds in wastewater using liquid chromatography–tandem mass spectrometry with electrospray and atmospheric pressure photoionization following desalting extraction. *Chemosphere* 74: 508–514.

Huerta-Fontela, M., Galceran, M.T., Ventura, F. 2011. Occurrence and removal of pharmaceuticals and hormones through drinking water treatment. *Water Research* 45: 1432–1442.

Hur, J., Schlautman, M.A. 2004. Influence of humic substance adsorptive fractionation on pyrene partitioning to dissolved and mineral-associated humic substances. *Environmental Science & Technology* 38: 5871–5877.

Huyghe, E., Matsuda, T., Thonneau, P. 2003. Increasing incidence of testicular cancer worldwide: A review. *Journal of Urology* 170: 5–11.

IARC. 1984. *IARC Monographs on the Evaluation of the Carcinogenic Risk of Chemicals to Humans: Polynuclear Aromatic Hydrocarbons, Part 2: Carbon Blacks, Mineral Oils (Lubricant Base Oils And Derived Products) and Some Nitroarenes*. World Health Organization, International Agency for Research on Cancer, Lyon, France. Vol 33.

Ibarluzea, J.M., Fernández, M.F., Santa-Marina, L., Olea-Serranon, M.F., Rivas, A.M. et al. 2004. Breast cancer risk and the combined effect of environmental estrogens. *Cancer Cause Control* 15: 591–600.

Jackson, J., Sutton, R. 2008. Sources of endocrine-disrupting chemicals in urban wastewater, Oakland, CA. *Science of the Total Environment* 405: 153–160.

Jeannot, R., Sabik, H., Sauvard, E., Dagnac, T., Dohrendorf, K. 2002. Determination of endocrine-disrupting compounds in environmental samples using gas and liquid chromatography with mass spectrometry. *Journal of Chromatography A* 974: 143–159.

Jenkins, B.M., Jones, A.D., Turn, S.Q., Williams, R.B. 1996. Emission factors for polycyclic aromatic hydrocarbons from biomass burning. *Environmental Science & Technology* 30: 2462–2469.

Jia, F., Cui, X., Wang, W., Delgado-Moreno, L., Gan, J. 2012. Using disposable solid-phase microextraction (SPME) to determine the freely dissolved concentration of polybrominated diphenyl ethers (PBDEs) in sediments. *Environmental Pollution* 167: 34–40.

Kim, S.D., Cho, J., Kim, I.S., Vanderford, B.J., Snyder, S.A. 2007. Occurrence and removal of pharmaceuticals and endocrine disruptors in South Korean surface, drinking, and waste waters. *Water Research* 41: 1013–1021.

Koh, Y.K.K., Chiu, T.Y., Boobis, A., Cartmell, E., Lester, J.N. et al. 2007. Determination of steroid estrogens in wastewater by high performance liquid chromatography–tandem mass spectrometry. *Journal of Chromatography A* 1173: 81–87.

Kolpin, D.W., Furlong, E.T., Meyer, M., Thurman, E.M., Zaugg, S. 2002. Pharmaceuticals, hormones, and other organic wastewater contaminants in U.S. streams, 1999–2000: A national reconnaissance. *Environmental Science & Technology* 36: 1202–1211.

Koutsouba, V., Heberer, T., Fuhrmann, B., Schmidt-Baumler, K., Tsipi, D. 2003. Determination of polar pharmaceuticals in sewage water of Greece by gas chromatography-mass spectrometry. *Chemosphere* 51: 69–75.

Kusk, K.O., Krüger, T., Long, M., Taxvig, C., Lykkesfeldt, A.E. et al. 2011. Endocrine potency of wastewater: Contents of endocrine disrupting chemicals and effects measured by in vivo and in vitro assays. *Environmental Toxicology and Chemistry* 30: 413–426.

Kuster, M., Lopez de Alda, M.J., Barcelo, D. 2004a. Estrogens and progestogens in wastewater, sludge, sediments, and soil. In *Water Pollution: Emerging Organic Pollutants in Waste Waters and Sludge*, ed. Barcelo, D. Springer, Berlin, Vol. 2, pp. 1–24.

Kuster, M., Lopez de Alda, M.J., Barcelo, D. 2004b. Analysis and distribution of estrogens and progestogens in sewage sludge, soils and sediments. *TrAC Trends in Analytical Chemistry* 23: 790–798.

Langdon, K.A., Warne, M.St.J., Smernik, R.J., Shareef, A., Kookana, R.S. 2012. Field dissipation of 4-nonylphenol, 4-*t*-octylphenol, triclosan and bisphenol A following land application of biosolids. *Chemosphere* 86: 1050–1058.

Lee, K.E., Barber, L.B., Furlong, E.T., Cahill, J.D., Kolpin, D.W. et al. 2004. Presence and Distribution of Organic Wastewater Compounds in Wastewater, Surface, Ground, and Drinking Waters, Minnesota, 2000–02. US Geological Survey Scientific Investigations. Report 2004-5138. USGS, Reston, VA. http://water.usgs.gov/pubs/sir/2004/5138/#pdf (accessed March 19, 2013).

Lima-Gomes, P.C.F., Barnes, B.B., Santos-Neto, Á.J., Lancas, F.M., Snow, N.H. 2013. Determination of steroids, caffeine and methylparaben in water using solid phase microextraction-comprehensive two dimensional gas chromatography–time of flight mass spectrometry. *Journal of Chromatography A* 1299: 126–130.

Lin, W.C., Chen, H.C., Ding, W.H. 2005. Determination of pharmaceutical residues in waters by solid-phase extraction and large-volume on-line derivatization with gas chromatography–mass spectrometry. *Journal of Chromatography A* 1065: 279–285.

Liu, R., Zhou, J.L., Wilding, A. 2004. Simultaneous determination of endocrine disrupting phenolic compounds and steroids in water by solid-phase extraction–gas chromatography–mass spectrometry. *Journal of Chromatography A* 1022: 179–189.

Llorca-Pórcel, J., Martínez-Parreño, M., Martínez-Soriano, E., Valor, I. 2009. Analysis of chlorophenols, bisphenol-A, 4-*tert*-octylphenol and 4-nonylphenols in soil by means of ultrasonic solvent extraction and stir bar sorptive extraction with in situ derivatisation. *Journal of Chromatography A* 1216: 5955–5961.

Ma, L. 2009. Endocrine disruptors in female reproductive tract development and carcinogenesis. *Trends in Endocrinology & Metabolism* 20: 357–363.

Mansilha, C., Melo, A., Rebelo, H., Ferreira, I.M.P.L.V.O., Pinho, O. et al. 2010. Quantification of endocrine disruptors and pesticides in water by gas chromatography–tandem mass spectrometry: Method validation using weighted linear regression schemes. *Journal of Chromatography A* 1217: 6681–6691.

Mao, L., Lu, J., Gao, S., Huang, Q. 2010. Transformation of 17ß-estradiol mediated by lignin peroxidase: The role of veratryl alcohol. *Archives of Environmental Contamination and Toxicology* 59:13–19.

Martínez-Vidal, J.L., Pablos-Espada, M.C., Garrido-Frenich, A., Arrebola, F.J. 2000. Pesticide trace analysis using solid-phase extraction and gas chromatography with electron-capture and tandem mass spectrometric detection in water samples. *Journal of Chromatography A* 867: 235–245.

Matamoros, V., Jover, E., Bayona, J.M. 2010. Part-per-trillion determination of pharmaceuticals, pesticides, and related organic contaminants in river water by solid-phase extraction followed by comprehensive two-dimensional gas chromatography time-of-flight mass spectrometry. *Analytical Chemistry* 82: 699–706.

Maurício, R., Diniz, M., Petrovic, M., Amaral, L., Peres, I. et al. 2006. A characterization of selected endocrine disruptor compounds in a Portuguese wastewater treatment plant. *Environmental Monitoring and Assessment* 118: 75–87.

McLachlan, J.A., Simpson, E., Martin, M. 2006. Endocrine disrupters and female reproductive health. *Best Practice and Research Clinical Endocrinology and Metabolism* 20: 63–75.

Meeker, J.D. 2010. Review: Exposure to environmental endocrine disrupting compounds and men's health. *Maturitas* 66: 236–241.

Miao, X., Yang, J., Metcalfe, C.D. 2005. Carbamazepine and its metabolites in wastewater and in biosolids in a municipal wastewater treatment plant. *Environmental Science & Technology* 39: 7469–7475.

Migowska, N., Caban, M., Stepnowski, P., Kumirska, J. 2012. Simultaneous analysis of non-steroidal anti-inflammatory drugs and estrogenic hormones in water and wastewater samples using gas chromatography–mass spectrometry and gas chromatography with electron capture detection. *Science of the Total Environment* 441: 77–88.

Mitra, S. 2003. *Sample Preparation Techniques in Analytical Chemistry*, ed. Mitra, S. Wiley, New York.

Mitra, A.K., Faruque, F.S., Avis, A.L. 2004. Breast cancer and environmental risks: Where is the link? *Journal of Environmental Health* 66: 24–32.

Mol, H.G.J., Sunarto, S., Steijger, O.M. 2000. Determination of endocrine disruptors in water after derivatization with *N*-methyl-*N*-(*tert*-butyldimethyltrifluoroacetamide) using gas chromatography with mass spectrometric detection. *Journal of Chromatography A* 879. 97–112.

Morales-Suárez-Varela, M.M., Toft, G.V., Jensen, M.S., Ramlau-Hansen, C., Kaerlev, L. et al. 2011. Parental occupational exposure to endocrine disrupting chemicals and male genital malformations: A study in the Danish National Birth Cohort study. *Environmental Health* 10: 1–9.

Moring, J.B. 2012. Water quality and the composition of fish and macroinvertebrate communities in the Devils and Pecos Rivers within and upstream from the Amistad National Recreation Area, Texas, 2005–7. US Geological Survey Scientific Investigations Report 2012–5038. http://pubs.usgs.gov/sir/2012/5038/SIR12-5038.pdf (accessed December 9, 2013).

Mul, D., Fredriks, A.M., van Buuren, S., Oostdijk, W., Verloove-Vanhorick, S.P. et al. 2001. Pubertal development in The Netherlands 1965–1997. *Pediatric Research* 50: 479–486.

Nakada, N., Tanishima, T., Shinohara, H., Kiri, K., Takada, H. 2006. Pharmaceutical chemicals and endocrine disrupters in municipal wastewater in Tokyo and their removal during activated sludge treatment. *Water Research* 40: 3297–3303.

Ni, Y., Zhang, H., Fan, S., Zhang, X., Zhang, Q. et al. 2009. Emissions of PCDD/Fs from municipal solid waste incinerators in China. *Chemosphere* 75: 1153–1158.

Nie, Y., Qiang, Z., Zhang, H., Adams, C. 2009. Determination of endocrine-disrupting chemicals in the liquid and solid phases of activated sludge by solid phase extraction and gas chromatography–mass spectrometry. *Journal of Chromatography A* 1216: 7071–7080.

Novikova, S.I., He, F., Bai, J., Cutrufello, N.J., Lidow, M.S. et al. 2008. Maternal cocaine administration in mice alters DNA methylation and gene expression in hippocampal neurons of neonatal and prepubertal offspring. *PLoS One* 3: 1919.

Öllers, S., Singer, H.P., Fässler, P., Müller, S.R. 2001. Simultaneous quantification of neutral and acidic pharmaceuticals and pesticides at the low-ng/l level in surface and wastewater. *Journal of Chromatography A* 911: 225–234.

Padilla-Sánchez, J.A., Plaza-Bolaños, P., Romero-González, R., Garrido-Frenich, A., Martínez Vidal, J.L. 2010. Application of a quick, easy, cheap, effective, rugged and safe-based method for the simultaneous extraction of chlorophenols, alkylphenols, nitrophenols and cresols in agricultural soils, analyzed by using gas chromatography-triple quadrupole-mass spectrometry/mass spectrometry. *Journal of Chromatography A* 1217: 5724–5731.

Parent, A.S., Teilmann, G., Juul, A., Skakkebaek, N.E., Toppari, J. et al. 2003. The timing of normal puberty and the age limits of sexual precocity: Variations around the world, secular trends, and changes after migration. *Endocrine Reviews* 24: 668–693.

Pawliszyn, J. 1997. *Solid Phase Microextraction: Theory and Practice*, Wiley-VCH, New York.

Peñalver, A., Pocurull, E., Borrull, F., Marcé, R.M. 2002. Solid-phase microextraction coupled to high-performance liquid chromatography to determine phenolic compounds in water samples. *Journal of Chromatography A* 953: 79–87.

Perera, F., Tang, W.Y., Herbstman, J., Tang, D., Levin, L. et al. 2009. Relation of DNA methylation of 5′-CpG island of ACSL3 to transplacental exposure to airborne polycyclic aromatic hydrocarbons and childhood asthma. *PLoS One* 4: 4488.

Petrovic, M., Eljarrat, E., Lopez de Alba, M.J., Barceló, D. 2004. Endocrine disruptors compounds and other emerging contaminants in the environment: A survey on new monitoring strategies and occurrence data. *Analytical and Bioanalytical Chemistry* 378: 549–562.

Planas, C., Puig, A., Rivera, J., Caixach, J. 2006. Analysis of pesticides and metabolites in Spanish surface waters by isotope dilution gas chromatography/mass spectrometry with previous automated solid-phase extraction: Estimation of the uncertainty of the analytical results. *Journal of Chromatography A* 1131: 242–252.

Prieto, A., Schrader, S., Moeder, M. 2010. Determination of organic priority pollutants and emerging compounds in wastewater and snow samples using multiresidue protocols on the basis of microextraction by packed sorbents coupled to large volume injection gas chromatography–mass spectrometry analysis. *Journal of Chromatography A* 1217: 6002–6011.

Prieto, A., Vallejo, A., Zuloaga, O., Paschke, A., Sellergen, B. et al. 2011. Selective determination of estrogenic compounds in water by microextraction by packed sorbents and a molecularly imprinted polymer coupled with large volume injection-in-port-derivatization gas chromatography–mass spectrometry. *Analytica Chimica Acta* 703: 41–51.

Quintana, J.B., Carpinteiro, J., Rodríguez, I., Lorenzo, R.A., Carro, A.M. et al. 2004. Determination of natural and synthetic estrogens in water by gas chromatography with mass spectrometric detection. *Journal of Chromatography A* 1024: 177–185.

Reid, A.M., Brougham, C.A., Fogarty, A.M., Roche, J.J. 2009. Accelerated solvent-based extraction and enrichment of selected plasticisers and 4-nonylphenol, and extraction of tin from organotin sources in sediments, sludges and leachate soils. *Analytica Chimica Acta* 634: 197–204.

Rhind, S.M., Kyle, C.E., Ruffie, H., Calmettes, E., Osprey, M. et al. 2013. Short- and long-term temporal changes in soil concentrations of selected endocrine disrupting compounds (EDCs) following single or multiple applications of sewage sludge to pastures. *Environmental Pollution* 181: 262–270.

Rodríguez, I., Quintana, J.B., Carpinteiro, J., Carro, A.M., Lorenzo, R.A. 2003. Determination of acidic drugs in sewage water by gas chromatography–mass spectrometry as *tert*-butyldimethysilyl derivatives. *Journal of Chromatography A* 985: 265–274.

Rodriguez-Mozaz, S., López de Alda, M.J., Barceló, D. 2004. Monitoring of estrogens, pesticides and bisphenol A in natural waters and drinking water treatment plants by solid-phase extraction–liquid chromatography–mass spectrometry. *Journal of Chromatography A* 1045: 85–92.

Rogers, J.A., Metz, L., Yong, V.W. 2013. Review: Endocrine disrupting chemicals and immune responses: A focus on bisphenol-A and its potential mechanisms. *Molecular Immunology* 53: 421–430.

Safe, S. 2005. Clinical correlates of environmental endocrine disruptors. *Trends in Endocrinology and Metabolism* 16: 139–144.

Salapasidou, M., Samara, C., Voutsa, D. 2011. Endocrine disrupting compounds in the atmosphere of the urban area of Thessaloniki, Greece. *Atmospheric Environment* 45: 3720–3729.

Samaras, V.G., Thomaidis, N.S., Stasinakis, A.S., Lekkas, T.D. 2011. An analytical method for the simultaneous trace determination of acidic pharmaceuticals and phenolic endocrine disrupting chemicals in wastewater and sewage sludge by gas chromatography-mass spectrometry. *Analytical and Bioanalytical Chemistry* 399: 2549–2561.

Santos, J.L., Aparicio, I., Alonso, E., Callejón, M. 2005. Simultaneous determination of pharmaceutically active compounds in wastewater samples by solid phase extraction and high-performance liquid chromatography with diode array and fluorescence detectors. *Analytica Chimica Acta* 550: 116–122.

Sarmah, A.K., Northcott, G.L., Leusch, F.D.L., Tremblay, L.A. 2006. A survey of endocrine disrupting chemicals (EDCs) in municipal sewage and animal waste effluents in the Waikato region of New Zealand. *Science of the Total Environment* 355: 135–144.

Schreiber, A., Kern, R. 2010. Quantitation and identification of pharmaceuticals and personal care products (PPCP) in water samples. AB SCIEX, Concord, Ontario, and AB SCIEX, Foster City, CA. http://www.absciex.com/Documents/Downloads/Literature/mass-spectrometry-PPCP-QuantandID-1790210.pdf (accessed April 25, 2013).

Schug, T.T., Janesick, A., Blumberg, B., Heindela, J.J. 2011. Review: Endocrine disrupting chemicals and disease susceptibility. *Journal of Steroid Biochemistry* 127: 204–215.

Schymanski, E.L., Bataineh, M., Goss, K.U., Brack, W. 2009. Integrated analytical and computer tools for structure elucidation in effect-directed analysis. *TrAC Trends in Analytical Chemistry* 28: 550–561.

Sharma, V.K. 2008. Oxidative transformations of environmental pharmaceuticals by Cl_2, ClO_2, O_3, and Fe (VI): Kinetics assessment. *Chemosphere* 73: 1379–1386.

Sharma, V.K., Anquandah, G.A.K., Yngard, R.A., Kim, H., Fekete, J. et al. 2009. Nonylphenol, octylphenol, and bisphenol-A in the aquatic environment: A review on occurrence, fate, and treatment. *Journal of Environmental Science and Health Part A* 44: 423–442.

Shen, G. 2005. Development of an analytical method to determine phenolic endocrine disrupting chemicals in sewage and sludge by GC/MS. *Chinese Science Bulletin* 50: 2681.

Sidhu, S., Gullett, B., Striebich, R., Klosterman, J., Contreras, J. 2005. Endocrine disrupting chemical emissions from combustion sources: Diesel particulate emissions and domestic waste open burn emissions. *Atmospheric Environment* 39: 801–811.

Spark, K.M., Swift, R.S. 2002. Effect of soil composition and dissolved organic matter on pesticide sorption. *Science of the Total Environment* 298: 147–161.

Stasinakis, A.S., Gatidou, G., Mamais, D., Thomaidis, N.S., Lekkas, T.D. 2008. Occurrence and fate of endocrine disrupters in Greek sewage treatment plants. *Water Research* 42: 1796–1804.

Suárez, B., Santos, B., Simonet, B.M., Cárdenas, S., Valcárcel, M. 2007. Solid-phase extraction–capillary electrophoresis–mass spectrometry for the determination of tetracyclines residues in surface water by using carbon nanotubes as sorbent material. *Journal of Chromatography A* 1175: 127–132.

Swan, S.H., Kruse, R.L., Liu, F., Barr, D.B., Drobnis, E.Z. et al. 2003. Semen quality in relation to biomarkers of pesticide exposure. *Environmental Health Perspective* 11: 1478–1484.

Tan, B.L.L., Hawker, D.W., Müller, J.F., Tremblay, L.A., Chapman, H.F. 2008. Stir bar sorptive extraction and trace analysis of selected endocrine disruptors in water, biosolids and sludge samples by thermal desorption with gas chromatography–mass spectrometry. *Water Research* 42: 404–412.

TEDX. 2011. TEDX list of potential endocrine disruptors. The Endocrine Disruption Exchange, Inc. http://www.endocrinedisruption.org/endocrine.TEDXList.overview.php (accessed October 12, 2013).

Thurman, E.M., Mills, M.S. 1998. *Solid-Phase Extraction: Principles and Practices*. Wiley, New York.

Tian, Q., Zhou, Z., Lv, C., Huang, Y., Ren, L. 2010. Simultaneous determination of paclobutrazol and myclobutanil enantiomers in water and soil using enantioselective reversed-phase liquid chromatography. *Analytical Methods* 2: 617–622.

Travison, T.G., Araujo, A.B., O'Donnell, A.B., Kupelian, V., McKinlay, J. 2007. A population level decline in serum testosterone levels in American men. *Journal of Clinical Endocrinology & Metabolism* 92: 196–202.

Trenholm, R.A., Vanderford, B.J., Holady, J.C., Rexing, D.J., Snyder, S.A. 2006. Broad range analysis of endocrine disrupters and pharmaceuticals using gas chromatography and liquid chromatography tandem mass spectrometry. *Chemosphere* 65: 1991–1998.

US EPA. 1997. Special Report on Environmental Endocrine Disruption: An Effects Assessment and Analysis. Report No. EPA/630/R-96/012. EPA, Washington, DC.

US EPA. 2003. Compendium Method TO-9A: Method for the Determination of Polychlorinated, Polybrominated and Brominated/Chlorinated Dibenzo-*p*-Dioxins and Dibenzofurans in Ambient Air. EPA, Washington, DC.

US EPA. 2008. Contaminant Candidate List 3, Chemicals: Screening to a PCCL. EPA 815-R-08-003. Office of Water, EPA, Washington, DC. http://www.epa.gov/safewater (accessed August 17, 2013).

US EPA. 2009. Final Contaminant Candidate List 3, Chemicals: Identifying the Universe. EPA 815-R-09-006. Office of Water, EPA, Washington, DC. http://www.epa.gov/safewater (accessed August 17, 2013).

US EPA. 2013. Endocrine Disruptor Screening Program (EDSP): What are endocrine disruptors? http://www.epa.gov/endo/pubs/edspoverview/whatare.htm (accessed November 28, 2013).

USGS. 2008. Research projects, emerging contaminants, toxics substances hydrology program. http://toxics.usgs.gov/regional/emc/index.html (accessed June 21, 2013).

Vallejo-Rodríguez, R., López-López, A. 2011a. Endocrine disrupting compounds: Identification and analysis in surface waters and their degradation by advanced oxidation process with ozone. In *Water Resources in Mexico. Scarcity, Degradation, Stress, Conflicts, Management, and Policy*, ed. Oswald Spring, U. Springer, Heidelberg-Dordrecht-London-New York, pp. 279–297.

Vallejo-Rodríguez, R., López-López, A., Saldarriaga-Noreña, H., Murillo-Tovar, M., Hernández-Mena, L. 2011b. Optimization of analytical conditions to determine steroids and pharmaceuticals drugs in water samples using solid phase-extraction and HPLC. *American Journal of Analytical Chemistry* 2: 863–870.

Vega-Morales, T., Sosa-Ferrera, Z., Santana-Rodríguez, J.J. 2010. Determination of alkylphenol polyethoxylates, bisphenol-A, 17α-ethynylestradiol and 17β-estradiol and its metabolites in sewage samples by SPE and LC/MS/MS. *Journal of Hazardous Materials* 183: 701–711.

Vega-Morales, T., Sosa-Ferrera, Z., Santana-Rodríguez, J.J. 2011. Determination of various estradiol mimicking-compounds in sewage sludge by the combination of microwave-assisted extraction and LC-MS/MS. *Talanta* 85: 1825–1834.

Vom-Saal, F.S., Nagel, S.C., Coe, B.L., Angle, B.M., Taylor, J.A. 2012. Review: The estrogenic endocrine disrupting chemical bisphenol A (BPA) and obesity. *Molecular and Cellular Endocrinology* 354: 74–84.

Waye, A., Trudeau, V.I. 2011. Neuroendocrine disruption: More than hormones are upset. *Journal of Toxicology and Environmental Health Sciences B* 14: 270–291.

Weiss, J.M., Hamers, T., Thomas, K.V., Van der Linden, S., Leonards, P.E.G. et al. 2009. Masking effect of anti-androgens on androgenic activity in European river sediment unveiled by effect-directed analysis. *Analytical and Bioanalytical Chemistry* 394: 1385–1397.

Weiss, J.M., Simon, E., Stroomberg, G.J., de Boer, R., de Boer, J. et al. 2011. Identification strategy for unknown pollutants using high-resolution mass spectrometry: Androgen-disrupting compounds identified through effect-directed analysis. *Analytical and Bioanalytical Chemistry* 400: 3141–3149.

Westerholm, R.N., Almén, J., Li, H., Rannung, J., Egebäck, K. et al. 1991. Chemical and biological characterization of particulate, semi-volatile and gas-phase-associated compounds in diluted heavy-duty diesel exhausts: A comparison of three different semi-volatile-phase samplers. *Environmental Science & Technology* 25: 332–338.

Wu, T., Wang, W.Y., Jiang, L.M., Chu, Q.C., Delaire, J. et al. 2008. Determination of natural and synthetic endocrine-disrupting chemicals (EDCs) in sewage based on SPE and MEKC with amperometric detection. *Chromatographia* 68: 339–344.

Xie, Z., Selzer, J., Ebinghaus, R., Caba, A., Ruck, W. 2006. Development and validation of a method for the determination of trace alkylphenols and phthalates in the atmosphere. *Analytica Chimica Acta* 565: 198–207.

Xu, J., Wu, L., Chen, W., Chang, A.C. 2008. Simultaneous determination of pharmaceuticals, endocrine disrupting compounds and hormone in soils by gas chromatography-mass spectrometry. *Journal of Chromatography A* 1202: 189–195.

Xu, N., Xu, Y.F., Xu, S., Li, J., Hu-Chun, T. 2012. Removal of estrogens in municipal wastewater treatment plants: A Chinese perspective. *Environmental Pollution* 165: 215–224.

Yang, L., Luan, T., Lan, C. 2006. Solid-phase microextraction with on-fiber silylation for simultaneous determinations of endocrine disrupting chemicals and steroid hormones by gas chromatography–mass spectrometry. *Journal of Chromatography A* 1104: 23–32.

Yaoi, T., Itoh, K., Nakamura, K., Ogi, H., Fujiwara, Y. et al. 2008. Genome-wide analysis of epigenomic alterations in fetal mouse forebrain after exposure to low doses of bisphenol A. *Biochemical and Biophysical Research Communications* 376: 563–567.

Yu, Y., Wu, L. 2012. Analysis of endocrine disrupting compounds, pharmaceuticals and personal care products in sewage sludge by gas chromatography-mass spectrometry. *Talanta* 89: 258–263.

16 Analysis of Plasticizers in Food and Environment

Leo M. L. Nollet

CONTENTS

16.1 INTRODUCTION

Phthalates (or phthalic diesters) are chemicals that have the common structure of 1,2-benzenedicarboxylic acid (Figure 16.1). The differences among various phthalates are the structures of the two hydrocarbon chains (R_1 and R_2) attached to the two carboxylic acid functional groups. The smallest chain is the methyl group forming dimethyl phthalate ($R_1 = R_2 = -CH_3$, DMP). Longer chains such as C12 (di-dodecyl phthalate) and C13 (di-tridecyl phthalate) have also been reported. Because of the large differences in the length of the side chains, wide ranges of physical properties such as vapor pressure and octanol–water partition coefficient are observed (Cousins and Mackay, 2000). Some of the physical properties are summarized in Table 16.1 for six phthalates.

Toxicology and epidemiology studies indicated that in animals and humans, some phthalates can mimic hormones and have endocrine-disrupting properties (Jobling et al., 1995; Wypych, 2013). Studies on rodents have shown that phthalates are estrogenic and are associated with adverse reproductive effects (Lamb et al., 1987; Poon et al., 1997; Fukuwatari et al., 2002; Higuchi et al., 2003). Significantly high levels of phthalates including di(2-ethylhexyl) phthalate (DEHP) and its metabolite were observed in very young Puerto Rican girls with premature thelarche (Colón et al., 2000). There was an inverse linear association between the levels of some phthalate metabolites (monobutyl phthalate and monobenzyl phthalate) in urine and observed mobility, concentration, and normal morphology of sperm in American men (Duty et al., 2003).

Phthalates are widely used in modern society. For example, DEHP is added to polyvinyl chloride, a thermoplastic polymer, to make it softer and more flexible. DEHP is also used in food packaging materials, medical products such as intravenous tubing, plastic toys, vinyl upholstery, shower curtains, adhesives, and coatings. Phthalates with smaller side chains such as diethyl phthalate (DEP) and di-*n*-butyl phthalate (DnBP) are used as solvents in perfumes.

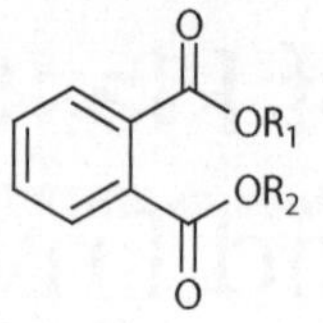

FIGURE 16.1 General structure of phthalates.

TABLE 16.1
Physical Properties of Six Commonly Monitored Phthalates

	Dimethyl Phthalate	Diethyl Phthalate	Di-*n*-Butyl Phthalate	Butyl Benzyl Phthalate	Di(2-Ethylhexyl) Phthalate	Di(*n*-Octyl) Phthalate
Abbreviation	DMP	DEP	DnBP	BBP	DEHP	DnOP
CAS no.	131-11-3	84-66-2	84-74-2	85-68-7	117-81-7	117-84-0
Structural formula	$C_{10}H_{10}O_4$	$C_{12}H_{14}O_4$	$C_{16}H_{22}O_4$	$C_{19}H_{20}O_4$	$C_{24}H_{38}O_4$	$C_{24}H_{38}O_4$
Molecular weight	194.2	222.2	278.4	312.4	390.6	390.6
Melting point (°C)	5.5	−40	−35	−35	−47	−25
Boiling point (°C)	282	298	340	370	385	428
VP(Pa) (25°C)	2.63×10^{-1}	2.83×10^{-3}	2.17×10^{-4}	1.17×10^{-4}	1.19×10^{-6}	1.00×10^{-7}
Specific gravity (20°C)	1.192	1.118	1.042	1.111	0.986	0.978
log K_{ow} (25°C)[a]	1.61	2.54	4.27	4.7	7.73	9.46
log K_{oa} (25°C)[a]	7.01	7.55	8.54	8.78	10.53	11.52
log K_{aw} (25°C)[a]	−5.4	−5.01	−4.27	−4.08	−2.8	−2.06

[a] Values cited from Cousins and Mackay (2000).

Phthalates have been reported to be present in the environment including water and soil (Penalver et al., 2000; Luke-Betlej et al., 2001; Suzuki et al., 2001; Fauser and Thomsen, 2002; do Nascimento Filho et al., 2003), consumer products (Aurela et al., 1999; Wilkinson and Lamb, 1999; Uhde et al., 2001; Bouma and Schakel, 2002), medical devices (Inoue et al., 2003), marine ecosystems (Lin et al., 2003), indoor air (Otake et al., 2001; Zhu et al., 2003), and indoor dust (Butte and Heinzow, 2002; Rudel et al., 2003). Phthalates are also widely present in dairy food (Zhu et al., 2010) and food in general (Cao, 2010). The presence of phthalates in food is attributed to the contamination of phthalates from both food processing and migration of phthalates from food packaging materials into food (Cao, 2010).

Adipates such as di-2-ethylhexyl adipate (DEHA) have a core structure of 1,6-hexadicarboxylic acid (−$HOCOC_4H_8COOH$−) instead of 1,2-benzenedicarboxylic acid (−$HOCOC_6H_4COOH$−). Adipate is used in many food processing products including food wraps and often detected in food (Page and Lacroix, 1992) (Table 16.2).

Because of the complex nature of different matrices, analysis of phthalates involves sample preparation and extraction, further cleanup in some cases, and instrumental separation and detection. Extraction and cleanup are the most challenging parts in the analysis of phthalate in foods and are often the critical steps in deciding the levels of detection limits of the overall methods.

Most of the information in this chapter deals with analysis methods of plasticizers in food matrices. Sample preparation and analysis methods are very likely as for environmental samples.

A great number of papers on plasticizers were published in the past several years. Only a selection is discussed here in this chapter.

TABLE 16.2
Selection of Sample Preparation and Cleanup Methods

	Extraction	Cleanup	Analysis	Analytes	Detection Limit	Reference
		(a) Water				
Water	SPME. Fiber: polyacrylate. Extraction time: 90 min. Extraction temperature: 45°C	n/a	GC-MS	DMP, DEP, DnBP, BBP, DEHA, DEHP, DnOP	7–170 μg L^{-1}	Peñalver et al. (2000)
Water	SPME. Fiber: PDMS-DVB. Extraction time: 30 min. Extraction temperature: 80°C	n/a	GC-MS	DMP, DEP, DnBP, BBP, DEHA, DEHP, DnOP	2–27 ng L^{-1}	Peñalver et al. (2001)
Water	SPME. Fiber: PDMS-DVB. Extraction time: 20 min. Extraction temperature: 100°C	n/a	GC-MS	DMP, DEP, DnBP, BBP, DEHP, DnOP	2–103 ng L^{-1}	Polo et al. (2005)
Drinking water	SPME. Fiber: PDMS-DVB. Extraction time: 60 min. Extraction temperature: 25°C	n/a	GC-MS	DEP, DnBP, BBP, DEHP	5–40 μg L^{-1}	Luke-Betlej et al. (2001)
Drinking water	Stir bar (coated with PDMS) sorptive extraction followed by desorption of phthalates from stir bar by organic solvents	n/a	GC-MS	DMP, DEP, DnBP, BBP, DEHA, DEHP, DnOP	3–40 ng L^{-1}	Serodio and Nogueira (2006)
Bottled water	SPE. Eluted with 5 mL of DCM/hexane (4:1) and 5 mL of methanol/DCM (9:1) at 1 mL min^{-1}	n/a	GC-MS	DMP, DEP, DnBP, BBP, DEHP	2–4 μg L^{-1}	Casajuana and Lacorte (2003)
Bottled water	SPME. Fiber: PDMS-DVB. Extraction time: 20 min. Extraction temperature: 25°C	n/a	GC-MS	DMP, DEP, DiBP, DnBP, DEHP	20 μg L^{-1} for DEHP	Montuori et al. (2008)
Bottled water	HS-SPME. Fiber: PDMS-DVB. Extraction time: 60 min. Extraction temperature: 90°C	n/a	GC-MS	DMP, DEP, DiBP, DnBP, BBP, DHP, DEHA, DEHP, DnOP	3–85 μg L^{-1}	Cao (2008)
		(b) Beverages and Wine				
Beverages	LLE. Extraction with hexane–DCM (10:1), twice. Extract reconstituted to ACN for analysis	n/a	HPLC-UV	DnBP, DEHP	4 mg kg^{-1}	Yano et al. (2002)

(Continued)

TABLE 16.2 (CONTINUED)
Selection of Sample Preparation and Cleanup Methods

	Extraction	Cleanup	Analysis	Analytes	Detection Limit	Reference
Beverages	Simultaneous steam distillation and extraction	n/a	GC-MS	DEP, DnBP, BBP, DEHP, DEHA	2–5 μg L^{-1}	Kato et al. (2002)
Wines	HS-SPME. Fiber: PDMS-DVB. Extraction time: 60 min. Extraction temperature: 70°C	n/a	GC-MS	DMP, DEP, DnBP, DEHP, BBP, DnOP	150–2200 μg L^{-1}	Carrillo et al. (2007)
		(c) Oils				
Citrus essential oils	Samples analyzed directly without extraction and cleanup	n/a	GC-MS	DMP, DEP, DPP, DnBP, DiBP, BBP, DEHP, DnOP	3–40 μg L^{-1}	di Bella et al. (1999)
Edible oil	Sample (4 g) dissolved in heptane (20 mL) and analyzed directly by GC	n/a	GC-FID	DEHP, DiOP, DEHA	30–70 mg kg^{-1}	Kozyrod and Ziaziaris (1989)
Vegetable oil	HS-SPME. Fiber: PDMS. Extraction time: 60 min. Extraction temperature: 40°C	n/a	GC-ECD	DMP, DEP, DnBP, BBP, DEHP, DnOP	0.2–0.5 mg kg^{-1}	Holadova et al. (2007)
		(d) Other Nonfatty Foods				
Nonfatty foods (fruits, vegetables, fruit juices and drinks, wines, beers, maple syrup, cereal grains)	LLE or SLE. Sample blended with ACN and filtered. Filtrate diluted with water and extracted with hexane and DCM (10:1). Organic layer collected, washed with water, dried through sodium sulfate, evaporated to dryness, and redissolved in hexane for analysis	n/a	GC-MS	DEP, DiBP, DnBP, BBP, DEHP, DnOP, DEHA	50–500 mg kg^{-1}	Page and Lacroix (1995)
Fruit juices, soft drinks, fruit	LLE or SLE. Sample extracted with heptane	n/a	GC-FID	DEHP, DiOP, DEHA	30–70 mg kg^{-1}	Kozyrod and Ziaziaris (1989)
Fruits of *Benincasa hispida*	SLE. Powdered sample extracted with chloroform	Column cleanup with a silica gel column. Eluted with hexane–chloroform	GC-MS	DEHP		Du et al. (2006)

(*Continued*)

TABLE 16.2 (CONTINUED)
Selection of Sample Preparation and Cleanup Methods

	Extraction	Cleanup	Analysis	Analytes	Detection Limit	Reference
		(e) Milk				
Milk	LLE. Milk added with sodium chloride and water. Extracted with pentane–acetone followed by hexane–MTBE. Acetone washed away with water	GPC with a PLgel column. Sample eluted with pentane–MTBE (1:1)	GC-MS	DEP, DnBP, BBP, DEHP, DnOP	0.12–3.0 μg L^{-1}	Hogberg et al. (2008)
Milk	LLE. Milk added with potassium hydroxide. Extracted with methanol–hexane (5:3). Centrifuged for phase separation. Organic layer reconstituted to dichlormethane–cyclohexane (1:1) for cleanup	GPC with a Bio-beads SX3 column. Sample eluted with DCM–cyclohexane (1:1)	GC-MS	DEHP	5 mg kg^{-1}	Castle et al. (1990)
Milk	LLE. Extracted with ethanol–diethyl ether–pentane (2:1:1), followed by diethylether–pentane (1:1). Organic phase washed with 2% NaCl solution, dried, and redissolved in DCM–cyclohexane (1:1) for cleanup	GPC with a Bio-beads SX3 column. Sample eluted with DCM–cyclohexane (1:1). The eluate changed to isooctane for analysis	GC-ECD	DEHP	0.05–0.2 mg L^{-1}	Petersen (1991)
Milk	LLE. Extracted twice with methanol–hexane–MTBE (1.5:2:2). Organic phase dried, evaporated, and redissolved hexane for cleanup	(a) For DnBP, BBP, and DEHP, the extract was mixed with ACN. Hexane phase was discarded, and ACN phase washed with hexane (b) For DiNP and DiDP, extract was loaded to a deactivated silica column and eluted with 0.7% ethyl acetate in hexane	LC-MS/MS	DnBP, BBP, DEHP, DiNP, DiDP	4–9 μg kg^{-1}	Soerensen (2006)
Cow milk and human milk	HS-SPME. Fiber: PDMS. Extraction time: 60 min. Extraction temperature: 90°C	n/a	GC-MS	DMP, DEP, DnBP, BBP, DEHP, DnOP	0.12–1.8 μg kg^{-1} for 3.25% milk	Feng et al. (2005); Zhu et al. (2006)

(*Continued*)

TABLE 16.2 (CONTINUED)
Selection of Sample Preparation and Cleanup Methods

	Extraction	Cleanup	Analysis	Analytes	Detection Limit	Reference
		(f) Milk Products				
Milk, cream, butter, cheese	SLE. Sample mixed with methanol–hexane and potassium hydroxide, shaken, and centrifuged. The organic phase collected and the lower phase extracted twice with hexane. The combined hexane extracts evaporated and redissolved in DCM–cyclohexane (1:1) for cleanup	GPC with a Biobeads SX3 column. Eluted with DCM–cyclohexane (1:1)	GC-MS	DEHP	0.01 mg kg^{-1}	Sharman et al. (1994)
Milk and cream	SLE. Sample blended with acetone and hexane and centrifuged. Hexane removed and extraction with hexane repeated. Hexane extract dried and made to 5 mL of hexane	Phthalates isolated from lipid by sweep co-distillation, Florisil trapping, and selective elution	GC-MS	DEP, DiBP, DnBP, BBP, DEHP, DnOP, DEHA	0.05–0.5 mg g^{-1}	Page and Lacroix (1995)
Powdered infant formula	SLE. Sample mixed with hexane-saturated ACN, sonicated, and centrifuged. Sample filtered and organic solution washed with hexane saturated with ACN-saturated hexane, and volume reduced for analysis	n/a	GC-MS	DnBP, DEHP		Yano et al. (2005)
		(g) Other Solid Foods				
Fatty foods (animal tissues, fats, cheese)	SLE. Sample blended with sodium sulfate and dichloromethane and filtered. DCM was changed to 5 mL hexane for cleanup	Phthalates isolated from lipid by sweep co-distillation, Florisil trapping and selective elution	GC-MS	DEP, DiBP, DnBP, BBP, DEHP, DnOP, DEHA	0.05–0.5 mg g^{-1}	Page and Lacroix (1995)
Baby food, infant formulae, total diet samples	SLE. Sample extracted with pentane. Supernatant reduced to dryness. Residue dissolved in 5 mL ethylacetate–cyclohexane (1:1) for cleanup	GPC with a Bio-beads SX3 column. Eluted with ethyl acetate–cyclohexane (1:1). Elutate fraction reduced to dryness and redissolved in 0.5 mL isooctane	GC-MS	DnBP, BBP, DEHP, DEHA	0.015–0.35 mg kg^{-1}	Petersen and Breindahl (2000)

(Continued)

TABLE 16.2 (CONTINUED)
Selection of Sample Preparation and Cleanup Methods

	Extraction	Cleanup	Analysis	Analytes	Detection Limit	Reference
Candy, chocolate, meat pasty, pork pie, chicken pie, sandwiches	SLE. Sample extracted twice with acetone–hexane (1:1). The extract dried with sodium sulfate, evaporated to dryness, and redissolved in DCM–cyclohexane (1:1) for cleanup	GPC with a Bio-beads SX3 column. Eluted with DCM–cyclohexane (1:1)	GC-MS	DnBP, DCHP, BBP	0.05 mg kg^{-1}	Castle et al. (1988a, 1988b)
Cheese, meats, fish, sandwiches, cakes, fruit, vegetables, cooked meals	SLE. Sample blended and extracted with acetone–hexane (1:1). The extract dried over sodium sulfate and evaporated to dryness, and the residue redissolved in DCM–cyclohexane (1:1) for cleanup	GPC with a Bio-beads SX3 column. Eluted with DCM–cyclohexane (1:1)	GC-MS	DEHA	<0.1 mg kg^{-1}	Startin et al. (1987)
Duplicate diet samples	SLE. Sample extracted with ACN twice. ACN layer mixed with sodium chloride and extracted with ACN-saturated hexane. ACN collected, evaporated, and redissolved in hexane	Column cleanup with a Florisil and Bondesil PSA dual-layer column, eluted with 5% of acetone in hexane. Eluate evaporated and redissolved in hexane	GC-MS	DEP, DPP, DnBP, DPeP, DHP, BBP, DCHP, DEHP, DiOP, DnOP, DiNP, DEHA	0.2–25.8 μg kg^{-1}	Tsumura et al. (2001, 2003)

Note: DiDP, diisodecyl phthalate; DiNP, diisononyl phthalate; DiOP, di-iso-octyl phthalate; DPeP, dipentyl phthalate; DPP, di-*n*-propyl phthalate.

16.2 CONTAMINATION ISSUES IN THE ANALYSIS OF PHTHALATES

Phthalates are common contaminants in the laboratory environment. They are present in air and were found in many of solvents, reagents, and common laboratory materials, such as tubing, cork, glass wool, filter paper, alumina, Florisil, sodium sulfate, and sodium chloride. The potential contamination of phthalates during sample preparation and instrument analysis has been mentioned in a number of studies. Among the phthalates, DEHP is usually the predominant one found in blanks, while DnBP and DEP are also commonly present. Every precaution must be taken to minimize the contamination. Common precautious measures include rinsing all glassware coming into contact with samples by using organic solvents such as methanol, acetone, and hexane (Sharman et al., 1994; Page and Lacroix, 1995; Petersen and Breindahl, 2000; Soerensen, 2006), heating the glassware to a high temperature (Bruns-Weller and Pfordt, 2000; Yano et al., 2005), or having the combination of heat treatment followed by solvent rinse (Gruber et al., 1998; Tsumura et al., 2001). In addition, septa, caps, and sample vials (Sharman et al., 1994), glass wool (Tsumura et al., 2001), and utensils for sample preparation (Page and Lacroix, 1995) are generally solvent rinsed before use.

Other measures to reduce phthalate contamination include careful selection and prescreening of solvents (Soerensen, 2006), distillation of purchased solvents in an all-glass device (Page and Lacroix, 1992; Page and Lacroix, 1995), or purification with aluminum oxide (Fankhauser-Noti and Grob, 2007), and avoiding using latex or vinyl gloves when handling samples (Page and Lacroix, 1992). When adsorbent is to be used in sample cleanup, it should be decontaminated by heating prior to use (Page and Lacroix, 1992; Soerensen, 2006). In some cases, sodium chloride for sample handling was heat treated prior to use as well (Tsumura et al., 2001).

16.3 SAMPLE PRETREATMENT, EXTRACTION, AND CLEANUP

16.3.1 WATER

Water is a relatively simple matrix. SPME has been widely used in the extraction of phthalates from water (Peñalver et al., 2000; Peñalver et al., 2001; Polo et al., 2005), drinking water (Luke-Betlej et al., 2001), and bottled water (Montuori et al., 2008). In SPME, a polymer-coated fiber is immersed into the water and the target analytes are extracted onto the fiber. This is also often referred to as direct SPME, as the fiber is in direct contact with the sample matrix. Because the water molecule is polar and has a minimal partition capability in the fiber selected for extracting nonpolar and lipophilic compounds such as phthalates, the target phthalates can therefore be concentrated effectively from water onto the fiber with the right choice of SPME fiber. PDMS–Carbowax–DVB-coated fiber was reported to be the most common one. The extraction temperature ranged from 25°C to 100°C with various extraction times among various studies. The exposed fiber is inserted immediately after SPME into a GC-MS instrument for analysis without a need for further cleanup. One of the advantages of the SPME method is the elimination of the use of organic solvents, which often are contaminated with phthalates to some degree, thus achieving low blank levels and good method detection limits of nanogram per liter (Peñalver et al., 2001; Polo et al., 2005) to microgram per liter (Peñalver et al., 2000; Luke-Betlej et al., 2001; Montuori et al., 2008) (Table 16.2).

Other methods reported for the measurement of phthalates in water include stir bar sorptive extraction followed by liquid desorption (Serodio and Nogueira, 2006) and traditional SPE methods (Casajuana and Lacorte, 2003).

Besides direct SPME, headspace (HS)-SPME is also used for the measurement of phthalates in bottled water with a detection limit at low microgram per liter (Cao, 2008). In HS-SPME, the fiber is suspended right above the sample in a closed system such as a vial, for example, instead of being immersed into the sample. Because phthalates are semivolatile compounds with low vapor pressures, the liquid sample usually is heated to aid the release of phthalates into the HS, where the extraction occurs. The performance of three fibers for the HS-SPME of DEHA and eight phthalates in water was investigated systematically under different extraction conditions (Cao, 2008). Good responses on the 65 μm PDMS/DVB SPME fiber were observed for DEHA and all phthalates. The PDMS SPME fiber had very poor responses for the lighter and slightly polar phthalates, DMP, and DEP, while the divinylbenzene/carboxen/polydimethylsiloxane (DVB/CAR/PDMS) SPME fiber had very poor responses for the heavier and nonpolar adipate and phthalates. The salt (NaCl) was found to increase the partitioning of DMP, DEP, diisobutyl phthalate (DiBP), DnBP, and butyl benzyl phthalate (BBP) from water into the HS, while partitioning of heavier adipate and phthalates from water into HS was suppressed when the concentration of NaCl was above 10%. The automated HS-SPME methods were developed and validated under two different salting conditions (30% NaCl for DMP, DEP, and BBP and 10% for DEHA, DiBP, DnBP, di-*n*-hexyl phthalate [DHP], DEHP, and di-*n*-octyl phthalate [DnOP]). Linearity with R^2 values better than 0.9949 was observed for DEHA and eight phthalates over the range from 0.1 to 20 μg L^{-1}. Method detection limits ranged from 0.003 μg L^{-1} for DnOP to 0.085 μg L^{-1} for BBP. Good repeatability was observed for DEHA and most phthalates with RSD values less than 10%. The methods were used to analyze bottled water samples for DEHA and eight phthalates. DMP, DHP, BBP, DEHA, and DnOP were not detected in any samples. Concentrations of the other phthalates were low (around sub-ppb) except for DnBP in the water from a polycarbonate bottle at 1.72 μg L^{-1}.

16.3.2 Food Commodities

A number of extraction methods have been reported for the measurement of phthalates in beverages and wines. They include LLE with hexane–dichloromethane (DCM) (Yano et al., 2002), simultaneous steam distillation and extraction (Kato et al., 2002), and HS-SPME (Carrillo et al., 2007). All samples are analyzed without further cleanup. The extract from LLE was concentrated and reconstituted to ACN for the analysis by HPLC. The method detection limit at milligram per kilogram level in this method (Yano et al., 2002) was higher than those in the other two methods, likely due to the use of a UV detector instead of an MS detector.

Oils are analyzed either directly without any sample treatment (di Bella et al., 1999) or diluted in heptane (Kozyrod and Ziaziaris, 1989). An HS-SPME method was developed for the analysis of phthalates in vegetable oil (Holadova et al., 2007). An unmodified PDMS fiber instead of PDMS/PVB fiber was used, and the extraction was conducted at a lower temperature of 40°C. The high detection limits of milligram per kilogram in the latter two cases were largely attributed to the use of an FID (Kozyrod and Ziaziaris, 1989) or an ECD (Holadova et al., 2007) instead of MS as a detector.

Only a few studies on the measurements of phthalates in solid nonfatty food have been reported. Solvent extraction method, either LLE or solid–liquid extraction (SLE) depending on the types of foods, was employed using heptane (Kozyrod and Ziaziaris, 1989) or chloroform (Du et al., 2006) as extraction solvents. Page and Lacroix employed ACN for the initial extraction of phthalates from various nonfatty foods, both liquid and solid, and the ACN extract was further extracted with other water-insoluble organic solvents hexane and DCM (1995). High detection limits of several dozen (Kozyrod and Ziaziaris, 1989) to several hundred (Page and Lacroix, 1995) milligram per kilogram were reported for these nonfatty foods.

There are a number of reports on the measurement of phthalates in milk. Powdered milk sometimes was reconstituted by mixing with water according to the manufacturer's preparation instructions. For example, Casajuana and Lacorte (2004) reported mixing milk powder with water in a 1:10 ratio (3 g of powdered milk in 30 mL of water) prior to sample extraction.

Several analytical methods have been developed for the analysis of phthalates in milk as well. LLE using different combinations of organic solvents was commonly employed for the extraction of phthalates from milk. Because of the water content in the milk sample, generally, a mixture of water-soluble solvents (methanol, ACN) and water-insoluble solvents (hexane, DCM, cyclohexane) was used for the initial extraction (Castle et al., 1990; Petersen, 1991; Soerensen, 2006; Hogberg et al., 2008). Such organic solvent systems include pentane–acetone followed by hexane–methyl *t*-butyl ether (MTBE) (Hogberg et al., 2008), methanol–hexane in the presence of potassium hydroxide (Castle et al., 1990), ethanol–diethyl ether–pentane (Petersen, 1991), and methanol–hexane–MTBE (Soerensen, 2006). Sodium chloride is added sometimes to aid the phase separation (Petersen, 1991; Hogberg et al., 2008).

An HS-SPME extraction method was used for the extraction of phthalates from raw milk (Feng et al., 2005) and human milk (Zhu et al., 2006) (Table 16.3).

Milk products such as cream, cheese, butter, and milk powder are solid materials, and therefore, SLE methods are often used by blending the samples with extraction solvents. Such solvent systems include methanol–hexane in the presence of potassium hydroxide (Sharman et al., 1994), acetone–hexane (Page and Lacroix, 1995), and ACN–hexane (Petersen and Breindahl, 2000). These methods are also applicable to milk samples (Sharman et al., 1994; Page and Lacroix, 1995). Because of the solid content in the extraction mixture, a centrifuge was employed to aid the phase separation and collection of organic extracts. Detection limits of sub-milligram per kilogram could be achieved using these methods.

Beside milk and milk products, phthalates have been measured in other solid foods in various studies involving duplicate diet and total diet samples. These samples contain various degrees of solid and fat content. Again, SLE with different solvent systems is the most common approach to extracting phthalates from food into organic solvents. The organic solvent systems used in SLE included pentane (Petersen and Breindahl, 2000) and acetone–hexane (Castle et al., 1988a, 1988b). Tsumura et al. (2001, 2003) used only ACN for the initial extraction followed by LLE to extract phthalates from ACN into hexane for the duplicate diet samples. For some fatty foods such as animal tissues, fats, and cheese, samples were mixed with sodium sulfate and extracted with DCM. DCM was then collected through filtration for further cleanup (Page and Lacroix, 1995). The detection limit for these solid foods was reported at sub-milligram per kilogram or lower.

To date, the largest problem in PAE analysis is the high blank value because PAEs are widely used in various applications and products (He et al., 2015). To overcome this shortcoming, gas purge microsyringe extraction (GP-MSE) was applied, which established a new and low-blank-value

TABLE 16.3
GC Retention Times of Some Commonly Measured Phthalates along with Quantifying Ions and Characteristic Ions

Abbr.	Full Name	MW	RT (min)	Q-Ion	Characteristic Ions
DMP	Dimethyl phthalate	194	14.34	163	194
DEP	Diethyl phthalate	222	15.5	149	177
DiBP	Diisobutyl phthalate	278	17.66	149	223, 167
DnBP	Di-*n*-butyl phthalate	278	18.47	149	223, 205
DPeP	Dipentyl phthalate	306	20.16	149	237
DnHP	Di-*n*-hexyl phthalate	334	21.94	149	251
BBP	Butyl benzyl phthalate	312	22.05	149	206, 91
DEHA	Bis(2-ethylhexyl) adipate	370	22.42	129	259, 241, 147
DCHP	Dicyclohexyl phthalate	330	23.63	149	167, 249
DEHP	Di(2-ethylhexyl) phthalate	390	23.71	149	167, 279
DOP	Dioctyl phthalate	390	25.21	149	279
DDcP	Didecyl phthalate	446	27.97	149	307
DUP	Diundecyl phthalate	474	29.97	149	321

Note: Injection port temperature: 280°C; column: DB-5MS 30 m × 0.25 mm i.d. × 0.25 μm film thickness and oven temperature: 45°C (5 min), 15°C min^{-1} to 210°C, 8°C min^{-1} to 270°C, and 30°C min^{-1} to 310°C (25 min).

analytical method for PAE analysis to analyze PAEs in foodstuff. GP-MSE was used as a cleanup method, and the overall recoveries ranged from 85.7% to 102.6%, and the RSD was less than 10%.

16.4 CLEANUP OF EXTRACTS

In general, nonfatty liquid foods such as water, juice, beverage, and wine do not go through further cleanup procedures. The only exception is the employment of a cleanup procedure using a silica gel column for fruit samples as reported by Du et al. (2006). Oil is analyzed either directly (di Bella et al., 1999) or diluted (Kozyrod and Ziaziaris, 1989) in organic solvents or using HS-SPME (Holadova et al., 2007) without further cleanup as well.

Further cleanup of the extracts is, however, almost always necessary for foods such as milk, milk products, or solid foods containing various degrees of fat and solid content as the matrix of these foods is much more complex. Isolation of phthalates from fat in fatty foods is often performed by size exclusion chromatography (SEC). Biobeads SX3 is the most commonly used SEC column with DCM–cyclohexane as the eluting solvents (Startin et al., 1987; Castle et al., 1988a, 1988b; Castle et al., 1990; Sharman et al., 1994), while ethylacetate–cyclohexane system is also mentioned (Petersen and Breindahl, 2000). The other SEC column reported is the PLgel column with pentane–MTBE as the eluting solvents (Hogberg et al., 2008). Other less popular approaches for cleanup of the extracts to isolate phthalates are by employing columns packed with silica gel (Soerensen, 2006), Florisil (Page and Lacroix, 1995), or dual-layer column of Florisil–Bondesil PSA (Tsumura et al., 2001, 2003).

A gel permeation chromatography (GPC) cleanup step with 7:3 cyclohexane/dichloromethane as the mobile phase was employed to remove the high-molecular mass species present in oil (Cavaliere et al., 2008).

16.5 INSTRUMENT ANALYSIS

Because of the semivolatile, nonpolar nature of phthalates, instruments used in the analysis are mostly GC, while HPLC-MS (e.g., Soerensen, 2006) or HPLC/UV (Yano et al., 2002) is also employed. For GC analysis, MS was almost a nominal detector nowadays for the measurements of phthalates. Other detectors, such as FIDs (Kozyrod and Ziaziaris, 1989) or ECDs (Petersen, 1991; Holadova et al., 2007), are used occasionally only.

In Tables 16.4 through 16.6, the reader can find a list of standardized analysis methodologies for plasticizers (Lambert et al., 2010).

16.5.1 Gas Chromatography–Mass Spectrometry

For GC-MS analysis, samples are injected into the GC-MS at an injection temperature of around 250°C (240–260°C reported from several studies). The commonly used GC capillary column for the separation of phthalates is a fused-silica capillary column containing 5% phenyl and 95% methyl polysiloxane with a column length of either 30 or 60 m. The oven temperature programs vary greatly depending on the complexity of the samples and number of target analytes monitored. For example, the rise of oven temperature ranged from 6°C min^{-1} in one study to 15°C min^{-1} in another. Usually, the phthalates can be well separated under these oven temperature programs. The elution orders of commonly monitored phthalates and DEHA are listed in Table 16.3 as an example, along with the target ions for quantification and several characteristic ions for the identification purposes.

The separated phthalates are detected by a mass spectrometer, which is operated either in full-scan mode or in SIM mode. Mass spectra generated through EI of six most commonly monitored phthalates are presented in Figure 16.2. For phthalates, the base peak of the mass spectrum is always ion *m*/*z* 149, which is also serving as the quantifying ion, except for DMP, which has a base peak of ion *m*/*z* 163. For DEHA, the base peak is ion *m*/*z* 129.

TABLE 16.4
US National Environmental Methods Index: Standardized Methodologies

Method: EPA-EAD 1625	Procedure: GC/MS; analyte: BBP; matrix: water; detection limit: not provided.
Method: EPA-EAD 606	Procedure: GC with ECD; analyte: BBP; matrix: wastewater and other waters; detection limit: 0.34 μg L^{-1}.
Method: EPA-NERL 506	Procedure: GC with PID; analyte: BBP; matrix: drinking water; detection limit: 2.67 μg L^{-1}.
Method: EPA-NERL 525.2	Procedure: GC/MS; analyte: BBP; matrix: finished drinking water, source water, or drinking water in any treatment stage; detection limit: 0.025 μg L^{-1}.
Method: EPA-NERL 625	Procedure: GC/MS; analyte: BBP; matrix: municipal and industrial discharges; detection limit: 2.5 μg L^{-1}.
Method: EPA-OSW 8061A	Procedure: GC with ECD; analyte: BBP; matrix: groundwater, leachate, soil, sludge, and sediment; detection limit: 0.042 μg L^{-1}.
Method: EPA-OSW 8270D	Procedure: GC/MS; analyte: BBP; matrix: solid waste matrices, soils, air sampling media and water samples; detection limit: 10 μg L^{-1}.
EPA: Method 8060	*Phthalate Esters*. This method provides GC conditions for the detection of ppb levels. For di-*n*-octyl phthalate, the method detection limit for ECD is 3.0 μg L^{-1} and for the average recovery range for four measurements is D50.0 μg L^{-1}, and the limit for the standard deviation is 13.4 μg L^{-1}.
EPA: Method 8250	*Gas Chromatography/Mass Spectrometry for Semivolatile Organics: Packed Column Technique*. This GC/MS method is used to determine the concentration of semivolatile organic compounds in extracts prepared from all types of solid waste matrices, soils, and groundwater. The practical quantitation limit for determining an individual compound is approximately 1 mg kg^{-1} (wet weight) for soil/sediment samples, 1–200 mg kg^{-1} for wastes, and 10 μg L^{-1} for groundwater samples. Under the prescribed conditions, di-*n*-octyl phthalate has a range for the average recovery of four measurements of 18.6–131.8 μg L^{-1} and a limit for the standard deviation of 31.4 μg L^{-1}.
Method: EPA-EAD 1625	Procedure: GC/MS; analyte: di-*n*-octyl phthalate; matrix: water; detection limit: 10 μg L^{-1}.
Method: EPA-EAD 606	Procedure: GC with ECD; analyte: di-*n*-octyl phthalate; matrix: wastewater and other waters; detection limit: 3 μg L^{-1}.
Method: EPA-NERL 506	Procedure: GC with PID; analyte: di-*n*-octyl phthalate; matrix: drinking water; detection limit: 6.42 μg L^{-1}.
Method: EPA-NERL 625	Procedure: GC/MS; analyte: di-*n*-octyl phthalate; matrix: municipal and industrial discharges; detection limit: 2.5 μg L^{-1}.
Method: EPA-OSW 8061A	Procedure: GC with ECD; analyte: di-*n*-octyl phthalate; matrix: groundwater, leachate, soil, sludge, and sediment; detection limit: 0.049 μg L^{-1}.
Method: EPA-OSW 8270D	Procedure: GC/MS; analyte: di-*n*-octyl phthalate; matrix: solid waste matrices, soils, air sampling media, and water samples; detection limit: 10 μg L^{-1}.
EPA: EMSLC Method 506	*Determination of Phthalate and Adipate Esters in Drinking Water by Liquid–Liquid Extraction or Liquid–Solid Extraction and Gas Chromatography with Photoionization Detection*. Detection limit: 12.000 μg L^{-1}.
EPA: EMSLC Method 525.1	*Determination of Organic Compounds in Drinking Water by Liquid–Solid Extraction and Capillary Column Gas Chromatography and Mass Spectrometry*. Revision 2.2. Detection limit: 0.60 μg L^{-1}.
EPA: EMSLC Method 525.2	*Determination of Organic Compounds in Drinking Water by Liquid–Solid Extraction and Capillary Column Gas Chromatography and Mass Spectrometry*. Revision 1.0.

(*Continued*)

TABLE 16.4 (CONTINUED)
US National Environmental Methods Index: Standardized Methodologies

EPA: Method 8060	*Phthalate Esters.* This method provides GC conditions for phthalate esters. This method provides GC conditions for the detection of ppb levels. For BBP, the method detection limit for ECD is 0.34 μg L^{-1} and that for FID is 15 μg L^{-1}, the average recovery range for four measurements is 5.7–11.0 μg L^{-1}, and the limit for the standard deviation is 4.2 μg L^{-1}.
EPA: Method 8250	Packed column GC/MS technique for the determination of semivolatile organic compounds in extracts prepared from all types of solid waste matrices, soil, and groundwater. This method is applicable to quantifying most neutral, acidic, and basic organic compounds that are soluble in methylene chloride and capable of being eluted with derivatization as sharp peaks from a GC packed column. Under the prescribed conditions, pentachlorophenol has a detection limit of 3.6 μg L^{-1}. Precision and method accuracy were found to be directly related to the concentration of the analyte and essentially independent of the sample matrix.
EPA: EMSLC Method 506	*Determination of Phthalate and Adipate Esters in Drinking Water by Liquid–Liquid Extraction or Liquid–Solid Extraction and Gas Chromatography with Photoionization Detection.* Detection limit: 12.000 μg L^{-1}.
EPA: EMSLC Method 525.1	*Determination of Organic Compounds in Drinking Water by Liquid–Solid Extraction and Capillary Column Gas Chromatography and Mass Spectrometry.* Revision 2.2. Detection limit: 0.60 μg L^{-1}.
EPA: EMSLC Method 525.2	*Determination of Organic Compounds in Drinking Water by Liquid–Solid Extraction and Capillary Column Gas Chromatography and Mass Spectrometry.* Revision 1.0.
EPA: Method 8060	*Phthalate Esters.* This method provides GC conditions for phthalate esters This method provides GC conditions for the detection of ppb levels. For BBP, the method detection limit for ECD is 0.34 μg L^{-1} and that for FID is 15 μg L^{-1}, the average recovery range for four measurements is 5.7–11.0 μg L^{-1}, and the limit for the standard deviation is 4.2 μg L^{-1}.
EPA: Method 8060	GC using solvent flush technique and a ECD or an FID for the detection of ppb levels of phthalate esters in solid waste. Groundwater samples should be determined by ECD.
EPA: Method 8250	Packed column GC/MS technique for the determination of semivolatile organic compounds in extracts prepared from all types of solid waste matrices, soil, and groundwater. This method is applicable to quantifying most neutral, acidic, and basic organic compounds that are soluble in methylene chloride.
Method: EPA-EAD 1625	Procedure: GC/MS; analyte: bis(2-ethylhexyl) phthalate; matrix: water; detection limit: 10 μg L^{-1}.
Method: EPA-EAD 606	Procedure: GC with ECD; analyte: bis(2-ethylhexyl) phthalate; matrix: wastewater and other waters; detection limit: 2 μg L^{-1}.
Method: EPA-NERL 506	Procedure: GC with PID; analyte: bis(2-ethylhexyl) phthalate; matrix: drinking water; detection limit: 2.25 μg L^{-1}.
Method: EPA-NERL 525.2	Procedure: GC/MS; analyte: bis(2-ethylhexyl) phthalate; matrix: finished drinking water, source water, or drinking water in any treatment stage; detection limit: 0.46 μg L^{-1}.
Method: EPA-NERL 625	Procedure: GC/MS; analyte: bis(2-ethylhexyl) phthalate; matrix: municipal and industrial discharges; detection limit: 2.5 μg L^{-1}.
Method: EPA-OSW 8061A	Procedure: GC with ECD; analyte: bis(2-ethylhexyl) phthalate; matrix: groundwater, leachate, soil, sludge, and sediment; detection limit: 0.27 μg L^{-1}.

(Continued)

TABLE 16.4 (CONTINUED)
US National Environmental Methods Index: Standardized Methodologies

Method: EPA-OSW 8270D	Procedure: GC/MS; analyte: bis(2-ethylhexyl) phthalate; matrix: solid waste matrices, soils, air sampling media, and water samples; detection limit: not provided.
EPA: Method 8060	*Phthalate Esters*. This method provides GC conditions for the detection of ppb levels. A 2–5 μg aliquot of the extract is injected into a gas chromatograph by using the solvent flush technique, and compounds in the gas chromatograph effluent are detected by an ECD or an FID. Groundwater samples should be determined by using an ECD.
EPA: Method 8250	*Gas Chromatography/Mass Spectrometry for Semivolatile Organics: Packed Column Technique*.
Method: EPA-EAD 1625	Procedure: GC/MS.
Method: EPA-EAD 606	Procedure: GC with ECD; analyte: di-*n*-butyl phthalate; matrix: wastewater and other waters; detection limit: 0.36 μg L^{-1}.
Method: EPA-NERL 506;	Procedure: GC with photoionization detection; analyte: di-*n*-butyl phthalate; matrix: drinking water; detection limit: 1.23 μg L^{-1}.
Method: EPA-NERL 525.2;	Procedure: GC/MS; analyte: di-*n*-butyl phthalate; matrix: finished drinking water, source water, or drinking water in any treatment stage; detection limit: 0.59 μg L^{-1}.
Method: EPA-NERL 625	Procedure: GC/MS; analyte: di-*n*-butyl phthalate; matrix: municipal and industrial discharges; detection limit: 2.5 μg L^{-1}.
Method: EPA-OSW 8061A	Procedure: GC with ECD; analyte: di-*n*-butyl phthalate; matrix: groundwater, leachate, soil, sludge, and sediment; detection limit: 0.33 μg L^{-1}.
Method: GPA-OSW 8270D	Procedure: GC/MS; analyte: di-*n*-butyl phthalate; matrix: solid waste matrices, soils, air sampling media, and water samples; detection limit: 10 μg L^{-1}.
Method: Standard Methods 6410 B	Procedure: GC/MS; analyte: butyl benzyl phthalate; matrix: municipal and industrial discharges; detection limit: 2.5 μg L^{-1}.
Method: USGS-NWQL 0-3118-83	Procedure: GC using an FID or an MS detector; analyte: BBP; matrix: water and water–suspended sediment mixtures; detection limit: not provided.
Method: USGS-NWQL 0-5130-913	Procedure: high-performance GPC, capillary-column GC/MS; analyte: BBP; matrix: soils and sediment; detection limit: 33.2 μg kg^{-1}.
Method: USGS-NWQL 0-3118-83	Procedure: GC using an FID or an MS detector; analyte: di-*n*-butyl phthalate; matrix: water and water–suspended sediment mixtures; detection limit: not provided.
Method: USGS-NWQL 0-5130-95	Procedure: high-performance GPC, capillary-column GC/MS; analyte: di-*n*-butyl phthalate; matrix: soils and sediment; detection limit: 31.6 μg kg^{-1}.
Method: USGS-NWQL 0-3118-83	Procedure: GC using an FID or an MS detector; analyte: bis(2-ethylhexyl) phthalate; matrix: water and water–suspended sediment mixtures; detection limit: not provided.
Method: USGS-NWQL O-S 130-95	Procedure: high-performance GPC, capillary-column GC/MS; analyte: bis(2-ethylhexyl) phthalate; matrix: soils and sediment; detection limit: 41.4 μg kg^{-1}.
Method: USGS-NWQL 0-5130-95	Procedure: high-performance GPC, capillary-column GC/MS; analyte: di-*n*-octyl phthalate; matrix: soils and sediment; detection limit: 22 μg kg^{-1}.
Method: DOE OM 1OOR	Procedure: GC with MS ion trap detector; analyte: di-*n*-octyl phthalate; matrix: solid waste matrices, soils, and groundwater; detection limit: 160 μg L^{-1}.
Method: Standard Methods 6410 B	Procedure: GC/MS; analyte: di-*n*-octyl phthalate; matrix: municipal and industrial discharges; detection limit: 2.5 μg L^{-1}.
Method: DOE OM 100R	Procedure: GC with MS ion trap detector; analyte: BBP; matrix: solid waste matrices, soils, and groundwater; detection limit: 76 μg L^{-1}.
Method: DOE OM 100R	Procedure: GC with MS ion trap detector; analyte: bis(2-ethylhexyl) phthalate; matrix: solid waste matrices, soils, and groundwater; detection limit: 260 μg L^{-1}.
Method: Standard Methods 6410 B	Procedure: GC/MS; analyte: bis(2-ethylhexyl) phthalate; matrix: municipal and industrial discharges; detection limit: 2.5 μg L^{-1}.

(*Continued*)

TABLE 16.4 (CONTINUED)
US National Environmental Methods Index: Standardized Methodologies

Method: DOE OM 100R	Procedure: GC with MS ion trap detector; analyte: di-*n*-butyl phthalate; matrix: solid waste matrices, soils, and groundwater; detection limit: 44 μg L^{-1}.
Method: Standard Methods 6410 B	Procedure: GC/MS; analyte: di-*n*-butyl phthalate; matrix: municipal and industrial discharges; detection limit: 2.5 μg L^{-1}.

Note: EAD, Environmental Appeals Decisions; EMSLC, Environmental Monitoring and Support Laboratory-Cincinnati; NERL, National Exposure Research Laboratory; OSW, Office of Solid Waste; PID, photoionization detector.

TABLE 16.5
US Occupational Safety and Health Administration Standardized Methodologies

Method: OSHA 104	Procedure: GC with FID; analyte: di(2-ethylhexyl) phthalate; matrix: air; detection limit: 0.09 μg L^{-1}
Method: OSHA 48	Procedure: GC using FID; analyte: petroleum distillate fractions; matrix: air; detection limit: 0.77 mg/sample (260 mg m^{-3})
Method: OSHA 104	Procedure: GC with FID; analyte: dibutyl phthalate; matrix: air; detection limit: 0.10 μg L^{-1}
Method: OSHA 104	Procedure: GC with FID; analyte: di-*n*-octyl phthalate; matrix: air; detection limit: 0.10 μg L^{-1}

Source: US Occupational Safety and Health Administration, Alphabetic Index of Sampling and Analytical Methods, http://www.osha.gov/dts/sltc/methods/toc.html, US Occupational Safety and Health Administration, Washington, DC, 2016.

TABLE 16.6
US Centers for Disease Control and Prevention Standardized Methodologies

Method: NIOSH 5020, Issue 2	Procedure: GC with FID; analyte: di(2-ethylhexyl) phthalate; matrix: air; detection limit: 10 μg per sample
Method: NIOSH 5020, Issue 2	Procedure: GC with FID; analyte: dibutyl phthalate; matrix: air
Method: NIOSH 1550, Issue 2	Naphthas using GC and FID; analyte: naphtha hydrocarbons; matrix: air; estimated level of detection: 0.1 mg per air sample

Note: NIOSH, National Institute for Occupational Safety and Health.

A GC-MS/MS method for the detection of six phthalates in olive oil was developed. Two ionization methodologies, that is, electron (EI) and isobutane–chemical ionization (CI), were compared, in MS/MS mode, to achieve better analytical performances. An overall evaluation of all analytical parameters shows that the EI-MS/MS approach provides satisfactory results and is to be preferred to CI-MS/MS, at least in the case of the examined analytes. The observed accuracies, ranging from 71.7% to 112.2%, and the RSD values of less than 9.7% confirm the effectiveness of the proposed method in the assay of phthalate content in such a complex matrix as olive oil. The proposed protocol for the identification and assay of phthalates in olive oil might be of interest for the implementation of the quality assurance scheme for residue monitoring in food safety assessment.

16.5.2 Mass Spectrometry Fragmentation Pathways

The fragmentation pathways of phthalates under EI ionization are illustrated in Figure 16.3. There are two fragmentation pathways, and both lead to the formation of ion of *m*/*z* 149, except for DMP. DMP looses a methoxy group (–CH_3O) under EI to form a stable ion of *m*/*z* 163. The ion of *m*/*z* 177

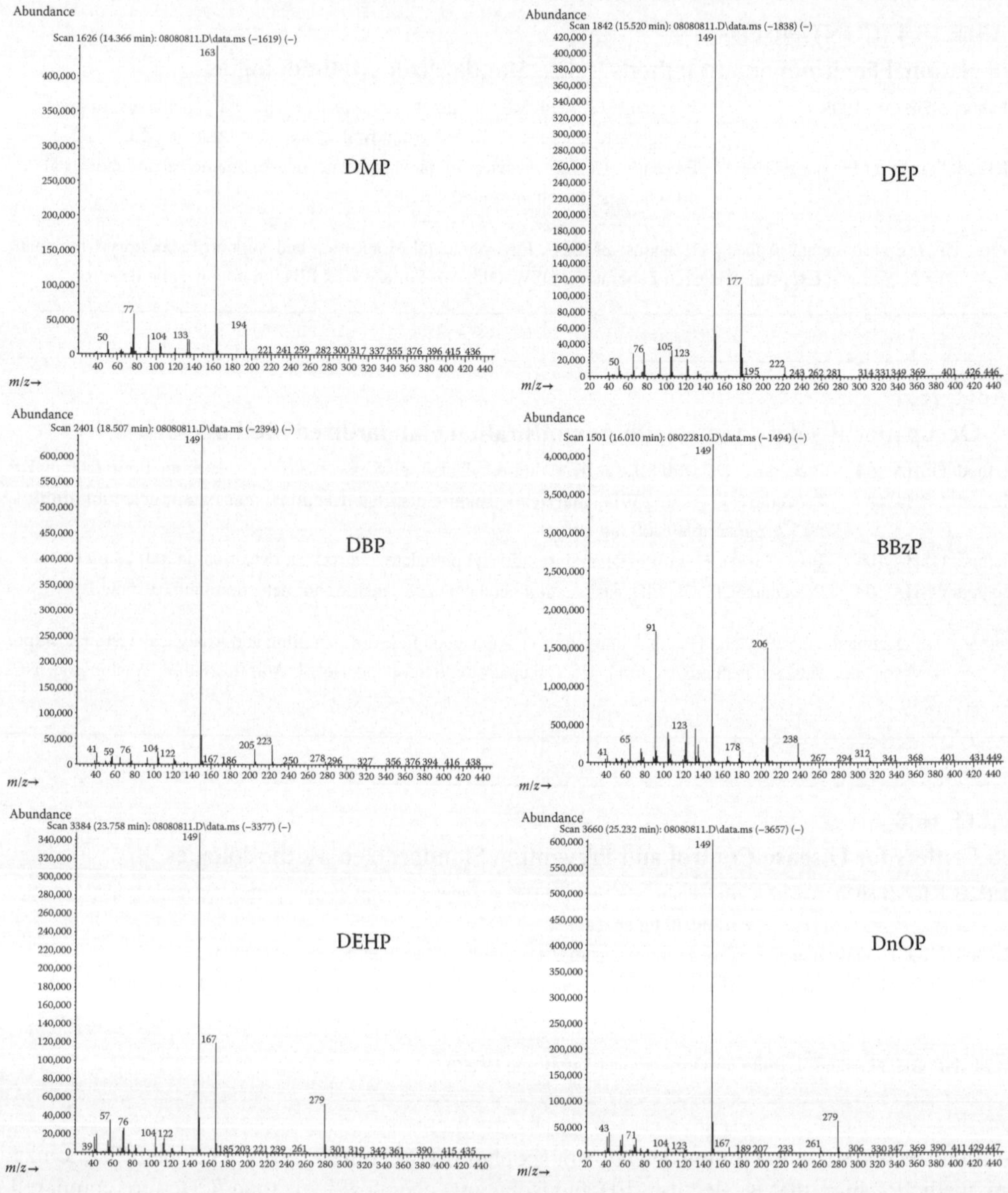

FIGURE 16.2 Electron impact mass spectra of six commonly monitored phthalates.

in DEP spectrum is also formed through the loss of an ethoxy group. For phthalates with long side chains such as DEHP, the major fragmentation pathway seems to be the loss of an alkyl side chain to form an [M−R+2H] ion (*m/z* 279 in the spectrum of DEHP and DnOP). This ion is further fragmented to the ion of *m/z* 167, which ultimately loses H_2O to form the stable ion of *m/z* 149. The mass spectrum of BBP is dominated by ions *m/z* 206 and *m/z* 91 besides the base peak of ion *m/z* 149. Ion *m/z* 206 is formed by the loss of neutral group of C_6H_5CHO, while ion *m/z* 91 corresponds to benzyl ion of $[C_6H_5CH_2]$.

The characteristic ions of phthalates resulting from the fragmentations under EI mode are summarized in Table 16.3. In practice, one has to consider the relative abundance of these fragments in selecting qualifier (Q-) ions. There are some variations in selecting Q-ions for phthalates among

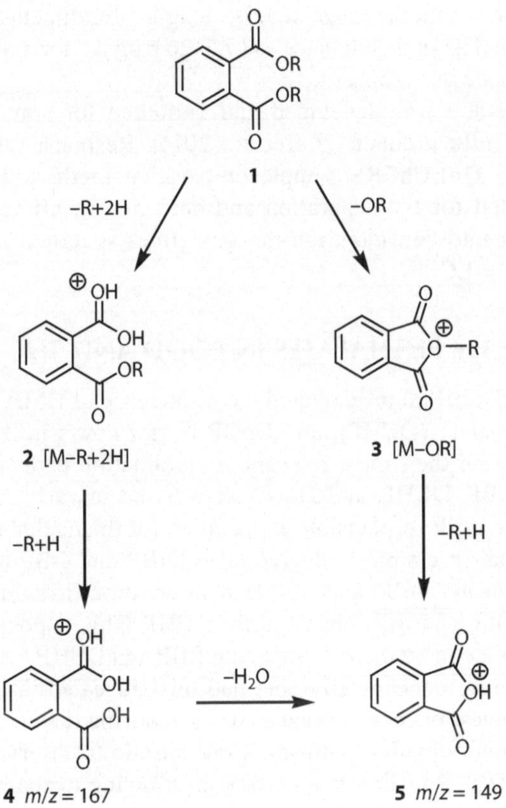

FIGURE 16.3 Possible fragmentation pathways of phthalates under electron impact ionization condition.

various studies. For example, Feng et al. (2005) selected Q-ions of *m/z* 77 and 194 for DMP, *m/z* 177 and 104 for DEP, *m/z* 223 and 104 for DnBP, *m/z* 91 and 206 for BBP, *m/z* 167 and 279 for DEHP, and *m/z* 279 and 104 for DnOP, while Casajuana and Lacorte (2004) selected Q-ions of 77 and 135 for DMP, *m/z* 177 and 105 for DEP, *m/z* 223 and 76 for DnBP, *m/z* 91 and 206 for BBP, and *m/z* 167 and 279 for DEHP.

16.5.3 High-Performance Liquid Chromatography–Mass Spectrometry

In his paper, Hayasaka (2014) describes the development and application of a novel method for the analysis of phthalates in wine by using HPLC-MS/MS combined with a hold-back column. Phthalates are ubiquitous contaminants in the environment and can be widely found in laboratory materials and equipment. An HPLC system is no exception and can be the source of contamination, affecting the accuracy and precision of analytical results. The new method successfully separates phthalates from the different sources, a wine sample, and an HPLC system by a simple technique using an additional HPLC column (a hold-back column) placed upstream of the injection valve. The hold-back column effectively retains the HPLC-derived contaminants during column equilibrium time and delays their elution times from an analytical column. Consequently, a phthalate from a wine sample can be baseline separated as it elutes sufficiently earlier than the same phthalate from the HPLC system. HPLC-MS/MS analysis combined with the hold-back column demonstrated virtually no influence of the HPLC contaminants on the quantification of phthalates present in wine. Together with a simple and rapid sample preparation and the use of labeled internal standards, the method was confirmed to be robust and reliable to determine concentrations of phthalates in wine.

Quantification limits were within the range of 1.6–9.8 μg L^{-1} for dimethyl, diethyl, dibutyl, benzyl butyl, bis(2-ethylhexyl), and dioctyl phthalates and 7.5–26.6 μg L^{-1} for multiple isomeric phthalates, di-iso-nonyl, and di-iso-dodecyl phthalates.

A new analytical method was developed and validated for simultaneous analysis of 27 phthalates in milk and milk products (Jia et al., 2014). Response surface methodology was employed to optimize a QuEChERS sample preparation method. UHPLC/ESI quadrupole Orbitrap HRMS was used for the separation and detection of all the analytes. The method was validated by taking into consideration the guidelines specified in Commission Decision 2002/657/EC and 2007/19/EC.

16.6 OCCURRENCE OF PHTHALATES IN FOOD AND THE ENVIRONMENT

The study of Fierens et al. (2012a) investigated the occurrence of DMP, DEP, DiBP, DnBP, BBP, DEHP, dicyclohexyl phthalate (DCHP), and DnOP in raw cow's milk and feed from Belgian farms in order to determine their most relevant contamination pathways in milk. Measurable levels of DMP, DEP, DnBP, DCHP, and DnOP were found in various feed samples, although they were not observed in milk. A plausible explanation for this is that they are rapidly metabolized in cows. DEHP and, in a smaller degree, also DiBP and BBP levels in milk seemed to vary across seasons and farms. DiBP and BBP levels were lower in summer than in winter milk, which was in contrast with what was observed for DEHP. This is possibly due to another feed composition during summer and winter. Comparing BBP and DEHP concentrations in manually obtained milk with those in mechanically obtained milk revealed that, besides environmental contamination via feed ingestion, contact materials used during the mechanical milking process are another important contamination pathway. Concentrations observed in this study confirm the decreasing trend of DEHP in European cow's milk owing to the substitution of DEHP by other plasticizers.

Food products can be contaminated with toxic compounds via the environment (Fierens et al., 2012b). Another possibility of food contamination is that toxicants are generated in foods or that chemicals migrate from food contact materials into foods during processing. The effect of cooking at home on the levels of phthalates in various food types (starchy products, vegetables, and meat and fish) was examined. Eight compounds were considered, namely DMP, DEP, DiBP, DnBP, BBP, DEHP, DCHP, and DnOP. Food products were analyzed before as well as after cooking (boiling, steaming, [deep-]frying, or grilling). In general, phthalate concentrations in foods declined after cooking, except in vegetables, where almost no effect was seen. Several factors influenced the degree of this decline (e.g., weight difference, fat uptake). Of all phthalates, DEHP, DiBP, and BBP were affected the most. In conclusion, cooking at home definitely affected phthalate concentrations in foods and thus needs to be considered in order to correctly assess humans' dietary exposure to these contaminants.

The presence of eight phthalate compounds—DMP, DEP, DiBP, DnBP, BBP, DEHP, DCHP, and DnOP—was investigated in 400 food products, divided over 11 groups, and packages sold on the Belgian market (Fierens et al., 2012c). For this purpose, suitable extraction techniques were developed and validated for four different matrices, namely, high-fat foods, low-fat food products, aqueous-based beverages, and packaging materials. The instrumental analysis was performed by means of GC-EI-low-resolution MS. A wide variety of phthalate concentrations was observed in the different groups. DEHP was found in the highest concentration in almost every group. Moreover, DEHP was the most abundant phthalate compound, followed by DiBP, DnBP, and BBP.

A study was conducted with a population of 25 German subjects aged between 15 and 21 months. Overall, 16 phthalates and DEHA were measured by GC-MS in a total of 171 duplicate diet samples collected over seven consecutive days, and 20 phthalate metabolites were analyzed in the urine samples collected over seven consecutive days by using an LC-MS/MS method (Fromme et al., 2013).

The median "high" daily dietary intake based on 95th percentiles was 4.66 μg kg^{-1} body weight (b.w.) for DEHP, 1.03 μg kg^{-1} b.w. for DiBP, 0.70 μg kg^{-1} b.w. for DnBP, and 1.0 μg kg^{-1} b.w. for DEHA. The high daily total intake from biomonitoring data was 4.9 μg kg^{-1} b.w. for DEHP, 2.2 μg kg^{-1} b.w. for DnBP, 3.9 μg kg^{-1} b.w. for DiBP, and 2.6 μg kg^{-1} b.w. for di-iso-nonyl phthalate.

The comparison of the two intake estimates indicates that the dominant intake source of DEHP was food ingestion, whereas other sources considerably contributed to the total intake of other phthalates. Using our high-intake scenario, none of the analyzed phthalates reached the recommended tolerable daily intake levels.

Sixty-eight sediment samples collected from Dongjiang River, Xijiang River, Beijiang River, and Zhujiang River in the Pearl River delta (PRD) region, Southern China, were analyzed for 16 PAEs (Liu et al., 2014).

PAEs were detected in all riverine sediments analyzed, which indicates that PAEs are ubiquitous environmental contaminants. Elevated PAE concentrations in riverine sediments in the PRD region were found in the highly urbanized and industrialized areas. Of the 16 PAEs, DiBP, DnBP, and DEHP dominated the PAEs. Influenced by local sources and the properties of PAEs, a gradient trend of concentrations and a fractionation of composition from more to less industrialized and urbanized areas were discovered.

The extensive utilization of phthalate-containing products has lead to ubiquitous contamination of PAEs in various matrices. However, comprehensive knowledge of their pollution in Chinese farmland and associated risks is still limited. In this study, 15 PAEs were determined in soils from agricultural fields throughout mainland China (Niu, 2014). The major source of PAEs in arable soils was associated with the application of agricultural plastic films, followed by the activities for soil fertility.

Takeuchi et al. (2014) measured 59 compounds, including plasticizers (phthalates, adipates, and others) and flame retardants (organophosphates and brominated compounds), from indoor air samples from six houses in Sapporo. These compounds were measured separately in the gas phase and the particle phase using a two-stage cartridge equipped with a quartz fiber filter (1 μm mesh) and C18 SPE disk for sampling and analyzed by GC-MS and LC-MS/MS (for the detection of brominated flame retardants). Among the 59 compounds measured in this study, 34 compounds were detected from the indoor air of the six houses. The highest concentration among the 34 compounds found in a newly built house was 2,2,4-trimethyl-1,3-pentanediol monoisobutyrate at 20.8 μg m^{-3}. Di(2-ethyl-1-hexyl)terephthalate (DEHT), which has been used in recent years as an alternative to di(2-ethyl-1-hexyl) phthalate (DEHP), was found in all six houses, although at low concentrations ranging from 0.005 to 0.027 μg m^{-3}.

A better understanding of the occurrence and degradation in the environment and during wastewater treatment processes of phthalates will facilitate the development of strategies to reduce these pollutants and to bioremediate contaminated freshwater and soil (Gao and Wen, 2016). PAEs occur at measurable levels in different environments worldwide. For example, the concentrations of DMP in atmospheric particulate matter; freshwater; and sediments, soil, and landfills are not detected (ND)–10.4 ng m^{-3}, ND–31.7 μg L^{-1}, ND–316 μg kg^{-1} dry weight, and ND–200 μg kg^{-1} dry weight, and ND–43.27 μg L^{-1}, respectively. DEHP and DnBP are primary PAE pollutants. Urbanization has increased the discharge of PAEs to atmospheric and aquatic environments, and the use of agricultural plastics has exacerbated soil contamination by PAEs in rural areas. Aerobic biodegradation is the primary manner of PAE mineralization in the environment, and this process has been widely studied. PAEs can be removed during wastewater treatment processes.

16.7 NONVOLATILE AND ALTERNATIVE COMPOUNDS

The objective of the study of Canellas et al. (2010) was to identify nonvolatile compounds as potential migrants from adhesives used in food packaging. A number of the current acrylic adhesive formulations were extracted and prepared for analysis. The extracts were screened using a UPLC-TOFMS detector. This approach allowed the identification of several components by a combination of exact mass and in-source collision-induced dissociation. Due to the lack of freely available information on adhesive formulations, further analyses were undertaken using UPLC coupled to high-definition MS. Using the MassFragment™ tool to interrogate fragmentation data, a wide series of compounds was identified, demonstrating the usefulness and importance of these tools for difficult problems. Moreover, using several packaging materials containing adhesives, qualitative migration tests were performed with Tenax® as a food simulant. Several nonvolatile compounds were identified as well in the Tenax, which emphasizes the importance of this work and demonstrates that even the nonvolatile compounds have the potential to migrate into food which is in contact with packaging materials. The main characteristics of the screening study and the results obtained are shown and discussed.

Alternative plasticizers to PAEs have been used for over a decade, but data regarding emissions, human exposure, and health effects are limited. Bui et al. (2016) reviewed 20 alternative plasticizers in current use and their human exposure, hazard, and risk. Most substances are hydrophobic with low volatility and are produced in high volumes for use in multiple applications. There is an increasing trend in the total use of alternative plasticizers in Sweden compared to common PAEs in the last 10 years, especially for di-iso-nonyl-1,2-cyclohexanedicarboxylate (DINCH). Evaluative indoor fate modeling reveals that most alternatives are distributed to vertical surfaces (e.g., walls or ceilings). Only 2,2,4-trimethyl-1,3-pentanediol diisobutyrate and glyceryl triacetate are predicted to be predominantly distributed to indoor air. Human exposure data are lacking, and clear evidence for human exposure exists only for DEHT and DINCH, which show increasing trends in body burdens. Human intake rates are collected and compared with limit values with resulting risk ratios below 1 except for infant's exposure to epoxidized soybean oil. Polybutylene terephthalate properties of the alternatives indicate mostly no reasons for concern, except that tri(2-ethylhexyl) phosphate is estimated to be persistent and tricresyl phosphate to be toxic.

REFERENCES

Arthur, C.L., Pawliszyn, J. 1990. Solid-phase microextraction with thermal desorption using fused silica optical fibers, *Analytical Chemistry* 62: 2145.

Aurela, B., Kulmala, H., Söderhjelm, L. 1999. Phthalates in paper and board packaging and their migration into Tenax and sugar. *Food Additives and Contaminants* 16: 571.

Bouma, K., Schakel, D.J. 2002. Migration of phthalates from PVC toys into saliva simulant by dynamic extraction. *Food Additives and Contaminants* 19: 602.

Bruns-Weller, E., Pfordt, J. 2000. Determination of phthalic acid esters in foods, mother's milk, dust and textiles. *Umweltwissenschaften und Schadstoff-Forschung* 12: 125.

Bui, T.T., Giovanoulis, G., Palm Cousins, A., Magnér, J., Cousins, I.T. et al. 2016. Human exposure, hazard and risk of alternative plasticizers to phthalate esters. *Science of the Total Environment* 541: 451–467.

Butte, W., Heinzow, B. 2002. Pollutants in house dust as indicators of indoor contamination. *Reviews of Environmental Contamination and Toxicology* 175: 1.

Canellas, E., Nerín, C., Moore, R., Silcock, P. 2010. New UPLC coupled to mass spectrometry approaches for screening of non-volatile compounds as potential migrants from adhesives used in food packaging materials. *Analytica Chimica Acta* 666: 62–69.

Cao, L. 2008. Determination of phthalates and adipate in bottled water by headspace solid-phase microextraction and gas chromatography/mass spectrometry. *Journal of Chromatography A* 1178: 231–238.

Cao, X.-L. 2010. Phthalate esters in foods: Sources, occurrence, and analytical methods. *Comprehensive Reviews in Food Science and Food Safety* 9: 21–43.

Carrillo, J.D., Salazar, C., Moreta, C., Tena, M.T. 2007. Determination of phthalates in wine by headspace solid-phase microextraction followed by gas chromatography-mass spectrometry: Fibre comparison and selection. *Journal of Chromatography A* 1164(1–2): 248–261.

Casajuana, N., Lacorte, S. 2003. Presence and release of phthalic esters and other endocrine disrupting compounds in drinking water. *Chromatogrgraphia* 57: 649.

Casajuana, N., Lacorte, S. 2004. New methodology for the determination of phthalate esters, bisphenol A, bisphenol A diglycidyl ether, and nonylphenol in commercial whole milk samples. *Journal of Agricultural and Food Chemistry* 52: 3702.

Castle, L., Mercer, A.J., Gilbert, J. 1988a. Migration from plasticized films into foods: 4. Use of polymeric plasticizers and lower levels of di-(2-ethylhexyl)adipate plasticizer in PVC films to reduce migration into foods. *Food Additives and Contaminants* 5: 277.

Castle, L., Mercer, A.J., Startin J.R., Gilbert, J. 1988b. Migration from plasticized films into foods: 3. Migration of phthalate, sebacate, citrate and phosphate esters from films used for retail food packaging. *Food Additives and Contaminants* 5(1): 9–20.

Castle, L., Gibert, J., Eklund, T. 1990. Migration of plasticizer from poly(vinyl chloride) milk tubing. *Food Additives and Contaminants* 7: 591.

Cavaliere, B., Macchione, B., Sindona, G., Tagarelli, A. 2008. Tandem mass spectrometry in food safety assessment: The determination of phthalates in olive oil. *Journal of Chromatography A* 1205: 137–143.

CDC. 1994. *NIOSH Manual of Analytical Methods*, 4th ed. US Government Printing Office, Washington, DC.

Colón, I., Caro, D., Bourdony, C.J., Rosario, O. 2000. Identification of phthalate esters in the serum of young Puerto Rican girls with premature breast development. *Environmental Health Perspectives* 108(9): 895–900.

Cousins, I., Mackay, D. 2000. Correlating the physical-chemical properties of phthalate esters using the "three solubility" approach. *Chemosphere* 41: 1389.

di Bella, G., Saitta, M., Pellegrino, M., Salvo, F., Dugo, G. 1999. Contamination of Italian citrus essential oils: Presence of phthalate esters. *Journal of Agricultural and Food Chemistry* 47(3): 1009–1012.

do Nascimento Filho, I., von Mühlen, C., Schossler, P., Caramão, E.B. 2003. Identification of some plasticizers compounds in landfill leachate. *Chemosphere* 50(5): 657–663.

Du, Q., Shen, L., Xiu, L., Jerz, G., Winterhalter, P. 2006. Di-2-ethylhexyl phthalate in the fruits of *Benincasa hispida*. *Food Additives and Contaminants* 23(6): 552–555.

Duty, S.M., Silva, M.J., Barr, D.B., Brock, J. 2003. Phthalate exposure and human semen parameters. *Epidemiology* 14(3): 269–277.

Fankhauser-Noti, A., Grob, K. 2007. Blank problem in trace analysis of diethyhexyl and dibutyl phthalate: Investigation of the sources, tips and tricks. *Analytica Chimica Acta* 582: 353–360.

Fauser, P., Thomsen, M. 2002. Sensitivity analysis of calculated exposure concentrations and dissipation of DEHP in a topsoil compartment: The influence of the third phase effect and dissolved organic matter (DOM). *Science of the Total Environment* 296: 89.

Feng, Y.-L., Zhu, J., Sensenstein, R. 2005. Development of a headspace solid phase microextraction method combined with gas chromatography mass spectrometry for the determination of phthalate esters in cow milk. *Analytica Chimica Acta* 538: 41.

Fierens, T., Van Holderbeke, M., Willems, H., De Henauw, S., Sioen, I. 2012a. Phthalates in Belgian cow's milk and the role of feed and other contamination pathways at farm level. *Food and Chemical Toxicology* 50: 2945–2953.

Fierens, T., Vanermen, G., Van Holderbeke, M., De Henauw, S., Sioen, I. 2012b. Effect of cooking at home on the levels of eight phthalates in foods. *Food and Chemical Toxicology* 50: 4428–4435.

Fierens, T., Servaes, K., Van Holderbeke, M., Geerts, L., De Henauw, S. et al. 2012c. Analysis of phthalates in food products and packaging materials sold on the Belgian market. *Food and Chemical Toxicology* 50: 2575–2583.

Fromme, H., Gruber, L., Schuster, R., Schlummer, M., Kiranoglu, M. et al. 2013. Phthalate and di-(2-ethylhexyl) adipate (DEHA) intake by German infants based on the results of a duplicate diet study and biomonitoring data (INES 2). *Food and Chemical Toxicology* 53: 272–280.

Fukuwatari, T., Suzuki, Y., Sugimoto, E., Shibata, K. 2002. Elucidation of the toxic mechanism of the plasticizers, phthalic acid esters, putative endocrine disrupters: Effects of dietary di(2-ethylhexyl)phthalate on the metabolism of tryptophan to niacin in rats. *Bioscience, Biotechnology, and Biochemistry* 66(4): 705–710.

Gao, D.-W., Wen, Z.-D. 2016. Phthalate esters in the environment: A critical review of their occurrence, biodegradation, and removal during wastewater treatment processes. *Science of the Total Environment* 541: 986–1001.

Gruber, L., Wolz, G., Piringer, O. 1998. Analysis of phthalates in baby food. *Deutsche Lebensmittel-Rundashau* 94: 177.

Hayasaka, Y. 2014. Analysis of phthalates in wine using liquid chromatography tandem mass spectrometry combined with a hold-back column: Chromatographic strategy to avoid the influence of pre-existing phthalate contamination in a liquid chromatography system. *Journal of Chromatography A* 1372: 120–127.

He, M., Yang, C., Geng, R., Zhao, X., Hong, L. et al. 2015. Monitoring of phthalates in foodstuffs using gas purge microsyringe extraction coupled with GC-MS. *Analytica Chimica Acta* 879: 63–68.

Higuchi, T.T., Palmer, J.S., Gray Jr, L.E., Rao Veeramachaneni, D.N. 2003. Effects of dibutyl phthalate in male rabbits following in utero, adolescent, or postpubertal exposure. *Toxicological Sciences* 72(2): 301–313.

Hogberg, J., Hanberg, A., Berglund, M., Skerfving, S., Remberger, M. et al. 2008. Phthalate diesters and their metabolites in human breast milk, blood or serum, and urine as biomarkers of exposure in vulnerable populations. *Environmental Health Perspectives* 116(3): 334–339.

Holadova, K.G., Prokůpková, G., Hajšlová, J., Poustka, J. 2007. Headspace solid-phase microextraction of phthalic acid esters from vegetable oil employing solvent based matrix modification. *Analytica Chimica Acta* 582(1): 24–33.

Inoue, K., Higuchi, T., Okada, F., Iguchi, H., Yoshimura, Y. et al. 2003. The validation of column-switching LC/MS as a high-throughput approach for direct analysis of di(2-ethylhexyl) phthalate released from PVC medical devices in intravenous solution. *Journal of Pharmaceutical and Biomedical Analysis* 31(6): 1145–1152.

Jia, W., Chu, X., Ling, Y., Huang, J., Chang, J. 2014. Analysis of phthalates in milk and milk products by liquid chromatography coupled to quadrupole Orbitrap high-resolution mass spectrometry. *Journal of Chromatography A* 1362: 110–118.

Jobling, S., Reynolds, T., White, R., Parker, M.G., Sumpter, J.P. 1995. A variety of environmentally persistent chemicals, including some phthalate plasticizers, are weakly estrogenic. *Environmental Health Perspectives* 103(6): 582–587.

Kato, K., Yoshimura, Y., Nakazawa, H. 2002. Preparation of samples for gas chromatography/mass spectrometry analysis of phthalate and adipate esters in plasma and beverages by steam distillation and extraction. *Journal of AOAC International* 85: 719.

Kozyrod, R.P., Ziaziaris, J. 1989. A survey of plasticizer migration into foods. *Journal of Food Protection* 52: 578.

Lamb IV, J.C., Chapin, R.E., Teague, J., Lawton, A.D., Reel, J.R. 1987. Reproductive effects of four phthalic acid esters in the mouse. *Toxicology and Applied Pharmacology* 88(2): 255–269.

Lambert, N., Rostock, C., Hetland, S. 2010. *Environmental Risks Associated to Plasticizers and Sweeteners in the Nordic Countries*. The Nordic Screening Group, Oslo.

Lin, Z.-P., Ikonomou, M.G., Jing, H., Mackintosh, C., Gobas, F.A.P.C. 2003. Determination of phthalate ester congeners and mixtures by LC/ESI-MS in sediments and biota of an urbanized marine inlet. *Environmental Science & Technology* 37(10): 2100–2108.

Liu, H., Cui, K., Zeng, F., Chen, L., Cheng, Y. et al. 2014. Occurrence and distribution of phthalate esters in riverine sediments from the Pearl River Delta region, South China. *Marine Pollution Bulletin* 83: 358–365.

Luks-Betlej, K., Popp, P., Janoszka, B., Paschke, H., 2001. Solid-phase microextraction of phthalates from water. *Journal of Chromatography A* 938(1–2): 93–101.

Montuori, P., Jover, E., Morgantini, M., Bayona, J.M., Triassi, M. 2008. Assessing human exposure to phthalic acid and phthalate esters from mineral water stored in polyethylene terephthalate and glass bottles. *Food Additives and Contaminants* 25: 511–518.

National Environmental Methods Index; Analytical, test and sampling methods. http://www.nemi.gov.

Niu, L., Xu, Y., Xu, C., Yun, L., Liu, W. 2014. Status of phthalate esters contamination in agricultural soils across China and associated health risks. *Environmental Pollution* 195: 16–23.

Otake, T., Yoshinaga, J., Yanagisawa, Y. 2001. Analysis of organic esters of plasticizer in indoor air by GC-MS and GC-FPD. *Environmental Science & Technology* 35: 3099.

Page, B.D., Lacroix, G.M. 1992. Studies into the transfer and migration of phthalate esters from aluminium foil paper laminates to butter and margarine. *Food Additives and Contaminants* 9: 197.

Page, B.D., Lacroix, G.M. 1995. The occurrence of phthalate ester and di-2-ethylhexyl adipate plasticizers in Canadian packaging and food sampled in 1985–1989: A survey. *Food Additives and Contaminants* 2: 129.

Peñalver, A., Pocurull, E., Borrull, F., Marcé, R.M. 2000. Determination of phthalate esters in water samples by solid-phase microextraction and gas chromatography with mass spectrometric detection. *Journal of Chromatography A* 872(1–2): 191–201.

Peñalver, A., Pocurull, E., Borrull, F., Marcé, R.M. 2001. Comparison of different fibres for the solid-phase microextraction of phthalate esters from water. *Journal of Chromatography A* 922(1–2): 377–384.

Petersen, J.H. 1991. Survey of di-(2-ethylhexyl)phthalate plasticizer contamination of retail Danish milks. *Food Additives and Contaminants* 8: 701.

Petersen, J.H., Breindahl, T. 2000. Plasticizers in total diet samples, baby food and infant formulae. *Food Additives and Contaminants* 17: 133.

Polo, M., Llompart, M., Garcia-Jares, C., Cela, R. 2005. Multivariate optimization of a solid-phase microextraction method for the analysis of phthalate esters in environmental waters. *Journal of Chromatography A* 1072(1): 63–72.

Poon, R., Lecavalier, P., Mueller, R., Valli, V.E., Procter, B.G., Chu, I. 1997. Subchronic oral toxicity of di-n-octyl phthalate and di(2-ethylhexyl) phthalate in the rat. *Food and Chemical Toxicology* 35: 225.

Rudel, R.A., Camann, D.E., Spengler, J.D., Korn, L.R., Brody, J.G. 2003. Phthalates, alkylphenols, pesticides, polybrominated diphenyl ethers, and other endocrine-disrupting compounds in indoor air and dust. *Environmental Science & Technology* 37(20): 4543–4553.

Serodio, P., Nogueira, J.M.F. 2006. Consideration on ultra-trace analysis of phthalates in drinking water. *Water Research* 40: 2572.

Sharman, M., Read, W.A., Castle, L., Gilbert, J. 1994. Levels of di-(2-ethylhexyl) phthalate and total phthalate ester in milk, cream, butter and cheese. *Food Additives and Contaminants* 11: 357.

Soerensen, L. 2006. Determination of phthalate in milk and milk products by liquid chromatography/tandem mass spectrometry. *Rapid Communications in Mass Spectrometry* 20: 1135.

Startin, J.R., Sharman, M., Rose, M.D., Parker, Ian., Mercer, A.J. et al. 1987. Migration from plasticized films into foods: 1. Migration of di-(2-ethylhexyl)adipate from PVC films during home-use and microwave cooking. *Food Additives and Contaminants* 4: 385–398.

Suzuki, T., Yaguchi, K., Suzuki, S., Suga, T. 2001. Monitoring of phthalic acid monoesters in river water by solid-phase extraction and GC-MS determination. *Environmental Science & Technology* 35(18): 3757–3763.

Takeuchi, S., Kojima, H., Saito, I., Jin, K., Kobayashi, S. et al. 2014. Detection of 34 plasticizers and 25 flame retardants in indoor air from houses in Sapporo, Japan. *Science of the Total Environment* 491–492: 28–33.

Tsumura, Y., Ishimitsu, S., Saito, I., Sakai, H., Kobayashi, Y. et al. 2001. Eleven phthalate esters and di(2-ethylhexyl) adipate in one-week duplicate diet samples obtained from hospitals and their estimated daily intake. *Food Additives and Contaminants* 18: 449–460.

Tsumura, Y., Ishimitsu, S., Saito, I., Sakai, H., Tsuchida, Y. et al. 2003. Estimated daily intake of plasticizers in 1-week duplicate diet samples following regulation of DEHP-containing PVC gloves in Japan. *Food Additives and Contaminants* 20: 317–324.

Uhde, E., Bednarek, M., Fuhrmann, F., Salthammer, T. 2001. Phthalic esters in the indoor environment—Test chamber studies on PVC-coated wallcoverings. *Indoor Air* 11: 150.

US Department of Labor/Occupational Safety and Health Administration's Index of Sampling and Analytical Methods. http://www.osha.gov/dts/sltc/methods/toc.html.

Wilkinson, C.F., Lamb IV, J.C. 1999. The potential health effects of phthalate esters in children's toys: A review and risk assessment. *Regulatory Toxicology and Pharmacology* 30: 140.

Wypych, G. 2013. *Handbook of Plasticizers*, 2nd ed. Chemtec, Ontario.

Yano, K., Hirosawa, N., Sakamoto, Y., Katayama, H., Moriguchi, T. et al. 2002. Phthalate levels in beverages in Japan and Korea. *Bulletin of Environmental Contamination and Toxicology* 68(4): 463–469.

Yano, K., Hirosawa, N., Sakamoto, Y., Katayama, H., Moriguchi, T. et al. 2005. Phthalates levels in baby milk powder sold in several countries. *Bulletin of Environmental Contamination and Toxicology* 74(2): 373–379.

Zhu, J., Feng Y.L., MacDonald, S., Newhook, R., Marro, L. 2003. Phthalates in indoor air of Canadian residences. *Proceedings of the ISIAQ 7th International Conference*. Vol. 1, pp. 542–547.

Zhu, J., Phillips, S.P., Feng, Y.-L., Yang, X. 2006. Phthalate esters in human milk: Concentration variations over a six-month postpartum time. *Environmental Science & Technology* 40(17): 5276–5281.

Zhu, J., Phillips, S.P., Cao, X.-L. 2010. Chemical contaminants: Phthalates. In: *Handbook of Dairy Foods Analysis*, ed. Nollet, L.M.L., Toldrá, F. CRC Press, Boca Raton, FL, pp. 771–800.

17 Instrumental Analysis of Brominated Flame Retardants

Mohamed Abou-Elwafa Abdallah,
Alin C. Dirtu, and Adrian Covaci

CONTENTS

17.1 INTRODUCTION

The extensive use of brominated flame retardants (BFRs) in various materials (furniture, plastics, electronics equipment, textiles, etc.) has led to the widespread and substantial contamination of the indoor environment, for example, indoor air and dust (Rudel and Perovich, 2009; Harrad et al., 2010; Covaci et al., 2011a). Ingestion and inhalation of indoor dust together with dietary intake are the most important exposure pathways for humans to BFRs, for which adverse health effects have been reported (Harrad et al., 2010). There is still limited knowledge about the mechanisms via which such chemicals incorporated in consumer goods and construction materials migrate into and behave in the indoor environment.

To obtain reliable data regarding the presence, relevance, and contribution of BFRs in consumer goods (plastic materials, textiles, etc.), in indoor matrices (air and dust), in food, and in biological matrices, it is important to employ innovative, suitable, and specific analytical methods. The relatively low levels of BFRs combined with the complexity of the matrices make, sometimes, their identification/quantification processes a challenging task.

Analytical methods for BFRs have shown a rapid development, and they were, in many cases, based on protocols previously established for persistent organic pollutants (POPs), such as PCBs. Although different properties of BFRs (e.g., polarity or vapor pressure) suggest that different procedures should be applied for their analysis, some common approaches can be found depending on the analyte, type of sample, or detection method (de Boer et al., 2001; Covaci et al., 2003, 2007; de Boer and Wells, 2006; Stapleton, 2006). Some BFRs, such as individual hexabromocyclododecane (HBCD) isomers and tetrabromobisphenol A (TBBPA), may require specific analytical approaches due to their particular properties (discussed in detail in Section 17.4). Since the sample preparation has recently been reviewed by the same authors (Dirtu et al., 2013), the present chapter will focus exclusively on the analysis of BFRs by using chromatographic techniques coupled to MS.

This chapter highlights the technical and methodological improvements in the analysis of BFRs with emphasis on the critical factors influencing their detection and accurate measurement. It also presents detailed information on QA issues.

17.2 STATE OF THE ART ON THE INSTRUMENTAL ANALYSIS OF BROMINATED FLAME RETARDANTS

Generally, BFRs, such as PBDEs, have been most commonly analyzed using GC-MS in electron capture negative ionization mode (GC-ECNI-MS) (Covaci et al., 2007, 2011b). GC-MS operated in electron impact ionization mode (GC-EI-MS) might also be used, although with a decreased sensitivity when compared to GC-ECNI-MS. However, due to their physicochemical properties (e.g., vapor pressure or polarity) and to their thermal instability, specific PBDE congeners need special precautions during GC analysis. It was shown that thermal degradation may occur during GC analysis of higher brominated PBDE congeners (especially BDE 209) (Björklund et al., 2004; Dirtu et al., 2008). Therefore, the characteristics of the GC system should carefully be optimized in agreement with the properties of the analytes. These characteristics include column brand, stationary phase, and column length and internal diameter, and the injection technique may have a very strong influence on the accuracy of PBDE analysis (Björklund et al., 2004; Covaci et al., 2007).

The recent restrictions and bans on the use of some PBDE formulations have created an increasing market for the use of novel brominated flame retardants (NBFRs) (Covaci et al., 2011a). Since methods were focused on the analysis of PBDEs, additional data on other BFRs present in samples were initially generated as collateral information (Covaci et al., 2011a). While some NBFRs (e.g., hexachlorocyclopentadienyl-dibromocyclooctane and 1,2-bis(2,4,6-tribromophenoxy)ethane) can be analyzed without major problems together with PBDEs, other NBFRs, such as decabromodiphenylethane (DBDPE), 2-ethylhexyl-2,3,4,5-tetrabromobenzoate (TBB)/di(2-ethylhexyl)-2,3,4,5-tetrabromophthalate (TBPH), and TBBPA-bis(2,3-dibromopropyl ether) (TBBPA-DBPE), can present several difficulties in their analysis, such as thermal decomposition or poor sensitivity (Covaci et al., 2011a; Papachlimitzou et al., 2012).

When GC-MS-based methods could not solve the previously mentioned issues or the need for the measurement of isomers (e.g., HBCDs or 1,2,5,6-tetrabromocyclooctane), the development of LC-MS methods was addressed (Guerra et al., 2011a). Although LC has lower chromatographic resolving power than that of GC, LC-MS and LC-MS-MS provide a good combination for the analysis of HBCDs and TBBPA, in addition to some NBFRs, such as TBB or TBPH (Zhou et al., 2010a, 2010b). LC-based analytical methods allow for isomer-specific determination of HBCDs, as opposed to the total HBCD analysis by GC-MS. Additionally, HBCD enantiomers can be separated and determined using chiral LC-MS methods (de Boer and Wells, 2006). As a consequence, LC-MS not only offers a reliable instrumental technique for accurate measurement of these compounds but also delivers important information for such flame retardants (FRs) and helps to better explain their environmental behavior.

17.3 GAS CHROMATOGRAPHY–MASS SPECTROMETRY ANALYSIS OF BROMINATED FLAME RETARDANTS

17.3.1 Gas Chromatographic Separation of Brominated Flame Retardants

While various injection methods can be used for the analysis of PBDEs, the most common injectors are splitless, on-column, and PTV injection systems. Their advantages and disadvantages stem primarily from their availability, price, detection limits, extract constitution, and discrimination of PBDE congeners on the basis of molecular weight.

Due to the relatively low levels of PBDEs in environmental samples, splitless injection is the preferred technique, but special precautions should be taken in order to minimize thermal degradation and discrimination of higher-molecular weight PBDEs. Therefore, the optimal injector temperature and splitless time should be kept as high as possible (e.g., 325°C and 4 min, respectively) to obtain an increased response factor especially for fully brominated PBDEs (e.g., BDE 209) (Björklund et al., 2004). An attractive way to reduce thermal degradation of PBDEs in an injection system is to apply on-column injection. Its main advantage is that it shortcuts the injector interface between extract and column by delivering the extract directly to the capillary column. As a result, a higher precision for the analysis of BDE 209 was observed when such injection was applied (Björklund et al., 2004). One of the main disadvantages is the need for very clean sample extracts so that potential matrix residues do not reach the capillary column, giving rise to increased noise, peak tailing, retention time shifts, more frequent column trimming, and reduced column lifetime. Both splitless and on-column injection methods are limited to volumes from 1 to 3 μL. In comparison, PTV injection may accommodate up to 125 μL after thorough optimization (Covaci et al., 2002; Björklund et al., 2003a; Tollbäck et al., 2003). Although splitless is the most common injection technique, injections in a properly used PTV system may have some advantages, such as minimal degradation of labile PBDEs (Covaci et al., 2007). Recently, it was shown that especially for the analysis of BDE 209, the use of PTV injector in solvent vent mode may not only positively influence the response factor of BDE 209 but may also reduce the formation of its main thermal degradation product, BDE 207 (Dirtu et al., 2008).

In general, capillary GC columns offer adequate resolution to determine individual PBDE congeners. In order to achieve adequate separation of all PBDE congeners present in the sample and possible interferences, there is a need for sufficiently long columns (30–50 m) with small internal diameters (≤0.25 mm). In case of clean extracts, the use of short narrow bore capillary columns (internal diameter ≤0.1 mm) may also provide satisfactory resolution (Covaci et al., 2002). In order to facilitate the selection of the most suitable GC columns for developing a congener-specific analysis, the elution order of 126 PBDEs was tested on seven different GC stationary phases (Korytár et al., 2005a). Firstly, less coelutions for PBDE congeners with different bromine numbers were observed compared to PCBs. Secondly, a stationary phase-dependent degradation was seen, indicating that column equivalency is not always a suitable criterion for column selection (Björklund et al., 2004; Korytár et al., 2005a). The most efficient column for PBDE congener-specific separation was found to be a DB-XLB (J&W Scientific) column, with a DB-1 (J&W Scientific) column as the second best. However, the latter is preferred for routine analysis due to reduced degradation of hepta- and higher PBDE congeners.

In another study (Wei et al., 2010), the GC retention times of 180 PBDEs were obtained under different operational conditions on two types of commonly used capillary columns, Restek Rtx-1614 and J&W DB-5MS, of different dimensions. This newly created database was used for the identification of mono- to hepta-PBDEs from the retention times of 39 congeners under various GC conditions to facilitate the identification of unknown PBDE peaks for which chemical standards are not available. Later, D'Archivio et al. (2013) predicted the retention of PBDEs in GC by using quantitative structure–retention relationships of 126 PBDEs in six stationary phases of different polarities. The predictive performance of the model was tested on a large external set consisting of retention data of 180 PBDEs collected in four separation conditions different from those considered in model calibration.

Even on the most efficient stationary phases, single-dimension GC cannot separate all PBDE congeners. Of the 126 PBDE congeners that were tested by Korytár et al. (2005a), 55 congeners were involved in 22 coelutions for DB-XLB, which is the most efficient stationary phase (Korytár et al., 2005a). In another study, six column combinations were evaluated for GC×GC separation of PBDEs coupled to μ-ECD or TOFMS detectors (Korytár et al., 2005b). It was concluded that a DB-1 × 007-65HT (Quadrex) combination was the most suitable due to (i) the highest number of PBDE congeners separated, (ii) less decomposition of higher brominated congeners, and (iii) the most suitable maximum operating temperature. With this setup, there were only 17 coeluting pairs involving 35 congeners.

Special precautions are needed for BDE 209 analysis due to its thermal sensitivity and, thus, its higher susceptibility for degradation in the GC system. Therefore, the GC column should be relatively short (<15 m) to reduce the residence time in the system (Covaci et al., 2007). Based on the ability of the low-pressure gas chromatography (LP-GC) technique to elute compounds at lower temperatures compared to those of conventional GC techniques (Ravindra et al., 2008), BDE 209 was analyzed at temperatures below the degradability limit. It was shown that the low elution temperature of BDE 209 (<295°C), combined with its short residence time in GC, leads to minimal thermal degradation (Dirtu et al., 2008). Besides minimal thermal degradation, the analysis of BDE 209 by using the LP-GC system resulted in an analysis time of merely 6.5 min. Baseline separation of 22 PBDE congeners (major components of PBDE technical mixtures) was possible in less than 12 min when using this LP-GC system (Figure 17.1).

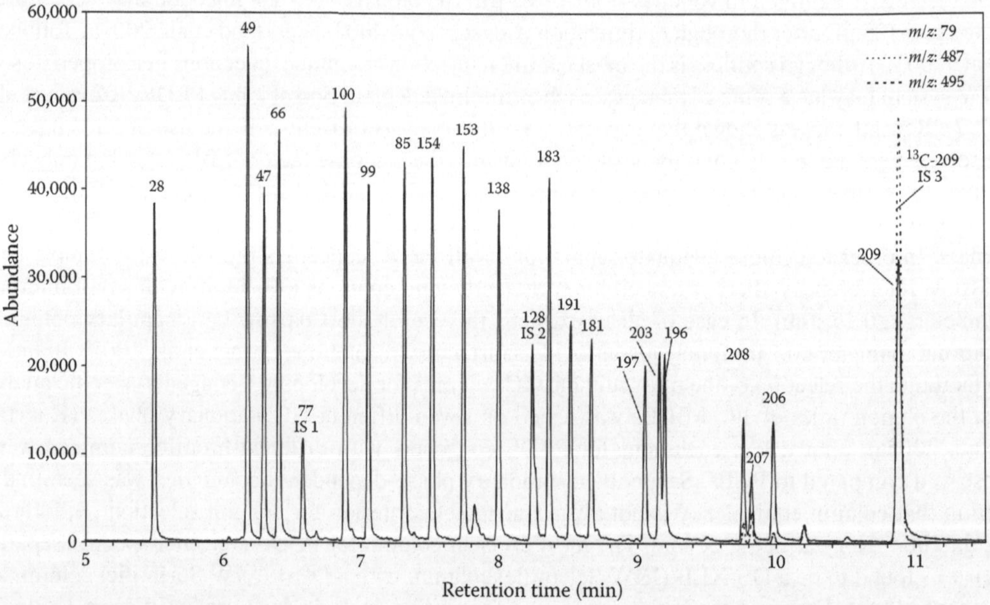

FIGURE 17.1 Mass chromatogram (m/z = 79, 487, and 495) of a standard mixture of 22 PBDE congeners (major components of penta-, octa-, and deca-BDE technical mixtures) injected in the LP-GC-ECNI-MS system (analytical column: 10 m × 0.53 mm internal diameter, df = 0.15 μm AT-5). Peak numbers correspond to PBDE congener nomenclature. (Reprinted from *Journal of Chromatography A*, 1186, Dirtu, A.C., K. Ravindra, L. Roosens, R. van Grieken, H. Neels et al., Fast analysis of decabrominated diphenyl ether using low-pressure gas chromatography—Electron-capture negative ionization mass spectrometry, 295–301, Copyright (2008), with permission from Elsevier.)

17.3.2 Mass Spectrometric Detection for Brominated Flame Retardants

The most widely used detectors for PBDE determination are mass spectrometers, classified into low-resolution (LR) or high-resolution (HR) MS instruments. LRMS instruments are operated in either EI or ECNI mode.

Using EI-MS for PBDE analysis, the major ions formed are $[M]^+$ and $[M{-}2Br]^+$, which can be used for their identification and quantification (Covaci et al., 2003). The main potential interferences that may influence the accuracy originate from chlorinated compounds, such as PCBs. For example, the nominal masses corresponding to ions monitored for di-BDEs and penta-chlorinated biphenyls (CBs) (m/z = 326) and also for tetra-BDEs and hepta-CBs (m/z = 396) are the same, and therefore, a resolution power of 24,000 is needed to separate them if they coelute on the GC column (Covaci et al., 2003). However, this is not recommended due to the significant loss of sensitivity at this elevated resolution (Eljarrat et al., 2003).

LR-EI-MS provides a higher selectivity compared to LR-ECNI-MS, since for the latter, only the bromine trace can be monitored (m/z 79 and 81). However, LR-EI-MS (with quadrupole as mass analyzer) is not routinely used for PBDE analysis due to its relatively low sensitivity, especially when measuring PBDE congeners with more than six bromine atoms. Determination of mono- to deca-PBDE congeners together with polybrominated biphenyls in electrical and electronic equipment was also achieved by GC-EI-MS (Li et al., 2009). Method LODs ranged between 0.2 and 8 ng g^{-1} for mono- to hepta-BDE and 70 ng g^{-1} for BDE 209, respectively.

In contrast to EI, the low-energy electrons (thermal electrons) generated in ECNI by interactions between a high-energy electron beam and a moderating gas (e.g., methane) react with the analytes to form negative ions. The electron energy should be low to facilitate electron capture, and the specific energy required for electron capture depends on the molecular structure of the analyte. PBDEs do not normally form molecular ions or other diagnostic ions under ECNI conditions (except BDE 209); thus, only bromide ions (m/z 79 and 81) may be monitored.

Compared to EI-MS, ECNI-MS is less selective because nonspecific bromide ions $[Br]^-$ at m/z 79 and 81 are used for the quantification of all homologue groups, except BDE 209, which gives rise to specific fragments ($[C_6Br_5O]^-$; m/z 486.7) (Björklund et al., 2003b). However, ECNI-MS is a very sensitive method with LODs that are one order of magnitude lower than those for LR-EI-MS. Therefore, this technique proves to be suitable for the analysis of low-contamination samples such as human serum and milk. Reliable results have also been obtained for the quantification of selected PBDEs in river water and sediment samples. Detection limits ranged from 3 to 160 pg L^{-1} and 5 to 145 pg g^{-1} for river water and sediment, respectively (Labadie et al., 2010).

Selectivity of LR-ECNI-MS can be improved under optimized conditions. By optimizing the electron energy, emission current, source temperature, and system pressure, the relative abundances of the larger molecular fragments $[M{-}xH{-}yBr]^-$ increases, and therefore, the technique can be used for the monitoring of each homologue group instead of the nonspecific bromide ions (Ackerman et al., 2005). Monitoring high-mass fragments under optimized ion source conditions was successfully applied for the determination of PBDEs in environmental and biological samples at concentration levels <0.01 pg in one single instrumental run, which was recently described (Vizcaino et al., 2009). Good repeatability (1.7–9.1%) and reproducibility (4.1–20%) and low LODs (1.5–15 pg mL^{-1}) were obtained, allowing PBDE quantification in snow and human serum samples.

Besides MS operated in either EI or ECNI modes, other MS-type detectors were used to analyze PBDEs. PCBs and PBDEs were simultaneously determined in soil by GC-TOFMS. Method detection limits ranged between 0.1 and 0.4 mg kg^{-1} for PCBs and 0.1 and 0.6 mg kg^{-1} for PBDEs with RSD <7.3% for PCBs and <6.3% for PBDEs (Park et al., 2009). Another multiresidue method for the quantification of 30 organohalogenated compounds in human breast tissue by using GC-QqQ MS has recently been developed. Analyses were performed in both EI and ECNI modes, and the acquisition of two transitions (in EI) or two ions (in ECNI) per analyte was acquired. This approach allowed confirmation through matching of ion ratios between the quantification and the confirmation transitions (EI) or ions (ECNI) (Medina et al., 2009). All PBDE homologues showed a common MS/MS transition consisting of the elimination of two Br atoms. Instrumental LOQs ranged from 0.05 pg μL^{-1} (BDE 47) to 0.5 pg μL^{-1} (BDE 154) for an injection volume of 2 μL.

17.4 LIQUID CHROMATOGRAPHY–MASS SPECTROMETRY ANALYSIS OF BROMINATED FLAME RETARDANTS

The inherent problems encountered with GC-MS analysis due to the high temperatures applied resulted in several difficulties in the analysis of some BFRs, particularly HBCDs, where isomeric interconversion takes place at temperatures >160°C, rendering isomeric separation impossible on GC columns (Marvin et al., 2011). Another problem encountered with GC-MS analysis of BFRs is the thermal decomposition of high-molecular weight compounds (e.g., BDE 209, DBDPE, and TBBPA-DBPE) (Ali et al., 2011). While the use of ^{13}C-BDE 209 as surrogate standard can—to some extent—account for the inevitable on-column degradation of BDE 209 during GC-ECNI/MS analysis, similar approaches could not be achieved for DBDPE and TBBPA-DBPE due to the lack of stable molecular fragments other than $m/z = 81$ in the NCI source (Kierkegaard et al., 2009; Ali et al., 2011). Furthermore, the increased interest in relatively polar compounds, such as TBBPA and hydroxylated PBDE metabolites, meant that derivatization is required prior to their GC-MS analysis, which may result in significant analyte loss and reduced recoveries (Covaci et al., 2009; Lupton et al., 2010). Therefore, LC/MS analysis emerged as an alternative technique to avoid the problems encountered during the analysis of thermolabile and relatively polar compounds by GC-MS.

17.4.1 Hexabromocyclododecane

HBCD is produced via bromination of cyclododeca-1,5,9-triene, resulting in the creation of six stereo centers at positions 1, 2, 5, 6, 9, and 10 of the formed product. This can give rise to a total of 16 possible optical isomers, 6 pairs of enantiomers, and 4 *meso* forms. To date, only three diastereomers—labeled α-, β-, and γ-HBCD—were detected in the technical formulations and environmental samples with minor contributions (up to 0.5%) of two *meso* forms named δ- and ε-HBCDs (Law et al., 2005).

17.4.1.1 Diastereomer-Specific Separation

17.4.1.1.1 Stationary Phase

Tomy et al. (2005) reported on baseline separation of HBCD diastereomers on a C18 reversed-phase column. While a 5 μm particle size column (Vydac 218MS, Mandel Scientific, Guelph, Ontario, Canada) packing is sufficient for baseline separation of α-, β-, and γ-HBCDs, better resolution with sharper peaks was reported using 3 μm particle size (Pursuit XRS3, Agilent, California, United States) (Abdallah et al., 2008a). Shorter retention times (<8 min) with narrower peaks could be achieved with C18 UPLC columns with 1.8 μm particles (Acquity HSS T3, Waters, Massachusetts, United States) (Lankova et al., 2013). Moreover, separation of α-, β-, γ-, δ-, and ε-HBCDs was achieved on a 1.7 μm UPLC column (Acquity UPLC BEH, Waters) (Arsenault et al., 2007).

17.4.1.1.2 Mobile Phase

Several mobile phase gradients using different combinations of methanol/acetonitrile (ACN)/water were reported for separation of HBCD diastereomers (Marvin et al., 2011). While the MS response of an ESI source for 200 pg μL^{-1} solution of HBCDs in methanol was reported to be slightly higher than that in ACN (Budakowski and Tomy, 2003), addition of ACN to the mobile phase (up to 20%) resulted in improved resolution of the three main HBCD diastereomers (mainly between β- and γ-isomers), which is recommended when the two minor *meso* forms (δ- and ε-HBCD) are to be monitored (Arsenault et al., 2007). Different mobile phase modifiers (e.g., ammonium acetate [Abdallah et al., 2008a], ammonium chloride [Gomara et al., 2007], and acetic acid [Cariou et al., 2005]) were reported to produce sharper peaks and improve separation efficiency of HBCDs.

17.4.1.2 Enantiomer-Specific Separation

17.4.1.2.1 Stationary Phase

Only one chiral stationary phase was reported in literature for efficient separation of HBCD enantiomers. Baseline resolution of the six enantiomers from an α-, β-, and γ-HBCD mixture was achieved on β-permethylated cyclodextrin-bonded (NUCLEODEX, Macherey-Nagel, GmbH, Düren, Germany) chiral LC column (4 mm × 200 mm, 5 μm) (Janak et al., 2005). It was observed that (−)-α- and (−)-β-HBCD eluted before their corresponding (+)-α- and (+)-β-HBCD enantiomers (Figure 17.2). These were followed by the γ-enantiomers, with (+)-γ-eluting ahead of (−)-γ-HBCD (Heeb et al., 2005). While the chiral column is sufficient for baseline separation of HBCD enantiomers, Yu et al. (2008a) connected a C18 achiral column to the β-permethylated chiral stationary phase in order to separate the HBCD diastereomers prior to enantiomeric resolution. This provided clear distinction between the respective enantiomers of each HBCD diastereomer in the resulting chromatograms (Figure 17.2).

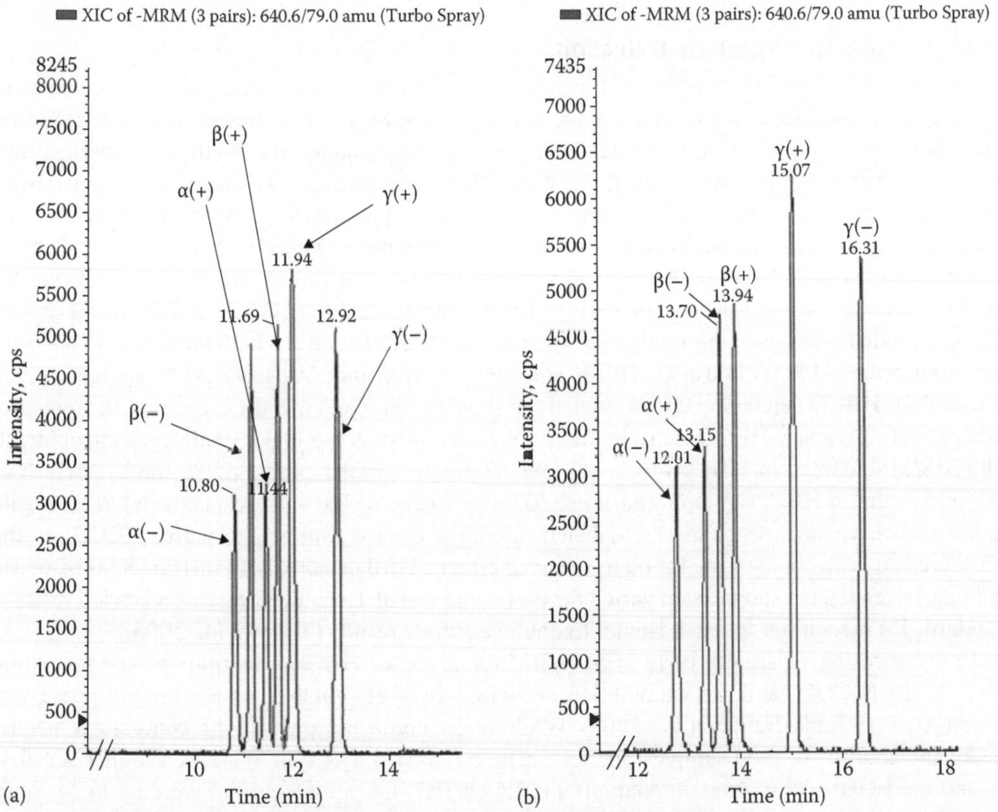

FIGURE 17.2 LC-MS/MS chromatograms showing chiral separation of 50 pg μL^{-1} α-, β-, and γ-HBCDs using (a) NUCLEODEX β-PM chiral LC column and (b) Pursuit XRS3 C18 column followed by NUCLEODEX β-PM chiral LC column. (Reprinted from *Chemosphere*, 61, Heeb, N.V., W.B. Schweizer, M. Kohler, and A.C. Gerecke, Structure elucidation of hexabromocyclododecanes—A class of compounds with a complex stereochemistry, 65–73, Copyright (2005), with permission from Elsevier.)

17.4.1.2.2 Mobile Phase

A combination of methanol/ACN/water in the mobile phase is mandatory for separation of HBCD enantiomers (Harrad et al., 2009a). Interestingly, Marvin et al. (2007) found that both mobile phase composition and column bleed could affect the MS response for different HBCD enantiomers. Dodder et al. (2006) observed that the MS response changed between the elutions of two enantiomers due to the extracted matrix component. In order to avoid such effects on the estimated enantiomeric fraction (EF), Marvin et al. (2007) introduced a mathematical formula for calculation of corrected EF values (Equation 17.1). This correction is based on the use of isotopic-labeled standards (e.g., d_{18}-HBCDs) since the labeled enantiomeric analogues behave identically to their native counterparts in the MS source (Marvin et al., 2007).

$$EF_{corrected} = \frac{\left[\left(A^+/A^+_{labeled}\right)\times\left(pgA^+_{labeled}\right)\right]}{\left[\left(A^+/A^+_{labeled}\right)\times\left(pgA^+_{labeled}\right)\right]+\left[\left(A^-/A^-_{labeled}\right)\times\left(pgA^-_{labeled}\right)\right]}, \quad (17.1)$$

where A^+ is the peak area of the (+) enantiomer, $A^+_{labeled}$ is the peak area of the labeled (+) enantiomer, $pgA^+_{labeled}$ is the mass of labeled isomer added in picograms, A^- is the peak area of the (–) enantiomer, $A^+_{labeled}$ is the peak area of the labeled (–) enantiomer, and $pgA^+_{labeled}$ is the mass of labeled isomer added in picogram.

17.4.1.3 Mass Spectrometric Detection

Several MS techniques were reported for detection of HBCDs. Morris et al. (2006) applied both single quadrupole MS and ion trap MS for detection of HBCDs in sediment and biota samples. Although HBCD molecular ion ($[M-H]^-$; *m/z* = 640.7) was monitored in both techniques, differences in instrumental response to the three studied HBCD diastereomers were observed. α-HBCD recorded the highest response when using the single quadrupole MS, while the ion trap MS was most sensitive to γ-HBCD. Nevertheless, the use of tandem mass (MS/MS) detection in QpQ mass spectrometers provided high sensitivity and very low LODs (≤1 pg on column) for all HBCD diastereomers by using the ion transition $[M-H]^- \rightarrow Br^-$ (Covaci et al., 2007). While an ESI source in negative ion mode is the most commonly used interface for HBCD analysis, both atmospheric pressure photoionization (APPI) (Zhou et al., 2010b) and APCI (Suzuki and Hasegawa, 2006) sources proved as useful for HBCD detection. Matrix-related ion suppression issues and differences in the response factors to α-, β-, and γ-HBCD diastereomers were identified as the major challenges encountered with MS/MS analysis of HBCDs by using ESI. These drawbacks can be overcome by the use of ^{13}C- or d_{18}-labeled HBCDs (monitored at *m/z* 652.4 → 79 and 657.6 → 79, respectively) as surrogate and/or recovery standards. The mass-labeled standards behave similarly to native HBCDs in the ion source and can compensate for matrix-related effects. Furthermore, both instrumental response and matrix-related ion suppression varied for α-, β-, and γ-HBCDs, indicating that a labeled internal standard (IS) is required for each isomer to obtain accurate results (Tomy et al., 2005).

To enhance the sensitivity of LC-ESI-MS/MS analysis of HBCD enantiomers, the formation of Cl^- and CH_3COO^- adducts via addition of NH_4Cl and CH_3COONH_4 to the mobile phase was investigated (Galindo-Iranzo et al., 2009). While both approaches presented a comparable behavior for the analysis of food samples, the Cl^- method (*m/z* 676.6 → 640.6) showed higher sensitivity and the LODs (0.2–0.4 pg on column) and LOQs (0.7–1.4 pg on column) were up to 14 times lower than those obtained applying the CH_3COO^- method (*m/z* 700.6 → 640.6). Another interesting approach involves the use of anion attachment atmospheric pressure photoionization (AA-APPI), with 1,4-dibromobutane in toluene as a bromide source for analysis of HBCDs (*m/z* 722.6 → 79) in sediment samples. This method offered increased sensitivity and lower LODs than those of APPI. Furthermore, minimal matrix effects were found with AA-APPI in sediment extracts, providing a major advantage over ESI-based methods (Ross and Wong, 2010).

17.4.2 Tetrabromobisphenol A

Avoiding the derivatization step of phenolic OH groups required prior to GC-MS analysis of TBBPA was not the only advantage gained by using LC/MS for determination of this BFR. Another advantage was the possible use of ^{13}C-labeled TBBPA as an IS, which greatly improves the quality of analytical data obtained via compensation for matrix-related effects that can affect analyte ion intensity (Covaci et al., 2009).

17.4.2.1 Liquid Chromatography Separation

17.4.2.1.1 Stationary Phase

Several studies have reported the use of C18 reversed-phase (RP) columns with various dimensions and particle sizes for analysis of TBBPA (Covaci et al., 2009). In general, TBBPA is rarely measured alone and is usually included in multiresidue analytical methodology for analysis of various BFRs (Guerra et al., 2011a). Guerra et al. (2008) applied a Symmetry C18 column (2.1 mm × 150 mm, 5 μm) preceded by a C18 guard column (2.1 mm × 10 mm) supplied by Waters (Milford, MA, USA) for baseline separation of TBBPA and related compounds bisphenol A (BPA), monobromobisphenol A, dibromobisphenol A, and tribromobisphenol A in sewage sludge and sediment samples. Application of UPLC columns (Acquity HSS T3, 100 mm × 2.1 mm, 1.8 μm, Waters) resulted in a shorter retention time (6.5 min) than that of HPLC columns (Lankova et al., 2013).

17.4.2.1.2 Mobile Phase

Chu et al. (2005) reported that by using methanol as the mobile phase, the LC-ESI-MS/MS response factor for TBBPA was ~1/3 higher than when ACN was used due to a more stable detector baseline. Similar results were recently reported by Lankova et al. (2013) using the Turbo V ion source for UPLC-MS/MS analysis of TBBPA in fish samples. Therefore, multiresidue analytical methods for determination of TBBPA with other BFRs applied only methanol/water mobile phase gradients (Abdallah et al., 2008b; Guerra et al., 2008; Abdallah and Harrad, 2011; Lankova et al., 2013).

17.4.2.2 Mass Spectrometric Detection

Unlike for HBCDs, an ESI source in negative ion mode is the most suitable interface for TBBPA analysis, with LC-MS providing 30–40 times lower LODs than those obtained by APCI (Tollbäck et al., 2006). Therefore, LC-ESI-MS/MS in negative ion mode was widely reported for determination of TBBPA concentrations in various environmental matrices via monitoring of the mass transitions corresponding to $[M-H]^- \rightarrow Br^-$ (*m/z* 542.6 → 79 and 552.6 → 79 for native and ^{13}C-TBBPA, respectively) (Covaci et al., 2009).

Although early studies were focused on the use of QpQ mass spectrometers, the high selectivity of ion trap MS was applied for the determination of TBBPA in sediment and sewage sludge, scanning the range from *m/z* 145–543 after LC separation (Saint-Louis and Pelletier, 2004). Guerra et al. (2008) described a method based on LC-QqLIT-MS for separation and quantification of TBBPA and other brominated bisphenols together with α-, β-, and γ-HBCD diastereomers in sewage sludge and sediment samples. The reported method displayed excellent LODs in SRM mode (0.1–1.8 pg), but even better results were obtained in enhanced product ion mode (0.01–0.5 pg). Interestingly, desorption atmospheric pressure photoionization–MS in negative ion mode was applied successfully for analysis of TBBPA in circuit board and orange peel samples by using anisole as spraying solvent. This method displayed the advantages of minimal sample treatment and low LOD (0.3 ng g^{-1}) (Luosujarvi et al., 2010). Recently, a different approach was adopted for analysis of TBBPA in plasma and serum samples using LC-ESI(+)-MS/MS. The method is based on derivatization of TBBPA in the extracts by using dansyl chloride reagent. The dansylated derivatives are then monitored at *m/z* 505.9 → 171.1 and 512.9 → 171.1 for native and ^{13}C-TBBPA with a method LOQ as low as 0.03 ng g^{-1} (Chu and Letcher, 2013).

17.4.3 Polybrominated Diphenyl Ethers and Novel Brominated Flame Retardants

Fewer studies have reported on the analysis of PBDEs and NBFRs using LC/MS techniques. This may be attributed to the presence of well-established, sensitive, and efficient protocols for analysis of these hydrophobic compounds using GC-MS techniques. However, LC/MS analysis can provide a major advantage for analysis of heavy-molecular weight BFRs (e.g., BDE 209 and DBDPE), which may undergo thermal degradation and/or extensive fragmentation during the course of GC-MS analysis (Covaci et al., 2007). LC/MS methods in ESI mode may have limited use for PBDEs due to poor ionization in this source (Marteau et al., 2012). Abdallah et al. (2009) reported an isotope dilution method using ^{13}C-labeled ISs for quantification of 14 major tetra- to deca-PBDEs using LC-negative ion (NI)-APPI/MS/MS. The 14 PBDEs were baseline separated on a C18 RP column (Pursuit XRS3, 250 mm × 4.6 mm, 3 μm, Agilent) by using mobile phase gradient of methanol/toluene and water (Figure 17.3). The method applied the soft photoionization technique to obtain stable pseudo-molecular ions $[M{-}Br{+}O]^-$ and $[M{-}2Br{+}O]^-$ in Q1, which enabled the use of isotopically labeled ISs for quantification. The method was then successfully applied for analysis of PBDEs in dust (Abdallah et al., 2009), air (Abdallah and Harrad, 2010), and human milk samples (Abdallah and Harrad, 2014). Zhou et al. (2010b) developed a sensitive and high-throughput LC-NI-APPI-MS/MS method for the analysis of 36 BFRs including PBDEs, HBCDs, TBBPA, and several NBFRs in fish samples. The method used an Ultra II C18 column (100 mm × 2.1 mm, 2.2 μm, Restek, Pennsylvania, United States) operated at 25°C for separation of target compounds with a methanol/water mobile phase gradient at a flow rate of 400 μL min^{-1}. In comparison with acetone, toluene provided around 10% higher ion intensity for less hydrophobic compounds. Three categories of precursor ions were observed in the APPI source: (1) displacement products, for example, $[M{-}Br{+}O]^-$ and $[M{-}HBr{-}Br{+}O_2]^-$; (2) elimination products, for example, $[M{-}H]^-$; and (3) association product, for example, $[M{+}O_2]^-$. The dominant precursor ion used for quantification of the studied BFRs was $[M{-}Br{+}O]^-$ (Zhou et al., 2010b). In another study, an APCI source was investigated by the same authors for determination of 38 BFRs in wastewater samples. For MS/MS detection, relatively high collision energy was required to produce abundant Br^- product ions, and the authors suggested that increasing the collision gas pressure may generate more of these ions. The method was simple, sensitive, and applicable to compounds with a wide range of physicochemical properties (Zhou et al., 2010a). Maşcolo et al. used a C18 BEH column (150 mm × 2.1 mm, 1.7 μm) for separation of 11 tetra- to deca-PBDEs. The column was kept at 40°C, while separation was achieved using a methanol/water gradient. Method LODs as low as 3–198 and 4–380 pg g^{-1} were reported for APCI and APPI (toluene as dopant) sources, respectively. Depending on the PBDE congener, the APCI source was two to eight times more sensitive than APPI (Mascolo et al., 2010).

Letcher and Chu reported the application of LC-NI-APPI-MS/MS for quantification of TBBPA-S-DBPE, TBBPA-AE, and TBBPA-DBPE in herring gull eggs. Target analytes were separated on a ZOBRAX SB-C18 column (2.1 mm × 30 mm, 3.5 μm). The method depends on the use of acetone as both the organic solvent in the mobile phase and the doping agent for the APPI source. The studied compounds were quantified via monitoring *m/z* 997.4 → 79, 655.8 → 79, and 975.5 → 79 for TBBPA-S-DBPE, TBBPA-AE, and TBBPA-DBPE, respectively, corresponding to the $[M{+}O_2]^- \rightarrow Br^-$ transition (Letcher and Chu, 2010). More recently, TBBPA-AE and TBBPA-DBPE were analyzed by APCI-MS/MS after separation on a C18 column (150 mm × 2.1 mm, 5 μm) by using a methanol/water mobile phase. The studied compounds were monitored at *m/z* 582.9 → 526.5 and 742.7 → 526.5 for TBBPA-AE and TBBPA-DBPE, respectively. Method LOD ranged from 10 to 30 pg g^{-1} in various environmental samples (Qu et al., 2013). Finally, Arsenault et al. (2008) reported an LC-ESI-MS method for analysis of tetrabromoethylcyclohexane (TBECH) isomers. TBECH is a novel BFR which has four thermolabile diastereomers that can interconvert at temperatures ≥125°C. Incomplete separation of the four isomers was performed on a UPLC BEH C18 column (2.1 mm × 100 mm, 1.7 μm) with a

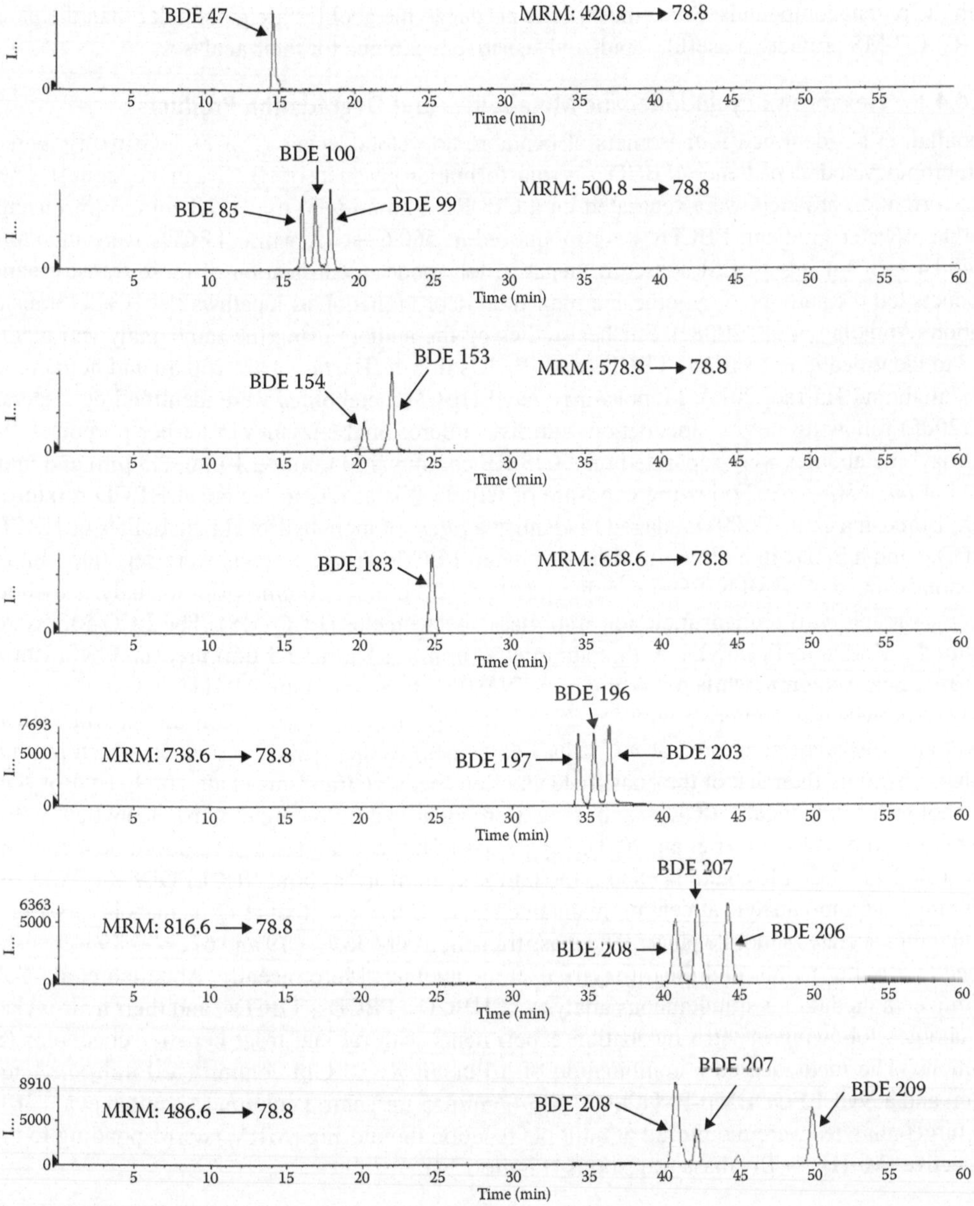

FIGURE 17.3 Separation of tetra- to deca-PBDEs on RP-C18 column (250 mm × 4.6 mm; 3 μm). (Reprinted with permission from Abdallah et al. (2009), 7460–7467. Copyright 2009 American Chemical Society.)

methanol/ACN/water gradient. Analytes were detected via monitoring Br^- ions in SIM mode due to the lack of molecular ion formation in an ESI source.

17.4.4 Brominated Flame Retardant Metabolites

The mounting scientific interest in BFRs in the past few years has resulted in an increasing number of studies on their fate and behavior in the environment and humans. This leads to the development of analytical methodologies to monitor BFR metabolites and transformation products together with the parent compounds. Since most of the produced metabolites are more polar than the parent BFRs, LC-MS provides a useful, rapid, and sensitive technique for their analyses.

17.4.4.1 Hexabromocyclododecane Metabolites and Degradation Products

Abdallah et al. identified four isomers of pentabromocyclododecene (PBCD) and two isomers of tetrabromocyclododecadienes (TBCDs) as transformation products of HBCDs in indoor dust. These transformation products were separated on a C18 RP column (150 mm × 2.1 mm, 3 μm) using a methanol/water gradient. PBCDs were monitored at 560.6 → 79, while TBCDs were monitored at 480.4 → 79 using an ESI source in negative ion mode. Identification of these transformation products led the authors to hypothesize sequential debromination as a pathway of HBCD transformation (Abdallah et al., 2008c). Further studies by the authors using the same analytical method lead to identification of various TBCDs and PBCDs in fish (Harrad et al., 2009b) and human milk (Abdallah and Harrad, 2011). Monohydroxylated HBCD metabolites were identified by Zegers et al. (2005) following in vitro incubation with liver microsomal enzymes of harbor porpoises. The hydroxyl metabolites were separated on a C18 RP column (150 mm × 2.1 mm, 3.5 μm) and monitored at *m/z* 656 → 79. Following exposure of female Wistar rats to technical HBCD mixture in feed, Brandsma et al. (2009) managed to identify a range of monohydroxyl metabolites of HBCDs, PBCDs, and TBCDs in addition to dihydroxylated PBCD. Tissue extracts were separated into 17 fractions using a μPorasil NP-HPLC column (10 μm, 7.8 mm × 300 mm) prior to analysis by liquid chromatograph with a quadrupole ion trap mass spectrometer (LCQ-MS). The LCQ-MS system featured a Zorbax eclipse XDB-C18 column (150 mm × 2.1 mm, 3.5 μm) preceded by a Zorbax XDB-C_8 guard column, while a mixture of ACN/0.01 mM ammonium chloride was used as mobile phase. The MS was equipped with an ESI source operated in negative ion mode, while all the target analytes were monitored at *m/z* values equivalent to their $[M+Cl]^-$ adduct, which provided higher sensitivity than that of the quasimolecular ion species (Brandsma et al., 2009). Hydroxylated metabolites of individual HBCD enantiomers were identified following in vitro incubation with rat liver microsomes (Esslinger et al., 2011). Separation of target analytes was achieved on a combination of Zorbax XDB-C18 column (1.8 μm, 150 mm × 4.6 mm) and a chiral NUCLEODEX β-PM (5 μm, 200 mm × 4.6 mm) analytical column maintained at 15°C using a mixture of methanol/ACN/10 mM ammonium acetate as mobile phase. The mass transitions of 656.7 → 79 and 672.6 → 79 were established to monitor mono- and dihydroxy-HBCD metabolites. More recently, Abdallah et al. (2014) optimized a method for simultaneous analysis of HBCDs, PBCDs, TBCDs, and their hydroxylated metabolites following in vitro incubation experiments with rat and trout hepatic subcellular (S9) fractions. The method used a combination of a Pursuit XRS3 C18 column (150 mm × 2.1 mm, 3 μm) and a NUCLEODEX β-PM (200 mm × 4.6 mm, 5 μm) chiral column (Abdallah et al., 2014). All target analytes were monitored at multiple reaction monitoring (MRM) corresponding to their respective $[M-H]^- \rightarrow Br^-$ mass transitions (Figure 17.4).

17.4.4.2 Polybrominated Diphenyl Ether Metabolites

Introduction of one or more hydroxyl groups to PBDEs can result in the formation of more toxic metabolites due to close structural similarity to the thyroid hormones (Song et al., 2008). Hydroxyl PBDE metabolites are nonvolatile, relatively polar compounds which require derivatization prior to GC-MS analysis. GC-MS methods for analysis of OH-PBDE metabolites must include a derivatization step with diazomethane, which needs to be handled with care due to its explosive characteristics. Furthermore, the efficiency of the derivatization step varies from sample to sample, since the reaction may give a yield less than 100%. Finally, additional sample preparation or cleanup steps could introduce errors and lengthen analysis time (Lupton et al., 2010). Therefore, LC-MS is the method of choice for rapid, fast, and sensitive analysis of these compounds (Guerra et al., 2011b). It was reported that ionization of PBDEs and their metabolites by an ESI source is poor (Yu et al., 2008b). Therefore, focus has shifted to the application of APCI and APPI sources for their ionization. Hydroxylated and methoxylated metabolites of tetra-PBDEs were analyzed in marine

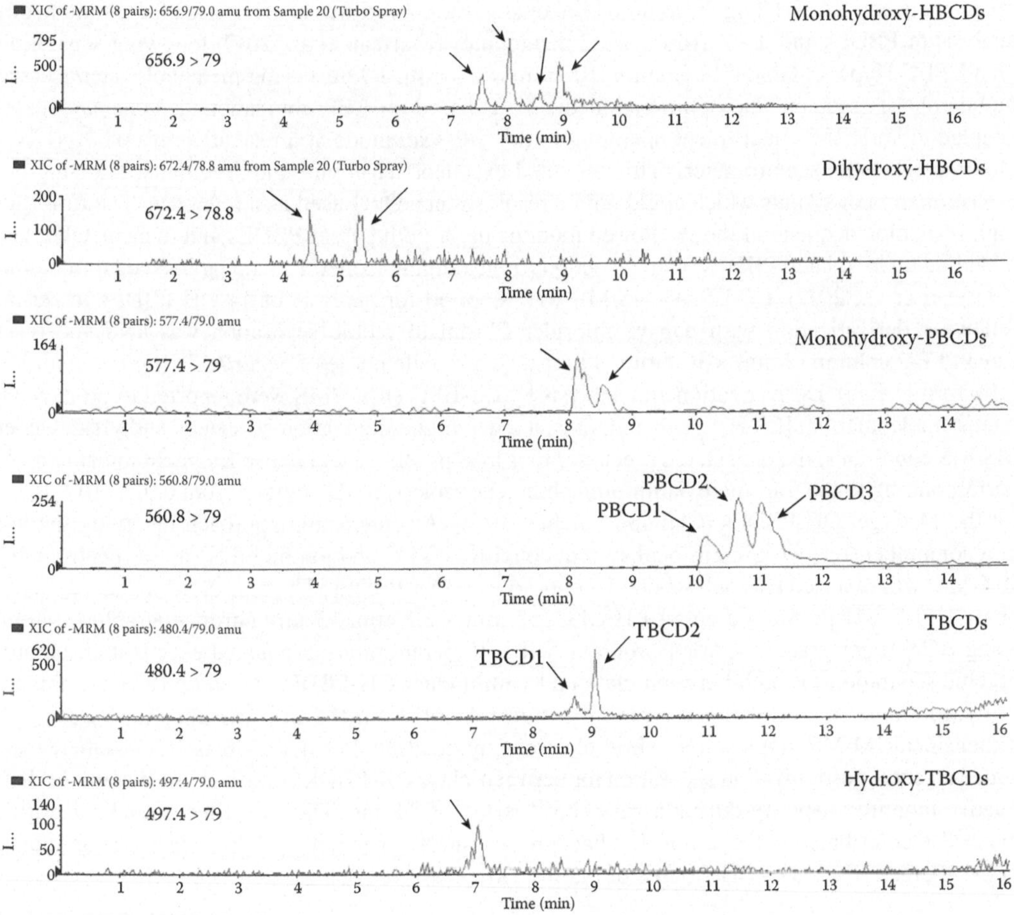

FIGURE 17.4 Selected LC-ESI-MS/MS chromatograms of HBCD metabolites in rat S9 fractions following exposure to 10 μM of α-, β-, and γ-HBCDs for 60 min. (Reprinted with permission from Abdallah et al. (2014), 2732–2740. Copyright 2014 American Chemical Society.)

biota using LC-APCI-MS/MS in negative ion mode. Chromatographic separation was performed on a C18 analytical column (150 mm × 4.6 mm, 3 μm) with ACN/water mobile phase. MRM was performed using the precursor $[M-H]^-$ ion for hydroxylated analogues and the $[M-Br+O]^-$ ion for tetra-PBDEs and their methoxylated analogues. Method LOQs ranged from 0.11 to 43 ng g^{-1} lipid weight (Kato et al., 2009). Nine OH-PBDEs, ranging from tri- to hexabrominated, were separated and quantified using a similar LC-APCI-MS/MS method. Notably, a significant decrease in ionization was observed in 6-OH-substituted PBDE metabolites with ortho-substituted bromine, relative to the other hydroxylated metabolites. This was attributed to the formation of dioxins as a result of high-temperature conditions in the APCI source, which prevented ionization by hydrogen abstraction. The MS/MS experiments also provided evidence of the neutral losses of HBr and Br_2, indicating the possible use of neutral loss scanning and SRM for screening of brominated metabolites (Lupton et al., 2010). LC-ESI-QqLIT-MS-MS in negative mode method was developed for the determination of 11 OH-tri- to OH-hexa-PBDEs (Feo et al., 2013). The optimal conditions for proper chromatographic separation of the studied OH-PBDE congeners were the following: Purospher® STAR RP-18 endcapped column (125 mm × 2 mm, 5 μm) working at pH = 10, using ACN and 3:2 water/methanol as the mobile phase. SRM was used in order to increase sensitivity using transitions corresponding to $[M-H]^- \rightarrow Br^-$ for all target metabolites. Instrumental LOQs ranged between 0.6 and 2 pg on column (Feo et al., 2013). APPI was also reported for simultaneous analysis of PBDEs and their hydroxylated metabolites (Marteau et al., 2012) following separation on a UPLC Hypersil Gold C18 column (100 mm × 2.1 mm, 1.9 μm) using methanol/water/acetone mobile phase gradient. The optimized method was based on APPI ionization (acetone as dopant) coupled to an HRMS instrument operating in the full-scan mode at a resolution of 60,000 (LTQ-Orbitrap XL mass spectrometer). This provided excellent sensitivity and specificity, allowing the discrimination of signals which could not be resolved on a QpQ used as a reference. The full-scan high-resolution acquisition mode allowed monitoring of both parent PBDEs and their metabolites, including hydroxylated PBDEs, with detection limits ranging from 0.1 to 4.5 pg injected on-column (Marteau et al., 2012). LC-ESI(+)-MS/MS was reported for analysis of 14 OH-PBDEs in serum following derivatization with dansyl chloride. Chromatographic separation was achieved on a Luna PFP-2 column (2 mm × 100 mm, 3 μm) with a mobile phase of water/ACN (both containing 0.1% formic acid). Derivatization and analysis by LC-ESI (+)-MS/MS were reported to produce an intense molecular ion $[M+H]^+$ peak and, thus, a much higher ionization efficiency and yield. Under MS/MS conditions, the dansylated precursor ions also produced an intense fragment ion at *m/z* 171 corresponding to the 5-(dimethylamino)-naphthalene moiety. LODs ranged from 0.01 to 014 ng g^{-1} for the 14 target OH-PBDEs (Chu and Letcher, 2013). An interesting approach involving the use of a comprehensive two-dimensional system coupling UPLC and ion mobility–mass spectrometry (IM-MS) was reported for analysis of 23 mono- to octa-OH-PBDEs. The first-dimensional reversed-phase UPLC was performed on a BEH C18 (150 mm × 2.1 mm, 1.7 μm) chromatographic column using ACN/water gradient elution program with a flow rate ramp. It enabled excellent chromatographic separation for both between-class and within-class OH-PBDEs based on their differences in hydrophobicity. Following the preionization resolution in the first dimension, the second-dimensional IM-MS employed a hybrid electrospray quadrupole IM-TOF mass spectrometer and added an extra postionization separation for between-class OH-PBDE congeners on account of their relative mobility disparity during a very short period of 8.8 ms. The two-dimensional separation plane also contributed to the removal of background interference ions and the enhanced confidence in the characterization of OH-PBDEs of interest (Ma et al., 2011).

17.5 PARAMETERS AFFECTING ACCURACY IN THE ANALYSIS OF BROMINATED FLAME RETARDANTS: QUALITY ASSURANCE/QUALITY CONTROL

One of the most important aspects influencing the instrumental measurement of BFRs is the commercial availability of the carbon-labeled standards for a range of these compounds. While labeled ISs are limited to be used for PBDE analysis when GC-ECNI-MS is applied (except for the case of ^{13}C-BDE 209) since generally only $[Br]^-$ ions are monitored (*m/z* = 79 and 81), the use of GC-EI-MS allows the use of multiple labeled ISs, albeit with lower sensitivity (Covaci et al., 2007). Sample preparation protocols might be responsible for a large percentage of the total method's uncertainty. The use of labeled ISs for the target compounds would be of a significant help in reducing potential problems, which might occur during both sample preparation (in order to correct for recovery losses) and instrumental analysis (to avoid potential interferences due to the matrix influence). This becomes of paramount importance when addressing the LC-MS methods using an ESI source for BFR analysis. Since an ESI source is prone to sample matrix effects, the use of unlabeled ISs might seriously affect the analysis results (Dodder et al., 2006; Marvin et al., 2007). In some cases, extensive sample preparation is needed for BFR analysis. This requires the use of both ISs and syringe (or recovery) standards in order to compensate for losses of the target analytes during sample preparation and correct for interinjection fluctuations (Covaci et al., 2011b). Appropriate standards should be chosen especially when labeled ISs are not commercially available. They should be absent from standard mixtures and environmental samples. Moreover, unlabelled ISs should also be checked for potential overlap (retention time and mass spectrum) with native target compounds.

Although sample preparation techniques for BFR analysis are not discussed in this chapter, another aspect which might have a significant influence on the accurate measurements of these contaminants is related to the use of properly cleaned glassware. Besides blank problems which might be generated, rather old glassware presenting surface active sites could influence the recoveries and, finally, the measurement of these compounds. Additionally, an audit to evaluate the analytical protocol in terms of the number of glassware involved in analysis should be performed in order to keep it to a minimum, thus reducing the potential influences on the method validation parameters. For example, BDE 209 might strongly absorb to glassware while it may not be redissolved completely, especially when using rather old glassware containing surface active sites. Addition of a solvent keeper (e.g., *iso*-octane) before final concentration of the extract/solution might significantly improve the recovery factor for the analytes, especially when addressing the analysis of more volatile BFRs when evaporation to complete dryness of the extracts should be avoided.

Besides contamination problems which might occur in each of the analysis steps, other processes may also have a strong influence on the method accuracy, namely, the capacity of some FRs to undergo photolytic degradations, especially in the UV region. It was proven that BDE 209 results might be affected by the light conditions used in the laboratory due to photolytic debromination processes in which it may be involved. In addition, the intensity of such processes might also be dependent on the matrix to be analyzed, on the solvent used during sample preparation procedures, or on the nature of the sunlight (continuous or discontinuous) to which the samples or the extracts are exposed (Söderström et al., 2004). As a consequence, the use of UV filters at laboratory windows or fluorescent lightings combined with amber glassware or even glass covered with aluminum foil would significantly reduce the chances of the analyte to undergo photolytic degradation.

The accurate measurement of some BFRs might also be influenced by several potential chromatographic coelutions already reported in the literature. For example, when performing GC-MS methods operated in EI mode, potential interferences originate from the copresence of the chlorinated compounds (namely, PCBs) in the extract for the analysis of PBDEs. Therefore, the nominal masses corresponding to di-BDEs are similar to those corresponding to penta-CBs ($m/z = 326$) and also masses for tetra-BDEs are similar to the ones for hepta-CBs. Special attention should also be given to the coelution reported for 30 m DB-5 capillary columns (5% phenyl–dimethyl polysiloxane) between BDE 47 and CB 180, BDE 100 and CB 194, or BDE 28 and BDE 33, although some of these congener separations are achievable by the use of an MS-type detector (Alaee et al., 2001; Ikonomou et al., 2002). Similarly, on DB-5 or CP-Sil 8 capillary columns, BDE 154 coelutes with BB 153 (Covaci et al., 2008). While this is resolved by the use of an HT-8 (8% phenyl polycarborane siloxane) column, another partial coelution is present, BDE 85 and BDE 154 (Covaci et al., 2003). Chromatographic interferences might also be encountered between PBDEs and some naturally produced brominated compounds (Vetter et al., 2002), whereas BDE 47 and BDE 99 may coelute with breakdown products of HBCDs (Voorspoels, 2006).

For the BFRs more susceptible to analysis by LC-based methods (MS- or MS/MS-type detectors), several pitfalls can also hamper results of good quality, such as coelution between target analytes and matrix components, leading finally to ion suppression (Tomy et al., 2005). Therefore, quantitative evaluation of the matrix effects should specially be addressed when applying LC-ESI/MS-based methodologies for the analysis of BFRs. In order to minimize the impact of such effects to be encountered, then some precautions should be taken, such as the use of the standard addition method, dilution of the sample, improved sample cleanup, improved chromatographic separation, use of a matrix-matched external standards, or spiking the sample with labeled ISs before injection.

Independent of the laboratory type (accredited, research, etc.) which performs analysis of BFR levels in various environmental matrices, there should be no differences concerning the QA/QC measures implemented. As a general rule that should be kept by regular research laboratories, between 10% and 20% of the analysis time is spent to ensure good quality of the performed analyses

(Päpke et al., 2004). Due to relatively low BFR levels measured sometimes in biological samples, contamination of the procedural blank samples became very important. Not only keeping blank values to a minimum is important, but blank consistency is also essential, generally with an RSD of 30% as a maximum. Properly validated protocols for BFR analysis should therefore be a standard practice in any laboratory dealing with their measurement. While the best procedure to evaluate the methods' accuracy remains the analysis of standard/certified reference materials, it should be noted that only a limited number of such materials are commercially available for BFRs. However, when reference materials are not available, the standard addition method (performed using a control sample spiked at least at two different levels) may still be accepted as a valid approach to evaluating the trueness and precision of the analytical procedures (Päpke et al., 2004; Covaci et al., 2008).

17.6 CURRENT CHALLENGES AND FUTURE PERSPECTIVES

As evidenced by previous experimental work in this research field, there are sufficient similarities in the sample preparation steps as well as in the instrumental setups reported for the analysis of both "traditional" BFRs and NBFRs. As a consequence, the available experience gained originally for the analysis of PBDEs will bring significant help in order to facilitate in the future the analysis of some other BFRs.

Following the environmental chemistry of NBFRs, it is expected that they will share physicochemical properties close to those of some traditional BFRs, such as PBDEs. It is therefore explainable that most of the protocols developed for the instrumental analysis of NBFRs were originally derived from the protocols applied for the PBDE analysis, while sample preparation protocols for the determination of each class of compounds in different environmental or biological matrices are rather similar (Covaci et al., 2011a, 2011b; Papachlimitzou et al., 2012). However, when screening the literature for NBFRs, surprisingly large differences were observed for the reported values of log K_{OW}, vapor pressure, solubility, and boiling points of some compounds (Dirtu et al., 2013). As a consequence, the solubility of NBFRs in organic solvents might differ from those of PBDEs, and therefore, that would necessitate significant alteration of either the sample preparation protocols or the instrumental setup concerning their accurate measurement. The absence of the harmonized literature on the physicochemical characterization of these FRs will lead to a continuous challenge for environmental chemists aiming to develop multiresidue analytical methods (Dirtu et al., 2013). With the increasing number of BFRs on the market sharing such a variability of their characteristics (from nonpolar to rather polar, from low-volatility to rather semivolatile, or from relatively low- to high-molecular weight compounds), complications associated with both sample preparation and instrumental analysis are expected to occur.

Additionally, the limited number of reference materials commercially available for the analysis of BFRs makes the task of their measurement even more difficult. Although spiking experiments are generally accepted as important validation parameters in order to prove the methods' accuracy, organizing interlaboratory comparison exercises would better help laboratories to evaluate the performance of their methodologies.

ACKNOWLEDGMENTS

Dr. Alin C. Dirtu acknowledges financial support from the University of Antwerp, Belgium, and the Research Scientific Foundation—Flanders (FWO). Dr. Adrian Covaci was financially supported by the Research Council of UA and from FWO.

LIST OF ACRONYMS

AA-APPI anion attachment atmospheric pressure chemical ionization
ACN acetonitrile

AE	allyl ether
APCI	atmospheric pressure chemical ionization
APPI	atmospheric pressure photoionization
BFRs	brominated flame retardants
BPA	bisphenol A
C18	octadecylsilane
CBs	chlorinated biphenyls
DBDPE	decabromodiphenylethane
ECNI	electron chemical negative ionization
EF	enantiomeric fraction
EI	electron impact ionization
ESI	electrospray ionization
FRs	flame retardants
GC	gas chromatography
HBCD	hexabromocyclododecane
HR	high resolution
IM-MS	ion mobility–mass spectrometry
IS	internal standard
LC	liquid chromatography
LCQ	liquid chromatograph with a quadrupole ion trap mass spectrometer
LOD	limit of detection
LOQ	limit of quantitation
LP-GC	low-pressure gas chromatography
LR	low resolution
MRM	multiple reaction monitoring
MS	mass spectrometry
NBFRs	novel brominated flame retardants
PBCD	pentabromocyclododecene
PBDEs	polybrominated diphenyl ethers
PCBs	polychlorinated biphenyls
POPs	persistent organic pollutants
PTV	programmed temperature vaporization
QqLIT-MS	quadrupole linear ion trap mass spectrometry
RP	reversed phase
TBB	2-ethylhexyl-2,3,4,5-tetrabromobenzoate
TBBPA	tetrabromobisphenol A
TBCO	1,2,5,6-tetrabromocyclooctane
TBECH	tetrabromoethylcyclohexane
TBPH	di(2-ethylhexyl)-2,3,4,5-tetrabromophthalate

REFERENCES

Abdallah, M.A.E., Harrad, S. 2010. Modification and calibration of a passive air sampler for monitoring vapor and particulate phase brominated flame retardants in indoor air: Application to car interiors. *Environmental Science & Technology* 44: 3059–3065.

Abdallah, M.A.E., Harrad, S. 2011. Tetrabromobisphenol-A, hexabromocyclododecane and its degradation products in UK human milk: Relationship to external exposure. *Environment International* 37: 443–448.

Abdallah, M.A.E., Harrad, S. 2014. Polybrominated diphenyl ethers in UK human milk: Implications for infant exposure and relationship to external exposure. *Environment International* 63: 130–136.

Abdallah, M.A.E., Harrad, S., Ibarra, C., Diamond, M., Melymuk, L. et al. 2008a. Hexabromocyclododecanes in indoor dust from Canada, the United Kingdom, and the United States. *Environmental Science & Technology* 42: 459–464.

Abdallah, M.A.E., Harrad, S., Covaci, A. 2008b. Hexabromocyclododecanes and tetrabromobisphenol-A in indoor air and dust in Birmingham, UK: Implications for human exposure. *Environmental Science & Technology* 42: 6855–6861.

Abdallah, M.A.E., Ibarra, C., Neels, H., Harrad, S., Covaci, A. 2008c. Comparative evaluation of liquid chromatography–mass spectrometry versus gas chromatography–mass spectrometry for the determination of hexabromocyclododecanes and their degradation products in indoor dust. *Journal of Chromatography A* 1190: 333–341.

Abdallah, M.A.E., Harrad, S., Covaci, A. 2009. Isotope dilution method for determination of polybrominated diphenyl ethers using liquid chromatography coupled to negative ionization atmospheric pressure photoionization tandem mass spectrometry: Validation and application to house dust. *Analytical Chemistry* 81: 7460–7467.

Abdallah, M.A.E., Uchea, C., Chipman, J.K., Harrad, S. 2014. Enantioselective biotransformation of hexabromocyclododecane by in vitro rat and trout hepatic sub-cellular fractions. *Environmental Science & Technology* 48: 2732–2740.

Ackerman, L.K., Wilson, G.R., Simonich, S.L. 2005. Quantitative analysis of 39 polybrominated diphenyl ethers by isotope dilution GC/low-resolution MS. *Analytical Chemistry* 77: 1979–1987.

Alaee, M., Backus, S., Cannon, C. 2001. Potential interference of PBDEs in the determination of PCBs and other organochlorine contaminants using electron capture detection. *Journal of Separation Sciences* 24: 465–469.

Ali, N., Harrad, S., Muenhor, D., Neels, H., Covaci, A. 2011. Analytical characteristics and determination of major novel brominated flame retardants (NBFRs) in indoor dust. *Analytical and Bioanalytical Chemistry* 400: 3073–3083.

Arsenault, G., Konstantinov, A., Marvin, C.H., Macinnis, G., McAlees, A. et al. 2007. Synthesis of the two minor isomers, delta- and epsilon-1,2,5,6,9,10-hexabromocyclododecane, present in commercial hexabromocyclododecane. *Chemosphere* 68: 887–892.

Arsenault, G., Lough, A., Marvin, C., McAlees, A., McCrindle, R. et al. 2008. Structure characterization and thermal stabilities of the isomers of the brominated flame retardant 1,2-dibromo-4-(1,2-dibromoethyl) cyclohexane. *Chemosphere* 72: 1163–1170.

Björklund, J., Tollbäck, P., Östman, C. 2003a. Large volume injection GC-MS in electron capture negative ion mode utilizing isotopic dilution for the determination of polybrominated diphenyl ethers in air. *Journal of Separation Sciences* 26: 1104–1110.

Björklund, J., Tollbäck, P., Östman, C. 2003b. Mass spectrometric characteristics of decabromodiphenyl ether and the application of isotopic dilution in the electron capture negative ionization mode for the analysis of polybrominated diphenyl ethers. *Journal of Mass Spectrometry* 38: 394–400.

Björklund, J., Tollbäck, P., Hiarne, C., Dyremarck, E., Östman, C. 2004. Influence of the injection technique and the column system on gas chromatographic determination of polybrominated diphenyl ethers. *Journal of Chromatography A* 1041: 201–210.

Brandsma, S.H., Van Der Ven, L.T.M., de Boer, J., Leonards, P.E.G. 2009. Identification of hydroxylated metabolites of hexabromocyclododecane in wildlife and 28-days exposed Wistar rats. *Environmental Science & Technology* 43: 6058–6063.

Budakowski, W., Tomy, G. 2003. Congener-specific analysis of hexabromocyclododecane by high-performance liquid chromatography/electrospray tandem mass spectrometry. *Rapid Communications in Mass Spectrometry* 17: 1399–1404.

Cariou, R., Antignac, J.P., Marchand, P., Berrebi, A., Zalko, D. et al. 2005. New multiresidue analytical method dedicated to trace level measurement of brominated flame retardants in human biological matrices. *Journal of Chromatography A* 1100: 144–152.

Chu, S., Letcher, R.J. 2013. Halogenated phenolic compound determination in plasma and serum by solid phase extraction, dansylation derivatization and liquid chromatography-positive electrospray ionization–tandem quadrupole mass spectrometry. *Journal of Chromatography A* 1320: 111–117.

Chu, S., Haffner, G.D., Letcher, R.J. 2005. Simultaneous determination of tetrabromobisphenol A, tetrachlorobisphenol A, bisphenol A and other halogenated analogues in sediment and sludge by high performance liquid chromatography–electrospray tandem mass spectrometry. *Journal of Chromatography A* 1097: 25–32.

Covaci, A., de Boer, J., Ryan, J.J., Voorspoels, S., Schepens, P. 2002. Determination of polybrominated diphenyl ethers and polychlorinated biphenyls in human adipose tissue by large-volume injection—Narrow-bore capillary gas chromatography/electron impact low-resolution mass spectrometry. *Analytical Chemistry* 74: 790–798.

Covaci, A., Voorspoels, S., de Boer, J. 2003. Determination of brominated flame retardants, with emphasis on polybrominated diphenyl ethers (PBDEs) in environmental and human samples—A review. *Environment International* 29: 735–756.

Covaci, A., Voorspoels, S., Ramos, L., Neels, H. Blust, R. 2007. Recent developments in the analysis of brominated flame retardants and brominated natural compounds. *Journal of Chromatography A* 1153: 145–171.
Covaci, A., Voorspoels, S., D'Silva, K., Huwe, J., Harrad, S. 2008. Brominated flame retardants as food contaminants. In *Comprehensive Analytical Chemistry*, ed. de la Guardia, M., Armenta, S. Elsevier, Amsterdam, Vol. 51, pp. 507–568.
Covaci, A., Voorspoels, S., Abdallah, M.A., Geens, T., Harrad, S. et al. 2009. Analytical and environmental aspects of the flame retardant tetrabromobisphenol-A and its derivatives. *Journal of Chromatography A* 1216: 346–363.
Covaci, A., Abdallah, M.A.E., Nadeem, A., Harrad, S., Herzke, D. et al. 2011a. Novel brominated flame retardants: A review of their analysis, environmental fate and behaviour. *Environment International* 37: 532–556.
Covaci, A., Dirtu, A.C., Voorspoels, S., Roosens, L., Lepom, P. 2011b. Brominated Flame Retardants, Handbook of Environmental Chemistry 16: 55–94.
D'Archivio, A.A., Giannitto, A., Maggi, M.A. 2013. Cross-column prediction of gas-chromatographic retention of polybrominated diphenyl ethers. *Journal of Chromatography A* 1298: 118–131.
de Boer, J., Allchin, C., Law, R., Zegers, B., Booij, J.P. 2001. Method for the analysis of polybrominated diphenyl ethers in sediments and biota. *Trends in Analytical Chemistry* 20: 591–599.
de Boer, J., Wells, D.E. 2006. Pitfalls in the analysis of brominated flame retardants in environmental, human and food samples—Including results of three international inter-laboratory studies. *Trends in Analytical Chemistry* 25: 364–372.
Dirtu, A.C., Ravindra, K., Roosens, L., van Grieken, R., Neels, H. et al. 2008. Fast analysis of decabrominated diphenyl ether using low-pressure gas chromatography—Electron-capture negative ionization mass spectrometry. *Journal of Chromatography A* 1186: 295–301.
Dirtu, A.C., Abdallah, M.A.E., Covaci, A. 2013. Advances in the sample preparation of brominated flame retardants and other brominated compounds. *Trends in Analytical Chemistry* 43: 189–203.
Dodder, N.G., Peck, A.M., Kucklick, J.R., Sander, L.C. 2006. Analysis of hexabromocyclododecane diastereomers and enantiomers by liquid chromatography/tandem mass spectrometry: Chromatographic selectivity and ionization matrix effects. *Journal of Chromatography A* 1135: 36–42.
Eljarrat, E., de la Cal, A., Barcelo, D. 2003. Potential chlorinated and brominated interferences on the polybrominated diphenyl ether determinations by gas chromatography-mass spectrometry. *Journal of Chromatography A* 1008: 181–192.
Esslinger, S., Becker, R., Maul, R., Nehls, I. 2011. Hexabromocyclododecane enantiomers: Microsomal degradation and patterns of hydroxylated metabolites. *Environmental Science & Technology* 45: 3938–3944.
Feo, M.L., Baron, E., Aga, D.S., Eljarrat, E., Barcelo, D. 2013. Development of a liquid chromatography–electrospray chemical ionization tandem mass spectrometry analytical method for analysis of eleven hydroxylated polybrominated diphenyl ethers. *Journal of Chromatography A* 1301: 80–87.
Galindo-Iranzo, P., Quintanilla-Lopez, J.E., Lebron-Aguilar, R. Gomara, B. 2009. Improving the sensitivity of liquid chromatography–tandem mass spectrometry analysis of hexabromocyclododecanes by chlorine adduct generation. *Journal of Chromatography A* 1216: 3919–3926.
Gomara, B., Lebron-Aguilar, R., Quintanilla-Lopez, J.E., Gonzalez, M.J. 2007. Development of a new method for the enantiomer specific determination of HBCD using an ion trap mass spectrometer. *Analytica Chimica Acta* 605: 53–60.
Guerra, P., De La Torre, A., Martinez, M.A., Eljarrat, E., Barcelo, D. 2008. Identification and trace level determination of brominated flame retardants by liquid chromatography/quadrupole linear ion trap mass spectrometry. *Rapid Communications in Mass Spectrometry* 22: 916–924.
Guerra, P., Covaci, A., Eljarrat, E., Barceló, D. 2011a. Recent methodologies for brominated flame retardant determinations by means of liquid chromatography–mass spectrometry. In *Brominated Flame Retardants, Handbook of Environmental Chemistry*, eds. Eljarrat, E., Barceló, D. Springer, Berlin, Vol. 16, pp. 95–122.
Guerra, P., Eljarrat, E., Barcelo, D. 2011b. Determination of halogenated flame retardants by liquid chromatography coupled to mass spectrometry. *TrAC Trends in Analytical Chemistry* 30: 842–855.
Harrad, S., Abdalla, M.A.E., Covaci, A. 2009a. Causes of variability in concentrations and diastereomer patterns of hexabromocyclododecanes in indoor dust. *Environment International* 35: 573–579.
Harrad, S., Abdallah, M.A.E., Rose, N.L., Turner, S.D., Davidson, T.A. 2009b. Current-use brominated flame retardants in water, sediment, and fish from English lakes. *Environmental Science & Technology* 43: 9077–9083.
Harrad, S., de Wit, C., Abdallah, M.A.E., Bergh, C., Björklund, J. et al. 2010. Indoor contamination with hexabromocyclodecanes, polybrominated diphenyl ethers, and perfluoroalkyl compounds: An important exposure pathway for people? *Environmental Science & Technology* 44: 3231–3231.

Heeb, N.V., Schweizer, W.B., Kohler, M., Gerecke, A.C. 2005. Structure elucidation of hexabromocyclododecanes—A class of compounds with a complex stereochemistry. *Chemosphere* 61: 65–73.

Ikonomou, M.G., Rayne, S., Fischer, M., Fernandez, M.P., Cretney, W. 2002. Occurrence and congener profiles of polybrominated diphenyl ethers (PBDEs) in environmental samples from coastal British Columbia, Canada. *Chemosphere* 46: 649–663.

Janak, K., Covaci, A., Voorspoels, S., Becher, G. 2005. Hexabromocyclododecane in marine species from the Western Scheldt Estuary: Diastereoisomer- and enantiomer-specific accumulation. *Environmental Science & Technology* 39: 1987–1994.

Kato, Y., Okada, S., Atobe, K., Endo, T., Matsubara, F. et al. 2009. Simultaneous determination by APCI-LC/MS/MS of hydroxylated and methoxylated polybrominated diphenyl ethers found in marine biota. *Analytical Chemistry* 81: 5942–5948.

Kierkegaard, A., Sellström, U., McLachlan, M.S. 2009. Environmental analysis of higher brominated diphenyl ethers and decabromodiphenyl ethane. *Journal of Chromatography A* 1216: 364–375.

Korytár, P., Covaci, A., de Boer, J., Gelbin, A., Brinkman, U.A. 2005a. Retention-time database of 126 polybrominated diphenyl ether congeners and two bromkal technical mixtures on seven capillary gas chromatographic columns. *Journal of Chromatography A* 1065: 239–251.

Korytár, P., Covaci, A., Leonards, P.E.G., de Boer, J., Brinkman, U.A. 2005b. Comprehensive two-dimensional gas chromatography of polybrominated diphenyl ethers. *Journal of Chromatography A* 1100: 20–31.

Labadie, P., Tlili, K., Alliot, F., Bourges, C., Desportes, A. et al. 2010. Development of analytical procedures for trace-level determination of polybrominated diphenyl ethers and tetrabromobisphenol A in river water and sediment. *Analytical and Bioanalytical Chemistry* 396: 865–875.

Lankova, D., Kockovska, M., Lacina, O., Kalachva, K., Pulkrabova, J. et al. 2013. Rapid and simple method for determination of hexabromocyclododecanes and other LC-MS-MS-amenable brominated flame retardants in fish. *Analytical and Bioanalytical Chemistry* 405: 7829–7839.

Law, R.J., Kohler, M., Heeb, N.V., Gerecke, A.C., Schmid, P. et al. 2005. Hexabromocyclododecane challenges scientists and regulators. *Environmental Science & Technology* 39: 281A–287A.

Letcher, R.J., Chu, S. 2010. High-sensitivity method for determination of tetrabromobisphenol-S and tetrabromobisphenol-A derivative flame retardants in great lakes herring gull eggs by liquid chromatography–atmospheric pressure photoionization–tandem mass spectrometry. *Environmental Science & Technology* 44: 8615–8621.

Li, Y., Wang, T., Hashi, Y., Li, H., Lin, J.M. 2009. Determination of brominated flame retardants in electrical and electronic equipment with microwave-assisted extraction and gas chromatography-mass spectrometry. *Talanta* 78: 1429–1435.

Luosujarvi, L., Kanerva, S., Saarela, V., Franssila, S., Kostiainen, R. et al. 2010. Environmental and food analysis by desorption atmospheric pressure photoionization–mass spectrometry. *Rapid Communications in Mass Spectrometry* 24: 1343–1350.

Lupton, S.J., McGarrigle, B.P., Olson, J.R., Wood, T.D., Aga, D.S. 2010. Analysis of hydroxylated polybrominated diphenyl ether metabolites by liquid chromatography/atmospheric pressure chemical ionization tandem mass spectrometry. *Rapid Communications in Mass Spectrometry* 24: 2227–2235.

Ma, Q., Wang, C., Bai, H., Xi, H.W., Xi, G.C. et al. 2011. Comprehensive two-dimensional separation of hydroxylated polybrominated diphenyl ethers by ultra-performance liquid chromatography coupled with ion mobility-mass spectrometry. *Journal of the American Society for Mass Spectrometry* 22: 1851–1861.

Marteau, C., Chevolleau, S., Jouanin, I., Perdu, E., De Sousa, G. et al. 2012. Development of a liquid chromatography/atmospheric pressure photo-ionization high-resolution mass spectrometry analytical method for the simultaneous determination of polybrominated diphenyl ethers and their metabolites: Application to BDE-47 metabolism in human hepatocytes. *Rapid Communications in Mass Spectrometry* 26: 599–610.

Marvin, C.H., Macinnis, G., Alaee, M., Arsenault, G., Tomy, G.T. 2007. Factors influencing enantiomeric fractions of hexabromocyclododecane measured using liquid chromatography/tandem mass spectrometry. *Rapid Communications in Mass Spectrometry* 21: 1925–1930.

Marvin, C.H., Tomy, G.T., Armitage, J.M., Arnot, J.A., McCarty, L. et al. 2011. Hexabromocyclododecane: Current understanding of chemistry, environmental fate and toxicology and implications for global management. *Environmental Science & Technology* 45: 8613–8623.

Mascolo, G., Locaputo, V., Mininni, G. 2010. New perspective on the determination of flame retardants in sewage sludge by using ultrahigh pressure liquid chromatography-tandem mass spectrometry with different ion sources. *Journal of Chromatography A* 1217: 4601–4611.

Medina, C.M., Pitarch, E., Portolés, T., López, F.J., Hernández, F. 2009. GC-MS/MS multi-residue method for the determination of organochlorine pesticides, polychlorinated biphenyls and polybrominated diphenyl ethers in human breast tissues. *Journal of Separation Sciences* 32: 2090–2102.

Morris, S., Bersuder, P., Allchin, C.R., Zegers, B., Boon, J.P. et al. 2006. Determination of the brominated flame retardant, hexabromocyclodocane, in sediments and biota by liquid chromatography-electrospray ionisation mass spectrometry. *TrAC Trends in Analytical Chemistry* 25: 343–349.

Papachlimitzou, A., Barber, J.L., Losada, S., Bersuder, P., Law, R.J. 2012. A review of the analysis of novel brominated flame retardants. *Journal of Chromatography A* 1219: 15–28.

Päpke, O., Furst, P., Herrmann, T. 2004. Determination of polybrominated diphenylethers (PBDEs) in biological tissues with special emphasis on QC/QA measures. *Talanta* 63: 1203–1211.

Park, H.M., Hong, S.M., Agustin-Camacho, M.R., Dirwono, W., Lee, K.B. 2009. Pressurized liquid extraction for the simultaneous analysis of polychlorinated biphenyls and polybrominated diphenyl ethers from soil by GC-TOF-MS detection. *Journal of Chromatographic Sciences* 47: 681–688.

Qu, G., Liu, A., Wang, T., Zhang, C., Fu, J. et al. 2013. Identification of tetrabromobisphenol A allyl ether and tetrabromobisphenol A 2,3-dibromopropyl ether in the ambient environment near a manufacturing site and in mollusks at a coastal region. *Environmental Science & Technology* 47: 4760–4767.

Ravindra, K., Dirtu, A.C., Covaci, A. 2008. Low-pressure gas chromatography: Recent trends and developments. *Trends in Analytical Chemistry* 27: 291–303.

Ross, M.S., Wong, C.S. 2010. Comparison of electrospray ionization, atmospheric pressure photoionization, and anion attachment atmospheric pressure photoionization for the analysis of hexabromocyclododecane enantiomers in environmental samples. *Journal of Chromatography A* 1217: 7855–7863.

Rudel, R.A., Perovich, L.J. 2009. Endocrine disrupting chemicals in indoor and outdoor air. *Atmospheric Environment* 43: 170–181.

Saint-Louis, R., Pelletier, E. 2004. LC-ESI-MS-MS method for the analysis of tetrabromobisphenol A in sediment and sewage sludge. *Analyst* 129: 724–730.

Söderström, G., Sellström, U., de Wit, C.A., Tysklind, M. 2004. Photolytic debromination of decabromodiphenyl ether (BDE 209). *Environmental Science & Technology* 38: 127–132.

Song, R.F., He, Y.H., Murphy, M.B., Yeung, L.W.Y., Yu, R.M.K. et al. 2008. Effects of fifteen PBDE metabolites, DE71, DE79 and TBBPA on steroidogenesis in the H295R cell line. *Chemosphere* 71: 1888–1894.

Stapleton, H.M. 2006. Instrumental methods and challenges in quantifying polybrominated diphenyl ethers in environmental extracts: A review. *Analytical and Bioanalytical Chemistry* 386: 807–817.

Suzuki, S., Hasegawa, A. 2006. Determination of hexabromocyclododecane diastereoisomers and tetrabromobisphenol A in water and sediment by liquid chromatography/mass spectrometry. *Analytical Sciences* 22: 469–474.

Tollbäck, P., Björklund, J., Östman, C. 2003. Large-volume programmed-temperature vaporiser injection for fast gas chromatography with electron capture and mass spectrometric detection of polybrominated diphenyl ethers. *Journal of Chromatography A* 991: 241–253.

Tollbäck, J., Crescenzi, C., Dyremark, E. 2006. Determination of the flame retardant tetrabromobisphenol A in air samples by liquid chromatography–mass spectrometry. *Journal of Chromatography A* 1104: 106–112.

Tomy, G.T., Halldorson, T., Danell, R., Law, K., Arsenault, G. et al. 2005. Refinements to the diastereoisomer-specific method for the analysis of hexabromocyclododecane. *Rapid Communication in Mass Spectrometry* 19: 2819–2826.

Vetter, W., Stoll E., Garson, M.J., Fahey, S.J., Gaus, C. et al. 2002. Sponge halogenated natural products found at parts-per-million levels in marine mammals. *Environmental Toxicology and Chemistry* 21: 2014–2019.

Vizcaino, E., Arellano, L., Fernandez, P., Grimalt, J.O. 2009. Analysis of whole congener mixtures of polybromodiphenyl ethers by gas chromatography-mass spectrometry in both environmental and biological samples at femtogram levels. *Journal of Chromatography A* 1216: 5045–5051.

Voorspoels, S. 2006. Environmental distribution of brominated flame retardants in Belgium. PhD thesis, University of Antwerp, Antwerp, Belgium.

Wei, H., Yang, R., Li, A., Christensen, E.R., Rockne, K.J. 2010. Gas chromatographic retention of 180 polybrominated diphenyl ethers and prediction of relative retention under various operational conditions. *Journal of Chromatography A* 1217: 2964–2972.

Yu, Z., Chen, L., Mai, B., Wu, M., Sheng, G. et al. 2008a. Diastereoisomer and enantiomer-specific profiles of hexabromocyclododecane in the atmosphere of an urban city in South China. *Environmental Science & Technology* 42: 3996–4001.

Yu, Y.J., Yang, W.H., Gao, Z.S., Lam, M.H.W., Liu, X.H. et al. 2008b. RP-HPLC measurement and quantitative structure–property relationship analysis of the *n*-octanol–water partitioning coefficients of selected metabolites of polybrominated diphenyl ethers. *Environmental Chemistry* 5: 332–339.

Zegers, B.N., Mets, A., Van Bommel, R., Minkenberg, C., Hamers, T. et al. 2005. Levels of hexabromocyclododecane in harbor porpoises and common dolphins from western European seas, with evidence for stereoisomer-specific biotransformation by cytochrome p450. *Environmental Science & Technology* 39: 2095–2100.

Zhou, S.N., Reiner, E.J., Marvin, C., Helm, P., Riddell, N. et al. 2010a. Development of liquid chromatography atmospheric pressure chemical ionization tandem mass spectrometry for analysis of halogenated flame retardants in wastewater. *Analytical and Bioanalytical Chemistry* 396: 1311–1320.

Zhou, S.N., Reiner, E.J., Marvin, C., Koilic, T., Riddell, N. et al. 2010b. Liquid chromatography–atmospheric pressure photoionization tandem mass spectrometry for analysis of 36 halogenated flame retardants in fish. *Journal of Chromatography A* 1217: 633–641.

18 Naphthenic Acids
Environmental Occurrence and Chromatographic Analysis

Adeola Adenugba, John V. Headley, Kerry Peru, and Dena McMartin

CONTENTS

18.1 INTRODUCTION

Naphthenic acids occur naturally in the environment in oil sands and crude oil deposits. In Canada, the Athabasca deposit forms the world's largest accumulation of hydrocarbon deposit totaling 600 billion barrels, which translates to about four times the size of the largest conventional oil field in Saudi Arabia (Schramm et al., 2000) (Figure 18.1). For this reason, the discussion herein will focus on the Canadian oil sands, to typify the operations and processes involved in the oil sand extractions and their impacts on the environment.

The IUPAC recently defined naphthenic acids (NAs) as mainly monocarboxylic acids which are cyloalkanes, especially cyclopentane, cyclohexane, and their alkyl derivatives (Grewer et al., 2010). However, in recent times, the generic term *naphthenic* has been used to include the complexity of other heteratomic compounds containing sulfur and nitrogen atoms that are present in tailing pond water, as discussed later in this chapter (see Grewer et al., 2010; Headley et al., 2013). The Athabasca,

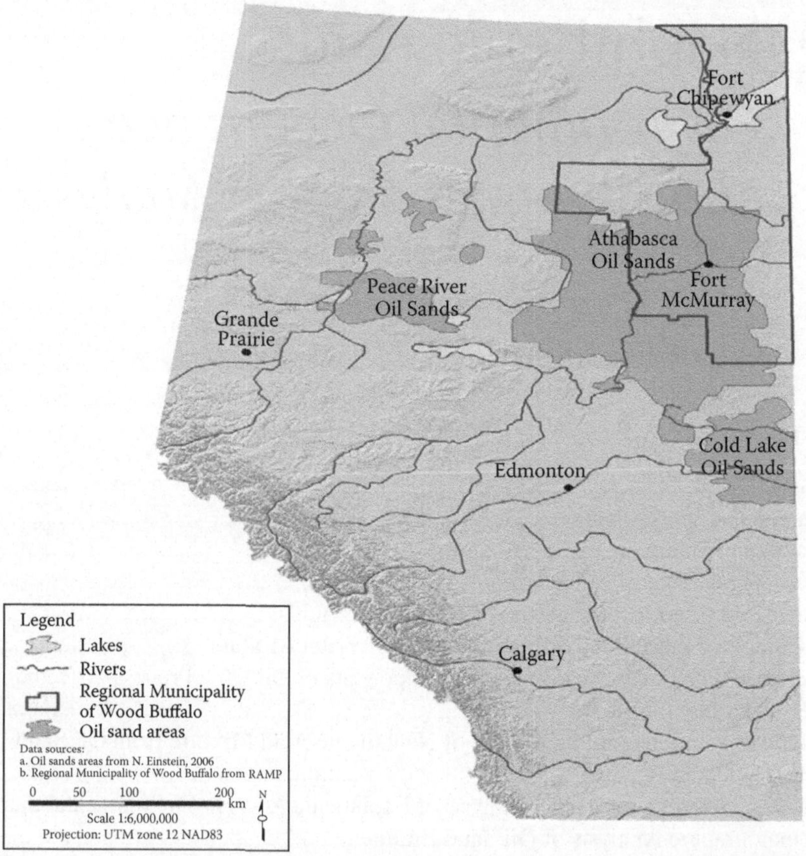

FIGURE 18.1 A map of the oil sand regions of Alberta. (From Einstein, N., Pembina Institute, Calgary, http://www.ramp-alberta.org/resources/gallery.aspx, 2006. With permission.)

Peace River, and Cold Lake oil sands are the three major deposits in Alberta, Canada. The river basin in the lower Athabasca region of Northern Alberta is abundant in natural oil sands, covering an area of 42,000 km^2 in the McMurray Formation (Gentes, 2006; Zhou et al., 2008). Combined, these three deposits cover a total of 140,000 km^2 of the boreal forest in the region (Gentes, 2006; Dimitriu et al., 2010). Other geographic regions of Alberta known as the oil sand deposits of the Western Canada Sedimentary Basin comprise nearly 85% of the total bitumen reservoir available around the globe. These deposits are highly dense in concentration such that they are economically feasible for recovery and transformation into crude oil (Zhou et al., 2008).

Over the past decades, there have been efforts among researchers to understand the actual molecular structure and group type of naphthenic acids. There remain challenges in characterization and identification of this complex group of compounds (Rudzinski et al., 2002; Grewer et al., 2010) and their naphthanate salts (Mapolelo et al., 2009) (Table 18.1) in environmental biological samples. This chapter discusses the environmental occurrence and persistence of NAs, the impacts and effects of their exposure on different biota, and their chemistry and how this relates to advancement in chromatographic analysis.

18.2 CANADIAN OIL SAND NAPHTHENIC ACIDS

Compositionally, oil sands are composed primarily of quartz sand, silt, and clay, which are present within an envelope of water about 10 nm in thickness, and a layer of viscous bitumen (Lo et al., 2003). The sand grains within oil sands are hydrophilic and water wetted. There is about 10 nm thickness of water separating each sand grain from another (Figure 18.2). Other components of bitumen include

TABLE 18.1
Commercial Uses of Naphthenic Acids

Naphthenic Acid Metal Salt and Soap	Commercial Applications
Na salt	Emulsifying agent for agricultural insecticides
	Additive for cutting oils
	Emulsion breaker in oil industry
Pb naphthenate	An extreme pressure agent for lubricating oil
Zn naphthenate	An antioxidant
Ca naphthenate	Lubricating oil detergent additive
Fe and Mn naphthenates	Fuel additives for improved combustion
Pb, Zn, and Ba salts	Wetting agent in pigment grinding
	Dispersion agents for paint
	Catalysts for oil-based paints
Zn, Co, Pb naphthenates	Driers in oil-based paint before they were phased out
	Possess good heat stability and noncorrosive
Cu naphthenates	Effective fungicide for sandbags, rope for use at sea, wood, cotton, jute product
Al naphthenates	Gelling agent for gasoline flame throwers
Co naphthenates	Curing agent in rubbers and resins
	Adhesion promoter of steel cord to rubber
Mn, Pb, Co, and Ca soaps	Oxidative catalysts

Source: McMartin, D.W., "Persistence and Fate of Acidic Hydrocarbons in Aquatic Environments: Naphthenic Acids and Resin Acids," PhD thesis, University of Saskatchewan, Saskatchewan, Canada, 2003.

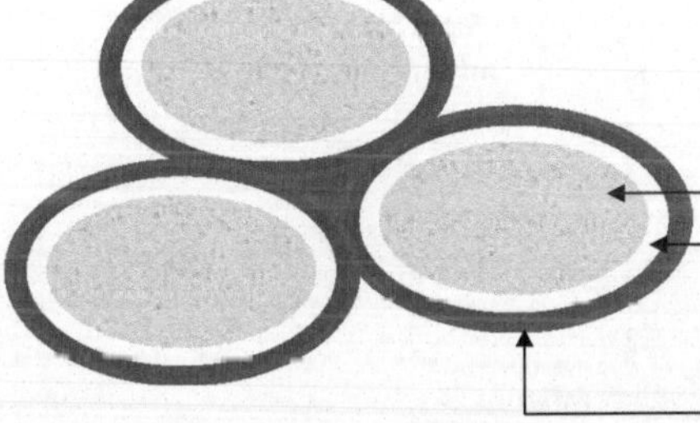

FIGURE 18.2 A structure of oil sand grains. (Reprinted with permission from Lo et al. (2003), 6394–6400. Copyright 2003 American Chemical Society; adapted from Headley, J.V. et al., *Journal of Environmental Science and Health Part A*, 48, 1145–1163, 2013.)

large amounts of chemical impurities such as S, N, Ni, and V, which are mainly encompassed in complex asphaltene composites (Headley et al., 2013).

18.3 EXTRACTION OF BITUMEN

A diagrammatic section of a mineable oil sand ore is provided in Figure 18.3. The extraction of bitumen begins by mining an oil sand (ore) body, which is approximately 60 m thick, and removing about 30 m of overburden material from the landscape. Large draglines and mobile power shovels are employed to dig the oil sand from an open pit created (Schramm et al., 2000).

Presently, there are two main methods of extracting bitumen from oil sands ore: surface mining and in situ (Frank, 2008; Thorley, 2012). Approximately 95% by mass of all ore mined in Canada

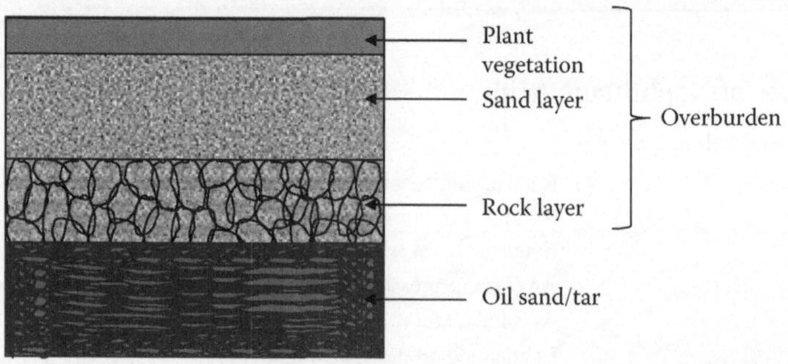

FIGURE 18.3 A diagrammatic representation of a mineable oil sand. (Adapted from Schramm, L.L. et al., *Surfactants: Fundamentals and Applications in the Petroleum Industry*, Cambridge University Press, Cambridge, 2000.)

is by surface mining, and this figure has seen an increase in the last decade (Thorley, 2012). In the early 1990s, the Suncor and Syncrude operations were the only oil sand projects in Alberta. Both of these projects entailed surface mining of bitumen. By mid-1980s, Imperial Oil Cold Lake operations and a few others were involved in in situ extraction or drilling of bitumen (Plourde, 2009). Currently, Cenovus, Connacher, Devon, Husky Energy, Shell, and MEG Energy, to mention but a

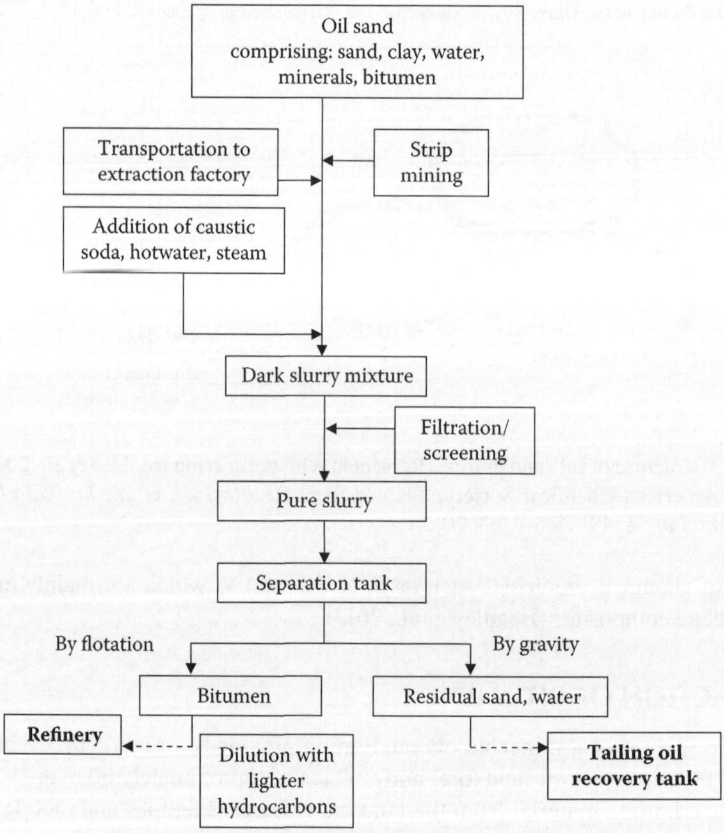

FIGURE 18.4 Extraction process of bitumen from oil sand ore. (Reprinted with permission from Lo et al. (2003), 6394–6400. Copyright 2003 American Chemical Society.)

few, are engaged in in situ projects (Heikkinen and Chipman, 2010). The in situ methods described by Heikkinen and Chipman (2010) mainly utilize cyclic steam stimulation (CSS) and steam-assisted gravity drainage systems. In CSS, steam is injected down a wellbore into the reservoir, heating the bitumen to a temperature of 130°C. This chapter focuses on wastes produced during the surface mining processes.

In order to extract 2 t of oil sand, half a ton of overburden must be evacuated to yield a barrel of oil (Schramm et al., 2000). After digging the boreal ground, the oil sand ore is transported to an extraction plant for further processing (Figure 18.4). Clark caustic hot water extraction process is employed in most surface mining operations, where the oil sand ore is digested with warm water (50–60°C) and sodium hydroxide (NaOH). The bitumen is separated from sand as froth and later upgraded by refinery processes to generate light, sweet crude oil (Schramm et al., 2000; Holowenko et al., 2002). Tailings consisting of a slurry of sand, silt, clay, and residual bitumen are placed in large holding basins (tailing ponds [TPs]), following the extraction process, to allow the solids to settle. The resulting overlaying water generated in the TPs, referred to as oil sand process-affected water (OSPW), is recycled back into the extraction process (Grewer et al., 2010). NA concentrations in the OSPW range from 40 to 120 mg L^{-1} (Holowenko et al., 2002).

A process known as dewatering, whereby a suitable polymeric additive is incorporated into the OSPW to alter the characteristics of clay present in the mixture, enhances coagulation and formation of solid deposits for future reclamation of the affected and disturbed landscape. Such additives are usually high-molecular weight, medium-charge density polyacrylamides. There is a lack of information about the fate and transport of such additives in aquatic habitats (Headley et al., 2013).

18.4 ENVIRONMENTAL FATE OF NAPHTHENIC ACIDS

The environmental implications of surface mining cannot be overemphasized. According to Flanagan and Grant (2013), TPs hold between 30% and 50% of a mine's total waste. In Alberta, Canada, these TPs currently contain up to 830 million cubic meters and cover approximately 176 km^2 of the landscape. In addition, it was estimated that two to four barrels (bbl) of fresh water are used per barrel of bitumen extracted through mining. Thus, a total volume of 2.4 million bbl per day of tailing waste is produced at the present mining rate. Up to 300 million cubic meters of tailings was held in the containments of the two major refineries, Syncrude and Suncor (Figure 18.5a and b), in 1993 (Smith, 2006).

(a) (b)

FIGURE 18.5 (a) Tailing pond north of Syncrude upgrader and (b) Suncor upgrader plant and related facilities. (From Dodge, D., Pembina Institute, Calgary, http://www.oilsandswatch.org, 2005; Evans, C., Pembina Institute, http://www.oilsandswatch.org, 2005. With permission.)

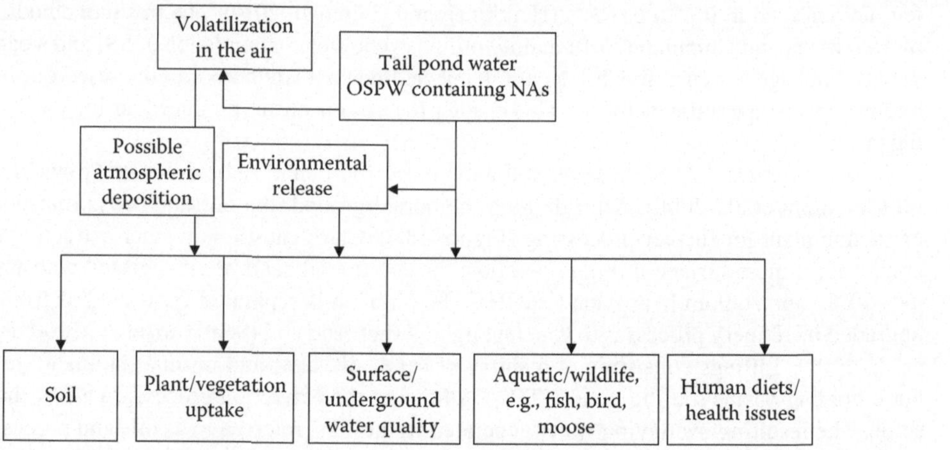

FIGURE 18.6 Potential environmental pathway of NAs to different biota. (Adapted from McMartin, D.W., "Persistence and Fate of Acidic Hydrocarbons in Aquatic Environments: Naphthenic Acids and Resin Acids," PhD thesis, University of Saskatchewan, Saskatchewan, Canada, 2003; Tenenbaum, D., *Environmental Health Perspectives*, 117, 150–156, 2009.)

An independent report on the magnitude and impact of oil sand mining on the environment was published in November 2005 by the Pembina Institute (Alberta, Canada) (Smith, 2006).

Barker et al. (2007) reported on a study by researchers at the University of Waterloo that estimated TP leaks at Suncor Energy's Tar Island pond in the range of 6 million liters per day into the Athabasca River. Critics of oil sand development are thus concerned about the potential for adverse environmental effects (Tenenbaum, 2009). Figure 18.6 provides a summary of the potential distribution of NAs released to the environment following the extraction of bitumen from oil sand deposits.

Background concentrations of NAs differ depending on the geological characteristics of the underlying area, but concentrations for waterways in the Athabasca region are generally less than 1 mg L^{-1}. In comparison, the NA concentrations in OSPW are as great as 110–120 mg L^{-1}, while natural groundwater levels in the region maybe as low as <4 mg L^{-1} in near-surface aquifers and high as >55 mg L^{-1} in basal and limestone aquifers (CEATAG, 1998; Clemente and Fedorak, 2005).

18.5 PHYSICAL AND CHEMICAL PROPERTIES OF NAPHTHENIC ACIDS

In the environment, NAs exhibit different characteristics depending on the geographical location (Headley et al., 2009). NAs are viscous liquids at room temperature. They differ in color from pale yellow to dark amber. They have low water solubilities (<50 mg L^{-1} at pH 7) but are readily soluble in organic solvents and oils. Their dissociation constants range from 10^{-5} to 10^{-6}, comparable to those of higher-molecular weight fatty acids (Brient et al., 1995). As the molecular weight of NAs increases, so do their polarity and nonvolatility, giving each individual compound unique physical and chemical properties (Headley and McMartin, 2004) as shown in Table 18.2.

Naphthenic acids are surfactant-like compounds, which dissolve well in organic solvents, but their solubility in water is pH dependent. They are readily soluble in basic media, dissociating as their salts; in acidic media, they are completely insoluble (Grbović et al., 2012). The general formula of NAs is $C_nH_{2n+z}O_2$, where n is the carbon number and Z ($Z = 0, -2, -4$, and -6 families) refers to the hydrogen deficiency due to a ring formation (Lo et al., 2003; Clemente and Fedorak, 2005) (Figure 18.7). The z number generally ranges from 0 to −12, and n is greater than or equal to 0. Naphthenic acids in which Z is greater than −12 constitute a very minor proportion of the acid content of compounds extracted in the caustic refinery process (Headley and McMartin, 2004; Clemente and Fedorak, 2005).

TABLE 18.2
Physical and Chemical Properties of Naphthenic Acids

Specifications	General Attributes
Boiling point	Ranges 250–300°C
Color	Varies from pale yellow, dark amber, yellowish brown, black
Density	Between 0.97 and 0.99 g cm^{-3}
Odor	Primarily imparted by the presence of phenol and sulfur impurities; musty hydrocarbon smell
Molecular weight	Generally between 140 and 450 amu
Refractive index	Approximately 1.5
Solubility	<50 mg L^{-1} at pH 7 in water
	Completely miscible in organic solvents
State	At room temperature is a viscous liquid
pK_a	Between 5 and 6
Log K_{ow} (octanol–water partition coefficient)	Approximately 4 at pH 1
	Approximately 2.4 at pH 7
	Approximately 2 at pH 10

Source: Headley, J.V., and D.W. McMartin, *Journal of Environmental Science and Health Part A*, 39, 1989–2010, 2004.

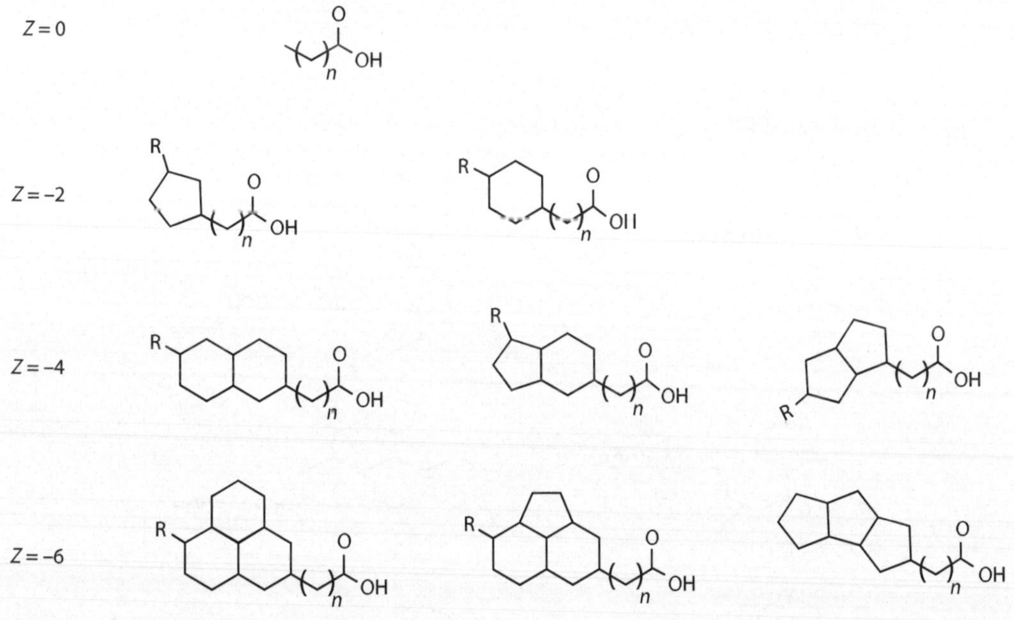

FIGURE 18.7 Structures of "classic naphthenic" acids in the Z = 0, −2, −4, and −6. (Reprinted with permission from Lo et al. (2003), 6394–6400. Copyright 2003 American Chemical Society.)

18.6 THE COMPLEXITY OF OIL SAND PROCESS-AFFECTED WATER MIXTURE

The extraction, processing, upgrading, and refining of bitumen generate different components which are xenophobic in nature and are persistent in the environment (Bauer, 2013). The presence of such components including polycyclic aromatic compounds, coupled with high-salinity and solubilized NAs, has made OSPW an extremely complex mixture, posing a hindrance to the characterization of individual components (Bauer, 2013). The relationship between *z* series, *n* families, and molecular weights is shown in Table 18.3 (McMartin, 2003). Numerous polar organic compounds are present

TABLE 18.3
Molecular Weights (MW) of Different *z* Series and *m* Families of "Classic Naphthenic Acids ($C_mH_{2m+z}O_2$)"

Number of Carbon Atoms	MW z = 0 (Open Chain)	MW z = −2 (1 Ring)	MW z = −4 (2 Rings)	MW z = −6 (3 Rings)
10	172	170	168	166
11	186	184	182	180
12	200	198	196	194
13	214	212	210	208
14	228	226	224	222
15	242	240	238	236
16	256	254	252	250
17	270	268	266	264
18	284	282	280	278
19	298	296	294	292
20	312	310	308	306

Source: McMartin, D.W., "Persistence and Fate of Acidic Hydrocarbons in Aquatic Environments: Naphthenic Acids and Resin Acids," PhD thesis, University of Saskatchewan, Saskatchewan, Canada, 2003.

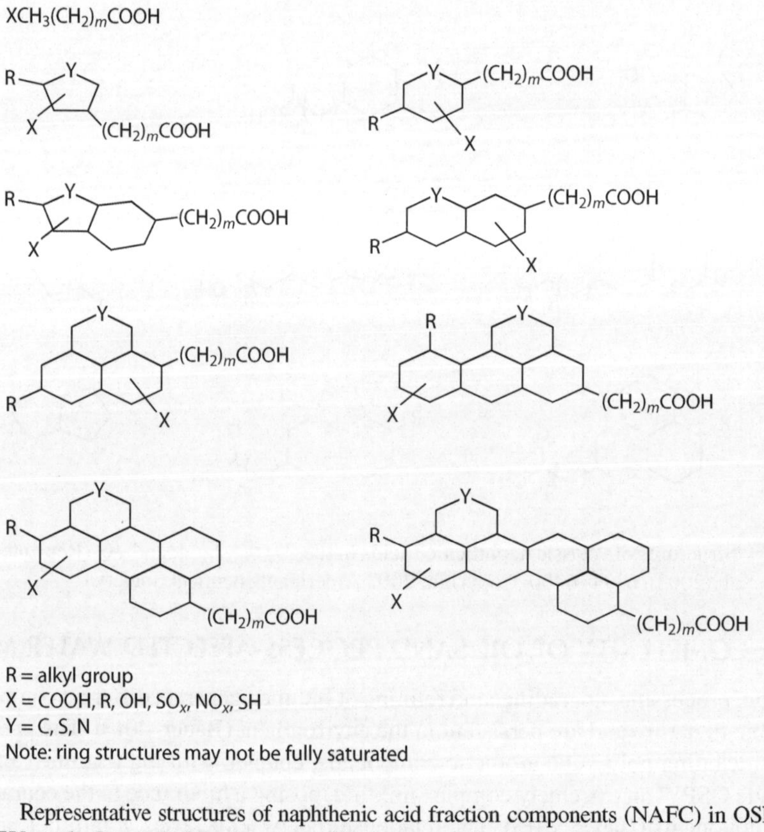

R = alkyl group
X = COOH, R, OH, SO_x, NO_x, SH
Y = C, S, N
Note: ring structures may not be fully saturated

FIGURE 18.8 Representative structures of naphthenic acid fraction components (NAFC) in OSPW. (Adapted from Headley J.V. et al., *Journal of Environmental Science and Health Part A*, 48, 1145–1163, 2013.)

in OSPW. Some of these compounds can now be characterized due to recent advances using ultra-high-resolution MS analyses of these mixtures (Bauer, 2013; Headley et al., 2013), to be discussed later in this chapter. Some groups of naphthenic acid fraction components (NAFCs) having the formula $C_nH_{2n+z}O_x$, where $x = 2–5$, and termed *oxy-naphthenic acids* (oxy-NAs) are formed after a mild oxidation process of the acids (Barrow et al., 2009; Grewer et al., 2010). Han et al. (2008) reported mono- and dihydroxy-naphthenic acids (i.e., $C_nH_{2n+z}O_3$ and $C_nH_{2n+z}O_4$) in extracts from Syncrude OSPW. Some of these oil sand NAFC compounds (Headley et al., 2013) are shown in Figure 18.8. Similar to the classic NA counterparts, the environmental fate and distribution of these NAFCs and OSPW compounds will be governed by factors such as geology and weathering processes, potential sorption to aquifer minerals, relative adsorption to TP sediments, and underground soils (Headley et al., 2013).

18.7 TOXICITY OF NAPHTHENIC ACID FRACTION COMPONENTS OF OIL SAND PROCESS-AFFECTED WATER MIXTURES

The toxicity of NAFCs spans across an array of organisms within the ecosystem and food webs including rats, birds, fish, plants, zooplankton, and bacteria (Bauer, 2013). Found that at high doses leading to chronic exposure, NAFCs are toxic to mammals, causing liver and heart damage, brain hemorrhage, weight loss, and liver enlargement. Peters et al. (2007) observed some deformity and lower birth size in yellow perch eggs exposed to NAFCs as opposed to the control population. Further, OSPW from oil sand processes was shown to be acutely toxic to rainbow trout and other mammals (Headley et al., 2002). Gentes et al. (2006) found that nestling tree swallows which were exposed to oil sand tailings containing NAFCs show lower weight and, thus, less resistance to stress.

The toxicity of NAFCs has been attributed to their surfactant properties, which facilitate their penetration of the cell wall, through the mechanism of osmotic stress. This results in membrane disruption and cytotoxicity as demonstrated in a molecular study specifically with *Escherichia coli* (Schramm, 2000; Quagraine et al., 2005). Leung et al. (2003) found the structure of aquatic microorganisms to be negatively impacted when the NAFC concentration was between 6 and 20 mg L^{-1}. To this end, NAFCs having $n \le 22$ have been suggested to be the probable cause of OSPW toxicity (Holowenko et al., 2002). To corroborate this finding, Frank (2008) conducted a quick assessment (Microtox assay) of the acute toxicity of NAFCs on a marine bacterium, *Vibrio fischeri*, and found that oil sand-derived NA fraction toxicity displayed an inverse relationship with the molecular weight of each fraction. In a follow-up study, Frank et al. (2009) determined that a decrease in acute toxicity to *V. fischeri* observed in NAFCs of greater molecular weight was possibly due to an increase in carboxylic acid content. Most recently, Jones et al. (2011) discovered that the toxicity to *V. fischeri* of an individual carboxylic compound was related to an increase in alkyl chain length and, hence, high molecular weight.

The probable mode of action by which NAFCs exert acute toxicity on the cell membrane of organisms is narcosis (Bauer, 2013). NAFCs are suggested endocrine disruptors and potential carcinogens, accounting for 65% of the estrogenic receptor activities in OSPW but not mutagens (Bauer, 2013; Wang et al., 2013).

A microbial community composition has been reported to be impacted by NAFCs (Hadwin et al., 2006). Indigenous microbial communities present within an oil sand mining region have the ability to adapt to and degrade NAFCs present in TPs and, as such, efficiently lower the toxicity of the OSPW mixture (Bauer, 2013).

18.8 ANALYTICAL ISOLATION AND PURIFICATION OF NAPHTHENIC ACID MIXTURE FROM OIL SAND PROCESS-AFFECTED WATER

Prior to instrumental analysis of NAs in any environmental sample, a series of analytical procedures must be performed, including extraction of the NA mixture from the matrices, purification,

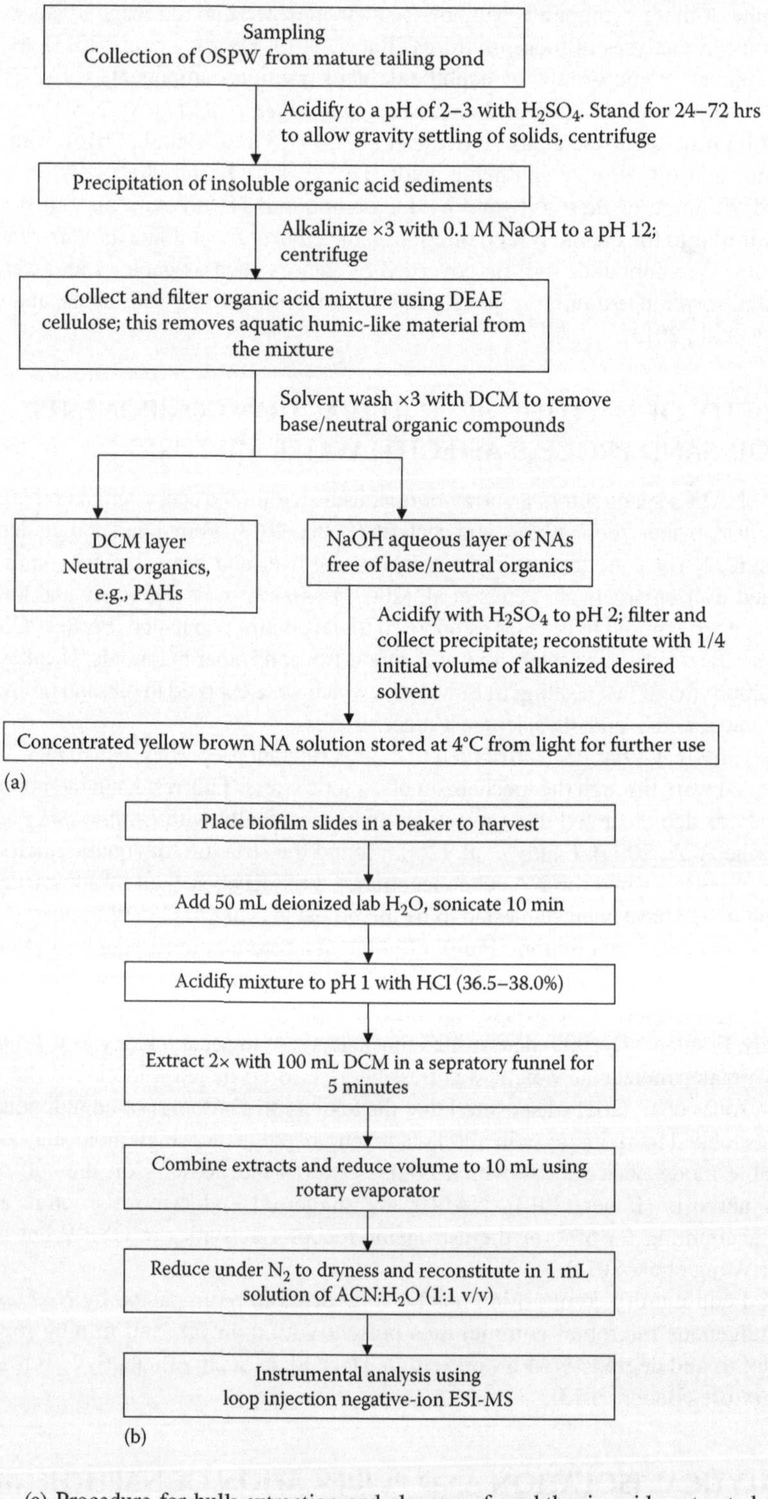

FIGURE 18.9 (a) Procedure for bulk extraction and cleanup of naphthenic acids extracted from oil sand tailing pond water. (b) Extraction procedure for microbial biofilm. (Headley, J.V. et al.: Characterization of oil sands naphthenic acids treated with ultraviolet and microwave radiation by negative ion electrospray Fourier transform ion cyclotron resonance mass spectrometry. *Rapid Communications in Mass Spectrometry*. 2010. 24. 3121–3126. Copyright Wiley-VCH Verlag GmbH & Co. KGaA. Reproduced with permission.) *(Continued)*

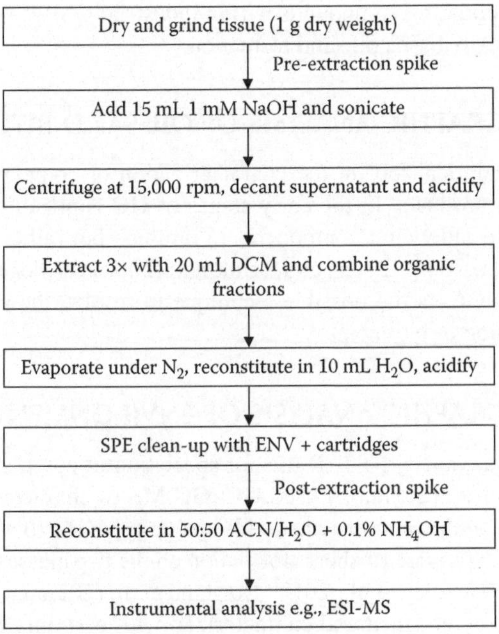

(c)

FIGURE 18.9 (CONTINUED) (c) Extraction procedure for plant tissue. (Headley, J.V. et al.: Characterization of oil sands naphthenic acids treated with ultraviolet and microwave radiation by negative ion electrospray Fourier transform ion cyclotron resonance mass spectrometry. *Rapid Communications in Mass Spectrometry*. 2011, 25: 459–462. Copyright Wiley-VCH Verlag GmbH & Co. KGaA. Reproduced with permission.)

cleanup, and, depending on the analytical instrumentation used, derivatization, such as methylation of the extracted NA mixture (Headley et al., 2002; McMartin, 2003; Yen et al., 2004; Young et al., 2007; Armstrong et al., 2009; Grewer et al., 2010; Headley et al., 2010; Woudneh et al., 2013). Figure 18.9a through c displays summaries of various analytical procedures available. These figures serve as general guidelines and are subject to modification to suit the intended purposes.

18.9 ADVANCES IN CHROMATOGRAPHIC ANALYSIS OF NAPHTHENIC ACIDS IN ENVIRONMENTAL SAMPLES

The complex nature of NAs and NAFCs in OSPW mixture has always been a limitation to the identification and characterization of each NA group in such mixtures (McMartin, 2003; Giesy et al., 2010). This problem is compounded by factors such as lack of individual standards, polarity of the compounds, lack of structural understanding of the compounds, and the presence of other coextractives which interfere with instrumental detection (McMartin, 2003).

The analytical techniques employed for the determination of NAs in OSPW have evolved over the last decade toward more robust procedures for accurate isolation, identification, and quantification of individual NA compounds in oil sand tailing mixtures. Such analytical methods include procedures such as MS analysis including ESI-MS, APCI-MS, GC-MS with derivatization, Fourier transform ion cyclotron resonance (FT-ICR-MS), LC-ESI-MS, positive-ion chemical ionization using isobutane reagent gas, and EI-MS following GC and/or LC/HPLC isolation techniques (McMartin, 2003).

In 2011, a workshop on naphthenic acid strategies was held in Saskatoon, Canada, to review routine methods for analyses of NA and NAFC mixtures and the growing need to develop chemical methods capable of differentiating compounds within industrial OSPW from those released during natural weathering of the Athabasca oil sand formation.

18.10 CHROMATOGRAPHIC ANALYSIS OF OIL SAND BITUMEN

GC has been a useful analytical tool in the characterization of complex mixtures, such as bitumen and other petroleum products. In the early stages of GC applications, it was difficult, if not impossible, to differentiate individual components in bitumen but rather was completed by group type (Beens and Brinkman, 2000). Beens and Brinkman (2000) further reported a comprehensive review on the use of GC×GC as the possible technique to resolve the ambiguity involved in the compositional differentiation of bitumen.

18.11 CHROMATOGRAPHIC ANALYSIS OF ENVIRONMENTAL SAMPLES

Barrow et al. (2010) used advanced FT-ICR-MS for speciation of components according to classes of compounds. Rowland et al. (2011a,b,c) used GC×GC-MS to characterize and confirm specific compounds, which do not conform to the classic NA group type. Barrow et al. (2010) utilized both ESI and APPI techniques to provide further information on the two classes (classic and oxy) of compounds present in OSPW (Headley et al., 2013). Rowland et al. (2011a,b,c) also reported on studies where classical NAs were determined and characterized with certainty into individual structured types, including aromatic carboxylic acids, tri- to pentacyclic diamondoid acids, highly branched isomers such as diamantane methyl and dimethyl diamantine acids, those acids with ethanoic acid side chains, and estrogen-like steroid acids.

18.12 CHROMATOGRAPHIC TECHNIQUES TO CHARACTERIZE THE ORIGIN OF NAPHTHENIC ACID FRACTION COMPONENT MIXTURES

In earlier studies by Holowenko et al. (2002), GC-EI-MS was employed for the characterization of NAs in water samples. GC-MS, although limited in its applicability, was capable of determining the origin of NAFCs (Headley et al., 2013). Headley et al. (2013) commented that it is necessary to use more advanced procedures, such as tandem LC (LC×LC), GC (GC×GC), and ion mobility, coupled with HRMS for more distinct separation of and definitive information on the source of oil sand NAFC mixtures (Headley et al., 2013). To this end, a combination of high-resolution mass accuracy analyzers such as FT-ICR and Orbitrap mass spectrometers coupled with either LC×LC or GC×GC would improve the quality of the characterization studies (Headley et al., 2013).

18.13 HIGH-PERFORMANCE LIQUID CHROMATOGRAPHY ANALYSIS OF OIL SAND PROCESS-AFFECTED WATER AND NAPHTHENIC ACID FRACTION COMPONENT MIXTURES

Martin et al. (2008) carried out a comparative study using ESI-MS and HPLC/HRMS to characterize commercially produced NA mixture and those derived from OSPW. HPLC-ESI QTOF-MS methods have been developed to characterize some individual groups of NAs in complex OSPW mixtures (West et al., 2013). By the same token, some oxy-type species containing O_3 compounds, such as alicyclic hydroxyl carboxylic acids, were identified in OSPW extracts by not only HPLC/QTOF-MS and FT-ICR-MS but also tandem GC×GC/TOF-MS (West et al., 2013).

Total oxy-NAs and each component present in environmental water samples have been quantitatively analyzed with HPLC technique coupled to a high-resolution accurate mass analyzer TOF-MS (Hindle et al., 2013). There was a fivefold reduction in ion suppression when using HPLC because a direct injection was utilized as opposed to flow injection methods in other procedures. This was the first accredited international standard method (ISO 17025) to analyze NAs in water samples; they ascertained the oxy-NAs in Sigma technical mixture at a level as low as 1 mg L^{-1}. Prior to the mentioned study, Noestheden (2011) also reported a 30–80% reduction in ion suppression through optimization of parameters such as sample dilution and elution time in the separation of oil sand NAFCs using HPLC/TOF-MS.

For future advances, there is a necessity for the development of authentic standards to quantify NAFCs expressed as total NAs, which is currently semiquantitative (Headley et al., 2013). Contributing to this concerted effort, Rowland et al. (2011a,b,c) have synthesized standards for a limited number of individual NAFCs to aid in their identification and characterization. The combination of high-resolution and chromatographic separation of HPLC/TOF-MS is a further development in the analysis of oil sand components in environmental samples (Headley et al., 2013).

18.14 LIQUID CHROMATOGRAPHY TANDEM MASS SPECTROMETRY OF NAPHTHENIC ACIDS IN AQUEOUS AND OIL SAND PROCESS-AFFECTED WATER SAMPLES

LC/ESI-MS has been applied by several groups to characterize NAFCs (Rudzinski et al., 2002; Wang and Kasperski, 2010; Headley et al., 2013). Other studies have been documented which utilized LC×LC-MS to separate OSPW into different isomers (Martin et al., 2010) and determine the recalcitrant properties of ringed and alkyl-substituted structured NAs (Bataineh et al., 2006; Han et al., 2008; Headley et al., 2013).

18.15 CONCLUSION

Advancements in instrumentation and analytical power have relieved some of the uncertainty about what precisely had been measured in OSPW and NAFCs. Not only have advancements in chromatographic analyses eased characterization and quantification challenges, but these methods are also more rapid and robust than those previously available. It is possible to now differentiate between the classic NAs, oxy-NAs, and N and/or S cocontaminants present in acid extractable fractions of OSPW and related environmental samples.

LIST OF ACRONYMS

APPI	atmospheric pressure photoionization
CSS	cyclic steam simulation
EI	electron impact ionization
ESI	electrospray ionization
ESI-FT-ICR-MS	electrospray ionization Fourier transform ion cyclotron resonance mass spectrometry
GC	gas chromatography
GC-MS	gas chromatography–mass spectrometry
HPLC	high-performance liquid chromatography
HRMS	high-resolution mass spectrometry
IUPAC	International Union of Pure and Applied Chemistry
LC	liquid chromatography
LC/ESI-MS	liquid chromatography/electrospray ionization multistage mass spectrometry
MS	mass spectrometry
m/z	mass-to-charge ratio
NA	naphthenic acid
NAFCs	naphthenic acid fraction components

OSPW	oil sand process-affected water
Q/q	quadrupole
TOF	time of flight
TPs	tailing ponds

REFERENCES

Armstrong, S., Headley, J.V., Peru, K.M., Germida, J.J. 2009. Differences in phytotoxicity and dissipation between ionized and non-ionized oil sands naphthenic acids in wetland plants. *Environmental Toxicology and Chemistry* 28: 2167–2174.

Barker, J., Rudolph, D., Tompkins, T. 2007. Attenuation of contaminants in groundwater impacted by surface mining in oil sands, Alberta, Canada. University of Waterloo, Waterloo, Canada.

Barrow, M.P., Headley, J.V., Peru, K.M., Derrick, P.J. 2009. Data visualization for the characterization of naphthenic acids within petroleum samples. *Energy & Fuels* 23: 2592–2599.

Barrow, M.P., Witt, M., Headley, J.V., Peru, K.M. 2010. Athabasca oil sands process water: Characterization by atmospheric pressure photoionization and electrospray ionization Fourier transform ion cyclotron resonance mass spectrometry. *Analytical Chemistry* 82: 3727–3735.

Bataineh, M., Scott, A.C., Fedorak, P.M., Martin, J.W. 2006. Capillary HPLC/QTOF-MS for characterizing complex naphthenic acid mixtures and their microbial transformation. *Analytical Chemistry* 78: 8354–8361.

Bauer, A. 2013. Identification of oil sands naphthenic acid structures and their associated toxicity to *Pimephales promelas* and *Oryzias latipes*. PhD thesis, University of Waterloo, Waterloo, Canada.

Beens, J., Brinkman, U.A.T. 2000. The role of gas chromatography in compositional analyses in the petroleum industry. *TrAC Trends in Analytical Chemistry* 19: 260–275.

Brient, J.A., Wessner, P.J., Doyle, M.N. 1995. Naphthenic acids. In *Encyclopedia of Chemical Technology*, ed. Kroschwitz, J.I. Wiley, New York, Vol. 16, pp. 1017–1029.

Clemente, J.S., Fedorak, P.M. 2005. A review of the occurrence, analyses, toxicity, and biodegradation of naphthenic acids. *Chemosphere* 60: 585–600.

Conrad Environmental Aquatic Technical Advisory Group (CEATAG). 1998. *Naphthenic Acids Background Information Discussion Report*. Alberta Department of Energy, Edmonton, Canada.

Dimitriu, P.A., Prescott, C.E., Quideau, S.A., Grayston, S.J. 2010. Impact of reclamation of surface-mined boreal forest soils on microbial community composition and function. *Soil Biology & Biochemistry* 42: 2289–2297.

Dodge, D. 2005. Figure 5A. Pembina Institute. http://www.oilsandswatch.org (accessed September 12, 2013).

Einstein, N. 2006. Figure 1. Pembina Institute, Calgary. http://www.ramp-alberta.org/resources/gallery.aspx (accessed September 12, 2013).

Evans, C. 2005. Figure 5B. Pembina Institute, Calgary. http://www.oilsandswatch.org (accessed September 12, 2013).

Flanagan, E., Grant, J. 2013. Losing ground, why the problem of oilsands tailings keep growing. Pembina Institute. http://www.pembina.org/pub/2470 (accessed February 14, 2014).

Frank, R.A. 2008. Naphthenic acids: Identification of structural properties that influence acute toxicity. PhD thesis, University of Guelph, Guelph, Canada.

Frank, R.A., Fischer, K., Kavanagh, R. et al. 2009. Effect of carboxylic acid content on the acute toxicity of oil sands naphthenic acids. *Environmental Science & Technology* 43: 266–271.

Gentes, M.-L. 2006. Health assessment of tree swallows (*Tachycineta bicolor*) nesting on the Athabasca oil sands, Alberta. MSc thesis, University of Saskatchewan, Saskatchewan, Canada.

Giesy, J.P., Anderson, J.C., Wiseman, S.B. 2010. Alberta oil sands development. *Proceedings of the National Academy of Sciences of the United States of America* 107: 951–952.

Grbović, L., Pavlović, K., Prekodravac, B. 2012. Fractionation of complex mixtures of naphthenic acids their characterization and biological activity. *Journal of the Serbian Chemical Society* 77: 147–157.

Grewer, D.M., Young, R.F., Whittal, R.M., Federok, P.M. 2010. Naphthenic acids and other acid-extractable in water samples from Alberta: What is being measured? *Science of the Total Environment* 408: 5997–6010.

Hadwin, K.M., Del Rio, L.F., Pinto, L.J., Painter, M., Routledge, R. et al. 2006. Microbial communities in wetlands of the Athabasca oil sands: Genetic and metabolic characterization. *FEMS Microbiology Ecology* 55: 68–78.

Han, X., Scott, A.C., Fedorak, P.M., Bataineh, M., Martin, J.W. 2008. Influence of molecular structure on the biodegradability of naphthenic acids. *Environmental Science & Technology* 42: 1290–1295.

Headley, J.V., McMartin, D.W. 2004. A review of the occurrence and fate of naphthenic acids in aquatic environments. *Journal of Environmental Science and Health Part A* 39: 1989–2010.

Headley, J.V., Peru, K.M., McMartin, D.W., Winkler, M. 2002. Determination of dissolved naphthenic acids in natural waters by using negative-ion electrospray mass spectrometry. *Journal of AOAC International* 85: 182–187.

Headley, J.V., Peru, K.M., Barrow, M.P. 2009. Mass spectrometric characterization of naphthenic acids in environmental samples: A review. *Mass Spectrometry Reviews* 28: 121–134.

Headley, J.V., Peru, K.M., Mishra, S., Meda, V., Dalai, A.K. et al. 2010. Characterization of oil sands naphthenic acids treated with ultraviolet and microwave radiation by negative ion electrospray Fourier transform ion cyclotron resonance mass spectrometry. *Rapid Communications in Mass Spectrometry* 24: 3121–3126.

Headley, J.V., Peru, K.M., Mohamed, M.H. et al. 2013. Chemical fingerprinting of naphthenic acids and oil sands process waters—A review of analytical methods for environmental samples. *Journal of Environmental Science and Health Part A* 48: 1145–1163.

Heikkinen, D., Chipman, J. 2010. Athabasca oil sands: Taking a bitumen break, Houston Texas. http://www.tudorpickering.com (accessed November 19, 2013).

Hindle, R., Noestheden, M., Peru, K., Headley, J. 2013. Quantitative analysis of naphthenic acids in water by liquid chromatography-accurate mass time-of-flight mass spectrometry. *Journal of Chromatography A* 1286: 166–174.

Holowenko, F.M., MacKinnon, M.D., Fedorak, P.M. 2002. Characterization of naphthenic acids in oil sands wastewaters by gas chromatography–mass spectrometry. *Water Research* 36: 2843–2855.

Jones, D., Scarlett, A.G., West, C.E., Rowland, S.J. 2011. Toxicity of individual naphthenic acids to *Vibrio fischeri*. *Environmental Science & Technology* 45: 9776–9782.

Leung, S.S., MacKinnon, M.D., Smith, R.E.H. 2003. The ecological effects of naphthenic acids and salt on phytoplankton from the Athabasca oil sands region. *Aquatic Toxicology* 62: 11–26.

Lo, C.C., Brownlee, B.G., Bunce, N.J. 2003. Electrospray-mass spectrometric analysis of reference carboxylic acids and Athabasca oil sands naphthenic acids. *Analytical Chemistry* 75: 6394–6400.

Mapolelo, M.M., Stanford, L.A., Rodgers, R.P., Marshall, A.G. 2009. Chemical speciation of calcium and sodium naphthenate deposits by electrospray ionization FT-ICR mass spectrometry. *Energy & Fuels* 23: 349–355.

Martin, J.W., Han, X., Peru, K.M., Headley, J.V. 2008. Comparison of high- and low-resolution electrospray ionization mass spectrometry for the analysis of naphthenic acid mixtures in oil sands process water. *Rapid Communications in Mass Spectrometry* 22: 1919–1924.

Martin, J., Wang, Y., Perez-Estrada, L., Gamal El-Din, M. 2010. Fingerprinting complex organic acid mixtures in oil sands process affected water. *Petroinformatics*. December 15–20, 2010, Pacifichem, Honolulu.

McMartin, D.W. 2003. Persistence and fate of acidic hydrocarbons in aquatic environments: Naphthenic acids and resin acids. PhD thesis, University of Saskatchewan, Saskatchewan, Canada.

Noestheden, M. 2011. Improving quantification of naphthenic acids mass accuracy and ion suppression. In *Trace Organics*, ed., Thompson, T., Johnson, S., Humphries, D., Chau, D. Western Canada Trace Organics Workshop, May 8–11, 2011, Edmonton, Canada, WCTOW, Edmonton, Canada.

Peters, L.E., Mackinnon, M., Van Meer, T., van den Heuvel, M.R., Dixon, D.G. 2007. Effects of oil sands process-affected waters and naphthenic acids on yellow perch (*Perca flavescens*) and Japanese medaka (*Orizias latipes*) embryonic development. *Chemosphere* 67: 2177–2183.

Plourde, A. 2009. Oil sands royalties and taxes in Alberta: An assessment of key developments since the mid-1990s. *Energy Journal* 30: 111–139.

Quagraine, E.K., Peterson, H.G., Headley, J.V. 2005. In situ bioremediation of naphthenic acids contaminated tailing pond waters in the Athabasca oil sands region—Demonstrated field studies and plausible options: A review. *Journal of Environmental Science and Health, Part A* 40: 685–722.

Rowland, S.J., Scarlett, A.G., Jones, D., West, C.E., Frank, R.A. 2011a. Diamonds in the rough: Identification of individual naphthenic acids in oil sands process water. *Environmental Science & Technology* 45: 3154–3159.

Rowland, S.J., West, C.E., Jones, D., Scarlett, A.G., Frank, R.A. et al. 2011b. Steroidal aromatic "naphthenic acids" in oil sands process-affected water: Structural comparisons with environmental estrogens. *Environmental Science & Technology* 45: 9806–9815.

Rowland, S.J., West, C.E., Jones, D. et al. 2011c. Monocyclic and monoaromatic naphthenic acids: Synthesis and characterization. *Environmental Chemistry Letters* 45: 525–533.

Rudzinski, W.E., Oehlers, L., Zhang, Y., Najera, B. 2002. Tandem mass spectrometric characterization of commercial naphthenic acids and a Maya crude oil. *Energy & Fuels* 16: 1178–1185.

Schramm, L.L., Stasiuk, E.N., MacKinnon, M.D. 2000. Surfactants in Athabasca oil sands slurry conditioning, flotation recovery, and tailings processes. In *Surfactants: Fundamentals and Applications in the Petroleum Industry*, ed. Schramm, L.L. Cambridge University Press, Cambridge, pp. 365–430.

Smith, B.E. 2006. Naphthenic acids: Synthesis, characterization and factors influencing environmental fate. PhD thesis, University of Plymouth, Plymouth, UK.

Tenenbaum, D.D. 2009. Oil sands development: A health risk worth taking? *Environmental Health Perspectives* 117: 150–156.

Thorley, U. 2012. Open pit mine planning: Analysis and system modeling of conventional and oil sands applications. PhD thesis, Queen's University, Kingston, Canada.

Wang, X., Kasperski, K.L. 2010. Analysis of naphthenic acids in aqueous solution using HPLC-MS/MS. *Analytical Methods* 2: 715–1722.

Wang, N., Chelme-Ayala, P., Perez-Estrada, L., 2013. Impact of ozonation on naphthenic acids speciation and toxicity of oil sands process-affected water to *Vibrio fischeri* and mammalian immune system. *Environmental Science & Technology* 47: 6518–6526.

West, C.E., Scarlett, A.G., Pureveen, J., Tegelaar, E.W., Rowland, S.J. 2013. Abundant naphthenic acids in oil sands process-affected water: Studies by synthesis, derivatization and two-dimensional gas chromatography/high-resolution mass spectrometry. *Rapid Communications in Mass Spectrometry* 27: 357–365.

Woudneh, M., Hamilton, C., Wang, G., McEachern, P. 2013. LC/MS/MS method for analysis of naphthenic acids in aqueous samples. In *Thursday Platform Abstracts*. November 7–11, 2013, Society of Environmental Toxicology and Chemistry (SETAC), Portland, OR. http://www.oilsands.ee/publications /historv.asp (accessed February 2007).

Yen, T.-W., Marsh, W.P., MacKinnon, M.D., Fedorak, P.M. 2004. Measuring naphthenic acids concentrations in aqueous environmental samples by liquid chromatography. *Journal of Chromatography A* 1033: 83–90.

Young, R.F., Orr, E.A., Goss, G.G., Fedorak, P.M. 2007. Detection of naphthenic acids in fish exposed to commercial naphthenic acids and oil sands process-affected water. *Chemosphere* 68: 518–527.

Zhou, S., Huang, H., Liu, Y. 2008. Biodegradation and origin of oil sands in the Western Canada Sedimentary Basin. *Petroleum Science* 5: 87–94.

Section VI

Transformation Products

19 Occurrence of Transformation Products of Pharmaceutical and Personal Care Products in the Aquatic Environment

Myrsini Papageorgiou, Eleni Evgenidou, and Dimitra A. Lambropoulou

CONTENTS

19.1 INTRODUCTION

Pharmaceuticals and personal care products (PPCPs) are widely consumed chemicals which have been designed for prevention, diagnosis, treatment, and cosmetic reasons. However, in recent years, PPCPs are considered as "emerging contaminants" and have been in the spotlight of the scientific community for their adverse effects on the aquatic biota and humans (Hedgespeth et al., 2012; Salgueiro-González et al., 2015). Despite their low concentrations, the ubiquity of pharmaceuticals in the aquatic environment together with their persistent biological activities explains the concern over this specific group of water contaminants.

Due to the use and release of the wide range of PPCPs, they end up in wastewaters and are transported in wastewater treatment plants (WWTPs) or disposed untreated in surface water and groundwater. However, the occurrence of PPCPs in effluents of the WWTPs reflects their partial elimination and WWTP malfunctions (Petrović et al., 2003; Mompelat et al., 2009). Physicochemical and biological properties of the compounds, operation systems, and treatment technologies (Barceló, 2012) in combination can explain this effect, which affects surface, ground, and drinking water. The circulation of PPCPs from wastewaters to drinking water has been depicted (Postigo and Richardson, 2014). However, the removal of such compounds is not commonly investigated because few or no regulations exist for their release in water bodies (Pasquini et al., 2013).

The past year's concerns, which can be reflected on the amount of increasing data (Radović et al., 2015), are the formation of metabolites and biotic and abiotic transformation products (TPs) of parent PPCPs. Through human metabolism, TPs can be formed, following two pathways, in phase I activation, for example, oxidation, or phase II, for example, conjugation (de Jongh et al., 2012), abiotic factors (physical and chemical factors, e.g., temperature, soil composition, amount, intensity, and wavelength of sunlight, sanity, and pH) and biotic factors (the living parts of an ecosystem with which an organism may interact) can have a significant contribution to the transformation of PPCPs and/or their TPs in the environment (Fatta-Kassinos et al., 2011). de Jongh et al. (2012) underline that TPs can also be formed during the wastewater treatment processes or in the environment with similar processes as in WWTPs including hydrolysis, oxidation, deconjugation, and photodegradation. Additionally, the TPs can be formed by particular oxidative techniques, such as ozonation; chlorination; or the use of chloramines, chlorine dioxide, UV, or UV/H_2O_2, during the drinking water treatment (Postigo and Richardson, 2014).

These products can be more toxic, estrogenic, and persistent than the parent compound (Agrawal et al., 2014; Postigo and Richardson, 2014; Yazdi and Es'haghi, 2005). The difference in toxicity of TPs from the parent compounds can be explained by their bioconcentration (toxicokinetics) and/or their mode of toxic action (toxicodynamics). In rare cases, transformation can create new toxicophores of higher toxicity by a similar or dissimilar mode of action (Cleuvers, 2004). Another field of research is the joint toxicity of multicomponent pharmaceutical mixtures in combination with multitransformation products (Ferrari et al., 2004). All the more, the independent, additive, antagonistic, or synergistic interactions in combination with the complexity of TPs could enhance the overall toxicity (Escher and Fenner, 2011). However, little is known about the environmental risk assessment in flora and fauna (Baumann et al., 2015; Radović et al., 2015). Escher and Fenner (2011) introduced a model-based framework to prioritize TPs for more detailed investigations in the context of chemical risk assessment and water quality assessment. Ideally, it would be necessary to assess the impact and toxicity of all the possible TPs released by urban WWTPs, many of which are already considered to be a possible threat to environmental health and safety. Therefore, it becomes important to determine their concentrations in wastewater treatment systems in order to make safe conclusions about their fate and the possible threat level.

In recent years, an increasing number of research groups are involved on these subjects and more data and broader knowledge have become available (Lambropoulou and Nollet, 2014). In this light, several studies were performed under laboratory-, pilot-, and full-scale conditions in order to assess their occurrence and elimination efficiency in WWTPs as well as to detect their presence in surface waters, in groundwater, and in drinking water treatment plants (DWTPs) and to evaluate the environmental and human health risks (Reddy et al., 2005; Jones et al., 2005; Castiglioni et al., 2006; Lajeunesse et al., 2008). Up to now, few review articles exist concerning exclusively the occurrence of TPs of PPCPs in wastewaters, while no intergraded knowledge is provided for their occurrence in wastewaters and other environmental compartments (Evgenidou et al., 2015; Haddad et al., 2015).

The aim of this chapter is to provide only the latest information (from 2010 up to now) about the occurrence of TPs of different therapeutic classes of PPCPs (e.g., analgesics, antibiotics, lipid regulators, psychiatric drugs, antidiabetics, β-blockers, antihistamines, anticonvulsants, antivirals, and cytostatics) in different environmental matrices (surface waters, groundwater, wastewaters, and drinking water) with a geographical perspective, attempting to summarize the recent research activity worldwide. However, since no recent available information exists concerning the occurrence of TPs of PPCPs in drinking water, earlier studies are included in this chapter in order to avoid exclusion of this water matrix from discussion.

19.2 SAMPLE PREPARATION METHODS FOR DETERMINATION OF TRANSFORMATION PRODUCTS

The detection and determination of TPs in environmental matrices are very complex and challenging. The most common analytical methodology for the trace determination of TPs in the aquatic environment was based on SPE followed by means of advanced chromatography and MS. In the last decade, several different SPE procedures were described thoroughly in worldwide literature (Bendz et al., 2005; Fatta et al., 2007; Barnes et al., 2008; Buerge et al., 2008; Prasse et al., 2010; Barclay et al., 2012; Subedi and Kannan, 2015). Nowadays, silica-based particles surface bound with C18 are rarely used. More popular are the new polymeric sorbents with both lipophilic (divinylbenzene) and hydrophilic monomers (*N*-vinylpyrrolidone)—mix-mode polymeric sorbent. Among them, Oasis HLB (Waters) is the most commonly used cartridge (Salgado et al., 2010; Schultz et al., 2010; del Mar Gómez-Ramos et al., 2011; Gracia-Lor et al., 2011; Huerta-Fontela et al., 2011; de Jongh et al., 2012; Cruz-Morató et al., 2013; Kosma et al., 2015; Radović et al., 2015; Subedi and Kannan, 2015). Other cartridges such as Isolute ENV+ (Prasse et al., 2010), Oasis MCX (Kostich et al., 2014), Strata X (Patrolecco et al., 2015), Strata-XC (Borova et al., 2014), and Ostro™ (Huerta et al., 2015) have been also selected. The extraction process has been improved with different extraction solvents (methanol or acetonitrile) and/or with pH adjustment or with ion pair reagents, for example, SDS for highly polar TPs (Kosma et al., 2015).

With respect to microextraction techniques, a number of sorbent and solvent microextraction techniques such as SPME, SBSE, LPME, and dispersive liquid–liquid microextraction have been explored for PPCP analysis in recent years, offering many advantages in terms of rapidity and simplicity and accompanied by good analytical results, which ensure that all these novel approaches will remain very bright in environmental research. However, it is clear that the number of microextraction applications focusing on TPs is substantially less than those with SPE technique, despite their potential. For example, a novel dSPE with MIPs has been recently proposed for the quantification of guanylurea in waters and wastewaters (a TP of metformin) (Kyzas et al., 2015), while a three-phase HF-LPME was applied for the determination of four sulfonamide metabolites (the corresponding N^4-acetyl derivatives: N^4-acetyl-sulfadiazine, N^4-acetyl-sulfamerazine, N^4-acetyl-sulfamethazine, N^4-acetyl-sulfamethoxazole) in environmental water samples (surface water, tap water, and wastewater) (Payán et al., 2011).

Currently, microextraction techniques coupled to GC-MS or LC-MS have not been well established in TP analysis as evident from the lack of publications in the open literature during the review period. However, the applicability of these approaches to the determination of parent pharmaceutical compounds to environmental samples such as waters and wastewaters indicates that, perhaps, this trend will be reversed in the next few years.

19.3 APPROACHES TO IDENTIFICATION AND DETERMINATION OF TRANSFORMATION PRODUCTS

19.3.1 Detection and Separation

Separation techniques such as GC and LC, hyphenated to MS detectors, are powerful tools for complex mixture analysis, since they can provide information regarding molecular weight and isotopic and fragmentation patterns, for precise identification of unknown compounds. Due to the high polar nature of TPs, however, GC is impossible in many cases, and therefore, only few studies reported the application of GC-MS studying the TPs of PPCPs (Calza and Fabbri, 2014). Hence, at present, LC coupled with ESI or APCI in positive or negative ion mode, is the most promising technique for the TPs of PPCP analysis due to its specificity, versatility, and simultaneous multianalyte quantification capability (Kim, 2005; Fatta et al., 2007; Radović et al., 2015). With respect to MS analyzers, various designs, performance, and functions have been used including single quadrupole (Q), QqQ, quadrupole ion trap, QqLIT, TOF, and QTOF. Based on the number of LC-MS applications being published in the field of TP/PPCP environmental analysis, it is apparent that the applicability of a single quadruple for detection and quantification has been less favored. On the contrary, HRMS, using TOF, QTOF (del Mar Gómez-Ramos et al., 2011), or Orbitrap mass spectrometers (Kosma et al., 2015), which provide higher selectivity with reference standards (actual m/z measurements, retention time) or with theoretically compiled libraries of elemental compositions applied without experimental accurate mass data, is dominating (Cruz-Morató et al., 2013; Postigo and Richardson, 2014; Robles-Molina et al., 2014; Baumann et al., 2015).

Often, however, MS alone is not enough to differentiate isomers or to provide the precise structure of unusual and/or unstable TPs. In addition, the substances present in environmental samples can suppress ionization, complicating TP identification. In such cases, LC with nuclear magnetic resonance (NMR) spectroscopy, chemical derivatization, and hydrogen/deuterium exchange combined with MS are used for the characterization of TPs, facilitating the MS structural assignment of these products (Calza and Fabbri, 2014; Postigo and Richardson, 2014).

19.3.2 Identification of Transformation Products

Undoubtedly, TP identification remains one of the greatest challenges in environmental analysis. In general, the analytical approaches to the determination of the TPs were differentiated according to the reference standards and the prior information. Knowledge about the environmental transformation processes is hereby favorable, since possible compounds arising from side or transformation reactions can be verified. With the knowledge about the possible TPs, present target analysis can be conducted, while for unknown substances, the analysis is more challenging.

Overall, three key analytical approaches are usually performed for TP determination: *target analysis* (with reference standards), *suspect screening* (with suspected substances based on prior information but no reference standards), and finally *nontarget screening* (no prior information, no reference standards) (Schymanski et al., 2014). The first approach simplifies the search for the identification of the TPs and guarantees their undoubted confirmation, using low-resolution tandem mass spectrometry systems (LRMS) or high-resolution mass analyzers (HRMS) with full-scan-based methods. Calculated TP masses can be systematically searched for in recorded mass spectra, and the substances can be identified via their masses. The use of HRMS analyzers such as QTOF or ion trap time of flight (IT-TOF) allows collecting full-scan spectra with accurate mass measurements and resolutions between 10,000 and 50,000 full width at half maximum (FWHM), whereas with hybrid Orbitrap instrumentation, even resolutions of 200,000 and higher FWHM can be achieved. In addition to the information of the precursor mass, further fragmentation experiments by recording mass spectra at increased collision energy are usually performed, in order to gain sufficient data for functional groups, and enhancing information for secure structure elucidation of the compounds. It is, however, worth noticing that the confirmation of the exact structure can only be achieved by NMR experiments or comparison of retention time and fragmentation behavior with that of reference standards.

The second approach, called suspect screening analysis, is a midway in which identification of compounds relies on specific available information such as their molecular formula and isotopic pattern. Accurate MS is commonly used in this approach. Identification of compounds relies on raw data processing with specific algorithms, which provide molecular formulas on the basis of (a) the monoisotopic mass measured and (b) relative abundances and *m/z* distances (spacing) of isotopic patterns. Hence, suspect screening is limited by the creation of databases as large as possible, with retention times and accurate HR mass spectra of both molecular and fragment ions related to TPs. "Suspects" can be screened using the exact mass of their expected ions, calculated from the molecular formula. Confidence in compound identification based on accurate mass and isotope pattern was expressed by "overall identification score," computed as a weighted average of the isotopic pattern signals of compound, such as exact masses, relative abundances, and *m/z* distances (spacing).

Despite their well-known advantages, the mentioned approaches have a significant drawback as they always will miss all TPs which were not selected at the start of the analysis, which means normally all unknowns or other untargeted compounds even in high concentrations or with severe toxic potential. To fill this knowledge gap, nontarget/unknown screening methods are very important tools for environmental chemistry. The last approach begins without any a priori information on the underlying compounds (Krauss et al., 2010; Schymanski et al., 2014; Evgenidou et al., 2015). Nontarget analysis would require visual inspection of the total ion chromatogram to find potential TPs present in the sample. Low-abundance compounds might not be apparent by visual inspection. On the contrary, intense peaks would be easily observed in the chromatogram, but they might not be necessarily associated with a single component. Under these circumstances, it is necessary to use powerful software with chromatographic peak deconvolution capabilities to identify the presence of multiple components and to produce pure spectra for each individual component. Experimental spectra can be compared against a homemade mass spectral library (either empirical or theoretical), so automating an efficient screening for many different compounds. When a match is unsatisfactory, the deconvoluted MS spectra can be used to investigate the identity of unknown compounds not present in the library. Based on this information, identification by HRMS/MS alone can often be considered as conclusive for this approach. A number of excellent works (Krauss et al., 2010; Calza and Fabbri, 2014; Chiaia et al., 2008; Schymanski et al., 2014; Zonja et al., 2014) giving a comprehensive theoretical background for the mentioned approaches are available in the literature and would be consulted for further information.

19.4 OCCURRENCE OF TRANSFORMATION PRODUCTS IN DIFFERENT WATER MATRICES

Due to their widespread and extensive production and consumption, the presence of PPCPs along with their TPs in different environmental compartments is ubiquitous. Numerous studies have identified several TPs that stem from biotic or abiotic transformation processes that take place during the route that PPCPs follow in the environment. After consumption, PPCPs are metabolized to some extent in the human (or animal) body and the unaltered parent compounds along with the produced metabolites are released though urine and feces. Entering the WWTPs, these compounds can undergo more transformations and structural changes that are attributed to biotic (biodegradation by bacteria and fungi) or abiotic processes such as hydrolysis or photolysis (Michael et al., 2014).

However, as aforementioned, WWTPs are designed to remove oxygen demand, suspended solids, pathogens, nutrients from wastewater, and easily or moderately biodegradable carbon (Verlicchi et al., 2012a). Thus, WWTPs could not always efficiently remove PPCPs or their TPs while there is a need for an upgrade technology (Verlicchi et al., 2012a). Subsequently, due to the direct influence of raw and treated waters, small rivers and creeks can be highly contaminated (Ternes et al., 2002). The released compounds can undergo further transformation or degradation in the aquatic environment such as direct or indirect photolysis or hydrolysis, giving rise to more TPs.

It is well known that groundwater is characterized as the largest and the most sensitive freshwater in the European Union (2006/118/EC). However, WWTP discharges, industrial discharges, accidental spills and landfills, agriculture, or leakages from sewage systems are point or nonpoint sources of pollution (Jurado et al., 2012), and as a result, a wide range of organic pollutants can be found in aquifers, posing a risk to groundwater quality. Moreover, the organic compounds traveling through the soil zone, before they reach groundwater, go through high microbiological populations, which are responsible for generating new TPs. Thus, a proper assessment of groundwater quality requires the identification of such pollutants.

Notwithstanding that groundwater is the main source of drinking water, numerous DWTPs use water sources impacted by wastewater, and therefore, the scientific community is strongly worried about the level of risk related to the presence of pharmaceuticals in drinking water (Benotti et al., 2009). Although, until now and based on precautionary principles, the low concentration of these substances (ng L^{-1}) does not cause health damage, it is of great importance to effectively remove the anthropogenic contaminants from drinking water (Ternes et al., 2002). Fortunately, the detection of pharmaceuticals in drinking water treatment effluent and tap water is uncommon due to the additional physical treatment (e.g., flocculation, sedimentation, filtration) and chemical disinfection (e.g., with chlorine, ozone, chloramines, or chlorine dioxide). Pharmaceuticals and TPs are generally removed by these processes, while few of them escape after these additional treatments remain in treated drinking water. However, growing interest is drawn to the formation of disinfection by-products (DBPs). Concerning DBPs, they can be easily produced through the reaction between the active aromatic rings of pharmaceuticals and oxidants such as chlorine, chloramines, chlorine dioxide, and ozone (Postigo and Richardson, 2014). Several studies have dealt with the monitoring of pharmaceuticals (Ternes et al., 2002; Petrović et al., 2003; Loraine and Pettigrove, 2006; Benotti et al., 2009; Mompelat et al., 2009; Huerta-Fontela et al., 2011), but there is relatively sparse information regarding TPs occurrence in drinking water (Heberer et al., 2001; Stackelberg et al., 2004; Jones et al., 2005; Chen, 2006; Vulliet et al., 2011; Gros et al., 2012; Kaiser et al., 2014).

The following section provides more detailed information on the occurrence of most commonly detected TPs of PPCPs in the aquatic environment. Tables 19.1 through 19.4 depict the concentrations of TPs, which are divided into the therapeutic classes of the parent compounds in WWTPs (Table 19.1), surface waters (Table 19.2), groundwater (Table 19.3), and DWTPs after finishing drinking water treatment processes (Table 19.4). Moreover, they underline the positive findings in each country worldwide.

TABLE 19.1
Occurrence of Most Frequent TPs in Influents and Effluents of WWTPs Worldwide

Therapeutic Class/ Parent Compound	Transformation Products	Matrix	Concentrations ($ng\ L^{-1}$) Influent Min	Influent Max	Influent Mean ± SD	Effluent Min	Effluent Max	Effluent Mean ± SD	Country	Reference
					Antibiotics					
Chlortetracycline	EACTC	6 WWTPs	–	–	<75[a]	–	–	–	Canada	Guerra et al. (2014)
	ECTC	6 WWTPs	–	–	<76[a]	–	–	<23[a]	Canada	Guerra et al. (2014)
	ACTC	6 WWTPs	–	–	<200[a]	–	–	–	Canada	Guerra et al. (2014)
Clarithromycin	14-OH-CLR	7 WWTPs	–	–	–	100	240	210[a]	Germany	Baumann et al. (2015)
Erythromycin	ERY-H_2O	3 SF	–	–	–	–	–	n.d. –22.5 ± 3.14	China	Zhou et al. (2013)
	ERY-H_2O	4 WWTPs	–	–	–	27	159	–	Spain	del Mar Gómez-Ramos et al. (2011)
	ERY-H_2O	ICW	–	–	44.5 ± 2.78	–	–	6.64 ± 1.80	South China	Chen et al. (2015)
	ERY-H_2O	6 WWTPs	14	600	92[a]	27	270	96[a]	Canada	Guerra et al. (2014)
Metronidazole	OH-MTR	WWTP	n.d.	145	62.9 ± 69	64.7	158	102 ± 33	Portugal	Santos et al. (2013)
	OH-MTR	4 HWWTPs	–	–	–	n.d.	11.344	121 ± 191– 1604 ± 3690	Portugal	Santos et al. (2013)
	OH-MTR	3 WWTPs	n.d.	454	–	n.d.	177	–	Spain	Gros et al. (2013)
	OH-MTR	HWWTPs	–	–	–	150	887	–	Spain	Gros et al. (2013)
Oxytetracycline	EOTC	6 WWTPs	–	–	<38[a]	–	–	<15[a]	Canada	Guerra et al. (2014)
Sulfadiazine	N^4-acetyl-SDZ	18 WWTPs	–	–	–	3.66	25.3	14.2	Germany	Garcia-Galan et al. (2012)
Sulfamethoxazole	N^4-acetyl-SMX	5 WWTPS	460	640	–	40	320	–	China	Qi et al. (2015)
	N^4-acetyl-SMX	WWTP	–	–	390–445	–	–	5–66	Australia	Le-Minh et al. (2012)

(Continued)

TABLE 19.1 (CONTINUED)
Occurrence of Most Frequent TPs in Influents and Effluents of WWTPs Worldwide

Therapeutic Class/ Parent Compound	Transformation Products	Matrix	Concentrations (ng L^{-1}) Influent Min	Influent Max	Influent Mean ± SD	Effluent Min	Effluent Max	Effluent Mean ± SD	Country	Reference
	N^4-acetyl-SMX	3 WWTPs	152.6	6224.2		4.9	72.4		Korea	Lee et al. (2013)
	N^4-acetyl-SMX	18 WWTPs	–	–	–	2.8	190	44.3	Germany	Garcia-Galan et al. (2012)
Sulfapyridine	N^4-acetyl-SFP	18 WWTPs	–	–	–	11	415	134.7	Germany	Garcia-Galan et al. (2012)
Tetracycline	ETC	6 WWTPs	26	120	65[a]	7.6	46	20[a]	Canada	Guerra et al. (2014)
	ATC	6 WWTPs	–	–	<91[a]	–	–	–	Canada	Guerra et al. (2014)
	EATC	6 WWTPs	–	–	<400[a]	–	–	–	Canada	Guerra et al. (2014)
Analgesics										
Acetylsalicylic acid	SA	8 WWTPs	n.d.	89,135	–	n.d.	431.9	–	Greece	Kosma et al. (2014)
	SA	WWTP	576	1673	1157	46.8	224	120	Greece	Stamatis and Konstantinou (2013)
	SA	WWTP	n.d.	11,920	4852	n.d.	570	110	Greece	Papageorgiou et al. (2016)
	SA	3 WWTPs	10,900	63,700	38,100	n.d.	n.d.	–	Spain	Gracia-Lor et al. (2012)
	SA	2 WWTPs	3224	13,640	–	<LOQ	114	–	Spain	Gros et al. (2012)
Diclofenac	4′-OH-DCF	3 WWTPs	212	530.6	–	40.2	582.1	–	Korea	Lee et al. (2013)
	4′-OH-DCF	2 WWTPs	–	–	–	–	–	1844–3032	Norway	Langford and Thomas (2011)
	5′-OH-DCF	2 WWTPs	–	–	–	–	–	551–675	Norway	Langford and Thomas (2011)
Ibuprofen	1-OH-IBU	WWTP	2550	5000	–	110	1410	–	Spain	Ferrando-Climent et al. (2012)
	2-OH-IBU	Sewer system	–	–	5100	–	–	5000	Spain	Jelic et al. (2015)
	2-OH-IBU	6 WWTPs	5700	130,000	24,000[a]	42	15,000	4200[a]	Canada	Guerra et al. (2014)
	2-OH-IBU	WWTP	1210	93,980	–	390	5870	–	Spain	Ferrando-Climent et al. (2012)

(Continued)

TABLE 19.1 (CONTINUED)
Occurrence of Most Frequent TPs in Influents and Effluents of WWTPs Worldwide

Therapeutic Class/ Parent Compound	Transformation Products	Matrix	Concentrations (ng L^{-1}) Influent Min	Influent Max	Influent Mean ± SD	Effluent Min	Effluent Max	Effluent Mean ± SD	Country	Reference
	CBX-IBU	WWTP	<LOQ	38,400	–	<LOQ	10,650	–	Spain	Ferrando-Climent et al. (2012)
MTZ	4-AAA	5 WWTPs	1520	2520	–	1300	2240	–	China	Qi et al. (2015)
	4-AAA	WWTP	1760	22,200	8333	<100	6745	4489	Spain	Rosal et al. (2010)
	4-AAA	WWTP	730	2340	–	40	1190	–	Hungary	Gyenge-Szabó et al. (2014)
	4-AAA	5 WWTPs	–	–	–	–	–	159–5343	Spain	López et al. (2014)
	4-AA	WWTP	262	3325	1517	127	2253	676	Spain	Rosal et al. (2010)
	4-AA	WWTP	28	150	–	20	340	–	Hungary	Gyenge-Szabó et al. (2014)
	4-AA	3 WWTPs	900	3200	2300	560	950	740	Spain	Gracia-Lor et al. (2012)
	4-FAA	WWTP	1005	71,000	17,579	<LOQ	27,444	559	Spain	Rosal et al. (2010)
	4-FAA	WWTP	410	1460	–	440	1950	–	Hungary	Gyenge-Szabó et al. (2014)
	4-FAA	5 WWTPS	–	–	–	–	–	62–2087	Spain	López et al. (2014)
	4-MAA	WWTP	314	1894	880	34	1098	291	Spain	Rosal et al. (2010)
	4-MAA	WWTP	10	890	–	n.d.	100	–	Hungary	Gyenge-Szabó et al. (2014)
Naproxen	*O*-Des-NPX	3 WWTPs	125	245	–	9.3	31.7	–	Korea	Lee et al. (2013)
Lipid Regulators										
Clofibrate	CLFA	WWTP	<LOQ	127	26	<LOQ	91	12	Spain	Rosal et al. (2010)
	CLFA	8 WWTPs	n.d.	265.9	–	n.d.	70.8	–	Greece	Kosma et al. (2014)
	CLFA	WWTP	308	740	527	86.5	258	184	Greece	Stamatis and Konstantinou (2013)
	CLFA	WWTP	n.d.	41,428	–	–	–	–	Spain	Salgado et al. (2011)
	CLFA	5 WWTPs	n.d.	65	28	n.d.	6	2	Korea	Behera et al. (2011)
Fenofibrate	FNBA	WWTP	<LOQ	117	79	<8	129	78	Spain	Rosal et al. (2010)

(*Continued*)

TABLE 19.1 (CONTINUED)
Occurrence of Most Frequent TPs in Influents and Effluents of WWTPs Worldwide

Therapeutic Class/ Parent Compound	Transformation Products	Matrix	Concentrations (ng L^{-1}) Influent Min	Influent Max	Influent Mean ± SD	Effluent Min	Effluent Max	Effluent Mean ± SD	Country	Reference
	FNBA	5 WWTPS	–	–	–	–	–	61–336	Spain	López et al. (2014)
Atorvastatin	2-OH-ATT	2 WWTPs	–	–	–	–	–	60–169	Norway	Langford and Thomas (2011)
	p-OH-ATT	2 WWTPs	–	–	–	–	–	83–233	Norway	Langford and Thomas (2011)
Psychiatric Drugs										
Alprazolam	AHA	2 WWTPs	5.86	28.5	17.4–21.4	n.d.	46.6	9.2–12.7	United States	Subedi and Kannan (2015)
Bupropion	TRE-BPP	24 WTTPs	–	–	–	<10	5700	–	United States	Writer et al. (2013)
	OH-BPP	24 WTTPs	–	–	–	<10	2000	–	United States	Writer et al. (2013)
Carbamazepine	CBZ-Ep	WWTP	n.d.	63	–	–	–	–	Spain	Rosal et al. (2010)
	CBZ-Ep	6 WWTPs	880	4026	–	69	2377	–	Spain	Huerta-Fontela et al. (2010)
	CBZ-Ep	6 WWTPs	–	–	59[a]	–	–	87[a]	Germany	Bahlmann et al. (2014)
	CBZ-Ep	2 WWTPs	–	–	48[a]	–	–	50[a]	Portugal	Bahlmann et al. (2014)
	CBZ-Ep	WWTP	–	–	16,208	–	–	–	Serbia	Petrović et al. (2014)
	CBZ-Ep	2 WWTPs	n.d.	10,240	–	n.d.	7480	–	Spain	Gros et al. (2012)
	Di-OH-CBZ	6 WWTPs	–	–	4000[a]	–	–	3400[a]	Germany	Bahlmann et al. (2014)
	Di-OH-CBZ	2 WWTPs	–	–	1900[a]	–	–	2000[a]	Portugal	Bahlmann et al. (2014)
	Di-OH-CBZ	WWTP	–	–	4960 ± 286	–	–	3380 ± 20	Germany	Kaiser et al. (2014)
	Di-OH-CBZ	24 WTTPs	–	–	–	<10	400	–	United States	Writer et al. (2013)

(Continued)

TABLE 19.1 (CONTINUED)
Occurrence of Most Frequent TPs in Influents and Effluents of WWTPs Worldwide

Therapeutic Class/ Parent Compound	Transformation Products	Matrix	Concentrations (ng L^{-1}) Influent Min	Influent Max	Influent Mean ± SD	Effluent Min	Effluent Max	Effluent Mean ± SD	Country	Reference
	Di-OH-CBZ	5 WWTPS	650	1020	–	450	560	–	China	Qi et al. (2015)
	2-OH-CBZ	WWTP	–	–	15,939	–	–	–	Serbia	Petrović et al. (2014)
	2-OH-CBZ	2 WWTPs	n.d.	928	–	n.d.	676	–	Spain	Gros et al. (2012)
	3-OH-CBZ	6 WWTPs	–	–	150[a]	–	–	140[a]	Germany	Bahlmann et al. (2014)
	3-OH-CBZ	2 WWTPs	–	–	92[a]	–	–	95[a]	Portugal	Bahlmann et al. (2014)
Carbamazepine/	10-OH-CBZ	24 WTTPs	–	–	–	<10	1900	–	United States	Writer et al. (2013)
Oxcarbamazepine	10-OH-CBZ	6 WWTPs	–	–	490[a]	–	–	500[a]	Germany	Bahlmann et al. (2014)
	10-OH-CBZ	2 WWTPs	–	–	270[a]	–	–	270[a]	Portugal	Bahlmann et al. (2014)
	10-OH-CBZ	WWTP	–	–	1590 ± 40	–	–	1430 ± 40	Germany	Kaiser et al. (2014)
	1/2-OH-CBZ	6 WWTPs	–	–	170[a]	–	–	140[a]	Germany	Bahlmann et al. (2014)
	1/2-OH-CBZ	2 WWTPs	–	–	90[a]	–	–	97[a]	Portugal	Bahlmann et al. (2014)
	ADON	WWTP	–	–	<LOQ	–	–	<LOQ	Germany	Kaiser et al. (2014)
	9-CA-ADIN	WWTP	–	–	700 ± 50	–	–	920 ± 50	Germany	Kaiser et al. (2014)
	BAQD	WWTP	–	–	162 ± 6	–	–	135 ± 15	Germany	Kaiser et al. (2014)
Citalopram	Des-Me-CT	24 WTTPs	–	–	–	<10	200	–	United States	Writer et al. (2013)
	Des-Me-CT	WWTP	–	133	104–133	n.d.	–	111	Canada	Metcalfe et al. (2010)
DZP	Des-DZP	5 WWTPs	0.6	3.1		2.6	4.7		Greece	Borova et al. (2014)
	Des-DZP	2 WWTPs	1.68	9.73	4.04–5.3	3.05	6.56	3.69–4.53	United States	Subedi and Kannan (2015)
	OXZ	5 WWTPs	19.7	62.7		5.8	74.1		Greece	Borova et al. (2014)
	OXZ	2 WWTPs	2.3	16.9	6.52–8.43	4.38	14.4	6.76–9.87	United States	Subedi and Kannan (2015)

(Continued)

TABLE 19.1 (CONTINUED)
Occurrence of Most Frequent TPs in Influents and Effluents of WWTPs Worldwide

Therapeutic Class/ Parent Compound	Transformation Products	Matrix	Concentrations (ng L^{-1}) Influent Min	Influent Max	Influent Mean ± SD	Effluent Min	Effluent Max	Effluent Mean ± SD	Country	Reference
	OXZ	90 WWTPs				–	1766	162	Europe	Loos et al. (2013)
Flunitrazepam	7-Amino-FLU	5 WWTPs	n.d.	15.6	–	n.d.	–	–	Greece	Borova et al. (2014)
FLU	NOR-FLX	8 WWTPs	–	118	62.8	–	20.1	13.2	UK	Baker and Kasprzyk-Hordern (2011)
	NOR-FLX	WWTP	45	78	–	20	26	–	United States	Hedgespeth et al. (2012)
	NOR-FLX	50 WWTPs	–	–	–	15	7.7	–	United States	Kostich et al. (2014)
	NOR-FLX	WWTP	45.1	226	112 ± 67	n.d.	99.6	36.1 ± 38.4	Spain	Santos et al. (2013)
	NOR-FLX	WWTP	n.d.	11	11	n.d.	5	5	Canada	Metcalfe et al. (2010)
Lamotrigine	2-*N*-gluc-LMTG	24 WWTPs	–	–	–	<100	455	–	United States	Writer et al. (2013)
Methylphenidate	RTLA	WWTP	<50	270	–	<50	170	–	Germany	Letzel et al. (2010)
Risperidone	9-OH-Ris	5 WWTPs	0.7	11.8		0.4	6.6		Greece	Borova et al. (2014)
SER	Des-Me-SER	WWTP	n.d.	21	5–21	–	–	<13	Canada	Metcalfe et al. (2010)
	Des-Me-SER	WWTPs	–	–	–	–	24	9.9	United States	Kostich et al. (2014)
VNF	*O*-Des-Me-VNF	24 WWTPs	–	–	–	<10	5500	–	United States	Writer et al. (2013)
	O-Des-Me-VNF	WWTP	n.d.	2600	2600	n.d.	–	1637	Canada	Metcalfe et al. (2010)
	N-Des-Me-VNF	WWTP	n.d.	259	134–259	n.d.	–	217	Canada	Metcalfe et al. (2010)
	Des-Me-VNF	WWTP	n.d.	5	5	n.d.	2	2	Spain	Huerta-Fontela et al. (2010)

(Continued)

TABLE 19.1 (CONTINUED)
Occurrence of Most Frequent TPs in Influents and Effluents of WWTPs Worldwide

Therapeutic Class/ Parent Compound	Transformation Products	Matrix	Influent Min (ng L^{-1})	Influent Max	Influent Mean ± SD	Effluent Min	Effluent Max	Effluent Mean ± SD	Country	Reference
Beta-Blockers										
Metoprolol/atenolol	MTPA	5 WWTPs	940	1380	–	280	1240	–	China	Qi et al. (2015)
	MTPA	WWTP	119	298	–	1107	2506	–	Spain	Rubirola et al. (2014)
	MTPA	WWTP	–	–	–	1500	5400	–	Norway	Reid et al. (2011)
Metoprolol	α-HMTP	WWTP	<26	36.1	–	n.d.	–	–	Spain	Rubirola et al. (2014)
Antidiabetics										
Metformin	GUA	7 WWTPs	n.d.	59	–	n.d.	328	–	Greece	Kosma et al. (2015)
	GUA	HWWTP	n.d.	84	29	<LOQ	627	188	Greece	Kosma et al. (2015)
	GUA	5 WWTPs	<250	3000	–	18,000	99,000	–	Germany	Scheurer et al. (2012)
	GUA	2 WWTPs	–	–	–	–	48,010	–	Netherlands	Oosterhuis et al. (2013)
	GUA	WWTP	–	–	–	28,000	67,000	–	Germany	Trautwein et al. (2014)
Antivirals										
Oseltamivir	OC	2 WWTPs	n.d.	42.7	42.7	n.d.	17.3	17.3	Germany	Prasse et al. (2010)
	OC	WWTP	–	–	–	–	293.3	–	Japan	Ghosh et al. (2010)
	OC	WWTP	n.d.	~1200	–	<LOQ	<1400	–	Norway	Leknes et al. (2012)
Cytostatics										
Paclitaxel	OH-PAC	WWTP	n.d.	4.4	–	–	–	–	Spain	Negreira et al. (2013)
	OH-PAC	12 WWTPs	n.d.	18.5	5.5[a]	n.d.	3.7	–	Spain	Negreira et al. (2014a)
Tamoxifen	OH-D-TAM	WWTP	–	–	–	n.d.	91.6	–	Spain	Negreira et al. (2014b)
	OH-TAM	12 WWTPs	n.d.	–	n.d.	n.d.	5.8	–	Spain	Negreira et al. (2014a)
	OH-TAM	WWTP	n.d.	–	n.d.	n.d.	164	100.7[a]	Spain	Negreira et al. (2014b)
Etoposide	3′-*O*-Des-ETP	WWTP	–	33	–	n.d.	<LOQ	–	Spain	Negreira et al. (2015)

(Continued)

TABLE 19.1 (CONTINUED)
Occurrence of Most Frequent TPs in Influents and Effluents of WWTPs Worldwide

Therapeutic Class/ Parent Compound	Transformation Products	Matrix	Concentrations (ng L^{-1}) Influent Min	Influent Max	Influent Mean ± SD	Effluent Min	Effluent Max	Effluent Mean ± SD	Country	Reference
Antihypertensives										
Enalapril	ENAT	50 WWTPs	–	–	–	–	150	14	United States	Kostich et al. (2014)
Valsartan	VALA	WWTP	–	320	101	–	–	–	Germany	Nödler et al. (2013)
	VALA	WWTP	–	–	136	–	–	1053	Switzerland	Kern et al. (2010)
Stimulants										
Caffeine	PARA-XNT	WWTP	–	–	–	n.d.	18,214	–	Canada	Ghoshdastidar et al. (2015)
	PARA-XNT	3 WWTPs	3222.3	10,442.1	–	<27.9	33.9	–	Korea	Lee et al. (2013)
	PARA-XNT	WWTP	4547	98,500	26,722	<LOQ	1796	836	Spain	Rosal et al. (2010)
	PARA-XNT	WWTP	–	–	–	–	–	2715	Spain	López et al. (2014)
Nicotine	COT	WWTP	–	–	–	n.d.	3095	–	Canada	Ghoshdastidar et al. (2015)
Hormones										
Estrone (E1)	E1-3S	WWTP	7.9	8.7	8.2	n.d.	n.d.	n.d.	Japan	Liu et al. (2010)
Estrone (E1)	E1-3G	WWTP	2.5	3.7	3.2	n.d.	3.7	1.4	Japan	Liu et al. (2010)
Estriol (E3)	E3-3S	WWTP	19.3	25.6	21.9	n.d.	2.2	1.3	Japan	Liu et al. (2010)
Calcium Channel Blockers										
Verapamil	NVP	WWTP	n.d.	0.908	n.d.	n.d.	n.d.	n.d.	Portugal	Santos et al. (2013)
Verapamil	NVP	4 HWWTPs	–	–	–	n.d.	8.63	–	Portugal	Santos et al. (2013)
Verapamil	NVP	50 WWTPs				–	20	5.8	United States	Kostich et al. (2014)
Verapamil	NVP	2 WWTPs	1.11	33.5	–	1.87	39.9	–	United States	Subedi and Kannan (2015)

(Continued)

TABLE 19.1 (CONTINUED)
Occurrence of Most Frequent TPs in Influents and Effluents of WWTPs Worldwide

Therapeutic Class/ Parent Compound	Transformation Products	Matrix	Concentrations ($ng\ L^{-1}$) Influent Min	Influent Max	Influent Mean ± SD	Effluent Min	Effluent Max	Effluent Mean ± SD	Country	Reference
Personal Care Products										
Paraben	ClMeP	WWTP	12	61	40	n.d.	6.9	–	Germany	González-Mariño et al. (2011)
	Cl_2MeP	WWTP	8	90	46	–	–	–	Germany	González-Mariño et al. (2011)
Triclosan	4-Cl-TCS	WWTP	2	17	–	n.d.	4	–	United States	Buth et al. (2011)
	6-Cl-TCS	WWTP	3	98	–	n.d.	9	–	United States	Buth et al. (2011)
	4,6-Cl-TCS	WWTP	8	29	–	n.d.	22	–	United States	Buth et al. (2011)

Note: n.d., not detected; <LOQ, below limit of quantification; ICW, integrated constructed wetlands; PF, production facility; SF, swine farms.

[a] Median concentration.

TABLE 19.2
Occurrence of Most Frequent TPs in Surface Water (River, Lake, Sea) Worldwide

Therapeutic Class/ Parent Compound	Transformation Products	Matrix	Concentrations (ng L^{-1}) Min	Max	Mean	Country	Reference
			Antibiotics				
Clarithromycin	14-OH-CLR	7 Rivers	2	80	24[a]	Germany	Baumann et al. (2015)
Erythromycin	ERY-H_2O	River	–	35	10	Netherlands	de Jongh et al. (2012)
	ERY-H_2O	River	–	–	62	United States	Ferrer and Thurman (2012)
	ERY-H_2O	Tributaries	n.d.	9.5	0.7	Spain	López-Serna et al. (2012)
Sulfamethazine	*N*[4]-acetyl-SMZ	River	7	85	18	China	Heeb et al. (2012)
Sulfamethoxazole	*N*[4]-acetyl-SMX	River	22	300	130	China	Heeb et al. (2012)
	N[4]-acetyl-SMX	Tributaries	n.d.	73.8	5.21	Spain	López-Serna et al. (2012)
	4-Nitro-SMX	Karst spring	3.4	5.4	–	Germany	Nödler et al. (2012)
	Des-SMX	Karst spring	2.9	7.7	–	Germany	Nödler et al. (2012)
			Analgesics				
Acetylsalicylic acid	SA	River	4	76	–	Spain	Gros et al. (2012)
	SA	Sea	–	2	–	Spain	Gros et al. (2012)
	SA	River	<LOQ	73.1	35.6	Spain	López-Serna et al. (2012)
	SA	Tributaries	<LOQ	89.1	42.5	Spain	López-Serna et al. (2012)
	SA	River-Canal-Lake	–	2.7	–	Serbia	Petrović et al. (2014)
	SA	River	86.7	184	119	Greece	Stamatis and Konstantinou (2013)
Diclofenac	4′-OH-DCF	River	n.d.	7.76	1.13	Spain	López-Serna et al. (2012)
	4′-OH-DCF	Tributaries	n.d.	48.2	5.87	Spain	López-Serna et al. (2012)
Dimethylaminophenazone	AMPH	River	–	66	16	Netherlands	de Jongh et al. (2012)
Ibuprofen	2-OH-IBU	River	<LOQ	3000	–	Spain	Ferrando-Climent et al. (2012)
	1-OH-IBU	River	<LOQ	560	–	Spain	Ferrando-Climent et al. (2012)
	CBX-IBU	River	<LOQ	3950	–	Spain	Ferrando-Climent et al. (2012)
	2-OH-IBU	Tributaries	–	573	–	Spain	Huerta et al. (2015)
MTZ	4-AAA	River	–	176	76	Netherlands	de Jongh et al. (2012)
	4-FAA	River	–	164	49	Netherlands	de Jongh et al. (2012)

(*Continued*)

TABLE 19.2 (CONTINUED)
Occurrence of Most Frequent TPs in Surface Water (River, Lake, Sea) Worldwide

Therapeutic Class/ Parent Compound	Transformation Products	Matrix	Concentrations (ng L^{-1}) Min	Max	Mean	Country	Reference
	4-AAA	River	270	2200	1000	China	Heeb et al. (2012)
	4-AA	River	4	31	16	China	Heeb et al. (2012)
	4-AAA	5 Rivers	31	512	–	Serbia	Radović et al. (2015)
	4-FAA	5 Rivers	9	186	–	Serbia	Radović et al. (2015)
Tramadol	*O*-Des-TRM	River	–	78	17	Netherlands	de Jongh et al. (2012)
			Lipid Regulators				
Clofibrate	CLFA	Tributaries	n.d.	14.4	2.25	Spain	López-Serna et al. (2012)
	CLFA	River	105	168	142	Greece	Stamatis and Konstantinou (2013)
Simvastatin	SVA	–	<LOD	66	–	Norway	Langford and Thomas (2011)
			Psychiatric Drugs				
Bupropion	TRE-BPP	Rivers	–	–	180	United States	Ferrer and Thurman (2012)
	OH-BPP	Rivers	–	–	150	United States	Ferrer and Thurman (2012)
Carbamazepine	CBZ-ep	River	–	35	17	Netherlands	de Jongh et al. (2012)
	CBZ-ep	River	2	25	11	China	Heeb et al. (2012)
	CBZ-ep	River	n.d.	214	76.5	Spain	López-Serna et al. (2012)
	CBZ-ep	Tributaries	n.d.	1667	212	Spain	López-Serna et al. (2012)
	CBZ-ep	-	14	77	–	Norway	Langford and Thomas (2011)
	CBZ-ep	River, canal, little lake	<LOQ	932	–	Serbia	Petrović et al. (2014)
	CBZ-ep	River	n.d.	105	–	Spain	Gros et al. (2012)
	CBZ-ep	Sea	–	21	–	Spain	Gros et al. (2012)
	Di-OH-CBZ	River	53	680	260	China	Heeb et al. (2012)
	Di-OH-CBZ	Rivers	–	–	80	United States	Ferrer and Thurman (2012)
	Di-OH-CBZ	River/small streams	–	–	131–2320	Germany	Kaiser et al. (2014)
	2-OH-CBZ	River	n.d.	5.38	1.27	Spain	López-Serna et al. (2012)

(Continued)

TABLE 19.2 (CONTINUED)
Occurrence of Most Frequent TPs in Surface Water (River, Lake, Sea) Worldwide

Therapeutic Class/ Parent Compound	Transformation Products	Matrix	Concentrations (ng L^{-1}) Min	Max	Mean	Country	Reference
	2-OH-CBZ	Tributaries	n.d.	61.7	7.46	Spain	López-Serna et al. (2012)
	2-OH-CBZ	River, canal, little lake	–	160	–	Serbia	Petrović et al. (2014)
	2-OH-CBZ	River	n.d.	52	–	Spain	Gros et al. (2012)
	3-OH-CBZ	River	n.d.	4.57	1.45	Spain	López-Serna et al. (2012)
	3-OH-CBZ	Tributaries	n.d.	70	9.17	Spain	López-Serna et al. (2012)
Carbamazepine/	10-OH-CBZ	Rivers	–	–	255	United States	Ferrer and Thurman (2012)
Oxcarbamazepine	10-OH-CBZ	River/small streams	–	–	29–819	Germany	Kaiser et al. (2014)
	ADON	River	n.d.	17.5	1.94	Spain	López-Serna et al. (2012)
	ADON	River/small streams	–	–	<LOQ-27	Germany	Kaiser et al. (2014)
	ADIN	River	n.d.	9.64	2.11	Spain	López-Serna et al. (2012)
	ADIN	Tributaries	n.d.	16.8	3.29	Spain	López-Serna et al. (2012)
	9-CA-ADIN	River/small streams	–	–	<LOQ-304	Germany	Kaiser et al. (2014)
	BAQD	River/small streams	–	–	<LOQ-74	Germany	Kaiser et al. (2014)
DZP	Des-DZP	River	n.d.	1.36	0.15	Spain	López-Serna et al. (2012)
	Des-DZP	Tributaries	n.d.	3.99	0.27	Spain	López-Serna et al. (2012)
	Des-DZP	River	–	5.5	3.2	UK	Baker and Kasprzyk-Hordern (2011)
	OXZ	River	<LOQ	51	–	France	Idder et al. (2013)
	OXZ-Glu	River	n.d.	48.8	22.7	Spain	López-Serna et al. (2012)
	OXZ-Glu	Tributaries	n.d.	70.9	12.9	Spain	López-Serna et al. (2012)
Lamotrigine	2-*N*-gluc-LMTG	Rivers	–	–	95	United States	Ferrer and Thurman (2012)
Methylphenidate	RTLA	3 Rivers	4	23	9.1	Germany	Letzel et al. (2010)
VNF	*O*-Des-Me-VNF	River	–	112	32	Netherlands	de Jongh et al. (2012)

(*Continued*)

TABLE 19.2 (CONTINUED)
Occurrence of Most Frequent TPs in Surface Water (River, Lake, Sea) Worldwide

Therapeutic Class/ Parent Compound	Transformation Products	Matrix	Concentrations (ng L^{-1}) Min	Max	Mean	Country	Reference
	O-Des-Me-VNF	River	25	109	–	Canada	Metcalfe et al. (2010)
	N-Des-Me-VNF	River	4	24	–	Canada	Metcalfe et al. (2010)
	Des-VNF	Rivers	–	–	84	United States	Ferrer and Thurman (2012)
			Beta-Blockers				
Metoprolol/Atenolol	MTPA	River	43	730	370	China	Heeb et al. (2012)
	MTPA	River, lake	–	197	32	Germany	Nödler et al. (2013)
	MTPA	Rivers	–	–	74	United States	Ferrer and Thurman (2012)
Metoprolol	α-HMTP		25	108	–	Norway	Langford and Thomas (2011)
Propranolol	PRP-β-D-Glu	River	n.d.	n.d.	n.d.	Spain	López-Serna et al. (2012)
	PRP-β-D-Glu	Tributaries	n.d.	n.d.	n.d.	Spain	López-Serna et al. (2012)
			Antidiabetics				
Glyburide	4-OH-GLY	River	n.d.	n.d.	n.d.	Spain	López-Serna et al. (2012)
	4-OH-GLY	Tributaries	n.d.	n.d.	n.d.	Spain	López-Serna et al. (2012)
Metformin	GUA	Lake	<LOQ	93	–	Germany	Trautwein et al. (2014)
	GUA	River	n.d.	391	–	Germany	Trautwein et al. (2014)
	GUA	Sea	n.d.	32	–	Germany	Trautwein et al. (2014)
	GUA	Rivers, creeks	–	–	25 ± 16	Germany	Scheurer et al. (2012)
			Antivirals				
Oseltavir	OC	River	<LOQ	2.4	–	Germany	Prasse et al. (2010)
	OC	River	6.6	190.2	–	Japan	Ghosh et al. (2010)
			Cytostatic				
Etoposide	3′-*O*-Des-ETP	River	n.d.	31	–	Spain	Negreira et al. (2015)
			Antihypertensives				
Valsartan	VALA	River, lake	–	2119	69	Germany	Nödler et al. (2013)

(Continued)

TABLE 19.2 (CONTINUED)
Occurrence of Most Frequent TPs in Surface Water (River, Lake, Sea) Worldwide

Therapeutic Class/ Parent Compound	Transformation Products	Matrix	Concentrations ($ng\ L^{-1}$) Min	Max	Mean	Country	Reference
Enalapril	ENAT	River	<LOQ	11.5	7.32	Spain	López-Serna et al. (2012)
	ENAT	Tributaries	n.d.	12.9	6.23	Spain	López-Serna et al. (2012)
			Stimulants				
Caffeine	PARA-XNT	Rivers	–	–	110	United States	Ferrer and Thurman (2012)
	PARA-XNT	River	–	421.4	277	UK	Baker and Kasprzyk-Hordern (2011)
	PARA-XNT	River	28	81	48	Spain	Martínez Bueno et al. (2012)
Nicotine	COT	River	20	80	39	Spain	Martínez Bueno et al. (2012)
			Antihistamines				
Ranitidine	RAN-*N*	River	n.d.	89.5	17.1	Spain	López-Serna et al. (2012)
	RAN-*N*	Tributaries	n.d.	196	32	Spain	López-Serna et al. (2012)

Note: n.d., no detected; <LOQ, below limit of quantification.
[a] Median concentration.

TABLE 19.3
Occurrence of Most Frequent TPs in Groundwater Worldwide

Therapeutic Class/ Parent Compound	Transformation Products	Concentrations (ng L^{-1}) Min	Max	Mean	Country	Reference
			Antibiotics			
Erythromycin	ERY-H_2O	n.d.	0.607	0.316	Spain	López-Serna et al. (2013)
	ERY-H_2O	n.d.	2.26	0.675	Spain	López-Serna et al. (2013)
	ERY-H_2O	n.d.	8.51	4.2	Spain	López-Serna et al. (2013)
Sulfadiazine	N^4-acetyl-SDZ	n.d.	37.2	8.2	Spain	López-Serna et al. (2013)
	N^4-acetyl-SDZ	–	1	0.85	Spain	García-Galán et al. (2011)
Sulfamerazine	N^4-acetyl-SMR	–	18	5	Spain	García-Galán et al. (2011)
Sulfamethazine	N^4-acetyl-SMZ	–	4.4	1.8	Spain	García-Galán et al. (2011)
Sulfamethoxazole	N^4-acetyl-SMX	–	5.5	1.4	Spain	García-Galán et al. (2011)
	4-Nitro-SMX	–	4.1	–	Germany	Reh et al. (2013)
	Des-SMX	–	6	–	Germany	Reh et al. (2013)
Sulfapyridine	N^4-acetyl-SFP	–	6	1.6	Spain	García-Galán et al. (2011)
			Analgesics			
Acetylsalicylic acid	SA	50.2	147	102	Spain	López-Serna et al. (2013)
	SA	26.6	98.2	58.9	Spain	López-Serna et al. (2013)
	SA	33.4	620	110	Spain	López-Serna et al. (2013)
Diclofenac	4′-OH-DCF	n.d.	147	45.9	Spain	López-Serna et al. (2013)
MTZ	4-AAA	20	105	–	Serbia	Radović et al. (2015)
	4-AAA	–	–	62.33–119.6	Spain	Cabeza et al. (2012)
	4-FAA	22	150	–	Serbia	Radović et al. (2015)
	4-FAA	–	–	63.2–98.75	Spain	Cabeza et al. (2012)
			Lipid Regulators			
Clofibrate	CLFA	n.d.	7.57	1.36	Spain	López-Serna et al. (2013)
			Psychiatric Drugs			
Carbamazepine	CBZ-ep	n.d.	1.8	0.36	Spain	López-Serna et al. (2013)
	CBZ-ep	n.d.	1.42	0.109	Spain	López-Serna et al. (2013)
	CBZ-ep	n.d.	8.36	2.91	Spain	López-Serna et al. (2013)
	CBZ-ep	–	–	7.25–33.66	Spain	Cabeza et al. (2012)
	2-OH-CBZ	<LOQ	47.9	27.8	Spain	López-Serna et al. (2013)
	3-OH-CBZ	<LOQ	39.9	18.8	Spain	López-Serna et al. (2013)
	ADON	n.d.	1.4	0.18	Spain	López-Serna et al. (2013)
	ADON	n.d.	8.22	2.99	Spain	López-Serna et al. (2013)
	ADIN	n.d.	5.39	1.08	Spain	López-Serna et al. (2013)
	ADIN	n.d.	15.8	2.33	Spain	López-Serna et al. (2013)
	ADIN	n.d.	10.7	1.17	Spain	López-Serna et al. (2013)
DZP	Des-DZP	n.d.	12.9	5.51	Spain	López-Serna et al. (2013)
VNF	N-Des-Me-VNF	1.5	3.9	–	Switzerland	Huntscha et al. (2012)
			Beta-Blockers			
Atenolol	ATN-Des	n.d.	6.4	–	Switzerland	Huntscha et al. (2012)
Propranolol	4-OH-PRP	5	21.4	–	Spain	López-Serna et al. (2013)
			Antihypertensive			
Enalapril	ENAT	2.9	8.5	4.78	Spain	López-Serna et al. (2013)

(Continued)

TABLE 19.3 (CONTINUED)
Occurrence of Most Frequent TPs in Groundwater Worldwide

Therapeutic Class/ Parent Compound	Transformation Products	Concentrations ($ng\ L^{-1}$) Min	Max	Mean	Country	Reference
	ENAT	n.d.	12.5	4.75	Spain	López-Serna et al. (2013)
	ENAT	n.d.	12.3	2.42	Spain	López-Serna et al. (2013)
			Stimulants			
Nicotine	COT	–	–	6.66–18.33	Spain	Cabeza et al. (2012)
	COT	6	320	27.2	Germany	Reh et al. (2013)
			Hormones			
Estradiol (E2)	E2-3G	–	–	425	United States	Shrestha et al. (2012)

Note: n.d., not detected; <LOQ, below limit of quantification.

19.4.1 Antibiotics

19.4.1.1 Wastewaters

Antibiotics are the most commonly detected compounds in the monitoring campaigns, while the adverse effects in human and aquatic organisms have been reported (Chen et al., 2015). The inefficient removal of parent compounds in combination with concerns about the mode of action of metabolites and TPs has triggered the identification of their TPs in different environmental compartments, however, to a small extent.

The most commonly detected TP is anhydroerythromycin (ERY-H_2O), a TP of erythromycin (ERY), belonging in the class of macrolide antibiotics. This dehydrated metabolite constitutes a necessary tool for the detection of ERY since the latter is rapidly converted to ERY-H_2O after administration (Hassanzadeh et al., 2007). According to the literature, ERY-H_2O no longer exhibits antibiotic properties (Roth and Fenner, 1994). The concentrations of ERY-H_2O in the influents of WWTPs were observed at levels up to 600 ng L^{-1} (Guerra et al., 2014), while earlier studies exhibited higher concentration levels in the influents (up to 1978 ng L^{-1}) (Xu et al., 2007).

Lower concentrations of ERY-H_2O were detected in the effluents ranging up to 270 ng L^{-1} (Guerra et al., 2014). However, Guerra et al. (2014) pointed out that in some cases, the concentrations of the specific TP were higher in the effluents compared to the influents. This is in accordance with previous studies (Gulkowska et al., 2008) that attributed this diversification either due to deconjugation of conjugated metabolites during the treatment process (Miao et al., 2002) since conjugate compounds are not analyzed in the influents and can be retransformed into the original compound during the treatment or due to adsorption of ERY-H_2O on particulate matter and, consequently, underestimation of the total content entering the WWTP since the TP is filtered out during sample preparation (Lindberg et al., 2005; Gulkowska et al., 2008). ERY-H_2O was also detected in integrated constructed wetlands in South China at lower concentrations ranging up to 44.5 ng L^{-1} in the influents and 6.4 ng L^{-1} in the effluents (Chen et al., 2015). As regards the removal efficiency, in general, high removal rates were reported (up to 85%) after secondary treatment (Karthikeyan and Meyer, 2006; Chen et al., 2015). Apart from ERY-H_2O, a hydrolysis product of ERY, called erythralosamine, has also been identified in wastewater effluents (Pieper et al., 2010). With regard to other TPs of macrolide antibiotics, concentrations up to 240 ng L^{-1} were reported for 14-hydroxy(*R*)-clarithromycin, the main metabolite of clarithromycin (Baumann et al., 2015). According to the authors' knowledge, this work is the only survey that described the occurrence of clarithromycin's TPs in WWTPs.

TABLE 19.4
Occurrence of Most Frequent TPs in Drinking Water Worldwide

Therapeutic Class/ Parent Compound	Transformation Products	Concentrations (ng L^{-1}) Min	Max	Mean ± SD	Country	Reference
Analgesics						
Acetylsalicylic acid	SA	3	19	–	France	Vulliet et al. (2011)
	SA	<LOQ	1.4	–	Serbia	Petrović et al. (2014)
	SA	–	2	–	Spain	Gros et al. (2012)
Dimethylaminophenazone	AMDOPH	–	900	–	Germany	Reddersen et al. (2002)
Lipid Regulators						
Clofibrate	CLFA	–	70	–	Germany	Stumpf (1996)
	CLFA	–	165	–	Germany	Stan (1994)
	CLFA	–	270	–	Germany	Heberer (2002)
	CLFA	–	170	–	Germany	Heberer and Stan (1997)
	CLFA	3.2	5.3	–	Italy	Zuccato et al. (2000)
	CLFA	0.9	1.1	–	Canada	Chen (2006)
Fenofibrate	FNBA	0.2	1	–	France	Vulliet et al. (2011)
Psychiatric Drugs						
Carbamazepine	Di-OH-CBZ	–	–	322 ± 8	Germany	Kaiser et al. (2014)
	CBZ-ep	–	128	–	Serbia	Petrović et al. (2014)
	CBZ-ep	–	2	1	Spain	Huerta-Fontela et al. (2011)
Carbamazepine/ oxcarbamazepine	10-OH-CBZ	–	–	21 ± 2	Germany	Kaiser et al. (2014)
	ADON	–	–	<LOQ	Germany	Kaiser et al. (2014)
	9-CA-ADIN	–	–	189 ± 3	Germany	Kaiser et al. (2014)
	BAQD	–	–	26 ± 2	Germany	Kaiser et al. (2014)
DZP	OXZ	n.d.	2.5		France	Vulliet et al. (2011)
Antihypertensive						
Valsartan	VALA	–	72	57	Germany	Nödler et al. (2013)
Stimulants						
Nicotine	COT	–	25	–	United States	Stackelberg et al. (2004)
Calcium Channel Blocker						
Nifedipine	NFD-H_2O	–	4	–	United States	Stackelberg et al. (2004)

Note: n.d., no detected; <LOQ, below limit of quatification.

Concerning tetracyclines, pH, redox, and light conditions can lead to abiotic transformation of these compounds, which can produce TPs via epimerization, dehydration, and proton transfer pathways (Halling-Sørensen et al., 2003). Among the possible TPs, anhydrotetracycline (ATC) and anhydrochlorotetracycline (ACTC), along with their epi-analogues, 4-epitetracycline (ETC), 4-epioxytetracycline (EOTC), 4-epichlorotetracycline (ECTC), 4-epianhydrotetracycline (EATC), and 4-epianhydrochlorotetracycline (EACTC), were detected in WWTP influents and effluents with mean concentrations up to 400 and 23 ng L^{-1}, respectively. Mean removals were calculated by

Guerra et al. (2014) for ETC ranging from 31% (WWTP with advanced biological nutrient removal) to 95% (WWTP with secondary activated sludge system) depending on the type of treatment.

Moreover, hydroxy-metronidazole (METR-OH), one of the major metabolites of METR which is biologically active, was detected in influents of WWTPs in Spain at concentrations ranging up to 454 ng L^{-1} (Gros et al., 2013). Slightly lower or higher concentrations were detected in WWTP effluents by Gros et al. (2013) and Santos et al. (2013), respectively, indicating the poor removal of the specific TP. Much higher concentrations up to 11 μg L^{-1} have been reported for METR-OH in hospital effluents in Portugal (Santos et al., 2013).

Despite the fact that sulfonamides are among the most used antibiotic compounds worldwide, sparse information is found about the occurrence of their TPs (Le-Minh et al., 2012; Qi et al., 2015). N^4-acetylsulfamethoxazole, a metabolite of sulfamethoxazole (SMX), has been identified in wastewater influents in China (Qi et al., 2015) at concentrations up to 640 ng L^{-1}, while lower concentrations were observed in the effluents (up to 320 ng L^{-1}). Removal efficiencies have ranged between 50% and 92% (Le-Minh et al., 2012; Qi et al., 2015). Concerning other TPs of sulfonamides, two TPs of the antibiotic trimethoprim, assigned as 2,4-diaminopyrimidin-5-yl) (3,4,5-trimethoxyphenyl) methanone and 2,4-diaminopyrimidine-5-carbaldehyde, were identified in the wastewater effluents in Greece (Kosma et al., 2014).

19.4.1.2 Surface Waters, Groundwater, and Drinking Water

The presence of antibiotics in the effluents of WWTPs or hospital wastewater treatment plants (HWWTPs) raises concerns since their presence in surface waters can contribute to an increase in bacterial species with antibiotic resistance (Le-Minh et al., 2010). Moreover, many TPs of antibiotics preserve the pharmacological activity of the parent compounds, thus underlying the need for further investigation concerning their presence in surface waters. Table 19.2 reports the concentration levels of TPs found in surface waters.

The presence of ERY-H_2O in surface waters has been confirmed by several authors. Ferrer and Thurman (2012) identified ERY-H_2O in different surface water samples, all downstream from effluent discharge, with a mean concentration of 62 ng L^{-1}. However, the authors mentioned that the specific TP was not very frequently detected (35%). Other studies exhibited slightly lower concentrations, up to 35 ng L^{-1} (de Jongh et al., 2012; López-Serna et al., 2012). Previous studies demonstrated similar concentration levels (Kasprzyk-Hordern et al., 2007), with an exception of Lin and Tsai (2009), who detected ERY-H_2O in river water at concentrations up to 75,500 ng L^{-1}. However, the specific rivers flow through municipal regions in Taipei district and are receivers of urban drainages, which include effluents from residential and industrial facilities.

Regarding SMX TPs, N^4-acetylsulfamethoxazole has been detected in river waters at concentrations up to 300 ng L^{-1} (Heeb et al., 2012), underlining the good correlation between the effluents and the surface waters, in accordance with previous studies (Ashton et al., 2004). Moreover, Nödler et al. (2012) detected other two TPs of SMX, desamino-SMX and 4-nitro-sulfamethoxazole (4-nitro-SMX), at concentrations up to 7.7 and 5.4 ng L^{-1}, respectively. The two TPs were detected in spring waters in Germany, and they are considered product of SMX by denitrifying microbial activity. Actually, SMX, as an aromatic amine, can undergo abiotic transformations under denitrifying conditions and can form diazonium cations, which could form desamino-SMX and 4-nitro-SMX, depending on the reaction conditions. However, these reactions are microbially mediated, since the reactants NO and nitrite were also microbially formed.

Moreover, target analysis for other *N*-acetyl derivatives of sulfamides (Heeb et al., 2012; López-Serna et al., 2012) was also conducted. N^4-acetylsulfadiazine and N^4-acetylsulfamethazine are metabolites of sulfadiazine and sulfamethazine, respectively. All acetylated derivatives of sulfamides, including N^4-acetylsulfamethoxazole, although they do not present pharmacological activity, might exhibit exotoxicity (Eguchi et al., 2004). However, only Heeb et al. (2012) detected N^4-acetylsulfamethazine in river water in China (up to 85 ng L^{-1}), while López-Serna et al. (2012) have not reported positive detections in Spain neither for N^4-acetylsulfadiazine nor for N^4-acetylsulfamethazine.

Finally, Baumann et al. (2015) detected 14-hydroxy(*R*)-clarithromycin in river water in Germany with concentrations up to 80 ng L^{-1}.

In studies conducted on the groundwater, the presence of ERY-H_2O has also been reported with concentrations not exceeding 8.51 ng L^{-1} (López-Serna et al., 2013). López-Serna et al. (2013) detected also N^4-acetylsulfadiazine (up to 37.2 ng L^{-1}), while García-Galán et al. (2011) reported positive findings for *N*-acetyl derivatives of sulfamides. N^4-acetylsulfamethoxazole, N^4-acetylsulfadiazine, N^4-acetylsulfamerazine, and N^4-acetylsulfamethazine have been detected in groundwater samples in Spain with maximum concentrations of 5.5, 1, 5.5, and 4.4 ng L^{-1}, respectively.

19.4.2 Nonsteroidal Anti-Inflammatory Drugs

19.4.2.1 Wastewaters

Nonsteroidal anti-inflammatory drugs (NSAIDs) are a therapeutic class widely consumed all over the world and very frequently detected in WWTP influent–effluents and surface water, groundwater, and drinking water. The most detected TPs from this therapeutic group are the TPs of acetylsalicylic acid, ibuprofen (IBU), and metamizole (MTZ).

Acetylsalicylic acid is rapidly deacetylated in the human body and initially forms the active metabolite salicylic acid (SA). SA concentration in the influents ranged up to 89.1 µg L^{-1} (Kosma et al., 2010), while in effluents, it ranged at lower levels up to 431.9 ng L^{-1} (Kosma et al., 2014), indicating good removal efficiency. High (67–99%) to complete SA removal was reported in secondary (Gracia-Lor et al., 2012; Gros et al., 2012; Stamatis and Konstantinou, 2013; Kosma et al., 2014; Papageorgiou et al., 2016) and tertiary treatment (Stamatis and Konstantinou, 2013).

IBU, which is one of the most frequently detected NSAIDs, is metabolized after administration, giving rise to 1-hydroxy-ibuprofen (1-OH-IBU), 2-hydroxy-ibuprofen (2-OH-IBU), carboxy-ibuprofen (CBX-IBU), and carboxy-hydratropic acid, with CBX-IBU being the main excreted metabolite. However, studies have shown that OH-IBU and CBX-IBU can also be formed during treatment in WWTPs but only to a small extent, suggesting that the major amounts in sewage come from human excretion (Zwiener et al., 2002). Moreover, among the hydroxyl derivatives of IBU, 2-OH-IBU is believed to be the most dominant, while 1-OH-IBU or even 3-OH-IBU is found in small concentrations in urine (Evgenidou et al., 2015).

Consequently, 1-OH-IBU was detected in influents of WWTPs in Spain at levels up to 5000 ng L^{-1} (Weigel et al., 2004; Ferrando-Climent et al., 2012), while higher concentration levels were reported for 2-OH-IBU in Spain and Canada (Guerra et al., 2014; Ferrando-Climent et al., 2012) reaching up to 130,000 ng L^{-1}. High concentration levels in WWTP influents were also reposted for CBX-IBU up to 36,230 ng L^{-1} (Ferrando-Climent et al., 2012; Guerra et al., 2014). These findings are in contrast with the percentage expected from the human metabolism input described by several authors (Djukovic et al., 2008; La Farré et al., 2008), where CBX-IBU excreted is reported as more abundant than 2-OH-IBU. Nevertheless, Ferrando-Climent et al. (2012) supported that 2-OH-IBU appeared to be the dominant metabolite in biodegradation studies indicating that the presence of metabolites in influent wastewater can thus be attributed to a double contribution: the input from human excretion and the input of microorganism activity in wastewater, transforming IBU before entering the WWTPs.

Lower concentrations were detected in the effluents of WWTPs for all three aforementioned TPs. 1-OH-IBU in the effluents ranged up to 1410 ng L^{-1}, while 2-OH-IBU ranged up to 15,000 ng L^{-1}. CBX-IBU concentrations reached 10,650 ng L^{-1}, indicating good removal efficiencies for all IBU TPs. More specifically, Ferrando-Climent et al. (2012) reported that all IBU metabolites were significantly removed, with removal rates for CBX IBU, 2-OH IBU, and 1-OH IBU at 72–100%, 77–100%, and 58–100%, respectively. However, 1-OH IBU appeared to be the most recalcitrant metabolite, being even less biodegradable than IBU itself. The lower elimination of 1-OH-IBU has also been reported by Collado et al. (2012) in experimental and simulation studies of IBU and metabolites. It is worth mentioning that the hydroxylated derivatives of IBU appear to be more toxic than the parent compound, increasing the inhibition percentage of bioluminescence from *Vibrio fischeri* as indicated by Méndez-Arriaga et al. (2008).

Metabolism of diclofenac (DCF) involves the direct conjugation of unchanged DCF or oxidation followed by conjugation. In humans, DCF is typically excreted as conjugated DCF (15%), 4′-hydroxy-diclofenac 4′-OHD (30%), 4′,5-dihydroxydiclofenac (15%), and 5-hydroxy-diclofenac 5-OHD (10%). Langford and Thomas (2011) identified 4′-OHD in WWTP effluents in Norway with mean concentrations up to 3032 ng L^{-1}, while 5-OHD ranged up to 675 ng L^{-1}.

After the administration, MTZ is metabolized to produce 4-methylaminoantipyrine (4-MAA). In the liver, demethylation of the absorbed 4-MAA transforms it to 4-aminoantipyrine (4-AA) and further acetylation forms acetylaminoantipyrine (4-AAA). Another important metabolite, 4-formylaminoantipyrine (4-FAA), is generated by an uncharacterized oxidation of the *N*-methyl group (Gyenge-Szabó et al., 2014). 4-AAA and 4-FAA are pharmacologically inactive metabolites. Monitoring studies confirmed the presence of the metabolites of MTZ and its active product 4-MAA such as 4-AAA, 4-FAA, and 4-AA in WWTP influents and effluents in Spain (Rosal et al., 2010; López et al., 2014) and Hungary (Gyenge-Szabó et al., 2014). Higher concentrations were observed for 4-FAA in WWTP influents with values up to 71,000 ng L^{-1} (Rosal et al., 2010). Similarly, for all MTZ metabolites, very high concentrations up to microgram per liter levels were found. For example, 4-AAA, 4-MAA, and 4-AA ranged up to 22,200; 1894; and 3325 ng L^{-1}, respectively (Rosal et al., 2010; Gracia-Lor et al., 2012). The same pattern is also observed in WWTP effluents but with lower concentrations compared to the influents. For instance, high concentrations were observed for 4-FAA up to 27,444 ng L^{-1}, while the concentrations of 4-AAA, 4-MAA, and 4-AA reached 6745, 1098, and 2253 ng L^{-1}, respectively. Removal efficiencies after the secondary treatment were calculated by Rosal et al. (2010) and ranged for all MTZ metabolites between 46% (for 4-AAA) and 68% (for 4-FAA). Moreover, Gyenge-Szabó et al. (2014) observed a seasonal variation of removal rates of 4-AAA and 4-FAA. 4-AAA reached 65–77% removal in summer and early autumn and 95% in winter and early spring, while 4-FAA had low removal rates in summer (~5%) and increased to 20–30% in winter with a conventional treatment. Although this outcome seems contradictory since the chemical oxygen demand of the effluent wastewater is higher in the cold seasons than in spring and summer, the authors explained the observation by the fact that the microorganisms used in the activated sewage sludge are capable of growth and reproduction at temperatures ranging between −15°C and +10°C. Therefore, better removals are observed in winter and early spring since, in summer, higher temperatures prevail.

19.4.2.2 Surface Waters, Groundwater, and Drinking Water

TPs of analgesic compounds were also reported in surface water. The concentrations of SA in river water ranged between <LOQ–89.1 ng L^{-1} in Spain, while in the same study, 4′-hydroxy-diclofenac (4′-OH-DCF) was also reported in river water at concentrations up to 48.2 ng L^{-1} (López-Serna et al., 2012). Concerning IBU metabolites, only 2-OH-IBU has been observed in river water at concentrations up to 573 ng L^{-1} in Spain (Huerta et al., 2015). The authors reported that the specific metabolite occurred at higher concentrations than those of the parent compound (193 ng L^{-1}) and the increased concentration of this TP in downstream sites indicated that WWTP could be a source of contamination.

High concentration levels were also detected in surface water for MTZ metabolites. The most abundant metabolite appeared to be 4-AAA, with concentration ranging from 79 to 2200 ng L^{-1} (de Jongh et al., 2012; Radović et al., 2015). 4-FAA, 4-AA, and AAA were also detected at concentrations up to 186, 31, and 260 ng L^{-1}, respectively (Heeb et al., 2012; Radović et al., 2015). Also, de Jongh et al. (2012) detected 1-acetyl-1-methyl-2-phenylhydrazide (AMPH), an oxidation product of dimethylaminophenazone, in river water in The Netherlands at concentrations up to 66 ng L^{-1}.

Moreover, López-Serna et al. (2013) reported that out of 23 detected TPs and metabolites in groundwater samples in Spain, SA was most ubiquitous, being present in every sample of the monitoring survey, but its concentration did not exceed 98 ng L^{-1}. Moreover, 4′-OH-DCF was also observed at one out of three groundwater locations with concentration of 147 ng L^{-1}. Target analysis was also conducted for DCF-acyl-β-D-glucuronide; however, it was not detected in any sample (López-Serna et al., 2013). Radović et al. (2015) analyzed 44 groundwater samples in Serbia and detected MTZ metabolites at concentrations up to 150 ng L^{-1} for 4-FAA and up to 105 ng L^{-1} for 4-AAA. Although MTZ has been banned in some countries such as the United States or the United Kingdom, it remains a drug of choice in others. Its metabolites are present in environment due to the spontaneous hydrolysis of the parent compound into the main metabolite, 4-MAA, which is further metabolized, giving rise to other compounds. The final metabolites, 4-FAA and 4-AAA, are often used as indicators of sewage contamination of natural waters (Radović et al., 2015).

Finally, sparse information exists concerning the detection of analgesics/anti-inflammatory TPs in drinking water. SA has been detected in DWTP effluents by Vulliet et al. (2011), who underline that the high concentration (3–19 ng L^{-1}) observed for SA is due to possible presence of the carboxylic function in the molecule, an electron-withdrawing group depressing the reaction of the aromatic ring with the oxidant. However, Boleda et al. (2011) achieved high SA removal (>99%) after oxidation with chlorine dioxide in a DWTP in Spain.

Additionally, Reddersen et al. (2002) detected, in DWTP effluent in Germany, 1-acetyl-1-methyl-2-dimethyloxamoyl-2-phenylhydrazide (AMDOPH), another oxidation product of dimethylaminophenazone, with concentration of 900 ng L^{-1}. AMDOPH was the only metabolite that was not efficiently removed (25%) by the conventional drinking water treatment. Moreover, AMPH was detected in drinking water in the Netherlands, which was produced from riverbank filtrates. The maximum concentration of AMPH was up to 10 ng L^{-1} (de Jongh et al., 2012).

19.4.3 Lipid Regulators

19.4.3.1 Wastewaters

Another therapeutic class which is consumed and detected worldwide is lipid regulators. Regarding the detection of metabolites/TPs of this group of pharmaceuticals in the environment, several publications have reported the detection of clofibric acid (Kosma et al., 2010, 2014), but scarce information exists for fenofibric acid (Rosal et al., 2010) and other metabolites, such as *o*- and *p*-hydroxy-atorvastatin and simvastatin hydroxy acid (Vanderford and Snyder, 2006; Benotti et al., 2009).

Clofibric acid (CLFA) and fenofibric acid (FNBA) are the polar metabolites of etofibrate or clofibrate and fenofibrate, respectively (Monteiro and Boxall, 2010). Chlofibric acid is one of first metabolites which have been detected in the mid-1970s in the United States. Clofibrate and etofibrate are immediately metabolized in human body to CLFA (95–99%), which exhibits persistence and low biotransformation (Stamatis and Konstantinou, 2013). CLFA was frequently detected with concentrations ranging up to 740 ng L^{-1} in the influents of WWTPs. As an exception, the findings of Salgado et al. (2011) revealed very high concentrations in WWTP influents, scaling up to 41.4 μg L^{-1}. Lower concentrations were detected in the effluents at concentrations up to 258 ng L^{-1}. The removal rates of CLFA after secondary and tertiary treatments were 64.2% and 73.8%, respectively, as reported by Stamatis and Konstantinou (2013). Additionally, higher than 75% removal was observed by Salgado et al. (2012) after biological treatment, while Behera et al. (2011) reported 76% removal after the secondary treatment in S. Korea.

Rosal et al. (2010) detected FNBA up to 117 ng L^{-1} in the influents, while higher concentrations were observed in the effluents, reaching up to 336 ng L^{-1} (Vanderford and Snyder, 2006; López et al., 2014). Moreover, Muñoz et al. (2009) reported very high FNBA concentrations in the effluents up to 200,000 ng L^{-1} at two WWTPs in Spain. Rosal et al. (2010) also reported that FNBA had almost negligible removal efficiency (~1.3%) after the secondary treatment.

19.4.3.2 Surface Waters, Groundwater, and Drinking Water

The metabolites of lipid regulators have been reported in samples from surface water, groundwater, and drinking water. CLFA occurred in river water at concentrations up to 168 ng L^{-1} and in groundwater up to 17.6 ng L^{-1}. Moreover, a statin metabolite simvastatin hydroxy-carboxylic acid was also detected in surface water in Norway at concentrations up to 60 ng L^{-1} (Langford and Thomas, 2011). Target analysis was also conducted by López-Serna et al. (2012) for 2-OH-atorvastatin, a metabolite of atorvastatin, in river water in Spain, but it was not detected in any sample.

No recent studies exist concerning the presence of lipid regulator TPs in drinking water. However, in studies conducted in the previous decade, CLFA was found at maximum concentration of 270 ng L^{-1} in finished drinking water treatment in Germany (Heberer et al., 2001). This compound has been reported in five separate studies by Heberer and Stan (1997), Stan (1994), and Stumpf (1996) in Germany and Zuccato et al. (2000) in Italy. Also, FNBA was detected in drinking water treatment, at concentrations up to 11 ng L^{-1} in France (Vulliet et al., 2011).

19.4.4 Psychiatric Drugs

19.4.4.1 Wastewaters

Psychiatric drugs are one of the most known widely prescribed and widely consumed pharmaceuticals worldwide. The psychiatric class includes antidepressants, neuroactive pharmaceuticals, anticonvulsants, and antiepileptic compounds, such as carbamazepine (CBZ), which is one of the most frequently detected pharmaceuticals.

CBZ has been detected in influents, in effluents of WWTPs, and in other water matrices. After consumption, 70% is metabolized, leading to the formation of more than 30 compounds (Miao et al., 2005). The main metabolic pathway for CBZ is oxidation, leading to the formation of carbamazepine-10,11-epoxide (CBZ-Ep), which is a pharmacologically active compound with anticonvulsant properties. CBZ-Ep is further metabolized into 10,11-dihydro-10,11-*trans*-dihydroxycarbamazepine (DiOH-CBZ), which, in turn, leads to the formation of 9-hydroxy-methyl-10-carbamoylacridan. The DiOH-CBZ metabolite is not pharmaceutically active. Another metabolic route involves hydroxylation and production of the hydroxylated compounds 2-hydroxycarbamazepine (2-OH-CBZ) and 3-hydroxycarbamazepine (3-OH-CBZ), while 4-hydroxycarbamazepine (4-OH-CBZ) plays a minor role. Apart from these two major metabolic pathways, traces of acridine, acridone (ADON), iminostilbene, 2-hydroxy-iminostilbene, and 9-acridine-10-carboxaldehyde are formed in a minor pathway. Acridine has also been reported as genotoxic, mutagenic, and carcinogenic (Chiron et al., 2006).

DiOH-CBZ and 9-hydroxymethyl-10-carbamoylacridan are also generated through metabolism of oxcarbazepine (OxCZ), a keto-analogue of CBZ, which is also used as an anticonvulsant. However, OxCZ is mainly metabolized into 10-hydroxy-10,11-dihydro-carbamazepine (10-OH-CBZ), its therapeutically active metabolite. On the contrary, according to pharmacokinetics, the amount of 10-OH-CBZ excreted in human urine after metabolism of CBZ is lowered to 2-OH-CBZ, 3-OH-CBZ, and EP-CBZ. Also, in a recent study conducted by Kaiser et al. (2014), some more biotransformation products have been elucidated and identified as possible TPs such as DiOH-CBZ, 10-OH-CBZ, and OxCZ. 1-(2-benzoic acid)-(1*H*,3*H*)-quinazoline-2,4-dione (BAQD), 1-(2-benzoic acid)-(1*H*,3*H*)-quinazoline-2-one, 9-aldehyde-acridine, 9-carboxylicacid-acridine (9-CA-ADIN), hydroxyl 9-CA-ADIN, ADON, 11-keto-OxCZ, and 2,2′-(carbamoylazanediyl) dibenzoic acid.

Furthermore, Li et al. (2009) also demonstrated the formation of DiOH-CBZ and acridine as photoproducts of OxCZ. More specifically, the authors demonstrated that acridine was generated by DiOH-CBZ, which underwent a ring contraction process probably upon direct photolysis (Evgenidou et al., 2015).

Most of the aforementioned TPs were detected in WWTP influents or effluents. The concentrations of CBZ-EP in WWTPs ranged up to 16,208 ng L^{-1}, while 10,11-dihydro-10,11-dihydroxy-carbamazepine (Di-OH-CBZ) was found at lower concentrations in the influents, up to 4960 ng L^{-1}.

However, Bahlmann et al. (2014) detected much higher concentrations of Di-OH-CBZ compared to CBZ-EP. Although these findings are in agreement with the measurements conducted in human plasma, where CBZ-EP concentrations are threefold to fourfold lower than those of DiOH-CBZ; in some cases, it was detected in sewage at concentrations at least 50-fold lower than those of DiOH-CBZ (Leclercq et al., 2009). This is consistent with the known instability of epoxides in aquatic solutions and the possible conversion of CBZ-Ep to DiOH-CBZ in sewage. 2-OH-CBZ was also detected in WWTP influents at concentrations up to 15,939 ng L^{-1}, while 3-OH-CBZ and 10-OH-CBZ were observed at much lower levels scaling up to 150 and 490 ng L^{-1}, respectively (Bahlmann et al., 2014). Moreover, 4-OH-CBZ was also identified by Bahlmann et al. (2014) in WWTP influents and effluents in Germany and Portugal, but it was not quantified. Kaiser et al. (2014) detected 9-CA-ADIN and BAQD in WWTP influents with mean concentrations up to 700 and 160 ng L^{-1}, respectively. The last research group identified also ADON, but its concentrations were below the LOQ.

In many cases, slightly higher concentrations were observed in the effluents for most CBZ TPs, indicating their negligible removal in WWTPs. This observation is because CBZ and its metabolites have been found to be partially excreted as glucuronide conjugates and the activated sludge exhibits glucuronidase activity, thus releasing the parent compounds and metabolites in free form (Ternes et al., 1999).

Furthermore, ritalinic acid was detected in wastewaters as the metabolite of the drug methylphenidate (known as Ritalin) (Letzel et al., 2010). Ritalinic acid shows none or very low pharmaceutical activity. Ritalinic acid was frequently detected by Letzel et al. (2010) in influents and effluents of WWTP in Germany (13 of 16 samples) with concentrations ranging up to 270 and up to 170 ng L^{-1} in the influents and effluents, respectively. The removal of the specific TP fluctuated from no removal to 44.9% with a mean value of approximately 23% (Letzel et al., 2010).

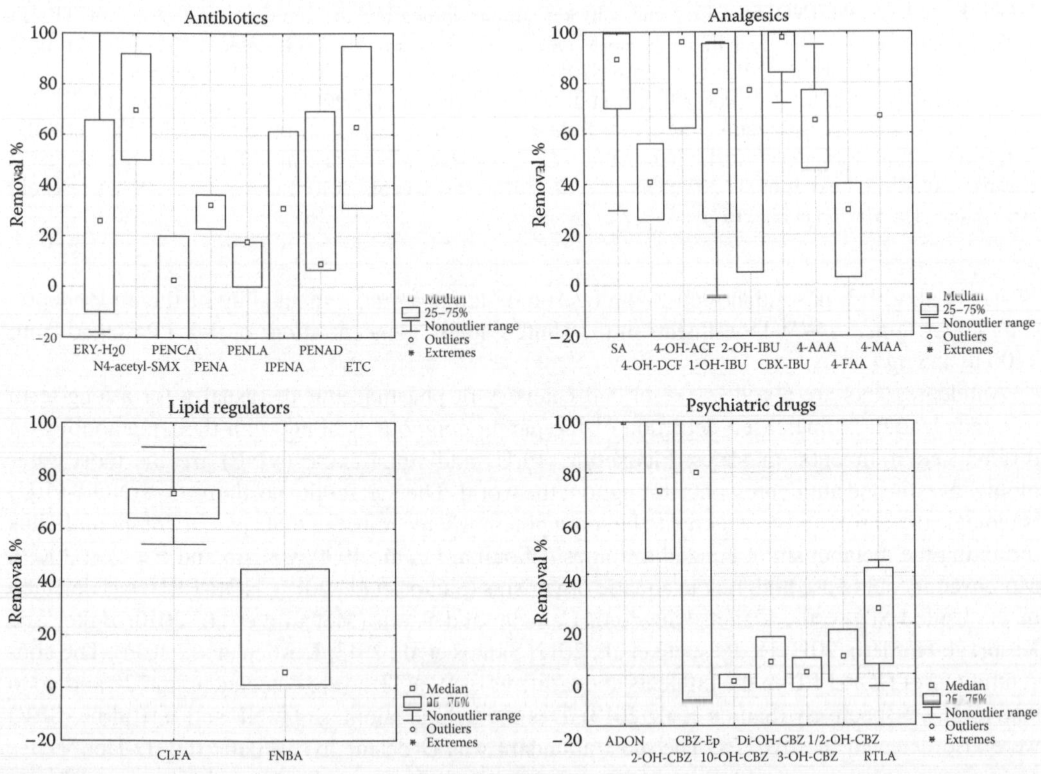

FIGURE 19.1 Box and whisker graphs indicating the removal efficiencies of TPs in WWTPs. *(Continued)*

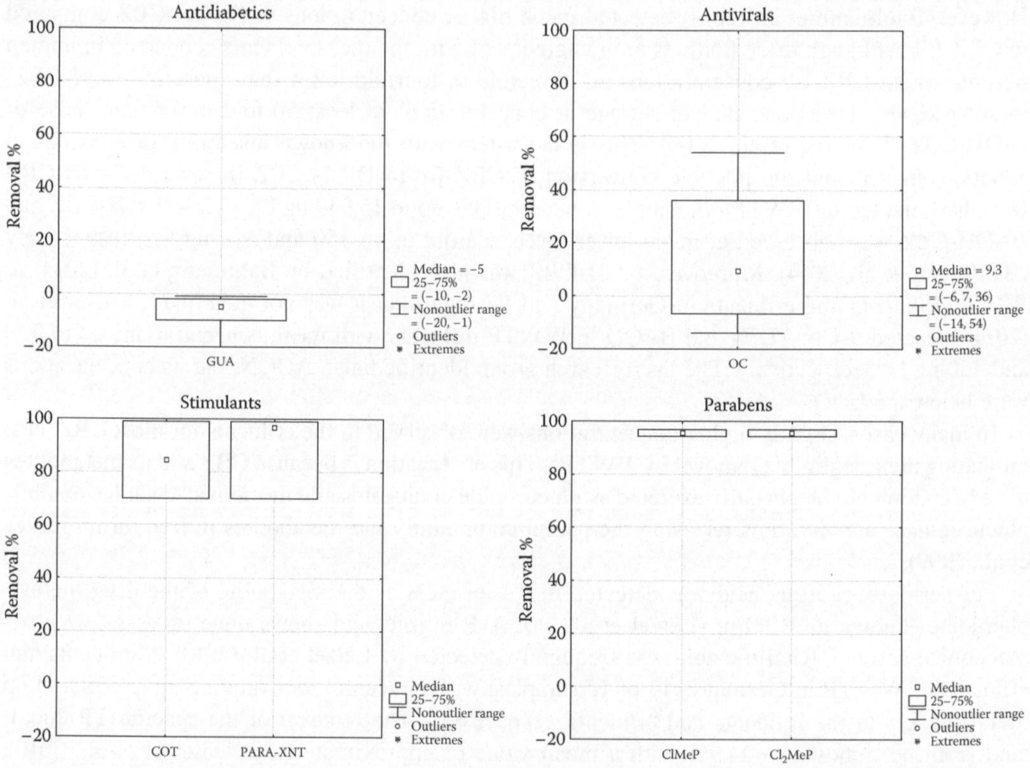

FIGURE 19.1 (CONTINUED) Box and whisker graphs indicating the removal efficiencies of TPs in WWTPs (Chen et al., 2011; Karthikeyan and Meyer, 2006; Guerra et al., 2014; Qi et al., 2015; Le-Minh et al., 2012; Gracia-Lor et al., 2012; Gros et al., 2012; Stamatis and Konstantinou, 2013; Kosma et al., 2014; Papageorgiou et al., 2016; Ferrando-Climent et al., 2012; Collado et al., 2012; Méndez-Arriaga et al., 2008; Rosal et al., 2010; Gyenge-Szabó et al., 2014; Boleda et al., 2011; Salgado et al., 2012; Behera et al., 2011; Letzel et al., 2010; Writer et al., 2013; Bahlmann et al., 2014; Huerta-Fontela et al., 2011; Scheurer et al., 2012; Kosma et al., 2015; Leknes et al., 2012; Ghosh et al., 2010; Prasse et al., 2010; Buerge et al., 2008; Lee et al., 2013; González-Mariño et al., 2011).

Writer et al. (2013) have identified 2-*N*-glucuronide-lamotrigine, a metabolite of the antiepileptic drug lamotrigine, in WWTP effluents in the United States. Concentrations of this TP ranged from <100 to 455 ng L^{-1}.

Antidepressants are a widely prescribed category of pharmaceuticals, usually for a long-term use, thus leading to increased detection of the parent compounds along with their metabolites in different environmental matrices. Fluoxetine (FLU) and venlafaxine (VNF) are the most commonly investigated antidepressants throughout the world. The metabolite norfluoxetine (NOR-FLU) is biologically active and considered to be more potent when compared to its parent compound, thus inhibiting the metabolism of benzodiazepines (Chouinard et al., 1999; Calisto and Esteves, 2009). Moreover, it could pose high risk to aquatic organisms (Santos et al., 2013). NOR-FLU was detected in the United States, the United Kingdom, Canada, and Spain (Metcalfe et al., 2010; Baker and Kasprzyk-Hordern, 2011; Hedgespeth et al., 2012; Santos et al., 2013; Kostich et al., 2014). The concentrations of NOR-FLU in the influents and effluents of WWTPs were detected up to 226 and up to 99.6 ng L^{-1}, respectively (Santos et al., 2013). Des-methyl metabolites of VNF and sertraline (SER) were also detected in WWTPs. The most abundant was *O*-desmethyl venlafaxine (*O*-Des-VNF), and its concentrations ranged up to 2600 ng L^{-1} in the influents and 1637 ng L^{-1} in the effluents (Metcalfe et al., 2010), indicating the poor removal in WWTPs.

Among benzodiazepines, oxazepam (OXZ), desmethyldiazepam (Des-DZP) (nordiazepam), 7-aminoflunitrazepam (7-amino-FLU), and α-amino-alprazolam are the active metabolites of diazepam, flunitrazepam, and alprazolam, respectively (Borova et al., 2014; Subedi and Kannan, 2015). Moreover, these metabolites were detected in the influents up to 62.7 ng L^{-1} and in the effluents up to 74.1 ng L^{-1}. Higher concentrations were reported for OXZ in 90 WWTPs up to 1766 ng L^{-1} (Loos et al., 2013). The removal rates of psychiatric drugs were depicted with box plots (Figure 19.1).

19.4.4.2 Surface Waters, Groundwater, and Drinking Water

TPs of psychiatric drugs were also identified in surface waters. Among these TPs, CBZ-EP appeared with higher concentrations in river water, scaling up to 1667 ng L^{-1} (López-Serna et al., 2012). The authors observed that CBZ-EP appeared with the highest individual and the highest mean concentration than other metabolites, and it was detected in 80% of the collected samples. Other studies also confirm its presence in river water in the Netherlands, China, and Serbia (Langford and Thomas, 2011; de Jongh et al., 2012; Heeb et al., 2012; López-Serna et al., 2012) or even lake and seawater (Gros et al., 2012; Petrović et al., 2014).

Moreover, Di-OH-CBZ (up to 889 ng L^{-1}), 10-OH-CBZ (up to 819 ng L^{-1}), 3-OH-CBZ (up to 70 ng L^{-1}), 2-OH-CBZ (up to 61.7 ng L^{-1}), acridine (up to 16.8 ng L^{-1}), ADON (up to 27 ng L^{-1}), 9-CA-ADIN (up to 304 ng L^{-1}), and BAQD (up to 74 ng L^{-1}) were also detected in river water (Table 19.2).

Metabolites of antidepressants were also detected in surface waters. *O*-Des-VNF, hydroxyl bupropion (OH-BPP), and eryhtrohydrobupropion were the most abundant with maximum concentrations of 112, 180 (mean), and 150 ng L^{-1} (mean), respectively. Lower concentrations were reported for ritalinic acid (RTLA) in surface water, up to 23 ng L^{-1} in tributaries in Spain (López-Serna et al., 2012).

Moreover, diazepam (DZP) metabolites, OXZ, OXZ glucoronide, and Des-DZP have been detected in surface water. Highest concentrations were observed for the former (up to 70.9 ng L^{-1}) since DZP is mainly excreted with its conjugated metabolite, OXZ glucuronide (33%), and just traces of Des-DZP and unchanged pharmaceutical (López-Serna et al., 2012).

The presence of these compounds raises great environmental concerns since the bioactivity of neuroactive metabolites has been demonstrated in vivo by the medical community (Howell et al., 1993; Haas et al., 2004). Moreover, metabolites of neuroactive compounds have also been shown to accumulate in the brain tissue of white suckers *Catostomus commersoni* (Schultz et al., 2010) and brook trout with evidence for negative biological effects (Lajeunesse et al., 2011).

In surface water, CBZ-EP was found at 15, 5, and 4 times higher concentrations than parent CBZ, while in groundwater, CBZ TPs were detected at much lower concentrations than the parent compound (Huerta-Fontela et al., 2011; Langford and Thomas, 2011; López-Serna et al., 2012). The concentrations of TPs increased in the order acridin < ADON < CBZ-EP < 3-OH-CBZ < 2-OH-CBZ in groundwater samples. Also, *N*-desmethylvenlafaxine was detected at concentrations up to 3.9 ng L^{-1} and Des-DZP up to 12.9 ng L^{-1} (Huntscha et al., 2012; López-Serna et al., 2013).

Finally, although CBZ, the parent compound, can be efficiently eliminated by treatment with the ozonation process, CBZ epoxide was found to persist in drinking water treatment processes, such as flocculation, sand filtration, sedimentation, ozonation, and chlorination. The different behavior of the metabolite of CBZ was explained due to the different reaction mechanism that takes place with ozone (Huerta-Fontela et al., 2011). Also, CBZ and its epoxy metabolite also survived chlorination with poor removals above 5% and 30%, respectively, yielding to almost unaltered concentrations after treatment. Petrović et al. (2014) detected this TP in drinking water with mean concentration of 128 ng L^{-1}.

Kaiser et al. (2014) conducted the identification of TPs in finished drinking water. The TPs DiOH-CBZ, 9-CA-ADIN, BAQD, and 10-OH-CBZ were detected at concentrations up to 322, 189, 26 and 21 ng L^{-1}, respectively, except ADON, which was below the LOQs. BAQD was present in drinking water because of its biotic formation. It should be noted that BAQD is also known to be present in drinking water as an ozonation product of CBZ as well as an (in)direct photolysis product of OxCZ and CBZ. Thus, its formation seems to be favored by several processes (Kaiser et al., 2014). Among others, psychiatric drug OXZ was detected up to 2.5 ng L^{-1} in drinking water (Vulliet et al., 2011).

19.4.5 Beta-Blockers

19.4.5.1 Wastewaters

Beta-blockers are among the top 200 drugs prescribed in the United States (Scheurer et al., 2010), used for the treatment of hypertension. Metoprolol and atenolol are pharmaceuticals of high consumption, and therefore, they have been widely detected in wastewaters and surface waters (Jelic et al., 2011; Petrović et al., 2014; Sun et al., 2014; Papageorgiou et al., 2016). The three main metabolites of metoprolol, metoprolol acid (MTPA), α-hydroxy-metoprolol (α-HMTP), and *O*-desmethylmetoprolol (O-DMTP), account for 85% of the urinary excretion, whereas around 10% is excreted in the unchanged form (Barclay et al., 2012). However, both atenolol and metoprolol can form MTPA as their major biotransformation product (Kern et al., 2010). MTPA is also referred to as atenolol acid (Qi et al., 2015).

Concentrations of MTPA in WWTP influents in China ranged up to 1380 ng L^{-1}, while slightly lower concentrations in effluents (up to 1240 ng L^{-1}) were identified in the same study, indicating poor removal (Qi et al., 2015). Moreover, Rubirola et al. (2014) identified much higher concentrations in the effluents (up to 2506 ng L^{-1}) compared to those measured in the influents (up to 298 ng L^{-1}). Additionally, concentrations up to 5400 ng L^{-1} in WWTP effluents were reported by Reid et al. (2011) in Norway, who observed that the MTPA was detected in every sample from the Oslo sewage stream effluent. These findings were explained by the fact that MTPA can also be formed from metoprolol or other β-blockers present in WWTP sewage (such as atenolol), by biotransformation during treatment (Rubirola et al., 2014).

On the contrary, α-HMTP was detected only in WWTP influents at concentrations up to 36 ng L^{-1} (Rubirola et al., 2014). Its complete removal was attributed by the authors to its biodegradation. Finally, the nondetection of O-DMTP in the influents and effluents of the selected WWTPs can be attributed to its rapid transformation to MTPA in human metabolism (Barclay et al., 2012). Therefore, although a small proportion of metoprolol is biodegraded to O-DMTP in the effluents, the latter is very fast removed according to batch experiments conducted by the authors. Toxicity tests with *V. fischeri* did not show any acute toxicity of α-HMTP and MTPA (Rubirola et al., 2014).

19.4.5.2 Surface Waters, Groundwater, and Drinking Water

Lower concentrations were observed for MTPA in surface water, not exceeding 197 ng L^{-1}. Similar concentrations were detected for α-HMTP (up to 108 ng L^{-1}). As an exception, Heeb et al. (2012) reported higher concentration levels for MTPA (up to 730 ng L^{-1}) in Haine River in China.

Concerning groundwater samples, Huntscha et al. (2012) detected atenolol-desisopropyl (ATN-Des), a TP of ATN, at concentrations up to 6.4 ng L^{-1}. Moreover, López-Serna et al. (2013) detected 4-hydroxy-propranolol (4-OH-PRP) up to 21.4 ng L^{-1}. Propranolol, another β-blocker, is extensively metabolized in the liver, giving rise to naphthyl acetic acid and 4-OH-PRP, which have been reported as the major metabolites in humans and animals (Partani et al., 2009). Moreover, PRP is also excreted as a glucuronic acid and sulfate conjugate. However, although target analysis was conducted for propranolol-β-D-glucuronide in surface waters, it was not detected in any sample (López-Serna et al., 2012).

19.4.6 Antidiabetics

19.4.6.1 Wastewaters

Antidiabetics are drugs which are widely used for the treatment of diabetes. This disease is chronic, and consequently, the specific drugs are continuously being released into the environment through wastewater discharges from wastewater treatment plants. Metformin (MTF) is the first-line drug of choice used for the treatment of diabetes mellitus type 2 and is excreted nonmetabolized in the urine. Recent studies (Scheurer et al., 2012; Trautwein et al., 2014) suggested that MTF is biologically transformed in activated sludge to the TP, known as guanylurea (GUA). GUA was detected in WWTPs in Greece (Kosma et al., 2015), in Germany (Scheurer et al., 2012; Trautwein et al., 2014), and in The Netherlands (Oosterhuis et al., 2013).

Kosma et al. (2015) detected GUA in seven WWTP influents in Greece at concentrations ranging up to 59 ng L^{-1}, while Scheurer et al. (2012) detected much higher concentrations of the TP, scaling up to nanogram per liter level (3000 ng L^{-1}). Moreover, higher concentrations were observed in the effluents, compared to the influents, in all studies. The higher values were observed by Scheurer et al. (2012), ranging from 18 to 99 μg L^{-1}. Negative removal efficiencies were presented by GUA since it is formed by biodegradation of MTF during treatment as aforementioned.

Consequently, because of their incomplete degradation and removal during wastewater treatment, both the compounds MTF and GUA are expected in considerable amounts into receiving waters. Furthermore, ter Laak and Baken (2014) reported that the concentrations of MTF and GUA together account for more than half of the total load of pharmaceuticals in surface waters. Trautwein et al. (2014) and Scheurer et al. (2012) detected GUA in different environmental matrices (lake, river, and seawater), with higher observed concentrations in Weser River (up to 391 ng L^{-1}).

Both MTF and its metabolite GUA are extremely polar compounds, and consequently, they are expected to be very mobile in the aquatic environment. Therefore, a certain degree of drinking water relevance cannot be excluded as they might be present in raw waters used for drinking water production (Scheurer et al., 2012).

19.4.7 Antivirals

19.4.7.1 Wastewaters

Antiviral drugs are among the most hazardous therapeutic classes with regard to their toxicity toward algae, daphnids, and fish (Sanderson et al., 2004). Additionally, the seasonal variations of antivirals, with peak emissions during influenza epidemics, have raised concerns about their potential environmental effects. Two groups of antivirals drugs have been extensively used for influenza treatment, the neuraminidase inhibitors and M2 ion channel inhibitors (Ghosh et al., 2010). Oseltamivir phosphate (OP) belongs to the first group, and it is a drug with common use as a first defense from a flu pandemic. After administration, OP can be extensively hydrolyzed to its active metabolite, oseltamivir carboxylate (OC). Its release into the environment might cause serious effects such as the generation of OC resistance in wildfowl, taking into account that birds are competent hosts of influenza viruses (Prasse et al., 2010).

Ghosh et al. (2010), Leknes et al. (2012), and Prasse et al. (2010) detected the concentrations of OC in wastewaters. Prasse et al. (2010) detected OC up to 42.7 ng L^{-1} in the influents, while higher concentrations (up to 1200 ng L^{-1}) were reported by Leknes et al. (2012) during an influenza pandemic in Norway. Higher concentrations were detected in the effluents in both studies, up to 293.3 ng L^{-1} in Germany (Prasse et al., 2010) and up to 1400 ng L^{-1} in the effluents of WWTP in Norway (Leknes et al., 2012). In the latter study, the authors demonstrated negative (−14%) to low (0.6%) removal rates after biological treatment. Additionally, Ghosh et al. (2010) and Prasse et al. (2010) reported also low (18%) to medium removal rates (54%) after secondary treatment in WWTPs.

19.4.7.2 Surface Waters, Groundwater, and Drinking Water

Few studies exist concerning the occurrence of OC in surface water samples. Ghosh et al. (2010) detected OC in river water in Japan with concentrations ranging from 6.6 to 190.2 ng L^{-1}, while lower concentrations were detected in Germany (up to 2.4 ng L^{-1}) (Prasse et al., 2010). The environmental fate of oseltamivir after release into water bodies has not been studied extensively; however, it has been shown that sun exposure and sediment content significantly reduce persistence of OC in surface waters, due to increased activity of OC-degrading bacteria (Goncalves et al., 2011).

19.4.8 Cytostatics (Anticancers)

19.4.8.1 Wastewaters, Surface Waters, Groundwater, and Drinking Water

Cytostatics (anticancers) are a major therapeutic class which is less studied worldwide. They are very toxic to microorganisms and harmful to crustaceans and fish (Carlsson et al., 2006). Due to their high pharmacological potency, such compounds often exhibit carcinogenic, mutagenic, or embryotoxic properties, even at low concentrations (Zounková et al., 2007; Haddad et al., 2015). Cytostatics and metabolites can be discharged in WWTPs, through hospital effluents and domestic wastewater and consequently in the environment, through excretion of urine and feces from patients who undergo chemotherapy. Although hospital wastewaters should be the main source of cytostatics, Ferrando-Climent et al. (2013) reported the equal discharge of cytostatics from HWWTPs and WWTPs, due to the fact that patients can leave the hospital after the administration of cytostatics. Moreover, Negreira et al. (2015) mentioned that hospital effluents are rarely pretreated before discharging in WWTPs.

Besides the extended use of cytostatics, scarce data exist concerning their TPs and their occurrence in the environment (Ferrando-Climent et al., 2013, 2014; Negreira et al., 2014a,b, 2015). Negreira et al. (2013) reported the occurrence of hydroxy-paclitaxel (OH-PAC), the metabolite of paclitaxel (Mahmood et al., 2014). In the influents, OH-PAC ranged up to 18.5 ng L^{-1} and in the effluents up to 3.7 ng L^{-1}. In the same survey, 4-hydroxy-*N*-desmethyl-tamoxifen (OH-D-TAM) and (*Z*)-4-hydroxy-tamoxifen (OH-TAM), the metabolites of tamoxifen (TAM), have been detected only in the effluents up to 164 ng L^{-1}. Studies also showed that OH-PAC and OH-TAM are rather labile compounds concerning their stability in wastewaters (Negreira et al., 2014a). Moreover, Ferrando-Climent et al. (2013) had detected these two metabolites of TAM and one metabolite of the cytostatic cyclophosphamide, carboxyphosphamide, in WWTP, but their detection was below the quantification limit.

Finally, Negreira et al. (2015) investigated the occurrence of 3′-*O*-desmethyl etoposide (3′-*O*-Des-ETP) in wastewater and surface water. Although this compound is a known metabolite of the drug etoposide (ETP), its urinary excretion has been reported to be minor. About 50% of an intravenous dose of ETP is recovered in urine; up to 17% is excreted as a glucuronide metabolite and less than 2% as the 3′-*O*-desmethyletoposide metabolite (Holthuis et al., 1986). The authors proved that 3′-*O*-Des-ETP can also be formed as a TP of the reaction of free chlorine with the parent compound ETP. 3′-*O*-Des-ETP has been quantified in raw wastewaters up to 33 ng L^{-1} and in similar concentrations (up to 31 ng L^{-1}) in surface water.

Taken together, the studies on the occurrence of cytostatics in WWTPs illustrate that further investigations on their occurrence and fate are needed regarding the concerns that metabolites can be more potent and active than the parent compound, as, for instance, TAM metabolites, which are 30- to 100-fold more potent and active than its precursor (Besse et al., 2012; Jager et al., 2012).

19.4.9 Antihypertensive Drugs

19.4.9.1 Wastewaters

The antihypertensive drugs are used to treat high blood pressure. Valsartan and enalapril have been detected in WWTPs worldwide (Huerta-Fontela et al., 2010; López-Serna et al., 2010; Salgado et al., 2010; Gros et al., 2012; Verlicchi et al., 2012a; Cruz-Morató et al., 2014; Jelic et al., 2015; Kostich et al., 2014; Petrović et al., 2014; Acuña et al., 2015; Moreno-González et al., 2015). Kern et al. (2010) reported that during activated sludge treatment, valsartan undergoes a series of sequential transformations. Firstly, the oxidative dealkylation of valsartan takes place followed by the hydrolysis of the dealkylated valsartan, giving rise to the amino-valsartan, which is finally oxidized to 2-(2*H*-tetrazol-5-yl)-[1,1-biphenyl]-4-carboxylic acid, referred to as valsartan acid (VALA). The latter appears to be the most stable of the three TPs. VALA was detected in the influents (up to 320 ng L^{-1}) and in the effluents (up to 1053 ng L^{-1}) of WWTPs in Germany and Switzerland (Kern et al., 2010; Nödler et al., 2013). Kern et al. (2010) reported that when 80% of valsartan was removed during treatment with activated sludge, the concentrations of its TP VALA increased about eightfold from 140 to over 1000 ng L^{-1} due to the formation of the specific TP during treatment, as aforementioned. However, due to structural similarities, this compound could potentially be formed from selected pharmaceuticals different from valsartan but belonging to the same compound class.

19.4.9.2 Surface Waters, Groundwater, and Drinking Water

Due to their short application period (<20 years), data concerning the fate and occurrence of sartans in water cycle is limited. However, VALA was detected in surface water, but at higher concentration (2199 ng L^{-1}) than wastewater, so further investigation needs to be considered. According to Nödler et al. (2013), the concentration of VALA in Berlin tap water (57–72 ng L^{-1}) might prove its persistence in bank filtration.

Furthermore, enalaprilat (ENAT), the active metabolite of the prodrug enalapril, was detected in WWTP effluents up to 150 ng L^{-1} (Kostich et al., 2014). Lower concentrations were detected in surface water and groundwater, not exceeding 7.32 and 4.78 ng L^{-1}, respectively.

19.4.10 Antihistamines

19.4.10.1 Surface Waters

Antihistamines are a therapeutic class administered for treatment and prevention of allergies. Moreover, seasonal variations have been presented, with peak emissions during pollen production (Petrie et al., 2015). Besides the extended use of antihistamines, scarce data exist for their TPs concerning their occurrence in the environment and fate. Ranitidine is an H_2 histamine receptor antagonist that works by blocking histamine and thus decreasing the amount of acid released by cells of the stomach. Urine is the main route of excretion for ranitidine. Approximately 30% of the orally administered dose is excreted as an unchanged drug in 24 h. The *N*-oxide (ranitidine *N*-oxide [RAN-*N*]) is the principal metabolite in the urine; however, this amounts to <4% of the dose. Other metabolites are the *S*-oxide (1%) and the desmethyl ranitidine (1%). However, only, RAN-*N* has been detected in river water up to 196 ng L^{-1} (López-Serna et al., 2012). Max concentrations of the parent compound ranitidine were also detected up to 136 ng L^{-1}. However, lower concentrations of the corresponding metabolite, ranitidine *N*-oxide, would be expected taking into account the data on excretion (4% vs. 30% of unchanged pharmaceutical). The difference in ratios may be attributed either to the variations of the theoretical percentages of excretions for elderly persons or patients with any other illness such as diabetes or high blood pressure since the metabolism may work differently in them (Tocco et al., 1982; Simon et al., 1988) or to the disposal of unused medicine (expired or no longer needed). However, other factors are the transformations in the WWTPs and/or in the environment that do not affect all compounds to the same extent.

19.4.11 Stimulants

19.4.11.1 Wastewaters, Surface Waters, and Groundwater

Nicotine and caffeine are two of the most widely consumed stimulants. The nicotine absorbed by a smoker is extensively metabolized in the human liver (70–80%) and excreted in the urine as cotinine (COT) (Bramer and Kallungal, 2003). COT concentrations have been detected in WWTP effluents up to 3095 ng L^{-1} (Buerge et al., 2008; Ghoshdastidar et al., 2015). Earlier studies demonstrated higher levels in WWTP influents and effluents (up to 42.3 and 17.9 μg L^{-1}, respectively)

(Huerta-Fontela et al., 2008). Due to the widespread consumption of tobacco, COT has been suggested as a potential chemical marker for domestic wastewater contamination of surface waters (Metcalfe et al., 2003; Glassmeyer et al., 2005; Benotti and Brownawell, 2007; Bradley et al., 2007). Consequently, the presence of COT in different environmental compartments was ubiquitous. COT was detected in river water (up to 80 ng L^{-1}) (Martínez Bueno et al., 2012) and in groundwater (up to 320 ng L^{-1}) (Barnes et al., 2008; Cabeza et al., 2012; Reh et al., 2013). Furthermore, Stackelberg et al. (2004) detected COT at 25 ng L^{-1} in finished drinking water.

The other stimulant, caffeine, is also extensively metabolized (~80%) through the catalytic action of CYP1A2, forming paraxanthine (PARA-XNT), theobromine, and theophylline (Gu et al., 1992). Ghoshdastidar et al. (2015) have reported high concentrations of PARA-XNT in WWTP effluents in Canada reaching 18.2 ng L^{-1}. PARA-XNT was also determined up to 421.4 ng L^{-1} in surface water (Baker and Kasprzyk-Hordern, 2011), while Barnes et al. (2008) detected PARA-XNT at 57 ng L^{-1} in groundwater.

19.4.12 Hormones

19.4.12.1 Wastewaters and Groundwater

Natural and synthetic hormones are present in wastewaters, surface waters, and groundwaters in measurable levels and could pose a risk in aquifers due to estrogenic effects, endocrine disruption, and interferences in reproduction (Evgenidou et al., 2015; Postigo and Barceló, 2015). Previous reviews report the presence of metabolites of estrone and estradiol such as 2-hydroxy-estrone, 4-hydroxy-estrone 16-α-hydroxy-estrone, and 2-hydroxy-estradiol in WWTP influents and effluents. Moreover, attention has been paid on estrogen conjugates, because, although they do not have affinity to estrogen receptors, evidence suggests that once they reach sewage treatment plants, they undergo chemical or enzymatic dissociation in bacterial sludge and reform active estrogens. From this perspective, β-estradiol-2-sulfate and estrone-3-sulfate (E1-3S) have also been identified in previous studies (Evgenidou et al., 2015). However, no recent studies exist concerning the occurrence of hormones in wastewaters. Moreover, conjugated estrogens were rarely detected in groundwater despite their high water solubility and they are likely to hydrolyze or deconjugate in the environment (Swartz et al., 2006; Shrestha et al., 2012; Postigo and Barceló, 2015). Shrestha et al. (2012) detected estradiol-3-glucuronide at an average concentration up to 425 ng L^{-1} in a well near an agricultural land in the United States. Furthermore, no available data occurred for these metabolites in surface water and drinking water.

19.4.13 Personal Care Products

19.4.13.1 Wastewaters

Personal care products (PCPs) are a diverse group of compounds, including fragrances, soaps, disinfectants, lotions, UV filters, and musks. Parabens, esters of 4-hydroxy-benzoic acid, are extensively employed as bactericides not only in a wide range of PCPs but also in beverages and in canned foods (Meyer et al., 2007; Lundov et al., 2009). Moreover, triclosan (5-chloro-2-(2,4-dichlorophenoxy) phenol), another member of PCPs, is an important bactericide used in various personal care (shampoo, toilet soap, deodorants, toothpaste) and consumer (footwear, plastic wear) products. As many personal care chemicals, they are continuously released in the aquatic media through domestic wastewater, and therefore, a growing concern has arisen in relation to their potential long-term effects on wildlife due to their estrogenic activity. Additionally, concerns may result from the possible formation of chlorinated derivatives since both parabens and triclosan are able to react with residual chlorine in tap water. Concerning parabens, although halogenation masks the apparent estrogenic activity of the parent compounds (Terasaki et al., 2009), the resulting chlorinated byproducts show higher acute toxicity responses in the *Daphnia magna* test (Terasaki et al., 2009).

Additionally, chlorinated triclosan derivatives (CTDs) such as 4,5-dichloro-2-(2,4-dichlorophenoxy) phenol (4-Cl-TCS), 5,6-dichloro-2-(2,4-dichlorophenoxy) phenol (6-Cl-TCS), and 4,5,6-trichloro-2-(2,4-dichlorophenoxy) phenol (4,6-Cl-TCS) exhibit antimicrobial and endocrine-disrupting properties (Canosa et al., 2005; Rule et al., 2005). Furthermore, additional concerns arise by the fact that 1,5,9-cyclododecatrienes (CDTs) can undergo photolysis in natural waters to form the respective dioxins, which are very toxic and their toxicity generally increases with chlorine substitution in the lateral positions (Goldstein and Safe, 1989).

Consequently, 3-chloro-methylparaben and 3,5-dichloro-methylparaben are some of the halogenated TPs of methylparaben that were detected in the influents of WWTPs in Spain by González-Mariño et al. (2011), at concentrations up to 90 ng L^{-1}. These chlorinated derivatives were also detected in some WWTP effluents but at lower concentrations (up to 12 ng L^{-1}). However, both TPs exhibited good removal efficiencies of 96.9% and 94.3%, respectively, although they are more persistent to biodegradation compared to the parent compound since halogenation decreases removal efficiency.

Additionally, all three chlorinated derivatives of triclosan have been identified in WWTP influents and effluents in the United States by Buth et al. (2011). Concentrations of CTDs in the influents were up to 98 ng L^{-1}, while lower concentrations ranged in the effluents up to 3 ng L^{-1}. Therefore, based on their hydrophobocity, CTDs are efficiently removed after secondary treatment due to their tendency to sorb on sludge. Nevertheless, it must be taken into account that disinfection with sodium hypochloride of secondary effluent can lead to the formation of CDTs in the final effluent, while, on the contrary, UV disinfection is a more "safe" choice since no chlorinated derivatives can be formed.

19.5 CONCLUSIONS

In this chapter, the occurrence of PPCP TPs in different environmental matrices is reported. The aim of this work was to provide an insight into the recent studies focused on the identification of PPCP TPs not only in wastewater but also in surface water, groundwater, and finally drinking water in order to give a more intergraded knowledge concerning the fate of these substances in the environment.

The most studied groups are antibiotics, analgesics, psychiatric drugs, and lipid regulators, with CBZ being the most extensively examined parent compound. Target analysis was conducted for more than 10 different TPs of the specific antiepileptic drug not only on wastewaters but also on surface waters, groundwaters, and drinking waters, thus providing a more thorough knowledge concerning its environmental fate.

Most investigations are conducted in Europe and more specifically in Germany and Spain. North American countries (United States and Canada) also provided scientific data, whereas data for other regions around the world are limited to China and Japan. Concentrations of the TPs in the wastewaters usually range in the nanogram per liter level; however, TPs such as ERY-H_2O, SA, OH-IBU, CX-IBU, 4-AAA, 4-MAA, 4-AA, 4-FAA, *O*-Des-Me-VNF, CBZ-Ep, DiOH-CBZ, MTPA, GUA, OC, and COT might reach the microgram per liter level. Lower levels of all TPs were observed in surface water (below μg L^{-1} levels), with 4-AAA and CBZ-EP as exceptions.

Most TPs detected are metabolites of PPCPs that stem from human excretion. Moreover, hydrolysis and photolysis products were also present in the different water matrices. It is worth mentioning that many of the identified compounds are TPs that arise from biotic or abiotic processes that take place during treatment. Biotransformation that takes place during secondary treatment is one route for possible generation of TPs, while photolysis, hydrolysis, or oxidation during the residence of the parent compounds in treatment plants is another. Nevertheless, disinfection techniques at the end of wastewater treatment may lead also to plausible toxic TPs in the final effluent.

Consequently, the clarification of the formation and fate of TPs in the cycle of water is a difficult task. First of all, excretion patterns must be taken into account in order to evaluate the load that enters WWTPs. However, the theoretical percentages of excretions may vary depending on the age, sex, or other diseases such as diabetes or high blood pressure that patients may suffer from.

In addition, pharmaceutical industries and the disposal of expired or no longer needed medicine contribute to increased concentrations of PPCPs in the wastewaters, thus changing the expected parent compound/metabolite ratio. Moreover, transformations of the parent compounds and/or their TPs, which take place during treatment in WWTPs, must be taken into account; however, these processes do not affect all compounds to the same extent. Finally, discharged effluents that carry unchanged parent compounds, metabolites, and TPs that arose from treatment can distribute these compounds in surface water or groundwater, where other abiotic (photolysis or hydrolysis) and biotic (by microorganisms) processes might take place and have to be considered. However, the environmental conditions (temperature, climate, meteorological conditions, physicochemical properties of the water matrix or soil, biota, etc.) affect these transformation processes; thus, it is not easy to predict the outcome. Finally, TPs that stem from all the aforementioned processes may arrive at DWTPs, where most of them might be eliminated or be subjected to new structural changes, thus giving rise to new TPs.

Therefore, since humans are the final recipient, including TPs in chemical risk assessment is highly warranted. However, only scarce data exist concerning the ecotoxicological effects of the TPs. Moreover, specific modes of toxic action of the TPs cannot be ruled out a priori based only on their physicochemical characteristics compared with those of the parent compounds (e.g., increased polarity). Hence, further investigation on the ecotoxicological properties of those TPs that exhibit a high exposure potential along with their mixtures is indispensable. In continuation of this aspect, approaches to prioritize TPs of PPCPs with regard to their toxicological significance should also be developed, integrating the existing research in this field.

ACKNOWLEDGMENTS

This research has been cofinanced by the European Union (European Social Fund [ESF]) and Greek national funds through the Operational Program "Education and Lifelong Learning" of the National Strategic Reference Framework (NSRF)—Research Funding Program "Excellence II (Aristeia II)," Research Grant No. 4199, which is gratefully acknowledged.

LIST OF ABBREVIATIONS

10-OH-CBZ	10-hydroxy-carbamazepine
14-OH-CLR	14-hydroxy(*R*)-clarithromycin
2-*N*-gluc-LMTG	2-*N*-glucuronide-lamotrigine
2-OH-ATT	*O*-hydroxy-atorvastatin
2-OH-IBU	2-hydroxy-ibuprofen
2-OH-CBZ	2-hydroxycarbamazepine
3-OH-CBZ	3-hydroxycarbamazepine
3′-*O*-Des-ETP	3′-*O*-desmethyl etoposide metabolite
3′-OH-DCF	3′-hydroxy-diclofenac
4′-OH-DCF	4′-hydroxy-diclofenac
4-AA	4-aminoantipyrine
4-AAA	acetylaminoantipyrine
4-FAA	4-formylaminoantipyrine
4-MAA	4-methylaminoantipyrine
4-nitro-SMX	4-nitro-sulfamethoxazole
4-OH-CBZ	4-hydroxycarbamazepine
4-OH-GLY	*trans*-4-OH-glyburide
4-OH-PRP	4-hydroxy-propranolol
5′-OH-DCF	5′-hydroxy-diclofenac
7-amino-FLU	7-aminoflunitrazepam

ACTC	anhydrochlortetracycline
ADIN	acridin
ADON	acridone
AHA	α-hydroxy-alprazolam
AMDOPH	1-acetyl-1-methyl-2-dimethyloxamoyl-2-phenylhydrazide
AMPH	1-acetyl-1-methyl-2-phenylhydrazide
ATC	anhydrotetracycline
ATN-Des	atenolol-desisopropyl
CBX-IBU	carboxy-ibuprofen
CBZ-Ep	carbamazepine-10,11-epoxide
CLFA	clofibric acid
COT	cotinine
DCF	diclofenac
Des-DZP	desmethyldiazepam
Des-Me-CT	*N*-desmethylcitalopram
Des-Me-SER	desmethylsertraline
Des-SMX	desamino-sulfamethoxazole
Di-OH-CBZ	10,11-dihydro-10,11-dihydroxy-carbamazepine
E1-3S	estrone-3-sulfate
E3-3S	estriol-3-sulfate
EACTC	epianhydrochlorotetracycline
ENAT	enalaprilat
EOTC	4-epioxytetracycline
ERY	erythromycin
ERY-H_2O	anhydroerythromycin
ETC	4-epitetracycline
FNBA	fenofibric acid
GUA	guanylurea
IBU	ibuprofen
MTPA	metoprolol acid
MTR-OH	hydroxy-metronidazole
MTZ	metamizole
***N*^{4}-acetyl-SFP**	*N*^{4}-acetylsulfapyridine
***N*^{4}-acetyl-SMR**	*N*^{4}-acetylsulfamerazine
***N*^{4}-acetyl-SMX**	*N*^{4}-acetylsulfamethoxazole
***N*^{4}-acetyl-SMZ**	*N*^{4}-acetylsulfamethazine
NFD-H_2O	dehydronifedipine
NOR-FLU	norfluoxetine
OC	oseltamivir carboxylate
***O*-Des-NPX**	*O*-desmethylnaproxen
***O*-Des-TRM**	*O*-desmethyltramadol
***O*-Des-VNF**	*O*-desmethyl venlafaxine
***O*-DMTP**	*O*-desmethylmetoprolol
OH-BPP	hydroxyl bupropion
OH-D-TAM	4-hydroxy-*N*-desmethyl-tamoxifen
OH-IBU	1-hydroxy-ibuprofen
OH-PAC	hydroxy-paclitaxel
OH-TAM	(*Z*)-4-hydroxy-tamoxifen
OxCZ	oxcarbazepine
OXZ	oxazepam
OXZ-Glu	oxazepam glucuronide

PARA-XNT	paraxanthine
***p*-OH-ATT**	*p*-hydroxy-atorvastatin
RAN-*N*	ranitidine *N*-oxide
RTLA	ritalinic acid
SA	salicylic acid
SMX	sulfamethoxazole
SVA	simvastatin hydroxyl carboxylic acid
TRE-BPP	threo/erythro-hydrobupropion
VALA	valsartan acid
α-HMTP	α-hydroxy-metoprolol

REFERENCES

Acuña, V., von Schiller, D., García-Galán, M.J., Rodríguez-Mozaz, S., Corominas, L. et al. 2015. Occurrence and in-stream attenuation of wastewater-derived pharmaceuticals in Iberian rivers. *Science of the Total Environment* 503: 133–141.

Agrawal, N., Esteve-Romero, J., Bose, D., Dubey, N.P., Peris-Vicente, J. et al. 2014. Determination of selective serotonin reuptake inhibitors in plasma and urine by micellar liquid chromatography coupled to fluorescence detection. *Journal of Chromatography B* 965: 142–149.

Ashton, D., Hilton, M., Thomas, K.V. 2004. Investigating the environmental transport of human pharmaceuticals to streams in the United Kingdom. *Science of the Total Environment* 333: 167–184.

Bahlmann, A., Brack, W., Schneider, R.J., Krauss, M. 2014. Carbamazepine and its metabolites in wastewater: Analytical pitfalls and occurrence in Germany and Portugal. *Water Research* 57: 104–114.

Baker, D.R., Kasprzyk-Hordern, B. 2011. Multi-residue analysis of drugs of abuse in wastewater and surface water by solid-phase extraction and liquid chromatography–positive electrospray ionisation tandem mass spectrometry. *Journal of Chromatography A* 1218: 1620–1631.

Barceló, D. ed. 2012. *Emerging Organic Contaminants and Human Health*. Springer, Berlin, Vol. 20.

Barclay, V.K., Tyrefors, N.L., Johansson, I.M., Pettersson, C.E. 2012. Chiral analysis of metoprolol and two of its metabolites, α-hydroxymetoprolol and deaminated metoprolol, in wastewater using liquid chromatography–tandem mass spectrometry. *Journal of Chromatography A* 1269: 208–217.

Barnes, K.K., Kolpin, D.W., Furlong, E.T. et al. 2008. A national reconnaissance of pharmaceuticals and other organic wastewater contaminants in the United States—I) Groundwater. *Science of the Total Environment* 402: 192–200.

Baumann, M., Weiss, K., Maletzki, D., Schüssler, W., Schudoma, D. et al. 2015. Aquatic toxicity of the macrolide antibiotic clarithromycin and its metabolites. *Chemosphere* 120: 192–198.

Behera, S.K., Kim, H.W., Oh, J.E., Park, H.S. 2011. Occurrence and removal of antibiotics, hormones and several other pharmaceuticals in wastewater treatment plants of the largest industrial city of Korea. *Science of the Total Environment* 409: 4351–4360.

Bendz, D., Paxeus, N.A., Ginn, T.R., Loge, F.J. 2005. Occurrence and fate of pharmaceutically active compounds in the environment, a case study: Höje River in Sweden. *Journal of Hazardous Materials* 122: 195–204.

Benotti, M.J., Brownawell, B.J. 2007. Distributions of pharmaceuticals in an urban estuary during both dry- and wet-weather conditions. *Environmental Science & Technology* 41: 5795–5802.

Benotti, M.J., Trenholm, R.A., Vanderford, B.J., Holady, J.C., Stanford, B.D. et al. 2009. Pharmaceuticals and endocrine disrupting compounds in US drinking water. *Environmental Science & Technology* 43: 597–603.

Besse, J.P., Latour, J.F., Garric, J. 2012. Anticancer drugs in surface waters: What can we say about the occurrence and environmental significance of cytotoxic, cytostatic and endocrine therapy drugs? *Environment International* 39: 73–86.

Boleda, M.A., Galceran, M.A., Ventura, F. 2011. Behavior of pharmaceuticals and drugs of abuse in a drinking water treatment plant (DWTP) using combined conventional and ultrafiltration and reverse osmosis (UF/RO) treatments. *Environmental Pollution* 159: 1584–1591.

Borova, V.L., Maragou, N.C., Gago-Ferrero, P., Pistos, C., Thomaidis, N.S. 2014. Highly sensitive determination of 68 psychoactive pharmaceuticals, illicit drugs, and related human metabolites in wastewater by liquid chromatography–tandem mass spectrometry. *Analytical and Bioanalytical Chemistry* 406: 4273–4285.

Bradley, P.M., Barber, L.B., Kolpin, D.W., McMahon, P.B., Chapelle, F.H. 2007. Biotransformation of caffeine, cotinine, and nicotine in stream sediments: Implications for use as wastewater indicators. *Environmental Toxicology and Chemistry* 26: 1116–1121.

Bramer, S.L., Kallungal, B.A. 2003. Clinical considerations in study designs that use cotinine as a biomarker. *Biomarkers* 8: 187–203.

Buerge, I.J., Kahle, M., Buser, H.R., Müller, M.D., Poiger, T. 2008. Nicotine derivatives in wastewater and surface waters: Application as chemical markers for domestic wastewater. *Environmental Science & Technology* 42: 6354–6360.

Buth, J.M., Ross, M.R., McNeill, K., Arnold, W.A. 2011. Reprint of: Removal and formation of chlorinated triclosan derivatives in wastewater treatment plants using chlorine and UV disinfection. *Chemosphere* 85: 284–289.

Cabeza, Y., Candela, L., Ronen, D., Teijon, G. 2012. Monitoring the occurrence of emerging contaminants in treated wastewater and groundwater between 2008 and 2010: The Baix Llobregat (Barcelona, Spain). *Journal of Hazardous Materials* 239: 32–39.

Calisto, V., Esteves, V.I. 2009. Psychiatric pharmaceuticals in the environment. *Chemosphere* 77: 1257–1274.

Calza, P., Fabbri, D. 2014. Advanced mass spectrometry-based techniques for the identification and structure elucidation of transformation products of emerging contaminants. *Transformation Products of Emerging Contaminants in the Environment: Analysis, Processes, Occurrence, Effects and Risks* 325–350.

Canosa, P., Morales, S., Rodriguez, I., Rubi, E., Cela, R. et al. 2005. Aquatic degradation of triclosan and formation of toxic chlorophenols in presence of low concentrations of free chlorine. *Analytical and Bioanalytical Chemistry* 383: 1119–1126.

Carlsson, C., Johansson, A.K., Alvan, G., Bergman, K., Kühler, T. 2006. Are pharmaceuticals potent environmental pollutants?: Part I: Environmental risk assessments of selected active pharmaceutical ingredients. *Science of the Total Environment* 364: 67–87.

Castiglioni, S., Zuccato, E., Crisci, E., Chiabrando, C., Fanelli, R. et al. 2006. Identification and measurement of illicit drugs and their metabolites in urban wastewater by liquid chromatography-tandem mass spectrometry. *Analytical Chemistry* 78: 8421–8429.

Chen, M., Ohman, K., Metcalfe, C., Ikonomou, M.G., Amatya, P.L. et al. 2006. Pharmaceuticals and endocrine disruptors in wastewater treatment effluents and in the water supply system of Calgary, Alberta, Canada. *Water Quality Research Journal of Canada* 41: 351–364.

Chen, J., Liu, Y.S., Su, H.C., Ying, G.G., Liu, F. et al. 2015. Removal of antibiotics and antibiotic resistance genes in rural wastewater by an integrated constructed wetland. *Environmental Science and Pollution Research* 22: 1794–1803.

Chiaia, A.C., Banta-Green, C., Field, J. 2008. Eliminating solid phase extraction with large-volume injection LC/MS/MS: Analysis of illicit and legal drugs and human urine indicators in US wastewaters. *Environmental Science & Technology* 42: 8841–8848.

Chiron, S., Minero, C., Vione, D. 2006. Photodegradation processes of the antiepileptic drug carbamazepine, relevant to estuarine waters. *Environmental Science & Technology* 40: 5977–5983.

Chouinard, G., Lefko-Singh, K., Teboul, E. 1999. Metabolism of anxiolytics and hypnotics: Benzodiazepines, buspirone, zoplicone, and zolpidem. *Cellular and Molecular Neurobiology* 19: 533–552.

Cleuvers, M. 2004. Mixture toxicity of the anti-inflammatory drugs diclofenac, ibuprofen, naproxen, and acetylsalicylic acid. *Ecotoxicology and Environmental Safety* 59: 309–315.

Collado, N., Buttiglieri, G., Ferrando-Climent, L., Rodriguez-Mozaz, S., Barceló, D. et al. 2012. Removal of ibuprofen and its transformation products: Experimental and simulation studies. *Science of the total environment* 433: 296–301.

Cruz-Morató, C., Ferrando-Climent, L., Rodriguez-Mozaz, S., Barceló, D., Marco-Urrea, E. et al. 2013. Degradation of pharmaceuticals in non-sterile urban wastewater by *Trametes versicolor* in a fluidized bed bioreactor. *Water Research* 47: 5200–5210.

Cruz-Morató, C., Lucas, D., Llorca, M., Rodriguez-Mozaz, S., Gorga, M. et al. 2014. Hospital wastewater treatment by fungal bioreactor: Removal efficiency for pharmaceuticals and endocrine disruptor compounds. *Science of the Total Environment* 493: 365–376.

de Jongh, C.M., Kooij, P.J., de Voogt, P., ter Laak, T.L. 2012. Screening and human health risk assessment of pharmaceuticals and their transformation products in Dutch surface waters and drinking water. *Science of the Total Environment* 427: 70–77.

del Mar Gómez-Ramos, M., Pérez-Parada, A., García-Reyes, J.F., Fernández-Alba, A.R., Agüera, A. 2011. Use of an accurate-mass database for the systematic identification of transformation products of organic contaminants in wastewater effluents. *Journal of Chromatography A* 1218: 8002–8012.

Djukovic, D., Appiah-Amponsah, E., Shanaiah, N. et al. 2008. Ibuprofen metabolite profiling using a combination of SPE/column-trapping and HPLC–micro-coil NMR. *Journal of Pharmaceutical and Biomedical Analysis* 47: 328–334.

Eguchi, K., Nagase, H., Ozawa, M. et al. 2004. Evaluation of antimicrobial agents for veterinary use in the ecotoxicity test using microalgae. *Chemosphere* 57: 1733–1738.

Escher, B.I., Fenner, K. 2011. Recent advances in environmental risk assessment of transformation products. *Environmental Science & Technology* 45: 3835–3847.

Evgenidou, E.N., Konstantinou, I.K., Lambropoulou, D.A. 2015. Occurrence and removal of transformation products of PPCPs and illicit drugs in wastewaters: A review. *Science of the Total Environment* 505: 905–926.

Fatta, D., Achilleos, A., Nikolaou, A., Meric, S. 2007. Analytical methods for tracing pharmaceutical residues in water and wastewater. *TrAC Trends in Analytical Chemistry* 26: 515–533.

Fatta-Kassinos, D., Vasquez, M.I., Kümmerer, K. 2011. Transformation products of pharmaceuticals in surface waters and wastewater formed during photolysis and advanced oxidation processes—Degradation, elucidation of byproducts and assessment of their biological potency. *Chemosphere* 85: 693–709.

Ferrando-Climent, L., Collado, N., Buttiglieri, G. et al. 2012. Comprehensive study of ibuprofen and its metabolites in activated sludge batch experiments and aquatic environment. *Science of the Total Environment* 438: 404–413.

Ferrando-Climent, L., Rodriguez-Mozaz, S., Barceló, D. 2013. Development of a UPLC-MS/MS method for the determination of ten anticancer drugs in hospital and urban wastewaters, and its application for the screening of human metabolites assisted by information-dependent acquisition tool (IDA) in sewage samples. *Analytical and Bioanalytical Chemistry* 405: 5937–5952.

Ferrando-Climent, L., Rodriguez-Mozaz, S., Barceló, D. 2014. Incidence of anticancer drugs in an aquatic urban system: From hospital effluents through urban wastewater to natural environment. *Environmental Pollution* 193: 216–223.

Ferrari, B., Mons, R., Vollat, B. et al. 2004. Environmental risk assessment of six human pharmaceuticals: Are the current environmental risk assessment procedures sufficient for the protection of the aquatic environment? *Environmental Toxicology and Chemistry* 23: 1344–1354.

Ferrer, I., Thurman, E.M. 2012. Analysis of 100 pharmaceuticals and their degradates in water samples by liquid chromatography/quadrupole time-of-flight mass spectrometry. *Journal of chromatography A* 1259: 148–157.

García-Galán, M.J., Garrido, T., Fraile, J., Ginebreda, A., Díaz-Cruz, M.S. et al. 2011. Application of fully automated online solid phase extraction-liquid chromatography-electrospray-tandem mass spectrometry for the determination of sulfonamides and their acetylated metabolites in groundwater. *Analytical and Bioanalytical Chemistry* 399: 795–806.

García-Galán, M.J., Frömel, T., Müller, J., Peschka, M., Knepper, T. et al. 2012. Biodegradation studies of N 4-acetylsulfapyridine and N 4-acetylsulfamethazine in environmental water by applying mass spectrometry techniques. *Analytical and bioanalytical chemistry* 402: 2885–2896.

Ghosh, G.C., Nakada, N., Yamashita, N., Tanaka, H. 2010. Oseltamivir carboxylate, the active metabolite of oseltamivir phosphate (Tamiflu), detected in sewage discharge and river water in Japan. *Environmental Health Perspectives* 103–107.

Ghoshdastidar, A.J., Fox, S., Tong, A.Z. 2015. The presence of the top prescribed pharmaceuticals in treated sewage effluents and receiving waters in Southwest Nova Scotia, Canada. *Environmental Science and Pollution Research* 22: 689–700.

Glassmeyer, S.T., Furlong, E.T., Kolpin, D.W. et al. 2005. Transport of chemical and microbial compounds from known wastewater discharges: Potential for use as indicators of human fecal contamination. *Environmental Science & Technology* 39: 5157–5169.

Goldstein, J.A., Safe, S. 1989. Mechanism of action and structure-activity relationships for the chlorinated dibenzo-p-dioxins and related compounds. *Halogenated Biphenyls, Naphthalenes, Dibenzodioxins and Related Compounds* 2: 239–293.

Goncalves, C., Perez, S., Osorio, V., Petrovic, M., Alpendurada, M.F. et al. 2011. Photofate of oseltamivir (Tamiflu) and oseltamivir carboxylate under natural and simulated solar irradiation: Kinetics, identification of the transformation products, and environmental occurrence. *Environmental Science & Technology* 45: 4307–4314.

González-Mariño, I., Quintana, J.B., Rodríguez, I., Cela, R. 2011. Evaluation of the occurrence and biodegradation of parabens and halogenated by-products in wastewater by accurate-mass liquid chromatography-quadrupole-time-of-flight-mass spectrometry (LC-QTOF-MS). *Water Research* 45: 6770–6780.

Gracia-Lor, E., Sancho, J.V., Hernández, F. 2011. Multi-class determination of around 50 pharmaceuticals, including 26 antibiotics, in environmental and wastewater samples by ultra-high performance liquid chromatography–tandem mass spectrometry. *Journal of Chromatography A* 1218: 2264–2275.

Gracia-Lor, E., Sancho, J.V., Serrano, R., Hernández, F. 2012. Occurrence and removal of pharmaceuticals in wastewater treatment plants at the Spanish Mediterranean area of Valencia. *Chemosphere* 87: 453–462.

Gros, M., Rodríguez-Mozaz, S., Barceló, D. 2012. Fast and comprehensive multi-residue analysis of a broad range of human and veterinary pharmaceuticals and some of their metabolites in surface and treated waters by ultra-high-performance liquid chromatography coupled to quadrupole-linear ion trap tandem mass spectrometry. *Journal of Chromatography A* 1248: 104–121.

Gros, M., Rodríguez-Mozaz, S., Barceló, D. 2013. Rapid analysis of multiclass antibiotic residues and some of their metabolites in hospital, urban wastewater and river water by ultra-high-performance liquid chromatography coupled to quadrupole-linear ion trap tandem mass spectrometry. *Journal of Chromatography A* 1292: 173–188.

Gu, L., Gonzalez, F.J., Kalow, W., Tang, B.K. 1992. Biotransformation of caffeine, paraxanthine, theobromine and theophylline by cDNA-expressed human CYP1A2 and CYP2E1. *Pharmacogenetics and Genomics* 2: 73–77.

Guerra, P., Kim, M., Shah, A., Alaee, M., Smyth, S.A. 2014. Occurrence and fate of antibiotic, analgesic/anti-inflammatory, and antifungal compounds in five wastewater treatment processes. *Science of the Total Environment* 473: 235–243.

Gulkowska, A., Leung, H.W., So, M.K. et al. 2008. Removal of antibiotics from wastewater by sewage treatment facilities in Hong Kong and Shenzhen, China. *Water Research* 42: 395–403.

Gyenge-Szabó, Z., Szoboszlai, N., Frigyes, D., Záray, G., Mihucz, V.G. 2014. Monitoring of four dipyrone metabolites in communal wastewater by solid phase extraction liquid chromatography electrospray ionization quadrupole time-of-flight mass spectrometry. *Journal of Pharmaceutical and Biomedical Analysis* 90: 58–63.

Haas, J.S., Kaplan, C.P., Barenboim, D., Jacob, P. III, Benowitz, N.L. 2004. Bupropion in breast milk: An exposure assessment for potential treatment to prevent post-partum tobacco use. *Tobacco Control* 13: 52–56.

Haddad, T., Baginska, E., Kümmerer, K. 2015. Transformation products of antibiotic and cytostatic drugs in the aquatic cycle that result from effluent treatment and abiotic/biotic reactions in the environment: An increasing challenge calling for higher emphasis on measures at the beginning of the pipe. *Water Research* 72: 75–126.

Halling-Sørensen, B., Lykkeberg, A., Ingerslev, F., Blackwell, P., Tjørnelund, J. 2003. Characterisation of the abiotic degradation pathways of oxytetracyclines in soil interstitial water using LC–MS–MS. *Chemosphere* 50: 1331–1342.

Hassanzadeh, A., Barber, J., Morris, G.A., Gorry, P.A. 2007. Mechanism for the degradation of erythromycin A and erythromycin A 2′-ethyl succinate in acidic aqueous solution. *The Journal of Physical Chemistry A* 111: 10098–10104.

Heberer, T. 2002. Tracking persistent pharmaceutical residues from municipal sewage to drinking water. *Journal of Hydrology* 266: 175–189.

Heberer, Th., Stan, H.-J. 1997. Determination of clofibric acid and *N*-(phenylsulfonyl)-sarcosine in sewage, river and drinking water. *International Journal of Environmental Analytical Chemistry* 67–4: 113–124.

Heberer, Th., Fuhrmann, B., Schmidt-Baumler, K., Tsipi, D., Koutsouba, V. et al. 2001. Occurrence of pharmaceutical residues in sewage, river, ground, and drinking water in Greece and Berlin (Germany).

Hedgespeth, M.L., Sapozhnikova, Y., Pennington, P., Clum, A., Fairey, A. et al. 2012. Pharmaceuticals and personal care products (PPCPs) in treated wastewater discharges into Charleston Harbor, South Carolina. *Science of the Total Environment* 437: 1–9.

Heeb, F., Singer, H., Pernet-Coudrier, B. et al. 2012. Organic micropollutants in rivers downstream of the megacity Beijing: Sources and mass fluxes in a large-scale wastewater irrigation system. *Environmental Science & Technology* 46: 8680–8688.

Holthuis, J.J.M., Postmus, P.E. et al. 1986. Pharmacokinetics of high dose etoposide (VP 16–213). *European Journal of Cancer and Clinical Oncology* 22: 1149–1155.

Howell, S.R., Husbands, G.E.M., Scatina, J.A., Sisenwine, S.F. 1993. Metabolic disposition of 14C-venlafaxine in mouse, rat, dog, rhesus monkey and man. *Xenobiotica* 23: 349–359.

Huerta, B., Jakimska, A., Llorca, M. et al. 2015. Development of an extraction and purification method for the determination of multi-class pharmaceuticals and endocrine disruptors in freshwater invertebrates. *Talanta* 132: 373–381.

Huerta-Fontela, M., Galceran, M.T., Martin-Alonso, J., Ventura, F. 2008. Occurrence of psychoactive stimulatory drugs in wastewaters in north-eastern Spain. *Science of the Total Environment* 397: 31–40.

Huerta-Fontela, M., Galceran, M.T., Ventura, F. 2010. Fast liquid chromatography–quadrupole-linear ion trap mass spectrometry for the analysis of pharmaceuticals and hormones in water resources. *Journal of Chromatography A* 1217: 4212–4222.

Huerta-Fontela, M., Galceran, M.T., Ventura, F. 2011. Occurrence and removal of pharmaceuticals and hormones through drinking water treatment. *Water Research* 45: 1432–1442.

Huntscha, S., Singer, H.P., McArdell, C.S., Frank, C.E., Hollender, J. 2012. Multiresidue analysis of 88 polar organic micropollutants in ground, surface and wastewater using online mixed-bed multilayer solid-phase extraction coupled to high performance liquid chromatography–tandem mass spectrometry. *Journal of Chromatography A* 1268: 74–83.

Idder, S., Ley, L., Mazellier, P., Budzinski, H. 2013. Quantitative on-line preconcentration-liquid chromatography coupled with tandem mass spectrometry method for the determination of pharmaceutical compounds in water. *Analytica Chimica Acta* 805: 107–115.

Jager, N.G.L., Rosing, H., Linn, S.C., Schellens, J.H.M., Beijnen, J.H. 2012. Importance of highly selective LC–MS/MS analysis for the accurate quantification of tamoxifen and its metabolites: Focus on endoxifen and 4-hydroxytamoxifen. *Breast Cancer Research and Treatment* 133: 793–798.

Jelic, A., Gros, M., Ginebreda, A. et al. 2011. Occurrence, partition and removal of pharmaceuticals in sewage water and sludge during wastewater treatment. *Water Research* 45: 1165–1176.

Jelic, A., Rodriguez-Mozaz, S., Barceló, D., Gutierrez, O. 2015. Impact of in-sewer transformation on 43 pharmaceuticals in a pressurized sewer under anaerobic conditions. *Water Research* 68: 98–108.

Jones, O.A., Lester, JN., Voulvoulis, N. 2005. Pharmaceuticals: A threat to drinking water? *Trends in Biotechnology* 23: 163–167.

Jurado, A., Vàzquez-Suñé, E., Carrera, J., López de Alda, M., Pujades, E. et al. 2012. Emerging organic contaminants in groundwater in Spain: A review of sources, recent occurrence and fate in a European context. *Science of the Total Environment* 440: 82–94.

Kaiser, E., Prasse, C., Wagner, M., Broder, K., Ternes, T.A. 2014. Transformation of oxcarbazepine and human metabolites of carbamazepine and oxcarbazepine in wastewater treatment and sand filters. *Environmental Science & Technology* 48: 10208–10216.

Karthikeyan, K.G., Meyer, M.T. 2006. Occurrence of antibiotics in wastewater treatment facilities in Wisconsin, USA. *Science of the Total Environment* 361: 196–207.

Kasprzyk-Hordern, B., Dinsdale, R.M., Guwy, A.J. 2007. Multi-residue method for the determination of basic/neutral pharmaceuticals and illicit drugs in surface water by solid-phase extraction and ultra performance liquid chromatography–positive electrospray ionisation tandem mass spectrometry. *Journal of Chromatography A* 1161: 132–145.

Kern, S., Baumgartner, R., Helbling, D.E. et al. 2010. A tiered procedure for assessing the formation of biotransformation products of pharmaceuticals and biocides during activated sludge treatment. *Journal of Environmental Monitoring* 12: 2100–2111.

Kim, S., Eichhorn, P., Jensen, J.N., Scott Weber, A., Aga, D.S. 2005. Removal of antibiotics in wastewater: Effect of hydraulic and solid retention times on the fate of tetracycline in the activated sludge process. *Environmental Science & Technology* 39: 5816–5823.

Kosma, C.I., Lambropoulou, D.A., Albanis, T.A. 2010. Occurrence and removal of PPCPs in municipal and hospital wastewaters in Greece. *Journal of Hazardous Materials* 179: 804–817.

Kosma, C.I., Lambropoulou, D.A., Albanis, T.A. 2014. Investigation of PPCPs in wastewater treatment plants in Greece: Occurrence, removal and environmental risk assessment. *Science of the Total Environment* 466: 421–438.

Kosma, C.I., Lambropoulou, D.A., Albanis, T.A. 2015. Comprehensive study of the antidiabetic drug metformin and its transformation product guanylurea in Greek wastewaters. *Water Research* 70: 436–448.

Kostich, M.S., Batt, A.L., Lazorchak, J.M. 2014. Concentrations of prioritized pharmaceuticals in effluents from 50 large wastewater treatment plants in the US and implications for risk estimation. *Environmental Pollution* 184: 354–359.

Krauss, M., Singer, H., Hollender, J. 2010. LC–high resolution MS in environmental analysis: From target screening to the identification of unknowns. *Analytical and Bioanalytical Chemistry* 397: 943–951.

Kyzas, G.Z., Nanaki, S.G., Koltsakidou, A. et al. 2015. Effectively designed molecularly imprinted polymers for selective isolation of the antidiabetic drug metformin and its transformation product guanylurea from aqueous media. *Analytica Chimica Acta* 866: 27–40.

La Farré, M., Pérez, S., Kantiani, L., Barceló, D. 2008. Fate and toxicity of emerging pollutants, their metabolites and transformation products in the aquatic environment. *TrAC Trends in Analytical Chemistry* 27: 991–1007.

Lajeunesse, A., Gagnon, C., Sauvé, S. 2008. Determination of basic antidepressants and their *N*-desmethyl metabolites in raw sewage and wastewater using solid-phase extraction and liquid chromatography-tandem mass spectrometry. *Analytical Chemistry* 80: 5325–5333.

Lajeunesse, A., Gagnon, C., Gagné, F., Louis, S., Čejka, P. et al. 2011. Distribution of antidepressants and their metabolites in brook trout exposed to municipal wastewaters before and after ozone treatment—Evidence of biological effects. *Chemosphere* 83: 564–571.

Lambropoulou, D.A., Nollet, L.M.L. eds. 2014. *Transformation Products of Emerging Contaminants in the Environment: Analysis, Processes, Occurrence, Effects and Risks*. Wiley, New York.

Langford, K., Thomas, K.V. 2011. Input of selected human pharmaceutical metabolites into the Norwegian aquatic environment. *Journal of Environmental Monitoring* 13: 416–421.

Le-Minh, N., Khan, S.J., Drewes, J.E., Stuetz, R.M. 2010. Fate of antibiotics during municipal water recycling treatment processes. *Water research* 44: 4295–4323.

Le-Minh, N., Stuetz, R.M., Khan, S.J. 2012. Determination of six sulfonamide antibiotics, two metabolites and trimethoprim in wastewater by isotope dilution liquid chromatography/tandem mass spectrometry. *Talanta* 89: 407–416.

Leclercq, M., Mathieu, O., Gomez, E., Casellas, C., Fenet, H. et al. 2009. Presence and fate of carbamazepine, oxcarbazepine, and seven of their metabolites at wastewater treatment plants. *Archives of Environmental Contamination and Toxicology* 56: 408–415.

Lee, E., Lee, S., Park, J., Kim, Y., Cho, J. 2013. Removal and transformation of pharmaceuticals in wastewater treatment plants and constructed wetlands. *Drinking Water Engineering and Science* 6: 89–98.

Leknes, H., Sturtzel, I.E., Dye, C. 2012. Environmental release of oseltamivir from a Norwegian sewage treatment plant during the 2009 influenza A (H1N1) pandemic. *Science of the Total Environment* 414: 632–638.

Letzel, M., Weiss, K., Schüssler, W., Sengl, M. 2010. Occurrence and fate of the human pharmaceutical metabolite ritalinic acid in the aquatic system. *Chemosphere* 81: 1416–1422.

Li, B., Zhang, T., Xu, Z., Fang, H.H.P. 2009. Rapid analysis of 21 antibiotics of multiple classes in municipal wastewater using ultra performance liquid chromatography-tandem mass spectrometry. *Analytica Chimica Acta* 645: 64–72.

Lin, A.Y.-C., Tsai, Y.-T. 2009. Occurrence of pharmaceuticals in Taiwan's surface waters: Impact of waste streams from hospitals and pharmaceutical production facilities. *Science of the Total Environment* 407: 3793–3802.

Lindberg, R.H., Wennberg, P., Johansson, M.I., Tysklind, M., Andersson, B.A.V. 2005. Screening of human antibiotic substances and determination of weekly mass flows in five sewage treatment plants in Sweden. *Environmental Science & Technology* 39: 3421–3429.

Liu, Z.-H., Hashimoto, T., Okumura, Y., Kanjo, Y., Mizutani, S. 2010. Simultaneous analysis of natural free estrogens and their conjugates in wastewater by GC-MS. *CLEAN—Soil, Air, Water* 38: 181–188.

Loos, R., Carvalho, R., Antonio, D.C. et al. 2013. EU-wide monitoring survey on emerging polar organic contaminants in wastewater treatment plant effluents. *Water Research* 47: 6475–6487.

López, S.H., Ulaszewska, M.M., Hernando, M.D., Martínez Bueno, MJ, Gómez, M.J. et al. 2014. Post-acquisition data processing for the screening of transformation products of different organic contaminants: Two-year monitoring of river water using LC-ESI-QTOF-MS and GC×GC-EI-TOF-MS. *Environmental Science and Pollution Research* 21: 12583–12604.

López-Serna, R., Pérez, S., Ginebreda, A., Petrović, M., Barceló, D. 2010. Fully automated determination of 74 pharmaceuticals in environmental and waste waters by online solid phase extraction–liquid chromatography-electrospray–tandem mass spectrometry. *Talanta* 83: 410–424.

López-Serna, R., Petrović, M., Barceló, D. 2012. Occurrence and distribution of multi-class pharmaceuticals and their active metabolites and transformation products in the Ebro River basin (NE Spain). *Science of the Total Environment* 440: 280–289.

López-Serna, R., Jurado, A., Vázquez-Suñé, E., Carrera, J., Petrović, M. et al. 2013. Occurrence of 95 pharmaceuticals and transformation products in urban groundwaters underlying the metropolis of Barcelona, Spain. *Environmental Pollution* 174: 305–315.

Loraine, G.A., Pettigrove, M.E. 2006. Seasonal variations in concentrations of pharmaceuticals and personal care products in drinking water and reclaimed wastewater in southern California. *Environmental Science & Technology* 40: 687–695.

Lundov, M.D., Moesby, L., Zachariae, C., Johansen, J.D. 2009. Contamination versus preservation of cosmetics: A review on legislation, usage, infections, and contact allergy. *Contact Dermatitis* 60: 70–78.

Mahmood, G., Mei, Z., Hojjat, H., Pace, E., Kallakuri, S. et al. 2014. Therapeutic effect of sildenafil on blast-induced tinnitus and auditory impairment. *Neuroscience* 269: 367–382.

Martínez Bueno, MJ, Ulaszewska, M.M., Gomez, M.J., Hernando, M.D., Fernández-Alba, A.R. 2012. Simultaneous measurement in mass and mass/mass mode for accurate qualitative and quantitative screening analysis of pharmaceuticals in river water. *Journal of Chromatography A* 1256: 80–88.

Méndez-Arriaga, F., Esplugas, S., Giménez, J. 2008. Photocatalytic degradation of non-steroidal anti-inflammatory drugs with TiO_2 and simulated solar irradiation. *Water Research* 42: 585–594.

Metcalfe, C.D., Miao, X.-S., Koenig, B.G., Struger, J. 2003. Distribution of acidic and neutral drugs in surface waters near sewage treatment plants in the lower Great Lakes, Canada. *Environmental Toxicology and Chemistry* 22: 2881–2889.

Metcalfe, C.D., Chu, S., Judt, C. et al. 2010. Antidepressants and their metabolites in municipal wastewater, and downstream exposure in an urban watershed. *Environmental Toxicology and Chemistry* 29: 79–89.

Meyer, B.K., Ni, A., Hu, B., Shi, L. 2007. Antimicrobial preservative use in parenteral products: Past and present. *Journal of Pharmaceutical Sciences* 96: 3155–3167.

Miao, X.-S., Koenig, B.G., Metcalfe, C.D. 2002. Analysis of acidic drugs in the effluents of sewage treatment plants using liquid chromatography–electrospray ionization tandem mass spectrometry. *Journal of Chromatography A* 952: 139–147.

Miao, X.-S., Yang, J.-J., Metcalfe, C.D. 2005. Carbamazepine and its metabolites in wastewater and in biosolids in a municipal wastewater treatment plant. *Environmental Science & Technology* 39: 7469–7475.

Michael, I., Vasquez, M.I., Hapeshi, E. et al. 2014. Metabolites and transformation products of pharmaceuticals in the aquatic environment as contaminants of emerging concern. *Advanced Mass Spectrometry-Based Techniques for the Identification and Structure Elucidation of Transformation Products of Emerging Contaminants*. Wiley, New York, pp. 413–459.

Mompelat, S., Le Bot, B., Thomas, O. 2009. Occurrence and fate of pharmaceutical products and by-products, from resource to drinking water. *Environment International* 35: 803–814.

Monteiro, S.C., Boxall, A.B.A. 2010. Occurrence and fate of human pharmaceuticals in the environment. In *Reviews of Environmental Contamination and Toxicology*. Springer, New York, pp. 53–154.

Moreno-González, R., Rodriguez-Mozaz, S., Gros, M., Barceló, D., León, V.M. 2015. Seasonal distribution of pharmaceuticals in marine water and sediment from a Mediterranean coastal lagoon (SE Spain). *Environmental Research* 138: 326–344.

Muñoz, I., Gómez-Ramos, M.J., Agüera, A., Fernández-Alba, A.R., García-Reyes, J.F. et al. 2009. Chemical evaluation of contaminants in wastewater effluents and the environmental risk of reusing effluents in agriculture. *TrAC Trends in Analytical Chemistry* 28: 676–694.

Negreira, N., López de Alda, M., Barceló, D. 2013. On-line solid phase extraction–liquid chromatography–tandem mass spectrometry for the determination of 17 cytostatics and metabolites in waste, surface and ground water samples. *Journal of Chromatography A* 1280: 64–74.

Negreira, N., López de Alda, M., Barceló, D. 2014a. Study of the stability of 26 cytostatic drugs and metabolites in wastewater under different conditions. *Science of the Total Environment* 482: 389–398.

Negreira, N., López de Alda, M., Barceló, D. 2014b. Cytostatic drugs and metabolites in municipal and hospital wastewaters in Spain: Filtration, occurrence, and environmental risk. *Science of the Total Environment* 497: 68–77.

Negreira, N., López de Alda, M, Barceló, D. 2015. Degradation of the cytostatic etoposide in chlorinated water by liquid chromatography coupled to quadrupole-Orbitrap mass spectrometry: Identification and quantification of by-products in real water samples. *Science of the Total Environment* 506: 36–45.

Nödler, K., Licha, T., Barbieri, M., Pérez, S. 2012. Evidence for the microbially mediated abiotic formation of reversible and non-reversible sulfamethoxazole transformation products during denitrification. *Water Research* 46: 2131–2139.

Nödler, K., Hillebrand, O., Idzik, K., Strathmann, M., Schiperski, F. et al. 2013. Occurrence and fate of the angiotensin II receptor antagonist transformation product valsartan acid in the water cycle—A comparative study with selected β-blockers and the persistent anthropogenic wastewater indicators carbamazepine and acesulfame. *Water Research* 47: 6650–6659.

Oosterhuis, M., Sacher, F., ter Laak, T.L. 2013. Prediction of concentration levels of metformin and other high consumption pharmaceuticals in wastewater and regional surface water based on sales data. *Science of the Total Environment* 442: 380–388.

Papageorgiou, M., Kosma, C., Lambropoulou, D. 2016. Seasonal occurrence, removal, mass loading and environmental risk assessment of 55 pharmaceuticals and personal care products in a municipal wastewater treatment plant in Central Greece. *Science of the Total Environment* 543: 547–569.

Partani, P., Modhave, Y., Gurule, S., Khuroo, A., Monif, T. 2009. Simultaneous determination of propranolol and 4-hydroxy propranolol in human plasma by solid phase extraction and liquid chromatography/electrospray tandem mass spectrometry. *Journal of Pharmaceutical and Biomedical Analysis* 50: 966–976.

Pasquini, L., Munoz, J.-F., Rimlinger, N., Dauchy, X., France, X. et al. 2013. Assessment of the fate of some household micropollutants in urban wastewater treatment plant. *Chemical Papers* 67: 601–612.

Patrolecco, L., Capri, S., Ademollo, N. 2015. Occurrence of selected pharmaceuticals in the principal sewage treatment plants in Rome (Italy) and in the receiving surface waters. *Environmental Science and Pollution Research* 22: 5864–5876.

Pieper, C., Risse, D., Schmidt, B., Braun, B., Szewzyk, U. et al. 2010. Investigation of the microbial degradation of phenazone-type drugs and their metabolites by natural biofilms derived from river water using liquid chromatography/tandem mass spectrometry (LC-MS/MS). *Water research* 44: 4559–4569.

Petrie, B., Barden, R., Kasprzyk-Hordern, B. 2015. A review on emerging contaminants in wastewaters and the environment: Current knowledge, understudied areas and recommendations for future monitoring. *Water Research* 72: 3–27.

Petrović, M., Gonzalez, S., Barceló, D. 2003. Analysis and removal of emerging contaminants in wastewater and drinking water. *TrAC Trends in Analytical Chemistry* 22: 685–696.

Petrović, M., Škrbić, B., Živančev, J., Ferrando-Climent, L., Barcelo, D. 2014. Determination of 81 pharmaceutical drugs by high performance liquid chromatography coupled to mass spectrometry with hybrid triple quadrupole–linear ion trap in different types of water in Serbia. *Science of the Total Environment* 468: 415–428.

Postigo, C., Richardson, S.D. 2014. Transformation of pharmaceuticals during oxidation/disinfection processes in drinking water treatment. *Journal of Hazardous Materials* 279: 461–475.

Postigo, C., Barceló, D. 2015. Synthetic organic compounds and their transformation products in groundwater: Occurrence, fate and mitigation. *Science of the Total Environment* 503: 32–47.

Prasse, C., Schlusener, M.P., Schulz, R., Ternes, T.A. 2010. Antiviral drugs in wastewater and surface waters: A new pharmaceutical class of environmental relevance? *Environmental Science & Technology* 44: 1728–1735.

Qi, W., Singer, H., Berg, M. et al. 2015. Elimination of polar micropollutants and anthropogenic markers by wastewater treatment in Beijing, China. *Chemosphere* 119: 1054–1061.

Radović, T., Grujić, S., Petković, A., Dimkić, M., Laušević, M. 2015. Determination of pharmaceuticals and pesticides in river sediments and corresponding surface and ground water in the Danube River and tributaries in Serbia. *Environmental Monitoring and Assessment* 187: 1–17.

Payán, M.R., López, M.Á.B., Fernández-Torres, R., Navarro, M.V., Callejón Mochón, M. 2011. Hollow fiber-based liquid phase microextraction (HF-LPME) for a highly sensitive HPLC determination of sulfonamides and their main metabolites. *Journal of Chromatography B* 879: 197–204.

Pieper, C., Risse, D., Schmidt, B., Braun, B., Szewzyk, U. et al. 2010. Investigation of the microbial degradation of phenazone-type drugs and their metabolites by natural biofilms derived from river water using liquid chromatography/tandem mass spectrometry (LC-MS/MS). *Water Research* 44: 4559–4569.

Reddersen, K., Heberer, T., Dünnbier, U. 2002. Identification and significance of phenazone drugs and their metabolites in ground-and drinking water. *Chemosphere* 49: 539–544.

Reddy, S., Iden, C.R., Brownawell, B.J. 2005. Analysis of steroid conjugates in sewage influent and effluent by liquid chromatography-tandem mass spectrometry. *Analytical Chemistry* 77: 7032–7038.

Reh, R., Licha, T., Geyer, T., Nödler, K., Sauter, M. 2013. Occurrence and spatial distribution of organic micropollutants in a complex hydrogeological karst system during low flow and high flow periods, results of a two-year study. *Science of the Total Environment* 443: 438–445.

Reid, M.J., Langford, K.H., Mørland, J., Thomas, K.V. 2011. Quantitative assessment of time dependent drug-use trends by the analysis of drugs and related metabolites in raw sewage. *Drug and Alcohol Dependence* 119: 179–186.

Robles-Molina, J., Lara-Ortega, F.J., Gilbert-López, B., García-Reyes, J.F., Molina-Díaz, A. 2014. Multi-residue method for the determination of over 400 priority and emerging pollutants in water and wastewater by solid-phase extraction and liquid chromatography-time-of-flight mass spectrometry. *Journal of Chromatography A* 1350: 30–43.

Rosal, R., Rodríguez, A., Perdigón-Melón, J.A. et al. 2010. Occurrence of emerging pollutants in urban wastewater and their removal through biological treatment followed by ozonation. *Water Research* 44: 578–588.

Roth, H.J., Fenner, H. 1994. *Arznelstoffe: Struktur-Bioreaktivität-wirkungsbezogene Eigenschaften; mit etwa 2500 Strukturformeln*. Deutscher Apotheke, Stuttgart.

Rubirola, A., Llorca, M., Rodriguez-Mozaz, S., Casas, N., Rodriguez-Roda, I. et al. 2014. Characterization of metoprolol biodegradation and its transformation products generated in activated sludge batch experiments and in full scale WWTPs. *Water Research* 63: 21–32.

Rule, K.L., Ebbett, V.R., Vikesland, P.J. 2005. Formation of chloroform and chlorinated organics by free-chlorine-mediated oxidation of triclosan. *Environmental Science & Technology* 39: 3176–3185.

Salgado, R., Noronha, J.P., Oehmen, A., Carvalho, G., Reis, M.A.M. 2010. Analysis of 65 pharmaceuticals and personal care products in 5 wastewater treatment plants in Portugal using a simplified analytical methodology. *Water Science and Technology* 62: 2862.

Salgado, R., Marques, R., Noronha, J.P. et al. 2011. Assessing the diurnal variability of pharmaceutical and personal care products in a full-scale activated sludge plant. *Environmental Pollution* 159: 2359–2367.

Salgado, R., Oehmen, A., Carvalho, G., Noronha, J.P., Reis, M.A.M. 2012. Biodegradation of clofibric acid and identification of its metabolites. *Journal of Hazardous Materials* 241: 182–189.

Salgueiro-González, N., López de Alda, M.J., Muniategui-Lorenzo, S., Prada-Rodríguez, D., Barceló, D. 2015. Analysis and occurrence of endocrine-disrupting chemicals in airborne particles. *TrAC Trends in Analytical Chemistry* 66: 45–52.

Sanderson, H., Johnson, D.J., Reitsma, T., Brain, R.A., Wilson, C.J. et al. 2004. Ranking and prioritization of environmental risks of pharmaceuticals in surface waters. *Regulatory Toxicology and Pharmacology* 39: 158–183.

Santos, L.H.M.L.M., Gros, M., Rodriguez-Mozaz, S., Delerue-Matos, C., Pena, A. et al. 2013. Contribution of hospital effluents to the load of pharmaceuticals in urban wastewaters: Identification of ecologically relevant pharmaceuticals. *Science of the Total Environment* 461: 302–316.

Scheurer, M., Ramil, M., Metcalfe, C.D., Groh, S., Ternes, T.A. 2010. The challenge of analyzing beta-blocker drugs in sludge and wastewater. *Analytical and Bioanalytical Chemistry* 396: 845–856.

Scheurer, M., Michel, A., Brauch, H.-J., Ruck, W., Sacher, F. 2012. Occurrence and fate of the antidiabetic drug metformin and its metabolite guanylurea in the environment and during drinking water treatment. *Water Research* 46: 4790–4802.

Schultz, M.M., Furlong, E.T., Kolpin, D.W. et al. 2010. Antidepressant pharmaceuticals in two US effluent-impacted streams: Occurrence and fate in water and sediment, and selective uptake in fish neural tissue. *Environmental Science & Technology* 44: 1918–1925.

Schymanski, E.L., Singer, H.P., Longrée, P. et al. 2014. Strategies to characterize polar organic contamination in wastewater: Exploring the capability of high resolution mass spectrometry. *Environmental Science & Technology* 48: 1811–1818.

Shrestha, S.L., Casey, F.X.M., Hakk, H., Smith, D.J., Padmanabhan, G. 2012. Fate and transformation of an estrogen conjugate and its metabolites in agricultural soils. *Environmental Science & Technology* 46: 11047–11053.

Simon, M.C., Chau, N.P., Levenson, J. 1988. Brachial artery hemodynamic response to acute converting enzyme inhibition by enalaprilat in essential hypertension. *Clinical Pharmacology and Therapeutics* 43: 49–54.

Stackelberg, P.E., Furlong, E.T., Meyer, M.T., Zaugg, S.D., Henderson, A.K. et al. 2004. Persistence of pharmaceutical compounds and other organic wastewater contaminants in a conventional drinking-water-treatment plant. *Science of the Total Environment* 329: 99–113.

Stamatis, N.K., Konstantinou, I.K. 2013. Occurrence and removal of emerging pharmaceutical, personal care compounds and caffeine tracer in municipal sewage treatment plant in Western Greece. *Journal of Environmental Science and Health, Part B* 48: 800–813.

Stan, H-J., Heberer, Th., Linkerhägner, M. 1994. Vorkommen von Clofibrinsäure in aquatischen System: Führt die therapeutische Anwendung zu einer Belastung von Oberflächen-, Grund-und Trinkwasser? *Vom Wasser* 83: 57–68.

Stumpf, M., Ternes, T.A., Haberer, K., Seel, P., Baumann, W. 1996. Nachweis von Arzneimittelrückständen in Kläranlagen und Fließgewässern. *Vom Wasser* 86: 291–303.

Subedi, B., Kannan, K. 2015. Occurrence and fate of select psychoactive pharmaceuticals and antihypertensives in two wastewater treatment plants in New York State, USA. *Science of the Total Environment* 514: 273–280.

Sun, Q., Lv, M., Hu, A., Yang, X., Yu, C.-P. 2014. Seasonal variation in the occurrence and removal of pharmaceuticals and personal care products in a wastewater treatment plant in Xiamen, China. *Journal of Hazardous Materials* 277: 69–75.

Swartz, C.H., Reddy, S., Benotti, M.J., Yin, H., Barber, L.B. et al. 2006. Steroid estrogens, nonylphenol ethoxylate metabolites, and other wastewater contaminants in groundwater affected by a residential septic system on Cape Cod, MA. *Environmental Science & Technology* 40: 4894–4902.

ter Laak, T., Baken, K. 2014. *The Occurrence, Fate and Ecological and Human Health Risks of Metformin and Guanylurea in the Water Cycle—A Literature Review*. The Water Research Institute, Nieuwegein, The Netherlands.

Terasaki, M., Kamata, R., Shiraishi, F., Makino, M. 2009. Evaluation of estrogenic activity of parabens and their chlorinated derivatives by using the yeast two-hybrid assay and the enzyme-linked immunosorbent assay. *Environmental Toxicology and Chemistry* 28: 204–208.

Ternes, T.A., Kreckel, P., Mueller, J. 1999. Behaviour and occurrence of estrogens in municipal sewage treatment plants—II. Aerobic batch experiments with activated sludge. *Science of the Total Environment* 225: 91–99.

Ternes, T.A., Meisenheimer, M., McDowell, D., Sacher, F., Brauch, H.-J. et al. 2002. Removal of pharmaceuticals during drinking water treatment. *Environmental Science & Technology* 36: 3855–3863.

Tocco, D.J., Duncan, A.E., Vassil, T.C., Ulm, E.H. 1982. The physiological disposition and metabolism of enalapril maleate in laboratory animals. *Drug Metabolism and Disposition* 10: 15–19.

Trautwein, C., Berset, J.-D., Wolschke, H., Kümmerer, K. 2014. Occurrence of the antidiabetic drug Metformin and its ultimate transformation product Guanylurea in several compartments of the aquatic cycle. *Environment International* 70: 203–212.

Vanderford, B.J., Snyder, S.A. 2006. Analysis of pharmaceuticals in water by isotope dilution liquid chromatography/tandem mass spectrometry. *Environmental Science & Technology* 40: 7312–7320.

Verlicchi, P., Al Aukidy, M., Zambello, E. 2012a. Occurrence of pharmaceutical compounds in urban wastewater: Removal, mass load and environmental risk after a secondary treatment—A review. *Science of the Total Environment* 429: 123–155.

Verlicchi, P., Al Aukidy, M., Galletti, A., Petrovic, M., Barcelo, D. 2012b. Hospital effluent: Investigation of the concentrations and distribution of pharmaceuticals and environmental risk assessment. *Science of the Total Environment* 430: 109–118.

Vulliet, E., Cren-Olivé, C., Grenier-Loustalot, M.-F. 2011. Occurrence of pharmaceuticals and hormones in drinking water treated from surface waters. *Environmental Chemistry Letters* 9: 103–114.

Weigel, S., Berger, U., Jensen, E., Kallenborn, R., Thoresen, H. et al. 2004. Determination of selected pharmaceuticals and caffeine in sewage and seawater from Tromsø/Norway with emphasis on ibuprofen and its metabolites. *Chemosphere* 56: 583–592.

Writer, J.H., Ferrer, I., Barber, L.B., Thurman, E.M. 2013. Widespread occurrence of neuro-active pharmaceuticals and metabolites in 24 Minnesota rivers and wastewaters. *Science of the Total Environment* 461: 519–527.

Xu, W., Zhang, G., Li, X. et al. 2007. Occurrence and elimination of antibiotics at four sewage treatment plants in the Pearl River Delta (PRD), South China. *Water Research* 41: 4526–4534.

Yazdi, A.S., Es'haghi, Z. 2005. Two-step hollow fiber-based, liquid-phase microextraction combined with high-performance liquid chromatography: A new approach to determination of aromatic amines in water. *Journal of Chromatography A* 1082: 136–142.

Zhou, L.-J., Ying, G.-G., Liu, S. et al. 2013. Excretion masses and environmental occurrence of antibiotics in typical swine and dairy cattle farms in China. *Science of the Total Environment* 444: 183–195.

Zonja, B., Aceña, J., Jelic, A., Petrovic, M., Pérez Solsona, S. et al. 2014. Transformation products of emerging contaminants: Analytical challenges and future needs, in *Transformation Products of Emerging Contaminants in the Environment: Analysis, Processes, Occurrence, Effects and Risks* (eds. D.A. Lambropoulou and L.M.L. Nollet). John Wiley & Sons: Chichester, UK.

Zounková, R., Odrášká, P., Doležalová, L., Hilscherová, K., Maršálek, B. et al. 2007. Ecotoxicity and genotoxicity assessment of cytostatic pharmaceuticals. *Environmental Toxicology and Chemistry* 26: 2208–2214.

Zuccato, E., Calamari, D., Natangelo, M., Fanelli, R. 2000. Presence of therapeutic drugs in the environment. *The Lancet* 355: 1789–1790.

Zwiener, C., Seeger, S., Glauner, T., Frimmel, F. 2002. Metabolites from the biodegradation of pharmaceutical residues of ibuprofen in biofilm reactors and batch experiments. *Analytical and Bioanalytical Chemistry* 372: 569–575.

Outlook

This book introduces fundamental principles and applications of different ionization methods and types of mass detectors and highlights the novel advances of recently developed MS methods in environmental analysis. The present contributions employed several chromatographic procedures and sophisticated instrumentation. These multidisciplinary approaches can elucidate analytical problems and provide holistic investigations in the field of environmental analysis. Relevant research studies from recent years have been summarized, the principles of the representative techniques have been highlighted, applications have been briefly introduced, and potential problems have been discussed.

Environmental analysis and quality is a very broad field that covers almost all environmental disciplines. A comprehensive review on such a broad topic is not feasible in a book due to limitations on space. Therefore, from a practical point of view, only some types of contaminants that are widely detected in different environmental compartments (e.g., pesticides, pharmaceuticals, plasticizers, persistent pollutants, emerging contaminants) have been discussed here, as examples to outline their MS analytical methods. Given the explosive growth in LC-MS research in environmental analysis, over the past decade, LC-MS techniques and studies have been overviewed in greater detail than the GC-MS ones.

Taking into account all the information included in this book, we have attempted to provide an overview of the major issues on the current MS-based methods used for the laboratory analysis and monitoring of pollutants:

- The continual innovations in MS technologies (e.g., new analyzers invented [e.g., Orbitrap], old ones improved [e.g., TOF], new configurations assembled [e.g., QIT-TOF and LIT-Orbitrap]) and the improvements in chromatography and interface techniques (e.g., UHPLC, comprehensive bidimensional or multidimensional separation techniques) have enabled new powerful strategies in environmental analysis. Nowadays, the utilization of MS-based techniques not only opens up new possibilities for overcoming the problems encountered in structure elucidation but also creates new opportunities for the use of these techniques as routine analysis methods to monitor different groups of pollutants.
- New trends in MS-based techniques tend, firstly, to improve the separation processes in terms of efficiency and fastness and, secondly, to incorporate effectively novel instrumentation in the analytical procedure. However, implementation of the MS-based methodologies (e.g., sample preparation, separation, ionization, data acquisition, and data analysis) differs depending on the complexity of the samples and the goals of the analysis.
- In order to obtain useful data, a variety of mass spectrometers are employed in MS-based research. Every instrument has special design, data system, and performance specifications, resulting in strengths and weaknesses for different types of experiments. In general, advanced analyzers such as QqQ-MS/MS, and QqTOF-MS is among the most widely used, although for a large quantity of compounds, the use of QqQ-MS/MS is recommended because of the possibility of simultaneous confirmation and quantification of the target compounds with excellent selectivity and sensitivity. ESI in positive and negative ion modes is the most commonly used ionization method in LC-based mass spectrometers.
- Quality assurance, quality control, and chemometrics are facing enormous challenges in developing robust and efficient methods for answering various analytical questions derived from measurement uncertainty. It is recommended to participate in proficiency testing schemes, especially for accredited laboratories. If such schemes are not available, then analysis of reference materials other than the ones used in the validation is also useful.

- The native binary data formats produced by each type of mass spectrometer are different and generally proprietary. The diverse character of the MS data structure causes difficulties in the integration of new instruments into established infrastructure; inhibits the analysis, exchange, evaluation, and publication of findings from different experiments and laboratories; and prevents the researchers from accessing data sets essential for the development of software tools. To date, research on this field entails much exercise of caution with regard to data acquisition, processing, and information interpretation because of the numerous limitations related to data processing.
- In recent years, much emphasis has been placed on the detection, occurrence, and fate of different groups of pollutants in environmental samples. Of particular concern are PPCPs, EDCs, plasticizers, BFRs, naphthenic acids, and air pollutants such as NO_x, SO_2, VOCs, and HAPs, many of which are ubiquitous contaminants in different environmental compartments. Apart from the parent compounds, an intensive research work has been performed in order to obtain information for their metabolites and TPs. Thus, in recent years, more data and broader knowledge have become available on the identification of TPs (e.g., degradation intermediates by biotic and abiotic processes). Despite this effort, a large variety of intermediates remain unknown and only a limited amount of information about the occurrence of a small number of metabolites and TPs has been produced.
- With respect to sampling techniques, conventional methods for monitoring pollutants in the aquatic environment mainly rely on grab and mechanical sampling. However, this method may not fully account for temporal variations in concentrations due to fluctuations in flow, precipitation, or episodic inputs (e.g., combined sewer overflows or sewage lagoon release). To overcome these drawbacks, passive sampling has been identified as a promising alternative tool. Many different devices have been developed and applied for determination of a wide variety of pollutants in freshwater, cave streams, marine waters, groundwater, and wastewater, with POCIS, Chemcatcher, and SPMDs being among the most widely used. Although the use of passive sampling in aquatic environment is recently introduced, for air quality studies, passive samplers (e.g., Radiello, PUF disks, and SPMDs) have been used for a long time in the short- and long-term monitoring at the local and global scales. Progress in monitoring device solutions has allowed their application in air monitoring, and nowadays, they are recognized as a suitable sampling technique for the determination of pollutants in the atmosphere. Much work is yet expected to be done in this challenging area, and sampling techniques will presumably have a paramount role in future developments.
- Proper sample preparation is critical for MS analysis, because the quality and reproducibility of sample extraction and preparation significantly impact results from MS instruments. A vast amount of research has been conducted, and a number of elegant methods have been proposed for analysis of contaminants belonging to different chemical groups over the past few years. Among the different sample preparation methods, SPE is by far the preferred approach for comprehensive chemical profiling in aqueous environmental samples prior to quantitative analysis of pollutants. In terms of autonomy, miniaturization, and the reduction in labor, microextraction techniques such as SPME and LPME show potential as sample extraction and preconcentration procedures; nevertheless, they still present limitations for routine methods and require significant investment in equipment for large sample throughputs. As part of this, the development of novel, more selective, robust, affordable, and easily prepared materials has become one of the most active research areas in this field. Thus, new materials in line with the concept of green sample preparation methods will attract more and more attention in the academia and industries, being an important and fruitful research area in the decades ahead.
- Regarding pollutant determination, UHPLC and MS/MS tend to be the ideal techniques because of the requirements of regulatory agencies that impose even lower limits. Another

noticeable trend is the development of multiresidue methods. The simultaneous analysis of several classes of organic pollutants in a single run appears to be the future direction. The use of HRMS, coupled to proper sample preparation techniques, databases, and data processing algorithms, could potentially allow the screening of thousands of compounds simultaneously in a given sample. The development of high-throughput and quantitative MS omics workflows within the last two decades has expanded the scope of what we know about the structure elucidation of analytes; hence, new substances generated under different physicochemical processes in the environment could be identified with a high degree of confidence.

To conclude, forecasting the future is always fraught with difficulties; however, given the enormous evolution of MS technology in recent years, the future of MS looks very bright and next-generation MS instruments (e.g., incorporating aspects of screening and targeted approaches by combining different MS techniques [e.g., QqQ and full-scan HRMS]) are anticipated very shortly.

We believe that this book can guide practitioners of MS and environmental analysis and provide insights into its present uses as well as new data processing applications.

Leo M. L. Nollet
Dimitra A. Lambropoulou

Index

Page numbers followed by f and t indicate figures and tables, respectively.

B

C

E

J

K

L

M

Q

R

W

X

Z